W9-BAR-104

MARTIN VAN BUREN

The Romantic Age of American Politics

Other books by John Niven

Connecticut for the Union
Years of Turmoil: Civil War and Reconstruction (Editor)
Gideon Welles, Lincoln's Secretary of the Navy
Connecticut Hero: Israel Putnam
Dynamic America (with Courtlandt Canby and Vernon M. Welsh)

Martin Van Buren

The Romantic Age of American Politics

John Niven

New York Oxford
OXFORD UNIVERSITY PRESS
1983

Library of Congress Cataloging in Publication Data

Niven, John.
Martin Van Buren: the romantic age
of American politics.

Includes index.
1. Van Buren, Martin, 1782–1862. 2. United States—
Politics and government—1837–1841. 3. Presidents—
United States—Biography. I. Title.
E387.N58 1983 973.5′7′0924 82-14528
ISBN 0-19-503238-1

Printing (last digit): 9 8 7 6 5 4 3 2 1

Printed in the United States of America

To

ELIZABETH *and* MARY ANN

Preface

Martin Van Buren has been a controversial figure in American history. By his contemporaries he was either extravagantly praised or roundly condemned. Distinguished writers like James Kirke Paulding and Washington Irving found Van Buren to be a charming person, a genial companion, a wise and able public man. Thurlow Weed, one of Van Buren's bitterest political opponents, admitted in retrospect that he was a man of ability, courage, and honesty, and strict personal integrity. Yet for all those who thought highly of Van Buren, there were many more who disliked or distrusted him, who considered him a mediocrity, conspicuous only for his political tricks and his appetite for the spoils of office. Disappointed but highly visible politicians whom Van Buren had worsted in partisan combat added their voices to the clamor of harsh criticism that frequently went beyond the limits of propriety, in days when libel and slander had little meaning in law and custom. Their success in picturing Van Buren as an unprincipled manipulator, a magician, a Talleyrand, who debased the pure coin of American democracy through the spoils system, however outrageous, has nevertheless made a lasting impression. The air of mystery and acrimony which surrounded Van Buren has to a considerable extent blinded succeeding generations of the informed public to his very real achievements.

In the management of politics Van Buren had few equals. The Albany Regency, which dominated the political life of New York for twenty years and was emulated in other states, was his personal creation. While a United States Senator, Van Buren introduced issue-oriented politics based on discipline, organization, communication, mission and financed through a judicious distribution of office to the faithful. This political instrument, which assisted mightily in the election of Andrew Jackson to the presidency, was the forerunner of what has come to be called the second American party system.

There is no doubt that Van Buren gave character and balance to the Jackson administration, especially in its foreign policy. As President he had the misfortune of being confronted with the first really devastating economic depression to face the nation. The policies his administration adopted to cope with the faltering economy were primarily corrective and aimed at curbing speculation. Though in line with the laissez-faire theory of the day, Van Buren's program was considered radical and destructive by the business community. Whether it was appropriate to the economic crisis or not, his actions were politically courageous, and did succeed in rescuing the public credit, his primary goal.

Van Buren was less successful in containing the divisive sectionalism between the free states and the slave states. His efforts to appease the South, to assure its spokesmen that abolition was not the threat they believed it to be, were viewed with suspicion and criticized in both the free and the slave states. Opposed to the expansion of slavery into the unsettled territories, Van Buren sought a middle ground that would preserve the Union and would be consistent with Jeffersonian views on states' rights. When southern leaders attacked both of these assumptions at the close of the Mexican War, Van Buren struck back with the first comprehensive articulation of the Free Soil stand. Though in retirement, and against all of his political instincts, Van Buren accepted the nomination of the Free-Soil party for President in the campaign of 1848. He knew he would be defeated, but he was willing to put his reputation on the line to administer what he felt would be a timely warning to southern extremists that the North would resist further expansion of slavery.

Lacking the charismatic qualities of Jackson and his well-dramatized military past, Van Buren nevertheless personified individual achievement so admired in early nineteenth-century America. Like Jackson, he came from humble beginnings but, unlike the Old Hero, he rose to the pinnacle of his profession at a time when his competitors were for the most part well-educated men from established families. Cautious and circumspect he certainly was in public life, but there was a strong moral fiber to the man, a cast of mind that could transcend the immediate and the practical and adopt the unpopular view because he thought it was right. As he remarked to a young Vermont Free-Soiler who visited him at Lindenwald in 1848, "Young man, you have chosen a good part. Persevere to the end . . . The recent aggressions of the slave power may destroy the old parties, but they will perpetuate the republic."

In the preparation of this volume I have accumulated many obligations. Robert V. Remini of the University of Illinois (Chicago Circle) read the entire manuscript and I profited much from his criticism. Similarly, Alfred Louch and Henry Gibbons of the Claremont Graduate School made invaluable comments, as did Charles Lofgren of Claremont McKenna College and John H. Kemble of Pomona College. I am indebted also to Sheldon Meyer, Leona Capeless, Kathy Antrim, and Melanie Miller of Oxford University Press, who helped me put the manuscript in manageable form.

Research for this book has taken me all over the country. Dozens of librarians and directors of special collections have given unstintingly of their time in ferreting out important documents. I want especially to thank John McDonough and Oliver Orr, of the Manuscript Division of the Library of Congress, who not only put their encyclopedic knowledge at my disposal but helped make my stay in Washington a pleasant one. Charles Dollar and James O'Neill of the National Archives also assisted me with their expertise in the vast holdings of the archives. George Franz, Director of the Van Buren Papers project at the Delaware County Campus of Pennsylvania State University, supplied me with valuable material. Mrs. Elizabeth Chittick, president of the National Woman's party, made available a pleasant room that I could afford at the Sewall Belmont House in Washington. To her and to other friends and colleagues in Washington, I am deeply grateful.

Paul Rugen and the staff of the New York Public Library's Manuscript and Archives Division could not have been more helpful. Thomas J. Dunnings, Curator of Manuscripts, New-York Historical Society, and Mary Jo Kline, Editor of the Aaron Burr Papers at the Society, went out of their way to make available valuable documents directly or indirectly relating to Van Buren and his times. I should also like to thank Mrs. Ruth Piwonka of the Columbia County Historical Society, John D. R. Platt, William Jackson, Bruce Stewart, and George D. Berndt of the National Park Service, James Corsaro, New York State Library, John D. Cushing, Librarian of the Massachusetts Historical Society, Nicholas Wainwright, Director of the Historical Society of Pennsylvania, Charles A. Ryskamp, Director of the J. Pierpont Morgan Library, Carolyn Wallace, Director of the Southern Historical Manuscripts Collection of the University of North Carolina (Chapel Hill), Peter Dzwonkoski, Head of the Department of Rare Books and Manuscripts, University of Rochester Library, Richard M. Ludwig, Rare Books and Special Collections, Princeton University Library, Robert M. Lunney, Director of the New Jersey Historical Society, Clyde Wilson, Editor, the Calhoun Papers at the University of South Carolina, E. L. Inabinett of the South Caroliniana Library, Dellman B. Sorrells, Special Collections Librarian, Clemson University, Sally Leach of the Texas Humanities Research Center, University of Texas (Austin), Kenneth L. Lohf, Librarian,

Rare Books and Manuscripts, Butler Library, Columbia University, William H. Runge of the University of Virginia Library, Mary Robertson of the Huntington Library, San Marino, California, Patrick Barkey, David Kuhner, Tania Rizzo, and Martha Smith, Honnold Library, Claremont, California.

I wish also to express my appreciation to the John Randolph Haynes and Dora Haynes Foundation which awarded me a research grant in the early stages of this work. Mrs. Marie Montgomery typed the manuscript and Mrs. Catherine Tramz typed the notes. Gerard A. Forlenza, Jr., a graduate student in history at Claremont, helped in checking various sources.

Claremont, California J. N.
June 1982

Contents

MARTIN VAN BUREN

The Romantic Age of American Politics

« CHAPTER 1 »

Beginnings

June days are apt to be hot in Sorrento. When the sun begins to set over the prowed headland of Capri, however, a mild breeze from the north generously lays the heat and sweeps away the mosquitos and flies.[1] In the Villa Farangola, overlooking Sorrento's plaza, an elderly gentleman could be found of an early evening in late June 1854. Even to the casual onlooker he would be marked as a person of distinguished appearance. Clean-shaven, broad of face and forehead, he was balding, with a ruff of white hair and white sidewhiskers. His deep-set penetrating eyes were the one striking aspect in otherwise bland features. He was Martin Van Buren, eighth President of the United States, now in his seventy-first year. Surrounded by books, documents, newspapers and notes, some in his own handwriting, others in the more controlled script of his son Martin, Jr., he was writing his memoirs.[2] Van Buren enjoyed the comfort, the gorgeous view from the terrace of Villa Farangola, perched as it was high above the Bay with the fabled gorge, Vallone die Mulini, dropping off to the right, its mountain stream rushing in a torrent to the sea 150 feet below. "I have hired a villa overlooking the town," he wrote his old friend Gouverneur Kemble on June 13, 1854, "roomy, exceedingly agreeable. Have a first rate cook, a commissaire."[3]

Van Buren had come to Sorrento two months earlier to enjoy picturesque surroundings, the daily salt water baths and the balmy climate he had read about in Fenimore Cooper's travel series, *Gleanings.* As an important foreign traveler, a former head of state, King Ferdinand of Naples accorded Van Buren every courtesy. He accepted these attentions as a matter of course, apparently choosing to ignore the unsavory reputation of his official host, a preposterous petty tyrant who represented everything he had opposed, indeed, despised, in his long political career.

While Van Buren's indifference to the oppressive government may be attributed to the infirmities of age, more probably it reflected concern for the health of his son Martin, seriously ill with tuberculosis, and anxiety over the political crisis developing at home.

Over several years Van Buren had been making notes for his memoirs. His son and namesake, who acted as his secretary, had helped him collect material and had copied pertinent abstracts from Jabez Hammond's three-volume political history of New York. Yet always, it seemed, something had come up that made it necessary for him to postpone the undertaking. Now, in this beautiful and quiet villa he delayed no longer. If for nothing else, work on the autobiography would divert his anxieties. A quest backward in time ranging over his own career might help him understand the causes for the present state of things at home.

From American newspapers and from careful reading of the speeches and debates in the *Congressional Globe,* Van Buren was well posted on current events in America. Old friends, associates, his sons—Abraham, Smith, but especially John, the most politically astute of his family—kept him informed of the intricate partisan maneuvers, the subtle shifts in public opinion, which were absent or muted in the party's public prints or in the self-serving debates and speeches of the politicians.

Van Buren's critical faculties had not lessened with age. He had watched the erratic course of President Franklin Pierce and had written him off as a second-rate man. For years he had mistrusted the young and drivingly ambitious Stephen A. Douglas, who repaid this distrust in equal measure. And he felt his worst fears to be justified as he followed the debates over the Kansas-Nebraska bill that Douglas was driving through Congress.

When Van Buren began to work seriously on his autobiography in Sorrento, he knew the bill, with its explicit repeal of the Missouri Compromise, had become the law of the land. He was fully aware of the staggering blow the Kansas-Nebraska Act had dealt his party, of the sectionalism it harbored like a deadly virus for invading and destroying the body politic. While he was listing the public offices he had held, glancing at Martin's abstracts from Hammond for points of reference, he must have wondered whether his political life had been worth the effort; whether, indeed, his role might not have contributed to the present crisis. As he recalled his failure to obtain the Democratic nomination in 1844 he blamed himself—his "long continued exercise of political power" which excited prejudice, envy and intrigue—for being the divisive factor that undermined the party structure. This weakening, he felt, provided the opportunity, quickly seized, to enlist "the cooperation of the slave power subsequently and adroitly brought to the assistance of designs already matured." Having written down these observations, he realized that he was digressing from his principal objective, the story of his life. With the remark that he would explain fully and completely the causes and the consequences of his defeat in 1844 (an explanation he never made), he brought his narrative back to the traditional canons of autobiography—his forebears and his beginnings in the little Dutch village of Kinderhook, New York.[4]

Van Buren's family was Dutch on both sides. No wandering Yankee, no enterprising Scot, none of the regulars of King William or King George who passed through Kinderhook to battle the French, had ever succumbed to the charms or succeeded in his suit with any of the winsome ladies of the Maessen-Van Buren clan. There had not been, said Van Buren, "a single intermarriage with one of different extraction from the time of the arrival of the first emigrant to that of the marriage of my eldest son, embracing a period of over two centuries and including six generations."[5]

This remark could have applied to most of the families in Kinderhook. Most cherished their Dutch heritage, spoke Dutch among themselves, attended the Dutch Reformed Church, and had some connection, by blood or by marriage, with each other. Washington Irving, who spent several months there in the home of Peter Van Ness, found Kinderhook a place that time had overlooked. There he discovered many of the characters who appear in his charmingly burlesque *Knickerbocker History of New York* and in his *Sketch Book,* the Wouter Van Twilers, the Rip Van Winkles, even the Ichabod Cranes.

Despite the kinship, the sense of belonging, the common culture, and perhaps because of it, class lines were closely drawn in Kinderhook and in the other predominantly Dutch villages of what was then Albany County. Van Buren may have been proud of his background, could even trace some sort of kinship with local notables, men of education and property—Van Shaacks, Van Nesses, Sylvesters—but he was always acutely aware of his and his immediate family's standing in the community. The Van Burens were of humble circumstances; they were insignificant in the affairs of the village, though not completely ruled out of its governance. Into this anomalous social and economic position Martin Van Buren was born on December 5, 1782, just five days before the American and British Commissioners signed a provisional peace treaty in Paris effectively ending the Revolutionary War. Martin was the third child of Abraham Van Buren and Maria Hoes; two daughters, Dirkie and Janetye, preceded him and two sons and one daughter would follow, Lawrence, Abraham and Maria. Besides the six Van Buren children, there were three young Val Alens in the family, for Maria was the widow of Johannes Van Alen by whom she had two sons and a daughter.

Abraham Van Buren was a thirty-nine-year-old bachelor at the time of his marriage to the widow Van Alen in 1776. In the accepted values of the community, especially among those of Abraham Van Buren's age and status, Maria could not have been a prudent choice. She had no dowry, no important connections but had that bane of all bachelors, three dependent children. Described by his son Martin as "an unassuming amiable man who was never known to have an enemy," his marriage to Maria was in character. She was a capable woman, and her condition, struggling to provide a home for her three fatherless children, appealed to the sentimental side of his nature.

Abraham Van Buren owned and operated a farm and a tavern that enjoyed a good location on the south side of the Post road to Albany. If its means were modest the family was always respectable. Though he had been a Whig and was an anti-Federalist in a Federalist town and county, Abraham held the post of town clerk for ten years, an office he was able to pass on to his step-son James Van Alen, who served from 1797 to 1801.

Martin Van Buren's first conscious recollections were of eight brothers and sisters crowded together in half of the upper story of his father's steep-roofed tavern. Food was prepared in a small cookhouse and taproom that adjoined the main house. The bedroom and parlor of Abraham and his wife occupied part of the ground floor of the building. They had just enough space to maintain a modicum of respectability. Abraham Van Buren owned six slaves; two were women who helped Maria in the kitchen and with the housework.[6]

Martin's early years were passed in what seems a large but at the same time cramped establishment, seventeen in family and slaves, without counting the tavern guests. Never one to press for debts owed him, always ready with a shilling or two for a neighbor or traveler in distress, Abraham gained friends, while his numerous family, despite the heavy labor of each and everyone in the tavern or in the fields, had a little less in worldly goods each year. Living conditions would have been worse had not Abraham received some trifling extra income in fees from his duties as town clerk, from renting his tavern rooms as a polling place and—with a fine lack of partisanship—as a meeting hall for Federalists and Jeffersonians alike when political parties began to emerge in the early 1790s.

James and John Van Alen, Maria's oldest children, were admitted to the bar when Martin was still a youngster and were able to help the family with small sums they were earning from legal business in Kinderhook and its surrounding towns. Martin's younger brothers, Lawrence and Abraham, were also intelligent, industrious young folk as were his sisters.

If the father of the family seemed content with his station in life, the mother, Maria, was prudent, shrewd and capable, given the restrictions imposed on a dutiful Dutch wife. She could not change her husband's habits; yet she contrived to manage a large household and maintain a certain respectability for the Van Burens. In her reserved way she ensured that her children loved and respected their father's fine human qualities but would not follow his example in practical affairs. Carefully concealed behind the reticent housewife and harried mother was a strong ambition for the success of her children. When times were grim she had seen to it that her older sons received sufficient formal education to be accepted as law clerks; and now when times were still hard for the family, her favorite, Martin, must somehow be rescued from the drudgery of a petty farmer or the declining fortunes of a small-time taverner.

Even as a very young child, there was something, some indefinable quality, that set Martin Van Buren apart from his brothers and sisters and from the other children of the town. A handsome child, small, rather delicate in appearance, his hair, worn long as was the custom, fell in fine reddish blond waves to his shoulders; bright blue deep-set eyes, a fair complexion; a merry disposition and an infectious smile made him popular among his friends. But it was his quickness of mind, his poise, astonishing for a mere stripling, that appealed to his elders. Although he did his share of farm work, his stint at tavern chores, he never seemed careless or dirty in appearance. Clothed in homespun like everyone else in his station, he was always a little neater, his clothes a little cleaner than those of his comrades. Inevitably he attracted the attention of the local gentry and they favored the bright young Van Buren who was so civil, whose address, while properly deferential, yet bespoke an independence of thought and of bearing.

Martin Van Buren attended the village school, referred to as the Kinderhook academy, but not to be confused with its successor, a private school of considerable distinction. The academy of Van Buren's day was a rundown, one-room, wooden building, not divided by class but by size and age of the students. Van Buren was fortunate, however, in having as a schoolmaster, David B. Warden, who managed to keep order and who had sufficient education to teach those who were responsive. He was fortunate also in having a family that kept him in school until he was fifteen, even though this meant loss of his labor for a part of the year, no small sacrifice. He could not hope for a college education, but he did learn a little Latin, enough grammar and rhetoric to form the basis of a flowing but prolix style of writing, and the rudiments of logic.

More than any formal education he might have received, his experience at Warden's school taught him that if he were to succeed in life he must rein in "a disposition ardent, hasty and impetuous." He recognized even at this early stage that he was capable of intellectual exertion provided he restrained a latent tendency to waste time on congenial but frivolous pursuits. No doubt his quiet mother reinforced this perception, and he had always before him the lesson of his good-natured, improvident father, whom he seemed to have in mind when he spoke of the "capricious and ill-sustained" aspect of his own temperament. Yet looking back on his youth after an extraordinarily rewarding career that had carried him to the pinnacle of success and power, Van Buren regretted his lack of a good systematic education. He had, he felt, always been forced to work harder than competitors far less intelligent than he but far better educated. "How often," he lamented, "have I felt the necessity of a regular course of reading to enable me to maintain the reputation I had acquired and to sustain me in my conflicts with able and better educated men."

However he might complain of his educational drawbacks, he was noticed by community leaders as a worthy lad, fit for higher things. The family still retained enough influence and connections to place him under the tutelage of one of the best lawyers in town, Francis Sylvester. As part of the understanding with the Van Burens, Sylvester provided a bedroom for Martin at his brother Cornelius's store in return for sweeping out the store and his law office each day, laying the fire, and dividing his time dispensing goods over the counter and clerking in the law office. He was expected, of course, to master the complexities of the New York legal system with its cumbersome mixture of statutory, common and chancery law, its elaborate and intricate course of pleadings; Van Buren also managed to involve himself in politics.[7] Somehow he found time to take part in the District convention of the Jeffersonian Republicans held at Troy in 1800. Obviously a country boy in his homespun, the charming manner and the shrewd but tactful suggestions made an impression on the Republican leaders of Columbia County. Though a political neophyte and only seventeen years old, he played an important role in securing the nomination of his fellow townsman John P. Van Ness for Congress.

The town of Kinderhook, and in fact Columbia County, was a Federalist bastion. Francis and Cornelius Sylvester, Van Buren's sponsors, numerous Van Schaacks, Van Rensselaers, and Ten Broecks, dominated the Federal party. Old Peter Van Ness, a self-made, tough-minded wealthy farmer, and his sons headed the minority Republicans in town. It was tempting fate for Van Buren, whose professional career at the time depended upon the good will of the Sylvesters, to associate himself with the Republicans; for family and politics were deemed one and inseparable. Yet Van Buren maintained good relations with the Sylvesters until he became actively involved with the Columbia County Republican organization. Only then, after he had read law for almost four years and had been admitted to the Columbia County bar, did he come to a parting of the ways with his Federalist mentors.

Francis Sylvester had done his best over the years of Van Buren's apprenticeship to make him a Federalist, but to no avail. The young law clerk had imbibed some Jeffersonian notions from his father and more from his reading of the New York City papers, enough to convince him that Jeffersonian principles and politics were about to sweep the county. To Van Buren, there was more to the world than Columbia County and more to his sensibilities than accepting patronage from the rich and the well-born. His repeated refusals to embrace the party of his teacher and benefactor led to "heart burnings with all and occasional tho' slight bickering between Mr. Sylvester and myself." He resolved to complete his education elsewhere, though prudently he did not make this decision until John P. Van Ness won the congressional election in the Jeffersonian landslide of 1800.

In keeping with his newly cultivated style and with genuine good feelings, Van Buren managed to break with the Sylvesters on a friendly basis. Only their uncle, the able and imperious Peter Van Schaack, who had early marked him out as a protégé, was hostile. "His prejudices against me early in life," Van Buren recalled, "were of the rankest kind."[8] This first brush with politics gave Van Buren invaluable experience even as it widened extensively his circle of acquaintances. More than anything else, the Troy convention banished whatever self-doubts Van Buren may have harbored about his future as a lawyer and as a public figure. He knew, however, he had much to learn if he were to make a mark beyond the confines of Kinderhook. He was already determined to try his hand in the perilous but exciting arena of New York City when John P. Van Ness offered him the means to make the move.

The Van Ness brothers were older, better educated and much more self-assured than their distant relative Martin Van Buren. The handsome John P. Van Ness owed his congressional seat largely to Van Buren's efforts and advice. It was now his turn to repay the obligation, which he did promptly. While attending the first session of the Seventh Congress, Van Ness met and married a Washington heiress. Returning to Kinderhook in March 1802, and now possessed of ample funds, he offered to lend Van Buren sufficient money to pay for his travel to New York and his room and board until he became settled. He also suggested that his brother William, who he said enjoyed a prosperous law practice in the city, would take him in "temporarily" until he could complete his clerkship. Anxious to broaden his horizons, and professionally, if not personally, estranged from the Sylvesters, Van Buren agreed to this arrangement. Van Ness was as good as his word. He advanced Van Buren the money before he returned to Washington, and stopping off in New York, prevailed upon his brother to accept the young man as his clerk.[9]

William P. Van Ness received Van Buren warmly, helping him to find inexpensive quarters at a boarding house on Catherine Street. Van Buren quickly discovered that the impetuous, attractive, young man practiced more politics than law. A devoted friend and supporter of Aaron Burr, Van Ness spent most of his time promoting Burr's political influence. Van Ness's inability to provide funds placed an unexpected financial strain on Van Buren. He wrote John Van Ness, explaining his predicament. The Congressman replied, enclosing twenty dollars and apologizing for the meager amount, promised he would send more money when he straightened out a real estate speculation.[10]

His immediate financial problems solved for the time being, Van Buren enjoyed himself thoroughly in New York, so much so that he received a lecture from Van Ness on the perils a young man faced in the city. "Temptations to vice are everywhere presenting themselves," he wrote. "I need not tell you that the first and often insensible step towards it is idleness." Lest he be misunderstood, Van Ness said that he was not opposed to what he called rational amusements; and high on the list was the theater, in moderation of course, because good plays like good reading would improve his social graces.

The Congressman need not have worried himself. Van Buren, once he had savored the entertainment that existed in the city of 60,000 inhabitants, found the political scene much more intriguing. He was careful to observe but not to participate; and he took the elder Van Ness's advice not to squander his leisure but to bear down and complete to his own satisfaction his reading in the law.[11]

Van Buren could not have arrived in New York City at a more exciting time of political maneuvering; nor could he have seen a more colorful, more interesting cast of characters. Dominating the Republican party was the rivalry of two men as unlike in appearance, temperament and style as they were alike in character and in ability, Aaron Burr and De Witt Clinton. Burr was a small, neat man of regular features, erect posture, always dressed at the height of fashion. Clearly, he would have the world take him for a gentleman. He worked hard on that image whether in conversation, or in pleading a case at the bar, or in his personal correspondence. A consummate flatterer when it served his purpose, whether it be to seduce a lady, to attract a useful man to his interest or to win over a judge or a jury—whatever his weaknesses, and they were many, he was a fascinating figure.

Where Burr would use compliment and maneuver to achieve his ends, the tall, generous-featured Clinton relied upon family connections, breadth of mind, and a driving ambition no less acute, no less scrupulous, than Burr's. If Burr maintained a fastidious distance from the broils of local politics, not so Clinton, who involved himself directly in the public affairs of his city and state. Both men blundered badly on occasions of critical import to themselves and their factions; but Clinton invariably rebounded while Burr, after his ambiguous stance in the election of 1800, never recovered his former stature. His duel with Hamilton in 1804 ruined him completely.

William P. Van Ness was absolutely committed to Burr, as were a small coterie of other well-placed young men, not just in the City but in every county of the state. When Van Buren arrived in New York, Burr was forty-five years old, his amazing career already faltering. He had had an enviable record for personal bravery and leadership during the Revolution. He had been attorney general of New York, and he vied with Hamilton as the leader of the New York bar. He had served in the Senate and was currently Vice-President of the United States. Unknown but to a few intimates, Burr had sensed the potential for creating a political power base in New York City through a Jeffersonian Republican patriotic club known as the Society of St. Tammany, or the Columbian Order.

Burr was not a member of the club himself; that was not his style. As befit his secretive ways and his aristocratic pretensions, he kept in the background. But just a few months before Van Buren left Sylvester's office to try his luck in New York, Burr began to infiltrate the Order with his lieutenants. William P. Van Ness was proposed for membership on May 11, 1801, along with a house carpenter, a shipwright, a grocer and two wheelwrights. As Burr calculated, were he to gain control of the Republican party in the City, he might drive a wedge between the great Republican families, the Clintons and the Livingstons, whose members or in-laws in an uneasy alliance ruled the City and the state. The irony that Burr must have relished was that none other than Governor George Clinton, De Witt's uncle, had proposed his man Van Ness for membership in Tammany. Burr himself placed in Tammany Jesse Hoyt, an energetic lawyer whose personal loyalty was beyond challenge, and Matthew L. Davis, an intimate friend for the past ten years. Two years later another of Burr's bright young men, Daniel D. Tompkins, became one of the Tammany tribe.[12]

The Society had come into being on May 12, 1789. William Mooney, a veteran of the Revolution and an upholsterer by trade, founded the organization. Fond of companionship, display and drink, this congenial patriot brought kindred spirits together in Tammany. At first the club was nonpartisan, its members, according to Mooney's fancy, dressing up as Indians for their meetings and organizing themselves as tribes with a hierarchy of titles deriving rather casually from Indian folklore—sachems, sagamores and wiskinkies. In 1801 Tammany met in a long wooden shed attached to the popular tavern of Abraham Martling. There they drank "the waters of the great spring," smoked the "calumet," sang patriotic ballads and ribald songs. On historic or other occasions of public observance they paraded in their outlandish Indian costumes, complete with war paint and feathers, no doubt to the fear and delight of the youngsters along the route of march.[13]

A few months earlier at the close of 1800 Burr had seemed to be headed for the Presidency, as much for his admitted ability and national reputation as for the significant if not crucial contribution he made in overthrowing the New York Federalists at the recent national election. Thomas Jefferson was the candidate of a decided majority in the Republican party for the Presidency and Burr for the Vice-Presidency. So disciplined were the Republican electors from the various states, however, that they voted for Jefferson and Burr without one of them casting a vote for some other person who was not a candidate, George Clinton for example. Thus, Jefferson and Burr were tied with 73 electoral votes each. President Adams and Charles C. Pickney, the Federalist candidates, received 65 and 64 votes respectively, one vote having been cast for John Jay.

The Constitution stated that the person receiving a majority and the highest number of electoral votes was elected President; the second highest number became Vice-President. Since there was no winner the issue would be decided in the House of Representatives, where each state cast one vote, with a majority necessary for election.

Whether Burr was involved in a complex intrigue with the Federalists or not, he removed himself from the contest, neither directly opposing nor supporting Jefferson. The immediate result was that a deep division occurred in Congress that required 36 ballots over a four-day period before Jefferson was elected to the Presidency, primarily through Alexander Hamilton's influence in the Federalist delegations. Burr immediately claimed that he had been a loyal supporter of Jefferson; and he may well have been, may well have simply let events take their course.

If so, he made the greatest mistake of his political career. Jefferson was quite certain that Burr had intrigued to cheat him out of the Presidency. Jefferson's ire went beyond mere personal pique. He believed that the Vice-President was guilty of a greater crime, that of attempting to set aside the clearly expressed will of the American people. And Jefferson could be merciless if he felt, as indeed he did about Burr, that any politician would place himself above party and people.[14]

Jefferson moved slowly and cautiously, however. The President had to enlist a lieutenant in New York who could match a man of Burr's mettle. His choice finally settled on De Witt Clinton, newly elected state senator, nephew of his old ally Governor George Clinton, a man who had behind him the money, the power and the influence not just of his own family but also the Livingstons, great manor lords in the southern counties of the state. Both families were of combined British-Dutch background, a formidable political asset in a state where each national group clung to its own unique distinctions. As might be expected, these families had competed with each other for supremacy. Aaron Burr had created a third faction whose astonishing success was the result of his very real organizing abilities and of the fact that his group was a neutral party standing between the two antagonists. Burr, sensing the opportunity to play the mediator, had managed to keep the Republican party together while, of course, advancing his own political career in the eight years from 1792 to 1800. His powers of judgment and persuasion had put together a mixed ticket of Clintons and Livingstons that, acting in harmony, carried the state for Jefferson in 1800. But there was an impermanence in his posture as mediator that became glaringly apparent after the election. Anxious to expand his power base outside of the City, Burr needed support throughout the state. Clinton recognized his strategy and was just as anxious to forestall him.

Both men looked to a revision of the 23rd article of the complicated New York constitution as the instrument that would determine the political future. This article established a five-man council of appointment consisting of four senators from the four great Districts of the state to be nominated by the lower house; the governor was the fifth member. It had complete power over all appointed officials from the attorney general of the state and the mayors of the cities to the lowest office of fence watcher. Until now the council had used its great powers sparingly, but on two occasions there had been heated arguments over whether the governor or his senatorial colleagues possessed exclusive or concurrent power. Hence the legislature in 1801 passed a law calling for a constitutional convention that would settle this and other questions.

Burr had himself and his two most able lieutenants, Daniel D.

Tompkins and William P. Van Ness, along with De Witt Clinton, elected as delegates to the convention where the Republicans would have an overwhelming majority. On October 13, 1801, when the convention met, the delegates unanimously elected Burr its president, a mark of high esteem. At this point Burr made a mistake as devastating to his future as his equivocating stand in the presidential contest. He let Clinton persuade the delegates to give the council of appointment concurring powers with the governor. He either trusted or he underrated Clinton; for he gave no sign that he recognized how vulnerable he would be if the Livingstons and the Clintons were to act in concert against him. But the potential weakness was seen and appreciated in Washington.

As early as April 1802, John P. Van Ness warned his brother William that Burr's "influence and weight with the Administration is . . . not such as I could wish." The President, he added, was covertly supporting De Witt Clinton, now a U.S. Senator, "but is considered one of the council of appointment." Van Ness had no illusions about Clinton, who "will doubtless be in direct opposition to you."

By June, Van Ness, writing from Washington, was deeply disturbed at the "unfavorable expressions [that] are here made against Burr. . . . If he must fall why should you be dragged along with him. As a friend he ought to receive a reasonable support; but be careful of yourself." William was not to be deterred by warnings from Washington. The Vice-President so dazzled him that he was beyond making any realistic appraisal of the maturing alliance between Jefferson and the Clintons and Livingstons. By now, Burr must have known something was amiss, especially when Jefferson, after appointing one of his nominees to an office in New York, refused him further patronage.[15]

The first blow at Burr's machine had come from the meeting of the new council of appointment the previous August. The commanding De Witt Clinton, a new member, brushed aside the remonstrances of the governor, his uncle George. With his crafty brother-in-law, Ambrose Spencer, at his side on the council, he began the largest proscription of public officers in the state's history. The deposed incumbents were primarily Federalists, a fact that may have justified Clinton's meat ax to the Burrites. But when the council began filling major posts with Livingstons and Clintons or their connections, Burr should have suspected that his budding organization was about to be pruned back.[16]

Martin Van Buren followed the contest with great interest. He heard only Burr's side, of course, from his conversations with William P. Van Ness and with Burr. Van Ness had introduced Van

Buren to Burr and had taken him several times to Burr's imposing mansion (just south of present-day Greenwich Village). "He treated me with much attention," Van Buren recalled, "and my sympathies were excited by his subsequent position."

Unlike William P. Van Ness, Van Buren was not an impressionable, impulsive man. The young law clerk from upstate had quickly and accurately assessed the power of the contending factions. As a native New Yorker of Dutch origin, he knew the clannishness of his own folk, their innate suspicion of an outsider, and one with Yankee forebears at that. He was a witness at first hand of the power and the weakness of the great Republican families. So too was Van Ness, but Burr's personality had so fascinated the young lawyer that he was incapable of making a rational judgment.

From what Van Buren had heard of De Witt Clinton's brilliance, and his use of the council, he thought Burr had a cloudy future among New York Republicans while the two families stood together. That Burr would maintain his Tammany machine Van Buren did not doubt; that eventually the Clintons and the Livingstons would revert to their age-old rivalry, his common sense told him. But for the present the right path was with Clinton, not with a faction that was being slowly but surely assailed on all sides. De Witt Clinton's election to the United States Senate in February 1802 furnished additional proof to Van Buren of Burr's waning fortunes. Though he kept his thoughts to himself, his silence was reinforced when Clinton resigned from the Senate to accept appointment as mayor of New York City in 1803.

Shortly after Clinton assumed his new office, all semblance of unity among the Republican factions dissolved, accompanied by a scurrilous, pamphleteering campaign. William P. Van Ness was in the middle of it. His attack on Clinton and his adherents under the pen name of Aristides was considered at the time to be the most venomous political invective ever seen in North America.

Van Buren's closemouthed attitude, his friendship with both of the Van Nesses, convinced many that he was a Burrite. Even the Van Nesses, with whom he maintained the most intimate ties, believed him a staunch supporter of the Vice-President.[17] Van Buren was anxious to terminate his clerkship and be admitted to the state bar. Close study with virtually no guidance left little time to spend in the political maelstrom of New York City—or so he claimed when pressed for an opinion or quizzed about his allegiance. In November 1803 he was admitted to the state bar as an attorney, which meant that he could practice law only before the lesser courts. Now a member of the legal profession, Van Buren made haste to leave the superheated political atmosphere in New York City and begin the required four years of practice before he became a counselor and could appear before the supreme court of the state.[18]

Burr may have been injured and weakened, but he was still very much alive politically. His organization, while small, counted among its loyal supporters some of the best political operators in New York. While he surely recognized the weight of power ranged against him, he also had no alternative but to make another challenge or retire into private life. And Burr thought the party was a fragile coalition precariously held together by patronage; there was a good chance Clinton would not be able to hold the coalition together. The Federalists, too, must be considered. Burr reckoned his chances as better than even that he could draw off those opposed to Hamilton, whose arrogance and burning ambition had alienated many among his own party. Burr might gather up these restive spirits and, with the promise of patronage dangled before them, forge a coalition that could win the governorship and a majority in the legislature. Accordingly, his "little band" of stalwarts nominated him for governor at the Central Coffee House in Albany, a nomination seconded by Tammany in New York City a week later.

Clinton was equally alert to the sensitive balance that existed within the majority faction. He would have preferred that his uncle, George Clinton, remain in the governor's chair. The Republican national caucus thought otherwise. On February 25, 1804, it nominated the Governor for the Vice-Presidency, Jefferson being the unanimous choice for a second term. Not one vote, not even one comment, was made at the caucus in favor of Aaron Burr. George Clinton's withdrawal from the contest in New York State posed a perplexing problem for his nephew. But he thought he had resolved the difficulty when he prevailed upon John Lansing to accept the nomination. Lansing was a distinguished lawyer, head of the chancery court, attractive to Clinton because he was not identified with any of the Republican factions.

When Lansing refused to accept Clinton's advice on patronage as a condition for his nomination and then removed himself from the race, Clinton and his faction reluctantly supported Morgan Lewis, chief justice of the state supreme court. Son-in-law of the revered first chancellor of the state, Robert R. Livingston, the fifty-year-old Lewis was averse to leave his secure, prestigious post for the uncertainties of elective office. Reluctantly he bowed to the family's decision that he was the only member of the Livingston faction with enough presence and energy to hold his own against Clinton, should that become necessary.

Clinton was uneasy about Lewis, about whether his own family would support a Livingston connection of such visibility to head the ticket. But with Burr in the field, he was able to rally all against the common enemy. Livingstons and Clintons particularly hearkened to his charge that Burr was again intriguing with the Federalists as he had in the presidential contest, a charge made in De Witt's paper the *New York Chronicle* and carried through the state in the form of handbills and broadsides. If Clinton tried to make any patronage preconditions for his support of Lewis as he had upon Lansing, he had not succeeded. Lewis may have been no match for Clinton or Burr in ability; he was, however, a man whose personal warmth masked a dogged perseverance. Clinton soon found he could neither manipulate nor intimidate Morgan Lewis, who defeated Burr in the election of 1804. Lewis's triumph began the process of unraveling the strands that tied the Livingstons and the Clintons together. Ironically, Burr had supplied the warp for the alliance, the binding fiber that held the party together.[19]

Meanwhile, Van Buren had found himself in an embarrassing position after he returned to Kinderhook and a law partnership with his half-brother James Van Alen. He had been home a matter of weeks when Burr's friends in the community importuned him to take an active part in the canvass. They assumed him to be one of them because of his close relations with the Van Ness brothers. Van Buren refused. He tried to follow a neutral course, which he soon found virtually impossible in such a vitriolic campaign. Seeking to temper some of the more outrageously ill-founded attacks on Burr, he brought on himself the Clintonians, who accused him of duplicity. The Burrites, equally intolerant, made what Van Buren called "illiberal and unmanly remarks with respect to me," simply because he was exercising what he deemed his freedom of choice. "Most men," he said, "are not scolded out of their opinions."

News of Van Buren's refusal to espouse Burr was taken to mean that he was a Clintonian and relayed on to New York City. The result was a letter from William P. Van Ness quizzing Van Buren about his politics. Earnestly, Van Ness pleaded that he support the embattled Vice-President. "I beseech you that you are not influenced by motives that will hereafter dishonor you," he wrote. Without saying so openly, Van Buren in his reply made it plain that he would support Lewis, leaving the issue open, however, should the situation favor Burr. While he claimed "strong personal prejudices" for Burr, he was convinced "under existing circumstances that it would not be expedient to support him."[20]

Van Buren's stand was in direct defiance of the Van Ness family to whom he was under a heavy obligation. The Van Nesses had not the lofty pretensions of the manor families. Peter Van Ness, the patriarch, had begun his career as a wheelwright, but he had prospered over the years in land speculation and was respected for his success. Socially the Van Nesses were secure, especially so in the immediate environs of Kinderhook. The Van Burens were considered part of the family. As might be expected of a self-made man, Peter Van Ness was overly conscious of his position and the obligations owed him. Like his sons, he backed the Burr cause. Dogmatic in his political opinions and accustomed to deference from poor relations, he did not spare Van Buren's feeling. The young man bore Van Ness's denunciations with outward composure, and though he was hurt he refused to back down.[21]

On election day in April 1804, Van Buren went to the polling place to cast his first ballot. There, with many of the leading men of the vicinity looking on, Peter Van Ness and the venerable Peter Van Schaack challenged his right to vote. Van Schaàck was the foremost lawyer in the town, one of the most eminent counsel in the state, a master of jurisprudence who had codified the colonial statutes of New York. During his long career he had been a Loyalist and an ardent Federalist. Along with his party he was supporting Burr, which brought him into a temporary alliance with the Van Nesses. Van Buren's explanation satisfied the inspectors at the polls but not Van Ness and Van Schaack, who demanded he take the oath prescribed where there was reason to believe the individual did not meet the suffrage requirements.

For a native of the town whose birth was recorded in the Dutch Reformed Church, this public challenge from the two leading men of the community was meant to be an affront, and was so taken. Van Buren had to swear that he was not a felon, that he was twenty-one years old, a white male who possessed a freehold estate of 200 dollars. Under the closest scrutiny he could not meet the property qualifications, a fact that all present knew, though a technicality because his half-brother would have come up with the sum. In the time-honored custom of Kinderhook the insistence upon an oath was an embarrassment that reflected on the character and the credibility of the person challenged. Yet Van Buren refused to be intimidated. He would risk the breaking of a close friendship, the damage to his social standing in the community, perhaps even his bread and butter, before he would allow himself to be coerced. He was his own man and no local squire would dictate whom he would support for public office.[22]

The complete rout of Burr in the state election removed the possibility of damage from the elder Van Ness and Van Schaack and actually improved the brothers' practice because they could expect some favors through Clinton's influence. Before Van Buren became a partner, James Van Alen had built up a small but steady practice among the farmers of Kinderhook and the neighboring villages of Valatie and Claverack. Since the bulk of his business came from small farmers and tenants, he opposed the Livingstons, and since he was an easy-going man, he would have sided with the Van Nesses and Van Schaacks had not his brother thought otherwise. With Van Alen, loyalty to his immediate family counted far more than his political affiliation or even business connections. He stood by his brother and cast his vote for Morgan Lewis and the Clintonians.

A significant part of the firm's legal business from the beginning had come out of defending actions for trespass, title searches and numerous small civil actions that arose out of the claims of the Livingston family to a large portion of Columbia County, including Van Buren's own town of Kinderhook. The original Livingston manor traced its origin to a Crown grant made in 1684. But a subsequent Livingston had expanded on the grant, creating a faulty title to much of their holdings; a title, strengthened, however, through the years by colonial statutes, orders-in-council of Royal governors and more recently by private acts of the state legislature. Farmers who thought they were freeholders fought back when agents of the Livingstons claimed rent. Tenants were restless, litigious and sometimes violent, especially after the social upheavals of the Revolutionary War. There had been no bloodshed, but there had been menacing crowds. Unrest was more or less endemic in the county, all of which profited Van Alen and Van Buren economically and politically.[23] It also involved Van Buren, who, in making the Livingston claim a specialty, became one of the more controversial political figures in Columbia County.

In establishing himself as a lawyer and politician, Van Buren knew he would be opposing the best counsel in the state. With so much at stake, the Livingstons could afford and would employ the best. Van Buren analyzed what he took to be his strengths and his weaknesses. His oratorical skills were modest; eloquence was not his suit. In the noisy public arena of the courtroom, his slight figure and light voice were serious detriments in pleading before judges and juries. Lawyers of large stature and of booming voice reinforced their arguments with the sheer weight of their physical presence. What he did possess was a careful, logical mind and a taste for research. Though naturally voluble, a good companion, he early recognized how dangerous a slip of the tongue could be in a court of law or in a political encounter. Even with close friends, Van Buren guided conversation away from sensitive topics or from those he was not completely prepared to discuss intelligently.

These rules of behavior he established early in his career, for his interest in the Livingston claim pitted him against the overbearing J.

Rutsen Van Rensselaer and Elisha Williams, considered the most brilliant trial lawyer in the state. Capable and sophisticated contemporaries all thought Williams the best advocate they had seen in a courtroom. Van Buren must have felt a deep sense of inadequacy when he prepared for their first encounter in court. But the test had to come if ever he were to be anything but a second-rate lawyer. He met it firmly, relying on preparation, knowing that he could never match Williams's delivery. From their first appearance and over the next decade, they clashed frequently in the courtroom. Self-assured as always, Williams, the confident Federalist of the old school, soon learned that his opponent was more than a match for his years of experience, his presence and his superior education. Once when Van Buren had soundly trounced Williams in a case heard before William W. Van Ness, another staunch Federalist, the judge remonstrated with the veteran lawyer. "How could you from want of a little industry, allow that little Democrat to get so much the advantage of you?" "Oh! Judge," replied Williams, "I relied upon you to supply my deficiencies."

Van Buren prepared his briefs with great care, toiling hours among the dusty records in Albany. After outlining his arguments, he then took the other side, trying to anticipate beforehand his opponents' line of attack. He studied their courtroom style so that he could exploit their weaknesses; and above all, he practiced memorizing and summarizing testimony, for there were no courtroom stenographers to refresh a lawyer's memory. Much depended upon the speed and the accuracy of note-taking during a trial and the readiness and precision of examination or rebuttal, the clarity of the argument.[24]

Van Buren's staunch defense of the Republican party, his courtroom offensive against the Livingstons, and his increasing reputation as a lawyer came to the attention of De Witt Clinton. Always interested in attracting able, enterprising young men to his personal standard, Clinton sent several lucrative cases Van Buren's way. More important than the increased legal business was that Clinton had seen how the young lawyer could be useful in his bid for political supremacy.

Clinton was in the process of concluding an alliance with a Tammany left leaderless after Burr's downfall. His next move was to gather up Burr's adherents throughout the state and then move against the Livingstons. To this ruthlessly ambitious man, what could be more logical than the enlistment of Martin Van Buren? Clinton was well aware that Van Buren still counted many Burrites among his personal friends.[25]

Although he had supported Morgan Lewis for governor, Van Buren had consistently opposed the Livingstons in principle and in his law practice. When Clinton began his campaign with some help from Washington, Van Buren watched the partisan battle carefully. But there was little doubt where his sympathy lay on both practical and ideological grounds.

Lewis played directly into Clinton's hands for immediate advantage when he managed temporarily to divide the Burrites in New York City and form a shaky alliance with the Federalists. The coalition achieved a narrow victory, still sufficient enough to capture the House and thus the all-powerful council of appointment. Prominent Clintonians, including De Witt himself, lost their posts. But, in trading with the common enemy, Governor Lewis gave Clinton a winning issue—the Livingstons were unprincipled in their hunger for office; Lewis himself, a crypto-Federalist. Ridiculous on the face, these charges were nevertheless effective with the rank and file of the Republican party.[26]

In 1807 at Clinton's instigation, a majority of Republican members of the legislature nominated for governor a former Burrite from Staten Island, Daniel D. Tompkins. Billed as "the farmer's son," Tompkins was an appealing candidate to all those small farmers and tenants who instinctively opposed what was known as the "manor interest." He was also a well-educated young man whose marriage to the daughter of New York merchant Mangle Minthorne brought him property and Dutch connections. Convivial, personable, a good speaker, he was a decided contrast, in Clintonian terminology, to that craggy apologist for the "greedy" Livingstons, Morgan Lewis.

Van Buren had met Tompkins during the year he spent in New York City. He liked the good-natured candidate, but he would have supported anyone the Clintonians put up, even old Peter Van Schaack, who had insulted him publicly. Van Buren had read the signs accurately. He was fascinated with what he regarded as the winning tactics of Clinton's campaign, while at the same time he believed deeply in the cause he espoused. He identified himself with the small farmers and tenants, not just for the business it brought his firm but because he felt himself one of them, at least in heritage.

Van Buren's father had been a Whig during the Revolution, an anti-Federalist and along with the Van Nesses a follower of old Governor George Clinton. Jeffersonian ideals came naturally to the Van Buren family; and Martin, struggling against haughty Federalists like Elisha Williams and the Van Rensselaers on the one side or the aristocratic pretensions and very real threat of the Republican Livingstons on the other, was faced with a political contradiction.[27] De Witt Clinton resolved this problem of allegiance. Now he could campaign with a will for Tompkins, Van Buren's conscience clear: he was for the purity of Republican principles that depended upon party regularity. He was against Federalists and Livingstons, both of whom it was clear would subvert the power of the state to their own selfish ends.[28] There would be no more coalitions. As Van Buren had foreseen, Tompkins and the Clintonians swept the state. On March 20, 1808, Van Buren was appointed surrogate of Columbia County, succeeding his half-brother and partner.

The office carried with it some prestige, less income. Far more impressive than these trifling emoluments was the visibility it gave Van Buren outside of his neighborhood. The surrogate was involved with the affairs of every citizen in the county who owned property and wished to devise it. The duties of the office, while not burdensome, made it possible for Van Buren to know a good deal about the personal worth of the voters, knowledge of no little importance to an aspiring politician. His brother James, whose political ambitions were concerned only with expanding the law practice of the firm, had willingly stepped aside. De Witt Clinton was happy to bestow the office on Van Buren. Clinton, too, recognized the political potential of the office, the political acumen of his young lieutenant. In such capable, energetic hands, the post would enhance his personal following where Federalists and Livingstons had long held sway.[29]

« CHAPTER 2 »

State Senator

After four years of law practice, Van Buren felt secure enough to marry and support a family. For some time he had been courting a young woman, Hannah Hoes, one year younger than he. On February 21, 1807, the young couple were married in Catskill, twelve miles southwest of Kinderhook. In keeping with their desire for privacy and frugality, the wedding was not held in their home town where it was customary for the bridegroom to entertain the entire village. Only their immediate families were present, Martin's father Abraham, mother, his brothers and sisters and a score or so of Van Alens, Hoeses and Cantines, another family connection. Martin and Hannah were distantly related; they had also known each other since childhood. Like her husband, Hannah Van Buren was small and slim. Fair of complexion, she was a very proper, very religious young lady of the same Dutch stock, the same social status as Van Buren. As a member of the household described her, she was "a woman of sweet nature but few intellectual gifts."[1]

Well before his marriage, Van Buren had decided that he must move from Kinderhook if he were to achieve the goals he had set for himself. It would not be to New York City, where competition among lawyers was intense and where he felt he lacked the necessary expertise in mercantile, admiralty and commercial law that then commanded the major share of litigation in the port. And at the time urban living did not appeal to him. For many of the same reasons he ruled out Albany. What he wanted was a growing community that combined some elements of city life with the clear air and the clean surroundings of the countryside. His choice was Hudson, a small city of about 4,000, on the east bank of the Hudson River, 15 miles from Kinderhook, 29 miles south of Albany, 116 miles north of New York City. Considering the kind of practice Van Buren was engaged in, his political interests and his need to enlarge his understanding of other branches of the law, Hudson was an ideal choice.

An additional incentive was a felt need to broaden his perspectives, to break the bonds that his Dutch heritage imposed, to associate with the sharp-witted Yankees who were beginning to pour into the Hudson River valley.

Hudson was a new city, almost exactly Van Buren's age. A group of Quaker merchants from New England planning to capitalize on the river trade had settled there in 1783. They had prospected the area thoroughly looking for a good port facility where they could also follow their former pursuits—shipbuilding, whaling, sealing and ocean commerce. The New Englanders had selected carefully and well—over the ages the river and tidal currents had scoured out a natural channel along the river banks at Hudson. They had also observed the richness of nearby farmlands. Produce carrying trade would, they decided, be a profitable enterprise there.

A testament to their good judgment, Hudson flourished from the beginning. When Van Buren bought a house on Warren Street—Hudson's main street—and moved there with Hannah in 1808, the city had been the county seat of Columbia for the past three years. Unlike the steep Dutch roofs and high stoops of Albany and Kinderhook, most of the dwellings were the sturdy clapboarded salt boxes of New England, some of which had been built in Nantucket or Providence, knocked down and shipped to be reassembled in Hudson. New England culture stamped itself squarely on the compact form of the city with its one main street and the seven side streets that crossed it to form right angle corners. The only exception to this grid pattern was Front Street, which conformed to the irregularities of the river bank. Van Buren could not have been blind to the natural beauties of the place, the great river flowing below the little city, the smokey Catskills in the distance.[2]

A further attraction to Van Buren were the schools that the energetic Yankees had built. Unlike his forebears, the more casual Dutch, the New Englanders had carried their traditional respect for education to their new home. The Hudson Academy, a three-story brick structure, was only four years old when the Van Burens took up residence, but it had already gained a statewide reputation for sound instruction. The school, which had a separate attached structure for teaching girls, was a far cry from the wretched little shed in Kinderhook where Van Buren received the only formal education he would ever have.[3]

Within two months of their marriage, Hannah Van Buren became pregnant and the couple's first child, Abraham, was born in Hudson. At regular intervals, Hannah had three more sons who would live to manhood, John, Smith Thompson, and Martin Junior. As his family increased so did his law practice and with it more frequent absences from home for extended periods.

He attended the supreme court sessions in Albany during the winter and the court for the correction of errors in New York City during the spring. The courts of common pleas and the courts of chancery carried him all over the state. Hannah Van Buren's lack of pretence, her unruffled disposition and her thoughtfulness were just what her husband needed after the constant stress of the courtroom, the miles and miles of riding the circuits in good weather and in bad, over execrable roads, through frontier wilderness or in the rutted, filthy streets of New York City rampant with yellow fever and cholera. Many were the nights he spent in dirty, disorderly taverns. The fastidious Van Buren had either the choice of sleeping in a chair before the fire or crowded in with two or sometimes three fully clothed, usually dirty bedmates. After such bouts, Van Buren appreciated the clean, comfortable, though modest household Hannah set up in Hudson. Perhaps more than anything else, she made no demands upon him.

While he extended his practice, Van Buren used whatever spare time was left to build up Clinton's faction. He had discovered that he had a peculiar knack for assessing a given political situation, merging men and motives and issues to develop a plan of action. To be sure, what successes he had in these early years of partisan apprenticeship were trivial ones, and he went astray as often as he found the right course. For his own purposes, De Witt Clinton induced Van Buren to work for Vice-President George Clinton's succession of Jefferson in the Presidency. But the aged Clinton had little or no chance; he did not even command the support of the party in his own state. Governor Tompkins, at the moment the most popular man among the Republicans, in opposing the nomination foreshadowed a rupture between himself and De Witt Clinton. Van Buren remained loyal to his patron and campaigned for George Clinton. In doing so Tompkins marked him down as a pleasant young man who was unable to see where his interest lay. Elsewhere, Van Buren's dearly bought influence, suffered along with that of Clinton from the desertion both of what was left of Burr's following in Columbia County and of those who in his Tammany organization were known as the Martling men—now combined with friends of former Governor Morgan Lewis.

Van Buren learned much from this temporary setback. He was quick to recognize the mistake of tying his fortunes to anyone and especially to one so controversial as De Witt Clinton. He also understood clearly a simple political fact that Clinton never learned—strict adherence to the decision of a party caucus. James Madison had received the presidential nomination in the usual way from the Republican congressional caucus. Allying himself with Clinton in opposition to a majority of the Republican party was construed as apostasy, an epithet that had blighted the career of many a young politician. Van Buren was lucky to escape relatively unscathed from the political fire-storm Clinton brought down on his own head.[4]

In fact, he eventually profited from the small part he played in the affair. He now saw that a break was in the making between Tompkins and Clinton. He could not, of course, predict when this event would happen, but the blunder he had made dictated a period of political quiescence on his part until the situation clarified. For the next four years Van Buren worked hard at his practice. He listened to the conversations on politics that went on incessantly among the lawyers on circuit, said little on that subject; and what little he said was carefully chosen so as not to associate him with any specific men or measures other than the tested principles of the Republican party.

By now Van Buren, though only twenty-six years old, had earned the reputation of being the best lawyer in Columbia County on the Republican side—a lawyer's politics were indistinguishable from his practice. Van Buren, ever sensitive about his poor education, early determined to master his profession, a dedication which left little time for the leisure of reading for pleasure. Every member of the bar in Columbia or in Albany County soon recognized that his knowledge of the law was correct and comprehensive, that he rarely left a loophole in his argument or missed one in his opponent's.

Beyond law books and statutory codes, however, Van Buren was virtually ignorant of literature. In polite company, should the conversation veer to the novels of Scott or to the classics, for instance, he maintained a discreet silence. Should a comment seem necessary, he had trained himself in the art of diverting the conversation to safer ground, a witticism if pertinent, or a neutral answer that seemed to cover the subject without revealing his cultural deficiencies. "His knowledge of books outside of his profession," said James A. Hamilton, a son of Alexander Hamilton, and at the time a close friend, "was more limited than that of any other public man I ever knew."

Van Buren, a sensitive man, was embarrassed at his lack of general knowledge. He deplored his failure to engage in vivid oratorial display so much in demand among the audiences in court or on the hustings. He compensated, perhaps overcompensated, for the lack of these skills through intense labor, through the careful cultivation of a charming manner, disciplining himself rigidly in never losing control of emotions no matter what the provocation. Possessed of a natural wit and fine sense of timing, Van Buren usually disarmed his foes and his rivals.

In due course, he was able to hold his own in small talk with polished men like Washington Irving and James K. Paulding, whom he delighted with his piquant style of expression, his quaint observations. And Van Buren also sought to make up for his presumed cultural deficiencies with his style of dress. Fastidious in his personal habits, he always dressed to the height of the fashion. When comparatively poor, he would deny himself necessities to maintain the wardrobe of a gentleman, well cut in the latest styles. He cultivated a rigidly erect posture which set off his neat, usually colorful attire and made him seem taller than he was, five feet, six inches.

Van Buren needed all the resources he could call upon during his years in Hudson. His living depended upon his ability to avoid the numerous political snares set up for any ambitious young Republican in a region where the Federalists were active, strong in numbers and influence—ever ready to put down any aspirant whose politics and background they did not approve. Nor were the remnants of the Burr organization to be discounted. Van Buren had patched up his differences with the Burrite leader in Columbia County, William P. Van Ness, but he could never be sure whether this impetuous, ambitious man would turn upon him again.

Van Ness's reputation, which had been in eclipse since the Burr-Hamilton duel, was now largely restored to its former state as the memory of his role (Burr's second) in that tragic event grew dim with the passing years. He was eager to try his hand again in politics and had his eye on a seat in the state senate that had fallen vacant in 1812. From that vantage point he could, he thought, organize the Martling men in New York City and the Lewisites in the river counties. Once he mobilized these disaffected elements he would move against Clinton, whom he personally detested and whose presidential ambitions he was determined to block. Van Ness, of course, knew that Van Buren was a Clintonian; though apart from his ill-fated support of George Clinton for the Presidency, he had not followed his patron into other more serious errors. Van Ness was also aware that Van Buren had supported Tompkins when as governor he had taken the unusual course of proroguing the legislature, a majority of whose members had been bribed by a corrupt group of speculators seeking a charter for the Bank of America. Had the legislature chartered this institution, it would have dwarfed all other banks in the state and threatened their very existence. Clinton, though opposed to the Bank, refused to line up behind Tompkins and would not speak out against the proposed charter because he felt it would injure his presidential campaign in the powerful New York business community. His silence on this crucial matter caused a break with his brother-in-law, Ambrose Spencer, the senior Republican in Columbia County, whose interest in another bank would have been jeopardized if the Bank of America received a charter.[5]

Van Buren had given no indication that he sought office beyond that of surrogate, which he had held for several years. In his planning against Clinton, Van Ness assumed he could utilize Van Buren's reputation and skills. He had no idea that the young lawyer had any desire for political advancement in mind, and he was right. When Van Ness sounded him out on the state senate seat, Van Buren declared that he had no interest, that his law practice absorbed all of his time and energy. Where Van Ness erred was that he agreed to support Van Buren's candidate for the senate, John C. Hogeboom, sheriff of Columbia County, though he had no intention of doing so. Hogeboom, a close friend of Van Buren and an able man, was much more popular than Van Ness. He had, in Van Buren's judgment, the best chance of winning the seat in a Federalist county. Having led Van Buren to believe he would support Hogeboom, Van Ness then began a secret intrigue to secure the nomination in such a way that Van Buren would be bound to support him as the caucus candidate. Van Ness overestimated his ability to mask his movements, and he underestimated Van Buren's reaction to his trap. In an offhand way, Van Ness had taken Van Buren's allusion to Hogeboom as simply a casual response. How could this warmhearted, young man, this boyhood friend who owed so much to him, be upset by such a trivial twisting of the truth? After all, political arrangements, as everyone knew, were not to be taken seriously if they proved inexpedient. And Van Ness had assured himself that Hogeboom's nomination was not expedient, while his was.

Van Buren and Hogeboom soon learned of Van Ness's moves which involved a deliberate distortion of their views. Both were angered and humiliated at his untruthful assertion that they were rival candidates at each other's throats over the nomination. Hogeboom promptly bowed out of the competition, pleading business commitments. He added, however, that his many friends, among whom he counted De Witt Clinton, wanted Van Buren to make the run. Clinton, said Hogeboom, was particularly anxious for Van Buren's nomination, as indeed he was because he had just broken with Ambrose Spencer.

Hogeboom spent several hours in an effort to convince Van Buren that he should be the candidate; but still Van Buren refused to commit himself claiming that he, too, had pressing business affairs, and even if nominated, defeat in the election seemed almost a certainty. Edward P. Livingston, son-in-law of Chancellor Robert R. Livingston, and a close associate of Governor Tompkins, was openly in the running. Livingston, a well-educated, articulate man, had the strong backing of his family; he would, in addition, command Federalist support. On the other hand, should Van Buren be nominated, Van Ness, with his motley of Burrites and Lewisites, would surely oppose him. Although Van Buren did not disclose all these doubts to Hogeboom, he suggested that a prominent citizen of Hudson, Robert Jenkins, deserved consideration; still he implied that if Jenkins declined he might enter the list.

The day following his conversation with Hogeboom, Van Buren had a long talk with Jenkins's brother and two of his closest friends. He told them in detail of the Van Ness intrigue, his conversation with Hogeboom and his own doubts about running. He added that Hudson deserved the honor of providing the candidate, and since he was but a newcomer, Robert Jenkins, an original settler, should have the preference.[6] Much to his surprise, the Jenkins group agreed with Hogeboom that Van Buren should be the candidate. While not directly removing Robert Jenkins from consideration, they inferred that he was not available. At least Van Buren made that assumption and reluctantly began making his preparations for the canvass. He drafted the call for the District convention and had it published in the local Republican papers. Soon after this public announcement, Van Buren learned that Jenkins was indeed a candidate. Had he been deliberately misled? He thought so and at once confronted the Jenkins group. They admitted sheepishly that Van Buren's recollection of their conversation was correct. They sought to mollify his feelings, declaring that circumstances had developed which they had not anticipated and could not control; nor could they at this time reveal what these circumstances were. Van Buren's reply was crisp. He was left with no other choice, he said, "than to obtain the nomination, if in my power, which I should assuredly do."

Commenting on this affair years later, Van Buren revealed a side of his character that belies the popular image of him as the devious manipulator, cunning, quick and furtive. "Mr. Jenkins had many of the good qualities of his race," Van Buren said, "but had besides an innate passion for political intrigue, and as I have almost always found to be the case with men subject to that infirmity, was neither skillful in his schemes or successful in their execution." When Jenkins learned that Van Buren was to be a candidate, he at once made a bargain with Van Ness. They would combine their forces with those of Livingston, deprive Van Buren of the nomination and contain, if not reverse, his rapidly growing political influence in the state. A Van Buren defeat would also injure Clinton, an important objective. All of these machinations came to naught. So widespread was Van Buren's popularity, fostered by a flurry of letter writing and appeals in person, that he won the nomination on the first ballot at the convention.[7]

Despite his quick victory, the outlook for his election appeared grim indeed. The state was divided into four great districts: the Southern District, which included New York City, Long Island, Staten Island, the populous county of Westchester and its environs; the Middle District, Van Buren's, contained all the river counties and the Catskills, one-quarter of the entire state and nearly as populous as the Southern District; the Eastern District ran to the Canadian border and west to Lake Ontario; while the Western District was a huge wilderness area sparsely populated, the scene of frenzied land speculation and rapid development. Senators and assemblymen represented each District according to its population based on a census taken every seven years. In 1812, the Southern and Middle Districts furnished twenty of the thirty-two senators and a similar majority of the lower house. Senators were elected to four-year terms, serving in rotation so that at least one senator would be chosen from each District at the annual election in April.

When Van Buren was nominated for the state senate, a presidential election was in the offing. Madison, candidate for a second term, was the majority choice of the party. There had been some grumbling among certain Republicans regarding his administration, which they held lacked vigor and consistency. War with Great Britain was a distinct possibility. Should it come to pass, argued those opposed to Madison, a stronger man was needed to meet the challenge effectively. Federalists and dubious Republicans resurrected the perennial argument that Virginians and southerners were monopolizing the executive branch. Among Republicans it was no secret that De Witt Clinton was planning to oppose Madison, while Governor Tompkins was openly supporting the President's reelection. These were perilous times for a novice politician to try to win his spurs; and no doubt the uncertain political climate had been one of the primary reasons why Van Buren initially chose not to be a candidate for the state senate. He had no desire to be caught again, as he had been in 1808, between his loyalty to Clinton and his loyalty to the congressional caucus.

Once he received the nomination, however, he campaigned vigorously throughout the Middle District, combining whenever possible legal business with political discussion. His supporters distributed handbills liberally, while the Clintonian press puffed his candidacy. Yet he did not expect to win once it became apparent that the Jenkins, Van Ness and Livingston men had allied themselves with the Federalists, who put up no candidate.[8] Nor could he look to Albany for any support. Governor Tompkins was conspicuously neutral in the contest. Van Buren had accepted what looked like a forlorn hope; yet he fought hard for a post he did not even want at the time because he felt Jenkins and Van Ness had acted in an underhanded fashion toward him. Now in concert with Livingston they were violating what he had come to consider a first principle of politics, party loyalty.

In forming this coalition with the Federalists, paradoxically, they were assisting Van Buren. Livingston would probably win, but only with the help of Federalist votes. Van Buren's reputation would, if anything, be enhanced in defeat; for he could charge his opponents with an utter lack of party principle, a willingness to barter and debase the pure, unalloyed Republican cause merely for place and profit. Far from snuffing out a promising political career, as Jenkins and Van Ness planned, it was they who would be entangled in their own machinations, bolters from the will of the party as expressed by its convention. Van Buren in defeat would stand as the upholder of Republican principles; Van Buren in victory would become instantly not just the fearless defender of the faith, but the undisputed leader of the Middle District—a mighty power in the legislature.

The election was held in the last week in April; and for the next two weeks returns trickled in, the result of the slow process of counting, authenticating, sealing and dispatching the voting results over the wretched roads and forest paths to designated collection points where they were opened, tallied and proclaimed. Returns from the river counties came in first and Livingston carried most of them, though by razor-thin majorities. Van Buren privately conceded that he had lost and made ready to leave for the spring session of the supreme court in New York City during the first week in May.

He gave no sign of concern to his family, except perhaps to be more solicitous than usual toward his wife Hannah, then two months' pregnant, frail and unwell. Yet he seethed inwardly at the laughter and drunken taunts of Livingston's partisans who were celebrating victory in the hotel directly across Warren Street from Van Buren's home. Though he knew that his reputation in the state had not suffered, and though he was pleasantly surprised to have received so many votes, he had to confess to himself that the defeat after all had hurt his pride. He steeled himself and met the revelers with his usual warm smile, his carefully feigned indifference, as he made his way through them to the steamboat for New York. Even on board he was not to be free of Livingston celebrants, but he did not have to put up with their jests for long. As the steamboat passed close to the town of Catskill, Van Buren saw the tall figure of his brother-in-law Moses Cantine among the usual crowd of idlers at the landing. Cantine, who was beyond earshot, gestured excitedly and pointed to a small craft that was moving toward them. When it came up alongside, the boatmen handed Van Buren a letter from Cantine that enclosed a certified copy of the vote in Delaware County. A glance at the figures showed him that his majority in Delaware was much larger than originally reported, large enough to give him a narrow victory of some 200 votes over Livingston in the more than 40,000 votes cast.[9]

The significance of this stunning success, over three distinct factions of his own party and the Federalists, was not lost on Van Buren. He attributed his victory to a lamentable lack of principle in his Republican opponents, to their ill-conceived opportunism, their trafficking with corruptionists, their contempt for the small freeholder. The unexpected outcome reinforced Van Buren's faith in what he was pleased to call correct principles from which one should never waver. That he was popular he was aware; but popularity was a fleeting thing, a mode of the moment that could be cultivated but was of no intrinsic value of itself and was peculiarly vulnerable to the hazards of fortune. One simply did what one could, did not confuse means with ends, prepared one's case for the people with the same care one employed for a judge, let the opposition make the mistakes, promptly recognize them and as promptly profit from them.

Van Buren also grasped at once the implications of his victory to his own career. At thirty years of age, he was suddenly catapulted to a position of equality among the leaders of his party, Clinton, Spencer and Tompkins, all college-bred, all proud and able men and all a generation older than he in years and in experience. But his position was precarious. One false step on these slippery heights could bring him down faster and farther than he had climbed. And almost at once he was put to the test, faced with a problem that if not met squarely yet adroitly, would have made a mockery of his professed principles.

Since 1812 was a presidential year, the New York legislature met in special session in November 2 to choose electors. Despite Madison's renomination by the congressional Republicans in caucus, Clinton went ahead with his plans to contest his party's nominee. Congress had declared war on Great Britain; and before Van Buren took his seat, Republican members of the legislature had selected Clinton for the Presidency.

As a state senator, Van Buren would be immediately faced with the dilemma of supporting either his party's candidate at home or that of the caucus in Washington. The decision was made doubly difficult for him because New York would be a major theater of operations, and he knew political unity was absolutely essential to the successful prosecution of the war.[10]

With these thoughts in mind, on the crisp morning of November 2, 1812, Van Buren walked from his headquarters at Baird's, an inexpensive boarding house on South Market Street, to the capitol. This stately building, which doubled as a meeting hall for the Albany city government, was only three years old. The senate chamber where he went to be sworn in and assigned a seat was a large, well-proportioned room on the second floor. There was a gallery for visitors, supported by rectangular columns with Ionic capitals. Tall windows were set off with velvet draperies and to the left of the senate president's platform was a huge oil painting of John Jay. Van Buren walked with his other colleagues through double doors, down the central aisle that divided the desks arranged in a semicircular pattern, to the low dais on which stood the irrascible Lt. Governor John Taylor, flanked by the judges of the supreme court.[11] After the swearing-in ceremony, Taylor assigned Van Buren and other newly elected senators their desks and called the session to order. Then, according to routine, the senate adjourned for one day. Republican senators and assemblymen separated from their Federalist colleagues, and both sides went into caucus behind closed doors.

Van Buren thought the Clinton nomination was inexpedient, but he was determined to vote for him. To his party colleagues he expressed his doubts that Clinton would win in the presidential election and his very decided feelings that whoever won, he would support his policies. Nor did he hesitate to voice his conviction that there must be no bargains with the Federalists. Clinton must run as a Republican on his own. If Federalists wanted to support him that was their privilege. All present agreed with Van Buren's views on avoiding any coalition with the Federalists, but not all with his avowed support of Clinton. They could do nothing, however, since the Clintonians had a majority over the Federalists and those electors favoring Madison. In the lower house the Federalists had a similar majority over the Madisonians and Clintonians.

At the conclusion of his remarks, Van Buren moved that Lt. Governor Taylor, a Madisonian, be made chairman of the caucus. When this carried, he proposed that the minority of Madisonians be given electors proportional to their strength on a joint ballot. After heated debate, during which it became evident that no compromise could be reached, Van Buren moved that the caucus bind itself to an exclusively Clintonian ticket. Over heated protests a Clinton ticket was rammed through, the angry Madisonians casting blanks. On the following day at the joint session in the chamber of the lower house, the Clintonians repeated their caucus performance. Republicans of New York State had to vote for De Witt Clinton or not vote at all. The Federalists offered no candidate.[12]

Immediately after the slate had been chosen, the legislature adjourned. Van Buren returned to his room at Baird's and made ready to catch the stage for Hudson. He was not surprised, however, when a servant announced De Witt Clinton; for it was a fair assumption that his conversations and the very important role he played in the session had been relayed to Clinton. As he expected, Clinton wanted to know why he thought Madison would win the election. Van Buren explained that the war spirit tied to the sectional pride of voters in the South and West would reelect Madison, who, in addition, would be expected to utilize the patronage at his disposal to reinforce potential waverers. Though the Federalists had no candidate, Van Buren did not believe that Clinton could count on their support; and even if they did vote purely on sectional prejudice, he did not believe they would bring out enough voters to make much of a difference unless there was an understanding, which, he reminded Clinton, the caucus had expressly forbidden.

Nettled, the self-confident Clinton disagreed and showed Van Buren some calculations a noted politician had made that proved he would win. As Van Buren recalled, "that did not change my opinion." During their rather tense conversation, Van Buren had made it plain that he had supported the electoral ticket for Clinton because he felt bound by the caucus action taken the previous spring.[13]

Had he known at the time how deeply involved Clinton was in trying to secure Federalist support; how he was willing to modify Jeffersonian-Republican principles to meet Federalist objections; how he would even head a peace movement; Van Buren would never have acted as the disinterested partisan in presenting him with New York's electoral votes.

What Van Buren did know was that Ambrose Spencer, now a justice of the state supreme court, Lt. Governor Taylor, Governor Tompkins and former U.S. Senator John Armstrong were close to an open break with Clinton. Men like Spencer and Tompkins held personal grudges against Clinton because of his neutrality in their unsuccessful struggle against the Bank of America, which had received its charter the previous May.[14] But now that war with Britain had come, they felt strongly that Clinton should not divide the party and possibly the nation by turning against President Madison. Considering the power of these individuals, Van Buren's management of the caucus had been a dazzling display of political virtuosity.

Rufus King, leader of the Federalists in the state, had early decided that his party should rebuff Clinton's overtures. Clinton was like any leader of a faction, King observed, "so long as he went on according to their views so long he wd. lead them, but as soon as he opposed their views, and more certainly so soon as he united with the rival Faction for any purpose he would be deserted by his own."

King was as anxious to preserve the identity and the principles of the Federalist party as Van Buren was to maintain those of the Republicans. In his opinion, Clinton's amalgamation tactics would never stand up against the leaders of his own party. He had not reckoned with Van Buren, an unknown quantity at the time; nor had he thought that so many battlefield disasters in quick succession should cast a relentless light upon the shortcomings of Madison's Administration.[15]

The claims of Clinton and his supporters that the President lacked the energy and the ability to conduct the war were proving correct. And whatever his numerous enemies might say about Clinton's duplicity, his chilly arrogance, his ruthless use of patronage, they all agreed that he was courageous, a worthy and able leader. In contrast to the little wizened man in the White House stood De Witt Clinton in all his majesty, the Magnus Apollo, whose administrative and executive talents had been on open display for a dozen years. Public opinion, which seemed to be setting against him at the onset of the War, was on the rise in his favor.

Van Buren had the sense to exploit this in the Republican caucus, though dubious about the eventual outcome. "The imbecility of Madison is daily more manifest," wrote Rufus King to his friend Christopher Gore, "still his friends and party in general adhere to him." Three days later, on September 12, 1812, King attended a secret caucus of Federalist leaders and over his vehement objections they agreed to give Clinton the electoral votes of all the Federalist states. King did not approve of Madison's administration, "but," as he said to his brother William, "between him and Mr. Clinton for reasons which in my judgement deeply concern the public liberties, I prefer the election of Mr. Madison."[16] King's preference was borne out in the election, though it had been a near thing, closer than he had thought possible, 128 electoral votes for Madison to 89 for Clinton.

The Federalists carried the lower house in the election of 1813 and with it the council of appointment. Hungrily following Clinton's precedent, the new council made a clean sweep of all major Republican officeholders except De Witt Clinton, who remained as mayor of New York. Had there been an understanding after all? Clinton's numerous enemies believed there had, but they could not be sure. A United States senator was to be chosen by the new legislature. The Republicans had a slight majority in joint session and assumed their candidate, James Wilkin, a Clintonian Republican Van Buren had selected as being acceptable to all factions, would be chosen.[17]

No sooner had the legislature convened in January 1813 than it was virtually invaded by Bank of America lobbyists attempting to have its charter amended so that the huge bonus the Bank was obligated to pay the state would be forgiven. Nothing so audacious in bank politics had been attempted since Aaron Burr smuggled a charter for the Bank of Manhattan through the legislature under the guise of forming a water company for the City of New York in 1799.[18]

Van Buren, though elected to the senate, was not yet a member when the legislature granted the Bank charter. But he was a member, and indeed one of the leading members, when the Bank of America sought this revision in its charter. The issue was joined on political lines. Federalist members were solidly for revision; Republicans generally opposed. In addition, Van Buren viewed the issue as a moral one, with the possibility of fraud—a conspiracy to alter a contract before it had been entered into. Clinton remained silent, a posture that increased Van Buren's suspicions; for a sizable group of legislators who he knew were partisans of the Bank were pledging their loyalty to Clinton. From sources he was bound to respect, Van Buren learned that the Federalists were confident King would be elected Senator. Now the connection seemed too obvious. Van Buren felt that he must see Clinton, who besides retaining his office of mayor of New York City had been elected lieutenant governor the previous year, and lay before him what appeared to be a concerted attempt by some of his supporters to vote for King and the amendment to the charter.

In his most portentous manner, Clinton gravely agreed that if Van Buren's inferences were correct then indeed the matter bordered on the criminal. But he could not believe that such a base conspiracy existed. He had been originally opposed to the Bank and was now opposed to a change in its charter. He maintained stoutly that since he was not a member of the legislature, he had no sanction to interfere, but as mayor of New York City, the Bank would be of immense help to the commercial and the business community. He had a public duty to his constituents whether Federal or Republican to advance their interests. However personally he felt about the Bank, as a mayor he must not take sides. Solemnly, Clinton promised he would look into the charge that some of his Republican supporters had pledged they would back King for the Senate in return for Federalist support of himself for the Presidency. Clinton spoke so earnestly, so openly, so sincerely, that Van Buren felt chastened. As he left the lieutenant governor's rooms at Cruttenden's Hotel, he regretted that he brought what after all was simply hearsay evidence that tended to impeach Clinton's integrity.

The next day the legislature met in joint session. Ruggles Hubbard, one of Clinton's friends and a supporter of the Bank, came to

Van Buren's desk. Would Van Buren write Wilkin's name on his ballot and accompany him to the ballot box where he would deposit it? Hubbard had been one of the men who Van Buren suspected of playing a double game. Reasonably sure that Clinton had inspired Hubbard's action to show his good faith, an embarrassed Van Buren disclaimed any need for such an open testimonial of party regularity. Hubbard persisted, leaving Van Buren no choice but to agree without being rude, and that he could never be. He did what was asked, then confidently he waited for the announcement that Wilkin had won. The tellers finished their count and the result—King was elected to the Senate by a majority of three votes. Van Buren was shocked. An elaborate charade had been played upon him. He had known that Hubbard was closely associated with the Bank Clintonians. Now it was clear that Hubbard's vote was not needed, hence his elaborate pretense, for that was what Van Buren immediately and correctly decided was part of the scheme. When many of his more gullible colleagues attributed the result to the three or four "Quids" (followers of Morgan Lewis) in the legislature attempting to discredit Clinton, Van Buren was certain that this rumor too was deliberate, a mere smokescreen.[19]

Even though Clinton's editor, Solomon Southwick of *The Albany Register,* poured editorial outrage on the traitorous Republicans who had voted for King, Van Buren concluded that Clinton was directly responsible. What contempt Clinton must have for him to play such a clumsy trick; what a consummate liar Clinton must be; what an ingrate to have scuttled his faithful supporter, Wilkin, who had presided over the caucus that nominated him for President. Van Buren, of course, said nothing. He was his usual amiable self when Clinton appeared at the senate chamber after the election and sought him out. "I hope," said Clinton, "you no longer entertain the suspicions you spoke of." Vexed at what he felt was Clinton's effrontery, Van Buren replied in measured terms, "Mr. Clinton, you misunderstand me. My suspicions have become convictions. I know that the men I pointed out to you have done this deed." An angry Clinton, implying personal prejudice, replied that Van Buren was wrong.

After this brief encounter, Van Buren did not speak with Clinton again for more than three weeks. On the afternoon of March 4, 1813, they had a few words together. The Republican caucus was to meet that evening. Van Buren told Clinton that he thought the caucus would seek to nominate Taylor for lieutenant governor in his place. He intended to name Clinton, but with such strong party leaders as Governor Tompkins and Supreme Court Justice Ambrose Spencer openly opposed to him, he was reasonably certain Taylor would be the choice of the caucus. Would Van Buren support Taylor in this event, was Clinton's query. "Certainly," said Van Buren, "if it is fairly made." The massive lieutenant governor, so accustomed to deference from his political subordinates, stared grimly at this impertinent young dandy for a moment or two, gave a slight bow and made his stately way to the presiding officer's chair.

Van Buren was concerned. He was certain that Clinton had duped him and betrayed the party for personal ends, but he feared a party rupture if the lieutenant governor was put down harshly. A war was being waged, and the American military, badly led and badly organized, had suffered reverse after reverse. New York State seemed about to be invaded by a large, well-trained army of British regulars. New York City lay exposed and defenseless to the overwhelming power of the Royal Navy. A separatist movement was gaining momentum in New England, and some Federalists like Gouverneur Morris were said to be trying to align New York with the New England secessionists. The national government, as Clinton had predicted so many times, was floundering in a morass of confusion. This, then, was no time for the Republican party to break up over real or imagined injuries. Nor was there any doubt even among his host of enemies about Clinton's patriotism or his great executive talents.[20]

At the Republican caucus that evening Van Buren was relieved that there was no disposition to humiliate Clinton, though a majority was opposed to his renomination as lieutenant governor. Van Buren saw the opportunity and promptly seized it. A motion had been made and seconded to name Tompkins for governor and John Taylor for lieutenant governor. No one, not even Erastus Root, bitter enemy of Clinton, rose to speak on the motion. After an awkward silence, Van Buren gained the floor and in his rapid, nervous manner of public speaking, alluded to the unsettled condition of the party following the recent election for senator. He shared that feeling, but considering the condition of the nation and of the state, considering also Clinton's long record of valuable public service, he thought it would be inexpedient to cast him off. Now for his clinching argument, Van Buren said that he knew Clinton would never accept the nomination unless he was sincere in acting with the party. Perhaps he had stumbled in the Senate election; perhaps his partisans had acted without his concurrence. In any event, the party was powerful and had always been magnanimous. He hoped the caucus would sustain him in substituting Clinton's name for Taylor's, but whatever it decided, he would support the nominee.

Van Buren's speech electrified the caucus. Members now saw the course to be pursued, the senator from the Middle District had shown the way. Other speakers followed Van Buren, some friendly to Clinton, others opponents. Party harmony and party discipline, Van

Buren's line, was the theme of their remarks. When the vote was finally taken, Taylor beat Clinton by a 16-vote plurality. The caucus was unanimous for Tompkins.[21]

Solomon Southwick, in the *Register,* after indulging in some acrimonious attacks on various well-known critics of Clinton, turned to other topics; and though he did not carry the Tompkins-Taylor slate in his paper, neither did he dwell on Republican divisiveness. Like Tompkins, Southwick was a handsome, friendly man with a common touch, treating starchy William James, Albany's only millionaire, in the same seemingly open and friendly manner he would the shabby, penniless printer's devil, Thurlow Weed. A natural promoter, a gifted persuader with that invaluable quality in any politician of rarely forgetting a name once he had met the individual, unfortunately, his character left something to be desired. His close connection with the Bank of America, the easily traceable bribes he dispensed to purchase legislators, was already injuring him severely; but he enjoyed the confidence of Clinton. The *Register,* in failing to mention Van Buren while it briefly castigated Tompkins, Taylor, Root and Spencer, meant that Clinton, too, had not cast away Martin Van Buren from his charmed circle. But Van Buren had already decided on a parting of the ways.

He harbored lingering doubts that Clinton was actually intriguing with the Federalists, but he found it hard to believe that he was completely innocent. Though not a resident of Albany, he lent his support to the establishment of a rival paper to Southwick's *Register.* Jesse Buel, for several years editor of the reliable Republican *Kingston Plebeian,* was brought to Albany for this purpose. The first issue of the new paper calling itself the *Albany Argus* appeared on Tuesday, January 26, 1813.[22] A triumvirate consisting of Van Buren, Spencer and Tompkins now ruled the Republican party, its avowed objective the overthrow of De Witt Clinton.[23]

« CHAPTER 3 »

War to the End

On a blustery evening in early February 1814, Van Buren sat up late drafting a document. Paper spilled over his table as he wrote rapidly in a sprawling hand. From time to time he paused, arranged the sheets in order and, sounding out phrases, crossed out sentences, substituted new ones, cut, added, polished, so that the pages were scarcely legible except to his own practiced eye. Careless of punctuation, uncertain in grammar and spelling, but with an ear for a ringing sentence, a startling expression and a continuity in argument, the manuscript had grown over the past several days to 3000 words, some sixty-four pages.[1] Van Buren was finishing the annual address to the Republican caucus, a vitally important political document. Conscious of the honor it conferred upon its author and the power it bestowed if the address were well and carefully done, he rewrote certain sections several times until he felt he had achieved just the right tone or emphasis. For he knew the annual charge to the party was not just a political guideline, or even a campaign platform, it must also serve other ends. Whether on the attack or on the defense, the language must inspire, must appeal to the emotions as well as to the reason of men whose ideas differed widely on public questions.

Others, Judge Spencer and Governor Tompkins certainly, would look over the draft when he had had it copied. Before he submitted the manuscript to those learned worthies, he would have close friends smooth over the rough edges, correct the punctuation and the spelling. But the direction of the argument, the telling phrases and comments that would soon be reechoed by friends, jeered at by foes, circulated in the party press, attacked by opposition journals, these were Van Buren's own.

The young senator had discovered to his own surprise when asked to write the legislative response to the governor's message the previous year (also a signal honor for a first-term senator) that he had a knack for an arresting phrase. His colloquial style, when polished, may have been devoid of literary merit, but it had a direct familiar quality better suited to a broad audience than the more learned, more ponderous prose of a De Witt Clinton or a Rufus King.

The manuscript, as he reread it that February night, satisfied him that he had made the points he had wanted to make. Not quite a manifesto, it was a ringing call to arms and a slashing attack on the Federalists, who were chastised for their unpatriotic behavior. Courageous and defiant, Van Buren called upon all Republicans and all patriotic citizens "to support with vigor this second struggle for freedom and independence." When he finished tinkering with the draft he felt certain he had made as strong a case as possible to his fellow New Yorkers that the War was a just one, that it had been forced upon the country and that it must be fought through to a successful conclusion.[2] As a political paper, he had aimed to put the Federalists on the defensive and he thought he had accomplished this goal. Of the public response he could not be sure. Would the argument convince the reading public? Would the rhetoric have any impact beyond party circles?

Republicans, divided among themselves, had but a slender hold on the state, the governorship and a small majority in the senate. Federalists controlled the lower house and the all-powerful council of appointment. The conduct of the war, thus far so lamentable, must have an impact on the election. Only two months earlier the British and their Indian allies had attacked all along the far western borders of New York, scattering the untrained militia that opposed them. Burning and pillaging as they went, they drove frontier refugee families in headlong flight as far south and east as Canandaigua and Geneva, 200 miles west of Albany. Had the conduct of the War damaged public morale to such an extent that the defeatist policies of the extreme Federalists so evident in New England crossed over the borders invading the mind as the British were invading the body of the state? Were the New York Federalists as separatist-minded as the Pickerings, the Daggetts and others of their ilk in Connecticut and Massachusetts?

Yet to the more perceptive among Republican leaders, not all Federalists, even in New England, agreed with the extremists, excepting always disgruntled old cynics like Gouverneur Morris. The Federalist leaders were less narrowly sectional than the leaders in New England. After all, those bulwarks of Federalism, Generals Stephen and Solomon Van Rensselaer, were loyal to the Union and had proven their loyalty on the battlefield; though that was about all. Stephen, who retained his ancient Dutch title of Patroon, had been a singularly inept commander among a veritable school of inept commanders. His cousin Solomon, brave to foolhardiness, had shown little or no military sense at Queenstown Heights. Rufus King, leader of his party in the state, had not spoken out on the War.[3]

As if all these political uncertainties were not worrisome enough, Van Buren was tired and concerned about his family. Hannah was pregnant again. He was acting as an associate prosecutor in the court-martial trial of William Hull, who had surrendered Detroit to an inferior force of British and Indians without firing a shot.[4] The legislative session was not half over and here he was almost single-handedly trying to cope with Federalist obstructionist tactics in the legislature, striving to energize Republicanism while worrying about Clinton's efforts to regain power at the expense of the party. He was uneasy, too, about Spencer's motives and Tompkins's presidential ambitions.

But at the moment the triumvirate was intact, working hard to sustain the national government and to mobilize the resources of the state at least in defense of its borders. Tompkins had taken hold of the state's military forces. He had appointed a former schoolmaster from Pennsylvania, Jacob Brown, to succeed Van Rensselaer as commander of the state's forces. Though untrained, Brown quickly demonstrated that he had a firm grasp of strategy coupled with organizational skills of the first order. Working with Tompkins, the new commander, within the legislative constraints imposed upon him, had made a promising start, equipping and drilling the militia.

The masterful Clinton was organizing the material resources of New York City and supplying most of the funds for this new war effort.[5] He had borrowed on his own signature well over a million dollars in hard cash, much of it from Federalist bankers. Through the efforts of the new secretary of war, John Armstrong, Brown had been commissioned a major general in the national army replacing the incompetent James Wilkinson on the northern border. Neither Tompkins nor Van Buren had much use for Armstrong, whom they regarded as an untrustworthy politician, close to Ambrose Spencer, a personal friend. Both were anxious to remove Armstrong from New York politics, and were satisfied that he was superior to the old time-server he replaced as head of the War Department, William Eustis, whose sole qualifications for the job were his experiences as a surgeon in the Revolutionary War.

At first Armstrong justified their confidence. Besides making Brown the commander of the northern army, the new secretary had the foresight to recognize the military talents of Winfield Scott. A lieutenant colonel in the United States army, Scott had attracted attention by his actions at the ill-fated battle of Queenstown Heights where he was taken prisoner. After his exchange Madison, at Armstrong's behest, nominated him a brigadier general. Before his Senate confirmation Scott had gone north to serve under Brown. "At the dark period of the war," recalled Scott, "Albany rather than Washington was the watchtower of the nation." The preparations going on in New York for a sustained conflict had impressed the young brigadier.

Van Buren, like Armstrong, was attracted to Scott; and the two men dined together frequently while Scott waited in Albany for his commission and his orders from Washington. When Hannah gave birth to another son, Van Buren named him Winfield Scott after the young general. The baby, Van Buren's fourth child, lived only for a few weeks.

However impressive Brown and Scott might have seemed to Van Buren, however vigorous Tompkins, Spencer and Clinton were in organizing and sustaining the war, he sensed that the Union had reached a critical point. Should there be further defeats on the frontier, followed by another defeat on the political front, who could say what the people of New York might do?[6]

The Republicans were fortunate that Sir James Yeo, the British commander on Lake Ontario, had been delayed a month in his successful attack on Oswego, a major American supply depot in western New York. For the fall of Oswego before the election would have certainly made a considerable difference in the outcome. As it was, Van Buren had his doubts whether the party could hold the senate, let alone gain the house, from the Federalists. He was agreeably surprised when the election returns came in.

The Republicans swept the state. Even New York City, where the Republican party was divided between Clinton's enemies, the Martling men—who had not forgotten how he had engineered Burr's defeat—and his equally vociferous defenders, the Federalists, did not elect one candidate to the house. The Republicans now had a decided majority in the house for the first time in three years, while all but one Federalist candidate for the senate was defeated.

There would be no more of those wearing conference clashes with Van Buren shouldering the major responsibility of maintaining Tompkins's war policies against an intransigent Federalist house. Van Buren may have gained public applause "by the vigor of his logic, his acuteness and dexterity in debate and in patriotic sentiments" he expressed, but it had been a fatiguing, thankless business. He and his rough, outspoken associate in the house, Erastus Root, frequently had to compromise on points that they deemed indispensable to the war effort. There was no doubt, however, that these well-publicized forays helped the Republican cause.

Nor should the value of Van Buren's "Address" as a campaign document be underestimated; but probably the most important factor was the organization of public war meetings all over the state. For these mass turnouts the Republicans mobilized their best orators, who made fiery emotional appeals to the patriotism of their audiences. Van Buren himself made an effective speech at a mass meeting in Albany on April 14, closing with the remark that Republicans must be "up and doing." As usual he had carefully outlined what he wanted to say in six pages of notes and then virtually memorized his speech.[7]

Scarcely had the harried Republican leaders finished celebrating their unexpected success in the election when they received further good news of victories along the northern and western borders. From June through September 1814, Brown and Scott won a series of little engagements, capturing Fort Erie, which had long menaced Buffalo, defeating the British in the battle of Chippewa, and fighting the British regulars to a standstill at Lundy's Lane. Thomas MacDonough, the pious Delaware-born commander of a small American squadron, utterly destroyed a superior British force on Lake Champlain, thereby securing Albany and the heart of the state from imminent attack. Relief and jubilation at the deliverance was quickly dashed as an even greater peril loomed.

The overworked triumvirate learned that a British fleet on the Chesapeake had landed a force on the Maryland shore. These troops had promptly scattered the hastily organized militia of Virginia and Maryland at Bladensburg, a village on the road between Baltimore and Washington. Van Buren was shocked when he learned of the rout at Bladensburg, the capture and burning of the capital city.[8] Even General Samuel Smith's successful defense of Baltimore gave scant comfort to the harassed New Yorkers. Everyone believed New York City would be next. Word had arrived that large units of the Royal Navy were at sea convoying British troops released by the surrender of France to the allies and the abdication of Napoleon. All the banks in the state suspended specie payments, bringing economic activity in the cities and towns to a standstill.

Governor Tompkins acted with dispatch. He summoned a special session of the legislature to meet on September 26, 1814. De Witt Clinton, suppressing his antipathy toward the triumvirate, for a second time in as many months managed to raise funds for the government and to instill a fighting spirit in the quaking business community. He would have been less successful in his efforts had it not been for the assistance of Rufus King, who cast off his mantle of silence and boldly called for energetic measures.

Yet the banks would not lend money on United States Treasury notes unless the governor endorsed them as the agent of the state. Tompkins hesitated in the crisis, fearing he had no legal authority to commit the state merely on his own authorization. Learning of his qualms, King visited Tompkins to express his complete support for any measures deemed necessary in the emergency. When the civilities were over, King came straight to the point. He told Tompkins that everyone must pledge his all to maintain the government. Tompkins replied that he would have to act on his own responsibility, assume powers that might well ruin him financially, as well as open him up to prosecution for exceeding his authority. "Ruin yourself if it becomes necessary to save your country, and I pledge you my honor that I will support you in whatever you do," said King.[9] Van Buren and Spencer had already urged Tompkins to adopt the same course.

With this assurance from the Federalist chief, one of the few leading figures of the Revolutionary generation still active in politics and government, Tompkins acted with dispatch. He endorsed the notes, which unlocked the coffers of New York's banking institutions. Ample funds were now available to support the war measures that Van Buren was even then drafting for the special session of the legislature. The proposed legislation was nothing less than the mobilization of the Empire State.[10]

Van Buren's first effort was the reply of the senate to Governor Tompkins's address to the special session. A forceful indictment of the enemy coupled with praise for American successes, it called for a united effort to secure victory and pledged the senate to support the governor's policies. More in sorrow than in anger, Van Buren alluded to the high Federalists whom he knew were considering secession from the Union. He did not mention them by party label but, reminding them of their Revolutionary forebears, urged them to avoid "criminal and unprofitable collisions."

No sooner had the senate voted to print the address, than Van Buren introduced a bill that would draft as a first installment 12,000 men from the militia. He had written the bill himself after consulting with his party associates in the legislature and with Aaron Burr, who made useful suggestions drawn from his military experience.[11]

The proposed legislation, known as the "Classification Bill," was bound to be unpopular, especially in those regions far removed from the conflict. Well aware that it might destroy him politically, might even ruin his prosperous law practice, he had not flinched from acknowledging his authorship or his defense of it on the floor of the senate. The bill came under immediate attack from men of means and from the Federalists generally. If Van Buren was not stigmatized as an evil man for writing the bill, he was thought a fool to sponsor it in the legislature. An old adversary, Elisha Williams, accosted him on the steps of the capitol. "Van Buren," he said, "my federal friends are such [damned] fools as to believe that you are in earnest with your conscription bill, and mean to carry it through, and I cannot convince them to the contrary." Van Buren promptly assured Williams that his friends were correct, momentarily astonishing one of the state's leading Federalists and an outspoken critic of the War.[12]

Never before had Van Buren's courage and his conviction of the right been more clearly displayed than in his battle for the "Classification Bill." At the outset his faithful associate in so many encounters with the Federalists, Erastus Root, deserted him. Root attacked Van Buren and his bill in his biting, uncompromising style. For a time Root's opposition jeopardized passage until, surprisingly, enough Federalists jumped party lines to offset Republican defections. Although the "Classification Bill" was the major accomplishment of the special session, additional legislation raising the pay of the militia, providing for a brigade of marines to serve on the waterways and harbors of the state and authorizing privateering under state auspices were driven through a legislature where the opposition was strong, its spokesman able and persistent.

By the end of 1814 Van Buren was acknowledged as one of the War leaders of the state. Largely due to his skill, his powers of persuasion and his untiring industry, New York had a comprehensive military organization ready for rapid implementation. In decided contrast were the New England states: a few weeks after the special session adjourned, representatives of Massachusetts, New Hampshire, Vermont, Connecticut and Rhode Island met at Hartford and debated whether they should seek a separate peace with Great Britain.

President Madison, his reputation at its lowest ebb, fearful that the Peace Commission he had sent abroad to treat with the British would fail, even more fearful that the Hartford Convention would take New England out of the War and perhaps the Union, was heartened by the vigorous support from Albany. The Empire State would stand by the Union. Republicans and most of the Federalists were opposing any extremist measures, and without New York the idea of an independent northern confederation lost much of its credibility.

Assuming the war program was the result of Tompkins's leadership, President Madison wanted such a strong and popular man near him in the highest levels of government. After the ludicrous affair at Bladensburg Secretary of War Armstrong resigned in disgrace. Madison persuaded Monroe, his secretary of state, to take over the War Department. The President then offered the State Department to Tompkins, a generous recognition of his worth as the nation's greatest War Governor. Madison's offer was tempting to the ambitious Tompkins; but he decided to remain where he was, calculating wrongly that the most direct route to the Presidency lay through Albany, not Washington. Like so many public men of his ardent temperament, he had visions of martial glory. As commander in chief of the state's military and naval forces, he would be the one personally to defend New York from an expected invasion of battle-seasoned British regulars in the spring of 1815 and then lead the offensive against Canada.[13]

In refusing Madison's offer, Tompkins made the greatest mistake of his long and successful political career. There is no doubt that his charm and his very real ability would have won over the impressionable President. Monroe, remote and laconic, would have been buried in the War Department with no war to wage. Just how wrong Tompkins had been became apparent a scant two months after his declination. On February 12, 1815, word reached Albany that a peace treaty had been signed between Great Britain and the United States at Ghent, Belgium, on December 24, 1814. Tompkins's dreams of martial glory had already received a rude shock a week earlier when Andrew Jackson's victory at New Orleans electrified the nation, producing an instant hero. Jackson's feat would certainly deter, if not completely alter, any full-scale attack on New York.[14]

A more direct weakening of Tompkins's position, and that of Van Buren too, was the abrupt change in the attitude of that master of intrigue, Ambrose Spencer. With the constraints of war now removed, the triumvirate showed signs of breaking up. Van Buren remained loyal to Tompkins and supported his presidential aspirations, but Spencer, who had ambitions of his own and had long been jealous of the Governor's popularity, would not sit idly by while a rival claimed all the power and the honors. Tompkins had had glory enough; it was time for him to step down. Before he made any move, the Judge had to test the relationship between Tompkins and Van Buren. Just how strong was it? Was there jealousy? And if any existed, how was it to be exploited? Where party principles had been little more than a convenient means of manipulating men to do his will, Van Buren was a potential rival of admitted skill, too independent for Spencer's personal plans.

Confident of his power, Spencer was determined that his old friend John Armstrong be elected to the United States Senate. Armstrong was still under a cloud because of his presumed mismanagement of the War Department, and he was unpopular with the party rank and file. On both scores he was completely unacceptable to Van Buren and Tompkins.

When Spencer asked Van Buren to push Armstrong through the

Republican caucus, Van Buren replied emphatically he would not do so. Unaccustomed to such a sharp rebuff, Spencer demanded to know who Van Buren's candidate was. Van Buren suggested that the judge himself be the candidate, but the imperious Spencer would have none of it. At the peak of his political power in New York State, planning to bring down Van Buren and Tompkins in due course, he was not about to have himself shelved in Washington. Van Buren then proposed Nathan Sanford, who he thought would make a good candidate. Spencer bridled. Sanford was the leader of the Martling men in New York City and could not be spared because he was an important curb on the political ambitions of De Witt Clinton. Van Buren had not bothered to assess Sanford in terms of New York City politics. He had thought of him as an upright Republican politician and a longtime resident of the Southern District, which had not been represented in the Senate since Clinton's short term in 1802. As far as Van Buren could see, Sanford would be popular in the legislature, a factor of no little importance to Van Buren's continued leadership. Sanford was duly elected to the Senate, though not before Spencer had speeded up his timetable for disposing of Van Buren and Tompkins.

Spencer saw to it that two of his henchmen were elected members of the council of appointment. Van Buren was a candidate for attorney general. Spencer's candidate was John Woodworth, many years Van Buren's senior, a prominent and capable lawyer and a well-seasoned Republican politician.

The council of appointment met on February 17, 1815, amid the general euphoria of peace. John Woodworth and Van Buren each received two votes, forcing Governor Tompkins to break the tie. Tompkins did so reluctantly because a vote for Van Buren was a direct challenge to Spencer; and Tompkins was not the kind of politician who thrived on controversy. He wished to avoid any trouble with the powerful judge; nor did he wish to alienate Van Buren, who had earned his respect and with whom he enjoyed an easy rapport. On more practical grounds, he could not afford to sacrifice this able supporter on whom he relied for steering administration policy so deftly through the rocks and shoals of the legislature. A worried Tompkins broke the tie by voting for Van Buren and hoped for the best.[15]

Van Buren knew his tenure as attorney general was uncertain, that another turn of the political wheel could spin him out. Still, he felt confident the Republican party would be in the majority for some years to come. Already the public was charging the Federalists with subversion and pointing to the Hartford Convention as a disloyal conspiracy. A minority party in the state since 1800, the occasional triumph of the Federalists at the polls had been more the result of factional strife among the Republicans than any sustained voter appeal.

In any event, Van Buren felt secure enough in his practice to move from Hudson to Albany. As Hudson had promised greater opportunities for himself and for his family than Kinderhook, so the bustling, growing capital now opened up more promising fields to cultivate, greater benefits to reap. For his family, too, Albany offered a more comfortable, more cosmopolitan environment. He was determined that his three young sons (a fourth had died in infancy) would have the best education then available in the state. Only New York City could boast better schools than Albany's, but New York City was out of the question. Prone to epidemics of cholera, typhoid and typhus, the great city was too dirty and too dangerous. He would never expose his children or his wife to the perils of the metropolis. His practice, still largely a country one, would not benefit from such a change, and the new post of attorney general demanded that much more time be spent in Albany and other upstate towns than in the City. As for politics, Albany was after all the capital and therefore the pivot on which the parties revolved.

Van Buren's financial affairs had improved substantially since his move to Hudson. His practice was flourishing to such an extent that he was able to conduct a profitable business in small loans and to speculate in state lands. He was a small operator when compared with Joseph Ellicott and others of the great Holland Land Company, Jonas Platt, a Federalist politician and one of the judges of the Supreme Court, or Peter B. Porter, war hawk Congressman, and one of Van Buren's links with politics in western New York and in Washington. Yet as a prominent politician and lawyer, he was privy to information on tax sales not easily accessible to the general public. For the past three years he had been picking up parcels of good land which because of location in prime areas of westward migration he felt were bound to appreciate.

As befit the attorney general and, of course, reflecting his easier circumstances, he rented a house on State Street large enough for him to maintain his law offices in one of the parlors off the spacious entryway. Besides Hannah, the children and three servants, the household included the twenty-one-year-old Benjamin F. Butler, who had been a member of the family for the past four years.

Butler's father, Medad, an early political supporter of Van Buren, had prevailed upon him to accept his son as a law clerk in 1811. The shy, sharp-featured, unobtrusive young man quickly won the affection of Van Buren and his wife. Industrious, thoughtful, with a precise analytical mind, he soon displayed a talent for the law that equaled that of his teacher. Under Van Buren's careful tutelage, Butler soon overcame his initial diffidence and while still a student began to develop latent powers for public speaking and for debate. Van Buren shared his political plans with Butler and encouraged the younger man to criticize them with candor. He did not always agree with Butler's comments, though he invariably profited from their discussions.

Butler had just completed his four-year clerkship at the time Van Buren was appointed attorney general and decided on the move to Albany. He offered Butler a full partnership and a room in his new Albany home, generous by any standards, particularly so to a young man who had just reached his majority and had just been admitted to the bar. Now that Van Buren had become one of the major public and political figures in the state, he had to have a confidential partner and he was absolutely certain of Butler's ability, discretion and loyalty. Butler accepted with alacrity, and for a time Van Buren had more leisure at his disposal for political affairs.[16]

Meanwhile, the struggle for control of the Republican party continued, with Spencer the aggressor and Van Buren and Tompkins on the defensive. In an effort to improve his position, Van Buren sought to have Samuel Young, one of his supporters, made secretary of state, a profitable post that carried with it considerable patronage. But Spencer had so arranged the council that it rejected Young. The Judge's candidate was Elisha Jenkins, a party hack from Hudson who, with William P. Van Ness, had intrigued against Van Buren when he first ran for the state senate. Though he anticipated countermeasures from Spencer, Van Buren had not thought he would be subjected to public humiliation besides. Jenkins was not just a weak man but one of the few Republicans in the state whose enmity for Van Buren was so vocal, so obsessive, that he had become a laughing stock in political circles.

When Van Buren heard what was proposed, he immediately saw the move as more than a power play, rather a concerted attempt to wound and to ridicule him. Briefly he was perplexed and then saw his way out of the procrustean bed Spencer had prepared for him. As one of the members of the council friendly to him was leaving for the meeting, he asked him to nominate Peter B. Porter of Buffalo in place of Young. Next to Tompkins, Porter was the most popular politician in the state. A wealthy landowner and developer of western New York, he had been an active and successful general officer in the War of 1812. Then serving his first term in Congress, he would obliterate Jenkins in any contest. Van Buren had turned the tables on Spencer. Neither the council nor the Judge himself dared oppose Porter with the likes of Jenkins.

Van Buren could not be sure that Porter would accept the post, and if he did, what political direction he would take. Temporarily his appointment would block Spencer and gain valuable time for organizing an opposition. Besides, Porter might well accept because he could use the patronage of the office to expand his political orbit and to further his ambitious projects. That he would pursue an independent course seemed the more likely given his prominence and his carefully groomed image as the spokesman of western New York. In Van Buren's opinion he would never knuckle under to Spencer or to Tompkins, for that matter. The only peril, and that seemed remote, was that he would come to an understanding with Clinton, for both men shared the same interests when it came to internal improvements. Van Buren could not have anticipated all these variables in the brief span of time he had to make a decision. Instinctively he knew that Porter was the one individual in the state who could best frustrate Spencer. The unknowns in the equation could not be that harmful to himself.

The primary reason Van Buren had discounted any Clinton connection was because he thought the council would retain the iron-willed politician. Spencer, his monumental dignity injured by his failure to put Van Buren down, had, along with the Martling men, insisted on Clinton's removal. Tompkins and Van Buren, though both had long since broken with Clinton, were most reluctant to play any part in his removal. If they disliked Clinton's cold and arrogant manner and distrusted his politics, they were wary about any public repudiation of a man with his family connections who was highly respected for his intellectual abilities and broad vision. But the vindictive Spencer must have his way; the council removed Clinton. Given their highly publicized differences, Tompkins had no choice but to go along with the majority. Van Buren was more fortunate. He was not a member and could avoid taking sides in the legislature. The deposed Clinton was rusticated to his home in Flushing, where for a time he drank excessively as he pondered what must have seemed the end of his public and political career.

Yet Clinton still retained an escape hatch, the post of canal commissioner to which he had been appointed in 1810.[17] While in retirement, with few distractions, he was able to devote his talents and his energies to his great canal project. Clinton may have seemed politically dead, but his fertile mind was busily engaged in preparing what would be his crowning achievement, a project that would dominate politics for the next decade and would shape the future growth of the Empire State—the Erie Canal.

Van Buren was reelected to the Senate in April 1816 over the covert opposition in his District from Spencer and the resurgent

Federalists who had made substantial gains over the debacle of the previous year because the party had persuaded a reluctant Rufus King to run for Governor. King's unexpected candidacy had forced a change in Republican strategy.

Tompkins had wanted to step down from the governorship and canvass actively for the presidential nomination. The Republicans feared, and rightly so, that the esteem King enjoyed throughout the state would secure his election. More important, he stood a good chance of bringing with him a Federalist majority in the legislature. Van Buren and others of his developing group of young politicians felt that Tompkins, their best man, must run again. But first, Van Buren felt he must see for himself what the Governor's chances were for the Presidency. In mid-December 1815, he visited Washington.

Next to Tompkins, William H. Crawford was the choice of New York, whose ambition for the Presidency had been so many times set aside by Virginia. Though himself a Virginian by birth, Crawford was a citizen of Georgia and he was emphasizing this geographical factor in his campaign. A large, handsome man of commanding presence, Crawford had behind him a long career of public service—U.S. senator, minister to France, secretary of war, and was currently secretary of the treasury. Considered a statesman of great ability, by no less a critic than John Quincy Adams, he had acquired a national reputation.[18] When Jabez Hammond, a representative from Otsego County, nominally a Clintonian acting as spokesman for a group of concerned congressmen, queried Van Buren, he replied, "We support Tompkins, of course." As he did not want to be drawn out further on the question he skillfully switched the conversation to a subject dear to the hearts of the group, gerrymandering congressional districts in the west which would strengthen the party at the expense of the Federalists.[19] Then, he made himself as scarce as possible visiting with congressmen from other states, conferring with Crawford, with Monroe, and visiting the President. His one cryptic comment had been enough, however, to line up the Republican members behind a public declaration of support for Tompkins. "Still," as Hammond observed, "the ways and means to accomplish the object of our wishes have not been exhibited to me . . . unless the Southern interest could be divided, all efforts to procure the nomination of a New York candidate would prove abortive." William W. Bibb, one of Georgia's Senators with whom Van Buren consulted, convinced him that Crawford would not be a candidate.[20]

Returning to Albany, Van Buren spread the word among close friends, and several newspapers announced it. No sooner had this information reached Washington than he received numerous letters controverting what he understood was Crawford's position. Promptly he asked for clarification from Bibb, who somewhat sharply replied that his understanding remained unchanged but emphasized that this was his personal opinion.[21]

Among the letters from Washington were several from an old friend whose opinion he respected, Samuel Betts, a congressman from the Ulster-Sullivan district. Betts had done a thorough job canvassing other delegations and concluded that Tompkins, while highly respected by many, really was not well enough known in the South and West and had no chance in the caucus. The congressman had also communicated with Tompkins, who decided reluctantly to withdraw. He now agreed to be a candidate for reelection to the governorship. In a hotly contested canvass Tompkins defeated King, though his majority was a slim one, and the Republicans had a reduced majority in the senate. The Federalists carried the lower house by one vote.[22]

Van Buren resigned himself to a Federalist council of appointment and the consequent loss of his attorney generalship after only one year. Then he espied a technicality that, if handled expeditiously, would ensure a Republican council. When he had worked out the parliamentary strategy, he quickly seized the initiative, though in doing so he was repudiating one of his fundamental political principles: he was rejecting the admitted will of the voters for partisan advantage. That he was using downright callous if not dishonorable means for political advantage, he knew. His defense—had the opposition been presented with the same situation they would have acted the same way—was at best an unworthy rationalization.

The point at issue involved the election of a member to the lower house from Ontario County. Henry Fellows, the Federalist candidate, was the victor by a majority of 30 votes over his Republican opponent, Peter Allen. In one of the towns, however, where Fellows had a majority, the election inspectors filed with the town clerk a certificate of the returns with Henry Fellows spelled out. On a duplicate certificate they sent to the county clerk as prescribed by law, they had unwittingly abbreviated his name to Hen. Fellows. The clerk promptly threw out the vote of the town, giving Allen a 15-vote majority in the county.

Meanwhile, the legislature had organized and because of the illness of two members, the Republicans managed to elect the speaker by one vote. As soon as Van Buren learned the result of this contest, he hastened to the assembly floor where, brazenly assuming command of the Republicans in the house, he managed to beat back the Federalists and seat Peter Allen until a Republican council had been chosen. While Van Buren was directing operations in an atmosphere charged with emotion, the Federalist leaders fought a hopeless battle as they sought parliamentary means to seat Fellows. Van Buren saw to it that every Republican member was in his seat and voting; even those who were ill were carried in. The Republican speaker instantly ruled the Federalists out of order and on their appeals from his ruling they were defeated by one or two votes.

Peter Allen represented Ontario County only long enough to vote for a Republican council; then the speaker permitted the committee on elections to make its report, which duly seated Henry Fellows. Summing up his role in this disgraceful proceeding, Van Buren said apologetically, "The case was in truth one of those abuses of power to which parties are subject, but which I am sure I could never again be induced to countenance."[23]

Unaware of the Governor's decision to withdraw from the presidential contest, the Republican members of the legislature recommended Tompkins as its candidate and copies of their resolutions were sent to the Republican members of the New York congressional delegation. This expression of party opinion at home had little impact; nor, considering the circumstances, had Van Buren expected it would. The delegation considered the resolutions, but debate centered on whether it was expedient to try forestalling a caucus nomination. Most felt that such a course was impossible at this late date. Their only hope, they finally concluded, was a last ditch effort to combine the Crawford and Tompkins forces in an effort to head off Monroe and make Crawford the nominee. Had Van Buren been in Washington backing Crawford, instead of in Albany, the strategy might have succeeded. But there was no member of the New York delegation or of other delegations for that matter who had the ability and the tact to bring it off.[24]

The caucus nominated Monroe on the first ballot with 60 votes to 54 for Crawford. As so frequently in the past, New York was recognized and presumably soothed after the caucus unanimously nominated Tompkins for the Vice-Presidency. When news of the nomination reached Albany, the *Argus* dutifully fell into line behind Monroe and Tompkins, following the lead of its detested rival, Southwick's *Register,* which had been backing Monroe for the past two months. Tompkins accepted the nomination, hopeful that it would extend his reputation sufficiently so that he could make another bid four years hence.[25]

As for Van Buren, he saw difficulties ahead when Tompkins would be elected Vice-President in the autumn. What candidate would the party put forward for the governorship if the Federalists renominated King? He was beginning to worry about Clinton again because various hints had been thrown out that Spencer might be about to reconcile himself with his brother-in-law.

Madison and Monroe regarded the caucus selection of Tompkins as a proper and indeed honorable assertion of New York's position in the Union. The vice-presidential nomination may not have suited Van Buren's particular plans, but it was a manifest of their good will, a peace offering to New York that enhanced rather than hurt his credibility with the state's delegation in Congress and the party at home. Spencer had the satisfaction of seeing the elimination of one rival in the departure of Tompkins. He still had to contend with Van Buren, however, whose popularity within the party seemed greater than ever.[26]

Spencer must now swallow his pride and bury the hatchet with Clinton, who still had his adherents about the state. As Rufus King wrote Oliver Wolcott, that shrewd observer of the passing scene, "Mr. Clinton has a more considerable party than those who wear his livery." Wolcott was reacting to the enthusiasm Clinton was generating throughout the state for his Canal project.[27] A great mass meeting favorable to Clinton had been held in New York City. This expression of support was amplified by a second mass meeting held in Albany, where leading members of both parties extolled Clinton and what had now become almost his personal property, the Erie Canal. The Albany meeting endorsed an elaborate memorial to the legislature that urged public funding for the great enterprise.

The memorial, comprehensive and persuasive as it was, simply set the stage for a report Clinton had prepared himself during his enforced retirement. In this elaborate document he outdid himself in presenting an astonishing array of facts, figures, and comments based on personal observation and on European experience, primarily British, to demonstrate beyond doubt the feasibility of the undertaking. He considered the financial resources of the state, calculated costs, showed clearly and concisely where the revenue for construction could be found and what financial means were necessary to tap this revenue. Needless to say, Clinton was well aware that increased taxation was universally unpopular. His proposals for funding were primarily to be laid on those who would benefit as the Canal progressed—travelers, shippers and the like.

The state salt monopoly, he estimated, could absorb initial funding. Anticipating objections from almost every source, he advanced arguments to refute them. And above all, he painted a golden picture of rapid economic growth for the state, which would open up the isolated west, and would bring immense wealth to the more settled east and south. He presented tables showing that the Canal would pay for itself out of tolls which would begin coming in as each section was completed. Eventually, he asserted, it would be so profitable that it would lower land taxes substantially while enhancing the value of lands generally. Nor were those counties along the proposed Erie Canal route the only ones to benefit. Clinton had foreseen the argument against local benefit. He envisaged a network of feeder canals that would reach into almost every area of the state, and a canal to link Lake Champlain with the Hudson, thus diverting much of the Canadian trade southward instead of eastward through the Saint Lawrence River. The feasibility of all these subsidiary canals he demonstrated with as many facts and figures as he used to justify the Erie Canal, which was the backbone of the system. Clinton's concept, like the man himself, was imperial in design, massive in scope, visionary, yet practical.

Clinton was on hand when the legislature met, making a great effort, at least for him, to be open and friendly with all. His report created such a sensation that his most bitter enemies and rivals dared not criticize it openly. Even the *Argus* spoke favorably, though cautiously, about the report. Governor Tompkins found it necessary to comment on the Canal, but not the report, in his address to the legislature. Van Buren helped write that politically delicate section in which the governor said that the matter was for the legislature to decide.

Van Buren had read Clinton's report and was impressed. Whatever his feelings may have been toward Clinton, he would not permit personalities to affect his judgment. Under his leadership, the legislature had created a permanent canal commission that was bipartisan in nature. Clinton, of course, was made a member and funds were appropriated so that the commissioners could survey the route, make up cost estimates and receive land donations from those great speculators and land companies that stood to profit directly from the Canal, as Clinton had proposed, and they had accepted beforehand.

Clinton was not only active in lobbying for the Canal, he was busy mending his political fences. The most significant of the repairs was not of his doing, however. It came from Ambrose Spencer, who extended a peace offering through his wife, Clinton's sister. During the legislative session and the summer of 1816, Spencer and Clinton met frequently. By November, when Tompkins called the legislature into special session for selection of electors for President and Vice-President, a complete reconciliation had been effected.[28]

Van Buren soon realized what was going on. From his knowledge of their political habits, he inferred a merger by which Clinton's sudden popularity would bolster Spencer's declining fortunes. As William L. Marcy, then a rising politician from the town of Troy, described the Clinton momentum; he found it "full of cunning and dextrous in intrigue [it] stimulates with hope and animates with zeal [its] profligate partisans and they go forth to proselytize among the unsuspecting Republicans."

Under the pretense of fostering harmony in the party, Spencer sounded out Van Buren on Clinton's being given a more active role—perhaps the gubernatorial nomination in the fall. He met with a courteous but decided negative coupled with a warning that the party would lose the confidence of the people if it was thought "that we were making a game of politics and playing it to serve our personal purposes." Both men lodged at the same boarding house in New York City during the May session of the Supreme Court. Spencer repeatedly sought to gain Van Buren's approval of the new alliance, or at least some sort of assurance that he would not interfere. Van Buren rejected, or rather turned aside, his requests.

Eventually, Spencer's patience, never a strong point, gave way and he exclaimed angrily to Van Buren, "Why you are a strange man! When I wanted to have Mr. Clinton removed, you were, in point of fact, opposed to it, and now that I want to bring him back you are opposed to that also!" Van Buren, accustomed to Spencer's outbursts, was prepared with a calm rejoinder that must have infuriated the choleric Judge. He would not oppose Clinton if he regained the confidence of the party through demonstrated loyalty and good works. But if either of them were to use their influence and power in the party for the sudden restoration, the equally sudden elevation, of a man who had been proscribed for his dealings with the Federalists, then it would cast doubt upon their own principles.

On that note they parted, Spencer angry, Van Buren composed, though concerned that this was not to be their last interview on the subject. Nor had his self-rectitude made any impression on Spencer except to reinforce an earlier decision that his merger with Clinton had been the only sensible way to finish off this dangerously precocious young politician.

The docket of cases before the court session in New York was completed in mid-June, and Van Buren found himself on the same boat to Albany with Spencer and Clinton. The brothers-in-law kept to themselves in the small after-cabin until the boat reached Newburgh, where Clinton disembarked. As Van Buren expected, Spencer sought him out as soon as the boat cast off. Would Van Buren have any objection to having Clinton made one of the two electors-at-large for the state, Chief Justice Smith Thompson being the other? He stressed the mere formality of being chosen an elector, at the same time declaring that Clinton would vote for Monroe and Tompkins. As a concession, Spencer said that Van Buren would have the option of selecting the elector-at-large who would head the ticket, a matter of some prestige, nothing more.

Van Buren promptly refused to accept Clinton on any terms. Spencer was taken aback and angrily accused him of excessive partisanship for opposing such a trivial formality as the appointment of an elector. Again Van Buren waited for Spencer to exhaust his temper and then calmly said that he would support Clinton for the post if he were certain there was no ulterior motive behind it. Looking directly at the Judge, Van Buren charged him with promoting Clinton for the governorship through the advance publicity he would gain as elector. Spencer wanted to know why he felt this way and then, before Van Buren could answer, offered his personal guarantee of Clinton's good conduct in patronage and in party affairs. Seeing his opening, Van Buren turned the Judge's offer against him. He knew there was bad blood between Spencer and Chief Justice Smith Thompson, so he replied with the same guarantees and suggested Thompson be the nominee for governor. Spencer scowled, rapped his snuffbox, a habit of his when excited, and responded melodramatically, "There, sir," he said, "you have touched a chord that vibrates in my heart!"

Van Buren and Spencer had several discussions in Albany without reaching any agreement. Spencer, concluding that he could not disarm his foe with gifts and deciding that open war was necessary, came up with an ingenious idea to cut at the very root of Van Buren's strength and simultaneously to make maximum use of Clinton's popularity. In the interest of democracy and proportional representation, neither of which Van Buren could oppose without compromising his vaunted principles, Spencer would urge at the proper time that counties with Republican minorities be represented at the party convention. What this proposal meant was that the Federalists would support delegates who favored Clinton, and they, together with Spencer's faction and those Republicans still loyal to Clinton, would be numerous enough to nominate him.[29]

Before Spencer matured his plan, Van Buren won a preliminary skirmish. The Spencer-Clinton faction lost out on the selection of electors-at-large, prompting the Judge to say bitterly, "Van Buren who is opposed to everything which does not promote his own views, organized all the opposition he could."[30] This minor success hurt rather than helped Van Buren. Spencer scored heavy gains for Clinton among the Republican legislators of the short session. Vice-President-elect Tompkins, his interest in local politics waning, much to Van Buren's disgust, was of little help in the buttonholing of members. By the end of the session, James Emmott, a Federalist member of the legislature, wrote Rufus King that the Republican party in the legislature was split into two factions of nearly equal strength. "One party," wrote Emmott, "calling itself Clintonian, but in truth headed by Spencer—the other professing to be the true Republican party, willing to support caucus nominations and to do all things necessary to promote the views of the Holy Father in Washington but in fact led by Van Buren."[31]

When the regular session of the legislature convened in January 1817, Spencer showed his power at the outset. He managed to gain control of the council of appointment, a heavy blow to Van Buren. For Spencer immediately used the power of the patronage to replace the legislative nominating caucus with a party convention that would have delegates from Federalist counties. Recognizing the danger, the Van Buren men tried to capture as many delegates as they could, though they knew this was a losing proposition. They must contend not only with the Federalists but also with that great engine of patronage, the council, whose levers Judge Spencer manipulated expertly.

After considering and rejecting various expedients, Van Buren turned to Peter Porter who at first declined but then reluctantly agreed to oppose Clinton for the governorship.[32] It was all to no avail. The Clinton-Spencer machine rolled over Porter with ease. When Clinton was nominated, his disgruntled opponents twice approached Van Buren to make a separate nomination, and twice Van Buren refused.[33] He would not countenance any break with the caucus nomination. The will of the majority must always prevail; if the party was to remain a cohesive unit it must have a set of fixed political principles that were enduring. Spencer had won, Clinton was riding high and Van Buren had few illusions about his future in politics if he crossed these all-powerful men. As he remarked ruefully, "My political candle was thus lighted at both ends."[34]

« CHAPTER 4 »

Mutiny in the Ranks

This low point in Van Buren's political career coincided with family problems and deepening worries about the health of his wife. Hannah gave birth to a fifth son on January 16, 1817, while Van Buren was fighting his losing battle with Spencer and Clinton. The couple named him Smith Thompson, in honor of Van Buren's friend and patron, the chief justice, who was just then trying to help him shore up the breaches in the legislative defenses.

It had been a difficult delivery and Hannah remained weak, feverish, confined to her room. The doctors suspected tuberculosis; their prognosis was not hopeful. Outwardly Van Buren presented his accustomed composure, his incisive judgment of men and measures, his unfailing good humor under more or less constant provocation. But he was deeply unhappy for the first time in his life, deeply concerned about the future. His political career seemed at a low ebb, his wife bedridden, and now he had four sons to provide for. Abraham, a sturdy ten-year old, was reserved like his mother; John at seven was just the opposite, a precocious, lively child, full of mischief and laughter, stubborn and willful too, his father's favorite, though he taxed his patience more often than not; Martin, Jr., two years younger than John, resembled his mother, small for his age and frail, a prey to frequent illnesses, and a source of concern to the family; and now the baby, Smith, as the family called him, added his tiny burden to the overworked Van Buren. He seemed healthy enough, but both parents were uneasy because he had to be left in the care of a nurse.

Van Buren reproached himself for his too frequent, lengthy absences from home on legal business, the many interruptions when he was at home that took him from the family circle. He saw to it that relatives and servants provided a comfortable and pleasant home environment. But he recognized that they could never make up for his own presence, the regular routine, the authority that he as head of the family instilled and represented. Yet he was so enmeshed, in politics, in his law practice and in his official duties as attorney general that he could find no way of extricating himself. To be sure, there was self-interest involved—Van Buren enjoyed the political arena. That he was good at it he knew, and he could not resist the temptation of matching wits with the movers and shakers of the state; nor could he turn away from the flattery and the dependence of lesser men. Always a little unsure of himself, he needed such testimonials. For one whose judgment of human motives was so penetrating and generally so accurate, he was curiously susceptible to self-serving blandishments.[1]

The soothing phrases of would-be admirers had prompted an overconfident Van Buren to challenge Spencer and Clinton when their increasing power in the state was obvious to the most casual observer. He had gone down to a defeat that he could have avoided had he thrust aside the cajolery and practiced his own maxim of noninvolvement when the factions were so nearly balanced.

Van Buren expected retribution from the triumphant Spencer-Clinton faction, and in fact, even welcomed what he assumed would be his removal from the attorney generalship. This post, a highly visible one, was just one more burden to carry, one more lien on his time. What he would give up in prestige and profit, he would gain in long-deferred attention to domestic affairs. But Spencer and Clinton were not ready to take such drastic action which they felt, and with some reason, might create a schism in the party.

For once, Clinton, his Canal project up for final approval in the legislature, did not listen to the sycophants who surrounded him and restrained his desire to dispatch the attorney general at once. Judge Spencer was assuredly a calming influence, pointing out that Van Buren could kill the Canal in the senate. With the fate of his project uppermost in his mind, Clinton sent word to Van Buren that he hoped they would continue to maintain cordial relations, a clear sign that he would retain his post. Clinton contented himself with the removal of several avowed Martling or Tammany men who held minor posts. As he had explained on the eve of his election, "the serpent is now torpid but if you warm him he will sting."

Though Clinton relished his victory over Van Buren and was simply biding his time, he would do nothing to disturb the apparent political calm. "We are all harmony," he wrote his friend Henry Post, and then went on to say that "an obscure painter of the Flemish school has made . . . a very ludicrous and grotesque representation of Jonah immediately after he was ejaculated from the whale's belly. He is represented as having a very bewildered and dismal physiognomy, not knowing from whence he came nor to what place bound.

Just so looks V.B., the leader of the oppugnation army." Clinton's gloating over a fallen enemy was a more accurate representation of his state of mind than the moderate policy he adopted toward the opposition. The new governor would, as he had written, support the Monroe Administration, "treat the V.P. [Tompkins] well; if this cannot be digested by all, not ill"; and though it must have cost him some mental anguish, "not to stir up strife with the Martling men . . ."[2]

In one of those impulsive comments meant to be derogatory, a Clintonian lumped all party opposition to the Governor under the rubric "Bucktails," a bit of Tammany regalia displayed on the hats of its members at certain celebrations. New York's elite had long sneered at the emblem and its rustic appearance, a slight well-known to the Martling men, the Tammany faction bitterly opposed to Clinton. As so often happens with an epithet, it was adopted as a badge of honor, not merely by Tammany but by all who opposed Clinton. While Van Buren never identified himself publicly as a "Bucktail," a term he considered rather vulgar, he saw the benefit to be gained and did not discourage its use. Henceforth, the combination of factions he headed, at first united only in its opposition to Clinton, would be known in New York politics as the Bucktails.

Jabez Hammond, the careful chronicler of New York politics, took his seat in the state senate just at this time. As a representative of the Western District, he had been elected as a Clintonian on the Canal issue. A retiring, reflective and somewhat naive politician, he had supposed that all was sweetness and light at Albany. He admired Clinton for his vision and intellectual prowess, and at the same time respected Tompkins and Van Buren. Hammond had not known the depth of Van Buren's break with Spencer and Clinton; nor had he reckoned with Clinton's confusion of the practical with the personal. "A very few days residence in Albany and attendance in the senate," he said, "convinced me of my error. I found on the one hand, that the Governor was cold, if not vindictive towards Mr. Van Buren and others who had opposed his elevation, and on the other, a determination of excited prejudices and jealousies against the Governor, to render him unpopular with the Republican members of the legislature and thwart him in his measures."[3]

Van Buren may have welcomed what he thought would be a period of relative calm, of watchful waiting, when he could slow down the killing pace he had set for himself over the past five years, but he could not resist the pressure of his colleagues in the legislature. When the anticipated removal from the attorney generalship did not occur, Van Buren found himself more committed to politics than ever before. A combination of circumstances—Spencer's ambivalent position, the cohesiveness of the Bucktails in the legislature, the demonstrable evidence that Clinton was still playing personal politics, all served to raise his spirits and to convince him that he had to rend the web of intrigue that seemed to have settled over the party, threatening its very existence.

"I have been in the furnace and am about the only one who has escaped unhurt," Van Buren said. "His excellency stands on a giddy eminence on which the difficulties of preservation are so great and distressing that many men would prefer political prostration as a choice of evils."[4] Spencer and Clinton had taken many tricks, he realized, but they had not won the game. As for himself, he recognized that he had made several gross errors. He had ignored his own political maxim—pursue a consistent course in party principle and practice, and leave the schemes, the inconsistencies that these begot to the opposition, until they tore themselves apart.

The great measure of the session was the Erie Canal bill. Van Buren's loose confederation of anti-Clintonians opposed it, especially the Tammany men from New York. In the lower house the measure passed by a narrow margin after a furious debate in which the Tammany men took turns attacking the project as impractical, one of Clinton's dark political intrigues, a ruinous expenditure that would bankrupt the state. All eyes, including those of the Governor, turned toward the senate, where Van Buren could easily defeat the measure and use his influence to frustrate any compromise with the house. Heavy pressure was exerted on Van Buren from most of his close political friends to defeat the measure, thus thwarting Clinton, whom they were certain was using the Canal project to consolidate his political fortunes. Van Buren was noncommital.

No one knew what to expect when the Attorney General claimed the floor to speak on the Canal bill. Clinton was present, standing with affected nonchalance on one side of the chamber. But his furrowed brow and his intense gaze upon Van Buren as he began to speak betrayed his anxiety. After a few moments, however, as he listened to Van Buren's rapid delivery, Clinton relaxed. For the next two hours, speaking from notes, Martin Van Buren gave one of the most powerful, yet most painful, speeches of his life. Employing facts and persuasion, he appealed for the support of his rival's great project. When Van Buren finished, Clinton strode over to his desk and shook hands with him, warmly praising his remarks.[5]

The bill passed in the senate, and as Van Buren expected, Tammany deserted him, while others jibed that he was playing politics, making a play for western votes. But he counted on the uproar to die down and if it did not, if he had committed political suicide, he was prepared. He had not acted rashly or on political grounds. The

Canal was right for the state if it were ever to fulfill its imperial destiny. He was ready to stake his political future on it.

Meanwhile, a plan had been slowly forming in his mind, an idea of organization that owed something to his assessment of the tactics of Aaron Burr years before, something more to his own convictions about party identity and unswerving loyalty to majority rule in political council—agreed-upon measures always, personal interest never. It had also dawned upon him that Clinton—through his policies, his personality and his contempt for anyone who disagreed with him—furnished a powerful adhesive that bound together Van Buren's enemies as well as his allies. Among this latter group he could count on the Administration in Washington, which had good reason to distrust Clinton.

Outside of the party there was, he was aware, a small but influential group of Federalists who not only disliked Clinton but thought him a dangerous man, a would-be dictator. Untainted by the heresy of the Hartford Convention, they had stood by the state and the nation at the time of its greatest peril. At their head was Rufus King. With all of these factors operating, Van Buren realized the time was ripe to try and unite the anti-Clintonians in a logical pattern of political behavior that was based upon consistency, complete loyalty once a decision was made and discipline not just in thought but in deed. If a crucial vote was to be taken in the legislature every member of his faction must be present, even the sick. That patronage was essential to party regularity Van Buren understood, though hardly a unique recognition on his part. If anyone had perceived and utilized patronage for political ends it had been Clinton, who may be said to have been the originator of the spoils system in New York, if not in the nation.

Newspapers in the two major cities of the state—Albany and New York—would set the party line which the country press would echo. They would also supply printed handbills and ballots for statewide distribution. The traditional committees of correspondence would supplement these means of communication and organization.

Frequent sessions of New York's complicated court system meant that lawyers were constantly on the move. They formed an excellent channel of communication and planning down to the villages and hamlets. The council-appointed officeholders formed a corps of political pressure groups in every locality of the state.

All that was necessary was to devise a program of political action for friendly lawyers and officeholders showing how it was in their interest to spread the word. None of these practices was original; none was confined to the Republican party. Van Buren's ideas were innovative in that he identified all of the parts and integrated them into a unified political structure which, once it came into being, supplied its own momentum, its own ideology, an organic mass responding to change, yet usually under control. Between 1817 and 1821 Van Buren created the modern political machine known to contemporaries as the Albany Regency. Despite his bold and comprehensive view of political structure, Van Buren recognized that it would not work without solid leadership. Fortunately he found most of his lieutenants in the 1818 state senate—the few exceptions included members of the lower house such as Michael Ulshoeffer and Erastus Root.

Of the thirty-two senators, twelve were so opposed to Clinton that they formed a unified group for Van Buren's wider purposes; of these, five shared his political views and broader vision of party formation. These five senators were also far more politically astute than their anti-Clintonian peers and indeed the entire body. Geographically, too, they represented the four Districts of the state, a practical consideration of great importance in the vast and largely undeveloped land mass of New York, where locality vied with the conflicting sentiments of a heterogenous population for political influence.

Henry Seymour, just beginning his second term in the senate, was the steadiest and most effective of the five. An emigrant from Connecticut, he had settled on the New York frontier in a village known as Pompey Hill, later to be called Utica. Van Buren had been attracted to this transplanted Yankee from western New York, or what was considered then the west, since his first appearance in Albany, some three years before. Seymour came from notable if not distinguished stock; yet he had made his own way and was now a wealthy man, a kind of personal achievement, always a matter of interest to Van Buren. Seymour spoke well and carried himself well. Though he was not a college man, he like Van Buren passed for a cultivated person who was at home in genteel society. Cautious, clever, guarded in conversation, Seymour was considered a schemer by his opponents. And certainly he was not above planning and executing a legislative coup if it seemed likely to succeed and promised a gain worthy of the effort.

From the beginning of their relationship, Seymour admired Van Buren and, as far as Seymour's frosty character would permit, became a personal friend as well as a valued associate. In so many ways the two men were alike—their distrust of Clinton's policies, their close adherence to Jeffersonian principles and firm belief in party loyalty. Henry Seymour deferred to Van Buren in political matters, acknowledged him as the leader, but he was no retainer, no flatterer, and quite as ready to criticize a Van Buren move as to agree with it.

A second member of the inner circle was Roger Skinner. He represented the far north, those counties largely populated by Vermonters who were steadily pushing westward along the Saint Lawrence and Lake Ontario. Hewing out farms from the wilderness along the fertile shores and hinterlands of those great waterways, they had borne the brunt of Indian and British raids during the War of 1812. Like their spokesman, Skinner, they disliked and distrusted Clinton because they considered his dealings with the Federalists in the election of 1812 unpatriotic and unprincipled. The Vermonters among them had, in addition, traditional scores they would repay, remembering the harsh treatment meted out to the Green Mountain boys by the Clintons, who regarded them as squatters on New York land. There was something about Skinner that appealed to Van Buren, though never in the way Henry Seymour appealed. Where Seymour was the epitome of caution, Skinner was apt to be impetuous, even reckless in political management.

Samuel Young, the most patently ambitious and least successful of the group, represented Saratoga County. An expansive lawyer from the Middle District, he had promoted his connections all over the state. Van Buren could count on him, though he remembered to be wary of this quick-tempered, hard-drinking man, who could antagonize as often as he could charm. The other two senators, Walter Bowne, a prosperous merchant from New York City, and Moses J. Cantine, Van Buren's brother-in-law, were solid, middle-of-the-road Republicans, alike in their opposition to Clinton and rather dazzled by Van Buren's manner, his quick perception, and his uncanny ability to size up events.[6]

As yet Van Buren had no solid links with Administration in Washington, though not for lack of trying. His only sources of information and influence were through Vice-President Tompkins, Senator

Nathan Sanford and indirectly through Smith Thompson, who had struck up a friendship with the President. "The truth is," Van Buren confessed, "I have so little influence with the folk at Washington and am reluctant to put myself under any obligation to them that I make it a point never to write them unless I feel strong confidence that my doing so will lead to success."[7] He was not concerned with weakness in that quarter, for he could count, he was certain, on the President's well-known dislike of Clinton to withold executive patronage from the Governor or to stand neutral in any impending contest.

Van Buren regarded himself as an orthodox party man, but he was certainly close to the Bucktails, who reluctantly accepted his restraining hand for the maintenance of harmony within their ranks. And never was restraint and careful management more essential than at the outset of the 1818 legislative session.

Clinton's popularity, and thus his political strength, had increased.

It was premature to oppose the Governor in politics or in public policy, though many of the Bucktails, still smarting over their defeat on the Canal, were clamoring for an open break. Van Buren managed to head off their separatist movement and line them up behind a clever plan he had devised for staffing the new council of appointment.

Had he listened to the Bucktails and used his influence in the lower house to appoint a council composed of Clinton's opponents, he reasoned, he and his supporters would have appeared before the public not just as factious politicians bent on revenge against a strong and powerful governor; they would also have to bear the burden of hasty appointments along with hastier removals that were sure to be made, many of them too openly inspired for personal rather than public benefit. Yet he feared the result of a council completely dominated by Clinton. For he believed Clinton would make a clean sweep of him and his Bucktail men and deliver what could be a staggering blow to his evolving organization. So his strategy was in the last analysis defensive, the fewer the removals the better. An ideal council, as he saw it, would be one made up of members, a majority of whom were not opposed to the Governor, yet could not be controlled by him. The enmities and the anguish, sure to be aroused when officeholders who thought of themselves as good Republicans were removed, would be directed at the "Clinton council." Van Buren may have played ducks and drakes with the appointments, but he was not altogether happy about this devious maneuver which depended on so many complex factors and which might well recoil back upon him.

Clinton wanted Jabez Hammond, the new senator from the Middle District, appointed. After Van Buren had sized him up as an honest but politically ineffectual Clintonian, he agreed to the selection in exchange for an outspoken, highly partisan Bucktail, Peter R. Livingston from the Southern District. Next came the appointment for the Middle District. Van Buren hit upon Henry Yates, an avowed Clintonian, with a number of importantly placed relatives. His brother, Joseph Yates, was a justice of the state supreme court and, as Van Buren had learned, harbored gubernatorial ambitions. More interesting to Van Buren was that Henry Yates heartily disliked Ambrose Spencer. If Clinton knew about Yates's connections, he dismissed any untoward influence; and even Spencer seems not to have thought that his loyalty was in question.

Yates was a calculated risk, but of all the Clintonians in the Middle District he best fit a council member whose family aspirations might clash with the Governor's will. Clinton's greatest strength lay in the west because of his Canal policy. And there Van Buren was able to make use of the bitter rivalry between two Clintonian candidates. While their respective supporters deadlocked in the house, Van Buren's faction proposed Henry Seymour, who was elected after tense and involved discussions in the various District caucuses. Now Van Buren was certain he had gotten a council that suited his original requirements. "All is safe," said he, "Seymour! Seymour! Seymour!"[7] and he was right. No sooner had the house made the appointments than Yates, though still considered a Clintonian, came out openly against Spencer.

Jabez Hammond, whom everyone thought a fumbling politician, acted as if he were a seasoned veteran. An innocent he certainly was when it came to understanding the motives and the schemes of intriguers and lobbyists who infested the capitol. But he soon observed, and quite correctly, that Van Buren and his far-flung allies had a decided majority over Clinton.[8]

The new council worked better than Van Buren had hoped. It protected almost all of his valued supporters from Clinton's political retribution, and on several important occasions, the Governor had to break ties that had been set up beforehand, bringing down upon himself the wrath of a defeated aspirant or a removed officeholder. Yet the Governor seemed unperturbed, even indifferent.[9]

Clinton's immense popularity enhanced an ego of epic proportions and further flawed a political judgment that always emphasized the natural rule of the rich and the well-born. Clinton's mind and energies were wrapped up in great projects of internal improvement, petty politics or the paltry intrigues of mean-spirited little men he had always borne with ill grace.[10]

While Clinton wrestled ineffectually with his unruly council, and with growing impatience cynically consigned all politics to the devil, Van Buren began to consolidate his organization. As the legislative session of 1819 approached, he initiated the same process that Hammond had earlier suggested to Clinton only to have it rejected out of hand, that of making his group the regular organization and have that of Clinton's stamped as a faction. The tactics Van Buren planned to use he had used before, and effectively. He would exploit Clinton's friendship with prominent Federalists who had opposed the War. And he would hammer on the old theme of Clinton's checkered past, where he had strayed from the organization when it suited his purposes. What happened once would happen again; Van Buren would emphasize publicly through the press, privately among active politicians—those committed to him and those not committed to either man.

Extreme caution was the rule. Understanding Clinton's foibles as well as he did, Van Buren waited for a favorable opportunity. What surprised him was that it came so quickly before the organization of the legislature. Van Buren's candidate for Speaker was William Thompson, a young lawyer from Seneca County in the Western District, who had had differences with Clinton in the past and personally disliked the man. Of mediocre ability, Thompson was not well-known in the state at large and could be counted on to follow directions from powerful figures like Henry Seymour of his own District, Samuel Young, or Van Buren himself.

Once Van Buren had decided upon Thompson, he and his close associates insisted that his candidacy be secret, a difficult proposition among garrulous politicians as they tippled in Albany taverns. Clinton and Spencer either seem not to have gotten wind of the decision on Thompson or, if they did, which is more likely, dismissed him as a mere cipher to be easily crushed. Those two worthies, overconfident of their prestige with the country members, especially those from the west, made no preparations for a nomination with any of their legislative leaders. Nor did they arrange, as Van Buren and his lieutenants had, that as full a complement of their supporters as possible be present at the party caucus to be held at the capitol on the evening before the legislature was to convene.

Thus far, the truly deplorable management of Spencer and Clinton had jeopardized their control of the party. They now proceeded to commit another egregious error by choosing as the candidate for Speaker an esoteric figure from the past, Obadiah German. It had been many years since he had been a member of the house and he was scarcely known by sight, though certainly by reputation, to the younger members; for German, a member of Congress in 1812, had voted against the Declaration of War. He had publicly opposed Governor Tompkins's energetic war measure and had formed alliances with the peace faction of the Federalists. If Clinton and Spencer had wanted a candidate more offensive to the party, they could not have found one in the state. German epitomized a small group of politicians whose conduct during the War had been more despicable to the Republicans than even the peace Federalists, who had the virtue at least of belonging to the opposing party. The selection of German and Clinton's lack of foresight in the caucus arrangements may have reflected his efforts to create a third party that the Federalists would be willing to join. If such was his design, it failed utterly.

When the caucus met on January 4, 1819, almost all of Van Buren's adherents were present, while seventeen Clintonians either had not arrived in Albany or, having no instructions, did not attend. The caucus nominated Thompson over German by a majority of nine votes. Had all the Clintonians been present, as they certainly could have been if the Governor and Spencer had reminded them of the vital importance of the caucus nomination, German would have been nominated by anywhere from 1 to 15 votes. Van Buren had taken a bold gamble and won. Had German been nominated, he would have been saddled with a caucus candidate whose dalliance with the Federalists mocked all of his political professions. The only alternative would have been to bolt a caucus nomination, a move that was wholly inconsistent with his oft-expressed political ideas.

Spencer and Clinton might have covered much of their losses had they acquiesced in the caucus nomination; but they refused, and in combining with orthodox Federalists to elect German Speaker, they committed partisan heresy. Now their faction, not Van Buren's, was the bolter; they had conspired with the Federalists; they had broken with the party will as expressed in caucus. Van Buren had seized the opportunity and as his opponents made one mistake, he set them up for another.

During the balloting for Speaker, the Van Buren men noted that the two most prominent Federalists in the House, Rufus King's son, John A. King and John Duer of Albany, voted against German. Van Buren himself had long admired Rufus King and had long known of his dislike for Clinton. As John King wrote his father, "I regret to say that there are some of the Federal gentlemen and influential ones too, who are deeply pledged to support the wavering fortunes of Mr. Clinton. On this point the Federal party must, if it has not already, deviate."[11]

The acknowledged leader of the Bucktails, Van Buren secured agreement that they would run Samuel Young for the United States Senate. As soon as the Bucktail caucus made this decision, powerful newspaper support rallied behind Young. The *Albany Argus* boldly asserted that Young was the best man for the post, while in New York City, the *Advocate,* edited by the talented Mordecai M. Noah, spokesman for Tammany, echoed the sentiments of the *Argus.* Both papers opened up a drum-fire attack on Clinton and Spencer and ridiculed the pretensions of their candidate, Spencer's son, John C. Spencer, from the Western District, then serving his first term in Congress. Apart from his support of the Canal, John C. Spencer was not a popular choice. Tall, slender, tense, humorless, he had no use for small talk and found it difficult to work with anyone, even his own father. In addition to these disagreeable personal traits Spencer, of course, was open to the charge of nepotism, a charge the *Argus* and the *Advocate* freely leveled against him.[12]

The elections of the previous year had given Clinton added strength in both houses of the legislature. He may not have paid attention to the details of party management, but he was a master at promoting the great Canal. Through regular reports of progress he emphasized the new machinery being developed to further the work, the tolls already coming in from completed sections of the Canal and interesting digressions on the geology, the fauna and flora along the route. If Clinton was doing nothing else, he was making the Canal a symbol for the age: an age of optimism, of expansion, of pride in native accomplishment. But the Canal reports were not the only means that ensured his increasing prestige. Clinton's addresses to the legislature, though on the pedantic side and bulging with foreign as well as domestic information, recommended reforms in education, in taxation, in fiscal affairs, prisons and insane asylums, in short the modernization of New York society. Clearly he was proud of his own state and his messages reflected a firm belief that through appropriate action, New York would become the richest, most powerful, most advanced state in the Union. Like his Canal reports, his addresses gained him adherents. One need not like the Governor, his "cold repulsive manner," his arrogance, but one was compelled to respect him, to consider him a great man.

When the legislature met in January 1819 the Bucktails found themselves in a minority despite their excellent attendance. Van Buren realized that Young could not receive a plurality much less the absolute majority the constitution required for election to the Senate. A Republican caucus, as usual, was set for the evening of January 4. Van Buren could not be present. Hannah's illness had reached a critical stage. For the past year she had slowly grown weaker; and Van Buren, deeply distressed, had sought what small measure of relief he could get by working out elaborate plans for the survival of the Bucktails in the face of the rapidly growing influence of Clinton and Spencer. He had also worked harder than usual on the senate committees he headed, Ways and Means, and Judiciary.

On the evening of the party caucus, Hannah was so ill that Van Buren spent the night at her side. The doctors held out little hope, but his wife, wasted by tuberculosis, still clung to life. In the morning Van Buren received the news that his lieutenants had managed the caucus every bit as well as he could have done. Again, they had taken advantage of poor leadership on the part of the Clintonian majority.

Several Bucktails began baiting German as soon as the caucus was organized. They knew him to be hot tempered and exceedingly argumentative. If they could turn the caucus into a shouting match, they might force an adjournment. After several hours of furious wrangling, a Bucktail made a motion for adjournment which was carried. Had the Clintonians stuck it out, they would have carried the caucus for Spencer; his would have been the regular nomination, and if the Bucktails refused to support him, they would have been the schismatics. The two factions eventually nominated their candidates at separate caucuses, the Bucktails claiming Young to be the regular candidate. The Republican party in New York for the time being, had ceased to exist.

Van Buren appeared in the senate on February 3 to vote for Young, who received a plurality over Spencer and King. In the house, Spencer had a plurality over King and Young. Since no one had a majority, and since the constitution prescribed the first Tuesday in February for election of senator, that was the end of it for the year.[13] Van Buren hurried home where he remained until Hannah died, on February 5, 1819. He left for Kinderhook the next day, accompanying Hannah's body; she was thirty-six years old, her husband a year older.

Van Buren had been upset, as any dutiful son would be, when his amiable and loving father died in 1817, to be followed within a year by the death of his indomitable mother. But they had lived out their time—Abraham was eighty-one years old, Maria just ten years younger. His wife's death affected him much more deeply, so young, so self-effacing, so responsive to his needs and to those of their four young sons. He had Darling and Company of Hudson, a noted local stonecutter, fashion a twenty-foot obelisk out of native granite on which he had Hannah's name and dates inscribed. Beneath them was a simple inscription which he composed. For a week after the funeral, he remained in Kinderhook, emotionally distraught, physically ill.[14]

Clinton and Spencer did not repeat the mistake they had made the previous year on the council of appointment. For once, they managed skillfully with the aid of Federalist votes in the lower house and secured an obedient council. Van Buren's days as attorney general would seem to be numbered. When the council met, it beheaded all Bucktail officeholders, yet the ax did not fall on Van Buren's neck. The restraining hand of Judge Spencer had granted the reprieve.[15] Spencer, and Clinton himself, had been deluged with letters and with pleas all in the interest of party harmony. Party officeholders and country editors especially knew not where to turn. Frightened about their uncertain future if they backed the losing faction, they wanted harmony.

Van Buren quickly regained his health and after putting his family affairs in order, plunged into the political swirl more deeply than ever before. Politics and work, he found, eased the psychic trauma he had suffered. In his casual way, he considered the olive branches that the Clintonians were extending, not rejecting them outright, nor even giving the appearance of evasiveness, but simply declaring that he had no quarrel with Clinton. Let the Governor follow proven

Republican principles, obey the rule of the caucus, place party over self, and all would be well. He knew that Clinton would never accept any terms except complete capitulation, that he was merely biding his time for the final reckoning.

Ever since Clinton's return to power he had all but wrecked the party and diminished New York's position in the nation. After exercising so much restraint upon his feelings, Van Buren opened himself up privately in a torrent of denunciation to his friend Gorham Worth:

> A man to be a sound politician and in any degree useful to his country must be guided by higher and steadier considerations than those of personal sympathy and private regard. . . . In the name of all that is holy, where is the evidence of that towering mind and those superior talents which it has been the business of myriads of puffers and toad eaters to attribute to him? . . . Mr. C. has not yet been governor two years and our state is distracted and our councils disturbed, to an extent beyond all precedent—speak not of the Tammany men, they are but a drop in the waters of bitterness

Van Buren was determined to restore, as he saw it, some semblance of good government and tried and true political practice in the place of public policy through personal whim and conscientious party management in place of Byzantine intrigue.[16]

The Canal was the core of Clinton's popularity, the source of his political strength. He presided over the board of commissioners who controlled all stages of construction. The board let contracts that were worth hundreds of thousands of dollars and through the contractor and subcontractors controlled thousands of jobs, which in turn meant patronage, and with patronage, votes. The Canal board at the moment vied with the council of appointment as the greatest patronage machine in the state. This project Van Buren had early realized must somehow be made palatable to those Bucktails who had opposed it from its beginning, had contemptuously labeled it "Clinton's Ditch."

Even as Van Buren was seeking some means of converting those of his faction who still opposed the Canal, Mordecai Noah was conducting a vigorous attack on the Canal in his *Advocate,* while the *Albany Register,* Clinton's paper, was replying. Most observers thought the *Register* was getting the better of the *Advocate,* and it was hinted that the Governor himself was the author of its rebuttals. Although opposition to the Canal came from isolated parts of the state that were not directly benefited, its primary source was in New York City and among the Tammany men, now Bucktails, who were against the Canal because the Canal meant Clinton to them.

Canal commissioners were elected on a joint ballot by the legislature. Over the past few years of political jockeying, the Bucktails had increased their representation on the board so that by 1818 they lacked but one member of having a majority. Some six months before the 1819 legislature convened, Joseph Ellicott, one of the commissioners and the agent of the Holland Land Company, resigned from the board.[17] Clinton promptly appointed Ephraim Hart, a senator from the Western District and an exceptionally keen businessman. He had given no thought to the political implications of the Hart appointment, which he simply made on the basis of his residence and his proven ability. Had he considered Hart's politics, he would have realized that he was a man who had openly and vociferously opposed the Federalists for years, among them Rufus King and those of his faction who sustained the War and opposed Clinton's policy of amalgamation. Hart held a recess appointment. His continuance as a commissioner had to be approved by the legislature. Seemingly oblivious to the patent fact that Hart was so controversial politically, Clinton submitted his name to the legislature for a permanent seat on the board.

Jabez Hammond, well aware of the partisan dangers surrounding the Hart nomination, sought to alert Clinton. The Bucktails nominated Henry Seymour, their best choice. Although opposed to Clinton, Seymour, ever cautious and ever courtly, had many friends among all the factions in the legislature. Hammond reckoned that with Hart as a candidate, and the Bucktails voting as a unit for Seymour, there were at least fifteen or so Rufus King Federalists who would vote with them. Quite possibly, their votes would put Seymour over, giving the Bucktails control of the governor's Canal board. Hammond explained all of this to Clinton who again scoffed at his reasoning.

Just as Hammond suspected, John A. King approached Van Buren and said that there were fifteen Federalists in both houses who could not vote for Hart. Depending on whether the Bucktails were backing a candidate and if the candidate were acceptable to them, they had enough votes to determine who would be the new commissioner. Van Buren replied that they were preparing to contest the Hart nomination and that their man was Henry Seymour. King thought him acceptable to his group. When the vote was taken at the joint session, enough Rufus King Federalists joined with a solid phalanx of Bucktails to elect Seymour to the board by a majority of one vote.[18] At one stroke, the Bucktails had seized Clinton's political power base, converting it to their own use.

The effect on the picturesque Noah was electric. His *Advocate* abruptly stopped attacking the Canal as a consummate folly, a financial disaster, a political trick, and edged slowly over a period of several months into a hearty support of the project. Walter Bowne and other Clinton-hating Tammany men needed little explanation from Van Buren to bring their supporters around. By the end of the legislative session, the Bucktails were as enthusiastic about the Canal as the Clintonians. To De Witt Clinton, the blow had been a heavy one. He had been incredulous that so many Federalists would join his foes and undercut him on what he regarded as his private preserve, but he reserved his anger and dismay for Van Buren. No longer would he listen to Spencer; Van Buren must go.

At the summer meeting of the council, Van Buren was removed from his lucrative post of attorney general, though not before that body had seriously considered appointing him to the state supreme court. The chief justice, Smith Thompson, had resigned to become secretary of the navy, and there was some talk of appointing two additional justices.[19] But the motive for the increase was not reform or even neccessity, it was strictly political—provide a judical post that would tempt Van Buren away from the senate and his leadership of the Bucktails.

When sounded out on the proposition, Van Buren indicated that he might accept the post. His domestic tragedies of the past three years, the uncertain state of his health, the constant bickering and maneuvering in the legislature, had taken its toll. If he had been appointed to fill the vacancy Thompson's resignation had created or one of the proposed new judgeships, he probably would have accepted. Missing another rare opportunity, Clinton vetoed Van Buren, "that prince of villains." The supreme court, itself, in a spasm of guilt by association, unanimously opposed any expansion.[20] Van Buren remained where he was in the senate, deftly captaining the legislative opposition to the state administration.

« CHAPTER 5 »

"The High-Minded Gentlemen"

Van Buren took his removal as attorney general and his rejection for a judgeship as a matter of course and began consulting with his friends about the 1820 election to which he had already given some thought. A United States Senator was to be elected. With the Canal patronage temporarily in Bucktail hands, it seemed a favorable time to topple Clinton from the governorship.

Van Buren now had his Washington connection in the person of Smith Thompson, who was working for the Bucktails in the cabinet and with the President.[1] If Monroe would not make removals, Presidential pressure on federal officeholders in New York would still be helpful. The basic question to be determined was the choice of a candidate to run against Clinton. Van Buren decided, as he had when Rufus King was the Federalist nominee in 1816, that Tompkins was the best choice. The Bucktails, even with the patronage in their hands, had to run their most popular man if they were to beat Clinton. Van Buren went ahead with the Tompkins nomination after he learned that the Vice-President would probably accept. Tompkins, always sensitive about his image, had found that his charm and his genial manner had proven no substitute for the superior talents and driving ambition of the men around the President.

As soon as the Tompkins candidacy became public information, the Clintonians immediately raked up a charge that had been smoldering for the past three years and had never been satisfactorily answered. Had Tompkins, when a War governor, defaulted on his accounts? The comptroller of the state, Archibald McIntyre, was a methodical, honest public servant, whose philosophy of accounting was narrowly rigid. He had examined Tompkins's accounts and found shortages. Neither of the factions wanted any public investigation at the time. Leading Republicans had differing interpretations of McIntyre's report and, out of respect for Tompkins's services and his present position as Vice-President of the United States, few were willing to impugn his integrity.[2] Now, as a candidate for the governorship, the Clintonians saw the issue in a different light. McIntyre made public his report that claimed a discrepancy between $110,000 and $120,000. Tompkins countered with a demand for compensation of $25,000.

Charges and countercharges rang through the nation's press. The Vice-President maintained his innocence, but the charges deeply disturbed him. A moderate drinker, Tompkins began drinking heavily, though he managed to conceal the extent of his intemperance. Van Buren first learned of Tompkins's problem when he visited the Vice-President at his home on Staten Island during the early fall of 1819 for a personal examination of his accounts. Troubled by this unexpected turn of events, Van Buren consulted with Smith Thompson, who was in New York City at the time and learned that Thompson had also become aware of the Vice-President's condition, which he thought was a temporary one brought on by the public controversy over the accounts. Somewhat relieved, but still skeptical, Van Buren was happy at least to find in Tompkins's accounts explanations that would, he thought, answer all the questions McIntyre had raised. When he explained the results of his examination to Tompkins he observed what he took to be "the renewal of his spirits and with them of his adroitness, tact and power." Under Van Buren's guidance, Tompkins drafted a six-page reply to McIntyre that promptly became a partisan issue, not just in New York but in the national press. The publicity did not concern Van Buren; in fact he counted on it because he was certain Tompkins, though careless in his accounting, was completely honest. If the Clintonians wanted to make Tompkins a martyr, let them do so.[3]

Van Buren was more concerned about the Vice-President's drinking problem. He shuddered at the repercussions that would affect him personally should Tompkins be elected and not be able to perform his duties. Van Buren knew he would be held accountable for placing a "drunkard" at the head of the state. Better perhaps to be defeated with Smith Thompson as the candidate than to win with the unstable Tompkins.

A few Bucktails had voiced opposition to the Vice-President because they thought the controversy with McIntyre would play into Clinton's hands. In similar situations Van Buren had brushed aside such reservations with some tactful but apt comments. This time, however, he relayed the doubts on to the Vice-President and to Thompson. In a masterly letter, he managed to convey a delicate but unmistakable hint that Tompkins withdraw, yet with language so carefully drawn that it balanced deference to the Vice-President with hope that Smith Thompson would stand. Neither man accepted his veiled suggestion, however. Tompkins wanted vindication, and Thompson did not want to tarnish his image in Washington with what seemed a certain defeat in New York. Chagrined at his failure, Van Buren put the best face on it he could and accepted the responsibility that he may have made a serious mistake.[4]

If he had acted rather precipitously in his early endorsement of Tompkins, he moved carefully and shrewdly in bringing the Bucktails behind his candidate for the United States Senate, Rufus King, to succeed himself. Over the past four years the Federalist party had been breaking up, a majority supporting Clinton, a minority claiming King its chief. King, needless to say, stood on his dignity and his record, even though he was known to oppose Clinton, blaming him for a variety of political sins, the worst being the destruction of the Federalist party.

The schism had first appeared openly in 1814 when a series of articles lampooning Clinton surfaced in a small two-page weekly, *The Corrector.* These were followed by another series in 1815 that humorously yet cuttingly put Clinton up for ridicule. His pedantic style, his pompous display of learning, his arrogance, even his physical bulk were caricatured in a pamphlet signed by one Abimelech Coody. The author was Gulian Verplanck, a young Federalist from New York City who harbored a personal grudge against Clinton that had quickly taken a political turn. From time to time Abimelech Coody published more pamphlets as other young Federalists joined in the sport of baiting Clinton. Van Buren and the Bucktails were delighted with the Coody pamphlets and laughed heartily at Clinton's discomfiture, which he did not conceal. That there was more than wit and sarcasm to the works of those amusing gadflies, collectively known as Coodies, politicians conceded.

Shortly after the publication of the Coody pamphlets, another series of amusing satires, this time in verse, entitled the *Bucktail Bards,* appeared in John A. King's *New York American.* The *Bards,* composed by Verplanck, John Duer and Rudolph Bunner, far surpassed Abimelech Coody's burlesque in trenchant, hilariously amusing commentary on Clinton's political pecadillos and his personal quiddities. There was no doubt now in Van Buren's mind what the Bucktails must do to defeat Clinton. They must be brought to an understanding with the young rebels, and Rufus King was the key.[5] Only management of the most tactful order and of the utmost discretion could bring about such a rapid change in the state of things. Van Buren decided that the move must be taken, but in such a way the alliance with the Federalists could not be interpreted as the apostasy he had so often charged upon Clinton.

Unlike the majority of the Federalist party, King had supported a vigorous prosecution of the War. Van Buren reasoned that King and his minority faction could be described as Federalist in name only. As he wrote Noah, whom he had brought into the scheme, "I should sorely regret to find any flagging in New York. We are committed to his support. It is both wise and honest and we must have no fluttering in our course."[6]

Another consideration which Van Buren found politically useful was King's forthright stand in the Senate against the admission of Missouri into the Union as a slave state. During the second session of the Fifteenth Congress three New Yorkers had precipitated and prolonged the first public debate in the nation's history over the issue of slavery in the territories. James Tallmadge, Jr., had begun it all when he attached an antislavery amendment to the bill admitting Missouri to the Union. Tallmadge's amendment passed the House after spirited debate, in which John W. Taylor, a Clintonian and the Speaker of the House, was an active participant.

Rufus King assumed the leadership of the antislavery forces, however, when the amended Missouri bill came before the Senate. Only twenty years had passed since the state began the phased abolition of slavery, only two years, since its total abolition. Van Buren and indeed most of the leading men in the state had grown up in households where slaves were part of the domestic scene. Most had long since decided that slavery, at least for their own state, did not accord with what they chose to identify as the progressive spirit of the times. The politicians among them, especially the Federalists and Clintonians, regarded slavery as one of the devices the South was using to perpetuate its hold on the national government.

Van Buren had quickly grasped the significance of King's position and its political importance to the Bucktails. King Federalists and, he hoped, eventually the bulk of Clinton's more extreme Federalist backers would fall away if the Bucktails accepted the position the old statesman staked out on slavery in the territories. He shared his thinking with Noah to whom he said, "The Missouri question conceals no plot, and we shall give it true direction." He also wrote, that "morally and politically speaking slavery is an evil of the first magnitude and whatever may be the consequences it is our duty to prohibit its progress in all cases where such prohibition is allowed by the Constitution. No evil can result from its inhibition more pernicious than its toleration."[7]

Van Buren was also a member of the legislature that unanimously passed resolutions strongly backing King's position. Previously he had agreed to have his name printed on the call for a public meeting in Albany that framed other resolutions condemning the spread of slavery. When first approached, he had been so busy that he explained he could not participate, and when asked to sign the resolutions, he refused because he had not been present at the meeting. There may have been something in the resolutions that he could not accept, but since they did not differ materially from the legislative resolutions that he did favor, it seems fair to take him at his word. Rufus King certainly never doubted Van Buren's stand. One of Van Buren's law clerks, John W. Edmonds, said that he was "overwhelmed with work," and Edmonds was drafting much of his private correspondence. Van Buren was at the time conducting delicate negotiations over the nominations for both governor and U.S. senator. Under the pseudonym *Leonidas,* he was drafting a long reply to a note Ambrose Spencer had written attacking the Bucktails and published in the *Albany Register,* and, of course, the law practice must go on. There were now four clerks under the tutelage of Van Buren and Butler, who had just resumed his partnership after a year's absence.[8]

Van Buren needed more help, not from his own circle, but a fresh face free of the complex intrigues that tainted the politics of New York City and Albany. For several years now, he had been watching the career of a young lawyer in Troy, William Learned Marcy, a Massachusetts emigrant of generous proportions, pugnacious temperament and an engaging style of writing editorial copy for a weekly newspaper, the Troy *Northern Budget.* Marcy was recorder of the Troy city council, a master in Chancery, a veteran of the War, and still an officer in the militia. He shared Van Buren's opposition to Clinton and did not bother to conceal it. The *Budget* bristled with Marcy's barbs and quips at the Governor's expense. As if this were not enough, he wrote many a letter excoriating Clinton in the *Albany Argus.* Clearly he was a talented man, full of drive, ambition and energy, yet always with a twinkle in his piercing eyes under beetling brows. Van Buren not only liked him but valued his political insights. Very soon the young man from Troy ranked with Roger Skinner as Van Buren's chief organizers in Rensselaer County.[9]

During December 1819 Van Buren had special need of Marcy's skills. He had drafted a carefully constructed argument in favor of Rufus King's reelection to the Senate. The major themes of patriotism, ability, probity, were of course developed, but the real meat of Van Buren's argument was his comparison of King's disinterested statesmanship with Clinton's checkered political career, where the Governor was held up as a monument of vanity and of personal partisanship. Van Buren divided the Federalists into two groups: those who opposed the War and who supported Clinton, and those who had supported the War and acted with the Tompkins and the Republican party. The forty-four-page manuscript was a subtle yet compelling document that, while urging Bucktail support for King, did so by skillfully avoiding any hint of amalgamation. The Bucktails remained distinctly Republican, King's Federalists remained Federalist. Uncertain of his style, needing more pith to his argument, more trenchant examples and telling phrases, Van Buren had Marcy go over the draft, which Marcy edited so heavily that the resulting "Considerations in Favor of the Appointment of Rufus King to the Senate of the U. States" can be considered a joint effort. A majority of the Bucktails accepted the Van Buren–Marcy document, though neither man could have predicted the outcome.

Clinton felt he could not risk his Federalist support by opposing King. The old statesman, who remained above partisan conflict, was reelected to the Senate with only three dissenting votes. King was grateful to Van Buren, who he recognized had risked his political leadership in supporting a Federalist candidate. When Van Buren was chided by his friend David Evans for rashness, he shrugged it off. "There is a radical difference between us," he said, "on the one hand [I] am over-sanguine & subsequently often err in my calculations, you on the other hand of a different temperament more dispassionate & therefore although you sometimes apprehend dangers to a greater extent than they really exist you generally make safe estimates."[10]

On the eve of the election, fifty anti-Clintonian Federalists published an "Address" to the people of New York severely criticizing the Governor for heading what they called a "personal party" that had no political principles and was devoted only to maintaining him in power. This cult of personality was "disgusting to the feelings of all truly high-minded and honorable men who entertained a decent self respect." The "Address" closed with an appeal that honest Federalists should support Tompkins or not vote at all. Everyone knew that the signers of the "Address" represented a minority faction, but what they lacked in numbers they made up in wealth, education and social standing. At their head were the sons of Alexander Hamilton and Rufus King, and the Duers of New York and Albany. The "Address" made more of a sensation than an impression.

Unfortunately the "high-minded gentlemen," as they would be called, came from Districts which were either solidly Bucktail or solidly ultra-Federalist. The "Address" was nevertheless welcome to Van Buren because it settled completely any residual criticism of his policy, and it would, he thought, be a useful document for future campaigns. As for the election itself, Van Buren was pleased at the outcome. Clinton, with the support of the majority of the state's Federalists, received a slim victory of 1,457 votes over Tompkins. The dread prospect of an alcoholic governor, which Van Buren had mentioned to Smith Thompson as "the fearful responsibility I was assuming in pressing his nomination . . ." was now forever set to rest. The election may have been a personal triumph for Clinton, but he emerged from the campaign without a party. The Bucktails had carried both branches of the legislature, effectively crippling him.[11]

Paradoxically, the architect of the triumph held no office. He had decided that eight years in the state senate were enough and declined nomination, but he was very much the undisputed leader of the Republican party. In August 1820 he became a major investor in the *Albany Argus*. Jesse Buel, its editor had offered to give up the *Argus* if suitable compensation were made. Van Buren quickly accepted. His brother-in-law Moses Cantine, hard hit by the depression, was in difficult financial straits and was more than willing to try his hand at editing.[12]

Though a lawyer with little editorial experience, Cantine proved to be a better newspaper man than Buel. The *Argus* was enlarged, new type purchased, a new flat-bed press installed. Editorial format remained largely the same. Articles, however, were more readable, and there were more national and international items, more exchanges with other papers to provide extensive coverage of political and feature material. With his brother-in-law as editor, the paper clearly reflected Van Buren's opinions, and in editorial policy hewed to a strict Bucktail line.

When Van Buren concluded his agreement with Buel he had decided to remain in Albany and to continue in politics. While temporarily shorn of his power, Clinton was quite capable of turning the tables on the Bucktails as he had done so often in the past. Before the spring election Van Buren had been seriously ill, which left him despondent and listless, unable to cope with the never-ceasing demands on his time and energy from the line of politicians that crowded in on him, staining the floors of his big front room with tobacco juice and filling the air with smoke from their long clay pipes. His lieutenants, capable, interested and industrious, somehow never seemed to make that difficult decision, never handled satisfactorily that earnest complaint from a country justice of the peace or a sheriff. And always and ever was the constant pressure from the horde of officeseekers bidding for the six thousand offices in the state. "The truth is," he complained to Rufus King just before the election, "I have scarcely had time to take my regular meals and am at this moment pressed by at least half a dozen unfinished concerns growing out of this intolerable political struggle in which we are involved." He was neglecting, he knew, his law practice despite the money pinch and the assistance of four clerks.[13]

Since February 1820 Van Buren had been thinking seriously of leaving it all, of moving to New York City and concentrating on his profession. On June 1, he poured his heart out to his old friend Gorham Worth in a long letter: "Until three or four weeks since I have labored with holy zeal and almost more than human industry to rid this devoted state of a Junta which sits like the nightmare upon her and which is hated and despised by all whom good opinion is worth having. . . . We have scotched the snake not killed it—one more campaign and all will be well." Had there been a complete victory over Clinton, or the reverse, had the Bucktails been routed, he would have left political life. But neither event happened. The Bucktails controlled the legislature and the ultra-Federalist Clintonian coalition remained intact.

The developing west still clung to the Governor though not his chosen representatives, variously denominated as the "Swiss Guard," the "Junta," "the mercenaries." Yet the political behavior of western leaders, even if they professed to be Bucktails, was unpredictable. If Van Buren left Albany now he would give Clinton and Ambrose Spencer a decided advantage. On the other hand, future prospects looked good if he continued to direct affairs with his usual tact, his restraint, his leadership on the attack. The Bucktails were nonetheless an unruly group, split up into local factions, divided by local interests, personal quarrels, bound together loosely with the Canal and council patronage at their disposal, but always before them the prospect of its loss at the hands of the cold, remorseless Governor and his hungry agents should there be a slip, a fall off in discipline. Van Buren, surveying his party with careful attention to its piebald composition, felt there had to be one or perhaps two more campaigns before Clinton and his faction were driven into private life. "For myself," Van Buren said, "I think we never had better prospects for another campaign, and notwithstanding my anxiety to get back to my profession I shall enlist again."

Clinton had become an obsession with Van Buren. He admired the man's accomplishments and his varied interests, at the same time believing sincerely enough that if unchallenged Clinton would smash up all parties in the state. Those Jeffersonian principles in which Van Buren had steeped himself meant little or nothing to De Witt Clinton. Should he prevail again in New York he would exploit its human and material resources for his own benefit and for those of his admiring coterie. Through a network of internal improvement at public expense, Van Buren saw an enormous and corrupt engine that would divert the property of the many to the enrichment of the few. He would never accuse Clinton of personal dishonesty, but he would and freely did accuse him of ignoring the corruption of others in his quest for power.[14]

Van Buren was doing a grave injustice to Clinton's character and his motives. That his image of Clinton as a public man was grossly distorted seems never to have occurred to him. Something in his intellectually deprived childhood, bred a hostility against the pretensions of men like Clinton. How could he forget the attitude of Elisha Williams, that mighty Federalist Lawyer, made all too clear when he was struggling so desperately to succeed? "Poor little Matty," Williams had once said, "What a blessing it is for one to think he is the greatest little fellow in the world. It would be cruel to compel this man to estimate himself correctly." Clinton, big, handsome, self-assured, college-bred, contemptuous of anyone who stood in his way, embodied a particularly wounding condescension.

The measure of Van Buren's anxiety may be seen in his manner, his emulation of their style of dress, their effusive urbanity, their studied courtesy, their amusing quips. Van Buren could not match Clinton's intellectual precocity, which in the eyes of educated young New Yorkers such as Gulian Verplanck may have been ludicrous pedantry; but to Van Buren it was a reminder of his inferior education, his uncertain social status. He could laugh with the rest of his friends at the caricatures of Clinton in the *Bucktail Boards,* but he still envied the Governor's vast display of learning, a wide-ranging intellectual breadth he could never hope to achieve.

So there was a personal as well as a political side to Van Buren's struggle against Clinton, a measure of vindictiveness that the Governor recognized and returned in full measure. To him, Van Buren was "intrinsically despicable." "I intend at the first convenient opportunity to cut him to the quick," he wrote an associate. Yet outwardly the two men preserved the civil amenities. When they met face to face they were reserved but courteous with one another, the one because he thought that was how gentlemen acted in public, the other because he knew how they acted.[15]

Once Van Buren had decided to remain where he was and continue his war against Clinton, he looked to the special session of the legislature which would meet in November to choose presidential electors. Clinton, in his annual message the previous winter, had finally responded to public opinion and recommended a call for a constitutional convention that would abolish the council of appointment and make other changes in the organic law the legislature might see fit to propose. He favored a convention whose powers would be limited to those in the call itself. His recommendation was a signal for the Clintonian majority in the legislature to stifle any action because members, whatever their faction, became hopelessly enmeshed in the specifics of the call. The only clear-cut division was that the Clintonians favored a convention of limited and specified powers; the Bucktails opposed any restraint.[16]

Clearly Clinton had planned on delaying tactics. Hopeful of victory in the coming election, he would recommend the same divisive procedure for the special session in November and the regular session that followed in January 1821. When the special session convened in November 1820, Clinton proposed that the call for a convention be submitted to the people, and if approved he proposed that there be a second election to accept or reject the results of the convention. The Bucktails were prepared for some sort of obstructive action, but they had not expected the form Clinton had given it. It was a shrewd political stroke that concealed a delaying tactic behind a façade of democratic procedure, a maximum of public participation in the convention process.

Weighing the risks, Van Buren advised his associates to push ahead and call for a convention with unlimited powers. After a brief debate, the special session approved such a call. A further and most formidable hurdle had to be overcome, however, before the call could be made. The council of revision, consisting of the governor, the chancellor and the judges of the supreme court, had the veto power. Most of the judges, and the chancellor, fearful of being "constitutionalized out of office," were expected to act against the bill, permitting Clinton to conceal his hand by voting with the minority. Of the judges, only one, Joseph C. Yates, had Bucktail leanings; Chancellor Kent was a Federalist; Chief Justice Spencer and a new appointee, John Woodworth, were Clintonians; and the remaining two justices, both Federalists, were absent on circuit. When the vote was taken, however, to the surprise of all and the intense anger of Clinton, Woodworth joined Yates in supporting the bill. Thus, the Governor was forced to break the tie and side with Spencer and Kent.[17]

Chancellor Kent prepared a veto message that was more remarkable for its special pleading than for any brilliant analysis or argument. Stripped of its legal verbiage, the veto argued simply that the people were as competent to vote on amendments or parts of the constitution as they were to consider a total revision. Further, the legislature had no power to propose any change in the constitution, even by amendment. Since the constitution was silent on this point, Kent simply stated that the omission had been purposeful, a judgment on his part that was surely open to an opposing view.

Kent's message infuriated the Bucktail majority and led to heated debate laced with acrimonious charges and countercharges. Van Buren, who held no public office but who was the acknowledged chief of the Bucktails, let the frenzy rage for a day or two. When emotion peaked, he had Michael Uhlshoeffer, the Tammany leader in the lower house, calm the troubled scene with a moderating speech followed by a motion for tabling the bill with the veto attached for consideration at the regular session of the legislature in January. Uhlshoeffer's motion prevailed in both houses. And as Van Buren planned, the Bucktail press went into action. For the next two months before the convening of the new legislature, the judges and Clinton were raked over unmercifully for their arbitrary behavior. Clinton had erred again by permitting such an ultra-Federalist as Kent, despite his legal erudition, to draft what was essentially a partisan document and then by signing the veto without filing a separate concurring opinion. A mystified John King wrote his father that "the opposition on the part of the Governor's friends has been inconsistent in the highest degree."[18]

Popular backlash, however, was simply incomprehensible to Clinton. "The objectives of the Council will be sustained by the people," he wrote his friend Post. "There is no inconsistency. I am in favor of a Convention properly and fairly called but not for one got up precipitately, for bad purposes under bad auspices and with a view to shake society to its foundations in order to sustain the predominance of bad men." Nor would he stop here in his headstrong, ill-conceived attempts to carry the war into the enemy's camp. As he wrongfully concluded, "The meekness of Quakerism will do in religion but not in politics." He had known for some time that Van Buren, through Smith Thompson and Tompkins, had solicited federal patronage in New York State and had received it. What he did not have at the time was direct evidence, and this he was determined to acquire. That he should have sought to injure Van Buren and the Bucktails by exposing a practice that he himself had indulged in early in his own political career, and which had become common usage, seems the height of folly. Perhaps Clinton believed that Van Buren's transgressions were so widespread and so flagrant compared with his own that the people would be shocked at the revelation. It is more likely he thought they would create an issue that would show that the Bucktails would barter state pride and states' rights to the Virginian dynasty and to the central government merely to maintain their political power at home. His ultimate objective, of course, was to discredit the Bucktail leadership in such a way that he could postpone indefinitely a constitutional convention. As usual, he acted without solid proof when he made his accusation in an address to the regular session on January 17, 1821.[19]

« CHAPTER 6 »

"The Tammany Horse"

As it was soon revealed, Van Buren's dealings with the federal government were neither corrupt nor in violation of either the federal Constitution or the state constitution. He had sought the removal of four Clintonian postmasters in the west, who he had reason to believe were holding up the transmission of Bucktail newspapers and pamphlets, and their replacement with members of his own party. He had also been instrumental in having his close collaborator, Roger Skinner, made the federal attorney for the Northern District. Skinner, in due course, had achieved a district judgeship and as he still remained a senator and the leading light of the council of appointment, he was the most striking example of a federal officeholder involving himself in state politics. Still there were many precedents, and Van Buren was not unduly concerned about Clinton's charges. Charles King put it neatly when he asked, "Is it criminal in the eyes of Mr. Clinton that a United States officer should be active against him and not for him?"

Van Buren suspected that Clinton did not have any hard evidence in hand. Why not call his bluff? At his suggestion, the Bucktails in the senate demanded that Clinton supply the legislature with more specific information. The Governor replied in a curt note of sixty words. Ironically he declared how much he appreciated the senate's "patriotic solicitude" and would in due course provide the evidence.[1]

Clinton again had given his enemies a means to embarrass and weaken him. Van Buren saw it at once and grasped the opportunity to turn his own words against him. Surely, if, as Clinton declared, an organized corps of federal officials in Washington were interfering with New York elections, he was making a most serious charge that impeached the integrity of the President and his cabinet. Since Clinton had not offered any evidence, the Bucktails in the state senate speaking through Peter R. Livingston stoutly maintained their "confidence in the patriotism and integrity of the general government, and will not change such opinion . . . but upon full and satisfactory testimony." The Clintonians had not anticipated this move and had no plans to oppose it by speech, debate or maneuver. Outgeneraled, they simply acquiesced to the resolution. Clinton had been caught in his own trap. His rage was olympian, his manner contemptuous, his reaction impulsively reckless.

The morning after the passage of the Livingston resolution the senate had before it the Governor's stinging reply expressing his "regret that any branch of the legislature should, in so unprecedented a manner, lose sight of the respect due to itself, and the courtesy due to a coordinate department of the government." A majority of the senate, including some Clintonians, were affronted at Clinton's brusque response. Led by Roger Skinner it voted not to receive the Governor's message. All of this rhetorical acrimony accomplished little but to place Clinton in a bad light. His tactic aimed at discrediting the Bucktails and thus further delaying any call for a constitutional convention had accomplished nothing.

Whatever delusion Clinton may have been laboring under about his part in the unseemly and rancorous exchange with the senate, he was realistic about the future of Van Buren who had seen to it that his own name was kept before the public by being named an elector on the Monroe-Tompkins ticket.[2] The Bucktails had lost no time in appointing their own council of appointment. "The Tammany horse rides over the Legislature like a wild ass's colt . . . thousands and tens of thousands of office leeches [appointed] under the new council," wrote Clinton with pardonable hyperbole. Referring to Van Buren, he said, "I am afraid he will beat Sanford for Senator. He will unless his friends stand out against a caucus decision." Even the Governor knew that would never happen.[3]

Van Buren had been thinking about entering the broader sphere of national politics since mid-summer, when he decided to remain in politics. He had held back for personal as well as political reasons. His practice was flourishing and he wanted to have his family comfortably settled with relatives and friends he could trust in looking out for their welfare. Now that his brother-in-law Moses Cantine was well established as editor of the *Argus* he was always available on short notice. His younger brother Lawrence in Kinderhook and Abraham in Ghent, a neighboring town, were nearby; and his sister Maria was running the household and providing in her way a parent's love to his motherless sons. Butler was on hand to help out if necessary. Van Buren had come to depend on this earnest, deeply religious young man, warm, friendly, transparently honest, valued in counsel, unstinting in his friendship. He was also fond of Butler's attractive young wife, Harriet, who he remembered as one of the belles of Hudson when he took in her husband as his first clerk.

The Butlers had rented a front room and parlor in Mrs. Jones's boarding house at 120 State Street just opposite Van Buren's law office and a short walk from the capitol. The partners were much together of an evening talking over business, politics and family affairs in one or the other's rooms. Before Mrs. Butler arrived from Sandy Hill, Van Buren often invited Butler, who was desperately lonely for his wife and family, to his house for a dish of hasty pudding and milk or some other light refreshment.

Socially, life in Albany was comfortable but hardly stimulating. Even the much more reserved, less discriminating Butler found it a dull place. In a letter to Jesse Hoyt, a Bucktail sachem in Tammany Hall, he said, "I like Albany about as little as you do—and with the exception of a few persons worthy of esteem have very little to say to the goodly inhabitants of this renowned metropolis." To Van Buren, who had lived there far longer than Butler and had pretensions to being a man of the world, the city had long since seemed a back water where society consisted largely of "bigots in politics . . . very full of prejudice and envy."

The special session in November had convinced Van Buren that Clinton was no longer a threat and that he could accept a seat in the United States Senate without prejudicing his own party's position in the state. He wrote at once to Rufus King and Smith Thompson asking their advice on whether he should be a candidate.[4] Both men responded favorably, so Van Buren went ahead and divulged his plans to his close associates, who now included Marcy and Benjamin Knower, a rich manufacturer in Albany and president of the new Farmers' and Mechanics' Bank, in which Van Buren owned an interest. All warmly backed his candidacy. By December the news was generally known and well received throughout the state, except, of course, from Clinton and William P. Van Ness, who still controlled a small faction in Tammany, and John W. Taylor, newly elected Speaker of the United States House of Representatives.[5]

Despite all the support Van Buren commanded in the party, he was still worried about Clinton. In early January 1821 he wrote Rufus King alluding to the editorial attack the Governor was making against King's son Charles and other "high-minded gentlemen" under the pen name of "Heraclitus" in the columns of his paper, the *Statesman.* Van Buren was also concerned about the fate of the convention bill, then before the legislature, and was most anxious that King speak about it to his close friend Stephen Van Rensselaer, the Patroon. He feared Clinton would persuade Van Rensselaer that a new constitution would injure property rights. "Although I have certainly not taken a lead in this business," Van Buren said, "being somewhat timid in all matters of innovation, still I am thoroughly convinced that temperate reform and that only is the motive of our friends who urge it most strenuously." In asking King to use his good offices with the influential Patroon, Van Buren was hoping also to reassure the conservative King that the convention would not get out of hand.[6]

Van Buren had another motive in conciliating King. Careful as always, he wanted to ensure the vote of the "high-minded" Federalists for his election to the Senate. Should Clinton be able to renew his alliance with most of the Federalists, a remote possibility he might with his own faction and disaffected splinter groups defeat him.

All these hypothetical situations were the imponderables of the political game Van Buren was playing. That they were unlikely to occur meant that they should be taken all the more seriously. Van Buren had learned, and learned well, that unpleasant surprises could confound the most carefully laid plans of the overconfident. Besides, he had his hands full controlling the unruly Bucktails, who were thirsty for office now that they controlled the council of appointment.

For every removal and new appointment from the humblest of offices, the fence watcher to the highest, the mayor of New York City, Van Buren knew there would be a dozen disappointed and disgruntled aspirants. From his lofty but lonely perch in the governor's office, Clinton waited confidently for the Bucktails to destroy themselves over the spoils. "The whole state is alive for office and next week will exhibit a scene of office hunting heretofore unknown in the annals of the community," he wrote Stephen Van Rensselaer. With Roger Skinner leading the way, the council could not be restrained even by Van Buren. "I hope the council will soon finish all they have to do," said Butler, "as the excitement produced by their labors is very great and the difficulty of pleasing everybody very strikingly illustrated. . . . The minor appointments for this city have given great dissatisfaction, and it is as much as we can do to keep the people from open rebellion."[7]

At its first meeting on January 12, 1821, the council removed eleven county sheriffs, the comptroller, the treasurer, the attorney general, secretary of state, all the chief officers of the militia, until then considered to be nonpartisan posts, the mayors and recorders of New York City and of Albany, and the superintendent of common schools, a record of proscription unparalleled in the history of the state. Over the next six weeks, the council systematically combed through the six thousand minor posts, removing the few Federalists yet remaining, all the Clintonians and even some Bucktails considered ineffective or doubtful politically. On the important state offices, Van Buren was as ruthless as Skinner and his fellow council members; for in his planning they were to act as the nexus of his personal power, directing the unruly party with a firm but tactful hand during his anticipated absence in Washington.

William L. Marcy was appointed adjutant general. Knower became treasurer and a very special Van Buren protégé, Samuel A. Talcott, attorney general. Talcott, a "high-minded" Federalist was an amiable, even-tempered, judicious man on whom Van Buren relied to balance Skinner's tendency to act impetuously in politics. Van Buren selected another stalwart supporter, Stephen Allen, one of the moderate Tammany men, to be mayor of New York City. Clinton's former friend and ally, now a political opponent, Richard Riker, was made recorder of the City, thus satisfying the members of Tammany whom he had managed to conciliate after his break with Clinton. Van Buren's friend and political ally in Albany County, Charles E. Dudley, a wealthy merchant, local philanthropist and the new mayor of Albany, would act as his alter ego in the capital. These men, including Roger Skinner and Walter Bowne of the council, were an efficient political combination[8] close to Van Buren, sharing his antipathy for Clinton and his firm belief in Jeffersonian principles. All of them were men of high spirit and independent thinking, however. Although they looked up to Van Buren and more often than not agreed with his reasoning, they did so because he had convinced them a particular course was the right one, not because of any subservience. Emphatically, he did not think of his group as disciplined subordinates; dissent and free discussion was the rule, not the exception. The only constraint, if it could be called one, was that majority opinion govern in political decisions; and once a consensus was reached, as in a party caucus, it was binding unless unforeseen events forced a change. Van Buren had been disturbed at Skinner's patronage policy, but when he found that he could not restrain the council he backed off, trusting that the bad temper stirred up over the minor officers would subside once the appointments had been made. As Butler put it, "I hope a few days of reflection will compose the angry elements."[9]

The disturbance Skinner's council had created among the Bucktails would have been more serious had not Governor Clinton unwittingly diverted attention from its doings by his message to the legislature supplying the proof of his previous allegations of Washington's interference in New York politics. Clinton had been collecting documents, letters and affidavits since November and by mid-January had several hundred in hand that demonstrated beyond doubt partisan activity on the part of federal officers. The Governor's collection was so voluminous that it was conveyed to the legislature in a large green bag, hence the *Albany Argus* instantly dubbed the message that accompanied the documents the "green bag message."[10]

As Van Buren had anticipated, the "green bag message" was a nine-day wonder. From Washington, Rufus King sent word that Clinton's "green bag has made no proselytes here; on the contrary it has dissatisfied those, or at least some of them, who were friends of his." The Bucktail legislature remained unshaken and simply referred it to a joint committee, where it remained until March 14, long after the charges excited any particular interest. The report from the joint committee, which bore traces of Skinner's vengeful hand, was a cynical document that proclaimed most of Clinton's affidavits and depositions fraudulent and then concluded with the preposterous assertion that there was not now, nor ever had been, any unwarranted interference by the federal government in New York politics.[11] Clinton's mighty bombshell had burst harmlessly above the legislature, lightened the chambers briefly, and drifted off in a billow of smoke.

While the "green bag message" languished in joint committee, the Bucktails and their "high-minded" allies met in caucus to nominate a United States Senator on February 2, 1821. With eighty-three members of the legislature present, Samuel Young called the meeting to order. He made some brief remarks, hoping that harmony and good fellowship would prevail. He also reminded his fellow members that the caucus decision must be binding. Young then recognized Samuel B. Romaine, a member from New York City, who nominated Sanford to succeed himself. A western member from Otsego followed, placing Van Buren's name in nomination. Several members spoke at length in support of Sanford. Young bore the burden of defending the Van Buren nomination while not offending Sanford's supporters. When he finished, a ballot was taken. Van Buren received 58 votes to Sanford's 24, with one abstention. Sanford's spokesman then rose and moved that the caucus make the nomination unanimous for Van Buren. His motion carried, and on February 6 the legislature elected him to the Senate for a six-year term. The Clintonians in a body with a few disgruntled Bucktails, associates of William P. Van Ness and Joseph Yates, voted for Sanford to the disgust of Jabez Hammond, who cast a blank, but thirty "high-minded" Federalists marking their ballots for Van Buren compensated for these losses.[12]

Van Buren's policy of separating patriotic Federalists from those who opposed the War had been a wise one. His support of Rufus

King for reelection to the Senate, his encouragement of the "high-minded" faction and selection of one of their respected group, Samuel Talcott, for attorney general, had all been clever political moves. Yet the great Governor was after all the reason for Van Buren's victory, not just in the Senate election, but in helping him keep the unruly Bucktails together and in making him the undisputed political leader of New York, an irony of fate and of circumstances. With all his intelligence, all his political experience, Clinton never understood how much he had helped his opponent.[13]

In Washington, Rufus King was delighted at the result of the election. He expressed a confidence that must have warmed Van Buren's heart. King would work to educate Stephen Van Rensselaer in what he termed "correct opinions" about New York politics. "Nobody better appreciates the character of Mr. Clinton and his chief supporters than the Patroon does . . ." said King. "He cannot be supposed to support his [Clinton's] sinking course."

Van Buren would have eleven months in which to settle his affairs, personal and political, before his term began in Washington. Uppermost in his mind was the forthcoming constitutional convention. The legislature had brushed aside Clintonian delaying tactics and passed a bill that met the principal objection of the council of revision by agreeing to a referendum on whether there would be a convention. The majority followed this action with a second bill that set no limits on the character of the convention. Public opinion was so obviously behind the legislative majority that the council dared not veto the bills. Voters approved the call by an overwhelming majority. As provided in the legislation, elections for delegates would take place in June and the convention itself would assemble at Albany on August 28, 1821. "A hot town, for a hot season," said Rufus King, who had been elected a delegate from Queens County, adding, "we must strive to keep ourselves cool, or the internal and external heat may lead to mischief."[14]

Van Buren had similar thoughts. In the forefront of the agitation for a convention since it first surfaced three years earlier, he sensed the time had come for a change, and he saw no other way of ridding the state of Clinton as well as the political meddling of Federalist judges.[15] He also felt intuitively a popular surge for reform, stimulated in part by constitutional changes in other states, and more especially by the rapid growth of New York. A prudent man, Van Buren was sensitively aware of the pitfalls the convention posed for the unwary. He had put together a loose and fragile coalition of eastern small farmers and tenants, radical westerners, urban machine politicians and anti-Clintonian Federalists, all of whom had distinct, social, regional and economic interests.

The convention posed the great political opportunity of his career, and he meant to capitalize on it, but at the same time he regarded it as the greatest challenge he had yet faced, a challenge that would engage all of his energies, all of his talents for careful management. Nor was this simply Van Buren the politician in his assessment; there was the Van Buren who had a deep sense of pride in his native state, its rapidly growing population, its expanding economy, and its dynamic mixture of cultures from the Irish immigrants crowding into New York City to the stolid Dutch farmers of his own stock, from the bustling, energetic Yankees to the more conservative native-born New Yorkers of British antecedents. He had long rejected the rigid class and caste of the closed Dutch and Palatine German communities of his boyhood, the cause of remembered humiliation and condescension. Outside of his immediate family, Van Buren's closest friends and associates were not of Dutch ancestry. And though he could still speak Dutch, he was more at home with the industrious, clever, well-educated Yankees than he was with his own people.

The conservative Rufus King found difficulties in the changing patterns of population, "nearly divided," as he said, "between the old and the new inhabitants. The latter are out of New England whose laws, customs and usages differ from those of New York." But Van Buren welcomed these changes, finding a kind of dynamism in which his style of life, business and politics could and did flourish. Reform was in the air, if it could be channeled into what he took to be the main currents; he could bring better government to New York while incidently paving the way for a new more democratic and more flexible party system. The risks were great, but he looked forward to August with a feeling of optimism for himself personally and for the state he served.

First, he had to be elected a delegate. Albany, where he lived, was solidly Federal-Clintonian. He had no chance there, nor any better chance in his home county of Columbia. While he was casting about for a solution to this problem, the nomination came completely unsolicited from Elisha Foot, a political acquaintance and local Bucktail boss for Otsego County, where Van Buren had extensive land holdings.[16]

The voters of New York showed unusually good sense in chooseing their delegates. With a few exceptions, they selected the most able men in the state, even though an overwhelming majority was Bucktail and there were rather more lawyers than farmers, merchants and artisans. Rufus King was the most distinguished, but Ambrose Spencer, Jonas Platt and Chancellor Kent had few peers in the nation as scholars of the law. Francis Sylvester, Van Buren's first teacher in the law, Abraham Van Vechten, Elisha Williams, William

W. Van Ness, Stephen Van Rensselaer and his cousin J. Rutsen Van Rensselaer, all men of wide experience and ability in the law and in public affairs, were chosen members. Federalists and Clintonians, these members of a repudiated aristocracy may have been a minority of the delegates, but they compensated for their numbers by their intelligence, their perception and in their conservative views, ably expressed, would act as a makeweight to the more radical, more voluble majority.[17]

The Bucktails had no such constellation of bright stars, but they did not lack for talent, for verbal skill, and for prestige. Vice-President Tompkins headed their list, along with ex-United States Senator Nathan Sanford and Senator-elect Van Buren, Colonel Samuel Young of Saratoga, Erastus Root, the vigorous champion of the west, and that violent partisan, Peter R. Livingston from Dutchess. Van Buren, Young, Root and Livingston, good speakers all, agile of mind, quick in debate, would prove a match for any member of the convention, and no delegate would be as adept at composing differences as Van Buren.

The temporary chairman, Erastus Root, opened the convention with 110 delegates present. Seating arrangements in the capitol did not follow political lines or importance. Rufus King sat in the first semicircle immediately to the right of the Speaker's dais. Chancellor Kent was placed far back in the hall, as was Van Buren on the far left and Ambrose Spencer middle left. Colonel Young was tucked in on the far right where he could not even see the presiding officer and thus would have to walk several steps to catch his eye if he wanted to speak.[18]

After the roll was called Root moved, and the convention accepted his motion, that a permanent president be elected by ballot. Tompkins was chosen. The convention, in a symbolic gesture of harmony, then selected Root and Kent, the one a radical from the west, the other a distinguished conservative from the east, to accompany the Vice-President to the chair. It took two days for the convention to organize itself, but finally on August 30, 1821, it was ready for business.

Appropriately, the only founding father of the United States Constitution present, Rufus King was the first speaker. In a voice so low that at times he could not be heard in the distant rows of the spacious chamber, he "urged the observance of moderation, of mutual confidence and the most exemplary prudence in our proceedings." At the end of his remarks, he proposed that the convention elect a committee on committees.

After a bit of inauspicious wrangling, the convention agreed with King and selected a committee of thirteen. The committee reported the next day, moving that ten select committees be appointed to examine and recommend changes in the articles of the constitution. Tompkins was given the authority to designate the chairmen and the membership of each of the seven-man committees. After the session adjourned, he asked Van Buren what committee he would like to head. "On the appointing power" was the prompt response. Tompkins smiled broadly and with a knowing look turned away.

When the convention resumed its deliberations, and Tompkins announced the membership of the committees, he named Van Buren to chair the committee of his choice. Delegates, whatever their political persuasion, assumed that the Bucktail leader would devise some clever scheme that would do away with the council but reserve the patronage for his and his party's purposes. Tompkins had assigned one Federalist, Philip Rhinelander, Jr., and one Clintonian, Victory Birdseye, to this committee, whose only distinguished member was Ogden Edwards, a Tammany sachem and a cousin of Aaron Burr. Van Buren realized what everyone in the convention was assuming when Tompkins named his committee members and as always brought to play his intuitive understanding of human nature in dealing with his colleagues. "I rather mischievously delayed calling my committee together until the suspicions . . . had time to mature," he recalled.

When the select committee met for the first time, Chairman Van Buren suggested each member give his opinion before he gave his. Ogden Edwards, assuming, and rightly so, that Van Buren had a purpose in mind, immediately objected, declaring that the chairman should give his views first. Van Buren blandly explained that as the presiding officer he regarded his role as being more of a mediator than a participant. Of course, he had his own opinions which may or may not coincide with those of his fellow members. Yet someone must summarize conflicting views without prejudice and try to reach an agreement; some member must take this responsibility, obviously the chairman. The committee, whatever its private misgivings, was unable to offer an alternative to this apparently sensible procedure. Each in turn gave his views and then all turned to the chairman, who promptly adjourned the meeting with the remark that he would make his comments on the morrow.

Van Buren knew that his ruling would be the subject of much speculation. He had counted on building up suspense and hoped to catch his committee off guard. He knew all of them wanted the council of appointment abolished and the appointing power lodged either in the legislature, or in the governor by and with the consent of the legislature, or through the direct election of all public officers. Two other important facts had emerged from the responses of the members. The committee was divided on the means of accomplishing reform, and it thought Van Buren wanted to preserve the council in some form. With this information, Van Buren decided to turn the tables on the committee by recommending that the council be abolished, that state officers and militia general officers be appointed by the governor with the concurrence of the senate, that all other militia officers be elected in ascending order of rank, and that mayors and other city officials be elected locally. Having done the unexpected, he hoped that he could slip through the committee a plan whereby the more than three thousand justices of the peace and as many more sheriffs would seem to be selected locally, but in effect would be partisan appointments still controlled from Albany. Justices of the peace in New York, through custom, precedent and law, had a great deal of power over local affairs in civil and criminal actions. They were a potent force in the community, the spine of any political organization.

As he had anticipated, the committee was baffled when he made his recommendations. The justices of the peace, he said, were civil officers combining judicial and executive functions, and thus county courts and the boards of supervisors, two local agencies, should each prepare lists of nominees. If the names were the same on both lists, they were appointed; if not, the governor would resolve the difference. Sheriffs would be selected in the same way.

Van Buren's plan was a clumsy, cumbersome attempt to centralize appointments of these politically important officers in Albany, yet preserve the semblance of local appointments. The voters in the towns did indeed elect their own boards of supervisors, but Judge Skinner's council had appointed the county judges then sitting who were uniformly Bucktail in their politics. Thus, the present justices of the peace were safe for their prescribed terms. After that, if the elected boards disagreed with the county judges appointed by the governor and the senate, Albany would dictate the appointment on partisan grounds.

As lackluster as the members of the committee were, as confused as they certainly were by Van Buren's unexpected turnabout, they immediately saw through his plan. Although heavily weighted with Bucktails, the committee showed considerable independence of spirit and made it quite plain that they were not creatures of their powerful chairman, who took their comments with good grace. In all other significant respects they agreed with Van Buren's proposal, but a majority would not go along with him on the justices and the sheriffs. Four members of the committee voted that these officers be elected by the people in their localities.[19]

The core of his plan defeated in committee, Van Buren fell back on the expedience of delay. He saw to it that his report would not be discussed in the committee of the whole until near the end of the session after the delegates had become weary of debate over such explosive issues as suffrage, the veto power of the council of revision, the powers of the governor and the judiciary.

It was not until October 1, after the convention had been in continuous session for over a month, that he submitted his report for debate, though it had been circulated since September 18. As in his select committee, discussion centered on the issue of the justices of the peace and the sheriffs. Van Buren admitted that he had opposed the election of these officers "at every stage of discussion; and it had been a source of sincere regret, that in that respect, [I have] been overruled by the Committee." In an effort to make his plan more palatable, he said that his committee had considered whether the county judges alone should appoint the justices. "But," he remarked, "on more mature deliberation it was concluded that it would be making political engines of them, and therefore it was abandoned." Stressing the fact that under his plan nominations would be made locally and underscoring the coequal status of the local supervisors in the nominating process, he won over many doubters, though not before he was forced to compromise on the appointment of sheriffs, next to the justices of the peace, the most important local political officers in the state. Sheriffs were made elective posts. And he lost ground with many of his closest supporters who wanted to retain the old council in some form. Michael Uhlshoeffer, his trusted lieutenant in Tammany, was harshly critical. "Some of your *good friends,*" he remarked, "say that you have quite dextrously made the city delegates assume the responsibility of a plan, instead of making the authority yourself and pointing out a plan to embrace the whole state."[20]

A completely opposite but not less critical view came from prominent western Bucktails. Samuel Beardsley, a rising young politician, wrote vigorously attacking Van Buren's plan for selection of justices of the peace. He wanted the voters in the various towns to elect their own justices. "These Justices," said Beardsley to Ela Collins, a member of Van Buren's committee, "although of trifling consequence individually, yet when arrayed together may need not shrink from contrasting their official power and their personal influence with any other class of officers in the state."

Nor could Van Buren's powers of persuasion convince the one man he most wanted to convince, Rufus King.[21] He expected the onslaught of Judges Van Ness and Spencer, and of Chancellor Kent, and was prepared to meet it, but he was severely hurt when King imputed his democratic principles. "Call a country town meeting an

Buren from the left and from the right, in a way, was a tribute to his management; Van Buren almost singlehandedly guided the delegates along a middle course. The completed document may have been imperfect in many details, but it represented a far more democratic and a far more realistic constitution than the one it replaced. It did not compare unfavorably with the constitutions of other states, and in fact, must be considered more representative of republican institutions than most.[26]

« CHAPTER 7 »

An Uncertain Regency

Van Buren was unwell when he bade goodbye to his family and friends in Albany, but his spirits were high. He felt a great sense of relief that the convention was over and the party organization, though strained, was still intact. As to the revised constitution now before the people, he remained somewhat skeptical. A cautious political leader, he had never been a zealous advocate of those sweeping reforms some of his closest associates urged. "On this," observed the old conservative Rufus King with obvious approval, "he has been not only sincere, but individually disinterested, and should the Constitution fail it will possibly create more joy than grief."[1]

There had been times when he despaired of composing what seemed to be irreconcilable differences between delegates from the established eastern counties and those of the new west. His role as chief mediator had not been a pleasant one. Old friends and close political colleagues whose vision was circumscribed by residence and particular interest had turned on him with a vehemence that startled, disturbed and hurt. For a man who prided himself on restraint, who had trained himself to divorce personal feelings from public issues, the arguments, the constant bickering, on the convention floor and behind the scenes had shaken his sense of detachment and even his confidence in himself. He had had a disagreement with Samuel Young and Erastus Root over the suffrage amendment which left a residue of mistrust. Far more serious was a bitter argument that developed between him and his faithful coadjutor, Roger Skinner, over several points that seemed to threaten centralized party control. Skinner was popular in the eastern counties, and permanent alienation would injure Van Buren's position. Yet, when he heard that Skinner was slandering him, he chose to believe it though the information was all hearsay and gossip. After he left Albany he resolved never to have any further political and personal friendship with Skinner. With several cases to argue before the fall session of the

Supreme Court in New York City before he went to Washington, he was anxious to consult other party members, lawyers like himself from all over the state who had business before the high tribunal.

Van Buren had scarcely settled himself in Mrs. Henderson's boarding house only to be dismayed when Skinner appeared with his valise.[2] The landlady had put them both in the same room. Van Buren at once decided "to cut all future intercourse with him" and would have sought other quarters "but found I could not without stating the reason explicitly." Skinner, unaware of Van Buren's burning resentment, ignoring or failing to recognize his "rigid reserve" toward him, was cordial, open and most friendly. When Van Buren's stubborn illness flared up again and he was confined to his bed for several days, Skinner treated him "with so much kindness that I was obliged in a great degree to alter my conduct." The deep-seated rancour remained, however, and continued to weigh on his mind after he reached Washington.[3]

It was foreign to Van Buren's character and foolish of him to bear a personal grudge over a length of time. Slander and libel were occupational hazards for any politician, as he well knew. He had taken more than his share of abuse with equanimity, rarely ever permitting vindictive outbursts to disturb his composure. To be sure, Van Buren was unwell, his nerves on edge, but his attitude toward Skinner bore unmistakable signs of concern about the state of the party in New York.

At a critical juncture in his career, he was determined to make New York the pivotal state in the Union; and if the two Clintons, Burr and Tompkins had all failed in their quest for the Presidency, he thought he knew why. None had a solid organization behind him. Magnetic, able, shrewd in their way, they had all relied overmuch upon their personal ties. Personality was one ingredient for success, but to Van Buren the least important. He would not make that mistake. They had all been impatient men. He would wait. Nor had the times been right to realize their ambition. Virginia and Pennsylvania were then more populous, more important states than New York. Now, however, his native state had outstripped them. When the great Canal was completed, opening up resources and land and putting New York in a position to exploit the trade of the Northwest, he was confident that villages like Buffalo and mere settlements near the Great Lakes and the Canal route like Rochester and Syracuse would vie with Albany for population, wealth and enterprise. As for the dirty, disorderly, expansive city of New York, with its magnificent harbor, its unsurpassed site and position for internal, coastwise and international trade, there seemed no limits to its economic growth. Time, Van Buren was certain, was on his side, provided always he

could control his highly independent followers, men like Skinner, Young, Root and Porter, who were apt to go their separate ways, apt to make mistakes, apt to pull him down.

Success for his grand design depended on this inherently unstable base at home. His jousts with Skinner over the fundamental aspects of political control during the convention had awakened suspicions first aroused when Skinner had overreached himself as the dominant member of the most proscriptive council of appointment in the history of state politics. Skinner's loose talk and the fire of his rhetoric had upset Van Buren. His state of mind led him to reassess Skinner's fitness, to question his loyalty, to wonder whether in his scheme of things at Albany the party could tolerate a leader so thoughtless of others' feelings.

Van Buren had a difficult and delicate job ahead of him, nothing less than establishing his primacy over New York's twenty-eight Congressmen as he had temporarily at least asserted his control in Albany. His friendship with his senior colleague, Rufus King, was secure, but the situation in the House was another and more difficult matter.

The informal leader of the New York delegation was John W. Taylor, now about to serve his fifth term in the House. Intelligent, hard-working, he had been chosen Speaker of the House, succeeding Henry Clay. His election was a distinct tribute to his qualities of mind and his industry because he was not a popular Congressman, and his forthright opposition to slavery in the territories had earned him few friends in the South. Yet such was the force of his leadership qualities and his admitted capacity that he had been able to unite the free states behind his candidacy. When the slave states divided their strength between two competing aspirants, Taylor was elected.

Taylor and Van Buren had been close friends and political allies during the War, but for the past six years they had gone their separate ways. So wise in the ways of politics, Taylor had made the mistake of identifying himself with De Witt Clinton. This attachment had alienated not just Van Buren's Bucktails but the "high-minded" Federalists as well.

Taylor had no illusions about Van Buren's intentions after his election to the Senate and the complete eclipse of the Clintonians in the spring elections of 1821.[4] Could he count on his well-cultivated friendships and alliances in the House to beat off opposition from his own state and from elsewhere? Calhoun blamed Taylor for the defeat of his program to reorganize the army; many of his committee appointments were considered hostile to other cabinet members.

Well before the convening of the Seventeenth Congress, Taylor learned that Van Buren was "drilling" the anti-Clintonians against him. But he was also advised that Tammany "radicals" were still chafing about their treatment at the convention. They had all been for universal suffrage and held Van Buren responsible for its defeat. His open alliance with the "high-minded" men in New York City had not set well with the mechanics and the tradesmen in the working-class wards Tammany men represented.

Despite the odds, Taylor remained a strong candidate for reelection and Van Buren knew that he would have a hard fight on his hands, that success was by no means certain.[5] As soon as he reached Washington, on November 5, 1821, and settled himself in temporary quarters at Strothers' Hotel at 14th Street and Pennsylvania Avenue, Van Buren had his first caller John C. Calhoun, secretary of war, one of those national celebrities he was anxious to cultivate. Tall, spare, almost exactly Van Buren's age, Calhoun was considered handsome among the ladies of the Washington social set, with his square jaw, rugged features, piercing brown eyes under heavy brows. Stiff dark hair shot through with gray added a touch of gravity to his otherwise youthful appearance. Calhoun bore a reputation for being imperious and aloof, but Van Buren found him to be "a very fascinating man" whose society he enjoyed. The mutual regard that for a time blossomed into friendship could not disguise the fact that as politicians each wanted something from the other.

Calhoun's residence was near Strothers', and during the next month he invited Van Buren to dinners and evening games of whist. Calhoun, who was just beginning his campaign for the Presidency, very much wanted Van Buren's support; and though he tried hard to mask his motives, he did not succeed. Van Buren had taken his measure at their first meeting. Though he feigned ignorance of Calhoun's intentions, he listened to the Washington gossip carefully and calculated his chances without committing himself in any way.[6]

Van Buren's reputation had preceded him to Washington. The mystifying character of New York politics baffled politicians from other states. How was it that such a pleasant man who seemed more interested in charming the ladies than in affairs of state, who heartily enjoyed the pleasures of the table and chatting about the merits of this year or that of madeira, could have destroyed the political career of such a giant as De Witt Clinton? Outside of New York most agreed that Clinton was not just a great man, but the greatest man in the Union, the tireless advocate of monumental projects, the father of the Erie Canal. There was not a newspaper worthy of the name in the Union that did not chart the progress of this titanic undertaking, this symbol of American enterprise, of all that Americans thought promising in their destiny. Clinton was its architect, its

promoter, its builder. Those Americans who counted were proud of Clinton, a mighty mover and shaker in national affairs for two decades. Those Americans who read *Niles Weekly Register* or purchased Longacre's lithographs, and that meant most of Washington society, were familiar with the generous features, the pensive brooding eye, the wide brow of the "deep" thinker. Yet this slight, average-appearing man who preferred small talk to graver matters, who occasionally stumbled in his grammar and who spoke so rapidly at times that his conversation could be difficult to follow, had pulled him down. True, Clinton was still governor of New York, but the knowledgeable ones in Washington, knew that his political career was in a shambles.

Calhoun had precise information on Clinton's fall and the extent of Van Buren's role in accomplishing it. His classmate at Yale and devoted friend, Micah Sterling, had kept him well informed, as had Peter B. Porter, a fellow "war hawk" in the Twelfth Congress and also a close friend. Clearly, any politician who could stand up to Clinton and defeat him on his own ground was one whose friendship was worth cultivating. As for Van Buren, he was aware of the differences that existed between Taylor and Calhoun and he wanted the Secretary's assistance in defeating Taylor's reelection to the Speakership. After the whist games were over he had had several private discussions on the matter. Without disclosing any of his reasons for wanting Taylor out of the way, he quickly obtained Calhoun's help.[7]

While he was partaking of Calhoun's hospitality and gaining his support, Van Buren was planning a flank attack on Taylor. Shortly after his arrival in Washington he had written John Van Ness Yates, New York's Bucktail secretary of state, for the addresses of the Republican members of the legislature and of the delegates to the constitutional convention. Obviously worried about the opposition to his moderate course there, he planned to head off any possible revolt should the referendum on the revised constitution fail and he be blamed for it. Since Congress did not assemble until December 3, Van Buren had almost a month to survey the scene, meet those of his colleagues he did not know and renew old acquaintances. Much of his time he devoted to the New York and Virginia members of the House. Enlisting Smith Thompson and the Bucktails who had come on with him to Washington, he swiftly organized his own campaign against Taylor. "Several of our northern and western members," complained the Speaker, "who left home favorable to my reelection were induced before their arrival here, but I know not what motive, to oppose me."[8]

Yet the battle was a stiff one. Taylor had not been idle either. He managed to make amends with John Quincy Adams for his conduct in the House on certain bills of special interest to the dogged New Englander, thus ensuring a bloc of votes from the Northeast. He also held on to thirteen of New York's twenty-eight-man delegation, and though his support from Maryland, Kentucky, Ohio and Indiana had been whittled away he still commanded substantial strength in those states. On the first day of balloting he missed a majority by a scant four votes, though his near victory was the result of a blunder on the part of his opponents rather than the success of his own campaign.

Van Buren and Thompson, sensitive to criticism for opposing a native son, made the mistake of settling on a first-term Congressman from Delaware, Caesar Rodney, whose principal attraction lay in the fact that his father had signed the Declaration of Independence. In all other respects he was a man of mediocre ability, unprepossessing in appearance and in manner. The scheme, like so many shrewd contrivances, had all the requisite political elements, but had neglected the most important of all, the capability of the candidate himself. Van Buren wanted an uncontroversial Speaker from an insignificant border state. Considered by many a self-serving nonentity, Rodney was not just uncontroversial, he was inconspicuous as well.

Calhoun had no candidate but sent word through a Pennsylvanian ally, Samuel Ingham, that he would back Rodney. When the House adjourned after seven ballots with Taylor leading the pack, it was all too clear that Rodney would not receive undivided support from any of the border states or from Virginia and Pennsylvania. Louis McLane, a Federalist from Rodney's state of Delaware, was next tried, but he proved weaker than Rodney, as did the hero of Fort McHenry, General Samuel Smith of Maryland.[9] Realizing that his policy to replace Taylor with someone from neutral ground between the great northern states of New York and Pennsylvania and the South had been a mistaken one, Van Buren could no longer conceal the responsibility of sacrificing the pretensions of his own state to his concept of party loyalty.

The next best tactic involved considerable risk to him personally, that of backing a candidate from Virginia who would not only receive the united vote of that state but of the entire South. The Bucktails caucused that night and, at Van Buren's behest, decided to support Philip P. Barbour, just beginning his fourth term in the House. In the Washington community Barbour was known as a competent man but a stickler for detail and fanatically devoted to the interests of his native state.

After the Bucktail caucus Van Buren sounded out other delegations on Barbour. He also consulted Calhoun, who offered no objection. By no means satisfied that Barbour would beat Taylor, Van Buren decided this candidacy was least likely to injure him if it failed. Luck was running with him when the House assembled the next day, December 3; and to the surprise of everyone, Van Buren included, Barbour won with a plurality of two votes over Taylor after five ballots. It had been a near thing. Had Taylor won, Van Buren's political standing would have been injured in Washington and New York. He might well have lost control over the Bucktails in Congress who Taylor would surely have disciplined in distributing committee assignments and in his control over debate.

The Speakership contest had a salutary effect on Van Buren. He had come to Washington expecting that the same tactics which worked so well in New York would work equally well in Washington. Personal relations among members of the various Messes, he discovered, could transcend party discipline, which was much looser than at home. Members of the executive branch, especially the cabinet members, exerted more influence than their counterparts in Albany, the President being the notable exception. He had not canvassed the ground thoroughly before he challenged Taylor, a mistake he would not make again. Yet he had gained much stature from the contest, and his role in backing Barbour had brought him the friendship of the most powerful Virginia politicians, including the valued support of the *Richmond Enquirer* and its erratic but influential editor Thomas Ritchie, a leading member of the so-called Richmond Junto of fastidious strict constructionists.[10]

With the Speakership contest now settled to his satisfaction, Van Buren turned his attention toward Albany and the campaign of 1822. One major concern, the referendum on the amended constitution turned out to be no problem at all; the electorate justified Van Buren's moderate course when it approved his handiwork with a majority of more than two to one.[11]

But before Van Buren made any decision about the campaign, he had to settle once and for all the problem of his personal and political relations with Skinner, which had been bothering him for the past two months. Could he have misjudged Skinner's remarks? Had Skinner actually slandered him or had his heated expressions been exaggerated in such a way that a difference of opinion became a premeditated impeachment of his integrity as alleged? Van Buren decided he needed counsel and wrote a long letter explaining his view of the affair to Charles Dudley, mayor of Albany, who had known of his displeasure at Skinner's remarks but had never dreamed that they had made a deep and lasting impression on Van Buren.

Dudley and Benjamin Knower, the state treasurer to whom he

showed the letter at Van Buren's suggestion, were aghast. Should the relationship between the two men be terminated as Van Buren plainly threatened, they saw a certain rupture ahead that could damage perhaps irreparably the party organization. After making some discreet inquiries they wrote Van Buren that he had been misled; "The judge [Skinner] had not spoken improperly of you or with malicious intent; and that if he did at any time during the session of the Convention indicate any dissatisfaction as to the course you were pursuing, it must have been owing to the great anxiety he had for the credit and honor of that body."[12] Dudley and Knower's comments had a soothing effect. Van Buren relented and his relationship with Skinner resumed its natural course. Having survived the tensions of the convention, Van Buren's Albany group now began to impose a kind of collective political leadership on the party that transcended local and regional interests.

Some months before in the remote and declining town of Manlius, an obscure weekly, the *Onondaga Republican,* made its appearance. This ephemeral sheet gave Van Buren's organization a name that caught the public fancy, the "Albany Regency." Its young, struggling printer-editor, Thurlow Weed, coined the expression that would be a household phrase for thirty years. Van Buren had learned of it during the convention and was soon using the term himself in his political correspondence.

Weed, a self-educated printer of remarkable foresight and editorial talent, had sensed and captured the essence of the new Republican organization just as it was taking shape. The fact that its leader was now a Senator in Washington gave an ironic twist to the expression. The Albany Regency was a tight little group of extraordinarily able editors, politicians and officeholders that Van Buren had gathered together in Albany. All of them had learned and were practicing his simple maxims of political behavior: party identity, party consensus, party loyalty, party principles and party organization down to the towns and villages as strict as uncertain communications could make them. All accepted his political strategy that a judicious mix of patronage, federal and state, bound together the organization which was informed and directed by a network of partisan newspapers. Thurlow Weed, who but for a mistake of Isaac Leake, business manager of the *Argus,* would have brought his talents to the Van Buren machine, had accurately described its principles and its operations in a phrase, save the fact that it was far more flexible than the accepted definition of a regency council and, though away from Albany for much of the time, Van Buren was no absentee monarch.[13]

During this formative stage, the Regency was vulnerable to personal friction and quarrels within the leadership, like that between

Van Buren and Skinner. Should ambitious party members sense any lack of cohesion, they were apt to go their separate ways, form their own blocs. Any setbacks to Regency leadership in Washington or in Albany could likewise have serious implications. Party members were fickle in their attachments, especially as so many were dependent for so many things on the power and the prestige of the Regency.[14] Patronage was an obvious adhesive, but beyond patronage were all those intricate social and business connections that flourished or withered according to the status of one's connections at any given time. Influence came in many subtle guises, but much depended on political attachment maintained through the local press, local office-holders, or in quasi-social events like militia musters, party caucuses and district conventions.

These activities demanded a constant nourishment of enthusiasm, a positive spirit bred of success that only the leaders in Albany and through them the congressional delegation in Washington could furnish. At the peak of this delicate structure stood Martin Van Buren, still threading his way through the treacherous paths and byways of Washington; hence his desperate need to overcome Taylor, his anxiety about a state ticket that would be popular, yet known to reflect views, that he felt best served the interests of the Empire State.

The Van Buren program emphasized the progressive development of a democratic social order where a presumed quality of opportunity would also bring an earned opportunity of equality. There were to be no personal politics like those De Witt Clinton practiced, no class and caste like that which the old Federalist oligarchy personified. It was all a felt need rather than any planned structure of a model state. Small farmers in the east and emigrant Yankees moving into the north and west, clearing away the forests and tapping the rich soil, were the constituents Van Buren and his lieutenants chose to represent. The only serious division in the ranks beyond the more or less constant tension between the established east and the emergent west had been Tammany's opposition to internal improvements—Clinton's network of canals. But Van Buren had thrown his influence behind these proposals in the legislature and had educated Clinton's enemies in New York City and elsewhere. The canals would enrich the farms, raise land values in the west, promote immigration, which meant more population and more prestige for the state in the Union, more prosperity because of increased trade for Albany, Hudson and New York City, more jobs for the tradesmen, the artisans, the unskilled porters, draymen and laborers, most of whom would eventually have the vote. In his systematic way Van Buren had analyzed Clinton's data and concluded there was a mutuality of interest between agriculture and enterprise providing a certain balance was maintained.

At this stage of party development, he felt that he could not afford further troubles over public questions like banking or constitutional reform. Except for the impending gubernatorial contest, the situation at home was stable. "In no state of the Union," said Van Buren, "was party discipline in so palmy a condition." But just when he was congratulating himself on the smooth workings of the Regency, the Monroe Administration inadvertently dealt him one of those serious blows to his prestige that reverberated all along the complex lines of political communication and jarred him out of his complacency. Solomon Southwick, the careless ex-editor, postmaster of Albany for the past six years, was about to be removed for misappropriating public funds. It never occurred to Van Buren that a Republican administration might appoint a Federalist to this lucrative post or that he, Vice-President Tompkins, Secretary Thompson and his Senatorial colleague, Rufus King, would be the last to know who would be appointed as Southwick's successor.[15]

Van Buren had made some tentative gestures regarding Southwick's replacement. Ex-Chancellor John Lansing, he thought, would be a good choice. An inveterate anti-Clintonian, a highly respected lawyer and judge, Lansing was in financial straits and needed the income. Not the least of Van Buren's motives was Lansing's relationship to the radical, outspoken politician from Dutchess County, Peter R. Livingston, Lansing's son-in-law. Van Buren never intended to have Livingston act as a surrogate postmaster, which Solomon Van Rensselaer, another aspirant for the post, charged. He indignantly denied the allegation. "I do not think," he wrote Knower, "Mr. Livingston has any such claim upon the party. And if the office was not desired and desirable to the Chancellor himself, I would not advocate it." Van Buren was being perfectly sincere, but what he did not mention was that he was preparing the way for the state campaign of 1822. The appointment of Lansing would be popular among all Republicans and a bridge through Livingston to the radical wing. Van Buren also counted on the friendship between Monroe and Lansing to secure the appointment.

Governor Clinton got wind of the imminent change a good ten days before any member of the Regency did. Promptly, he informed Solomon Van Rensselaer, lately deposed as adjutant general but still representing the Albany District in Congress. Van Rensselaer, a much-wounded war veteran, had been an orthodox Federalist since he came of age. Unlike his cousin, Stephen the Patroon, he did not possess extensive land holdings or any sizable property. The Albany post office was one of the more attractive political plums in the state since it carried a substantial salary, $2000 and patronage of its own, two deputies and two clerks. In the lax political morals of the day, the postmaster customarily doled out this patronage worth an additional $2000 a year to his own family. For Van Rensselaer this was not quite wealth beyond the dreams of avarice, but it was a very comfortable living for himself and a modest clutch of dependent adult children.[16]

Van Rensselaer lost no time in soliciting the office from his good friend, the postmaster general, Return Jonathan Meigs, Jr., a Yankee transplant from Ohio. For reasons known only to the principals, Van Rensselaer himself wrote out his own recommendation that he be appointed. This document he circulated among New York's Congressmen on December 21, 1821, but not its Senators. Twenty-two of the states' twenty-eight Congressmen signed it, at least ten of whom were Bucktails. Van Buren learned of their action about the same time Meigs presented the recommendation to Monroe, stating he would appoint Van Rensselaer if the President had no objection. Monroe approved but did ask Meigs to consult further with Smith Thompson, who was present and objected in no uncertain terms.

After their brief meeting, Thompson hurried to Strothers' where he informed Van Buren of Meigs's recommendation and Monroe's approval of it. Incredulous, angry and alarmed by turn, Van Buren immediately sought out Vice-President Tompkins, who had just arrived in Washington, and his venerable colleague, Rufus King.[17] That anyone should be considered for such an important post in the capital of his own state without consulting him was gall enough, but that the person should not even be a member of the Republican party and had won the support of most of the New York Congressmen was a heavy blow to his newly won, dearly bought prestige. "If I could have supposed," he said, "that our members could have recommended Mr. Van Rensselaer I would have expostulated with them on the subject. I have no doubt they are as much mortified about it now as I am."

Tompkins was angry, too, at what seemed a calculated affront to his position in the Administration and his status in New York. Even King was irked at the high-handed procedure and agreed to sign a note Van Buren drafted to Meigs asking if a vacancy existed at the Albany post office. When Meigs replied in the affirmative, Van Buren drafted a second note, which King also signed, as did Tompkins, requesting that no appointment be made for at least ten days, which they judged a minimum of time needed to gauge the feelings of Albany's citizens on the appointment. Meigs's reply to the second note, which he addressed to Tompkins, though courteous enough, was chilly in tone. The appointment could not be delayed; Van Rensselaer's commission was then before the President for his signature.[18]

King now retired from the field, but Van Buren and Tompkins

pressed on. While Van Buren was applying pressure on the President in a personal appeal, he was explaining the matter to his associates in Albany and instructing them how to assist him. "In the course you pursue," he wrote, "you should have regard to the circumstance that the Postmaster General is committed, etc. (was so before I knew that Southwick would be removed) . . . and nothing but the positive interference of the President can arrest the proceeding. . . . Let the Republicans of the city send a memorial to the President recommending the Chancellor, if that is agreeable to them."

Lest his associates be carried away in their effort to sway Monroe, Van Buren tactfully warned them that the memorial be written with "the utmost delicacy and respect. This is extremely important . . . let them speak with firmness, and as if conscious of their respect and regardful of their duties; and lastly, I would send a letter to those republican members who have subscribed V.R.'s petition . . . to be signed by all the members of the legislature setting forth the strong reasons against having a federalist in the office at Albany."[19]

Before the memorial arrived from New York, Monroe, reacting to Van Buren's appeal, judged the issue to be serious enough for him to convene a special cabinet meeting. When the cabinet assembled on January 5 the President submitted all the correspondence for perusal. Then he called in Meigs. Upon being asked his opinion, the Postmaster General said that he would approve Van Rensselaer unless the President objected. Citing his distinguished war record, the recommendation of the New York delegation, the incontrovertible fact that he was now representing the people of Albany in Congress and thus obviously enjoyed the confidence of a majority in his district, Meigs concluded his remarks by declaring there was no other candidate for the post. Thompson promptly inquired if Lansing was not a candidate. When Meigs denied it, Adams wondered whether the appointment could not be delayed a fortnight as Van Buren, Tompkins and King had asked. If that were done, replied Meigs, "Southwick would in the meantime be receiving the money of the public and the longer he remained in office the greater would be the amount of the delinquency."

After Meigs left, Monroe "thought it very questionable whether he ought to interfere in the case at all." A heated argument arose among the cabinet members. William Wirt, the attorney general, and Crawford were unsparing in their criticism of Meigs. Adams, while anxious to placate the Vice-President and the two New York Senators, saw no wrongdoing in Meigs's handling of the case. Calhoun, who had lent his support to Van Rensselaer earlier, said very little; but after the meeting broke up, confessed to Adams that he thought Meigs's conduct had been correct. "Why," the suspicious Adams mused, "did he not say so while the discussion was going on at the President's?"

Van Buren, who anticipated the President's position, had no complaint about it. For home consumption he put the best possible light upon his rebuff. He told Knower what he knew was true, if taken literally, that the law establishing the post office department gave the postmaster general control over appointments; but, practically, the President as chief executive had the last word. Meigs himself had admitted as much to the New York Senators. "If the President thinks it proper to have the appointment delayed, it will be so," he said.[20] Van Buren was not about to make any public issue with the President. He had only been in Washington for three months, and loyalty to one's party chief was a political imperative with him.

Meigs was a different matter; he could be cast as the villain in the piece. In this way Van Buren hoped to recoup any prestige he might have lost even though he had been caught napping and had been worsted in an important skirmish over patronage. He had already drafted a final letter to be sent to Meigs, recommending Lansing and planning "to put the question on such political ground that the people of the United States may distinctly understand what principles prevail in that department of the government, and, may take measures necessary to a wholesome reform. This point must be settled one time or other, and no time more opportune than the present." On January 7, 1822, Meigs replied that he would not delay the appointment; Monroe on the same day declined to intervene because, as he put it, the case "involves a principle in the administration of the Post Office Department."[21]

As soon as Tompkins and Van Buren read these notes, they sent off Van Buren's carefully prepared letter of eight pages nominating Lansing, or any other good Republican from the area deemed more suitable to the Postmaster General and acceptable to the party in Albany. Van Buren had composed a political indictment that unmistakably made its case yet was neither captious nor argumentative. "We forbear discussion on the matter," he said, choosing his words carefully, "and therefore content ourselves with observing that whatever might be the correct course as to removals from office at this time when the feelings of party are in some degree relaxed, we had flattered ourselves with the hope that for new appointments at least (all other matters equal) a preference would be given by every department of a republican administration to its republican supporters."[22]

That Van Buren meant business whenever his position was threatened, as in this case, quickly became apparent. The appropriate public meeting was held in Albany, adopting resolutions and a remonstrance along the lines he had suggested. Dozens of prominent Albanians signed the document. The *Argus* printed copies that were sent to all the members of the New York congressional delegation and to the President himself. On the same day, January 22, 1822, the *Argus* published all the correspondence dealing with the affair. There, spread for all to see, was Meigs's unwavering course, Monroe's disclaimer of all responsibility, Van Buren's indictment, and most important of all, public confirmation that many Bucktail Congressmen had supported Van Rensselaer. Van Buren had persuaded several of them to endorse another recommendation for postponing the decision, and this document, too, appeared in the same issue of the *Argus*. But in private correspondence, the Regency knew that of the twenty-two members of the House who previously endorsed Van Rensselaer, Van Buren had only been able to secure the names of eleven for his petition to Meigs.

The alarm bells began ringing in Albany and in New York City. Van Buren, it would appear, was on a collision course with many of the New York members of the House. Many prominent Bucktails also worried that he was cutting himself off and thus themselves from the Administration. "I have greatly feared, that resenting this course, it will be very unpleasant for you at Washington . . . with such a delegation. Will there be any harmony or any unity of action under such circumstances?" asked Ulshoeffer. He questioned also Van Buren's campaign against Meigs, stating that it would "perhaps entirely destroy all your influence with the heads of the departments at Washington, under the present administration—and with the president."

Much of what worried Ulshoeffer was in fact occurring. Van Buren's stand against Meigs was taken as an attack on the Administration for personal political gain. Monroe was slow to anger, but to be accused in public, even indirectly, of appointing Federalists in the place of faithful Republicans, was too much. Why the Federalists did not even exist as a distinct party. Years later Van Buren recalled how the White House castigated him, how he was denounced as a disloyal political upstart because he had wrung "this peal of the party tocsin in the ears of those who glorified the 'Era of Good Feelings.' "[23] Yet he was not concerned about punitive actions from the President or from the cabinet. He subscribed to Rufus King's view: "We have no administration. Mr. M. tho' not buried is dead as respects direction, or control: his cabinet is insincere of necessity; and some wanting in advice, cannot acquire it from men who from jealousy will unite in no opinion."

If Van Buren discounted any reprisals from the White House, he was concerned about the members of the House, the Bucktail members in particular, who had broken party discipline and might continue on their independent course. One House member, Churchill C. Cambreleng, was a highly intelligent, wealthy New York merchant, a man of judgment and character already marked out by Van Buren as the leader of the New York delegation. Cambreleng was a staunch Bucktail, a Tammany sachem, and a signer of the Van Rensselaer recommendation. Try as he might, Van Buren could not move him. Van Buren's charm, his suave manners, were met by equally charming and suave manners from this portly little man who retained the mellow accent of the North Carolina tidewater where he was born and grew to manhood.

Disturbed by the attitude of his colleagues in the House, Van Buren was more disturbed that party workers as important as Ulshoeffer should misunderstand what was happening within the Monroe Administration.[24] If Ulshoeffer was right, and he had every reason to believe that he was a good judge of public opinion, Van Buren's campaign against Meigs and indirectly against a majority of New York's congressional delegation could injure the Regency. When he decided to challenge Meigs, he had anticipated some of the problems in communication that Ulshoeffer had raised. Writing to Dudley, he laid the groundwork for diverting political opinion away from the post office imbroglio to the impending state election and the presidential question. After noting that Crawford, Adams and Calhoun were inspiring resolutions from Republican meetings and legislative caucuses in the state, he suggested that party members in New York might "express their regrets at the early agitation of this matter, and in a well drawn resolution express their convictions of the mischief which must inevitably flow from it, and their hopes that their members will not mingle in the fray." Further, he urged the Regency not to chastise the Bucktail Congressmen for their impulsive action. "They are very unhappy, and it would be unwise to mortify them more."[25]

After receiving Ulshoeffer's letter, however, he decided that he must be more explicit with his associates at home about conditions in Washington. He wrote long letters to Dudley and Knower, which had a calming effect when passed around within the Regency. He was also reassured that his campaign against Meigs, far from injuring him in New York and elsewhere, was being picked up and applauded by the Republican press all over the Union. By mid-February he had received sufficient proof to be certain he had been right in asserting a bold course on partisan grounds. "It is not in any degree to be questioned that the post office business has been in every point of view useful to us. The papers in every part of the Union come out loudly in our favor, the spirit that animates them is felt here," he wrote John A. King.

Within the cabinet he had at least two strong supporters, Smith Thompson and, more important, William H. Crawford. His brief infatuation with Calhoun had come to an end. John Quincy Adams was still hopeful that Van Buren's friendship with King, his northern birth and the well-known antagonism against Virginia in New York political circles would bring Van Buren behind his candidacy. What Adams had failed to understand was how traditionalist Van Buren had always been in political strategy, tactics and persuasion; what a realist he was in determining the sources of power. New England might well go as a unit for its illustrious son, but New York, Pennsylvania and Virginia would swing the contest; agriculture rather than commerce or manufacturing was still the major economic interest in the country.

All the daring moves, the publicity and the attention paid Van Buren in the national press shook much of the dullness and the lackluster politics out of the Monroe Administration. He suddenly found himself a social as well as a political lion in the community. The Washington scene was far more appealing than he had expected, the presidential succession, just being bruited about, an exciting power game that kindled his interest.[26] In letters to Erastus Root and to his old friend Gorham Worth, Van Buren declared frankly that he was happy with his new career. Even though less convenient, he decided to move out of Strothers' and join Rufus King in Georgetown. Van Buren admired his senior colleague and enjoyed his company. He also liked being with the educated, urbane Federalists who boarded with King, men like Henry Warfield, a Maryland Congressman who was secretly carrying on a correspondence with Henry Clay, and even that prime mover of the Hartford convention, the Massachusetts Senator Harrison Gray Otis.

The only straight-line Republican in the Mess was Charles F. Mercer, serving his first term as a Congressman from Virginia. When Stephen Van Rensselaer was elected to Congress in place of his cousin Solomon, he, too, joined the Georgetown group. At first the Patroon, who assumed Van Buren was still at Strothers', thought he would not be welcome because of the post office affair. But none of the boarders was more cordial or more genuinely desirous of having him join the Georgetown Mess than Van Buren.

Van Buren's new quarters and new associates were roundly criticized at home. To smaller minds, where politics and society were one and the same, Van Buren was acting the hypocrite and the social climber, boarding with "new fashioned aristocrat republicans." He was even accused of being the willing tool of Rufus King.[27] Van Buren soon made it clear that he was not a captive of New Englandism. When Johnston Verplanck, a "high-minded gentleman" and an editor of the *New York American,* publicly committed his paper to the presidential hopes of Adams, Van Buren took the opportunity of warning King's son, John, in a friendly way not to commit himself.[28]

Involved as he was with internal affairs of the party, and judging that a new member of the Senate should wait a decent interval before engaging in debate or in making a speech, Van Buren said nothing during the first four months of his tenure. He had been elected to two important committees, Finance and Judiciary, more a sign of respect for the state he represented than for any knowledge of his ability. He was faithful in his attendance at committee meetings, and his habit of thorough preparation made him such a valuable member that his colleagues soon elected him chairman of the Judiciary Committee.

Van Buren regularly attended also the sessions of the Senate, where he occupied himself in observing carefully the style and the procedure of its principal speakers. He made extensive notes on Senate rules which he committed to memory and finally felt himself prepared to make his debut. A bill to legitimize certain land claims in the Louisiana territory known as the *Maison Rouge* tract had been introduced to the Senate and referred to the Judiciary Committee, which found so many flaws in the bill that it made an adverse report. As chairman of the committee, Van Buren could make the report or delegate it to some other member. It was a minor matter and typically he chose to deal with it himself, though not before he had researched the claim carefully and thoroughly. When he rose to report on the bill and to state the committee's objections to it, much to his humiliation his mind went blank and after a few disconnected stuttering remarks he had to sit down in blushing embarrassment.

The advocate of the bill, James Brown, a veteran Senator and as a resident of New Orleans familiar with the civil code of Louisiana, seized the opportunity to score what he thought would be an easy victory. In a condescending manner he proceeded to lecture Van Buren as if he were a schoolboy, to the applause and laughter of the gallery. Brown's tone, his manner and the response from the gallery so irritated Van Buren that he overcame his temporary block. The slight, erect figure snapped up from his seat and rapidly, but logically and coherently demolished his argument. After he finished, Nathaniel Macon, the North Carolina spokesman of simple Jeffersonian Republicanism, came to Van Buren's desk and complimented him warmly on his rebuttal. Never again would Van Buren lose his poise in the Senate or in any other public assemblage.

In commenting on this initial failure Van Buren said, "that he had experienced the same thing before." Whenever placed in new and unfamiliar surroundings, he had always at first feared entering debate and had lost his poise, which, however, he soon regained. Deep within the genial, articulate man-of-the-world figure he cultivated there lurked an insecurity, a self-consciousness that only acute embarrassment could check and erase. As a public man, Van Buren tended to underrate his ability, to take safe refuge in expecting the worst and being inwardly surprised, yet pleased, that he won more often than he lost, that his judgment of his own capability turned out to be better than he thought when measured against others whom he had overrated.

He had no sooner felt at home in the Senate and was preparing himself for a major speech on the floor than he had to post-pone it and assist King in helping Tompkins. The Vice-President had again fallen into hopeless alcoholism and could not perform his duties as president of the Senate. Tompkins, now a physical wreck, was still a proud and sensitive man who had earned the deep obligation of the party and the nation. The Senators from New York exerted themselves to have Congress settle the nation's accounts for the Treasury bills and other obligations of the government Tompkins had assumed during the War, a slow, laborious process.[29]

Meanwhile, they took upon themselves the painful task of persuading him to relinquish his official duties and return home. At no time did they suggest he ought to resign. In late January 1822, Tompkins took leave of the Senate. Two months later the Vice-President left Washington, never to return. Completely broken in body and spirit he lingered on at Staten Island for three more years to die at the age of forty-eight.[30] Saddened at Tompkins's collapse, Van Buren decided he must get away from Washington for a short time. He wanted to visit Jefferson and Madison and, like any tourist from the North, view the sights of the Old Dominion—the Natural Bridge so celebrated in Jefferson's *Notes,* Richmond and the famed Sulphur Springs.

Van Buren had by now become accustomed to the rumors that followed him wherever he went, whatever he did. At several of the Washington parties he had been observed paying particular attention to one of Jefferson's granddaughters, Ellen Wayles Randolph. When the marine commandant Colonel Henderson and his wife gave a ball in the spring of 1822, Van Buren, Miss Randolph and oddly enough Rufus King, despite his years, were guests. The young lady asked the marine band to play a popular Scottish ballad, "The Yellow-Haired Laddie." King immediately made a connection and passed it on to his son John. Others must have also thought that there was more between Van Buren and Miss Randolph than mere friendship.

Likewise, as soon as it was known that Van Buren had gone to Virginia, gossip in Albany and New York City was divided between those who thought the trip was of a political nature and those who were certain it would result, as Ulshoeffer put rather grandly, "to consummate a matrimonial scheme." Whether Van Buren visited the intelligent and well-read Miss Randolph or a bevy of Virginia politicians or both is not known. All he would say about this trip was that he found Richmond interesting and entertaining, so much so that he spent ten days there and postponed his visit to Jefferson and Madison. "I have been treated," he wrote Gorham Worth, "with a degree of kindness and respect which I could not have anticipated and which has detained me longer than I contemplated."[31]

Van Buren visited Ritchie's cousin, the ailing Spencer Roane, who criticized unsparingly what he believed was Monroe's unmistakable drift toward Federalist views of centralization. We may safely assume he went on to Shokoe Hill in Richmond and was entertained by Dr. John Brockenbrough, another of Ritchie's innumerable cousins, whose mansion was as much the center for important Virginia politicians as the nearby state capitol. It would be strange indeed if the Senator from New York who had such a reputation for party management, and who had openly clashed with the Monroe Administration, had not talked politics and presidential candidates with Thomas Ritchie. If he did, he did not disclose his preference to the editor of the *Enquirer* or to anyone else in Richmond.[32]

« CHAPTER 8 »

Harp on the Willows

"Why the deuce is it that they have such an itching for abusing me?" asked Van Buren of Rufus King in September, 1822. "I try to be harmless and positively goodnatured, and a most decided friend of peace." He was alluding in this particular instance to an outrageous fabrication that a Clintonian weekly invented, which was promptly copied in other anti-Van Buren papers throughout the state. Van Buren was supposed to have been challenged to a duel by a worthy old Revolutionary War veteran when he would not give up his place in a coach. He was pictured as a craven who, after insulting the old, enfeebled patriot, refused his challenge. Spiteful people everywhere, even those like De Witt Clinton who should have known better, retailed this atrocious libel.

Van Buren was no stranger to personal attack in the extravagant rhetoric of the partisan, the slurs and whispering campaigns of disappointed officeholders. He had learned to tolerate political billingsgate with affected indifference, making it a rule never to reply. But inwardly Van Buren rankled, though he recognized it was an inevitable consequence of public life. As he gained in reputation, especially as he stepped on the national stage, the chorus of critics increased in numbers and in volume.[1] By many he was believed to be the absolute master of the Regency, which he was not, though the mistakes of those associates less gifted than he in Albany and in New York were always attributed to him. The tensions within the party that had always existed, and which Van Buren had with great difficulty managed to control, broke into open quarrels when he was away in Washington and his restraining hand, his extraordinary powers of tact and persuasion, were delayed at least a week or longer by the mail service. Van Buren, so lucid and so quick, so penetrating in conversation, wrote with extreme difficulty. Even after editing and rewriting, his advice and directions were frequently obscure and diffuse.

He had thought that he had left behind him a corps of able men who would be able to adjust differences, smooth ruffled feelings, keep up party discipline. He had not been away three months when Edward P. Livingston, long since a loyal Bucktail, remarked how the organization had suffered: "Van Buren's young tribe, that he has been training for the last eighteen months, thought they could rule the state, but he is too cunning for them. The party is in an unsettled state; we want a firm leader."[2]

When Livingston voiced his complaints, Van Buren had already made two decisions that involved considerable risk. The legislature met for its regular session in January. Should the customary caucus make nominations for governor and lieutenant governor or should it postpone any action for a special caucus to be held sometime in the summer closer to the date of election, which under the new constitution was to be held in November? Samuel Talcott fretted about nominating candidates who would be running for eight months "without getting out of breath." "So long an interval," he said, "between the nomination and election would afford too good an opportunity to organize cabals of every description; and God knows where all the undercurrents and countercurrents would drift us."

Despite these pertinent observations, Van Buren would risk it. He felt confident that the Regency would control things provided the candidates were loyal and above all willing to place the party and its principles over personal ambition. Under the revised constitution, the office of the governor had been granted greater powers to implement and to direct public policy. The governor appointed all judges, except justices of the peace, with the consent of the senate, and he could veto legislation, subject, of course, to an override of the legislature.[3] When Van Buren learned that a majority of the party seemed inclined to support the popular and impressive Samuel Young as its candidate for the governorship, he decided also to risk taking immediate steps to quash his nomination.

Young, like Skinner, had clashed with Van Buren at the convention, but for precisely opposite reasons. Where Skinner held Van Buren responsible for the abolition of the council of appointment, Young pressed for the popular election of judges and the extension of the suffrage.

There was no doubt in Van Buren's mind that if Young were elected he would use the expanded powers of the office to push for another constitutional revision that would range the radicals squarely against the moderates and the conservatives. Joseph Yates, Van Buren thought, would be reasonably immune to the wiles of disruptive elements. Even in physical appearance, Yates and Young were distinctly opposite types. Young was an angular, pugnacious trial lawyer; Yates was short, rather portly, whose bland features and habitual reticence seemed to bespeak a safe, conventional candidate, a steady governor.

At Van Buren's behest, Yates's cousin, John Van Ness Yates, was already preparing the way when the word went out to men like Skinner and Talcott that Van Buren had no objection to the nominations being made in February as usual, and that Yates was more acceptable than Young, particularly so to the conservatives and the "high-minded" men. The radicals, as Van Buren had foreseen, were divided between Young and Porter, between banking interests and land speculators. Even if they could compose their differences, they were in a distinct minority.

In the caucus held in the house chamber of the capitol at Albany on March 18, 1822, Yates won easily on the first ballot. Young took the result philosophically, but he was clearly disappointed and began to question whether his political future lay with the Regency. "I can do without the State of New York," Young said when he learned of Yates's nomination, "as well as the State can do without me." To ease over any regional and factional differences, the caucus nominated Erastus Root, a radical and a westerner, for lieutenant governor.[4]

The Clintonians, now leaderless, offered but token opposition, and the Regency candidates carried the state in November. All members of the senate for the first time since Independence belonged to the same party, whose acknowledged leader was Van Buren. No more than a dozen assemblymen avowed themselves Clintonians. The Federalists had completely disappeared, their identity gone; the once proud party of Washington, Hamilton and Jay had become a political artifact.

Of course, it was a stunning personal triumph for Van Buren with far-reaching implications in a Washington that had looked on with sardonic detachment at his rebuff in the Albany post office affair and his seemingly ineffective efforts to engage in presidential politics. Van Buren could congratulate himself that he had not fallen prey to the fears of his associates or given in to pressure from the various factions.[5] He was pleased at the successes of friends like Charles E. Dudley, Edward P. Livingston and Jacob Sutherland, who won seats in the senate from Albany, long the personal fief of the Van Rensselaers. Equally satisfying were the elections to the house of three protégés, Jesse Buel, former editor of the *Argus,* Azariah C. Flagg, an unusually astute printer-editor, and Jesse Hoyt, a growing power in Tammany.

Of these men, Van Buren had been especially attracted to Flagg, who, as editor of the Plattsburgh *Republican,* had spread the gospel of Jefferson through the northern counties of the state. What had particularly marked out Flagg for him was an insistence on party regularity that he dinned at his reading public in column after column of agate type. Much as Van Buren was impressed by Flagg's emphatic loyalty, he reserved his special admiration for the editor's grasp of facts and figures, his knack for converting public reports with their tedious columns of customs duties, ship clearances, appropriations, expenditures, and the like, into matters of interest and political consequence. Compared to the bluff, genial Jesse Hoyt, whose life-style inclined toward gambling, tippling, easy camaraderie and a rather broad construction of legal and political behavior, Flagg was frugal and industrious. Transparently honest, he belied the current sterotype of a native Vermonter.[6]

But no sooner had a confident Van Buren taken quarters at Mechanics Hall in New York City for the November session of the state supreme court before traveling to Washington, than the political calm he had left in Albany began to show signs of impending disturbance. Skinner and Marcy, who also had cases before the court, had come on with him and all had rooms in the same hotel. Naturally, politics were discussed after long hours in court. And thither came the Tammany sachems, the "high-minded" men of the City, the ubiquitous Noah.[7] There had been nothing secret or devious about their talks, though Van Buren made some unflattering reference in an offhand manner about some of the new governor's confidants; but political busybodies promptly enlarged on these informal conversations, relaying them to Yates. The uneasy governor, conscious of his shortcomings in the political management of his new office, was quick to believe and quicker to resent the imputation that the Regency would dictate his policies.

Van Buren had favored Yates precisely because he believed him to be an honest man of modest abilities who would exercise the greatly increased powers of the governor's office under the new constitution as the Regency thought best. But he was far too tactful ever to discuss such matters with anyone, including his closest friends; nor did he feel that Yates was to be a mere figurehead. There would be an experimental period during which the politicians would test the new powers of appointment, the new judicial system and other reforms that had been inaugurated. It would be a time for caution when the organic law would be shaped and proven in such a way that public policy and party harmony supported and strengthened each other. In all this he expected Yates would consult with Regency members as he himself had always done before making decisions in sensitive areas like appointments.

He had not been in Washington more than a week when Cantine, writing from Albany, disclosed that Yates believed the rumors that Van Buren had held a private meeting in New York City where steps had been taken to control the office of the governor. "There was, of course, not a word of truth in the story," said Van Buren, "and under ordinary circumstances, I would not have noticed it. But I knew the Governor's disposition, and that he was surrounded by men in whom I had little confidence." After discussing the news with King, Van Buren wrote a long letter to Yates emphatically denying any such "caucus" as he put it, and pledging his confidence in the new governor's administration. "With most men this would have been sufficient," said Van Buren, "but as to him the soil was too favorable to the rank growth of the seed I endeavored to eradicate and the sowers were too numerous and industrious to admit of any success to my efforts."

Instead of conciliating the radicals, in particular the embittered Samuel Young, Yates pointedly ignored them and the Regency when he prepared his annual message. The result was a rambling, wordy document in decided contrast to Clinton's powerful state papers. Even the *Albany Argus* could find little to praise. Coleman's *Evening Post* scorned the limping prose, and finding not one specific recommendation for legislative action, concluded its only merit was its brevity.[8]

The message irritated the Regency and embarrassed Van Buren, but no great harm had been done. Had Yates followed Van Buren's advice on appointments, the public exhibition of his deficiencies would have soon been forgotten. Unfortunately, he was determined to pursue the same independent course on appointments, the most important of which were the new justices. The irrepressible Noah did not help matters by suggesting in the *National Advocate* that the old justices be retained. Since Noah was a Van Buren spokesman, the radicals jumped to the conclusion that this was the official line. Peter R. Livingston, Young and Root were harshly critical and threatened dire results should Spencer and his discredited colleagues be confirmed. Van Buren took immediate steps to disavow Noah's ill-considered statements. But they had already poisoned the atmosphere and involved Van Buren, at least by implication, in a political debacle of the Governor's making[9]

To an incredulous legislature, Yates nominated all the old judges to the new supreme court. One of the more important reasons for framing the new constitutional amendments had been the removal of these politically minded men. In fact, so heated had the opposition been to the blatant partisanship of the three Yates nominees, that Van Buren was forced in the convention to exert heavy pressure for the preservation of an independent judiciary from the violent attacks of radicals like Peter R. Livingston, now Speaker of the house. The senate lost no time in refusing to confirm the nominations and in the process did not spare Yates's feelings. But the incident reflected also on Van Buren himself, whose judgment the radicals impugned, albeit indirectly. They dwelt on the lingering suspicions Noah had aroused, and they questioned how the party could nominate such a bumbler.

Chastened, a craven Yates withdrew the nominations the very next day without consulting the rejected judges or giving them any time to defend themselves, either in the public prints or before the Judiciary Committee of the senate. Young and other radicals who opposed the old judges now attacked the Governor roundly for his sudden turnabout, asserting vehemently that he was playing politics with the judiciary. Yates then submitted the names of the men Van Buren had suggested. The senate promptly confirmed two, but in a direct affront to the Governor rejected his third nomination. Yates now did what he should have done in the first place—consulted the Regency and accepted their recommendations for the third place.

Although Yates's ineptitude had created serious problems for Van Buren and had spurred the nascent factionalism in the party, the Governor was more a symptom of unrest than its basic cause. The root and branch of the gathering disturbance lay in the presidential question and the apparent indifference of Monroe to the succession. Van Buren had kept his own counsel and watched the maneuvers of Clay, Calhoun, Crawford, Adams and a new challenger, a fellow senator, Andrew Jackson. Even Clinton, ostensibly in retirement, was known to be angling for an opportune moment when his small band of adherents might go forth and create a favorable climate of opinion. Everyone knew that New York, and this meant Van Buren must have a leading say in who would be the next President.

Rumors that Van Buren would back Crawford had been circulating since the winter of 1822 when he criticized Johnston Verplanck, the editor of the *New York American,* the paper of the "high-minded men" in New York City, for openly advocating the election of John Quincy Adams. By April, when General Stephen Van Rensselaer arrived in Washington to take his seat in the House, he wrote his cousin Solomon that Van Buren was "in Crawford's interest." Shortly thereafter, when Noah came out for Crawford in the *National Advocate,* the rumors gained a semblance of reality because of Van Buren's close ties with this erratic spokesman for Tammany.[10] In fact, the explanation for Noah's eccentric behavior in the matter of the old judges was that they were partisans of Crawford. Van Buren had, indeed, made up his mind to back Crawford after his trip to Richmond the previous spring, but judged it expedient to delay any public pronouncement until he had been able to evaluate the mood at home and in Washington.

Just as he considered the time ripe for his public avowal in February of 1823, his plans were upset and his judgment clouded by the sudden death of his closest friend, the ever-faithful brother-in-law, Moses I. Cantine. Cantine's death left the *Albany Argus* without an editor at a critical juncture in his affairs when pressures were mounting for him to announce his presidential preference. But Van Buren's personal loss transcended all of these considerations. The Cantines had looked after his four boys, Cantine himself acting as a surrogate father. When all of his energies and faculties should have been concentrated on the rapidly maturing presidential campaign, Van Buren had to make arrangements for his family and to provide financial assistance for Cantine's widow.

Cantine had edited the party's paper and shared in the state's printing, but left almost nothing to support his large family. Van Buren enlisted his relative, Peter Hoes, Butler, Jesse Buel and William A. Duer to have someone "take the establishment [the *Argus*] off our hands so that we can get rid of ultimate responsibilities" and also make provision for the support of Cantine's family. In a letter to Jesse Hoyt he was more explicit. "I am so overwhelmed with the account of poor Cantine's death," he wrote, "I know that nothing from me can be necessary to secure your zealous attention to Mrs. Cantine's interest, if anything can be done for her. . . . Mr. Hoes and myself are responsible to Mr. Buel for $1500 of the last payment." Concluding his letter, Van Buren underscored the absolute necessity of quickly finding an able editor to conduct the *Argus*.

Significantly, he did not mention Noah or the state printing, which he knew the Tammany editor would try to secure. He could not have been more emphatic when he said, "without a paper thus edited at Albany we may hang our harps on the willows. With it the party can survive a thousand such convulsions as those which now agitate and probably alarm most of those around you." Ulshoeffer echoed Van Buren's concerns about the *Argus* and the state printing. With an obvious reference to Noah, he wrote Hoyt, "Do not make a state printer who will transfer the feuds of New York to Albany and throughout the state. Dullness would be preferable to indiscretion."[11]

Fortunately there was at hand a young printer-editor who was not only discreet but gave promise of considerable ability, Edwin Croswell of Catskill. The Croswells, proprietors of a struggling weekly, the *Catskill Recorder,* had been close friends of the Cantines, who had once lived in that small river town. Edwin had come up to Albany for Cantine's funeral, and there he met the group Van Buren had charged to find a new editor for the *Argus.* None of them knew Croswell personally, but they were familiar with his work for the *Recorder.* A seasoned printer, he had also helped his father, MacKay Croswell, edit the paper.

The elder Croswell had a fondness for drink and cards and a casual attitude toward publishing a newspaper. For several years, a large share of the editorial responsibilities had fallen upon his son, who in addition, set the type, ran the press, sought subscribers and advertising. Yet his political letters and articles signed variously, "Ed.," "Ned" or "Edwin," caught the attention of the politicians in Catskill, Hudson and Albany.

Strictly Republican in politics, Croswell commanded a clear, vigorous, somewhat scholarly style. A good debater in print, he relied more on logic and on common sense than on flamboyant rhetoric. Just the right stance, thought Butler and his associates, to provide direction and substance for the party press without involving the *Argus* in those profitless duels over careless or scurrilous language. As an experienced printer, the *Argus,* it was assumed, would be relatively free of annoying typographical errors or muddy, broken letters.

Still they were cautious and astute men. Croswell may have done well on a country weekly, but could he take on the responsibilities of a city daily that was the all-powerful communicator of the Regency's political line? Could he defend the Regency and its policies against opposing editors far more experienced and mature? Although this studious, young man of twenty-five had much to recommend him, he might be unable to provide essential political direction for the dozens of Regency papers. Was he alert enough to sift out the innumerable issues, state and national, commercial, financial, agricultural and manufacturing, that beset New York's political leadership, and then explain clearly those of current importance? Pressed as they were by Cantine's death, Croswell's appearance seemed a blessing. They decided to put him on a trial basis and scout the state for a more experienced editor. If, however, he proved he could handle the *Argus,* then the post would be his, and the money for him to purchase a half interest would be advanced on generous terms.

After Butler explained all this to a surprised Croswell, he at first doubted whether he should saddle himself with the unremitting labors that had hastened Cantine's death. He worried too about his family. Could his father do without his services? Evidently receiving assurances from home, assurances of editorial support from Butler and Flagg, and realizing that if he were successful, he would inevitably wield power and influence in the politics of the burgeoning Empire State, he agreed to take over temporarily.

After a few weeks, the Regency knew that it had made a lucky discovery. Croswell's editorials and anonymous letters were more varied and far more provocative than Cantine's had ever been. The appearance of the paper improved as the young editor set high standards in the composing room, doing his own proofreading and cleaning the type. Delighted with Croswell's work, a relieved Van Buren agreed to a permanent arrangement, which was accepted. Over the next twenty years, Croswell made the *Argus* not just the mouthpiece of Van Buren and his party, but a paper of national influence and prestige whose editorials were widely copied.[12]

With the *Argus* situation now settled and with satisfactory arrangements made for his family in Albany, Van Buren was able to concentrate on the presidential question. As he surveyed the scene in February and March 1823, he formed a relatively clear picture of the strength of the various candidates. His operating principle, with over twenty-three years of success behind it, was the union of New York, Pennsylvania and Virginia on one candidate. This arrangement, virtually unbeatable, was fraught with problems, not the least of which were those in his own state. Van Buren had satisfied himself that the ruling spirits in Virginia were, if anything, more purely agrarian-minded than he, more Jeffersonian than Jefferson himself. Calhoun's and Clay's nationalist views were anathema to them.

Van Buren agreed; but he was not as doctrinaire as the Richmond Junto. After all, he came from a state of much more diverse interests and resources, which must in some way be accommodated. He had supported Clinton's Canal, a scheme of internal improvements at great expense that shocked those supreme individualists, John Taylor and Spencer Roane. There was moreover a practical streak to Van Buren's political thinking, a capacity for rationalizing his principles to accommodate and to exploit a given situation. He was quite capable of bending his precepts to fit a particular course of action; yet when accused of deviating from principle, he brushed aside criticism of his motives, rarely bothering himself with means provided the ends he sought comported with the public welfare as he understood it.

A Jeffersonian by education, experience and conviction, he was no agrarian determinist like Taylor, no feverish anachronism like Randolph of Roanoke; rather he was an extremely perceptive politician who saw that Jeffersonian verities were supple enough to embrace changing realities. Van Buren may have been a conservative, an advocate of states rights, an agrarian, a party regular; he was above all a New Yorker, a northerner and a pragmatic politician.

He professed an early opposition to Calhoun's candidacy because he thought the Secretary's views on the federal Constitution "were latitudinous in the extreme." Doubtless Calhoun's nationalism, his support of internal improvements at public expense was a matter of real concern to Van Buren. A national program of canals and roads would rob New York of its edge and add to the state's tax burden, already substantial, for the construction of the Erie Canal by boosting imposts through a higher tariff that would fall disproportionately upon the merchants and farmers of New York. However important these factors may have been, they were certainly not the only reasons for Van Buren's refusal to be drawn into Calhoun's orbit. The South Carolinian's decided opposition to the "caucus system" as Van Buren put it, was a much more compelling argument for his decision to look elsewhere. Caucuses had been intrinsic to Van Buren's political success thus far. They had the force of tradition behind them and Van Buren was nothing if not a believer in precedent, especially when the precedents had spelled success for the party over the past twenty years.

Calhoun, as Van Buren knew all too well, had invaded his own domain, the state of New York, and had already set up—through social connections, the patronage of the War Department and Monroe's personal interest—a base of support centered in New York City with lines extending westward. Congressman Micah Sterling from Watertown, a college classmate of Calhoun, was a devoted supporter. Another classmate, Ogden Edwards, was working hard in New York City to bring the Tammany organization behind him. Monroe's son-in-law, Samuel L. Gouverneur, a "high-minded" man who was capable of some low-minded tricks, was Calhoun's entree to that small but influential group of ex-Federalists. Generals Joseph Swift and Winfield Scott and the engaging young officer, John A. Dix, were all stationed in New York and were all advocates of Calhoun. They were popular and had some patronage of their own in supplying their various commands. Calhoun was also actively seeking through Charles G. Haines and other Clintonians to make converts, much to Clinton's disgust.

Van Buren had problems enough in maintaining his ascendancy without tolerating yet another menacing faction under the auspices of the ambitious Secretary of War. A final consideration was Calhoun's organization in Pennsylvania under the leadership of George M. Dallas, who with his relatives and connections were dubbed the "family party." They controlled the persuasive *Franklin Gazette* of Philadelphia that took every opportunity to undercut Adams while it advanced the cause of Calhoun. Should the Calhoun organization capture Pennsylvania, Van Buren's grand alliance would come apart.

Another contender was building an organization in New York, the popular Henry Clay, self-appointed champion of the West. Despite the fact that Clay was more nationalist in his political doctrines than Calhoun, Van Buren had not let his "latitudinous views" interfere with their relationship. In fact, it was Van Buren who had made the first overture to Clay shortly after his arrival in Washington during the winter of 1821–22. At that time Van Buren told Clay he thought "the prospects" of John Quincy Adams and Calhoun "were altogether hopeless." The subtle Clay, while giving Van Buren no assurances, observed that "whilst he very positively asserted who would *not* be elected and thereby limited very much the field of competition, he abstained from mentioning (and I gave him no encouragement to say) who would be elected."[13] Van Buren certainly fell prey to Clay's captivating manner. Personalities aside, Clay did not at the time represent the threat Calhoun posed to Van Buren's New York machine. The Clay organization in the Empire State consisted of only two men, Samuel Young and Peter B. Porter. Thus far, Porter had moved so cautiously that Van Buren was unaware of his connections with Clay. As for Young, he knew his man too well to be concerned about any rupture in the party over Clay's candidacy.

Adams commanded substantial support in the state, from the old Federalists and the "high-minded" men headed by Rufus King's sons, the Duers in Albany and in New York City, and the Verplanck brothers, Gulian and Johnston. That Adams was the only prominent northern candidate in the field carried weight among men of all persuasions, numbering many from Van Buren's own party. Van Buren had tried several times without success to woo the *New York American* away from Adams. He was making greater headway, he thought, with the old-line Federalists through his carefully cultivated friendship with his Messmate, General Stephen Van Rensselaer. Rufus King, his colleague and friend, avoided any comment that might elicit a preference. And Van Buren was too respectful and tactful to press him, though he was quite certain King was for Adams, as indeed he was.

Everyone, even at this early date, thought Adams would carry all of New England, Delaware, possibly Maryland and might receive some votes from South Carolina and New York. Van Buren, who admired the Secretary and shared to some extent New York's bias for a northern President, did not think Adams could win. An overwhelming vote from the South and West would go against him. The campaigns of Clay, Calhoun and Crawford would weaken him in New York and in the middle states.

A fourth unannounced candidate, De Witt Clinton, bore watching. The former governor was living in active retirement, delivering lectures on scientific and philosophical topics, writing public letters on politics under assumed names that deceived few as to the identity of the author. Ever busy with projects for educational and penal reform, for the improvement of insane asylums and hospitals for the indigent, he kept his name before the public with his frequent reports on the progress of the Erie Canal. As unpaid president of the Canal Board, he was a careful yet imaginative manager of construction, an honest custodian of canal tolls that were accruing from finished portions of the Canal. Clinton had become obese with advancing years, but he had lost none of his vigor and to the people of the nation he was still the muscular giant he had always been, projector of great and good causes, the intellectual politician, the embodiment of statesmanship.

Ambrose Spencer had again fallen out with Clinton and was cooperating with Van Buren, but there remained a sizeable group of devoted adherents in the state, while outside New York Clinton's reputation was untarnished. Andrew Jackson, just now being spoken of as a candidate, thought Clinton one of the greatest men of the age. He was strong in the Northwest, which stood to benefit from the Canal. Clinton could not be factored out of the equation.[14]

Congressman Henry R. Warfield of Maryland gave a biased yet at the same time canny appraisal of Van Buren's stand on the election for Henry Clay's edification,

> As to Van Buren, with whom I was on the most social [*sic*] terms and who is a pleasant, good tempered facetious little fellow, is at the same time a deep observer, looks quite through the deeds of men—subtle and intriguing—I was afraid to trust him—he is searching, ambitious, indefatigable—no particular attachment to either of the candidates trying to find out whose chance is best—if he could be secured and I have no doubt he might, you would unquestionably take New York—he is now inclined for Crawford because he thinks his chance the best.[15]

Others of Clay's friends in New York, Porter and William B. Rochester, urged him to cultivate Van Buren. But they were wasting their ink and their paper, if not their breath. Rufus King had it right when he said that "V.B. *au fond,* is for Crawford" and followed up this assertion with the astute comment, "tho' he cannot be insensible of the difficulty of the labor."

Van Buren was well aware of the perils and obstacles ahead of him in lining up New York for Crawford. Apart from the Regency, he had but one important politician in the state who would go with him for Crawford, Lieutenant Governor Erastus Root. In Washington, however, the situation was decidedly better. Crawford far and away was the leading contender among Congressmen.[16] Despite his personal feelings about Clay, Calhoun and Adams, all of whom had publicly condemned congressional caucuses electing Presidents, Van

Buren saw them for what they were, using the cloak of popular election and the magic words of democracy, to disguise their weakness in Congress. Van Buren's faith in tradition, in party discipline and organization, in his own estimate of Crawford's strength, blinded him to the fragmenting effect of personal rivalry and the powerful appeal that all caucuses were essentially aristocratic. The people should choose their own electors, the argument went, in open elections free from the intrigues and the drills of party managers. Dismissing all these criticisms as so much chaff, Van Buren found his own way. Cautiously and carefully cementing his alliance with Ritchie, the Virginia political leader, with Randolph, Tazewell and Macon, he kept a weather eye on Pennsylvania and assumed that the Regency would soon wipe out the pockets of Calhoun, Clay and Adams resistance in the Empire State.[17]

Meanwhile, he found himself in an embarrassing situation with his old friend and political ally, Smith Thompson, the secretary of the navy. On February 2, 1823, Thompson visited Van Buren at his quarters in Georgetown. After some discussion on matters of mutual interest, Thompson and Van Buren went to King's rooms where he cordially welcomed them. The immediate topic of conversation was Yates's nomination of the old judges, which all regarded as a political blunder that would rekindle factionalism. Musing on this turn of events, Thompson suddenly asked if it were not time for the legislature to speak forth on the presidential question. Would not this new issue smother the clamor of contending interests? Van Buren thought not, giving as his opinion that it would have precisely the opposite effect. It was premature to stir up any legislative discussion when the election was more than a year and half away.

Obviously, he did not wish to discuss the campaign and left the room for a brief time. While he was absent, Thompson referred to Noah's abortive espousal of Crawford and asked King to keep him informed about Van Buren's presidential preference if and when he learned who the candidate would be, an odd query considering the close relations between Van Buren and Thompson and one that suggested ulterior motives to King. Just then Van Buren returned and King, who was as curious as Thompson, boldly explained what they had been talking about. Van Buren refused to be drawn out. As he often did in similar circumstances, he neatly parried King's remark diverting the conversation to other less sensitive topics. When he realized that Van Buren would not commit himself, Thompson gave up and left for his home.

As soon as the inquisitive little Secretary had gone, Van Buren remarked that there was yet another candidate for the Presidency. The evening's conversation had convinced him that Thompson hoped to challenge Adams as the northern candidate. King agreed, suggesting that since "he and the Secretary were friends it might be proper to seek an occasion of freely conferring with Mr. Thompson on the subject." Van Buren mulled over the matter for three weeks before he sought out Thompson, wondering whether he actually thought of himself as a candidate or was simply acting for Calhoun, his cabinet colleague and warm friend.

Again, without disclosing his own preference, he learned that Thompson was opposed to Crawford. Ten days later the two men dined with Rufus King and over some prime Madeira, Thompson frankly asked Van Buren to back him for the Presidency. Realizing that this was no time for sparring, Van Buren laid out the near impossibility of Thompson's securing the vote of New York. Radicals like P. R. Livingston and Samuel Young on one side and conservatives like Spencer, the old Federalists and the Clintonians on the other would never support him; nor could Van Buren force them into a coalition without risking a rupture in the party. The vice-Presidency was a possibility, he thought. Thompson was not interested, and well warmed with wine, made the extravagant statement that Calhoun was no longer a candidate. He believed the South Carolinian would throw his support to Adams, a suggestion Van Buren hotly disputed.[18]

That such a man of barely second magnitude among the stars in Washington should consider himself in the presidential galaxy, and that this man, a fellow New Yorker, should presume on the many obligations he owed him, strengthened Van Buren's convictions about Crawford. Clinton, now Thompson; what other aspiring New York politician might there be waiting to build up a personal party that would weaken an already distracted leadership in New York?[19]

The current was running strong in New York for a northern candidate; the Empire State deserved the honor. Popular feeling at home restrained Van Buren from any public or even private expression of support. He had met with Crawford frequently during the winter and spring of 1822–23, but maintained such a delphic stance that the Secretary himself was certain he wanted the Presidency for New York. Van Buren left Washington after the adjournment of the short session of Congress in March. He had not committed himself publicly to anyone or anything except his unwavering devotion to the caucus system.[20] The Regency had just turned back a bill that would have repealed the law giving the legislature the power to appoint presidential electors and would have turned it over to the people.[21] Van Buren quashed what had been a serious attempt to popularize the nomination and election process.

When he reached Albany, he received an annoying letter from

Smith Thompson. Apparently he had not been explicit enough in his discussions with this man because Thompson wanted himself nominated in the party caucus and wanted the Regency to push his campaign. By then Van Buren had learned that New York's representative on the Supreme Court, Brockholst Livingston, who had been sick with pleurisy since the end of February, was dead. He seized upon this event to deflect Thompson's presidential drive, suggesting that he might be an appropriate candidate to fill the vacancy on the bench. Much to his surprise, Thompson wrote back that the President had offered to nominate him, but that "my present impression is very strong against accepting it. My hesitation arises principally and indeed I may say solely from my state of health." Thompson wondered whether Van Buren might be willing to accept the nomination. "I am inclined," said Thompson, "to think I shall not accept the appointment."

Van Buren recognized Thompson's carefully phrased offer as a bribe for his support in the presidential contest. His first thought was to pen a polite refusal; but since he happened to be in New York City at the time, he decided to show the letter to Michael Ulshoeffer, who, in his deliberative way, considered both its personal and political implications. On the personal side, a seat in the Supreme Court was tempting, a place not just of honor and dignity but a secure one, which would enable him to establish a permanent home for his family whose welfare since Cantine's death had been a source of constant worry. At the time, Van Buren happened to be in one of his desponding moods where the future seemed an unending series of political trials and intrigues, his law practice ever more demanding. Politically, Thompson had presented him with a singular opportunity. He reasoned that Thompson would not give up a place on the Supreme Court without the requisite pledge of his support for the Presidency, a pledge he would never give. On the off-chance Thompson was sincere about bowing out, Van Buren was quite willing to accept the appointment. It also occurred to him that he could use Thompson's letter to weaken Adams's candidacy. The prospect of eliminating one contender and weakening another appealed to his political instincts.

He and Uhlshoeffer traveled over to Jamaica where Van Buren showed Thompson's letter to Rufus King. As he anticipated, King chose to ignore the carefully inserted caveats and assumed Thompson was stepping aside unequivocally. Van Buren knew that King was for Adams and would be most pleased if Crawford's manager was playing into his hands when he eagerly urged him to accept what he termed, "the proposal of the Secretary of the Navy." Solemn and vehement by turns, King declared:

> The office was very important, and in our system of great authority, dignity and independence; but that does not admit of any expectations of ulterior advancement, nor could it tolerate the interference of the Judge in party or personal politics; that he, V.B., had been deeply engaged in the party politics of the times, of this all are acquainted. To be a member of the Supreme Court he must be wholly and forever withdrawn and separated from these connexions. The dissolution must be absolute, and entering the judicial department, like taking the vow and veil in the Catholic Church, must forever divorce him from the political world.

Van Buren readily concurred, but asked him to write Secretary Adams and the President stressing that he was not an applicant and was available only if Thompson declined the office. He, too, would draft a letter of acceptance to Thompson which he would show King before he sent it. As soon as Van Buren left, King wrote the President and sent Adams a note explaining the circumstances, pointing out that "the appointment would not only be good, but better than any other which could be made from this state." Commented King in a memorandum on the subject, "If Mr. A. has a good nose, he can be at no loss for my object." On April 3, 1823, they met again in New York City; Van Buren presented the draft of his letter to Thompson, which King went over carefully, suggesting various changes which would clarify the acceptance in such a way Thompson would be open to the charge of misrepresentation if he did not decline positively and recommend Van Buren. Van Buren made little or no comment at the time, but the letter that went out over his signature bore only King's stylistic, none of his editorial, corrections. The Secretary would have ample room to change his mind if he so chose without being misunderstood.

Well satisfied with his part in the affair, King returned to Jamaica, believing that he had advanced Adams's prospects in New York. Adams did have "a good nose," and when he received King's note he gave a most flattering description of Van Buren's capabilities when he suggested to Monroe that if Livingston's successor be chosen from New York, he was the best candidate. But Van Buren had "a good nose" too, far keener than King's or Adams's. He knew that Monroe had distrusted him since his bold stand in the Albany post office affair. He also knew that if the President was backing any candidate to succeed himself, he was surely not pushing his cold, contentious secretary of state. The President had been at first perplexed by Thompson's reaction to his offer, but he was not deceived after he received fulsome letters of recommendation for Van Buren from King and Adams. Nor was Thompson when he read Van Buren's letter which made no reference to the all-important issue—his support for the Presidency, yet left loopholes for him to change his mind on the Supreme Court judgeship.[22]

Van Buren was not surprised when he heard from Thompson on April 6 that Monroe would delay the appointment for some time. Van Buren had taken the precaution of making certain the Secretary had not misunderstood him. Before acknowledging Thompson's letter, he sent a copy of it to Rufus King and followed this up with a letter to King in which he rather sharply criticized the President for being surrounded by "court parasites" and for being of "habitual indecision."

After a suitable passage of time he asked Thompson whether "your declension has been definitive and whether Mr. Monroe so understands it." Thompson's answer, a model of obfuscation, said that he had intended his declension to be definitive, but he was not certain the President so understood him, and in fact, he had just learned from Adams that the President believed he had not expressly declined the nomination. There the matter rested until July 1823, when Thompson asked whether he could "with propriety as it respects yourself" take the judgeship. Van Buren assented courteously enough with a certain studied distance that Thompson could not fail to recognize. Soon thereafter Monroe nominated Thompson to be an associate justice of the Supreme Court. Van Buren voted for his confirmation when the Senate convened in Executive Session in January 1824.

A friendship that went back to 1808, deep enough for Van Buren to name one of his sons for Thompson, had become an early casualty in the bitter campaign of 1824. Both men were to blame, the one for his ill-conceived ambitions, the other for seeking political capital in a comedy of errors. Thompson had never been a serious candidate; and even if Adams had drawn upon him the President's displeasure for his part in the contest over the judgeship, it was of no material significance. Nor were Rufus King's motives beyond reproach. He had sought political gain for Adams, and his unsparing criticism of Thompson contributed to dissension in the party. King's thinly disguised effort to remove Van Buren from state and national politics did not become the old-fashioned disinterested gentleman of honor. But such were the ways of politics during the "era of good feeling."[23]

« CHAPTER 9 »

A Broken Politician

A compelling, perhaps the paramount reason for Van Buren's willingness to accept a nomination to the Supreme bench was the disorderly state of the party. That stormy petrel Noah had now turned against Tammany, blaming it wrongly for depriving him of a share of the state printing. So caustic, so dictatorial, so personal had the *Advocate*'s attacks been on Ulshoeffer, Bowne and other powers in Tammany that they organized a faction against Noah and defeated his reelection for county sheriff. Noah paid them back in even more vicious diatribes that seemed to rule out any reconciliation. This was the situation when Van Buren arrived in New York City after Congress adjourned in March 1823. Somehow he managed to calm down Noah; quite probably he promised his support for a national office under the new administration. Next he persuaded the Tammany sachems to doff their war bonnets, no mean feat.

After arranging a truce in New York City, he went to Albany where party factionalism was at fever heat. As James Campbell, a Van Burenite, a Tammany man and the Surrogate of New York County, described it:

> True it is, there appears to be no direct or open opposition in this contest to the Republican Party, all candidates professing themselves to be pure Republicans; but if we examine the matter coldly, it will be found notwithstanding these appearances that the stability of the Republican Party was never more seriously threatened or endangered than at present. Our old foes are still arrayed against us; the mode of warfare is only changed. And they now hope to try insidious wiles and strategems that they could never achieve by open force.[1]

Van Buren took steps to toughen party organization and to reduce factionalism. His method was to have the party endorse caucus nominations that he felt would temporarily, at least, contain the combustible elements. The day before the adjournment of the legislature on April 22, 1823, ninety members under Regency direction passed the appropriate resolution Van Buren had drafted, which in due course would be forwarded to Congress.

Having smothered the hot spots in New York and Albany, having reassured the Regency, and having assured himself that Thompson would take the judgeship, Van Buren turned his attention to family affairs and long-deferred business commitments. He had arranged to have his second son, John, then a lively lad of thirteen, spend the summer in the Rufus King household while he attended school in New York City. His eldest son, Abraham, was bound for West Point. Father and sons left Albany on May 20, 1823. Smith and Martin, Jr., remained behind in Albany with their aunt Maria, and Mrs. Cantine and her family. After settling Abraham and John, Van Buren stayed ten days in the City before returning to Albany for the Columbia County circuit.[2]

Political affairs there were again taking on an ugly look, though this time because of outside pressure, not internal bickering. While Van Buren was in Albany, a new paper appeared in New York, the *Patriot.* In its opening editorial, the *Patriot* opposed the presidential candidacy of William H. Crawford, claiming he represented politicians in Washington, not the people at large. After a few issues, the paper broadened its campaign, launching vehement attacks on both Noah's *Advocate* and Tammany Hall. Now, everyone who interested himself in such things recognized the Calhoun bias of the *Patriot* and guessed the identity of its skillful corps of writers, most of whom were connected with the War Department or identified with the President. Charles K. Gardner, editor of the *Patriot,* had been an aide to Army Commander Major General Jacob Brown before serving as assistant postmaster general. Brigadier Generals Swift and Scott and Lieutenant John A. Dix saw no conflict of interest between their army careers and their partisan interests. Henry Wheaton, an erudite lawyer and an experienced journalist, and Samuel Gouverneur completed the editorial department.

With this array of skillful penmen, the *Patriot* was soon artfully squibbing away at Van Buren's popular support in New York City. The object of its editorial onslaught remained unperturbed, urging his supporters not to notice the *Patriot* until its policy became more specific. Not long after Van Buren left the City, however, it began denouncing the caucus system as un-American, undemocratic and unprincipled, the tool of intriguing managers. As the campaign gained momentum, the *Patriot* launched an anticaucus faction that it called the People's party. A rallying point for all those individuals and factions opposed to Van Buren, the People's party attracted

Clintonians, old-line Federalists, disappointed officeseekers and a sizeable number of floating voters.

Throughout the summer, the new organization gained in popularity and though it maintained its anti-Crawford line, it did not add many recruits to the Calhoun cause. If anyone was profiting from the strong current of opinion against a caucus nomination, that person was John Quincy Adams and to a lesser extent, Henry Clay and De Witt Clinton. Still, Van Buren refused to be swayed from his conviction that the *Patriot* and the People's party were temporary aberrations despite mounting evidence to the contrary.

Although Van Buren had confided in no one, he was now committed to Crawford. Unmoved by the ever-increasing clamor of the People's party, Van Buren placed his confidence in his ability to control the legislature that would be chosen in November.[3] For once Van Buren was out of tune with public opinion; he had misjudged the political climate. Every candidate for the legislature on the People's party ticket from New York County was elected over the Regency slate. Six anticaucus candidates won election to the senate by large majorities. Van Buren and the Regency controlled the legislature but the portents were unmistakable. The Regency must change its tactics or it would lose the state for Crawford, and with this loss possibly control of the party mechanism. Much of Van Buren's success, thus far, had come as a result of his opponents' errors and misjudgments, his own ability to recognize and exploit them. He chose to regard the election and the People's movement as just another mistake of the opposition. Any practical politician could see the new party was simply a cover for the campaigns of the competing presidential candidates, a blatant attempt of political enemies to injure him through misrepresentation. His obsession about Clinton, whose hand he saw in the maneuver, made him all the more certain that the traditional path of caucus nomination was not only correct in a tactical sense but worthy in a moral sense.[4]

Others were not so sure. A veteran local politician who had twice been elected Speaker of the house, canvassed the state in July. He found no agreement on the caucus question in the New York congressional delegation. In New York City he was shocked to find widespread and vociferous opposition to Van Buren from the voting public: "It may routinely be asked what has caused such a change? I answer that a very general belief has gone abroad (whether true or false is not for me to say) that he [Van Buren] has offered to barter this state for the purpose of furthering his own views." He concluded that Crawford had little or no chance of carrying New York; Adams and Clay were about equal. Jackson, who had been nominated by the Tennessee legislature and was gathering strength at

Calhoun's expense in Pennsylvania, was "not much talked of." Nor, despite the success of the *Patriot,* was Calhoun popular among the voters of the City.[5]

When Van Buren left for Washington, he still had confidence in the caucus system and in New York's eventual support of Crawford. But there were too many storm signals flying for such an experienced political navigator as he to be completely unconcerned. A major problem involved the governorship and an unrepentant Samuel Young, whose sullen attitude toward the Regency was being fostered by the able, thoroughly unscrupulous operator, John Cramer. Unknown to Van Buren, Cramer was also active in the People's party. A far graver matter for Van Buren was the health of his candidate. Crawford had become desperately ill; and while the extent of his sickness was kept secret, Van Buren could not have been reassured by the accounts he read in Crawford's own newspaper, the Washington *Daily National Intelligencer,*[6] or in the private information he received. Still he and his partisans plodded ahead with the Crawford candidacy.

Skinner and other members of the Regency, hoping to brake the momentum of the anticaucus forces, prevailed upon Yates to oppose election by general ticket in his annual message to the legislature. Anxious also to prepare the way for dropping Yates with a minimum of recriminations, they intimated to him that he could be nominated for Vice-President on the Crawford ticket. The weak-willed Governor, against the advice of his cousin, John Van Ness Yates, and other intimates, agreed.

Yates's second annual message arrived in Washington just after Van Buren and his Messmates had finished dinner on a raw evening in mid-January. Van Buren and most of those assembled were certain Yates would come out for a general election law. King was particularly anxious for such a declaration because he felt the Governor's influence would improve the chances of John Quincy Adams getting all or a substantial portion of New York's thirty-six electoral votes. King proposed that Yates's message be read aloud. All agreed. Andrew Stephenson of Virginia was designated for the task. "Mr. King," as Van Buren recalled, "folded his handkerchief on the table before him and resting his arms upon it, as was his habit, his complacent countenance indicated the confidence and satisfaction with which he prepared himself to hear the welcome tidings."

When Stephenson came to the crucial subject of the election law, there was an audible gasp in the room. Yates had thrown the weight of his office and his influence behind the present mode of choosing electors. King, who had a hot temper when aroused, scowled deeply and turning to Van Buren said in a tone heavy with irony, "I think

Mr. Van Buren, that Mr. Crawford's friends ought to send the Governor a drawing of the Vice-President's chair." In all innocence Van Buren asked why. "Because I presume they have promised its possession to him." Rather sharply, Van Buren said that he could not speak for his associates, but he was certain they had done nothing of the kind. "I hope so," said King angrily. "I know so," said Van Buren, who had also lost his temper, a rare occasion for him. He, of course, did not know so, but felt perhaps unjustly that King had singled him out as the culprit in an underhanded intrigue. Without another word, King left the room. He absented himself from the Mess for the next day, taking his meals in his chamber.

But King realized after reflection on the subject that he had no proof for his ill-tempered remarks at the expense of a fellow-Senator whom he liked personally and whose abilities he respected. King approached Van Buren and asked if he would join him at a reception to be given that evening at the French minister's. Van Buren, who likewise regretted his display of temper, was happy to accept the invitation. On the way, King apologized for his rudeness, as did Van Buren for his. Their friendship remained unimpaired, but the strains of the presidential contest were altering personal relationships in the small circle of official Washington. Van Buren saw little of Thompson, and his relations with Calhoun were correct but distinctly chilly.[7]

Van Buren had been right in his calculations that the state legislature would reject the general election bill Henry Wheaton introduced. But he must have been shaken at how slim the margin had been—three votes in the senate. Moreover the public clamor the opposition press raised when it printed the names of the Regency senators who voted against the general election bill in bold type set off with black borders was approaching a fever pitch.

Cramer, secretly pushing Samuel Young for governor on a People's ticket, fed innuendo, rumor, whatever seemed appropriate to his purpose, into the anti-Regency press. Through whispering campaigns, he shrewdly organized anti-Caucus sentiment; remarks made the rounds of public meetings, militia training days, patriotic celebrations and testimonial dinners. Van Buren, who was trying to rally his circle in Albany, was having more serious problems in Washington trying to preserve the caucus system and ensure the election of the ailing Crawford.

As soon as he had what he took to be reliable word that Crawford was on the mend, he sounded out friendly fellow Senators and Congressmen on putting together a nominating caucus toward the end of December 1823. He was anxious to get a substantial endorsement, which he believed would ensure the necessary momentum. To his chagrin, he found little support for an early caucus nomination.

And then Crawford, who had returned to Washington, became ill again. What started with an inflammation of one eye spread to the other and the condition gradually became worse, so that by the end of December the Secretary was very nearly blind and had to remain in a darkened room. Crawford's condition put an abrupt end to Van Buren's unproductive efforts until his eye infection slowly cleared up; by early February 1824, though still confined to his home, he was carrying on his Treasury responsibilities.[8] Van Buren, who was now acting as his chief manager, judged that the caucus must now be convened.

He was especially concerned about Pennsylvania, where Jackson was making rapid headway in the western part of the state and where support for Calhoun's "People's party" was thought to be losing ground. In discussion with Crawford, Van Buren and other Crawford supporters like Jonathan Russell (an anti-Adams Congressman from Massachusetts), Senator James Barbour from Virginia and Senator Jesse Thomas (a Marylander who had moved to Illinois) set February 14, 1824, as the date for the caucus. Having given up on Pennsylvania, Van Buren tried unsuccessfully to have Henry Clay accept the caucus nomination for the Vice-Presidency.[9] After Clay's refusal, he agreed with others of Crawford's managers to seek out the aged Swiss financier and Jeffersonian politician, Albert Gallatin, who it was thought might counteract Jackson and Calhoun sentiment in Pennsylvania. Grasping at any means of improving their position, the desperate managers also wanted Gallatin to visit Washington before the caucus on the outside chance he would bring some of the doubting Pennsylvanians around to Crawford. Gallatin reluctantly agreed to accept the caucus nomination for Vice-President but, pleading other engagements, remained at home.[10]

Van Buren knew several weeks before the caucus that the turnout would not be as high as he had hoped. He had been cautiously optimistic when a Calhoun-sponsored anticaucus meeting adjourned without making any recommendations. He and his associates counted on a split between Jackson and Calhoun supporters that would enable them to pick up some additional votes from Pennsylvania. But their hopes were unwarranted. Although the Calhoun and Jackson men were rivals in Pennsylvania, they were one in opposing the caucus.

Saturday evening, February 14, was cold. As the hacks, carriages and people on foot made way through the frost-hardened streets to the Capitol, snow flurries swirled about them. It was an unpleasant night and the hall of the House of Representatives where the caucus would soon take place was cold and dimly lit with candles that flickered in the drafts. The galleries, jammed with bundled-up spectators, looked down on a scene where all efforts to cultivate the dignity and the solemnity of a presidential nomination were lost in the scant attendance of Congressmen and Senators who seemed swallowed up in the huge chamber.

Van Buren was appalled. He had thought that nearly a hundred Congressmen would be present, a respectable one-third of the Congress; but instead there was a bare one-fourth, and of this number almost one-half were New Yorkers and Virginians. Crawford did not even have a solid Georgia delegation present and voting. If the galleries were any measure of public opinion, the caucus and its candidate were decidedly unpopular. After the clerk tallied the vote, Benjamin Ruggles, a Senator from Ohio who had been elected chairman, announced the vote in his high nasal voice with a New England twang. There were 62 votes cast for Crawford, 2 for Adams, 1 for Nathaniel Macon and 1 for Andrew Jackson.

Van Buren, in his correspondence with the Regency, and in his conversations with King, MacLane and others of his Mess, made the best of what he knew was a bad situation. Return mails from Albany were somewhat more consoling. A poll taken two weeks after the caucus revealed that the New York legislature stood 88 for Crawford, 36 for Adams, 11 for Calhoun, 6 for Clay. At the end of the list was Andrew Jackson with but 4 votes, heartening news to Van Buren who now judged Jackson to be Crawford's most formidable opponent.

On February 18 at a Philadelphia town meeting, George M. Dallas, head of the Family party, Calhoun's chief support in Pennsylvania, shifted to Jackson. Van Buren was certain now that Calhoun would either withdraw from the contest or accept nomination for the Vice-Presidency on the Jackson ticket. Ten days later this judgment was confirmed when the Pennsylvania State Convention at Harrisburg unanimously nominated Jackson for the Presidency, Calhoun for the Vice-Presidency.

Van Buren continued to encourage his friends in New York. "I have no time doubted," he wrote Jesse Hoyt on March 6, 1824, "of our complete success. . . . despondency is a weakness with which I am but little annoyed. On the assumption that New York will be firm and promptly explicit, we here consider the question of the election substantially settled. Neither Mr. Adams nor Mr. Clay can keep in the field after the course of New York is positively known."[11]

Meanwhile, in an attempt to soften the anticaucus sentiment in New York and to draw, if possible, some of the teeth from the People's movement, Van Buren had introduced in the Senate a resolution for a constitutional amendment that would require that nominations for President and Vice-President be made by general vote in the states. Van Buren's amendment was simply a gesture to show that he believed in the principle of direct and general selection of presidential electors. He was astonished when Governor Yates used his bill to justify a caucus recommendation.[12] The Adams and Clay newspapers quickly pointed out the absurdity of any federal constitutional amendment altering New York's election law; and the Crawford-aligned newspapers talked of "high-minded" reforms, skirting the issue that Van Buren's amendment bill would have no effect whatever on the presidential election, while ignoring Yates's comment upon it.[13]

Worried by popular unrest at home, Van Buren was facing ever more serious problems in Washington. A cloud still lingered over the political fortunes of his candidate, William H. Crawford. In early 1823 a series of letters signed "A.B." appearing in the Washington *Republican,* a Calhoun paper, had charged Crawford with mismanagement, if not downright abuse, of the public trust in his dealings as secretary of the treasury with certain speculative western banks. Despite the obvious political motivation of the "A.B. letters," they offered sufficient hard information to bring Crawford's policies under scrutiny. The House appointed a special Committee of Investigation in an effort to discover the author of the letters so that a deeper probe could be made. But the committee became hopelessly enmeshed in presidential politics. Unable to identify the author, it issued a report of little value, neither substantiating nor dismissing the charges.

Amid a confusion of voices, the House appointed a new select committee drawn from its leading members. Crawford obligingly supplied it with Treasury Department documents that refuted A.B.'s charges. There the matter stood when Crawford fell seriously ill in mid-September 1823. Concerned as he was about his candidate's health, Van Buren had also watched with his accustomed care the course of the A.B. affair. Early on he satisfied himself that it was a political foray expressly designed to injure Crawford. His faith in the Secretary's integrity unshaken, he had to admit to himself that the accusations and the investigations were injuring Crawford's candidacy.

But any worries he may have felt about the A.B. affair were trivial when compared with Crawford's illness. Van Buren found himself during the spring of 1824 publicly committed to a candidate who now seemed unable to assume the responsibilities of the Presidency. The trouble was that Crawford's marvelous vitality would bring fresh hopes of recovery before another attack ensued; and compounding the dilemma, he refused to take himself out of the race. While his attitude remained what it was, his chances for a complete recovery possible, Van Buren could do nothing but remain loyal to a man who was gradually losing his political as well as his physical strength.[14]

The nagging A.B. affair now made its last appearance, this time instigated by Crawford's friends who had long suspected a former Senator from Illinois, Ninian Edwards, of being the author of the letters. Edwards had denied authorship under oath during the first investigation. But the Crawford men, whom Calhoun and Clay denominated "Radicals," had been secretly collecting their own information and were satisfied that Edwards was the man. They chose to make their charges at an opportune time when it would cause the most damage to the Monroe Administration. Monroe had just appointed Edwards minister to Mexico, and the Senate confirmed his appointment. Then they struck. Edwards was actually on his way to his post when he received information that was so conclusive he readily admitted he had been the author of the letters.

The President was incensed and would have removed Edwards from office immediately, but was restrained by the more politic counsels of Adams and Calhoun, who argued, and rightly so, that now exposed as a perjurer, Edwards would resign. Meanwhile, the House sent its sergeant at arms to arrest Edwards and bring him back to Washington where he would be compelled to testify before a third committee composed of its leading members. The investigation, which dragged on for several months, could not substantiate any of the charges against Crawford and the matter faded away. Even though the House report exonerated Crawford, enough evidence and testimony was spread before the country to tarnish his towering image as the exemplary administrator, the wise and able public servant.[15]

Van Buren had been asked to remain in Washington after the adjournment of Congress in case his advice might be needed while the investigation went on. He took this opportunity for a short vacation where he could be free of his many cares. He had long wanted to visit Thomas Jefferson and imbibe political wisdom from the original source; he had another motive, too. He wanted Jefferson to come out publicly for Crawford as the only candidate who faithfully reflected his political philosophy. An endorsement from the demigod of Monticello he felt would bolster Crawford's flagging fortunes and might add as many as half a dozen Clay and Jackson states to his candidate's column.

Accompanied by Mahlon Dickerson, one of the "martyrs" of the Alien and Sedition Acts, Van Buren slipped out of the city on May 29, though their departure did not escape attention. The day before they left, Henry Clay wrote a friend, "I conjecture that a visit Mr. Van Buren and Gov. Dickerson of New Jersey are about to make to

Virginia is connected with . . . measures for a fresh campaign." What Clay did not know, nor anybody else in Washington, even Crawford, was that the two men planned to visit Jefferson. When they reached Fredericksburg, Virginia, Van Buren posted a hasty note to Crawford's confidant, Asbury Dickens, giving him notice of their intent. "Mention this to no one but Crawford," he enjoined.[16]

Jefferson, at eighty-one, seemed many years younger to the admiring Van Buren. "His imposing appearance, as he sat uncovered—never wearing his hat except when he left the carriage and often not then—and the earnest and impressive manner in which he spoke of men and things," fascinated the younger man. For the most part neither he nor Dickerson did much talking over the several days of their visit. As chairman of the Senate's Judiciary and Finance Committees, Van Buren did raise questions about the federal judiciary, the Second Bank of the United States and Chief Justice John Marshall's recent decision on the Bank in *McCulloch v. Maryland*.

These timely topics gave Jefferson an opportunity to range over the whole issue of judicial tenure, appointments and powers, on all of which he held strong opinions. They talked also of banks in general, internal improvement and the present state of the Republican party. Jefferson, as was his wont, reminisced about days gone by, the birth of the Republic, Washington, Adams, Hamilton, Patrick Henry, the extreme factionalism of the 1790s. The tall, gaunt man, his once red hair now sandy gray, his hazel eyes alert to questions and comments, his easy laugh and the stream of anecdotes, some to illustrate points, others prompted by nostalgia or elaborated on for the benefit and the interest of his guests, made such an impression that Van Buren remembered those few days with amazing clarity thirty-five years later.[17]

The visit was, of course, a pilgrimage for Van Buren because of his veneration for a man whom he considered the real founder of the Republic, the philosopher who had devised what Van Buren felt was the most perfect system of politics ever conceived. But try as he would, Van Buren failed to achieve his other purpose, a strong statement of support for Crawford. There had been some talk and many questions posed about the Monroe Administration. Was an incipient federalism shaping the President's policies? What of the nationalist ideas of Clay and Calhoun? Would not the personal ambitions of presidential contenders break up the Republican party into factions pursuing personal programs unless Monroe's "fusion" program was checked? And was not the ailing Crawford the only hope for a restoration of Jeffersonian principles, a wise and frugal central government, the sovereign rights of the states and the people, the curbing of an arrogant, centralist Supreme Court that was encroaching on state power and within the central government on the power of the executive and the legislative branches?

Jefferson was unable or unwilling to give specific answers at the time, or at least answers that satisfied Van Buren. He could not steer the conversation to Crawford except in a most general way. What he wanted most, the endorsement, the accolade, he did not get. A month later, Van Buren sent along a recent pamphlet authored by that brilliant but bitterly unforgiving Federalist, Timothy Pickering, that indicted both Jefferson and Adams for placing their own ambitions over the public welfare. In his covering letter he tried again. After raising the issue he and Jefferson had discussed, Monroe's amalgamation of parties, Van Buren spoke of Crawford's health and the A.B. investigation. "Mr. Crawford's condition is greatly ameliorated and I entertain no doubt of his speedy restoration to perfect health," he wrote. On the A.B. investigation, he was certain Crawford was the victim of a political plot and would be completely cleared.

If he thought his carefully worded letter would bring the necessary blessing, he was doomed to disappointment. Jefferson replied with a spirited defense that was in effect a history of the political relations among himself, Adams, Washington and Hamilton. But at the end of his five-thousand-word letter he wrote, almost as an afterthought, a single line alluding to Van Buren's request for his views on Monroe's political course. "Tories are Tories still," he commented, "by whatever name they may be called." He did not even mention Crawford's name.[18]

By now Van Buren was clutching at straws. Struggling with what appeared to be a losing cause in Washington, he was in serious trouble at home. The Regency without his careful hand on the tiller had become increasingly driven toward dangerous reefs. The anticaucus sentiment in the state and the rapid rise in factionalism as the presidential contenders sought to build up personal followings was destroying unity of action and weakening party leadership. Van Burenites, reacting to this clear and present danger, concocted a series of poorly considered maneuvers without waiting for Van Buren's advice. That such a group of able, experienced politicians, commanding such powerful support from press and officeholders, could have made so many mistakes in such a short time was little short of amazing.

Their first move was the unceremonious dropping of Governor Yates. Flagg and the newly elected senator from northern New York, Silas Wright, just beginning his political career under the protective eyes of Van Buren and Roger Skinner, at first stood by Yates. But a thoroughly frightened Regency, deeply worried about Samuel Young, then the acknowledged leader of the People's party that had the backing of Adams and Clay supporters in the state, sacrificed Yates to his ambition.[19]

After his initial falterings, Yates had proven a loyal Regency man, though quietly resentful of Van Buren's leadership. As early as February 1, a stream of letters went forth from Albany to Washington, voicing increasing fears that Young was gaining, and that if Yates were renominated, the party would go down to defeat. Van Buren urged restraint, pointing out that Young was leaning toward Clay and presumably opposed the caucus nomination. Members of the Regency countered that if Young stood for governor on a People's ticket, secured Clinton's support, a distinct possibility, and won the election, he would surely split the ticket and Clay might even make away with a majority of the electoral votes.

Six days before the legislative caucus was to meet and make nominations for governor and lieutenant governor, Butler wrote that Yates "can not obtain a nomination even if the caucus were exclusively composed of the regular republican members and that to defeat Young and harmonize conflicting interests and opinion, our friends will be obliged to give up Gov. Y." Butler was concerned about the ethics of the thing. "The abandonment of Y. looks ill and I fear will be thought a breach of good faith and a violation of political morality and honour."[20] Van Buren moved as quickly as he could to repair the damage his impetuous subordinates had done. He assured Yates that he would be recompensed for his sacrifice, probably with the Senate seat of Rufus King whose term would expire in a year's time and whose age and uncertain health precluded his seeking reelection.

On April 3, 1824, Young was nominated for governor at the legislative caucus and Erastus Root for lieutanant governor. The hand of the Regency was obvious in the proceedings. Young, who had been considered partial to Clay and vigorously supportive of the general election law, suddenly became silent on both subjects. So far, the Regency had managed to avoid a political explosion, but then it committed an irretrievable error. Under the constant prodding of Skinner, whose hatred of De Witt Clinton knew no bounds and who had an unfortunate tendency to mix his politics with his own personal feelings, it decided to depose Clinton from his unpaid post as president of the Canal Board.[21]

The Regency took this dangerous step only four days after the nomination of Young and Root. Considering themselves betrayed, vocal spokesmen of the People's party held an indignant meeting at which they denounced the caucus nominations and passed a resolution for a convention to be held in Utica on September 21 to choose the People's candidates. The prime mover in this movement was

James Tallmadge, Jr., a crusty conservative from Dutchess County, who, like so many other politicians in the state, had once been close to De Witt Clinton but was now a political opponent. It was soon learned that Tallmadge would be the gubernatorial nominee and, in all likelihood, would give Young a stiff fight in November. Hence the quick and foolish reaction of the Regency to divide the leadership of the People's party from its rank and file by forcing it to vote on Clinton's removal.

If Tallmadge and Wheaton voted to return Clinton, they risked the anger of their party's Tammany members, whose enmity for the ex-governor was of twenty years' standing. If they voted for removal, they would alienate their Clintonian following. The plan may have seemed appealing as an abstract political move, but politics, as surely these veterans must have known, was emphatically not practiced in the abstract. In striking down Clinton, they were striking at the symbol of the age, the Erie Canal. For the Canal and Clinton were synonymous in the popular mind. They were creating a martyr for a short-term and mean-spirited advantage.

There had been no time to consult with Van Buren, who would have saved them from their folly, for the legislature was due to adjourn on the morrow. The word went out to the Regency leaders in the house and senate to remove Clinton by resolution, forthwith. A witness to the event was Thurlow Weed, then a writer for the *Rochester Telegraph,* one of the few Adams papers in the state. Weed was in Albany lobbying for a bank charter. His open, genial manner, his talent for not forgetting a name, his readiness to stand any member of the legislature to a drink or a meal on funds supplied him by the businessmen of Rochester, earned him warm friends and the privilege as a reporter to be on the floor of both legislative chambers. Weed had noticed a hasty and excited conference in the house among the Regency men, one of whom soon left for the senate chamber. Behind Weed's *bomhomie* lay concealed a talent for political management that rivaled Van Buren's. Making himself as inconspicuous as possible, he followed the messenger, whom he saw hand a note to Silas Wright. Wright opened his desk drawer and removed a slip of paper on which had been written the resolution to depose Clinton. He handed the paper to another Regency senator who promptly offered it to the senate where, despite conflicting views on the caucus issue, a solid majority were long-term political enemies of Clinton.

In an instant, Weed's quick mind had grasped what was happening. Without waiting for the senate to vote, Weed hurried back to the house chamber and sought out Tallmadge, whom he knew was hostile to Clinton. Weed gave Tallmadge sound advice. Address the house he urged, and in a few brief remarks explain his long opposition to Clinton, the man and the politician, but insist that removal of the guiding spirit of the Canal when it was progressing so well would be construed by the people of the state as sacrificing the public interest for political ends. Weed could not move Tallmadge, who probably suspected him of advancing Adams. The editor-lobbyist went to others with the same counsel and with no more success than he had had with Tallmadge, though he finally persuaded an obscure assemblyman to denounce the resolution from the senate. Two other speakers followed with brief remarks and then the resolution passed almost unanimously.

Weed thought it a clever trap that had caught many of those opposed to Crawford and who supported a general election law. Although certainly a trap, the removal of Clinton caught those who set it, not the Tallmadges and the Wheatons, whom it was supposed to snare. As soon as the news was out, the uproar began. People everywhere were shocked at the political malice that was exposed. Even many of those who disliked Clinton considered him a great man, a statesman, the embodiment of the Empire State. Public meetings sprang up spontaneously all over the state denouncing the act and blaming the Regency for it. In the western counties and along the Canal route, the speeches and the editorials and the public letters in the weeklies demanded vengeance. From the political ashes where Van Buren had consigned him, Clinton rose almost over night to become a public hero and, for those whose political antennae vibrated in the storm of protest and adulation, the next governor of the state.

Van Buren received the news with something akin to despair. Instantly he knew that his associates had made the political error of their lives, and he realized that Clinton would seize power and ride roughshod over his carefully constructed organization. It had taken him ten years of hard political work to finish off his rival only to see it all vanish in less than an hour on the floor of the legislature. Never had Crawford's chances looked more dismal; never had Van Buren's political future been more clouded.

After his arrival in Albany came another shock. The People's party convention at Utica nominated Clinton for governor; and sufficient pressure was brought to bear on James Tallmadge, Jr., to accept the nomination for lieutenant governor.

When Van Buren appeared at the polling place in November, the crowd of idlers, petty politicians and pettifogging lawyers who lounged about such places on election day jeered at him with the taunting remark that became a chant, "Regency! Regency!" Over a dozen voters challenged his right to vote and, despite the best efforts of the local board of inspectors to spare him the humiliation, Van

Buren was required to take the oath prescribed; for the second and last time in his career he was forced to make a public acknowledgment that he was not a liar.

Two days after the election when half of the returns had come in, Van Buren knew that his party had been beaten badly. Skinner, the man most responsible for the debacle, brought the news to him while he was eating breakfast. As Van Buren scanned the totals on the tally sheets, a silent Skinner nervously tapped his fingers on the window. "I could not resist saying to him," recalled Van Buren, "I hope, Judge, you are now satisfied that there is such a thing in politics as *killing a man too dead.*" Skinner left the room immediately. Later Van Buren apologized for his hasty remark, but Skinner carried a deep sense of remorse for the damage he had done his party to his grave. He died within a few months, leaving a gap in the Regency ranks which at the time seemed irreplaceable.[22]

Skinner may have been overwhelmed by a defeat that saw Clinton brush Young aside by 16,000 votes, Tallmadge engulf Root by twice that number, the loss of six seats in the senate and a three-to-one majority against them in the house. But Van Buren was a more resilient politician, more realistic in his assessment of the results than his associate. The Regency still controlled the senate, and its minority in the house was a cohesive unit for Crawford, the majority being divided between Clay and Adams. Clinton, despite his overwhelming victory, was disliked by many in the legislature who had concealed their enmity in the face of popular enthusiasm for the governor-elect. According to law, the legislature must meet immediately after the election to select a ticket for President and Vice-President. On November 10, 1824, the special session convened. Van Buren was hopeful that he could salvage some Crawford electors out of the wreckage of the Republican party in the state.

Elections throughout the country made it appear virtually certain that if he could hold the line in New York, the House of Representatives would choose the next President, each state delegation casting one vote. Adams and Jackson already had enough votes to go into the House, the initial question then was who would be the third candidate? The federal Constitution prescribed that in case of no majority, the House would select from among the three highest candidates. New York was crucial. If Clay were to secure enough electoral votes from New York without a compensating loss in Adams's strength, Crawford could be ruled out. Van Buren and his lieutenants went to work at once.

New York law prescribed that only two electoral tickets containing the names of thirty six electors could come before the legislature, one from the senate and one from the house. After they had been made up, the tickets would be compared at a joint session. If they agreed completely, the election was over; if not, conflicting names and blanks were thrown out, another vote would be taken on a revised ticket and the sequence repeated until all electors matched. Since there were three active candidates with factions of their own before the legislature, a coalition ticket had to be designed that would involve careful and tactful management if it were to succeed, precisely the skills at which Van Buren excelled.

A trial vote in the two houses revealed two things: how much the recent election had shaken the faith of the legislature in the magic of Van Buren's leadership and how much the presidential contest had eroded party loyalty. Adams led the pack in the house with 50 votes, Crawford next with 43, Clay a poor third with 32. In the senate, which the Regency still controlled, Crawford received 17 votes, though not before Van Buren made a deal with Cramer, Young's manager, that he would place seven moderate Clay men on the Crawford ticket. In return, Young and Cramer promised 16 Clay votes from the house.

The Crawford ticket that finally came out of the senate reflected five days of bargaining, caucusing and not a little acrimony. A Clay-Crawford coalition was a shaky one at best; and with the cunning, unscrupulous John Cramer acting for Young, Van Buren should have been especially cautious. For the next several days, balloting went on in an unsuccessful effort to break the deadlock, the strength of the three candidates changing by two to three votes as deals were made and broken. But faction lines remained firm.

Van Buren was faced with a dilemma. Clearly the house was deadlocked; even if Cramer's promise held, a promise he was most reluctant to stake his hopes upon, the coalition would lack two votes of a majority with every one present and voting. Popular clamor was rising, the feeling against the legislative caucus, against the Regency, against him, was becoming more intense each day the session remained at an impasse. Something had to be done, but what? The only way out was to stake all on Adams in the house. Van Buren reasoned that this would reduce the candidates to Crawford and Adams.

The next day after another fruitless ballot, Flagg rose and announced that he was marking an Adams ballot, the signal for all the Crawford men to do likewise. Clay's supporters were outraged as they saw their man vote by vote being thrown out of the competition. With the house chamber in a complete uproar, a motion for adjournment was heard, seconded and carried. Van Buren and his Regency associates met with the Clay leadership that night, but they were angry, restless, confused and highly suspicious of his intentions. At this point, Van Buren ought to have realized he had failed. The Clay contingent, now stubbornly furious to a man at what it regarded as the base behavior of the Regency, would never vote for Crawford even if this meant the sacrifice of their electoral votes. Van Buren had not thought that the Adams faction might approach these angry men bearing gifts.

Wheaton, Tallmadge and Weed sought out Porter, Young and others of the Clay faction. They had a plan that would lead the Clay men out of their deadlock, save their votes on the Crawford ticket and avenge themselves on Van Buren. Weed had concocted the scheme that was simplicity itself, though of doubtful legality. A third ticket would be made up and the moderate Clay men on the Crawford ticket would be placed upon the new Adams ticket. If the utmost secrecy prevailed, if discipline could be maintained, and if the full Adams and Clay strength was present and voting at the joint session to be convened on the morrow, the third or Union ticket, as Weed called it, would win with two votes to spare. Weed himself would set the type and strike off the new ticket. He would do it that very night in the pressroom of the Albany *Daily Advertizer,* an anti-Regency paper whose editor had agreed to lend the premises. Amazingly, in rumor-laden Albany, the secret was kept, even though eighty-two voluble politicians and potential place seekers were in on it.

On the following day, the two houses assembled and the legislators in solemn procession cast their ballots. When all had voted, Lt. Governor Root opened the ballot box and took out the first ballot. His heavy, flushed face at first assumed a quizzical expression, followed quickly by a look of amazement. "A printed split ticket," he announced. For a moment the chamber was hushed, then as the Van Buren men realized what had happened, one of their senators leapt to his feet and exclaimed, "Treason by God."

At once pandemonium gripped the legislators, and the Regency contingent was unable to break up the proceedings by leaving the hall before the ballots had been counted, as the law prescribed, in the presence of the two houses. The tally proceeded and on the first ballot for the 36 electoral votes, the Union ticket garnered 32 votes; 25 for Adams, 7 for Clay, 4 were thrown out. On the second ballot, 6 supporters of Clinton, who had been acting with the Clay men, crossed over and voted the Crawford ticket, giving him the remaining 4 electoral votes.[23]

Poor timing, poor communication, faulty judgment of human nature, faulty calculation of the odds, a reckless disregard for the pressures of public opinion, all of these must be charged to Van Buren for the collapse of the Crawford cause in the legislature. The joint session of 1824 also represented more, much more than a series of tactical blunders. It was a rebuke, a mighty rebuke to the caucus system, and it spelled the beginning of the end of northern subservience to the South in the federal executive branch.

Slavery, unmentioned in the heated debates, in the public prints, in the caucuses and in the street, was nevertheless present. The Missouri debates and the state constitutional convention of 1821 were still fresh in the minds of the public, if not of the legislators. A new order of politics was emerging where the North, New York—its grand canal and its great statesman, De Witt Clinton—all combined in a popular image of the state as power, as future, as prototype of modernism.

Van Buren recognized these undercurrents and meant in due course to give them direction. His basic error as a politician was in relying upon tactics that he thought to be traditional but were in effect obsolete. Party machinery, however realistic it seemed to the Regency, was no substitute for the mounting pressures of changing attitudes.

As he agonized over his mistakes and misfortunes, Van Buren maintained a cheerful public face and kept the Regency together. Thurlow Weed, just as anxious as Van Buren to keep Clay out of the House, worked in his own subtle way to reduce the seven Clay electoral votes.[24] But there was no doubt in Van Buren's own mind that he had failed utterly. "I left Albany for Washington," he said, "as completely broken down a politician as my bitterest enemies could desire."[25]

« CHAPTER 10 »

A Quick Recovery

When Van Buren arrived in Washington, he found a city full of intrigue. In his present state of mind he had little patience with the sort of huckstering that was in full cry. After paying his election bets in New York City he had hurried south avoiding Jamaica and Rufus King, his customary traveling companion. He was still irritated at King's support of Adams; and though he knew the old Senator had a perfect right to his own opinion, as he later explained, "Since I was in no mood to form a very correct estimate . . . I indulged my feelings."

He would have preferred to journey alone but could not resist the entreaties of Stephen Van Rensselaer, whom he met in New York City. At Philadelphia, King overtook them and sought a reason for the change in Van Buren's itinerary. King's approach was sensitively indirect, infinitely courteous. "I muttered some civil explanation that explained nothing," said Van Buren, and when they reached Washington, King moved out of the Mess where for the past three years he had lived in the closest association with Van Buren. That he should have acted this way toward an old and distinguished man nearing the end of his public career shows the depths of Van Buren's despondency at the wreck of his plans and his ambitions.

Yet all was not dust and ashes; word from Albany confirmed that the Regency had picked up one more vote for Crawford at Adams's expense and that Clay received only 4 of 8 electoral votes the Adams men had promised. The final result in New York was Adams with 26 votes, Crawford next with 5, Clay with 4 and Jackson with 1. Calhoun had received 29 votes for the Vice-Presidency to ex-Senator Nathan Sanford's 7. The Clintonians working for Jackson had helped in reducing Clay's vote, thus indirectly assisting Crawford. "Several agents of Jackson attended," Roger Skinner reported, "and did all they could to effect a division amongst the Adams electors."[1]

Two weeks later, it was learned in Washington that Clay had lost

Louisiana and was thus counted out of the running. Van Buren's spirits picked up perceptibly with this news which placed Crawford in the House, albeit a very weak candidate against Jackson and Adams. Still, if the leading contenders deadlocked, Crawford might just be selected as a compromise candidate; or failing that, he and his Virginia friends could and would use the Crawford states as bargaining counters to rid the executive branch of what they chose to believe were Federalist influences.

Oddly enough, in this game of President-making, no Congressman was as inflexibly devoted to Crawford as Van Buren's close friend and fellow Messmate, Louis McLane, Delaware's one Congressman. Formerly a staunch Federalist, he still clung to the principles of his youth, though he could see no inconsistency in his ardent support for Crawford, the strict constructionist, Old Republican of the extreme Virginia school. The paradox, however, was more apparent than real. McLane who had a quick temper and agile mind was strongly attracted to Crawford's magnetism, his powerful will; he was also susceptible to Van Buren's good nature, his clarity of thought in political matters.

Even if McLane had a more accurate perception of public policy, he still would have supported Crawford over Adams, the Yankee apostate, or Clay, the reckless western gambler. As for Andrew Jackson, his bumbling career in Congress and his reputation for military adventuring did not appeal to McLane's quick mind and eastern ways. Ambition, too played an important part in McLane's scheme. He craved the position of kingmaker; and even though Crawford's chances for the Presidency were slim, he recognized that it was too late for him to switch candidates.

Like Van Buren he had no choice but to make a virtue out of necessity and support his crippled candidate. For there was no other place to go.

Van Buren, McLane, and their Virginia, North Carolina and Georgia supporters faced considerable risk in trying to bargain Crawford into the White House. Calhoun had already been elected Vice-President. Should Crawford become President and then die in office, or be unable to discharge his duties, Calhoun would succeed him. But Crawford's backers let no such possibility deter them. Van Buren wrote Butler on December 27 that "Mr. Crawford is substantially well. You would be surprised to see what fine spirits he is in; indeed he never appeared half so interesting or respectable to me as this winter." Crawford's prospects, he deemed "good" provided he could get Clay's support which he thought was "probable but not certain." "If only Crawford had all of New York's electoral votes," he lamented, "his election would have been virtually assured."[2]

When, as Van Buren hinted, the three Clay states joined Adams, and Clay himself announced his support, Adams had twelve votes. He needed only one more state for election, but his margin in at least two states (Louisiana and Maryland) was tenuous, a one-vote majority in each. "The push on the part of Mr. Adams' supporters will be to succeed on the first ballot," Van Buren concluded. "If they do not there is no such thing as forming a rational conjecture as to the results." New York was the key because the Crawford delegation trailed Adams by one vote, if the lines held firmly. Stephen Van Rensselaer, the only avowed Jackson man in the delegation, had it in his power either to elect Adams or to produce a tie that would deprive New York of its vote.

Whatever the result, the election of Crawford seemed unlikely except through some complex intrigue that would have thwarted the popular will. Van Buren excelled at persuasion and bargaining, but his good sense led him to back away from any deals. He urged strict neutrality with respect to Jackson and Adams, loyal support for Crawford to the end.[3] As he saw it, a reasonable solution, apart from a deadlock and the subsequent draft of his own candidate, was the eventual election of Adams. The Secretary of State was a northerner, the strongest candidate in New York, a brilliant man with whom he had maintained good relations. Van Buren had known Andrew Jackson as a fellow Senator, but had not thought much of his performance, though he was conscious of the man's immense popularity and was attracted to his nervous vigor and forceful personality. Van Buren had no way of knowing how Jackson stood on what he regarded as vital issues beyond the compelling fact that his revered political mentors, Jefferson and Madison, were not impressed. His arch-rival, Clinton, had already preempted the Jackson candidacy in New York, a powerful deterrent.

If Crawford could not win after a second or third ballot, then an Adams victory over Jackson would simplify Van Buren's position. Van Buren would be a central figure in building an opposition party to Adams who he suspected would continue or more likely enhance the centralizing tendencies he had observed in the Monroe Administration. He would support Crawford wholeheartedly but not be too distressed if Adams won. All of these factors, personal, political and public, were balanced in Van Buren's mind as the time approached for the election in the House and it became clear that the vote of New York was pivotal.

Elaborate mind-reading

Stephen Van Rensselaer, the Patroon, on whom New York's vote depended, was an uncomplicated, wealthy landowner and farmer. This adventurous, patriotic soul whose drive for military achievement had ended disastrously at the Battle of Queenstown Heights, still yearned vicariously for the glory that fate and the cowardly militia had withheld. Jackson's victory at New Orleans had inspired in him a kind of awe at the success of the frontier commander. As a Congressman, he had met Jackson and the two men soon became warm friends. Had it not been for Van Buren's influence Van Rensselaer would have held out for Jackson to the end. The flinty personality and pedantic manner of Adams repelled the easy-going Patroon, and his family had long harbored political differences with the older Adams.

Beyond these personal traits Van Rensselaer was an indecisive man upon whom responsibility rested heavily. He was also a deeply religious man who, in times of stress, sought and usually found divine guidance. Sixty-one years old, time had not dealt harshly with his physical appearance. Only the thatch of white hair and the set expression of his features lent a venerable touch to his ruddy unlined face, his clear blue eyes and his lean, erect figure.[4] Van Buren, so acute in his judgment of character, failed to assess Van Rensselaer's weaknesses as a public man despite ample opportunity to do so. Perhaps Van Buren's humble background blinded him to the flaws of New York's first citizen, whose easy manners and friendly, unaffected feelings toward his younger colleague were deeply appreciated. Van Rensselaer, like King, a fellow graduate of Harvard, represented that blend of wealth and social standing Van Buren could never resist.

Van Rensselaer had discussed his vote earnestly with both McLane and Van Buren, and up to the day of the election they had assumed he would go for Crawford. His increasing anxiety and the last-minute question he put to Van Buren about the consequences of his voting for Jackson should have alerted the two men that the Patroon was wavering. Van Buren had answered the query with the simple though misleading statement that "as his vote could not benefit Mr. Crawford it was of no importance to us whether it was given to him or General Jackson." He suggested in his casual way that the Patroon would be thought fickle "by a change which would produce no beneficial result to anyone."

Van Buren was not being candid. Had Van Rensselaer voted for Jackson he would not, of course, have helped his man, but he would have given Adams the majority. Whereas if he voted for Crawford the two Jackson-leaning men in the New York delegation would have joined him and forced a second ballot, which Van Buren, McLane and Alfred Cuthburt (an amiable young Congressman from Georgia and the third member of his Mess), so ardently desired. Van Rensselaer pondered Van Buren's comment briefly and left for the House, declaring he would vote for Crawford.[5] When he arrived at the chamber, Adams's men immediately besieged him. McLane, arriving an hour or so later, found the Patroon completely beside himself, certain the fate of the Presidency and perhaps the Union rested upon his shoulders. Neither McLane nor Van Buren knew at the time that Van Rensselaer had visited Adams some five days earlier at the suggestion of Daniel Webster and received assurances from the stiff New Englander that if elected he would not discriminate against former Federalists in the distribution of patronage. Adams, who with all his self-conscious rectitude was not above a timely political thrust, reminded the Patroon that Van Buren had attempted to proscribe his cousin Solomon purely because of his Federalist background.

No wonder McLane found Van Rensselaer in such a wretched state. Clay and Webster had already committed him to Adams; yet he was genuinely distraught at what he believed was a betrayal of his closest friends.[6] Honor and friendship meant much to Van Rensselaer; but the fate of his country, as he so believed it, meant more. He listened to McLane but would give no assurances, prompting that excitable Congressman to dash off a note to his wife in which he referred to Van Rensselaer as having "womanish fears and miserable wavering."[7]

At this point, the Senators filed into the House chamber and in joint session elected a Senator and a Representative to open and proclaim the electoral votes of the states. For over two hours this tedious formality, whose result everyone knew beforehand, went on, and more Adams men clustered around Van Rensselaer.

Van Buren, steadfast in his appearance of strict neutrality, had remained at his lodgings when an agitated William S. Archer appeared with distressing news from McLane. Archer, a Virginia Congressman and firm supporter of Crawford, described the situation in the House. Van Rensselaer was faltering; would Van Buren go to the capitol and talk with him? Van Buren would not. He said that it would be presumptuous to force himself on the older man. Their prior conversations had all been voluntary and initiated by Van Rensselaer. As a member of the Senate he thought it unseemly and provocative to meddle in the affairs of the House. His policy had been, as Archer well knew, strict neutrality; he meant to keep it so. But when Archer appeared again "with a request of the same character, and from the same sources, of increased urging," he reluctantly consented to go, casting some doubt on the sincerity of his previous remarks. He did set an express condition, however: Van Rensselaer must come to him for advice; he would not make any advances.

No sooner had Van Buren entered the House chamber than Alfred Cuthburt greeted him with the news that Van Rensselaer had a few moments before assured him and McLane that he would not vote for Adams on the first ballot. Since Van Buren was quite certain the Patroon would not vote for Jackson either, New York, it appeared, would have no vote on the first round, forcing another ballot which would weaken Adams and boost Crawford's chances.

By now the counting of the electoral votes was completed. Van Buren remained at the rear of the cavernous chamber to observe the final act, the voting by states. The scene was tense, though not unrelieved by a bit of whimsy in the election of the tellers, John Randolph of Roanoke and Daniel Webster. Randolph, a wraith of a man dressed in riding habit, his curiously small head and boyish face set upon his scarecrow of a body, had a piercing voice. Webster, shorter, with massive head, deep chest, and booming resonant voice, formed a counterpoint to Randolph which in any other situation would have been irresistibly amusing; and perhaps some of the spectators who crammed the galleries that snowy afternoon did chuckle as first Webster announced, "thirteen votes for Adams, seven for Jackson, four for Crawford. John Quincy Adams, having a majority, is elected President." No sooner had the solemn tones of Webster echoed through the Chamber, than the shrill voice of Randolph followed with a similar announcement; thirteen states, not votes, confirmed the count.[8]

Van Buren was as solemn as anyone on the crowded floor after Speaker Clay had cleared the galleries. He knew, of course, that Van Rensselaer had switched his vote. He felt sorry for the old gentleman as he had observed his distress and had seen him bow his head in prayer before he marked his ballot.

Van Buren had thought New York would be tied but he was not that chagrined at the unexpected result. Van Rensselaer would apologize and be overwhelmingly contrite for his behavior. Instead, it was McLane who would stand on principle, whose unthinking rage at what he regarded as betrayal would destroy what had been one of the few pleasant relationships Van Buren had known in the capitol.

When he reached home, he found Van Rensselaer and Cuthburt in the parlor sitting at each end of a long sofa, tense and studiously ignoring each other. Van Rensselaer broke the silence as soon as Van Buren entered the room. "Well Mr. Van Buren," he said, "you saw that I could not hold out!" Soothingly, Van Buren replied that he had no doubt Van Rensselaer had done what he conscientiously believed was right and that the question should not be allowed to break up their pleasant Mess and their warm social relationship. Leaving the strained atmosphere of the parlor, Van Buren went immediately to McLane's room. As he expected, the emotional Congressman was deeply upset, not just with the New York vote but also with Maryland's. That state had gone to Adams by one vote, too. McLane had just spurned an apology from Van Rensselaer and it took all of Van Buren's diplomatic skills to reconcile the two men. Finally, McLane agreed with Van Buren that Van Rensselaer's motives had been pure, his vote "was sheer weakness of mind." After the initial excitement of Adams's election had died down, the clubable atmosphere of the Mess went on as before.[9]

The three men had certainly enjoyed their comfortable quarters, a furnished house just behind the capitol. Throughout all of the tempestuous politics and the intrigues for the Presidency that had been going on since November, the Messmates had been kindred spirits. All were fond of entertaining, of good food, good wines and good company. They had made a specialty of giving small dinners with menus as elaborate as Washington's premier caterer could provide. For once McLane's chronic dyspepsia did not oppress him; and Van Buren, who also was prone to bouts of indigestion, remained healthy enough to consume his share of the many fish, meat and game courses, the jellies, the molds, the rich cakes and pies that graced their table along with the customary half-dozen sweet, dry and sparkling wines that accompanied the courses. In his frequent letters to his wife Kitty, McLane exulted in the comfort and ease of the Mess. Van Buren, too, enjoyed their arrangement and after the tensions of Adams's election was at great pains to keep the group together.

The dinner he and his associates had given Lafayette, who was then making a triumphal tour of the country, had been an unforgettable experience. And Van Buren had been able to reciprocate the many social debts he owed to politicians and public men, to comely widows like Mrs. Stephen Decatur whose company he enjoyed so much. McLane suspected the *bon vivant,* one of Washington's most eligible widowers, of "licentious" conduct; and possibly he was right, for Van Buren was in his early forties and his wife had been dead for many years. But if he did not lead the life of a strict celibate in Washington or elsewhere, he was certainly one of the most discreet of men in a city of gossip. McLane's references to Van Buren's conduct in his letters home were invariably accompanied with complaints of fancied snubs, political treachery, or some other personal problem that was tormenting him. He was not exactly a reliable witness to Van Buren's love life; and he and De Witt Clinton, scarcely an unbiased observer, are the only sources for Van Buren's affairs, if any. That he enjoyed the company of lovely ladies, especially those of charm and wit, is abundantly manifest; that he recognized they were worth cultivating for political or public reasons because of their influence on their husbands is certainly true. But

Van Buren was far too cautious, far too ambitious for that matter, to make any slips of an amorous nature that his numerous enemies would use against him with undisguised relish.[10]

Should the choice lay between love and power, Van Buren had no hesitation in deciding where his interest lay. Why he had not remarried remains a mystery, especially when he knew the welfare of his four boys would have been better served if his household had been the conventional one of its day. His numerous relatives in Albany and in Kinderhook acted as surrogates, an arrangement Van Buren found satisfactory. And busy as he always was, he found time to write, to instruct and to be with his family as much as possible. This kind of domestic life was scarcely suitable to the mental health and development of his children; yet none seemed to have suffered emotional deprivation.[11]

Apart from the purely social aspects of the Mess for Van Buren in Washington during the mild, wet winter of 1824–25, it held important political assets as well. Van Rensselaer was not just a revered figure among the elite and the articulate in New York, he was considered a true aristocrat in Washington. If his manners, democratic to a fault, fit the notion of the true republican, his lands and lineage placed him on the same social level as the most affected and most affluent tidewater planter or Charleston nabob. That Van Buren could claim the closest of ties with Van Rensselaer inescapably heightened his social status in the eyes of those southerners who counted. He was not little Mattie Van Buren, but Senator Van Buren, whose intimate friend was the greatest landowner in the Empire State. McLane, though not conferring the social distinction of a Van Rensselaer, was a much admired Congressman of ability and education, a border state southerner whose sympathies were more northern and entrepeneurial than most of his contemporaries from the slaveholding planter class. Both men disliked Adams, and for personal as well as public reasons would fit into the core of opposition to the new Administration that Van Buren was planning. He needed them within his circle for his social well-being and the enhancement of his political fortunes.[12]

An alert Van Buren listened to Adams's inaugural address searching for those telltale hints that signaled the direction of the new Administration. He found what he had anticipated, a continuation of the Monroe amalgamation policy—"Our political creed is without a dissenting voice"—what he chose to consider new Federalism—"to proceed in the great system of internal improvements . . ." and the clinching statement—"the extent and limitation of the powers of the general government in relation to this transcendently important interest will be settled and acknowledged to the common satisfaction of all." Van Buren had grasped the dimension of Adams's centralism, though at the time he had underestimated the depth and the boldness of the new President's committment to the planned development of the nation on a scale that dwarfed its Hamiltonian model.[13] Realist that he was by training, Jeffersonian by background and by fervent belief, Van Buren was now gratified that he had hewed unwaveringly to his principles during the campaign and had identified himself as the leader of the uncompromising Old Republicans.

His political strategy of opposition to the Administration could be a clear-cut one. There would be no fuzziness, no amalgamation of parties, as there had been under Monroe. On internal improvements alone he was certain that state, not nationally designed and funded development was what a country of diverse interests, of local pride in accomplishment, wanted and what it would accept. New York had taxed itself, had borne the full cost for its great Canal nearing completion; it would reap its own harvest. Nor would it share its wealth in the form of taxation to improve other distant, poorer, less enterprising states. Any thinking New Yorker could see the logic of this. Self-interest was common sense. Adams's proposals would cost him dearly in New York, in Pennsylvania, and in Virginia, all of which had embarked on their own projects of local improvement.[14]

Congress adjourned on March 3, 1825, but most members, including Van Buren, delayed their departure until the roads dried up sufficiently to make coach travel faster and more comfortable. In the Messes that catered to the Jackson men, the campaign to discredit the Adams Administration had already begun. Some weeks before the election when it became known that Clay would throw his influence into the balance for Adams, an idiosyncratic Pennsylvania Congressman, George Kremer, had accused Clay of engineering what he termed "a corrupt bargain" with Adams. For his support, claimed Kremer, Clay demanded he be made secretary of state in the new Administration, and the President-elect had agreed. Clay should have ignored the charge; rashly, he did not. He issued his own statement of heated denial and then injured himself further when he demanded that the House investigate Kremer's allegation.

Van Buren had not only recognized the Kremer charges for what they were, but at this stage he was still pursuing a course of strict neutrality. As he had always done when he was outlining a political strategy, he was letting the potential opposition make its own way, but waiting for the proper time when he judged his own course of action sufficiently matured to have a better than fair chance of success. The impetuous westerners who were hurling loose charges at the Adams Administration might do more harm than good. Nor was he, with his grand design to form a second two-party alignment, about to hitch his fortunes to the star of Jackson, whether ascendant or not. He knew that some of the more strident Jacksonians were referring to his "noncommitalism," that the Clintonian press in Albany was reiterating this presumed defect in his political character. But he kept his own counsel on the Kremer investigation and urged his fellow Crawford men to do likewise.[15]

The presidential election, so disastrous to his organization at home, was behind him. Optimist that he was, he had recovered his poise and was studying party alignments in the state. Thus, when he arrived in Albany he had formulated a campaign based on his personal evaluation of the party and a sure feel for Clinton's political failings. Several months before, Clinton's friends had urged him to make common cause with the People's party against the Regency. Most of the People's men had been ardent supporters of Adams, and Clinton had been for Jackson. But as it was presented to the Governor, the election was over and the Presidency should not remain a divisive force. Clinton coldly rejected this sound advice. They could come to him if they wished and in effect bow the knee. His policies, public and political, were inflexible. As for Van Buren and the Regency, he regarded them as completely finished in the state, capable only of annoyance, nothing more.[16] Van Buren had been kept informed of Clinton's imperious stance, and he meant to use it for the benefit of rebuilding his organization.

The largest faction was the People's party in whose ranks marched a considerable number of his own Bucktails. They were supporting Adams and were backing a popular election law. The Van Buren–Crawford faction Clinton had dismissed with contempt was far from prostrate. Its capable leadership was intact, indeed strengthened, through the recent admission to its ranks of two extraordinary men, Silas Wright, Jr., and Michael Hoffman of Herkimer, a lawyer and Congressman.

Wright and Hoffman were uncommonly sedulous workers who studied a subject or considered an issue until they had gained complete mastery of it in all its details. Hoffman was the more intense, the more impulsive. Above average height, he was thin, wiry and radiated nervous energy. His early training as a physician had provided a solid foundation for his powers of observation, discrimination and logical analysis. Legal studies and long practice at the bar had amplified these qualities. Like Wright, he made lasting friendships.

The more diffident Wright provided a striking contrast in appearance to Hoffman. Everything about Wright bespoke strength. Muscular, robust, with the powerful arms and the broad chest of one who was accustomed from an early age to hard physical labor, his large, ruddy face, dark hair and curiously gentle brown eyes, all marked him as an impressive man. Deliberate and earnest in speech or debate, Wright gave the appearance of force combined with a natural dignity; there was no pretense about the man. Like so many other New Yorkers, he had come from Vermont, from a hardworking farming family. Through family connections he began the study of law, completing his legal training with Roger Skinner, who introduced him to Van Buren.[17]

Wright may have been slower than his more energetic friend Hoffman at analyzing complex problems, but what he lacked in surface brilliance, he made up with the broad span of a highly disciplined mind. When Silas Wright rose in the legislative halls he would occupy with distinction for over twenty years, he commanded respect, attention, even admiration. But more than these powers of intellect, Wright had qualities of heart that would make him the most beloved public man of his generation. Loyal, generous to a fault, scrupulously honest, rarely ill-tempered whatever the provocation, he was as frank and friendly with a neighbor in his home village of Canton, New York, as he was with a truculent, self-important southern grandee. Wright's great fault, a weakness he shared with other public men of his generation, was his heavy drinking.[18] But in 1825 he was healthy, vigorous, definitely the Van Buren leader in the state senate, which was still under Regency control.

The way was clear for Van Buren. Presidential politics were to be strictly avoided. The Adams Administration was to be left free of any partisan attacks and Clinton permitted the luxury of his lofty isolation. Meanwhile, the Regency would move quickly and efficiently to regain what it had lost in the countryside, the villages and towns. It would select and elect sound men who would dominate the next legislature, purged of any presidential distractions.[19] Clinton seemed wrapped in his dreams of glory and ambition—he still thought of himself as a presidential candidate despite his public pronouncements for Jackson—and with some reason.

On November 2, 1825, the Erie Canal was completed and elaborate preparations were made to celebrate the event. The Governor went to Buffalo, accompanied by aidės, Canal commissioners, and a hive of politicians. He embarked for eastern New York on the *Seneca Chief,* a special canal boat decorated for the occasion. The rest of the throng crowded into a second canal boat, the *Young Lion of the West,* which followed. For the next several days that it took to make the journey of almost four hundred miles to Albany, the banks of the Canal were lined with farmers and townsmen who turned out in great numbers to watch the slow, majestic progress, punctuated at each lock and each landing with the barks of saluting cannon, the rolling thunder of eighteen pounders. Conspicuous as always among the passengers of the *Seneca Chief* was the tall, portly form of the great Governor, who to many observers seemed the very embodiment of national character and to not a few the creator of what they confidently expected would be their personal fortunes. John C. Calhoun in far and distant South Carolina charted Clinton's course in the press and confidently predicted he would be the next President. "Those who do not see that," he wrote, "know but little of the man, or the times. His is a situation of great power."[20]

Van Buren wisely chose to await the Clinton triumphal progress in Albany. There he attended the great public dinner the citizens of Albany gave for their hero. He absented himself from the official party that traveled down the Hudson to New York City where Clinton poured a keg of Erie water into the Atlantic, a bit of symbolic pageantry that moved Hezekiah Niles to ecstatic prose in the columns of his popular *Register*.[21] Van Buren could not afford to be associated with Clinton's hangers-on or seem to be clutching at his coattails. And he had much more important concerns, much more practical ones than giving up precious time and energy to witness Clinton's display. Let him bask in his richly deserved glory; Van Buren was after sound nominations for the legislature of 1826, only a fortnight away.

The Regency was working as it never had worked before. Each member was writing a dozen or more letters a day to leading men all over the state. The message was simple, direct and time-tested. Clinton was a great man, perhaps the greatest the state had ever produced. As governor, he had, and should continue, to protect lofty undertakings that would advance the material and moral worth of New York. As a politician, however, he could never be trusted. The old formula of Clinton's Federalist leanings was brought out and commented upon. If the addressee happened to have favored Adams or Clay on the last election, he was reminded that Clinton had been for Jackson, but then he was urged to put aside presidential politics, to concentrate on state issues, to remember the importance of discipline and loyalty in advancing Republican principles and the varied interests of the state. More important inducements were the hints of patronage in which they would share only if they stood together.

Van Buren and his friends had early weighed the reforms made by the revised constitution in which the legislature as a whole had a significant part in distributing the thousands of minor offices in the state. The *Argus,* in much more muted tones as became a public print, echoed the same sentiments while remaining conspicuously silent on the policies of the Adams Administration or on patronage, national or local. Noah's *Advocate* caught the party line and kept its indiscretions to a minimum. Quite likely Gulian Verplanck and the Kings, editors of the *New York American,* recognized Van Buren's tactics; but their hatred of Clinton and the seemingly tacit support the Regency was according the Adams Administration gave them little ground for complaint.[22] The result was a complete overthrow of the Clintonians in the November election, every bit as overwhelming as the Regency's had been the year before.

Until the new legislature convened in January 1826, the Regency kept a steady pressure on the members-elect that featured above all harmony within Republican ranks. The message constantly repeated was a clear one. Clinton may have been untrustworthy in the past, and through his rule-or-ruin politics he had disorganized the party in the state; but his wings had been clipped. He was not to be abused in the local papers and at political gatherings.[23]

Van Buren himself set the pattern when he did not oppose the election of the Clintonian, John W. Taylor, for Speakership of the House in the Nineteenth Congress. Taylor, who had been twice denied the powerful post largely through Van Buren's intervention, won on the second ballot. These tactics paid handsome dividends. The legislative session of 1825 had enacted a law making popular election of presidential electors by district. As the federal Constitution prescribed, the appointment of the United States Senators remained with the legislature. In control of both houses, the Regency selected the chancellor of the state's equity court, Nathan Sanford, former Senator, loyal, but not politically controversial. He received a caucus nomination and, after resigning the chancellorship, was elected to succeed King in the Senate. The next step in the Van Buren strategy was the most delicate and difficult for him personally, fraught as it was with political hazards and personal animosities of long standing. If he was to act the parliamentary leader in Congress, he had to abandon his policy of neutrality toward the Administration, or as his opponents would have it, his noncomittalism.[24]

Adams's first annual message in which he called for an unprecedented system of internal improvement at national expense was the kind of clear-cut issue Van Buren had been waiting for. And the President had gone even further when he announced that he accepted an invitation for the nation to be represented at a Congress of newly independent Latin American states to be held in Panama where members would discuss "objects important to the welfare of all."[25] Van Buren could now put together a congressional coalition on real issues of public policy and political ideology. Partisan effusions like "corrupt bargain," or abuse of patronage, he left to others who had not as yet comprehended the grave threat he imputed to the Adams Administration, the corruption he saw in its centralizing tendencies—none other than the corruption of the democratic political order Jefferson and his successors had established.

In preparing himself for his new and challenging function as parliamentary leader of the loyal opposition, he realized that the presidential election of 1828 would be of paramount importance and he had to be prominently identified with the successful candidate. All political signs pointed him toward Jackson. With Crawford now completely out of the picture, only Clinton stood in the way. Van Buren had effectively destroyed Clinton's home base and felt that in so doing he had ruled him out of presidential contention. But he was well aware that Clinton, when he assured himself he could not win the Presidency, would revert to his original position and support Andrew Jackson, if only to embarrass his rival.

After the debacle of 1824, when his spirits were very low, Van Buren chanced to meet Mrs. Clinton and her brother, James Jones, on the steamboat to New York. While they were at breakfast, Jones said, "Now is the time admirably fitted for a settlement of all difficulties between Mr. Clinton and yourself." Van Buren thanked Jones for his well-meant suggestion but declined, declaring that his "fortunes were at too low an ebb to be made the subject of a compromise and that when they improved a little I would remember his generous offer."[26]

Recalling that conversation, and confident he would best the Clintonians in the election of 1826, Van Buren judged the time ripe to make preliminary overtures. Though he knew from past experience Clinton would drive a hard bargain, Van Buren steeled himself for what he was sure would be a humiliating experience. Infinitely more difficult, as he considered the problem, would be the task of making satisfactory explanations to the party leaders in the legislature, many of whom had long cherished their resentment against the arrogant Governor. With Adams sentiment running so strong in the legislature, Van Buren ran the risk of sharing with Clinton political isolation in New York, scarcely an enviable position from which to develop parliamentary leadership.

On his way to Washington just after the state election of 1825, he had an intimation of just how high the price would be for Clinton's cooperation. Despite the defeat of his faction, Clinton suggested at a meeting with Van Buren that he be unopposed to his reelection as governor in 1826. A harsh demand, it placed Van Buren in a most perplexing position. He had no sooner reached Washington than he received a letter from Edward P. Livingston that was even more disturbing. Livingston outlined a conversation he had had with William James, the richest man in Albany and second to John Jacob Astor, the richest man in the state. James, long a close personal friend and confidant of Clinton, asked Livingston to tell Van Buren "that Mr. Clinton had given him a *carte blanche* to treat with you—he said that Mr. Clinton was determined to be a candidate for President at the next election—that he wished you to keep yourself this winter uncommitted on that question." For this service, Clinton, if elected President, would support Van Buren for the governorship, or if he preferred, he could have any cabinet post he desired.[27]

Instantly, Van Buren saw how damaging even the report of such a deal would be to his reputation. If Clinton were a strong candidate Van Buren would be subjected to the same sort of abuse the Jacksonians were heaping upon Henry Clay. "I never have and I never shall subject my course on a public question to the control of my personal interests," he replied to Livingston. "I have no reason to doubt that Mr. J.'s [James'] views are friendly and that he was unconscious of the impropriety of his suggestions. I have only to ask that the conversation be forgotten and that you be careful not to speak of it to *anyone*. You know what political gossips are apt to make of such things." Even with this timely warning, Van Buren decided to go ahead with his policy of conciliation. He had carefully tested the political climate in Washington during the second session of the Nineteenth Congress and returning home in the spring, resolved to meet, if at all possible, Clinton's demand that he be unopposed for the governorship.

The Regency, whose leaders shared Van Buren's confidence, understood and concurred in his long-range plans and was entrusted with the first formal approach to Clinton.[28] As they expected, Clinton at once demanded patronage in the appointment of an in-law, Samuel Jones, to the chancellorship Nathan Sanford had just vacated. From a partisan point of view, Jones's appointment could not have been worse for Van Buren's pretensions. A life-long Federalist, he was just the sort of man whom the Regency had repeatedly accused Clinton of courting to the detriment of the Republican party. But his professional qualifications were of the highest order; and emphasizing that the office was nonpolitical in character, the Regency-dominated senate promptly confirmed Jones's appointment. The *Argus* and other Regency papers throughout the state managed to dampen charges of inconsistency and, in fact, to narrow and soften deep-seated prejudice against Clinton within the Republican party.

When Van Buren returned to Albany in the spring of 1826, he felt he had no alternative but to convince his following that they should not oppose Clinton for the governorship. The Regency planned to hold its first state convention at Herkimer, a village near Utica which at the time roughly marked the center of the state. He was well aware that his following would object. Silas Wright had already written him a long letter on the subject saying that the Herkimer convention must nominate a candidate in opposition to Clinton.[29]

At a series of meetings in Albany, Van Buren explained his views to the Regency on the forthcoming political contest. He was unable to convince his associates that they should support Clinton at Herkimer, however. So vehement was the opposition that Van Buren, though tired and burdened down with family and business responsibilities, decided to make a trip throughout the state where he could see for himself whether the members of the Regency were correct in their views of the contest. He discovered that, if anything, they had underrated the opposition to Clinton's nomination. How could he resolve this dilemma? He was chary of meddling, as he had before, with those factions of the party which still clung to their presidential preferences. He now realized that the Herkimer convention must nominate a candidate to oppose Clinton, if the Regency were to endure. Yet he must be a weak candidate whom Clinton would defeat handily.

Van Buren's choice was William B. Rochester, an undistinguished state circuit court judge who had served one term in the state legislature and one term in Congress. Rochester's nomination posed some risks—he was an ardent backer of Henry Clay and of the Adams Administration.[30] In fact, he had been appointed one of the American commisioners to the Panama Conference, which Van Buren had attacked in an elaborate speech just a few months before. But as Van Buren saw it, the obvious weakness of Rochester as a candidate would reelect Clinton. An added incentive for the Rochester nomination was that he had opposed a Clintonian project to build an east-west turnpike at the state's expense through the southern tier of counties that were not benefiting from the Erie Canal.

When Van Buren presented his plan to the Regency at his home in Albany, there was again heated discussion and some division of opinion. Silas Wright and Perley Keyes, a state senator, one of the seventeen who had stood out against the election law endangering their political careers in defense of party loyalty, argued that Rochester's nomination would come as such a surprise to the party faithful that they would be demoralized. Rochester's well-known attachment to Clay, they thought, was dangerous and divisive.

Their arguments made no impression on Van Buren, who insisted that presidential politics were not of immediate issue in the state election. The party had to be overhauled; since all present agreed that Jackson would be supported three years hence, it was expedient to join with Clinton in their new endeavor. Eventually Wright and

Keyes came round and Van Buren, with an eye to weakening Rochester's candidacy, further saw to it that Nathaniel Pitcher, a resident of the southern tier and an avowed advocate of the turnpike Clinton had recommended, was nominated for lieutenant governor.

Peter B. Porter, the Administration's chief in the state, learned that Van Buren and Clinton were attempting a reconciliation, but as he wrote Henry Clay in March, 1826, "Whether they have any distinct and definite object, I know not." Porter was astonished and delighted when Rochester was nominated for governor at the Herkimer convention in late September 1826; it was the end of factional strife and the triumph of the Administration, he thought. What Van Buren had not considered was that seasoned politicians like Porter would be so enthusiastic about a pallid candidate like Rochester; or that Porter and others would immediately grasp the significance of the nomination as a means of identifying the Administration with party harmony in New York.

Through his small but compact network in the state and the numerous legislators sympathetic to Adams, Porter and Thurlow Weed strenuously urged the electorate to support Rochester. Despite the declarations of the *Argus* and Clinton's *Albany Register* that the election was a state matter exclusively, Porter and the Administration press discounted Clinton and persisted in minimizing the objectives of any coalition between Van Buren and his erstwhile political foe.[31]

Porter and Weed did such a good job in hammering on that theme that for a time it appeared a majority of the electorate believed a vote for Clinton was a vote for Andrew Jackson, that dangerous military chieftain, a demagogue from the West who was bent on subverting civil liberties. Van Buren's "high-minded" Federalists, who had become increasingly critical of his policies and suspicious of his intentions, joined the hue and cry. Small in numbers but influential, and sworn enemies of Clinton, they opened the columns of the *New York American* to the Porter campaign.[32]

As the autumn wore on, Van Buren became deeply concerned at the impetus of the anti-Clinton forces. Perhaps his analysis of the party structure had been faulty. He had assumed that the Clay, Adams and Calhoun men had voted for Clinton in the state election of the past year to bring down the Crawford candidacy and that as soon as this had been accomplished, "the hollow alliance" between Clinton and the other factions had fallen apart; they would resume their own special quarrels with each other over state patronage and state issues. But he had again miscalculated and to his consternation found himself and the Regency cast as loyal supporters of the Adams Administration. If Rochester should win, Van Buren's political career along with Clinton's would be in jeopardy. Hailed as an Administration triumph, it would in all likelihood make New York safe for Adams in 1828.

Croswell did a fine job in downplaying Rochester in the *Albany Argus* by contrasting his opposition to the turnpike through the southern counties with Pitcher's record of hearty support.[33] And it was this issue, which Van Buren had the foresight to anticipate, that won the election for Clinton by a scant 3,650 votes, while Pitcher triumphed over the Clintonian candidate for lieutenant governor by 4,188 votes. Silas Wright confessed that the result went far beyond his expectations. He wrote Flagg, "Nothing is more clear than that Rochester would have succeeded if any confidence of success had been felt. . . . Clinton would be whipped for certain." The assembly which would choose the next United States Senator was safe for Van Buren so long as the coalition held.[34]

« CHAPTER 11 »

"With Muffled Oars"

The election had been a near thing and there had been grumblings among the Administration Republicans that Van Buren had not played fair with Rochester. Rochester himself, however, was perfectly satisfied that Van Buren had loyally supported him. And when Noah suddenly made a savage attack on the Regency in the wake of the election, it merely served to dispel most of the doubts. Marcy exulted in the *Advocate*'s unexpected course, writing Van Buren that it would "secure you a unanimous vote from the Republicans" for reelection to the Senate.

Van Buren needed this cheering news. While the campaign he had planned was being put into operation and his future in politics seemed chancy at best, his son John fell desperately ill. The harried lawyer, politician and would-be statesman spent hours at his bedside, at times in utter despair as the son closest to him struggled for life. Loyal friends sought to ease his many burdens. Butler and Talcott took over most of his unfinished legal business. Knower, Marcy, Keyes, Wright, Dudley and Evans shouldered the political burdens. John finally recovered, but the ordeal had left Van Buren weak in body and dispirited when he was able to make the long, uncomfortable journey to Washington.[1]

Although he remained confident and optimistic, he was worried about his hold on the legislature and consequently his reelection to the Senate. He knew that it would take more than Rochester's belief in his good faith or Noah's animadversions to maintain a majority of the restless spirits in the capital.

Again he was faced with great opportunity and great peril, and again he had to gamble. Should he make a frontal assault against Administration policies in the Senate on constitutional grounds, in particular the program of internal improvements Adams had proposed? If he made the challenge before the legislature acted on his reelection and was triumphantly sustained, the effect would focus national attention on New York and especially upon him as the opposition leader. On the other hand, if he laid out a program of opposition and was defeated for reelection, the triumph would lie with the Administration and he could never survive the repudiation.

Anxiously he wrote the Regency for advice and was heartened when it responded through Marcy that "nothing has transpired at all calculated to impair our confidence in the success of your reelection." While waiting for the Regency reply, Van Buren had already asked his close friend, David Evans, one of the agents of the Holland Land Company and a power in western New York, to work for his reelection in the legislature. He had also requested John C. Hamilton, "a high-minded" Federalist with whom Van Buren maintained close ties, to persuade others of their faction that he had made no arrangements regarding state politics with their hated enemy, Clinton. Hamilton managed to curb what had threatened to be a bolt of the Kings and more especially their influential *New York American.* He had visited Noah in addition and secured assurances of support from that erratic editor. At length, satisfied that his seat was as safe as it could be, Van Buren went ahead with his plans to face down the Administration.[2]

The President and Clay were still hopeful they could detach the Crawford men from Jackson.[3] Van Buren, their leader, was the obvious target of Administration attention. "It seems," he wrote James A. Hamilton, "they are determined not to give me up. . . . there is no doubt that the folks here would do anything now that was desired of them, but it is out of the question."

Van Buren dispelled any lingering hope the Administration may have harbored of co-opting him, and as yet his publicly uncommitted following early during the second session of the Nineteenth Congress.[4] In the course of a long, well-researched speech on a federal bankruptcy bill, he emphasized repeatedly his devotion to states' rights. Now the Administration realized that his views were in such fundamental disagreement with its policies, that there could never be a basic agreement. Van Buren had used the bankruptcy bill as a pretext for formulating and presenting to the country his political creed, an eloquent and forceful declaration of Republicanism that earned praise from Jeffersonians everywhere, but especially from those in Virginia. His speech was anything but "noncommittal," his argument as clear and direct as the best effort of Nathaniel Macon. And his ideas, though scarcely original, were expressed in such terms that they formed the basis for a cohesive opposition to what was considered the centralizing objectives of the Administration. Looking back on that stirring moment when all of his uncertainties were resolved he wrote many years later, "I have always regarded [the speech] as the most successful of my senatorial efforts."[5]

Van Buren had always been a hardworking Senator. Since his election in 1821, he had held the chairmanship of two of the major committees, Finance and Judiciary, through three congressional sessions. After his first unfortunate experience in the Senate, he had gained poise and participated frequently in debate. He was not a powerful speaker and made few elaborate addresses; but when he did speak in his conversational way, he always went directly to the root of the matter and proceeded logically to his conclusion with a minimum of rhetorical flourish. All of his speeches bore evidence of extensive research and a lawyer's habit of anticipating anomalies that a skillful debater could exploit. This kind of exposition may have weakened the force of his argument; but what he lost in impassioned discourse, he usually gained by examining and rejecting views or arguments that ran counter to his position. In debate, Van Buren was more agile than most of the great orators who surrounded him. And he never spoke on any subject unless he had mastered it to his own satisfaction.

Although he took his legislative duties seriously, maintained a heavy correspondence with other politicians, and especially their wives, enjoyed an expansive social life that at times must have been fatiguing, Van Buren never lost sight of his primary objective, the restoration of a party system based on what he chose to regard as governing principles of political behavior. Personal politics had become abhorrent to him through painful experience, his long duel with Clinton and the damage the last presidential campaign had done to his own political organization. Where he was most effective was in working out informal caucuses of like-minded Senators who would agree to vote as a bloc against Administration measures that could be construed as centralizing.[6]

Reasonably confident of his reelection to the Senate, he began cautiously to make some preliminary maneuvers before he publicly announced his support of Jackson. Three major problems, all interrelated, must be resolved. He had to be reelected to the Senate, he had to deal with Clinton and with Calhoun. His Crawford contingent had opposed Calhoun for the Vice-Presidency in the recent election. Calhoun, as yet publicly uncommitted to a candidate for 1828, was one of the most prominent and influential power brokers in Washington. As Van Buren mused about Clinton in New York and Calhoun in Washington, he decided that his relationship with Calhoun had to change for the better.

Calhoun was a more adroit politician than Clinton, and if he did not possess the monumental image of the Magnus, he was obviously moving away from the Administration. Van Buren reconciled himself to a second term as Vice-President for Calhoun despite the additional influence and political leverage this would give the South Carolinian. Furthermore, he was certain that Calhoun had an understanding with Jackson.

More difficult for Van Buren when he finally made up his mind was the problem of swinging his Crawford contingent into line. Crawford himself detested Calhoun personally and was skeptical of his political views. Others of the Radicals, like Ritchie and his Richmond group, were distrustful of the Vice-President and suspicious of Jackson. They would have to be coaxed into the new arrangement, a delicate proposition with these sensitive but politically persuasive old Republicans where a promising relationship could be injured through a careless phrase or a thoughtless remark.

First, Van Buren sounded out the Vice-President. Where did he stand? Was an alliance possible and if so on what terms? The initial move was purely social. He, his Messmate Stephen Van Rensselaer and Calhoun were guests for the Christmas holiday at Ravensworth, the splendid home of William Henry Fitzhugh, a rich Virginia planter in nearby Fairfax County. Van Buren and Calhoun skirted any commitments at Ravensworth, but in general conversation each established for the other grounds of mutual interest and further discussion.[7]

For the next month, as the Vice-President gradually began to disassociate himself from the Administration party in Congress, Van Buren turned his attention toward Albany where the legislature was in session. A large number, perhaps a majority, of the members were supporters of the Administration; but the Regency preached party discipline and argued forcefully that Van Buren was not committed to any future presidential candidate. His differences with the Administration had not been partisan, they had been honest differences of opinion on public policy.

The Clintonians, most of whom opposed the Governor's preference for Jackson, were more than willing to support Van Buren on his record, on his apparent understanding with Clinton, and for the sake of party harmony. Still, Van Buren and his managers had to tread carefully, pull the right wires and somehow get across to the legislature that the Regency was not supporting Clinton and Jackson. They had to do this without calling forth any blazing rays from the Magnus in Albany. The young Hamiltons and Jesse Hoyt were helpful in accomplishing this project, though the course of the *Advocate* was still a burden to Van Buren, not just in New York politics, but in his delicate relations with Ritchie and the Old Republicans to whom Clinton was anathema and Jackson at best an enigma. The Regency, having learned from bitter experience, picked its way through all the potential hazards in the legislature with a finesse that matched its master. On February 6, 1827, Van Buren was reelected overwhelmingly in both chambers, 23 to 8 in the senate and 82 to 31 in the house.[8] The first problem, his reelection to the Senate, turned out to be no problem at all, but in fact a personal triumph. His stature in Washington soared and with it his relationship with an impressed Calhoun who came calling. Previously, the two men had agreed to pool their political resources behind the Jackson candidacy; and the Vice-President had little difficulty in persuading Van Buren of the benefits to be gained by having Jackson and presumably himself nominated at a national convention. After several conversations in which they thoroughly canvassed the attitude of the Crawford men, Van Buren agreed to write Ritchie of the *Richmond Enquirer* proposing the convention idea and disposing of any claims that the touchy editor might have seen in Noah's espousal of Jackson and Clinton. After writing an unusually long letter for him, Van Buren submitted the draft to Calhoun and his chief manager, Congressman Samuel D. Ingham of Pennsylvania. Whatever suggestions or editorial corrections they may have made are unknown, probably few, for Van Buren was skilled at sensing and avoiding any misunderstandings that might arise through careless usage or undue emphasis in the written or the spoken word. It was agreed at the outset that Calhoun's name would be mentioned only as a strong advocate of the convention idea. Jackson's name would be introduced in a casual way, deferring to the Virginia editor's qualms, yet deftly drawing him into the logic of the General's nomination.

Van Buren was "not tenacious whether we have a Congressional caucus or a general convention, so that we have either." He did not favor state nominations for several cogent reasons—they promoted personal politics, they gave Adams a better chance to blur the issues and to consolidate his strength among the crypto-Federalists of the Northeast, the middle Atlantic states and the Northwest. In balance he thought Jackson, as an anti-Administration candidate, would be most favorably presented through a convention that the Adams men would boycott, having one of their own in due course. He admitted he was as troubled as anyone by what he felt was a general ignorance of Jackson's position on public issues. "General Jackson has been so little in public life that it will be not a little difficult," he said, "to contrast his opinions on great questions with those of Mr. Adams." And in emphasizing this point, Van Buren said that if former differences were

> suppressed, geographical divisions founded on local interests, or what is worse, prejudices between free and slaveholding states will inevitably take their place. Party attachment in former times furnished a complete antidote for sectional prejudices by producing counteracting feelings. It is not until that defense had been broken down that the clamour about Southern influence and African slavery could be made effectual in the North.[9]

He stressed repeatedly that his aim was a restoration of the old Republican party and Federalism, too. "We must always have distinctions," he wrote, "and the old ones are the best of which the nature and the case admit." Despite his backhanded compliment of the Federalists, he left no doubt which party he favored and which party best represented the highest ideals and the most practical concerns of the nation. In words that must have warmed Ritchie's heart, he said, "Political combinations between the inhabitants of the different states are unavoidable and the most natural and beneficial to the country [is] that between the planters of the South and the plain Republicans of the North."[10]

After his resounding victory in New York, Van Buren's views were taken much more seriously in Richmond, as they were in Washington. But problems remained, reassurances had to be given, enthusiasm generated, suspicions lifted and personality conflicts submerged. The difficulties were not simply confined to Virginia but to North Carolina and Georgia as well. The Crawford contingent in North Carolina shared with the Richmond Junto the same dislike of Jackson, the same distrust of Calhoun. Nathaniel Macon, still the most influential politician of the Old Republicans in the South, thought that those favoring Jackson and those favoring Adams were cast in the same mold. They ignored principles while whoring after the strange gods of personality and section. There was even talk in North Carolina of supporting Clinton for the Presidency, most discomforting news to Van Buren.

Romulus Saunders a rough, gregarious lawyer from the Piedmont region of North Carolina, now serving his third term in the House, was Van Buren's source for that particular piece of information and at the same time his primary link with party opinion in North Carolina. Saunders was one of those neighborly souls who preferred the North Carolina legislature to the halls of Congress. He had been Speaker of the state's lower house for two terms and knew everybody of any consequence in the state. As such, he was just the sort of person who appealed to Van Buren, a virtual catalogue of public opinion, and public men, in North Carolina.

Through Saunders, Van Buren recommended to other North Carolinians that the wisest policy was to stay neutral and united on presidential candidates while the situation in Washington remained unstable. Van Buren was also in touch with Macon and through him, Bartlett Yancey, another local politician and lawyer, whose outlook was similar in many ways to that of Saunders. From what he could gather, the Crawford group was in disarray, not quite ready to espouse Jackson, still opposed to Calhoun on any national ticket, yet more violently opposed to Clay and to Adams. "Anything to get clear of this cursed union of 'puritan and blackleg' " said Saunders, echoing the unseemly epithet Randolph had hurled at the Adams Administration.[11]

Precisely attuned to these vibrations, Van Buren let no opportunity go by, however slight, to underscore his Jeffersonian convictions and carry the gospel of party definition southward. In his public letter of acceptance to the New York legislature, he declared that "it shall be my constant and zealous endeavor to protect the remaining rights reserved to the states by the Federal Constitution; to restore those of which they have been divested by construction."[12]

Van Buren and Churchill C. Cambreleng, now close collaborators, slowly and carefully drew Saunders into the Jackson cause. Once converted, the persuasive Saunders began suggesting to his legion of political acquaintances that the time for neutrality was over. Jackson was the only candidate who could beat John Quincy Adams and make a campaign on Old Republican principles against the New Englander's obvious Federalism.

Van Buren was now moving closer to the noisy Jacksonians in Congress who had begun their attack on the Administration almost immediately after Adams's inauguration. The most vituperative of the critics was the eccentric John Randolph of Roanoke; the most effective phrasemaker was the bombastic Thomas Hart Benton of Missouri. Van Buren had cultivated their friendship for some years. He was genuinely attracted to Benton's vigorous style and perceived his very definite ability behind the florid rhetoric, the self-important pose. He was, of course, not unaware that Benton was a powerful spokesman for the aims and aspirations of the rapidly growing West. As for Randolph, Van Buren rather looked up to the aristocrat he believed this Virginian to be, much in the same way he deferred to the Patroon.

Apart from Randolph's pretensions, Van Buren found him vastly entertaining, drunk or sober, with his gift for invective, his colorful language, his amusing remembrances of times past; though he could be exasperated, even bored at times, with Randolph's lengthy wandering speeches. And he worried about the propriety of Randolph's intemperate abuse of the Administration, the public spectacle he frequently made of himself in Congress when exhilarated with "toast water." If Randolph's life was a maze of bewildering inconsistencies, his utterances frequently verging on the ridiculous, no one doubted his devotion to Jeffersonian principles. Like Benton, behind the posturing, the ranting behavior in public, there was a shrewd, political mind that Van Buren soon detected. The Virginia squire had surveyed the political scene and had concluded that Jackson was the coming man, months before Van Buren had made up his mind. He readily permitted himself to be Van Buren's stalking horse for Jackson in the Old Dominion.[13]

On January 13, 1827, to the consternation of Ritchie, the Virginia Legislature elected ex-Governor John Tyler to the Senate in place of Randolph, who was completing James Barbour's term. Tyler's political beliefs were Jeffersonian, but he had supported the election of John Quincy Adams and was a good friend of Clay. Independent in his thinking and ambitious, he was a distinct threat to Ritchie's power base. Workers for the Administration had managed to forge a coalition between those western Virginians who favored internal improvements and those everywhere who disliked Randolph, who considered his antics had brought disgrace to their state.

Randolphs's departure from the Senate and the reaction to this in Richmond prompted Van Buren to make a trip through the Crawford states as soon as the Senate adjourned on March 3. With his usual fine sense of timing, he judged that if he were ever to bring Macon and Ritchie and their followers squarely into the Jackson camp, he must attempt it while Randolph's defeat was fresh in their minds. He also wanted to confer with the Virginians on the prospect of establishing another press in Washington.

Despite his new relationship with Calhoun, he was not satisfied with the newspaper, the *United States Telegraph,* which Duff Green had recently established in the capital. The *Telegraph* was out for Jackson and Calhoun, but its editorial page seemed more concerned with men rather than measures. Some months before Van Buren had sent Calhoun a letter of Cambreleng's suggesting a new paper to espouse Old Republican views and develop public issues. Calhoun had doubted whether another paper was needed. In defending the *Telegraph* he argued "that two on the same side must distract and excite jealousy. Each will have its partisans." Calhoun's argument had been reasonable enough, but Van Buren continued to distrust his motives and to question their expediency, at least among the Crawford bloc.

Before he set off for his trip South, Van Buren had several exploratory conversations with the senior Senator from Virginia, Littleton Tazewell, about forming a rival paper to the *Telegraph* that would emphasize measures and principles, not presumptive candidates. His choice for editor was Ritchie. Tazewell agreed and promptly relayed the substance of their conversations to Richmond. When Ritchie replied that he preferred to remain where he was rather than brave the uncertainties of Washington, Van Buren still remained hopeful he could persuade the reluctant editor to change his mind.[14] He was more successful in his campaign to make Calhoun palatable to the Crawford men. In late January 1827, Saunders wrote his friend Yancey, "You see the Legislature of Pennsylvania are on Monday next to go into nomination of Jackson—if he is nominated *without* Calhoun, then he goes down. I do not much regret it, damn him."[15]

With these uncertainties in mind, Van Buren set out on his difficult mission. For personal and political reasons he chose Cambreleng as his traveling companion. The stout little Congressman was a native of North Carolina. Though long a resident of New York City, he had maintained business and personal ties with his relatives and boyhood friends. And with his ready wit, his good nature and his keen mind, he provided pleasant and instructive conversation that eased the discomfort of the bad roads, the interruptions and delays that were ever the traveler's lot over the stretches of forest and swamp, the creeks and bays of the seaboard slave states. Van Buren had purchased horses from the Dutch minister, Baron Tuyl, which he described as "undoubtedly the handsomest in the United States." As part of the deal the Baron lent him his carriage, which Van Buren eventually purchased. The trip would be as comfortable and as private as possible. Many of the overnight stops would be made at the homes of congressional colleagues, thus avoiding for the most part bad accommodations and poor food.

In their separate carriages, two Charlestonians, Colonel William Drayton and James Hamilton, Jr., rounded out the group. Both had become fast friends of Van Buren and, though formerly political opponents, were now ardent supporters of the Andrew Jackson-Calhoun ticket. Drayton was a new friend, having just finished his first term in Congress, but he was a member of Van Buren's Mess and enjoyed as close a familiarity as Hamilton. A Crawford elector, he had not, like Hamilton, supported Calhoun for the Vice-Presidency, and on that score had been politically closer to Van Buren. Both men were good companions, well read, cultivated lawyers and plantation owners; both had extensive connections in South Carolina, Hamilton was a former mayor of Charleston and Drayton a former recorder of the city council.[16]

Van Buren and Cambreleng parted from their companions at Charleston and made straight for Georgia, the furthest south of their trip, to visit with Crawford at his comfortable but unpretentious plantation, Woodlawn. They met him in late April and were happy to find their former champion in much better health than when they had bade him farewell in Washington eighteen months before. The reunion had both practical and symbolic aspects. It was a fitting display of loyalty and affection for the fallen chief that went over well with Crawford's sentimental southern followers and became the subject of laudatory editorials in the southern press. On the political side, Van Buren wanted to learn about Crawford's political plans, if any. Would he be a candidate in 1828? Had he changed his opinion of Jackson? Was his hatred for Calhoun still as intense and who were his closest and most trusted political associates in the villages and towns from Savannah and Milledgeville to Richmond.

Crawford was not just the titular head and the spiritual leader of the Old Republicans, he maintained his political influence over a small but cohesive group of followers. His speech and vision still impaired, he had given up all hopes of the Presidency. Although it cost him some effort, he agreed with Van Buren that Jackson was the logical candidate. He disliked the man but he was willing to support him. Calhoun was another matter. The hulking, disease-ravaged Crawford had lost none of his prejudice and was determined, if he could, to finish off the Vice-President's political career.

Van Buren and Cambreleng handled the vice-presidential question gingerly, seeming to agree with Crawford but not committing themselves. Van Buren realized, however, that Crawford would soon learn where he stood and balk at Calhoun's renomination, but more than a year remained before binding decisions had to be made and much would happen before then.

The carriage jolted through Columbia, South Carolina, to Raleigh, North Carolina, the small rural capital of the state where Romulus Saunders had arranged meetings with the Crawford leaders. Van Buren, at his most engaging, suggested to the skeptical North Carolinians the political logic of a Jackson-Calhoun ticket. The General, he assured them, would maintain Old Republican principles; and Cambreleng's cogent analysis of the political situation, coupled with the favorable impression Calhoun had made on his February visit, converted some of the local politicians to the new order of things.

Apart from enlisting a few adherents, the real gain lay in the future, the new acquaintances and the willing correspondents; otherwise their visit did not add immediate strength to the budding organization Van Buren was so carefully nurturing. He had managed to wring out of Crawford a reluctant agreement of support for Jackson, but this was balanced by the Georgian's obvious displeasure with Calhoun. More important was the opportunity the trip gave Van Buren to educate the southern electorate about the Adams Administration and the threat it posed to concepts of Jeffersonian government. His warnings were made at small gatherings and in the course of conversation. But he also seized every opportunity of presenting this view publicly and to a wider audience. In declining to accept a dinner in his honor at Raleigh, he wrote, "The spirit of encroachment has assumed a new and far more seductive aspect and can only be resisted by the exercise of uncommon virtue."

The following week Van Buren and Cambreleng reached Richmond at the same time as Calhoun's emissary, the spare, tight-lipped George McDuffie, appeared. McDuffie worshipped Calhoun, who had recognized the keenness of his mind and had rescued him from a country blacksmith to whom he had been apprenticed. Now a Congressman from the Edgehill district of South Carolina, McDuffie had proven himself a tenacious advocate once convinced he was right. With Van Buren's suave manner and his nimble mind, ballasted by the arguments of McDuffie and the political insights of Cambreleng, the Virginians who were already half convinced that Calhoun would bring strength to the ticket came round completely.

South Carolina and Virginia were behind the movement; Van Buren could afford to let time and organization take its course with North Carolina and Georgia. He adjudged his southern mission a complete success.[17] But ever circumspect, ever mindful of placing the opposition on the defensive and curious as to what Adams might make of his trip, he visited the President the day after he returned on May 12, 1827. Adams was courteous, somewhat stiff as always; yet even he was not immune to Van Buren's charming manner.

He inferred accurately that Van Buren and Cambreleng had been on an electioneering trip. Adams had also assumed that his visitor had requested an interview to pick up, if possible, the current thinking of the Administration on public affairs and therefore was on his guard. He was so successful in concealing his position that he forced Van Buren to engage in small talk. With that sixth sense he always had of knowing when the well was dry, Van Buren soon left. In this case, he had simply provided the astute President with a reasonably accurate profile of his present role and future prospects rather than gaining any useful ideas about the plans of the Administration. "Van Buren," wrote Adams of the interview, "is now the great electioneering manager for General Jackson as he was before the last election for Mr. Crawford. He is now acting over the part in the affairs of the Union which Aaron Burr performed in 1799 and 1800; and there is some resemblance of characters, manners and even person between the two men." After making these comments, the President paid Van Buren the compliment of saying that he was a far abler politician than Burr—". . . has improved as much in the art of electioneering upon Burr as the State of New York has grown in relative strength and importance in the Union."[18]

Van Buren left Washington for New York the next day, but not as in previous years to a comfortable home in Albany. After his re-election to the Senate, he made a major decision affecting his family and professional life. Washington, not Albany, would be the site of his future activities. He disposed of his house in Albany and was considering the shipment of his household effects to Washington, "as I cannot sell my furniture and as my future business will be here . . ." He planned to board in Albany and if possible rent an unfurnished home in Washington where he would live with Cambreleng and possibly other members of his Mess, friends of old standing like McLane and Drayton or new acquaintances like the intelligent William C. Rives, a Virginia Congressman with close ties to the Ritchie circle in Richmond.

Van Buren was scaling down his law practice, though when he left Washington in May of 1827 he spent several weeks in New York City attending the spring session of the state supreme court. He had been engaged along with Daniel Webster and former Chancellor Kent as counsel for the state of New York in a complex proceeding that involved extensive land claims of John Jacob Astor. The case dragged on through May, and it was before one of the conferences with his colleagues that he first learned about the alcoholism of his brilliant protégé, Samuel Talcott, then attorney general of the state.

Van Buren was no abstainer, but felt that wines and liquors should be taken in moderation as an accompaniment to a good meal or a good conversation among friends. He had seen the able Tompkins die a drunkard's death; he had observed how excessive drinking was dimming the minds and undermining the health of such vigorous individuals as Erastus Root. He worried about signs of overindulgence in Silas Wright, whose great ability he was beginning to appreciate. Nor was intemperance a concern merely outside of his own family. His son John, now at Yale, was carousing more than Van Buren thought proper. But John had responded to his father's stern warnings; and for the time being under the eye of Edwin Croswell's brother Harry, a New Haven minister, he was applying himself to his studies.

As for his other sons, Abraham was now a lieutenant in the regular army. Martin, Jr., and Smith were at school and spent their holidays with relatives in Kinderhook, where Van Buren maintained a farm and owned extensive properties picked up at sheriff's sales for taxes over the years. Comptroller Marcy, for instance, made it possible for Van Buren to purchase large sections of valuable land in Otsego County at a mere fraction of their value. By the summer of

1827 Van Buren's income from his law practice, his investments, especially his land holdings, made him a comparatively wealthy man. From habit, if from nothing else, he was careful about money, though he always lived well and paid attention to appearances.[19]

After the "Astor Cause" as he termed it (the case he, Webster and Talcott were prosecuting for the state) was settled in June 1827, Van Buren went to Kinderhook where he spent a week or so with his family and then left to make a personal canvass of the state. His correspondence had indicated that Adams sentiment was strong, particularly among the Yankee settlers in the western counties. The trip confirmed for him that Adams was as popular as represented, but much more important was the excitement over the abduction and presumed murder of William Morgan for allegedly publishing Masonic secrets. Morgan had disappeared the previous September under mysterious circumstances and sufficient publicity had been generated to provide local excitement. The Regency, preoccupied by Van Buren's campaign for reelection to the Senate, had not paid close attention to the Morgan episode, though one of its western lieutenants, Van Buren's good friend David Evans, was a member of a local committee raised to investigate Morgan's disappearance. The committee's letter to Governor Clinton charging obstruction of justice on the part of the Masons was passed over without much comment. Even Clinton's response virtually admitting that a crime had been committed did not ruffle the calm of the Regency or alert it to the political implications of anti-Masonry.

Van Buren, who was an omniverous reader of newspapers and subscribed to many that were published in western New York, was aware of the excitement, but he too was so completely involved in putting together the Jackson campaign that he had not given the anti-Masonic feeling the attention it deserved. He himself was not a Mason, nor was President Adams; but Jackson and Clay were, and De Witt Clinton was the head of the order in the nation. As far as Van Buren knew, it was a social and philanthropic organization that had counted among its members Washington, LaFayette and Benjamin Franklin.

As soon as he reached Canadaigua County, the Morgan affair rudely wrenched Van Buren from his complacency. Several Adams papers were drumming up the excitement of what was not only a much deeper and wider movement than he had thought possible, they were also making the obvious political connections. Thurlow Weed, whom he had ample reason to consider a clever politician and a hard-hitting editor was out in full cry, crowding the columns of his *Rochester Daily Telegraph* with sensational matter that had no particular regard for the truth. Disturbing also was the revivalist fervor that lent a unique shape, style and momentum to an incident that was obviously being diverted into secular channels.[20]

Unlike the Regency, unlike the eastern Bucktail press, unlike even Evans, who was in the midst of the controversy, Van Buren quickly saw it for what it was becoming, an organized political opposition to himself, to Jackson and to the party. Anti-Masonry was no resurgence of the Federalist party, though there were Federalists in its ranks, nor was it a sectional phenomenon, though presently surging through the western counties; emphatically it was not a personal faction, though anti-Clinton and anti-Jackson because both men were prominent Masons.

Yet it was this aspect of the uproar that was most surely the political part of its appeal. The *Telegraph* was rapidly broadening the definition of Masonry as an elitist conspiracy which sought nothing less than absolute control of the community for the extension of its vast accumulation of economic and social privileges. Anyone like Morgan who threatened its insidious control was to be removed. Murder, threats, kidnapping, torture, were commonly used not just to protect its secrets and its position, but to force compliance with its undemocratic, un-American, hypocritical program.

Van Buren was a troubled political leader when he returned to Albany in early 1827. He was concerned at the course of a new Jackson paper he had helped establish a few months before in New York City, the *Morning Courier.* After a brief tutelage from Croswell of the *Argus* and from Cambreleng, the *Courier* had gotten off to a good start. But Van Buren became increasingly worried about its refusal to take the anti-Masonic movement seriously. "I am sorry to see that a paper of so much early promise as the *Courier,*" he wrote Cambreleng, "should fall into so great a mistake as to speak lightly of the Morgan affair. Depend upon it, this course may do us much injury." He urged Cambreleng to have a talk with the editors, and at the same time the *Argus* began a series on Morgan, stoutly condemning any outrage, if any had occurred, and demanding the most scrupulous inquiry. The *Courier,* responding to Cambreleng and to Croswell, followed suit.

Noah, who was now editing the *Enquirer,* a new morning daily, pursued his highly personal, and to Van Buren, perilous course. "I am heartsick at his reckless indiscretion upon that subject," said Van Buren. "It is passing strange that a man so capable can commit so great a blunder as unnecessarily and unwisely to run in the face of so unmistakable a current of public feeling upon this subject in many of the western counties." At least once a week Noah ridiculed the anti-Masons, and with his gift for the ironic paragraph, mocked them as credulous creatures, victims of a monstrous fraud.[21]

The anti-Masonic problem was a thorny one for Van Buren, but with some luck and his skill in management, he felt he could turn the threat aside or at least blunt it by having the Regency's communication network identify Henry Clay with Masonry. In this way, the Adams Administration would be shown to be as closely affiliated with the Masonic Lodge as any of its opponents.

But devising a strategy to meet the anti-Masonic movement was not the only serious political problem that beset Van Buren. The tariff issue which he had managed to avoid the previous session of Congress had become far too pressing for any further delay in taking a public stand. A convention of manufacturers was about to meet in Harrisburg, Pennsylvania. Henry Clay, speaking for the Administration, had given it his blessing. Hezekiah Niles, in his influential weekly, was promoting the convention and for the past year insisting that a protective tariff for manufacturers and producers of commodities like raw wool could no longer be delayed without disastrous consequences.

Times were difficult for all American manufacturers, and those who produced woolen cloth were experiencing particularly severe economic hardships. The depression of 1819 still lingered; changes in economic policy which reduced the duty on imported woolens had given British manufacturers of all grades of woolen cloth an edge in the American market, even after absorbing the costs of transportation, insurance, warehousing and auctioning. Quite apart from the fact that British mills were more efficient than American, that British labor was cheaper and its product in all grades better, a large capital investment in New England and in New York was seriously threatened with extinction. Politicians who may not have worried much about the plight of Yankee or Yorker entrepreneurs, were susceptible to the argument of national self-interest. Even those consuming sections of the nation that benefited from the sale of cheaper and better woolen cloth could be moved by arguments that the loss of the American woolen industry would be a broad retreat to colonial status.

More trenchant than patriotic appeal was the declining European market for all American agricultural staples. Along with diminished international trade, agricultural land values had not recovered from the panic and were still depressed. A tariff on agricultural products—wheat, corn, flax, hemp—could raise domestic prices, it was argued, and ease the plight of the farmer. Exceptions were the seaboard slave states, which had flooded the raw cotton market and were blaming everyone but themselves for their plight. Any tariff that made the manufactured articles they depended upon for plantation agriculture more expensive was looked upon as a special tax levied upon them to subsidize northern manufacturers.

Besides sensational tales of secret Masonic activities, there came from the western counties reports of farmers and manufacturers demanding tariff relief. Marcy, returning from a visit to Detroit, wrote Van Buren that his political enemies were busy circulating reports that he had killed the "woolens bill" in the past congressional session. The Administration was in favor of protection, and Marcy believed Van Buren must take a public stand on the issue before the Albany meeting to be held on July 10, 1827. Marcy and others were worried that the Albany meeting, which they knew would be packed with partisans of the tariff, would attack Van Buren's record. The cause of Jackson, and indeed Van Buren himself, might be set back seriously. Characteristically, Van Buren kept his own counsel until the very morning of the meeting when he told Benjamin Knower and Charles Dudley he intended to be present and speak out. Both men were horrified. Certain the meeting was being held not simply to choose delegates for the Harrisburg convention, they thought it would censure Van Buren's conduct on the woolens bill. Should he be present, they argued, his enemies would have an opportunity to humiliate him publicly. Van Buren shrugged off the risk. Stephen Van Rensselaer would be chairing the meeting and would never permit such an action. Although unconvinced, they walked with him up State Street to the capitol.

The morning was sultry and it was a business day; yet the representatives' hall was packed, as was the gallery, with farmers, manufacturers, politicians and the idle curious. When the well-known, slight figure appeared, a sudden hush came over the noisy assemblage. Unperturbed, Van Buren, nodding at acquaintances, made his way to the Speaker's rostrum where Van Rensselaer greeted him warmly, offering him the clerk's chair beside him. But Van Buren declined and a seat was found for him in the chamber. Knower and Dudley found places, too, directly in front of him.

Despite the stifling atmosphere in the overcrowded hall, attention did not flag as the lengthy Harrisburg roll was read and a half dozen speakers droned on, every one of them making frequent caustic allusions to Van Buren's absence (visiting the congressional cemetery) when the woolens bill had been defeated with Vice-President Calhoun casting the deciding vote. As each orator concluded his remarks, eyes were turned toward Van Buren, but he remained imperturbable, impervious to the insinuations of bad faith on his part, of political cowardice, of low cunning. Finally, when it became obvious that the oratory had come to an end, Van Buren rose and with a slight smile acknowledged recognition from the chair.

He spoke rapidly though clearly for two hours, explaining his absence from the Senate when the crucial vote was taken and then leading his audience through the complexities of tariff legislation.

He avowed himself a wool producer and acknowledged that the market for American wool was so bad he had kept back his last two shearings. Yet he sympathized with the plight of the manufacturer, who, he claimed, was a victim of a conspiracy on the part of European interests and their American agents to evade the present revenue laws. False invoices understating the value of the imported materials were in general use, he charged. Foreign raw wool was being imported "in the skin" at a much lower tariff schedule than if it had been fleece. Finally he came to the crux of the matter—was he in favor of protection or not?

Any answer was bound to be difficult because it involved many more interests than simply those of protection for the producers of raw wool and the manufacturers of woolen cloth. He alluded to the iron masters of Pennsylvania, the hemp of Kentucky, flax in many states. Commerce and shipbuilding had to be considered, too. Tariff legislation was of national concern and must, he warned his audience, be kept clear of any partisanship. As for himself, he would extend encouragement and protection where it was manifestly called for to woolen producers, woolen manufacturers and "to every branch of domestic production and industry that may request it" and could prove its need.

Never once in a speech that bristled with statistics and knowledgeable statements did he allude to any conflict of interest between woolen producers and manufacturers. Knower, who had recently invested heavily in woolen mills, was well satisfied that Van Buren was in favor of protection for the manufacturer. Marcy agreed with his father-in-law. "There was last spring a more than half formed opinion," he wrote Van Buren, "that you was [*sic*] hostile to the tariff; this opinion was settling down to a conviction . . . and was doing (or rather was about to) infinite mischief to the course of Gen'l. Jackson in this state. . . . You destroyed in the speech you made at the capitol all of the works which long premeditated mischief had contrived." As for Van Buren himself, looking back on his remarks, he could not resist making a droll comment: "I have frequently been told and have always believed that I rendered much service to the cause of truth by that speech, but . . . directness on all points had not [been] its most prominent feature."[22]

That he had straddled the issue for the time being was good politics for there were too many unresolved conflicts in New York and elsewhere. From South Carolina came the warnings of a new acquaintance, the elderly Oxford-educated president of South Carolina College, Thomas Cooper. "Meetings are going; and feelings are becoming hourly more exasperated," he wrote. Prophetically, Cooper declared that if a protective tariff were to be adopted, South Carolina would secede from the Union and declare Charleston a free port. Two weeks after his Albany speech in which he had been so careful not to have himself labeled either for or against any specific tariff, Cooper wrote again that Van Buren was being charged with being a protectionist in South Carolina. "The manufacturers will make the woolen bill a sectional question; you will be burned with the multitude," he said.

In the face of these dire warnings from a prominent political economist whose opinion he respected, Van Buren was already moving toward a strategy of compromise that would garner the most votes for Jackson and for himself. He was quite certain that if the tariff issue were handled equitably for manufacturers and farmers alike, Jackson would bury Adams under a landslide of popular and electoral votes.[23]

As self-appointed congressional manager for Jackson, Van Buren hoped to advance his own position. His temporary alliance with Calhoun blocked Clinton from the Vice-Presidency where he could be a formidable opponent to Van Buren's aspirations, which were no less than succeeding Jackson as President. Calhoun, of course, would have to be dealt with in time. Clinton, too, with whom he was also in temporary accord, might stand in the way, though there was little he could do about that except wait for the Magnus to make some fatal error such as challenging Jackson and Adams for the Presidency. But above all, Van Buren had to preserve his political hegemony in New York, and this meant a cautious position on the tariff and on anti-Masonry. The heterogeneous population and the manifold interests of his burgeoning state demanded the utmost discretion and the most delicate balancing.

When Jackson was a member of the Senate during the Eighteenth Congress, he had joined Van Buren in supporting most of the provisions of the mildly protective Tariff of 1824. Van Buren's desk at that session was on the aisle and directly behind Jackson's, but they rarely spoke except to express the usual civilities. Jackson had little use for the Senator from New York who he was assured was a sly, political trickster with inordinate ambition.[24] He continued to be suspicious of Van Buren, and though he was well aware how important New York and its Senator were to his presidential ambitions, he favored De Witt Clinton, whom he deeply admired. Nor was he particularly disturbed at evidence that cropped up now and then of Clinton's ambition for the presidential nomination in 1828, or when a group of Clintonians urged him to shelve Calhoun in favor of Clinton as his vice-presidential running mate. Jackson turned down the proposal, but he was quite ready to offer Clinton the State Department should he win the election.

Van Buren soon learned of this maneuver from Alfred Balch, a close friend of his and an intimate friend of Jackson. In his first direct communication with Jackson, he took notice of the Clinton movement. "Attempts will doubtless be made to entangle your friends in the Vice-Presidential and other questions but they will, I am persuaded, have good sense enough not to meddle in them," he wrote. "I have no other feelings in relation to the Vice-Presidency than as it may operate on the main question. Let it be left to the natural course of public sentiment and it will fare best." Jackson was being pressed from many sides to abandon Calhoun. Crawford, whose political strength was still substantial in Tennessee and very strong in the seaboard slave states, ceaselessly pursued his vendetta against the Vice-President. Jackson resisted them all, giving Van Buren an opportunity to strengthen his position with his own organization, the Nashville Committee, who were conducting his campaign.

Van Buren deftly played on Jackson's known friendship for Clinton while demonstrating that the Governor had little or no following in New York. "I have no evidence," he said, "that Gov. Clinton had any hand direct or indirect in the matter and I am perfectly satisfied that (unless it should take to an extent not to be anticipated) he can not be so mad as to enter the field under existing circumstances."

Van Buren had been careful not to presume on his acquaintance with the highly sensitive Jackson. He had, however, kept in touch through third parties with William B. Lewis and John H. Eaton, Jackson's closest advisers and the most prominent members of the Nashville Committee. His friend Balch, who for the past three years had been promoting Van Buren in Tennessee, kept Jackson constantly aware of the Senator's overwhelming influence in New York. Other important Radicals like James Hamilton, Jr., wrote glowingly of Van Buren as "the first man" in New York.[25]

By the fall of 1827, Van Buren was ready to display his might. He had informed Jackson that almost all of the fifty Republican papers in New York would come out in support of his candidacy. By the time his letter reached Nashville, Tammany Hall had held "a general Republican Committee" that warmly endorsed Jackson. The *Albany Argus,* which had followed a most circumspect course, now came out boldly for Jackson and Calhoun. As Van Buren had said, nearly every Clintonian and Bucktail paper quickly followed suit. Meetings all over the state convened and echoed in oratory and enthusiasm the extravagant rhetoric of the press. Jabez Hammond said "the effect was prodigious. All the machinery, the construction of which had for two years put in requisition the skill and ingenuity of Mr. Van Buren and his friends at Albany, was suddenly put in motion, and it performed to admiration."[26]

The events in New York excited the General's martial spirit, his feeling for a well-prepared, well-organized military campaign. Neither Jackson, the Nashville Committee, the Old Republicans nor the Administration in Washington had ever witnessed such a spectacular display of integrated political strength. The most disinterested observer could have predicted the outcome, a complete sweep of the Jackson ticket in the fall election. Even the anti-Masonic districts in western New York were caught up in the Jackson enthusiasm.

There was but one more important task for Van Buren to perform in this preliminary stage of the contest and that was securing a Speaker of the House who would bind together all the disparate elements in the new Congress and marshal them behind Jackson and Calhoun. Andrew Stevenson, in his judgment, was just such a man. A Virginian of good family, tall, vigorous and imposing in appearance, he had the good lawyer's gift of bringing about compromise between heated adversaries. More important than this personal quality was a shrewd understanding of motives and how to use them for advantage, a trait Van Buren particularly esteemed. His Crawford credentials were impeccable; yet for well over a year he had been counted a power for Jackson in Virginia. He was acceptable to Ritchie and his group, though John Randolph thought him completely self-centered and imbued with what he termed "pedantic arrogance." Louis McLane, who coveted the Speakership himself and was therefore scarcely a disinterested observer, called him "a pompous hypocrite." The conclusive argument for Stevenson, however, was Van Buren's firm belief that he was the best and strongest man available from the Old Dominion who could handle what was sure to be the most politically sensitive issue confronting the Twentieth Congress, the tariff.

On December 5, when Congress convened, Stevenson was handily elected over John W. Taylor, the Administration candidate. Van Buren had opened the sluice gates, and the New York delegation again set the mill stones revolving when it repudiated a native son. He had scored a victory every bit as important as his success in New York. In defeating the avowed Administration candidate, Taylor, Van Buren defeated the President. Silas Wright gloried in the rout of Taylor and writing his friend Flagg said, "I like to see the galled jade wince and I can never doubt the correctness of my course." One by one, Van Buren was splicing together the political strands of the Jackson campaign with what he regarded as clear-cut issues of public policy.[27]

« CHAPTER 12 »

An Uncalculated Risk

In Albany on the evening of February 11, 1828, De Witt Clinton died of a heart attack. The great Governor had been ill for several months, but careless of his health, had ignored the symptoms and followed his daily routine. Since few outside of the immediate family circle knew just how sick Clinton was, his sudden death shocked the nation. In New York even those who had bitterly disliked him, and who had made a career out of opposing him in politics, experienced a sense of loss as if in some way Clinton's passing was the end of a certain, distinct period in their lives. Nothing would be quite the same without the towering Governor whose imperial stance so long had shaped the attitudes, the accomplishments and the aspirations of his native state.[1]

To the practical politicians, sentimental feelings were but a momentary phase before they began to calculate what impact Clinton's death would have on the politics of New York, the direction of the Jackson campaign. From far off Shelbyville, Tennessee, one of Jackson's protégés, Archibald Yell, wrote Congressman James K. Polk that "A grate [*sic*] man indeed has 'fallen.' His loss to New York is irreparable and the U.S. will feel his loss. . . . If he had lived he was destined for higher honors than that of Govr of New Yourk. I fear *we* shall feel his loss in New Yourk tho we have a *Van Buren* left."

In Washington, Van Buren was deeply moved at the news. Though he had spent most of his political career in opposition to Clinton, he could now view the Governor's achievments in perspective, cast aside his personal dislike for the man and do honor to his memory. Van Buren was not well at the time, but he delivered the finest oratorical display of his life when he eulogized Clinton at the request of the New York delegation in the House chamber. On February 19, with Van Rensselaer in the chair, the floor and the galleries crowded,

Van Buren made a direct appeal to the sentiments of his audience and to the people of his state, closing with a long-remembered flight of rhetoric:

> I doubt not that we will, with one voice and one heart, yield to his memory the well deserved tribute of our respect of his name and our warmest gratitude for his great and signal services. For myself, sir, so strong, so sincere and so engrossing is that feeling that I, who whilst living never envied him anything; now that he has fallen, am greatly tempted to envy him his grave with its honors.[2]

Van Buren may have been sincere when he lauded Clinton in these emotion-packed words, but he was aiming his message directly to the eighty thousand or more Clintonian voters who preferred Adams and who had remained passively for Jackson simply because their leader had been one of the General's firmest supporters since 1824. At first glance, the removal of Clinton from the scene seemed to clear the way for Van Buren. His most formidable opponent over the years was gone; the troublesome paradox of Van Buren and Clinton both backing Jackson, though heads of rival factions and each uneasy about the intentions of the other, was now resolved. Wright, for one, was certain Clinton's disappearance from the political scene rolled away "the greatest obstacle with our best men in taking the side of Gen'l. Jackson."[3] William C. Rives, though a Virginian and as such mystified by the vagaries of New York politics, took the same position as Wright. But the veteran politician Van Buren viewed the consequences of Clinton's death far more accurately than these less-experienced younger men.

Orthodox Federalists who had followed Clinton's varying fortunes and their "high-minded" associates who had scorned him would now go for Adams along with sizeable numbers of disgruntled Bucktails and a majority of Clintonian Republicans. Only their long-time allegiance to the Governor and their belief that he would succeed Jackson if he were elected had kept these restive elements in line. In Van Buren's opinion the loose coalition in New York would break up once Clinton's forceful personality was counted out. The Yankee West would combine with the Federalist East in opposing Jackson, the military adventurer, the slaveholding autocrat. If the anti-Masons, now in the early stages of political organization, joined the coalition as seemed likely, the state would go for Adams and imperil Van Buren's organization. That these were no mere speculations came from Regency correspondence with Washington not a month after Clinton's death. Flagg dispelled Wright's initial optimism when he wrote him on March 7 that "Mr. V.B. should put nothing on paper as to the Gov. at present. Things are now in such a state of chaos that it is impossible to say what is best at this time. . . . it is a critical juncture."[4]

The Regency had tried to divert anti-Masons from Adams by playing up the influence of Clay, a Mason of long standing. The anti-Masons, however, had been but briefly receptive to its campaign. Adams, after all, was President; he was not a Mason; he was a northern man and a New Englander like themselves. Van Buren's stock rose dramatically in Nashville and in Washington when Jackson received reassurances from New York; yet the vacuum left by Clinton's death dramatically altered the political balance at home, a balance that had to be restored at all costs. Silas Wright had reflected accurately Van Buren's sentiments on the Jackson campaign before Clinton's death. "Save the state," he wrote Flagg, "and let the nation save itself."[5] Now with Clinton gone, Van Buren recognized all too clearly how much more important it was to have a tariff bill passed, a bill that satisfied as far as possible the many-sided interests of New York.

Since the convening of the Twentieth Congress, he had bent most of his energies toward framing a tariff bill that would strike an equitable balance among the many special interest groups clamoring for protection. With this idea in view, he had engineered the election of Andrew Stevenson to the Speakership of the House. A Virginian, a Jacksonian and a moderate on the tariff issue, Stevenson had justified Van Buren's faith in his selection. His appointments to the House Committee on Manufactures that would report a tariff bill could not have been improved upon. All were able, hardworking men representing a composite of political views and sectional interests. For chairman, Stevenson chose Rollin Mallory, an Adams man from Vermont and a decided spokesman for the woolen manufacturers. But the majority of the committee were Jacksonians primarily concerned with protecting agricultural products.

The keenest member was Silas Wright who was serving his first term in the House. One of Wright's initial moves was to convince the committee that it should ask the House for delay in reporting a bill and give it the power to hold hearings, a novel but eminently practical approach to weigh the economic and political issues involved. Chairman Mallory, who personally opposed the request, introduced it on the floor where the Adams men vehemently attacked the measure as a political tactic. The *Daily National Intelligencer* saw in it a typical maneuver that would play presidential politics with a matter of grave public interest. But neither criticism from the floor nor strictures of the *Intelligencer* and the *National Journal,* another Administration press in Washington, could convince the Jacksonian majority. The committee received all the time and powers it deemed necessary.[6]

Throughout January, Mallory and his colleagues heard testimony from manufacturers and from agricultural interests. Members like Wright studied pertinent reports and papers, both American and English, far into the night, a tumbler of whiskey or brandy near at hand when his spirits flagged. Wright and Van Buren went over the schedules state by state calculating as closely as possible the political effects on the Jackson campaign, on the rate of protection demanded in one section, that would be sustained in another. Coarse hemp for the manufacture of ordinary cordage, flax and cotton bagging were items of particular interest to the western states of Kentucky, Ohio and Missouri. Shipbuilders and merchants in New England, New York City, Philadelphia and Baltimore were bound to oppose high prices for the ordinary cordage they consumed. The cotton-growing states would likewise oppose any increase in bagging. Higher prices for flax would hurt home spinners and weavers of linen cloth everywhere. Pennsylvania and New Jersey iron-masters demanded protection for bar and strip iron, among the leading products of those important states. But again, consumers of pig and fabricated iron, primarily the shipbuilding industry, wanted cheaper iron products.

Despite the plethora of disparate interests, it was almost immediately apparent that the fundamental division between the manufacturers of woolen cloth and the producers of raw wool was to be the most significant, the most controversial and the most politically sensitive of any of the schedules. Yet the House Committee moved ahead rapidly with its hearings and reported a bill on January 31, 1828.

Woolen schedules were placed as high as Wright and Van Buren thought would be acceptable in Kentucky, Pennsylvania and Ohio. Molasses, which New Englanders had profitably distilled into rum since colonial times, competed with spirits distilled from western grain. Duties were raised sharply on molasses, thus favoring western farmers and acting as an inducement for them to support higher prices for woolens. The bill that emerged from the committee favored agricultural interests over manufacturing. As with any compromise measure, it was greeted with a storm of criticism which was most vocal from representatives of the Northeast and the cotton states.

Reports from New York were mixed. Jesse Buel, former editor of the *Argus,* and Benjamin Knower, both woolen manufacturers, were the leading critics within the Bucktail ranks. Most of the New York City merchants and shipbuilders were Adams men, and their anger at the increase in the rates for iron, molasses and cordage could be discounted. But the discontent of such staunch Van Buren supporters as Walter Bowne and Saul Alley, prominent members of Tammany Hall and of the City's mercantile community and big dealers in raw wool, was unsettling to say the least.

Pressures from the mercantile community forced the passage of an amendment that reduced the specific duties on raw wool by three cents a pound, a significant benefit to importers and to dealers in the staple. The Jacksonians from the West and the Middle Atlantic states were strong enough to beat down all the amendments the southern and New England members proposed. On April 22, the bill finally passed in the House by a vote of 105 to 94, a bare margin of 6 votes would have killed it. Reduction of rates on imported wool had probably done the trick, garnering additional votes from New England and the Middle Atlantic states.[7]

Complicating the problem was that many of the farmers, especially in New York State, had an interest in woolen manufacturing that was practiced extensively in the home under the putting-out system. Wright and Van Buren were made well aware of this vexing situation when they received a confidential circular written in Albany signed by such old political allies as Benjamin Knower and Jesse Buel. The Knower-Buel circular demanded the Harrisburg specific rates on woolens, which, if adopted, would have doubled the price of coarse woolen cloth and raised substantially the prices of higher quality products. Knower and his associates also wanted the least amount of protection for domestic wool, though the sheep population in New York stood at more than 5.5 million head.

Had Van Buren heeded Knower's circular, he would have run into heavy weather, not just from the slave states but also from other cloth-consuming states like Pennsylvania. Moreover, it was in direct conflict with the official Regency position as published in a previous circular and given wide distribution. Nor was the Knower-Buel position at all consistent with the bland resolutions the legislature had adopted on January 31 instructing the New York Senators and requesting its representatives to "afford sufficient protection to the growers of wool, hemp and flax and the manufacturers of iron, woolens and every other article so far as the same may be connected with the interests of the manufacturers, agriculture and commerce." The startling information of a break in the ranks at home, the evidence he had gleaned from the hearings, the advice he had received from trusted political friends, guided Van Buren's deliberations.

Clinton's death, occurring as it did while the House was considering the tariff bill, complicated the delicate balancing act Van Buren was attempting. One thing remained unchanged, the tariff must not injure his position in New York. As Silas Wright had reported to Flagg earlier, "even Mr. V.B. would rather jeopardize the presidential election itself than to risk a breaking up of our ranks at home, or of destroying our strength and harmony in the present Legislature." Van Buren managed to have Knower, Buel and other manufacturers back away from their extreme position yet still retain their good will and support, no mean feat. He was much disturbed, however, at the difficulties in Albany, even though they proved to be transient ones.[8]

When the bill came before the Senate, he decided that if it were to pass, and he was determined that it would, the rate on woolens must be raised enough to satisfy the interests of his own state and gain sufficient votes from New England to neutralize any sectional combination between that region and the Southeast. Somehow he prevailed upon his Pennsylvanian and western colleagues to accept higher rates on woolens which were to be compensated with the lower rates on foreign wool in the House bill. Obviously, he was thinking of those thousands of farmer-manufacturers not just in New York but in Ohio, Vermont, New Hampshire and in western Massachusetts. Anticipating a hard fight, he outlined a major speech he was prepared to give in defense of the bill with the woolens amendment. He never had to make his speech.

New Englanders, champions of southern cotton and other special interest groups tacked on amendments. In every instance, Van Buren voted with the majority against them. Only when the amendment on woolens came up, which he himself had worked out, did he deviate and vote in the affirmative. The amendment passed 24 to 22. Had Van Buren voted against it, the result would have been a tie that Vice-President Calhoun would have broken with a certain negative vote, thus killing the revised schedule and with it the entire bill. For the manufacturers, as Van Buren and Wright had feared, would have joined the cotton-producing states to defeat the measure.

Outraged southerners accused Van Buren of inconsistency then and later for his role in passing what would soon be stigmatized as the "Tariff of Abominations," but the charge was absolutely false. Whatever he may have said in private conversation to mislead critics of his course, he made abundantly clear in his conversations with Wright that he had always wanted the bill to pass. The upward adjustment of the woolen schedule in the Senate was essential to that end, just like the reduction of rates on raw wool had been necessary for its acceptance in the House. The amended bill went back to the House, which approved it without delay.[9]

It had been a hard fight during which Van Buren strained the political ties he had carefully cultivated with many of his southern friends like Tazewell, Randolph, now in the House, James Hamilton, Jr., and Drayton. He had few illusions about Calhoun who opposed him on the tariff issue and whose further ambitions were clear to anyone who read Duff Green's *United States Telegraph*; and Van Buren was a careful reader.

For the past two years each had needed the other. Calhoun had given up on Adams, had reconciled himself to Jackson's nomination and election. Jackson was old, his health precarious, and Calhoun, along with many other politicians, thought of him as a one-term President. Reelection to the Vice-Presidency was indispensable to Calhoun's plans. He would associate himself in the public mind with a popular figure, an authenic American hero. He would also be on national display for another four years, his claims to the succession strengthened immeasurably. But he needed the support of the Crawford bloc and this meant Van Buren, its titular leader. He also needed New York and again the amiable Senator, "the little magician" as he was being called, held the combination that would swing open its electoral votes.

Van Buren had seen in the Vice-President a necessary link for him and his faction with the Jacksonians. He had done all he could as manager of Crawford's candidacy to defeat the General. And Crawford himself had made no secret of his dislike for Jackson, his belief that the Old Hero was entirely unfit both by experience and by temperament for the Presidency. Van Buren knew that he could not escape sharing some of the personal antipathy that existed between the two men. Nor could he expect a hearty welcome from Jackson based simply on his tenuous alliance with De Witt Clinton along with his shaky leadership of the small group of restive Radicals. Calhoun was the logical man to pave the way, to assist with his not inconsiderable political strength in Congress.

Both men had adhered to the bargain while mutually beneficial and each secured the benefits he had sought. Van Buren convinced the dubious Radicals, though never Crawford, to forgive Calhoun's flirtation with Adams and back him for reelection. Calhoun worked successfully to make Van Buren acceptable to Jackson and the Nashville Committee.

But the sudden death of De Witt Clinton had removed a potential rival from Calhoun's path and had elevated Van Buren to that unenviable position. As Silas Wright wrote Flagg, "Let me, however, suggest to you that I think he has at this early period a very jealous *friend* in the Vice Pres't." The tariff debates provided Calhoun with an excuse, if one were needed, to break the partnership and work covertly but actively against any pretentions the New Yorker might have for higher office.

Van Buren, too, had by now little need of Calhoun's active support; yet he would back his reelection as he made abundantly clear to Crawford. The alliance of the Middle Atlantic states with most of the South and most of the West, which had taken such a great deal of effort to create, could not be jeopardized at this late hour without the weakness of confusion. Too many Jacksonian caucuses, though not New York's, had nominated a Jackson-Calhoun ticket. Nor did the tariff issue, then in delicate stage, augur any change in the arrangements already made.[10] He would bide his time, strengthen himself with the Nashville Committee and work to bolster the Regency at home. He was well satisfied that the tariff he successfully piloted through Congress had strengthened the Regency in New York and would ensure Jackson's election. The South had been opposed, but its politicians were completely circumscribed. As a section, it had to support either Jackson or Adams, and it would never back the "federalizing arch protectionist" New Englander on principle, if nothing else. In the first flush of his dearly bought victory, Van Buren had no idea how strong and how deep the currents of sectional resentment were already running against the tariff.

Another factor was claiming his undivided attention, the governorship of New York. Since Clinton's death Van Buren had to face the possibility that he might have to be a candidate, as the only means of eliminating the political vacuum in Albany. The Regency was composed of talented, ambitious men, but none of them had the experience or could command the popularity of Van Buren throughout the state. In fact, the tariff debates had disclosed internal strains within its ranks that required his sure touch. Only he had the prestige and the ability to redirect the party's energies against an evolving, highly menacing rival organization under the capable leadership of Thurlow Weed and Peter B. Porter. Even though he had a clear picture of the political landscape in New York, and his place in the coming contest, he resisted the mounting pressures upon him to risk his position, his ambitions, on the chancy race for the governorship.

Van Buren's open identification with the Jacksonians and his highly visible management of Jackson's cause in Washington had not at first set well with the Regency. His close associate, perhaps at this time his closest associate, Silas Wright, confessed to Flagg his misgivings about Jackson whose "political creed as he has published it to the world by no means suits me." Like the Southerners, Wright was supporting Jackson simply because Adams, the only alternative, seemed practically a Federalist in policy and politics.

Van Buren had wanted a caucus of the Republicans in the New York legislature to join Tennessee, Louisiana and Pennsylvania in nominating Jackson and Calhoun on January 8, the fourteenth anniversary of the Battle of New Orleans. Michael Hoffman sent the word from Washington, but it was not until the end of the month that the Regency could put together a legislative caucus that nominated Jackson but not Calhoun, bowing to the sentiment expressed forcibly that the North and particularly the Empire State was being discriminated against.[11]

Van Buren, perhaps alive to these divisions, perhaps listening to Wright's words of caution at making the nominations so early, arranged to have the facile, imposing James A. Hamilton, "a high-minded" Federalist who remained with the Bucktails, sent West to describe the intricacies of New York politics to the Nashville Committee in case, as he suspected, the legislative caucus would not make the nomination at the appropriate time. He also wanted to broaden communications between himself and the Tennesseans. The charming Hamilton accomplished this task admirably.[12]

But as Van Buren's stock rose with Jackson and as he pushed the tariff to its successful conclusion, the political weather in New York took on a more threatening cast. More than one astute observer regarded the Administration faction about equal to the Jacksonians, with the anti-Masonic organization holding the balance of power and leaning toward Adams. There was early evidence that a fusion was being planned when on March 7, anti-Masons from twelve counties met in convention and overwhelmingly endorsed Adams's reelection. At the same time, the convention issued a call for a state anti-Masonic convention in August to nominate an entire slate. A few days later, Thurlow Weed, who had organized the convention, brought out the name of Francis Granger for governor on an anti-Masonic ticket in the *Rochester Telegraph.*

Granger, a stately expatriate from Connecticut, the son of Jefferson's postmaster general, was a highly respected lawyer and public figure in western New York. As chairman of the legislative committee investigating the Morgan affair and the alleged activities of the Masons in the state, he had gained a reputation for fair dealing that satisfied the anti-Masons and those Adams Republicans who were either Masons or who belittled the excitement in western New York as a transient outburst of illiterate fanaticism. Since Granger had achieved the near impossible of being acceptable to Masons and anti-Masons alike, he was the obvious choice to lead a fusion ticket; so believed Thurlow Weed whose political instincts were as sound as Van Buren's.[13]

For Van Buren, these developments gave further credence to his own initial conclusions that the fall campaign in New York would be long and bitter, fraught on all sides with hazards to Regency control. Yet he had barely digested this information when to his annoyance several newspapers floated his name as a logical candidate to run for governor in the fall. The *Daily National Intelligencer* immediately called

Van Buren a blatant opportunist who was plotting covertly to strengthen himself further with the Jacksonians by standing for a popular election. Thomas Ritchie, who had recently returned from a trip to New York, viewed these press comments with some concern. While in Albany, the nervous editor had been exposed to the intensity of New York politics. He had come away convinced that agents of the Adams Administration were organizing the anti-Masons to assist them in carrying western New York. Van Buren would injure himself if he entered the political melee and stood for office. "It is presumptive of me to pretend to advise you," said Ritchie, choosing his words carefully, "but I should suppose you ought to reserve yourself for another destiny under the General Government."

By this time Van Buren had taken steps to quash any more talk of the governorship. The *Argus* distinctly omitted all reference to him when it called for a convention to nominate a state ticket and presidential electors for Jackson. Noah's new paper the *Enquirer* and the *Courier,* edited by the young, flamboyant James Watson Webb, were likewise silent upon the subject. The *Evening Post,* edited by William Coleman (though his young assistant, the precocious William Cullen Bryant, was now writing most of the editorials), was a violent antagonist of Adams. Like all Regency papers, it too avoided specific mention of nominees for governor, while berating the Administration. Bryant depended upon Gulian Verplanck and Cambreleng, personal friends, for his Washington news. Both Congressmen were confidants of Van Buren and reflected his political posture at any given time.[14]

By the end of March, however, the silence of the Regency press on Van Buren's intentions was becoming strained. The *New York American* and other Adams papers, when not abusing Jackson as a bloody-minded autocrat, a murderer of militia-men and an open adulterer, were taunting the Van Buren papers for their "noncommittal" stance in state politics, imputing all sorts of sinister motives.

If the editors were restive, the Regency had grown frantic as it observed the apparent coalition between the anti-Masons and the Adams Republicans, Federalists of all hues and Clintonians, a hybrid mixture, patently unstable but potentially devastating if they could act in unison. Flagg and Marcy kept up a stream of gloomy letters to Van Buren and Wright that all but insisted Van Buren accept the nomination for governor. By the end of March, while still deeply involved with the tariff bill, the pressure on Van Buren had become so intense that he knew he had to decide one way or another. The safe way, of course, was to heed Ritchie's advice, keep clear of the state campaign and support either Marcy, whose patronage as comptroller had given him widespread influence, or possibly even Wright.[15]

It was a reasonable assumption that Jackson, if elected, would reward New York and its preeminent spokesman who had in effect put together the winning coalition with one of the two top cabinet posts, State or Treasury. On the other hand, if Jackson should lose New York, a distinct possibility, while Van Buren sat safely ensconced in Washington, his ambitions might very well not be gratified. And he would certainly forfeit much of his personal power in the party. Should he accept the nomination and lose the state election, also a distinct posibility, the result would be even more disastrous. For his political organization would go down with him, quite possibly putting an end to his political career and certainly destroying his influence in Washington.

He talked over the political prospects with Wright and Hoffman, leaving them with the distinct impression that he would prefer to stand clear of the extremely volatile state of politics in New York. "I hope it may not become necessary to nominate M.V.B.," wrote Hoffman to Flagg, "but if it does, the party may safely rely that it can command the services of *any man* in the state." Van Buren revealed his indecision in a letter to his friend and land agent, David Evans:

> A crisis in my political life is approaching which must be very consequential in its results . . . the matter that plagues me is the office of Governor. My interests and the true interests of all require that I should be continued here. But I have already been greatly persecuted upon the point and expect to be more. . . . But suppose the worst comes to worst and I cannot avoid a nomination without a rupture with my party . . . is there a reasonable or any chance . . . that you will run with me as Lieut. Gov.

Although he had not made up his mind, he felt he had to prepare for all contingencies. Through Hoffman, he instructed Flagg to advise the papers it was too early "to name a candidate." A timely reminder from Albany emphasized how important it was for Van Buren to win a popular election. He had, of course, been popularly elected to the state senate but had never stood for an election before the people of the entire state. "Being a candidate at this time," a friend argued cogently, "would make justices, constables and all the numerous active men in the towns familiar with your name, and from them it would be scattered throughout. In fighting the battles of a man at election in a warm contest, we become insensibly attached to him, and when you descend to the very lower ranks, this attachment becomes enthusiasm."[16]

Whether he decided to enter the New York race or not, Van Buren wanted to set the record straight on his political beliefs. He took advantage of a debate then in its final stages on the floor of the Senate over the powers of the Vice-Presidency to make a rather elaborate historical analysis of parties in the nation. As usual, he had researched and rehearsed his subject well.

Van Buren's theme was one that he had developed over the past six years and emphasized clear party divisions between the centralists and the advocates of states' rights. To an attentive audience on the floor with the grim-visaged Calhoun in the chair, the faultlessly dressed Senator from New York, speaking rapidly but logically, based his argument on the conflicting opinions between Hamilton and Jefferson on the power of the state. There would always be, he was certain, a basic disagreement between these opposite views of the nature of government. This being so, argued Van Buren, sharp lines must be drawn in the form of organized parties that would strive to convince the people where their best interests lay. He left no doubt in his audience where his sympathies lay and where his energies as a public man would be directed. Should Van Buren remain where he was in the Senate, he gave notice that he would act as a parliamentary leader who would rally men of like ideas and range them against centralism in any form.

If he were to go before the people of New York State, he would carry the same message of states' rights and individual liberties as the best guarantee of progressive social change and economic growth. The speech was in part an earnest, and by and large, successful attempt of Van Buren's to sort out and articulate his political ideas. But there was also an opportunist and partisan side to his remarks when he cleverly linked the Adminstrations of the two Adamses and cast them both as aristocratic, potentially dangerous to the well being of all the states, but especially of New York where the party battles between the Federalists and the Jeffersonians were still fresh in the memory of his constituents, where the "Manor" interests were still entrenched, where the westward movement was in full tide and, finally, where the state had taxed itself and was still taxing itself to expand its own internal improvements.

Parts of the address would be valuable campaign documents, but its major importance was that it set the line for the state campaign. Croswell gave up the entire editorial section of the *Argus* on April 20 to reprinting the speech. The *Evening Post* devoted several columns to it and noticed it favorably in its editorial comments as did the other Van Buren papers in New York City.

Van Buren now had to make the fateful decision. Congress adjourned on May 26 and he hurried home, disregarding the unseasonably bad condition of the roads. He stopped in New York City where he conferred with the Tammany Sachems, now thoroughly Jacksonian but enmeshed in quarrels over local office and patronage in which Noah and Richard Riker, Clinton's former associate, were deeply embroiled. Van Buren managed to dampen the controversy before going north. When he reached Albany he found similar difficulties. Erastus Root, a confirmed alcoholic, but still commanding a large following in western New York, wanted the gubernatorial nomination. Marcy was equally demanding and at cross purposes with Root. So much so that word of their differences reached the ears of Henry Clay, who asked Porter if these differences could be exploited.

Other ambitious aspirants urged their claims, quarreled with each other, threatened to go their separate ways if not recognized for their part in the stunning victory of the Bucktails the previous year.[17] As he had in New York City, Van Buren, still silent, still undecided about his own plans, arranged a temporary truce before he traveled west to gauge political opinion, to estimate for himself the strength and direction of the anti-Masonic organization.

The great question in his mind was whether the anti-Masons would merge with the Adams men who were now calling themselves National Republicans and who had just held a convention in Albany. A majority of their delegates favored Francis Granger who had publically identified himself with the anti-Masonic organization. But no nominations were made as the National Republicans, like Van Buren, waited to see what the anti-Masons would do.[18]

Shortly after the Albany convention, Van Buren, his carriage well stocked with provisions, and in fact all those material comforts that could not be expected in the raw, new villages and towns of western New York, set forth on his journey. Ahead of him were hundreds of miles on the road, tropical heat and high humidity, sudden thunderstorms that turned the wretched roads to quagmires in a matter of hours, and swarms of insects that infested the wilderness areas he would transit. Van Buren did not let the hazards of travel interfere with the taxing schedule he had set for himself. It was not until August 21, almost seven weeks later, that he returned to Albany.

For a man of forty-five who had been sick a good deal of the time the past winter in Washington, who was again concerned about his son John, worried about the education of his younger sons, Smith and Martin Jr., Van Buren bore the hardship, the nervous strain of appearing composed in public regardless of provocation, with good humor and uncomplaining patience. Though exposed to all sorts of what were euphemistically termed "summer complaints," he was remarkably fit during his travels; and he never neglected his personal appearance.

When he attended church in Rochester, he was dressed in the latest fashion and in colors that complemented his fair complexion. The beige swallowtail broadcloth coat with matching velvet collar set off his bright reddish-blond hair. He wore an orange cravat with lace tips, white duck trousers, pearl gray vest and silk hose. As he entered the First Presbyterian Church, doffing his broad-brimmed, long-furred beaver hat, he held yellow kid gloves in his left hand while he waved and bowed to the crowd of curious onlookers. He must, however, have been happy to reach Enos Throop's comfortable home and enjoy his hospitality. Throop was an acquaintance of some standing with whom he wished to discuss the anti-Masonic problem in western New York.

The movement was more vigorous and widespread than he had supposed. And Throop, one of the most prominent lawyers of western New York, was the only leading Jacksonian enjoying the confidence of the Yankee farmers, who formed the backbone of anti-Masonry in the counties of the southern tier along the Pennsylvania border and those edging on Lakes Ontario and Erie.

As the circuit judge of the region Throop had held the first trial of the Morgan conspirators, four prominent Masons of Canadaigua who admitted helping to abduct Morgan but denied any foul play. Before imposing jail terms ranging from two years to one month, Throop had lectured the convicted men in a severe homily from the bench.

From that time until Van Buren's visit, Throop had maintained the one formidable Jacksonian outpost in the anti-Masonic wilderness. And the only prominent Van Buren Republican to attend the Harrisburg convention, he was also identified with the woolen manufacturers of western New York, primarily small farmers whose families spun and wove cloth for the dealers in raw wool and woolens.[19]

Physically, Throop was an imposing figure, tall, angular, with deep-set eyes and generous features, he combined a self-imposed sense of judicial dignity with a keen aptitude for business. There was no pretense about Throop, who treated Senator Van Buren no differently than he would treat any of his neighbors. Yet with it all, he was a kindly man, a just man of some ability, a believer in the economic potential of western New York and a committed Jeffersonian. Throop was an ideal candidate to run for lieutenant governor should Van Buren decide to head the ticket. For among these strengths, he had never been associated with the Regency, an important consideration to Van Buren for campaign purposes and for the internal unity of the party.

Well before Van Buren sat with his host outside Throop's one-room cottage office overlooking beautiful Lake Owasco he had made up his mind to run for governor. Of the many reasons for his decision, the most important was the obvious confusion of the National Republicans. Thurlow Weed had sought unsuccessfully to bind the eastern Adams men with the western anti-Masonic contingent. His candidate, Francis Granger, was unacceptable to the eastern delegates at the Utica convention. Primarily Clintonians, they would have none of him. Many of them who were Masons, many others who were not, refused to accept what they regarded as dictation from an absurd minority.

Weed tried his level best to convince them that the nomination of Granger would in no sense be an anti-Masonic victory but would capture anti-Masonic votes. Granger would win the state election and with it thirty-six electoral votes for Adams. Neither Weed's cogent explanations nor the entreaties of Peter B. Porter could prevail. Phineas Tracy, an insightful National Republican Congressman from Batavia, compared the new party with the Regency:

> Their leaders are adroit managers and expert politicians. They have a rallying point and a council of review. Our party is formed of discontent materials. We want organization, unity and energy. We are like an oligarchy without any one acknowledged head. All wish to be leaders. Van Buren is a host in himself; the idol and pride of his party, tho not over scrupulous in the means which he adopts to carry his points.[20]

The convention reached into the distant past when it nominated Smith Thompson for governor and Francis Granger for lieutenant governor. Thompson had never held an elective office. During the intervening years since 1818 when he went to Washington as Monroe's secretary of the navy, he had all but lost touch with state politics and state political leaders. Clearly, Thompson was a last-minute compromise candidate whose national stature it was hoped would elevate him above the anti-Masonic bias.

Before the convention adjourned, Thurlow Weed went west and for two days sought fruitlessly to convince the anti-Masonic leaders that they should back Thompson. All he received for his pains was a call for a separate anti-Masonic state convention from the county papers around Rochester. The zealous anti-Masons went ahead with their convention plans and promptly nominated Granger for governor along with a relatively unknown political hack, John Crary, for lieutenant governor. Obviously Granger had to make a choice and he opted for his National Republican nomination. The anti-Masons then chose Solomon Southwick in his place. Southwick's struggling weekly, the Albany *National Observer,* was the only paper east of Utica that endorsed the anti-Masonic party. Considering his checkered past,

Southwick's espousal of anti-Masonry was typical of a man who "was credulous to the excess and even superstitious . . . swindled by every knave and duped by every imposter he met on the road."[21]

When Van Buren learned of these events he was as certain as he had ever been that he would win the election along with a fair prospect of securing most of New York's electoral votes for Jackson. But still he could not afford to take any undue chances and he was concerned about the unity of his party. In all likelihood his term would be a short one. Jackson would defeat Adams and he would be offered State or Treasury in the new Administration, either of which he would accept. The post of lieutenant governor would be a crucial one in maintaining the power of the Regency. Pitcher lacked the confidence of party leaders and more than that the political insight and tact to run the organization while Van Buren was in Washington. He was also an easterner and there was a very good chance that Crary would bow out of the campaign in favor of Granger; should this event occur, Van Buren might face the anomaly of having a National Republican lieutenant governor who would take over when he went to Washington. And in fact he was anticipating Thurlow Weed, who had fallen back on this second line of defense and was exerting all the pressure he could bring to bear in anti-Masonic circles against the Crary nomination.

Throop, on all counts, was the proper choice in Van Buren's mind. After explaining the various options open to him, and the probable course of the election, Van Buren had no difficulty in persuading him to be his running mate.[22]

The over-all picture may have looked bright to Van Buren, but even with the split between the anti-Masonic party and the National Republicans, there were areas of weakness that were being exploited by a lavish use of campaign materials, peculiarly well adapted to the prejudices of New York and indeed of the entire Northeast.[23] Isaac Hill, editor of the Concord *New Hampshire Patriot,* allied with the cautiously shrewd Senator Levi Woodbury, the popular old Revolutionary War veteran Benjamin Pierce and his handsome eloquent son Franklin, had made a promising start. The caustic editor had opened a correspondence with Van Buren more than two years earlier and the past year had visited him for what he declared in the *Patriot* was a tour of the Revolutionary battlefields.

No doubt the battlefield was Saratoga, but the talk was surely not devoted to Benedict Arnold's frontal assault on Burgoyne's line while Gates sulked in his tent. The conversation was surely about politics, about the Regency organization and how a similar machine could be developed in New Hampshire. Van Buren's expertise, Hill's acumen, Woodbury's family connections and the popularity of the Pierces were simply not enough to encounter the avalanche of coffin handbills, the broadsides of crude woodcuts picturing Jackson as an immoral, bloody-minded tyrant, a gambler and a duelist, all of which made a profound impression on the suspicious yet gullible Yankees on the hard-scrabble farms and the isolated little communities where life was a constant struggle against a less than bountiful nature.

The same signals came out of Vermont, Massachusetts and Connecticut, where the Jacksonians had hoped to snatch some electoral votes from Adams. John M. Niles, whose flaming hair matched his fiery temperament and his nationalistic fervor, had traveled to Albany from Hartford to learn about politics from its acknowledged master and had gone home to initiate Gideon Welles, his editorial associate on the *Hartford Times* into the mysteries of the Albany Regency. And there were even correspondents in Adams's home state where Van Buren had cultivated ambitious editors like Nathaniel Greene of the *Boston Post* and through Benjamin Knower, restive wool wholesalers, independent manufacturers and disaffected merchants.[24] But the Adams campaign was smothering all these potential sparks of defiance under a thick coat of dampening sectionalism, selected patronage and partisan innuendo.

Weed's activities, editorial and otherwise, seldom rose above the level of scurrilous abuse, standard fare for the National Republican press in the Northeast. But he did refuse to distribute coffin handbills and other circulars more offensive to taste and intellect. He was not so scrupulous when it came to accusing Van Buren and what he termed his Masonic cohorts of flooding Genessee and Monroe Counties with money for the purchase of votes and of newspapers favorable to their cause. As with all the libelous campaign literature, Weed's phillipics were millet seeds of truth embedded in husks of fabrication.[25]

In other areas of the state, National Republican editors did not confine themselves, as Weed had done, to unproved assertions of political corruption. And they were making an impression. The Regency was particularly anxious about St. Lawrence County where Silas Wright was up for reelection to Congress. The "Manor Counties," Columbia, Van Buren's own, Albany and Rensselaer, centers of old Federalist sway, were all but given up; Saratoga to the north, was in the balance. Van Buren was concerned with the New York City Districts, even though he had secured a truce among the warring factions of Tammany the past summer. Still, factionalism gave every sign of infecting the congressional elections and losing the City for the Jacksonians. A local group of mechanics had organized themselves into an antiauction party and were attacking Cambreleng and Verplanck as toadies to the auctioneer monopoly which set prices on all imports. Distressed at this unexpected outbreak, Van Buren wrote Cambreleng, "It is as much as we can do to carry you and Verplanck and [Jeromus] Johnson."[26]

In the presidential contest, the Regency concentrated on impressing the public that the policies of the father and the son were virtually identical. Corruption and waste were also charged against the Administration. Busy Jacksonian politicians, Van Buren and his associates among them, had ransacked public records and had found evidence of misappropriation, though none of it was large and certainly no more flagrant than that exposed during the Monroe Administration. Yet the demand for reform in public policy as well as in political alignments was asserted and reasserted at boozy speeches after militia musters, at "spontaneous" meetings in villages and towns, in editorials and on handbills and broadsides.

In mid-September, Van Buren informed the Regency that he would accept the nomination for governor on a ticket that was now calling itself Jacksonian-Democrat. At the same time, he explained why Throop must replace Pitcher. Even with Southwick and Crary expected to draw off votes from Adams and Thompson, the race would be a close one.

Besides the war of words, the Jacksonians capitalized on local issues, concentrating on those that seemed most likely to attract votes. High on their list of priorities was the Chenango Canal, that had been proposed from time to time since the completion of the Erie and the Champlain Canals. The Chenango was the largest of the proposed lateral canals that would open up the rich farmlands of the southern tier. On his visit west, Van Buren had listened carefully to the arguments for the canal and decided that the issue was timely, popular and practical. The word went out to the local organizations through the counties that would be primarily affected, Broome, Chenango, Otsego, Delaware and Madison. So effective were the communication links and so enthusiastic were the farmers, speculators and local village folk, that well before the election it was clear the Van Buren ticket would sweep all those counties.[27]

Once the announcement was made to his inner circle, Van Buren began drafting the party address. From long experience he knew the beginning was the most important because it set the tone. And he always had trouble condensing his thoughts and striking just the right note at the onset of any of his important speeches or public letters. With the Herkimer convention only five days away, Van Buren wanted Butler to read what he had already written, edit it thoroughly and return the copy. James A. Hamilton had been with Van Buren and had helped him with the draft before returning to New York City. There he advised Cambreleng, who was a delegate to

Herkimer, that the document was completed. In his playful manner, Cambreleng dashed off a note to Van Buren asking for a copy to be sent by the next boat, "so that we can give it an *honest parentage,* if our board of censors here think it will do." He also wanted Van Buren to "tell Hamilton he must not long delay his letter on the subject of Jackson's personal character."[28]

Cambreleng was anticipating the letter, which Hamilton had already written, that was to be sent out to the Democratic press in the state testifying that Jackson was not only a religious man but regularly attended church. The Adams press had repeatedly cast doubts on Jackson's faith, and leaving nothing to chance, Van Buren had asked Hamilton to refute the charges. Hamilton complied. On September 16, Van Buren returned the draft with only slight alterations which he deemed of a political nature. In a postscript, he asked "Does the old gentleman have prayers in his own home? If so, mention it modestly."

Wisely, Van Buren decided not to have the party take cognizance of the anti-Masons though neither he nor Throop belonged to the order. Rather he would have it take a stand on national issues, on clear-cut political divisions, on the patriotism and the presumed merit of Andrew Jackson. A full three-fourths of the address was an arraignment of the Adams Administration in specific areas like the abortive Panama Conference, its policy on internal improvements and on tariffs exclusively for the protection of manufacturers. Ideologically, he accused the National Republicans of being Federalists in disguise bent on extinguishing states' rights and threatening individual liberties. Van Buren devoted only about 250 words to himself, being careful to sketch briefly his patriotic record during the War of 1812, his backing of the Erie Canal and his support of New York's interests in the United States Senate. The address was a typical Van Buren production, rather prolix, matter of fact in content, cautious in skirting potentially dangerous issues and modestly offering the candidates for governor and lieutenant governor on their records.

Nothing disturbed the orderly procession of events. Walter Bowne placed Van Buren and Throop in nomination with a minimum of oratory to the accompaniment of the cheering delegates who approved the ticket and the address by acclamation. The next day, before many of the delegates had reached their homes, Van Buren wrote his letter of acceptance, declaring that even though he was "entirely satisfied with my present situation in public service" he regarded it as "a duty to acquiesce" in the decision of his fellow citizens.[29]

Van Buren expected to remain out of public view during the campaign, but inevitably he was drawn into the fray. When he returned from his western trip, he found among his accumulated mail a letter from William B. Lewis. The Nashville Committee was fearful that a combination between the anti-Masons and the National Republicans would give most of New York's thirty-six electoral votes to Adams. Lewis wondered whether the legislature could meet in special session and repeal the electoral law. Van Buren was not a little irritated at what he took to be a foolish and unwarranted incursion into the internal affairs of New York politics where, if anyone was the acknowledged master, it was he. He scribbled his decided opposition to any such blatant maneuver on the letter and filed it among his papers.

There it would have remained had not someone given a garbled version of it to the editors of the *Daily National Intelligencer,* declaring that Van Buren was scheming to do just what Lewis had suggested. Gales and Seaton eagerly grasped the charge and began a series of editorial attacks on him and on Ritchie, "the master spirits" of a corrupt alliance that would destroy for personal gain the manufacturing interests of the country. Van Buren may not have noticed the *Intelligencer*'s abusive remarks; after all, it had been attacking him almost constantly for the past year. But this time the Adams press throughout the country took up the conspiracy charge with such a vengeance that it began to make an impression on public opinion. Ritchie's *Enquirer* and Croswell's *Argus* not only denied the charges explicitly but accused Gales and Seaton of gross libel.

The party's New York City press corps took up the challenge. Van Buren refrained from any comment until Richard Mentor Johnson, one of the Senators from Kentucky, a fervent Jackson supporter, wrote him in alarm and in some bewilderment. Johnson added that the Administration press in Kentucky and other western states had accepted the *Intelligencer*'s comments as truth and were injuring the Jackson compaign. Recognizing now the dangerous drift of the controversy, Van Buren wrote a stern denial in the form of a public letter not to the *Argus* or any paper connected with the Regency, but to the most prominent National Republican paper in New York, the Albany *Daily Advertizer.*[30]

Apart from his one public letter to the *Advertizer,* Van Buren maintained a discreet silence. Otherwise he was active and innovative as always. He had his lieutenants secure cartloads of cheap Jackson badges, broadsides, bucktails and other identifying marks for public distribution. Through Hamilton and others he raised a substantial war chest, primarily in New York City, in Albany and Hudson. He drafted Marcy and Flagg, both able writers, to assist Croswell with the *Argus.*

Had these extraordinary efforts not been made; had these novel campaign devices not been employed; had the Van Buren organization not strained itself to the utmost, identified and exploited local issues, Jackson would have gone down to defeat even with the anti-Masonic party in the field. Van Buren defeated Thompson by about 30,000 votes while Southwick polled well over 33,000 votes. Jackson's majority in the state was only about 5,000. He carried twenty of New York's thirty-six Districts, but in many of them the switch of less than a hundred votes would have spelled defeat. In New York City, Cambreleng and Verplanck managed to hold their seats, but for days it was not known whether Silas Wright would be returned to Congress. He finally won by the narrowest of margins, 45 votes.[31]

« CHAPTER 13 »

Governor Van Buren

Narrow as it was, it was victory all the same. Jackson received 18 electoral votes to Adams's 16, and since he had a majority the two at-large electors brought his vote to 20. The Jacksonians swept the house, electing 81 members to 33 for the National Republicans and 13 for the anti-Masons. The senate remained safely in Regency hands.[1] Disgruntled that Jackson had not taken more of New York's electoral votes, Lewis had the effrontery to scold Hamilton in a letter he knew would be passed on to Van Buren. Obliquely, he criticized the Regency for involving the presidential with the state election. He did, of course, take due note of the anti-Masonic vote as well he might have. Had the anti-Masons cooperated wholly with the National Republicans, Jackson's electoral vote in New York would have been halved. This kind of political arithmetic was not lost on Calhoun, whose many supporters from the South and the West were jockeying for position in the new Administration and who recognized Van Buren as the only northern man with the ability and the prominence to thwart their ambitions. Lewis's carping note reflected some of these fears. Others saw that the Nashville Committee had this report fresh from New York on Van Buren: "A mere politician not a statesman and has been too long regarded by the people of this state as a political jobber to retain either their confidence or esteem."[2]

But Van Buren was not unduly alarmed at the information that had come in from John Forsyth of Georgia, his friend Balch, and when Congress began its December session, from Verplanck, Cambreleng and Wright. In fact, Lewis's complaining letter to Hamilton admitted "a strong probability that the State Department will be offered to our friend Van Buren."[3] "Don't trouble yourself about the intrigues in Washington," he wrote Cambreleng, "but do me the favor to let me have as much of the gossip as your leisure will admit. . . . You might," he added jocularly, "as well turn the current of the

Niagara with a ladies fan as to prevent scheming and intriguing."[4]

Van Buren was simply too busy with a multitude of personal, political and public affairs that pressed upon him to worry about Nashville or Calhoun or Washington. As governor, he had to have a much larger house than the one that had been his home for so many years. New York expected its governors to live in a style that befit the Empire State, a pattern that had been set for the past twenty years. As chief executive and as party leader, he would have to entertain members of the legislature and other state officers at official functions and at more discreet meetings when he discussed policies, patronage and party organization. He needed a much larger household staff to handle these activities and minister to his own personal needs, those of his sister and his two younger sons who were still at home.

Van Buren believed in living well, if not expansively. Memories of the cramped quarters he knew as a child in Kinderhook were repellent to him. And he felt the politicians and the electorate would respect him the more if he acted the gentleman even if not to the manor born; his slight figure and undistinguished appearance would be lost in the throng unless he called attention to himself in his apparel, his studied courtesy, his readiness with a touch of wit, or an appropriate compliment, and an ease of manner but not of familiarity, except with his closest friends. Likewise, his residence had to reflect all these personal traits as well as serve the more practical functions of his new office. Since he might well be called to Washington to serve in the new Administration, he wanted a furnished home that he could rent on a monthly basis. Except for essential items, he would store his own household effects.

A few blocks from the capitol, at 92 State Street, stood an imposing mansion known as the Stevenson House. Built in 1780 by the wealthy Albany merchant John Stevenson, it was a three-story brick structure in the Georgian style with a gambrel roof. Van Buren made suitable arrangements with the owner and during December 1828 spent some of his precious time supervising the moving, engaging servants and looking after the education of Smith and Martin, Jr.[5] While he was settling into his new residence and beginning to draft his inaugural address, he had to donate much of his attention to pressing party demands.

The steps and hallways of 92 State Street were thronged with visitors, each one of whom claimed at least a few minutes of Van Buren's time even if it were some well wisher like a Bucktail farmer from a far distant county who had come to see the sights of Albany after the fall harvest was over. But certain hours each day were set aside for politics and making policy. Members of the Regency, though occupied with their own affairs, helped with the scheduling and, of course, the counseling. Still the days were long and chaotic and trying for the Governor-elect who had no appointments secretary or clerk save those in his own law office, all of whom were busy with work and study of their own. Van Buren's systematic mind, his amazingly strong physical constitution, his capacity for hard work, long hours at the desk without flagging, stood him in good stead.

Among the many problems that must be solved, appointments were the most perplexing and demanded the most skill. Long before he had decided to run for governor, Van Buren learned that Samuel Talcott was an alcoholic and would have to be replaced as attorney general. "You must be Attorney General," he wrote Butler. Until now Butler had lent his valuable counsel on political matters to the Regency but had steadfastly refused to follow a political career. His single term in the legislature where he brilliantly captained Van Buren's forces had shown him how ugly and self-seeking politics could be. An intensely moral man, Butler had no use for the cut and thrust of the political arena. All of this Van Buren knew, and he respected Butler the more for it. Such a man was sorely needed in public office. "Is it possible," he asked his diffident junior law partner, "to be anything in this country without being a politician . . .?" To Van Buren's relief, Butler reluctantly agreed to the appointment.

The next appointment seemed far more complicated—his own replacement in the United States Senate. At first, when it appeared that Wright had lost his congressional seat, Van Buren was insistent that he be elected to the Senate. "I fear," he said, " our people will put him at once into the Comptroller's office if that should become vacant. This ought not to be done. His post is a different and a higher one." Wright won his election, but Van Buren continued to press him for the Senate, a position he felt neither prepared nor willing to accept. He saw Van Buren in New York and urged him "plainly and frankly" not to consider him. Should he be ramrodded through the legislature, it would arouse all sorts of personal rancors that would injure party harmony. Fortunately for Wright's peace of mind, he did not have to make the painful decision.

Van Buren began to waver shortly after he spoke with Wright. Erastus Root, ex-Governor Yates, and several other prominent members of the party descended upon him, each citing his service, each making it plain that the organization would suffer if he were not appointed. While still preferring Wright, Van Buren acquainted him with the situation in Albany, especially mentioning Root's insistent demands. "Now I see that much trouble is to come upon our friends at Albany in making this appointment and I very much fear, if a disastrous result follows, some will feel that I am chargeable for it," said Wright.[6] Nor was the election of a United States Senator the only office that was arousing fierce competition.

William L. Marcy had served three terms as comptroller, the customary tenure for that important office. Flagg had his eye upon it and so did several other Regency figures. Even Wright hinted that he would not be averse. But where would Marcy go? Van Buren had moved him out of the governor's race and was not considering him for the Senate. Marcy was a valuable man and a loyal member of the Regency. He was, moreover, the son-in-law of that redoubtable elder statesman of the Regency, the rich and powerful Benjamin Knower, Van Buren's new neighbor on State Street. Surely the Governor-elect would arouse resentment in that quarter if he passed over Marcy's claims in favor of Root or Wright. Yet this is just what Van Buren was planning to do until he was rescued from what would have been a serious error by an intrigue among his close friends. That the alleged "great intriguer," should have been caught napping at his presumed game, was amazing to those of his circle who wrongly believed Van Buren never made a mistake in politics. Yet the success of the intrigue saved Van Buren from embarrassment, even if it did pique him and thus, oddly enough, add to "the magician" myth that was growing about him.

A sharp-eyed lawyer from Oneida, Greene C. Bronson, engineered the coup right under Van Buren's nose and in such a way that the Governor-elect's hands were tied. Bronson, a former Clintonian, but for the past six years a warm supporter of Van Buren, wanted the attorney general's office. He bargained with John Suydam and with others of Van Buren's personal and political friends who were opposed to the contenders for the vacant seat in the U.S. Senate. In return for their support, Bronson would use his influence in the legislature to back any candidate they put forward. Their candidate was one of Van Buren's oldest friends, Charles E. Dudley, a former mayor of Albany and former state senator but not conspicuous for his political or intellectual talents. A shy man in public, no one had ever heard him make a speech, either in the legislature or at political meetings.

When Bronson put together his coalition, John Suydam requested that Van Buren not interfere in the contest for attorney general. Suydam, a friend of twenty years' standing and also a power in the party, spoke for a legislature which, to the discerning Van Buren, gave unmistakable signs of becoming a cockpit of personal factions before it convened. Always the realist, he agreed to remain neutral if a contest developed. Butler, sensing opposition to his candidacy and no doubt privately relieved that Van Buren would not force the issue, readily agreed with his chief. Accordingly, when the legislature met, Bronson was elected attorney general. Then to Van Buren's astonishment, as he explained in a letter to Jesse Hoyt, "Bronson's friends had the address to push Dudley into the Senate."

After reading a seven-page letter from Silas Wright setting forth lucidly the reasons why he should not be considered a candidate, Van Buren had decided to back Marcy. But when Dudley overwhelmed Marcy on the first ballot in both houses, it was a heavy blow to Marcy's prestige and to his pride. Fortunately, there was a vacancy on the supreme court. As Van Buren said, "Marcy was so situated that I must make him a judge or ruin him." The latter alternative was unthinkable. For some years he had been one of the linchpins that held the Regency together. On occasion he had irked Van Buren with his independent course and his long, complaining letters, but on the whole he was a trusted counselor whose services were indispensable. Marcy was nominated to the court and readily confirmed. Although he could not engage in politics while on the bench, he was always available for advice.

Flagg was named secretary of state. Wright, who became comptroller, resigned his seat in Congress and took up his new position in Albany.[7] Meanwhile, Van Buren was wrestling with his annual message, accepting help from Butler, Marcy, Flagg, Croswell and Cambreleng, who reworked some of his awkward syntax and offered suggestions. "Don't let your politeness induce you to make a few immaterial remarks," he wrote Butler, "and then say it is so good that you can't make it better. I know better than that." Van Buren was determined to deliver a message superior to any of those much admired state papers of his illustrious predecessor and rival. Van Buren was painfully conscious that many people in high places regarded him as a mere political manipulator, an acute lawyer but no match for Clinton as a public man. He had made a start in brightening the statesman image he craved with his speeches in the Senate, but the real test he felt was in this address, his first as chief executive of the most populous, most powerful state in the Union. He had identified several problems he wished to consider—the banks of the state, the auction monopoly in New York City, internal improvements, a new election law that would abolish the district system and substitute a general ticket for presidential electors, legislation that would limit expenditures in elections. Having had a hand himself in purchasing political advantages, he recognized the evils that were developing and hoped to check them. Though he realized that on his past record he could and would be accused of hypocrisy, he also knew that popular attitudes were short-lived; that all parties and factions had engaged in disreputable practices; that it was time for a change and he wanted to be on the right side of a needful reform.

Van Buren's proverbial luck was with him. As he pondered various reforms in banking, consulted with Cambreleng, Knower and Thomas Olcott, the cashier of the Farmers' and Mechanics' Bank of Albany, all of whom were far more knowledgeable than he in the practices of trade, of credit and debt, he received a visit from a prosperous bustling Yankee who had made a fortune in land speculation. His name was Joshua Forman, now retired from active business and living in New York City.

Strangely enough, the two men had never met, even though Forman had been one of the more prominent figures in western New York. At one time he owned all of the land on which the town of Syracuse was situated. An ardent promoter of the Erie Canal, he saw his profits in land and other investments soar as the Canal enriched his holdings. He had come to the Governor-elect in early December with a plan for establishing a state-supervised banking system that he was certain would not just strengthen all banks and thus protect depositors and businessmen, but also provide a stable state currency.

Van Buren, who had been a director of the Bank of Hudson and was currently a director of the Farmers' and Mechanics' Bank, knew enough about banking practices to grasp the basics of Forman's plan. It was nothing less than a state-imposed system in which the banks remained independent yet at the same time supported each other through contributions to a common fund that any member bank could draw upon if subjected to extraordinary pressure from its depositors—a local crop failure, for instance, a maritime disaster or simply the fluctuations of the economy. Forman had anticipated many of the objections of the big city banks being forced to underwrite the credit of country banks where management was considered either self-serving or inept. A state board of bank commissioners, similiar to the Canal Board, would be appointed and charged with annual inspection of all member banks. Along with the law creating such a commission, Forman suggested that the state impose a reasonable limit or ratio of note issue to the specie deposits held in their vaults. Should the state adopt the system, which he dubbed the Safety Fund, it would eventually compel all banks to join as a condition of rechartering.

Van Buren was impressed. There were at present forty banks doing business in the state, thirty-one of which would come up for rechartering during the next three years, most in the second and third year. Thus, a small number of banks would be affected by the proposed Safety Fund, which appealed to his cautious nature because if enacted it would provide a clear test of its feasibility without endangering the economy of the state. Van Buren also liked the system of modest control in the public interest while the banks themselves would remain entirely free of restraint save that imposed by mutual benefit and prudent management. He suggested Forman consult with Thomas Olcott and with George Newbold, a prominent banker in New York City, whose judgment in banking affairs he respected.

Olcott was a small person, whose heavy eyebrows, his only distinguishing feature, nearly matched those of Marcy. If undistinguished in figure, he was not so in his grasp of business affairs and of banking procedures. Indeed, he was perhaps the most astute banker in a state where banking was fast becoming a major source of wealth. Olcott instantly grasped the crux of Forman's plan, contributed several technical improvements that he thought would make it more attractive to the big city bankers and bestowed upon it his hearty blessing. Forman then consulted with Knower, who was equally impressed, before returning to New York City and facing what he knew would be the hardest test, the support of the big banks. With Van Buren and Olcott behind him, Forman sought out Newbold and presented the Safety Fund as a choice of evils.

Clearly, the present situation, where scores of lobbyists for bank charters annually crowded the capitol and exerted corrupt means to achieve their ends, could no longer be tolerated. The incredibly rapid economic growth of the state was reflected in the mounting pressure for more banks. Some means had to be devised to protect the public and to reduce corruption in Albany. The Safety Fund, he maintained, would provide the means yet leave the banks free to regulate themselves within rational guidelines. After speaking with several other bankers, Newbold gave the plan his cautious approval "as altogether the best he had heard offered." Van Buren hesitated no longer but inserted it in his message and recommended it without reservation.[8]

The Safety Fund was the single most important part of Van Buren's message, but it also included a call for reform in elections and spoke out strongly on the influence of corrupt politicians in the appointment of auctioneers in New York City. Much of the foreign goods that came into the port of New York after release from the customs office would then be forwarded to the auctioneers who sold them to merchants pocketing rich fees for their services. Van Buren, denouncing the procedure as a source of graft and of monopoly, recommended that the state pass a general law under which anyone who had sufficient capital as the legislature determined could be a licensed auctioneer. As a whole, the message breathed a spirit of individualism and of reform buttressed with concrete recommendations.

Only on the subject of internal improvements was Van Buren ambiguous. All one could infer was that new projects be curtailed until the debt incurred for the Erie Canal and Champlain Canal was paid off. Citizens of Chenango, Broome and adjoining counties along the southern tier, who had taken his promises literally and voted for him, were enraged at what they promptly dubbed "noncommittalism." An angry resident of Chenango called him a "Proteus of many shapes" and his treatment of internal improvements, "a honeyed production and intended to please all men, the friends of internal improvements and the severe economists of the public treasures opposed to lay out one cent in opening facilities to those parts of the state which are now shut out by distance and bad roads from market."

Outside the southern tier most of Van Buren's contemporaries praised the message as the equal of any Clinton had written. Such an old Clintonian as Jabez Hammond said, "It is among the best, if not the best, executive message ever communicated to the Legislature of this State."[9] With able state officers assisting him, and those of the Regency like Butler, Knower and Croswell who held no official position always available for consultation, with loyal, though not outstanding leadership in the legislature, Van Buren's program began to be implemented with unprecedented speed and a minimum of contentious debate.

Van Buren could now turn his attention to the unfolding events in Washington. On the surface, everything seemed calm, almost tranquil. He, of course, had many warnings, if he needed them, of Calhoun's consuming ambition for the Presidency. His friend Thomas Cooper had described Calhoun's political anatomy with precision. John Forsyth, a Georgia Congressman and a close associate, reminded Van Buren of "my prophecy if Calhoun goes into the V. P'cy. as the Republican candidate at the next election he will be the southern candidate four years thereafter for the Presidency and will have the support of Virginia and Pennsylvania." Yet Van Buren took time out of a busy schedule to write Calhoun when he decided to run for governor and received a most cordial reply. In accents similar to those of Ritchie, he regretted that Van Buren would be in Albany when "in the present state of things, it seems to me desirable that there should be as much experience and talents as possible in the service of the nation." Calhoun's concern for his welfare did not deceive him one bit as he placed the letter in his file for future reference.

New York had not yet made a nomination for Vice-President, but as Calhoun knew, it would give him the Jacksonian electoral vote in the state. Still the calculating South Carolinian must have felt some uneasiness about the absence of his name on its ticket. As the second session of the Twentieth Congress convened, Edward Everett, a thoughtful outsider serving his second term as a National Republican Congressman from Massachusetts, detected three distinct factions in Washington, Calhoun's, Van Buren's and that of Adams's postmaster general, John McLean of Ohio, who had declared his neutrality but maintained his position through his control of post office patronage.

The Vice-President, he observed, "has been working South Carolina into a frenzy on the tariff question for the purpose of repopularizing himself in the southern Crawford states, and making it impossible for Van B. to get any strength there. Van B. is, of course, stronger in all the tariff states." William C. Rives disagreed, "The last summer's proceedings in S.C. have utterly destroyed Mr. Calhoun's prospects." Analyzing the gossip in the lobby of the Capitol, Everett thought Van Buren would be named secretary of state, but others were mentioned, in particular Edward Livingston and Littleton W. Tazewell, one of Virginia's Senators and a member of Ritchie's Richmond Junto.

By mid-December, while Van Buren was grappling with bankers and auctioneers and anti-Masons in New York, the conventional wisdom in Washington was that Jackson would serve one term and Calhoun would succeed him. Messages of alarm from Wright, Verplanck, Cambreleng and even John Randolph of Roanoke came into 92 State Street, Albany. "There are not a few men who professing great esteem for VB are anxious he should stay where he is," wrote Verplanck to Jesse Hoyt, who passed the letter on to Van Buren. The ever watchful Alfred Balch, close to Jackson's most intimate advisers, had been reporting from Nashville on Calhoun's efforts to undermine Van Buren with the President-elect. "We are also ready," he said. "The former friends of Crawford here neither few in numbers nor weak in talents have an acc't. to settle with Mr. Calhoun which must be settled."[10]

Van Buren remained ostensibly imperturbable. Although he could have wished a more resounding triumph for Jackson in New York, he had been responsible for the Democratic victory there. He was the undisputed leader of the majority party in New York and the chief executive of a state which had to be recognized for the most important positions in the cabinet, State and Treasury, the one a traditional stepping stone to the Presidency, the other commanding impressive patronage. Early on, he advised his friends in Washington to calm themselves; yet he indicated to Cambreleng on December 17 that he would not accept the Treasury, if offered. He made no comment when Silas Wright, obviously seeking direction, wrote that Duff Green was a spokesman for the Vice-President, but "it seems to me we cannot avoid making Duff printer without bringing on a quarrel now which I think should be delayed at all events for the present."

He had received no direct word from Jackson since the election, not even any response to his congratulatory letter of November 14, though he had enough information from other sources that he was virtually certain to be named secretary of state. In a fatherly way, he advised his unofficial aide James A. Hamilton not to be concerned about the absence of any direct word from Tennessee. "I would not write to Lewis," he said. "Let them worry and fret and intrigue at Washington. Six weeks hence they will find themselves as wise as they were when they began. If our friend Jackson wants admonishing and advising upon the point, it would in the end be better for me that he erred in the beginning."

In early January Hamilton received another letter from William B. Lewis that made it plain Van Buren would be offered the State Department.[11] Any apprehensions he may have felt about the appointment he now cast from his mind. He would be going to Washington and he would play the major role in the new Administration. Hamilton, as he had in the past, would act as his link with Jackson and the Nashville Committee. With his program being smoothly guided through the legislature, he began to yearn for the excitement of Washington. Social life in Albany, once so exhilarating, was now becoming stale and tedious. The routine of the governor's office he found burdensome; even state politics had lost much of its savor. "You have not the slightest idea of the pressures on me for favors and commissions. I have not a moment's leisure," he wrote Cambreleng. Repeatedly and plaintively he asked his friends for the gossip, not the political rumors and intrigues that were sweeping the Capitol, but the talk of the social world of Washington. "I should be ashamed if I did not," he confessed to Rives, "for these smaller concerns are among the real comforts of life." Cambreleng and Rives did supply Van Buren with the morsels he craved in the jocular style he found amusing. "Poor Eaton," wrote Cambreleng, "is to be married tonight to Mrs. T! There is a vulgar saying of some vulgar man, I believe Swift, on such unions—about using a certain household . . . and then putting it on one's head." Cambreleng may have been frivolous in his correspondence, as Van Buren desired, but he was all business when it came to cabinet making.[12]

Since Hamilton's visit to Nashville the previous year, it was understood that he was Van Buren's informal but accredited agent, even as Major Lewis was acting for Jackson outside the Tennessee circle. Van Buren was never the man to presume on Jackson, and he was too conscious of his position to initiate any correspondence with Jackson's lieutenant, though he responded when his advice was sought. As Jackson remained aloof in any personal exchange between the two but acted through Lewis, so Van Buren adopted the same stance, relying upon Hamilton to maintain communications. The urbane Hamilton had made a good impression on Jackson and particularly upon Rachel Jackson, a most important consideration. Except for one or two letters from Lewis before the election, all correspondence was addressed to Hamilton. And following Jackson's example, Van Buren made Hamilton a member of his official family with the rank of colonel in the state militia.

Hamilton's position had been understood for some time in Tennessee and in Washington. The New Yorkers also credited him, and rightly so, as having more influence with Jackson than any of themselves. Wright had never met "the Hero"; he had only a passing acquaintance with Senator John H. Eaton, Jackson's longtime friend and spokesman in Washington. Cambreleng knew Eaton well, but thought him a fool, while he scarcely knew Jackson and had never met any of the Nashville Committee. He was on familiar terms with Thomas Hart Benton, the burly Senator from Missouri, many of whose tastes in dress, in food, wine and good conversation he shared. Benton had known Jackson for years, though not always on friendly terms. Years before, a frontier brawling incident, punctuated by a shooting and stabbing affray nearly cost both of them their lives. Since the early twenties Benton, however, had been a devoted adherent, and one should add, a close friend of Van Buren. Yet he was as much in the dark as anyone in Washington about Jackson's plans. The New Yorkers on the scene were a closed group when it came to the fortunes of their chief. They did not take Benton into their confidence; nor, in fact, did he intrude himself either with them or with Van Buren personally. He had intimated that his services were available in his letter of congratulations to the Governor, but receiving no reply, he contented himself with touting Van Buren for the senior position in the new cabinet.[13]

Thus Cambreleng and other members of the New York delegation looked to Hamilton as the one person completely conversant with Van Buren's views on how to deal with what they perceived as a Calhoun move to dominate the new Administration before it fairly began. The letters that went forth to Hamilton throughout early January began to take on panic proportions as they urged him to come to Washington, presumably bearing the ideas and instructions of the Chief.

Hamilton would not move until Van Buren thought it proper, and he had no word from Albany until February 2, 1829. Van Buren had received a letter from James Hamilton, Jr., that spurred him to action. The South Carolinian submitted what he called a cabinet "project." Van Buren was to be secretary of state; Langdon Cheves, one of the best financial minds in the country, or Louis McLane, also considered most able and, as Hamilton knew, a close friend of Van Buren, for the Treasury Department. He also suggested McLane for navy and John McLean for war secretary; Samuel Ingham of Pennsylvania as postmaster general. For attorney general, Hamilton had three candidates; William T. Barry of Kentucky, ex-Speaker of the House Philip P. Barbour of Virginia, and Levi Woodbury of New Hampshire. As Van Buren cast his eye over the list, he was instantly alert to a possible power grab by the Vice-President. Of particular concern were the two posts with the most patronage—the Treasury and the Post Office.

Van Buren, too, respected Cheves's acumen, but even though he had moved to Pennsylvania, he was a South Carolinian, born and bred. A former president of the Bank of the United States, his political philosophy was not above suspicion. The most puzzling aspect, however, was that Cheves had not been active in politics for twenty years. Where and how would he distribute the patronage? Surely not in any consistent manner as Van Buren understood it, unless he was already enlisted in another cause, unless he was Calhoun's man.

Van Buren was more receptive to McLane, whom he knew was too independent to be anybody's cat's-paw, a personal friend, a fellow Radical and also a man of financial talent. Van Buren had a week or so earlier referred McLane's name to Cambreleng as a fit candidate for attorney general, a place he thought more befitting his status in the party; but the more he thought about McLane in the Treasury, the more he thought it would work out. Ingham he knew to be a creature of Calhoun, though apparently he acquiesced in giving up the Post Office patronage to his rival. Van Buren enclosed the South Carolinians's letter to James A. Hamilton with the query, "When do you go to Washington?" Hamilton went to Albany where Van Buren briefed him on his mission, and then left for the capital, arriving on the same day as the President-elect, February 12, 1829.[14]

Jackson made his appearance in Washington as unobtrusively as possible to avoid the place seekers, the politicians and the well wishers. He went straight to Gadsby's Hotel, which would be his residence and his headquarters until he moved into the White House.

As soon as Hamilton had settled himself at Mrs. Peyton's boarding house, he called upon the President-elect. When he was ushered into the presence, Jackson took both of his hands in his and in an excess of emotion or sheer weariness stood looking at Hamilton without saying a word, before he inquired about his family, and then reminded him that the last time they had seen each other his beloved Rachel was at his side. They reminisced briefly about the celebration in New Orleans the previous year. Jackson asked about Van Buren and made a few complimentary remarks about him before Hamilton took his leave.

At Mrs. Peyton's, Elias Kane, a New Yorker who had moved West and was now a Senator from Illinois, brought him up to date on the current crop of rumors. Then Major Lewis arrived with an invitation to attend a family dinner with Jackson at his suite of rooms in Gadsby's. Lewis had a private chat with Hamilton before he left, his primary objective to learn whether Van Buren would accept the State Department if offered. Hamilton thought he would and as he explained to Van Buren, "further than this (I hope it is not too far) I have not gone in your name."[15]

Hamilton hastened to Gadsby's at the appointed hour. After dinner, Jackson's personal entourage withdrew discreetly, leaving the two men alone. At the moment Jackson was preoccupied with what appeared to be an effort on the part of the Richmond Junto to place Littleton Tazewell in the State Department. Tazewell, the heavy, slouching Senator from Virginia, who was almost as eccentric as Randolph, seemed to think the honor of the Old Dominion was at stake if it and he were not accorded first place in the cabinet.[16]

Jackson was obviously distressed at the thought of excluding Virginia from the cabinet. He canvassed this subject freely, finally declaring he would offer Tazewell the War Department. Hamilton, who had reason to believe Tazewell would decline, instead of remaining silent or stating he had no opinion on the subject, made the mistake of agreeing. Jackson incorrectly assumed he spoke for Van Buren. Although Lewis had informed him of his private talk with Hamilton, Jackson seemed to need reassurance that Van Buren would accept the State Department; for he brought the subject up several times during their discussion. Finally, fixing his clear blue eyes on Hamilton, he asked him to write Van Buren and say the post would be offered.

The next day, the ubiquitous Lewis requested another interview, this time in Jackson's suite at Gadsby's which the newspapers were already calling "the wigwam." Lewis wanted his opinion on prospective cabinet appointments. This time Hamilton was more discreet than he had been with the President-elect. As he reported to Van Buren, "I answered his inquiries without reserve but did not in any measure authorize the opinion that you had confided with me,"

If at times he overestimated his influence, James A. Hamilton certainly kept Van Buren well informed. On February 13 he wrote him four letters, two describing his talks with Jackson, a third on what he had gleaned about the two cabinet posts of major interest to Van Buren. He made only a passing reference to Tazewell, though the

Virginia Senator was a member of Mrs. Peyton's mess and talked freely in her parlor.[17]

For Van Buren these were extremely trying days. Here he was trying to play a deep political game by remote control, depending primarily on his aide, with some assistance from Cambreleng, since he was unable to focus his entire attention on the complexities rapidly developing in Washington. His first duty after all was to his office and his state and the fulfillment of his ambitious legislative program. "I have never worked so much in six months as I have during the past thirty days," he complained. "My occupation is incessant and very irksome."

On February 14, 1829, Jackson wrote Van Buren inviting him to be his secretary of state, politely requesting an early answer and a hope that if he should accept, he could come to Washington as soon as he conveniently could manage so "that I may consult with you on many and various things purtaining [*sic*] to the general interest of the country." Jackson certainly had need of Van Buren's stature in the party, in the country at large and with the Congress. Apart from his informal talks with James A. Hamilton his only intimate adviser of capacity, experience and common sense was Hugh Lawson White, one of the Senators from Tennessee. Lewis and Donelson, with whom he confided too freely, were both neophytes in the jungle world of Washington politics. The other Tennessee Senator, John H. Eaton, was his closest personal friend on the scene and as such had Jackson's ear. But Eaton, a pleasant, pliable person who certainly was experienced in the intrigues of Washington, was too casual in his affairs to give sound advice. "Through his habitual carelessness about his letters," said Van Buren, "one runs the risk of having his highly confidential correspondence find its way into one of the Committee rooms folded up in a petition in behalf of some good fellow, who has no friend except the Major."[18] Harassed at the constant importunities from the rival cliques, worried about party harmony, sensitively alert to the slightest criticism, Jackson wanted and needed Van Buren's counsel.

On February 18, he entertained at dinner James A. Hamilton, Calhoun, Cambreleng, the Tennessee congressional delegation and Henry Baldwin, the candidate of "the Amalgamator" minority faction among the Pennsylvania Jacksonians. The Family faction, based in Philadelphia and temporarily in the ascendancy, was backing Ingham for the cabinet. From the table talk Baldwin gathered McLane and Ingham were the chief competitors for the Treasury, words that must have made him a rather solemn dinner companion. Jackson seemed preoccupied during the meal and as soon as the cloth was removed, took Hamilton aside, no doubt to Cambreleng's disgust, for a private chat. Computing the number of days it would take for a reply from Van Buren to his offer of the state department, the anxious General wondered whether it would be possible "for him [Van Buren] to be here for two days before the 4th." Hamilton replied in the negative and thought Van Buren "could not probably be here before the 10th or 15th."

Van Buren received Jackson's invitation on February 19, 1829. He replied promptly and affirmatively. In accepting, however, he explained he would need time to prepare the population of New York for his resignation, to finish up his legislative program, acquaint the lieutenant governor with the duties and the unfinished business of his office and wind up his domestic affairs. Nor could he resign the governorship before his confirmation by the Senate. For that would be presumptuous on his part and might well offend the dignity of that body. In short, he could not be in Washington until late March at the earliest. Meanwhile, he suggested that his aide, James A. Hamilton, carry on the duties of the office until he was able to take over. Jackson had apparently not considered Van Buren's obligations and had assumed he would be in Washington by March 4. He was irritated at the delay and not a little baffled on how he should proceed. But he also understood Van Buren's situation and agreed to have Hamilton act as an interim secretary.[19]

Well before then, official and social Washington had the cabinet lists and save for Van Buren, and McLean, an Adams holdover, the members were considered undistinguished and generally partial to Calhoun. Senator Kane was so disgusted with what he considered its mediocre character that he urgently advised Van Buren not to be associated with it. Another letter from Hamilton informed him that Louis McLane was railing against the cabinet, which was natural considering his disappointment at not being chosen.

Jackson, fed up with the constant importunities, had Duff Green publish the list in the *Telegraph* on February 26. There would be one more change, however, a change that must have heartened Van Buren. John McLean resigned as postmaster general to accept a nomination to the Supreme Court. William T. Barry, an amiable Kentuckian who was not identified with any faction, was appointed in his place. Van Buren, ever the realist and not as concerned with his self-image as some of his friends thought he ought to be, was not distressed at being associated with an undistinguished cabinet. Eaton, in the War Department, was clearly a personal choice of the General, one of his closest friends. That he was likeable and shared his position on the tariff was quite sufficient to recommend him as a colleague. John Branch of North Carolina, whom Jackson named for Navy, and John M. Berrien of Georgia for attorney general, he had known as supporters of Crawford. Branch was careless and indolent, Berrien fussy and pompous, but as far as Van Buren knew both were not in Calhoun's interest. If he had any worries on that score they were that both were southerners and both were malleable. Should sectional tendencies become more sharply defined over the tariff and other matters of policy, Calhoun could draw them into his orbit. At the moment these fears could have had little or no substance.

As Cambreleng wrote him on March 1, "this grumbling and letter writing has made an impression on the General not calculated to increase the influence of Virginia and South Carolina." Cambreleng, in fact, made a strong argument for mediocrity. If the Treasury were occupied by the immovable pertinacity of Cheves, and Navy by the vanity and eccentricity of Tazewell, "You would have had all leaders and no wheelhorses and the first hill you reached would have upset you all." Ritchie said much the same thing.[20]

Van Buren's major concern at this point was not the caliber of his cabinet colleagues or even his suspicion of Calhoun's motives; he wanted to wind up his governorship as rapidly as possible and hurry to Washington. On the whole, he had been reasonably satisfied with Hamilton as a surrogate, but his aide could only go so far without incurring enmity, jealousy and suspicion that could be damaging.

Already Hamilton had ruffled the feathers of Cambreleng and Hoffman, two of Van Buren's most trusted advisers. Hoffman, describing him as "a Federalist of the Old School and unknown to Republicans as a partisan, . . . [he] has come out here and taken charge of the State Department." Virtually accusing Van Buren of dodging responsibility, he said, "I do not think the General has any system in these matters because every one seems to get something out of him." As if this blast were not enough, Van Buren received probably at the same time a coldly curt letter from James Hamilton, Jr., whose tone verged on the insulting when he accused him of hiding in Albany. "Depend upon it," he wrote, "there is such a thing as pushing prudence too far."[21]

Yet as vexing as these importunate demands were, the interests of the state came before all. He would not leave until he was certain that his program was completed. The big banks of New York City, after first accepting the Safety Fund plan, had reneged and sought to kill it with amendments. Van Buren and his lieutenants successfully fought them off, but not before much intricate maneuvering in the legislature. The auctioneers, too, lashed back, requiring the Governor to divert energy and attention before he stemmed their assault on his general incorporation bill. Even the electoral reform attracted partisan attack and delaying actions from the National Republicans. And then the legislature came very near passing a resolution instructing New York's Senators to offer a bill that would abolish slavery in the District of Columbia. Van Buren headed off this explosive issue, too. While he was immersed in these last-minute problems, he heard from Hamilton that Jackson had named Tazewell as minister to England. He had not been consulted, and he did not believe the man was well suited to this important post.

While engaged in his executive duties, advising through Hamilton on cabinet matters, he had also borrowed whatever time remained to study and make notes on the current difficulties that existed between the United States and Europe. It was all so disheartening. Perhaps James Hamilton, Jr., had been right after all when he told Cambreleng that Van Buren would cut his own throat if he became secretary of state.[22]

On March 12, 1829, Van Buren sent in his resignation. By now, of course, the word that he would be going to Washington had penetrated even the most remote hamlet of the state. Public opinion was now prepared for the move without risking a wave of criticism that he was a blatant opportunist playing fast and loose with their suffrage for personal preferrment. The *New York American* and other National Republican papers had been exploiting this theme since definitive news of the make-up of the cabinet reached the state on March 1. His judgment had been correct. Except for the usual partisan comments, the legislature and a majority of the electorate seemed proud of Van Buren's elevation as a mark of respect to the state, rather than otherwise. Complimentary resolutions were passed, invitations to public dinners flooded 92 State Street (all declined with appropriate remarks), even political enemies came to call with their congratulations; his landlord made no difficulties about the termination of his rental agreement.

Van Buren was fortunate to have his son John, who had graduated from Yale and was now studying law in Albany, available to inventory, pack up and ship whatever furniture might be needed in Washington. Butler took on unfinished law business and looked after Van Buren's investments. His sister Maria would return to Kinderhook and live on a farm Van Buren had purchased years ago at a sheriff's sale—near his brother Lawrence.

On March 17, 1829, Van Buren, his sister, two youngest sons, and Benjamin Butler, left Albany for Kinderhook where he would spend one day seeing friends and relatives, and then on to Washington. He had left the state and the party organization in good shape and in good hands; his reform program, the equal of any Clinton had sponsored, would have a lasting beneficial effect.[23] Van Buren had served as governor exactly forty-three days, the shortest term ever in the history of New York.[24]

« CHAPTER 14 »

All the Coppers in the Mint

Van Buren was not in the best of spirits when he left Kinderhook for New York City. From what he had read in the newspapers and in his letters from Washington, the new administration seemed to lack direction. His brief tenure as governor with its ambitious program had left him physically weak and emotionally drained. The winter weather had been unusually severe, low temperatures, sudden thaws, sleet storms interspersed with heavy snowfalls, made traveling more difficult and unpleasant than usual.[1]

New York City was depressing. Extreme cold still lingered, along with unemployment even among skilled tradesmen. Many a family could not afford enough wood or coal to maintain a minimum of comfort. In the poorer areas, incidents were reported of death from exposure. The best efforts of richer inhabitants, the mayor and the common council to provide food and fuel for the desperate were insufficient.

Social unrest expressed itself in political unrest. Tammany Hall, as usual, was riven with factionalism of a sharper edge than ever before. Frances Wright and Robert Dale Owen, the British utopians, were inspiring the younger and liberal-minded men with their radical rhetoric and infuriating the older, more conservative groups of all political persuasions. Many of the city's mechanics, reacting against bad times and what appeared to be growing social inequalities, were showing distinct signs of concerted political action.

Van Buren was too worn out to explore the labyrinth of New York City politics. Nor did he have enough spare time for the kind of thorough investigation of all the conflicting interests and issues that he always made before committing himself or even advising a plan of action. But he could not escape the flood of office seekers that coursed around his coach and crowded into the barroom of the City Hotel, where he had taken rooms. He retired to his bedroom immediately, but he had the hotel clerk announce that he would accept cards and letters. After skimming through them, he decided he would have to address the multitude, which he did that evening. After this wearing incident, during which he confessed he could promise nothing but would try to be of assistance, he managed to close the meeting and go to bed.[2] He had just fallen asleep when he was awakened by Levi Woodbury. Woodbury expressed deep dissatisfaction, saying Van Buren's adherents in Washington were confused and upset by the cabinet and other executive appointments. Van Buren listened carefully to this uncharacteristic litany of complaints from the New Hampshire Senator; for Woodbury was a careful, unemotional person, tightlipped about political matters. Always the evaluator, the cool-tempered judge of men's motives, Van Buren attributed Woodbury's unique outburst to his chagrin at not being offered a cabinet position. For the first time since the founding of the Republic, the six New England states were not represented in the cabinet. Yet personal motives apart, Van Buren found Woodbury's comments disturbing.

More bad news awaited him in Philadelphia, where he met Edward Livingston and his wife, both in a somber mood. The Livingstons made much of Eaton's and Lewis's influence over the President. John M. Berrien, the new attorney general, and John Branch, the navy secretary, were both considered pawns of the secretary of war. Van Buren learned at first hand how Peggy Eaton, the vivacious widow of a navy purser and the recent bride of the war secretary, was churning up Washington society, involving it with cabinet politics in strange and ludicrous ways. The worst news of all was when Livingston announced that he had been appointed minister to France and was disposed to accept if he could settle his affairs, which he admitted would take some time.[3] Tazewell to England and now Livingston to France, the two most important missions involving the most difficult problems the United States faced in its foreign policy, had been made without as much as a note from the President to his secretary of state.

Livingston's qualifications were beyond exception. Although then a Senator from Louisiana, he was the senior member of the aristocratic Livingston lords of the manor, a firm Republican, a cultivated gentleman who spoke French fluently, an honorable person who had assumed the financial responsibility of a defaulting clerk when he had been United States district attorney for New York and had paid back the entire sum with interest to the government out of the proceeds of his law practice in New Orleans, where he went after the scandal came to light. Not the least of his many accomplishments, Livingston was a brilliant lawyer and scholar of the law; the nation's leading authority on jurisprudence, he was one of the few Americans at that time with an international reputation.

Despite Livingston's eminence and his fitness for the post, Van Buren had doubts about his age and his resilience in the give and take of a diplomatic effort that had defied the best efforts of such a supple mind as Albert Gallatin.[4] The President must have let personal considerations govern this appointment. Van Buren knew that the two men had been friends for years. Whatever Livingston's great merits or Tazewell's political importance, was the Secretary of State to be a mere cipher for the imperious old General and his inexperienced palace guard from Tennessee? Adept at concealing his feelings, he sympathized with the Livingstons, who dwelt upon the baneful effects the Eaton affair would have on the French court. But his pride was deeply injured, his future clouded with misgivings, his reputation at stake. After all, he, more than anyone else, had put together the coalition that elected Jackson. He had served as Senator and spokesman for New York in Washington, had just resigned as governor of the richest and most populous state in the Union, and was still the leader of its dominant party.

There was more to come. When his steamboat from Philadelphia pulled into the quay at New Castle, Delaware, he saw as he expected the figure of Louis McLane, anxiously awaiting him. Although tired, he forced himself to walk with the excited Delaware Senator and listen patiently as he complained of being overlooked by Jackson and he voiced his disdain for the people around the President. McLane's harangue, for he became more emotional as he warmed to his topic, was most depressing to Van Buren and suggested more emphatically that he made a bad mistake. Resignation might be the only way to recover his diminishing political capital before it was all lost.[5]

Van Buren's mood was not improved when he reached Mrs. Miller's boarding house after a six-hour cold and jolting ride from Baltimore to find a crowd of office seekers and office holders awaiting him. Bone-weary, travel-stained as he was, he courteously but firmly told the crowd that he would listen to them for only one hour because it was evening and he had to pay his respects to the President. If any had brought letters with them that pertained to his department, they should leave them for his examination which, he assured them, would be prompt and careful.

It was quite late when Van Buren arrived at the White House. The mansion was quiet, a single lamp glowed in the vestibule through which Major Lewis ushered him to Jackson's office where he saw the sharp-featured man silhouetted against the light of a fire and a single candle. The President arose immediately and greeted him warmly.

He could not help but notice how Jackson had aged since he last saw him in 1824. More obviously careworn and depressed than his guest, Jackson seemed so pleased to see his secretary of state and to make him comfortable that Van Buren felt better almost at once. After a brief conversation before the fire, the President noticed how tired Van Buren was and suggested he return to his lodgings for a good night's sleep.[6] They would defer all business until the morrow.

Most persons of Van Buren's stature in public life would have demanded an explanation of the President for the Tazewell and Livingston appointments. Quite probably he would have gone on and appealed to Jackson's highly developed sense of the military chain of command, pointing out that he had improperly interfered with the authority of a trusted field commander before he presented his plan of operations in his assigned sector. Van Buren was the exception, perhaps the only exception, to this approach which would have placed a heavy strain on the relationship between the two men at the outset. As he had done all his political life, he would not take the initiative, would not open their conversations on a complaining note.

Jackson had either been poorly advised or had acted on impulse. He knew the President as a man of native intelligence and extreme sensitivity. Inferring, and quite properly, that Jackson read the Washington papers which, except for Duff Green's *Telegraph,* were critical of his appointments thus far, he would be on the defensive and would sooner or later bring up the subject. Then Van Buren planned to explain his reasons why he thought the appointments were not appropriate to the tasks demanded. He would do this without any reference to the President's interference and he would emphasize the sterling worth of both individuals, especially their present importance in the Senate where the Jacksonian majority was slim. Never would he question the President's right to do what he had done, or its propriety.

Van Buren had judged his man well. No sooner had he raised the question of foreign policy in a general way than the President apologized for appointing Tazewell and Livingston without consulting his secretary of state. As Van Buren remembered the occasion years later, the President's explanation was "creditable to his heart" not his head. Given the opening, Van Buren, who was well prepared, carefully and logically laid out the thorny issues that existed between the United States and Great Britain and France, issues so perplexing that they had confounded the persistent efforts of three administrations over a period of fifteen years.

What was needed, he thought, was a new departure, the appointment of younger ambitious men who had yet to make their mark. He could have said that Tazewell had uniformly taken the British side on many issues that divided the two countries in his speeches and debates in the Senate. He might have been tempted to remind the President that he himself had sharply differed with Tazewell's recommendations from the Senate Committee on Foreign Relations of which he was chairman. He avoided any critical references or, indeed, any comment over differences in policy. As for Livingston, Van Buren had only words of praise, though the hint about age and established reputation was not lost on the President.

The President remarked that since he had made the appointments he had had serious misgivings and if he could withdraw them without creating any ill will or any reflection on his probity, he would do so. In the course of discussion, Jackson mentioned he had not as yet received any word from the two gentlemen, though it had been more than a month since he had offered the missions to them. Van Buren at once saw a possible way out. He agreed completely with the President that the invitations could not be withdrawn; but considering the importance to the public of a prompt settlement, he thought it would not be unreasonable for them to depart by August 1, four months hence. The President agreed, adding that the Spanish mission was vacant and he would welcome Van Buren's suggestion on a candidate. He named Woodbury. New England had no place in the cabinet or in the senior diplomatic appointments. Woodbury was able, tactful and one of the few influential Jacksonians in that section. Jackson warmly seconded the recommendation.

Van Buren was reasonably certain Livingston could not meet that deadline because of cases he had pending before various courts, and it soon became apparent that Tazewell had doubts about the success of any mission to England with the object of removing the restrictions to the West Indies trade. Obviously he was concerned about his reputation. Was there any possibility of success in having England repeal its corn laws that discriminated against American grain and meat production? Van Buren thought not, but advised Tazewell to make an informal inquiry of the British minister, Sir Charles Vaughan. If anything, Vaughan was more discouraging than Van Buren had been. After the interview, Tazewell declined the mission on the grounds that he could accomplish nothing of substance on American claims. Livingston, as Van Buren anticipated, also declined.[7]

All was not smooth, however. Van Buren wanted to strengthen his hand in the cabinet and then direct foreign relations personally. McLane, he knew, craved public recognition and the security of a cabinet post. After he switched to Jackson, he had lost control of the majority party in Delaware. His political future looked bleak when his term in the Senate expired. Van Buren had always thought well of McLane's abilities and his industry, but he was especially attuned to the respect the Washington community accorded the Senator. McLane in the cabinet would raise the prestige of the Administration with important figures like Ritchie, Woodbury (on whom rested his plans for building party strength in New England) and among his old Radical faction of Crawford supporters. Why not induce Berrien to forgo the post of attorney general and take the English mission? McLane could then go into the cabinet.

The new attorney general was still in Georgia preparing for his departure, and Van Buren was so sure of his powers of persuasion once Berrien arrived, that he sent James A. Hamilton, who was acting as his assistant in the State Department, to Wilmington with the conditional offer of attorney general. McLane accepted after some soul-searching because as he wrote Hamilton, "I am not, however—unfortunately—in a situation to consult all my feelings, much less to be fastidious." And he was not as certain as Van Buren that Berrien would agree to leave the cabinet. At the same time, Van Buren offered Woodbury the Spanish mission, and after consulting with the President, offered the Dutch mission to William Pitt Preble of Maine. Thus, two New Englanders would receive important foreign missions. After England and France, Spain was considered the next ranking mission. And since the king of Holland was acting as arbitrator between British and American claims over the long disputed northeastern boundary of Maine, Preble, a leading Jacksonian from that state, was a logical choice.[8]

But his calculations on the English and the Spanish mission and the attorney generalship went awry; both Woodbury and Berrien declined the offers. When Berrien reached Washington and resisted Van Buren's blandishments, there was nothing the Secretary could do but take advantage of McLane's initial reluctance, frankly state the reasons and offer him the English mission. Van Buren was embarrassed and nettled, especially by Woodbury's declension, which had all the earmarks of overweening self-esteem. Fortunately, the President placed no blame on Van Buren for what were undoubtedly misjudgments. McLane swallowed his pride and accepted the English mission. Van Buren selected an old acquaintance, Cornelius P. Van Ness, the Jacksonian governor of Vermont, for the Spanish mission.[9]

Even if Virginia need not be placated, Van Buren would have chosen William Cabell Rives in place of Livingston. Rives was articulate, intelligent and a younger member of the Ritchie circle. He had studied law with Jefferson after his retirement from the Presidency; James Madison considered him a foster son. He was just completing his third term in the House from a safe district and was eminently available. Rives, and particularly his attractive wife, had charmed Van Buren from their first meeting. The President, who had only a limited acquaintance with Rives, certainly saw the political need for recognition of the sensitive Virginians. He accepted Van Buren's recommendation, and the appointment was made.[10]

While he was concerning himself with these major appointments, he put his own house in Jacksonian order by removing six of his twenty-four clerks and replacing them with party stalwarts, though taking due care that they were honest and capable men. He retained Daniel Brent, Clay's chief clerk, temporarily because of his knowledge of the Department's affairs and his ability. There was nothing novel or corrupt, or unfeeling in rewarding one's friends as far as Van Buren was concerned. Patronage had been an axiom of politics in New York since De Witt Clinton had made the first great sweep when he gained control of the council of appointment twenty-five years before. Since then the "spoils system," or as it came to be called more euphemistically "rotation in office," had been modified in New York and elsewhere so that the public interest would not be subordinated to mere partisan gain. Although the President expressed on more than one occasion that any citizen of the country (free blacks and women excluded) was capable of holding any office, he demanded honesty, industry and, of course, loyalty.

Van Buren was completely comfortable with these generous guidelines, though he and other cabinet members came under spirited newspaper attacks for wielding an ax, or, as they responded, inaugurating reforms in their departments. Van Buren ignored the *Intelligencer* and other National Republican papers, but neither he nor the President felt they could afford to remain silent when the *Richmond Enquirer* and other conservative Jacksonian papers vehemently criticized the appointment of a score of partisan newspaper editors all over the country to federal offices. The posts in question ranged from the fourth auditorship in the Treasury Department, which went to Amos Kendall, formerly editor of the Frankfort *Argus of the Western World,* to the postmastership of Portsmouth, New Hampshire, where Abner Greenleaf of the *New Hampshire Gazette,* a Woodbury advocate, replaced the incumbent. In all, only about 10 percent of the federal officeholders were "rotated," but the Virginians were upset at what they deemed a deliberate policy to subvert the press. Speaker of the House Andrew Stevenson warned from Richmond that "this course, if persevered in by the Administration will not only arm their enemies but take away the moral energies of friends." Van Buren wondered whether Ritchie was simply displeased at the caliber of the cabinet because Virginia was not represented in it. Or was he sincere about his campaign against rewarding editors with federal jobs? Choosing the realistic course that every politician, whether he be a newspaper editor or a high-minded Virginian, had his price, Van Buren with Jackson's agreement offered the post of Treasurer of the United States to Major John Campbell, a close friend of Ritchie, and a member of Virginia's upper house, the executive council.

With his usual tact, Van Buren conducted the delicate negotiations through Ritchie so that the querulous editor could enjoy this manifest of his influence in Washington without any hint he was being placated or that the office of Treasurer, a sinecure of no particular prestige, was unworthy of Campbell and of Virginia. He succeeded with the aid of the Campbell appointment in convincing Ritchie that the issue of removals had been vastly exaggerated and that in itself, if wisely conducted according to the President's direction, would be beneficial to the country. What the public service needed was new blood, new ideas and a renewed sense of responsibility. Ritchie seemed satisfied, "but" he asked the Secretary, "in applying the Razor, ought you not to proceed with much caution not to take denunciation for proof?"[11] Van Buren had one more office of sufficient stature that he was confident would ease the conscience of Ritchie and his friends, the mission to Russia.

For this post, considered fourth in importance of the major foreign appointments, he proposed John Randolph, whose uncertain health had taken a turn for the better. There was, of course, a personal side to this decision. For many years Randolph and Van Buren had been friends. That Randolph had lost much of his physical and mental resources through age, indulgence and chronic bad health, that his appearance was ludicrous—"a forked radish" as a contemporary described him with his overlong legs, short body, tiny head and skeletal thinness—made no difference to Van Buren. He saw only the Randolph he remembered, and he knew his appointment would please the Virginians.

Yet he was sufficiently realistic about Randolph's infirmities and his eccentricities to play the waiting game. The present minister, Henry Middleton, had been in his post over four years, and if his recall was not especially in the public interest, the length of his stay furnished an appropriate excuse. Characteristically, Van Buren made an indirect approach to Jackson when he thought the time was ripe. He and Jackson were riding back to Washington from a pleasant afternoon visit with George Washington's adopted son, George W. P. Custis, and his wife. Custis had made his distinguished visitors feel at home in the spacious airy drawing room of Arlington where the conversation ran to agriculture and the theater, topics in which he was something of an expert. His play, "The Indian Prophecy" had

had a successful run in Washington's one small theater and in the larger theaters of Philadelphia and New York.

The Custises had put Jackson in a relaxed, mellow mood and as they rode, Randolph's name figured in the conversation. At this point, Van Buren judged it a proper time to broach the Russian mission. He had a suggestion for the President that would surprise him. In fact, if Van Buren were President he would not accept the suggestion, nor would he make the proposal to any other President except Jackson because to him alone it would make sense, "tho not without hazard." Jackson was completely baffled, as well he might be, and demanded an explanation. "It is to give John Randolph a foreign mission!" The President was astonished, but wanted Van Buren's reasons. The Secretary was ready. He outlined briefly Randolph's services as a distinguished chairman of the House Ways and Means Committee for many years; his intelligence, quick perceptions, marred unfortunately by his pugnacious conduct that led to quarrels with three Presidents because he thought they had strayed away from the pure and simple agrarian democracy both Jackson and Van Buren espoused. Randolph needed some recognition to reward his dedication to the Republican cause. And then Van Buren provided the conclusive argument that his appointment would be popular among the Virginians. Jackson would be pleased to honor "Jack Randolph" as he called him. What mission did Van Buren have in mind? "The Russian mission" was his prompt reply. "Our relations with that government being simple and friendly, little harm would be done if it should turn out that we had made a mistake." Jackson agreed to make the appointment but it would be delayed for six months in order to ease Middleton out of the post without offending him or his powerful relatives.

Randolph's acceptance, which Van Buren received the first week in May 1830, meant the Richmond Junto had placed three of its members in major positions. After the Randolph appointment, it was reasonable to expect there would be no more caustic letters from Ritchie or Stevenson, no more testy editorials in the *Enquirer.* In satisfying an old friend's exalted notion of his self-importance, Van Buren had brought the Virginians into harmony with the Administration, though it had taken a year to accomplish.[12]

Meanwhile, he had failed utterly in the federal patronage for his own state. The most significant position, the post with the most patronage and the fattest income, was collector of the port of New York. The incredible growth of commerce through the port of New York after the opening of the Erie Canal, and the high tariff schedules imposed in 1828, meant that millions of dollars worth of goods passed through customs each year. Customs collections alone were estimated in 1829 to be $15 million, an enormous sum for that day. Some forty or so tidewaiters, appraisers, deputy collectors and clerks gave the office powerful leverage over the quarreling factions in New York City. Van Buren's candidate was Jesse Hoyt.

Van Buren was well aware that several unsavory characters, whose greatest claims to preferment were that they had supported Jackson, were angling for the office. In particular, Samuel Swartwout, the youngest of three brothers who had cut a meandering path through the political thickets of New York City for the past twenty years, had been pushing hard in Washington and elsewhere. Swartwout, with the covert backing of Calhoun, had made the most of Van Buren's enforced absence from Washington to visit with Jackson. He had been busy flooding the President with recommendations of all sorts, some from Adams's men who felt that with him in the collector's chair they could expect some morsels from the table, others from men of no political stature, others completely spurious, like those who claimed the support of New York's two Senators.

The Swartwouts had all been close associates of Aaron Burr in his political heyday, swaggering young bloods, always with an eye on the main chance. But Samuel, the youngest, was considered an adventurer of no fixed habits, known throughout the City for his improvidence, his fondness for speculation, fast women and matched thoroughbreds. Like so many men of his temperament, Swartwout was a genial individual with a wealth of wit and anecdote in his conversation. He had met Jackson while in the West with Burr. Sensing in the General a man he should cultivate, Swartwout over the years had maintained their relationship.[13] Of Swartwout's friendship with Jackson, its extent and its longevity, Van Buren was unaware. He, of course, knew that Swartwout was an original Jackson man in 1824 and as a Tammany brave had pushed the General's cause in the City. Yet he had played a minor role in securing the large Jackson majority in the metropolis when compared with men of his controlling faction like Cambreleng, Verplanck, Saul Alley, Walter Bowne, Jesse Hoyt, all Sachems of long standing and high repute. When Van Buren first learned of Swartwout's candidacy for the collectorship, he erred gravely by dismissing it as having no possibility of success.

Jackson, too, had heard that Swartwout was not in good standing with Van Buren's faction of Tammany Hall; he determined nevertheless to appoint him collector. But he anticipated opposition from his secretary and resolved to give him two appointments that he naïvely thought would compensate for the loss of the collectorship. Since he had involved himself directly in New York patronage, he would allow Van Buren the appointment of the federal attorney in Nashville, Tennessee, and in addition select Van Buren's aide, James A. Hamilton, to be federal attorney in New York City.

Confusing the close-working relationship between Hamilton and Van Buren in the State Department with equally close political ties, Jackson blandly assumed Van Buren would be delighted. He could not have been farther from the mark. Hamilton was simply an assistant and nothing more—a personable and useful link with the Jackson circle, a skillful helpmeet in drafting or editing messages and correspondence, a confidential messenger. His Federalist background, even though of the "high-minded" variety, did not set well with the Regency and with Tammany. Hamilton was too close to Charles King and his hostile *New York American* for them to feel comfortable with his new access to Van Buren, which they did not properly understand. Van Buren had come under veiled criticism for seeming to promote a man they felt was a self-serving political renegade, at best an opportunist, an eleventh-hour Democrat, over hardworking party leaders who had been faithful over many years.

Swartwout, who had never been known to be reticent about his exploits, had increased Hoyt's indignation by boasting of his imminent appointment to the collectorship, his friendship with the President and the Vice-President. Hoyt dashed off a complaining letter to Van Buren that stung him to the very quick. "Some gentlemen," wrote Hoyt, "called on me this morning not office-seekers—but men who you respect and whose opinion you would respect and stated that your friends were falling off here, from the apprehension that either you had no influence in the Cabinet or that you were not disposed to exercise it for the benefit of your party." One week later, Van Buren received a six-page letter from Cambreleng, in part confirming Hoyt's allegations. The voluble Cambreleng was horrified at the thought of Swartwout presiding over the customs house. He reminded Van Buren of Swartwout's "unfortunate management of his own concerns and the speculative eccentricity of his career."

These communications irritated and alarmed Van Buren. For once, he dropped his genial mask and lectured Hoyt strongly for his rudeness in giving credence to rumors and idle boasts. "Here I am," he wrote, "engaged in the most intricate and important affairs which are new to me, and upon the successful conduct of which my reputation as well as the interests of the country depend, and which keep me occupied from early in the morning, until late at night, and you can think it kind or just to harass me under such circumstances with letters."

By now fully alert to the danger, Van Buren sought to sidetrack Swartwout, yet not give Jackson any offense. Without acknowledging that he had any prior information on the New York City appointments, he read Cambreleng's lengthy indictment of Swartwout to the President. Jackson kept his own counsel, though he did refer favorably to his "original friends in New York," an allusion not lost on Van Buren. On April 20, Jackson said that he had sent Swartwout's papers to Ingham and that both New York Senators "verbally supported" the appointment, astonishing information which seemed scarcely credible. Would Van Buren read them over and give his opinion?[14]

Van Buren immediately wrote New York and asked Cambreleng and Bowne to have their faction write remonstrances stating "unreservedly the effect that such an appointment must have upon the confidence of the people of the State—the objections to him . . ." At the same time he wrote Senator Dudley expressing doubts that he and Senator Sanford had recommended Swartwout as Jackson had said. "Do me the favor to see Sanford," he asked, "and to explain yours and my views upon the subject by *return mail* as the appt. will be delayed until you are heard from."

Armed with additional information from Cambreleng, but not from the Senators (Dudley and Sanford's letters of denial were not received until after the appointment was made), Van Buren composed a lengthy exposition of his own in which he showed that most of Swartwout's support was unjustified or based on hearsay, rumor and gossip. He followed up his letter with a visit to the White House where he gave his frank opinion of Swartwout's unfitness for the office. Jackson listened carefully, but reserved judgment, a sufficient hint for Van Buren to expect the worst.

Tired, his nerves on edge, still feeling out of sorts physically, he vented his chagrin on the hapless Cambreleng as he had earlier on Hoyt. "If the result should be adverse to the wishes of our friends in N. York," he said, "they will have to bear much of the blame themselves. I do not find among the papers (with the exception of one or two private letters) a single remonstrance ag't. the appointment." Before Van Buren sent off his letter to Cambreleng, the President replied, brushing off Van Buren's indictment and those of others. Jackson found no specific charges of dishonesty in Swartwout's papers. On the contrary, he stoutly upheld him as a "warm-hearted, zealous and generous man, strictly honest and correct in all his dealings and conduct," and gave notice that he would appoint him collector.[15]

As he had planned from the beginning, the President now offered Van Buren the appointment of federal attorney in Nashville and in another note said he had named James A. Hamilton district attorney in New York City. The Nashville appointment meant little or nothing even as a consolation prize to Van Buren. And the Hamilton appointment could cause almost as much trouble as Swartwout's.

Hamilton was sitting next to Van Buren at Mrs. Miller's when the President's note arrived in the evening of April 23, 1829. Although he made it clear to his assistant how his appointment would complicate matters further, he was unable to convince him he should withdraw. Apart from the all too obvious political mischief Hamilton's appointment would cause, Van Buren had become quite dependent on his ambitious assistant for the management of the State Department. And the President was fond of him, a most important consideration in keeping the channels of communication between Van Buren and Jackson clear and open.[16]

These developments had shaken Van Buren's self-confidence. However much he blamed his New York friends for being remiss in countering Swartwout, he could not banish from his mind the ugly doubt that he was much, if not more, to blame. After all, he had selected Hamilton; had he acted more promptly, he might well have checked Swartwout. But he was shocked at Jackson's conduct and questioned whether he could serve with a President who would deliberately interfere in the politics of his chief subordinate's home state to make a crucial partisan appointment on mere personal interest. How could one trust a man who professed to place the public interest over all, who sternly demanded reform and then refused to accept the best-informed opinion on a bad appointment, and who rejected the plain facts set before him? Could the offers of two unsolicited posts be bribes to purchase his silence? Could Calhoun have inspired them? Ingham would not show him his reply to the President. Was this part of the plot?

Careful to conceal his dismay from his ambitious assistant, Van Buren seized coat and hat and left the house. For several hours under the bright moonlight he walked along the muddy streets of Washington, turning over in his mind whether he ought to resign. With nothing resolved, he returned to Mrs. Miller's exhausted and went straight to bed. A sound night's sleep did wonders for Van Buren's morale and for his perceptions. In sorting out the events of the past three weeks in his mind, the Tazewell and Livingston appointments, and now those of Swartwout and Hamilton, three conclusions could be drawn. The President had acted on impulse, or by design, or was under the influence of a cabal out to destroy him. Jackson's actions on Tazewell, Livingston, Berrien, McLane and Randolph argued strongly for Jackson's innocence and against any premeditation or consort with others to injure his secretary of state. And when carefully evaluated, so did the Hamilton appointment which, however inept and troublesome, was really meant to conciliate, not to weaken.

Van Buren came round to the view that he was in a sense the guardian of the President; men like Ritchie and Randolph, Cambreleng, Woodbury, Jackson himself, and countless others looked to him as the steadying influence in the new Administration, as a sort of political balance wheel.[17] Finally, there was always the specter of Calhoun, his great ability paired with great ambition. His vaunted nationalism marred by his particularist stand on the tariff. If Van Buren abandoned Jackson now, would he not betray the people who elected him? Would he not be giving up the field to Calhoun and, as a result, abandoning a public trust to the unorthodox views, the political caprice, of the Vice-President?

It was his duty then to remain, to cope as best he could with the situation in New York. To that end he would do two things. He would write a courteous but rather reserved letter with an evident injured tone to the President. He would, in the same letter, ask Jackson if he might inform Cambreleng and a few others in New York that he had opposed Swartwout and send them copies of the President's letters to him on the subject. As to the post in Nashville, he pleaded complete ignorance, but in an offhand way suggested Alfred Balch, his longtime supporter in Tennessee. Jackson promptly agreed, and in the next mail to New York went all the pertinent correspondence between them. Cambreleng, disguising his emotions with pointed amusement, returned the letters with a brief note of mock congratulations on the resolution of the New York appointments: "If our collector is not a defaulter in four years I'll swallow the Treasury if it was all coined into coppers." Cambreleng was far more solemn about the political implications. "As it regards myself," he said, "I care not for these gentlemen. I can probably stand their attacks which are already openly threatened—but I care for others who are now in the corporation, in the legislature and various offices in our city. It cannot and it ought not to be disguised from the authority that the patronage of the Custom and the Post Office will be directed to overturn the very men who have been zealously engaged in gaining for us our elections in 1827 and 1828."[18]

Cambreleng had every right to be deeply concerned about the fate of the organization in New York City. Van Buren's handling of the affair had been adroit. He had shifted the burden of the appointments to the shoulders of the President; yet at the same time, he had shown the fault was not one of calculation but of Jackson's innocence in the disposal of patronage. Most important, he assured his worried lieutenants that they had nothing to fear from either Hamilton or Swartwout. He would see to it that their patronage went into the right hands. As he expected, Hoyt was furious but even he did not charge Van Buren with bad faith, reserving his barbs for the President.[19]

Although he had not made any open moves as yet, Calhoun had been one of the forces behind Swartwout's appointment. And the postmaster, Monroe's son-in-law Samuel Gouverneur, had long been one of the Vice-President's firm advocates in the North. Calhoun was steadily advancing his fortunes through a network of trusted friends. In Massachusetts, Marcus Morton, one of the justices of that state's supreme court, was already committed to the Calhoun succession as was the rich, abrasive druggist, David Henshaw, his candidate for collector of the port of Boston. Francis Baylies, one of the small band of original Jacksonians in Boston, employed some choice invective in describing the new collector as "mean, false, envious and treacherous . . . a permanent nuisance not unlike an eternal pigsty under a parlor window."[20] Ansel Sterling in Connecticut, his brother Micah in western New York, the majority "family party" in Pennsylvania and their representative in the cabinet, Samuel Ingham, were all moving to enhance the availability of the Vice-President.

Duff Green was skillfully promoting Calhoun in the capital city's only Administration paper and reaping large gains from the public printing. The Van Buren New Yorkers in Congress and and others who distrusted Green's motives had tried unsuccessfuly to establish a rival paper and had approached Amos Kendall on the subject. But that cautious veteran of many a political skirmish decided he would keep clear. Van Buren, who had marked Kendall as a coming man, likewise lent no support to the movement. He recognized how vain and opinionated Green was, and having made his assessment, waited for his man to commit the inevitable blunder that would help run Calhoun's candidacy on the rocks.[21]

Once Van Buren had decided to remain in the cabinet, he began making plans to move out of Mrs. Miller's into a home of his own that would befit the dignity of the secretary of state. He wanted a residence near the White House and the Congress of sufficient size to accommodate his family, and servants and the official entertainment expected of a secretary of state. Van Buren found just what he was looking for in the house the Clay family had just vacated.[22]

He charged his son John, now a tall handsome young man who towered over his father, with disposing of the family furniture. Van Buren planned to purchase all new furnishings, even down to the curtains and portieres, and in the latest, most elegant style. He instructed John to sell all, provided he received a fair price. Anything he could not dispose of was to be shipped to Washington, along with the linen and the silverware. He had fewer qualms about John's industry and strength of purpose and was now entrusting him with family business, listening to his advice about real estate holdings and encouraging him to report on local political news. John was also deputed to keep in touch with his younger brothers, Smith and Martin, Jr., both in school.[23] Van Buren's eldest son, Abraham, an army lieutenant, had returned from his tour in the Michigan territory and was then in Albany. Both he and John planned to join their father in Washington as soon as the furniture was sold and certain other business details in Kinderhook and Albany were cleared up. Abraham, who had been assigned to temporary duty in the capital, would remain near his father. John would return to New York for the completion of his law studies.

The careworn Van Buren looked forward to having half his family with him to ease the burden of his lonely, overworked existence.[24] For what amusement he may have had was not calculated to be relaxing. The tight social scene in Washington was being rocked with what any outsider would consider a ridiculously self-conscious little war of manners with the President and secretary of war ranged on one side, and the rest of the cabinet, the Vice-President, and many of the diplomatic corps and Congressmen on the other. It had all begun when Eaton married Peggy O'Neal Timberlake, the widowed daughter of William O'Neal, a local tavernkeeper in whose boarding house he had lived for some time, as had Jackson. The new Mrs. Eaton bore a reputation for a rather casual life-style, all knowledge of which was circumstantial, but largely believed to be true. Whether true or not, Mrs. Eaton was as snobbish and determined to be a great hostess as any of the cabinet and diplomatic ladies who passed tidbits of gossip about her alleged affairs behind their fans while they took their tea and nibbled on cakes during the afternoons at home or at the crowded, stiff evening receptions that made life bearable for them in the remote, rural capital.

Thus, *la bellona* (which meant in Italian a vulgar woman of generous proportions, and in Latin the Roman goddess of war) became her sobriquet among the ladies. And inevitably, the affair became a political one, a joust for influence over the President. When Van Buren arrived in Washington, John Eaton was still the cabinet member whose advice the President most valued. His counsel had been the most significant in shaping the cabinet. Even though one Calhoun partisan, Samuel Ingham, occupied the Treasury Department, this post had not been the one the Vice-President had wanted for him.

Calhoun, whose covert moves were becoming transparently obvious, had made the best of his bridgehead in the cabinet and was busy through Ingham's patronage building up his strength in the country—Swartwout and Henshaw being the most recent flagrant examples. He had lost the Post Office when McLean gratefully sought refuge in the Supreme Court, a move in which Eaton and Donelson had played major roles. The smaller patronage of war, navy and attorney general's office would be useful if it could be controlled. But the compelling fact in the Vice-President's mind was to win over those members of the cabinet who were uncommitted and through Duff Green fashion an alliance with the most widely read Jacksonian editors like Mordecai Noah of the *New York Enquirer,* Thomas Ritchie of the *Richmond Enquirer,* Nathaniel Greene of the *Boston Post,* Isaac Hill of the *Concord Patriot,* and Gideon Welles of the *Hartford Times.*[25]

Ordinarily, the ladies' rebellion against Peggy Eaton would have simmered down, despite *la bellona*'s thirst for recognition. But Calhoun partisans could not resist taking advantage of the break to enlarge their sphere of influence in the cabinet from which they hoped to drive out Eaton and Barry, and ultimately Van Buren, too, if he stood in the way of the succession. They had not reckoned with Jackson's fondness for Eaton and with the vivacious Peggy, who knew how to charm the General and play upon his sympathies. When Floride Calhoun refused to receive Mrs. Eaton and other cabinet wives followed suit, Jackson demanded to know the reason. When told of the rumors that impugned Peggy's virtue, Jackson demanded evidence; when nothing concrete could be produced he dismissed the charges. But he could not persuade his niece, Emily Donelson, who was acting as hostess at the White House, nor other female members of his family to receive Peggy. The situation was close to becoming an affair of state when Van Buren arrived.[26]

« CHAPTER 15 »

Intrigues and Intriguers

Hounded by officeseekers, correcting political mistakes the new Administration had made in his absence, familiarizing himself with the responsibilities of the State Department, and consulting daily with the President, "a pale and haggard" Van Buren found time to make a courtesy call upon ex-President Adams and leave his card at Mrs. Eaton's. Both individuals were gratified, though Adams felt Van Buren had an ulterior motive. "His principles," said the ex-President, "are all subordinate to his ambitions."[1]

When Van Buren learned how deeply the President felt about the snubbing of Peggy Eaton, he tried to bring about some compromise which would restore harmony in the cabinet and in Jackson's own family. Van Buren, himself, was attracted to Peggy's voluptuous charms, but he was far too shrewd not to miss her efforts through Eaton to meddle in politics and patronage. With nothing to lose himself, no wife, no daughters, he could play the conciliator, perhaps curb for a time Peggy's driving ambition. He counted also on an appeal to the better nature of Washington's *grandes dames* whose husbands he felt were overestimating Eaton's influence and grossly underestimating the President's independence.

Van Buren had never counted Eaton as much of a politician or an administrator for that matter. But he had quickly sensed Jackson's emotional dependence on his fellow Tennessean. If Eaton's continuance in the cabinet was important to the President's peace of mind, it was indispensable to Van Buren's political status. If he, as premier, could not hold the cabinet together, give it tone and leadership and cohesion, as the party leaders expected of him, he could go down with the unfortunate secretary of war. For all the sacrifices, and it must be said the gains he had made already, he was not about to forfeit them in a flurry of petticoats and lace. He could play the drawing room game to studied perfection. In this kind of affair, Van

Buren, the ultimate courtier, risked little or no harm to himself, yet posed a serious threat to his adversaries.

For the moment, more immediate problems claimed his attention. During the hot, humid summer of 1829 he labored in his airless office at the State Department and at home. Though he had some help from James Hamilton, he personally prepared instructions for the new ministers after asking and receiving advice from such old hands as Gallatin, Smith and even John Quincy Adams. The only exercise he got was horseback riding of an evening over the Virginia and Maryland countryside. Yet his health, which was poor when he first arrived in Washington, had steadily improved; a condition that he attributed to a "regular diet (or at least tolerably so) and some exercise and *mustard seed.*"

Jackson, when he felt well enough, frequently joined Van Buren on his evening rides. "We are getting along extremely well," he wrote Enos Throop. "I am upon terms of kindness and full and affectionate confidence with the President and every one of the heads of Department. The President proves to be in all respects a finer man than I anticipated. . . . he is intelligent, laborious and patient to a fault."[2]

But Jackson was not patient when it came to the Eatons and their exclusion from Washington society. His compulsive obsession with the defense of Peggy's chastity was now seen as a Clay-inspired plot against him through Eaton. Van Buren listened to the old gentleman fulminate and, agreeable always, came in with just the right comment at the conclusion of a tirade.

The test would come, he knew, when the Twenty-first Congress assembled in December. Thus he was not surprised when he received a letter from James Hamilton, Jr., in late November, a few weeks before Congress convened for its long session. Hamilton wrote about his "long and frequent conversations with Calhoun" during which the Vice-President assured him that when he arrived in Washington he wanted to establish "the most cordial and confidential relations with you." Van Buren may have been skeptical about Calhoun's good intentions, but he welcomed them as a possible way out of the Eaton imbroglio. Perhaps Calhoun would also exercise a restraining influence through Ingham on the Treasury officers in New England and New York, whom he suspected strongly of fomenting factionalism.[3]

Van Buren did manage to get away from Washington in early November, ostensibly to observe the opening sessions of the Virginia constitutional convention but actually to take stock of politics in the Old Dominion and to enjoy some much needed diversion in the congenial atmosphere of Shockoe Hill. He would have been less the thoughtful politician he was had he not noticed the bias among some of the Virginians for Calhoun as they sought to discredit Ritchie and establish a rival newspaper. The make-up of the convention itself presented a broad spectrum of Virginia society, ranging from the homespun and buckskin clad delegates of the west demanding tax reform and abolition of slavery to tidewater potentates just as determined to retain their privileges in landed wealth and slaves. Van Buren was an attentive listener at debates that featured notables like Randolph, Madison, Monroe, the venerable Chief Justice John Marshall, and other individuals, less-illustrious but more in tune with public opinion such as Dr. Brockenbrough or Ritchie's former partner, Claibourne W. Gooch, a veritable mine of information, whose opinions were at once accurate and perceptive.[4]

Refreshed, Van Buren returned to Washington where piles of letters and dispatches awaited him. The last two weeks of November were days of incessant activity as he drafted his report for the President's annual message, helped out with other portions of it, especially those dealing with internal improvements, the Bank of the United States, the tariff, the anticipated surplus of federal funds. By and large, his comments were aimed at moderating the language of the message on points that involved highly controversial issues. Substantively, he accepted the President's views, many of which they had discussed informally on their evening rides. Jackson had come round to Van Buren's opinions on the tariff and on internal improvements, both of which sharply diverged from those of Calhoun.

When the Twenty-first Congress convened on December 7, 1829, and its members digested the message, the Vice-President's partisans promptly and vehemently expressed their criticism, implying that the President had come under Van Buren's influence and was discriminating against the South. So heated were the discussions in the lobbies of the Capitol and in the debates on the floor that a cleavage of the Union along sectional lines seemed not improbable to some. By late December, Edward Everett remarked that the rivalry between Van Buren and Calhoun had reached a crisis.[5]

The sweetness and light Calhoun breathed to Van Buren through James Hamilton, Jr., never materialized. Jackson's message had been a warning Calhoun could not overlook; nor could the other southerners in the cabinet, Branch and Berrien. Though owing their seats to the influence of Eaton and the President's two secretaries, Donelson and Lewis, they had listened to their wives and daughters and to the advocates of the Vice-President, the tendentious arguments of Duff Green, the analysis of Robert Y. Hayne, and others in the exceptionally able South Carolina delegation.

Hard times persisted in their states, the world market for cotton had not recovered from the slump of 1826; planters everywhere in the seaboard slave states were blaming the tariff for their plight. The tariff, it was said, was a northern measure, favoring northern interests, authored by Van Buren through his creature Silas Wright.

Then there was the matter of internal improvements. The South and the West needed federal support if these regions were ever to get their staples to market rapidly and economically. Clay, Jackson and Calhoun had all been advocates of internal improvements at federal expense. Whatever other differences existed between them, Clay and Calhoun hewed to their original line on this question, but Jackson had changed, if his message was any indication. Van Buren, charged many angry southerners and westerners, including former friends and working partners, had been responsible, and for purely selfish reasons, in establishing the new policy. In fact, any part of the message, however pure the Jeffersonian accents, that seemed in some way injurious to southern interests, they attributed to the baleful influence of the Little Magician over an old, sick malleable President. That these allegations were simply untrue, were grossly unfair to Van Buren and demeaning to the President's intelligence and ability, meant little in the overheated partisan atmosphere of Washington, where the smoldering Eaton affair was flaring up again.[6]

The social season in Washington began with the convening of Congress in December. Jackson, as was customary, led off with a state dinner for his cabinet and others of the executive branch. The dinner was a dismal failure despite the best efforts of Van Buren to promote some gaiety and interest. Those of Washington society who came snubbed the Eatons. The Secretary of State, in order of precedence, gave the next dinner which turned out to be a stag affair, all the wives declining on one pretext or another. No more state dinners graced the social scene for that season, but Van Buren tried again by giving a large "unofficial" party at his own home. This time, many of the Washington wives did attend. The Eatons were conspicuous among the glittering assemblage, though Peggy's behavior left much to be desired. Either in an excess of youthful or of ardent spirits, she made a scene on the dance floor. Van Buren, resting on a sofa in one of the private rooms below passed off the incident, but a pall descended over the ballroom and the buffets. Guests left earlier than usual.

Van Buren made one last effort to harmonize the social life of the cabinet, to purge it of partisan intrigue. He arranged with Baron Krudener and Charles Vaughan, the Russian and British ministers, both single like himself, to have a ball. The Vice-President, cabinet members and their wives, including the Eatons, attended along with the entire diplomatic community, which had also taken sides. The grand ball and dinner, like all the earlier events Van Buren had arranged, was a failure. Most of Washington society again pointedly snubbed Peggy.[7] Thereafter, Branch, Berrien, Ingham and the Vice-President excluded not just the Eatons but also Van Buren and Barry from their list of guests. Van Buren and his group retaliated in like manner, though the Secretary was courtly and amiable, as only he could be, when he met any of the Calhoun group at social affairs on neutral ground.

The President had fully investigated Peggy's past and satisfied himself that she was innocent. Pronouncing her the victim of a vile slander, he warned the cabinet their tenure rested in harmony and trust as much, if not more than, in their administrative duties or their counsel on matters of policy.

Despite his chivalrous instincts in protecting the virtue of a fair lady, Jackson, and Van Buren, too, had always suspected that there was a partisan basis to the affair. Van Buren called it "a conspiracy of excited women and infuriated partisans."[8] In its early stages, like Jackson, he considered Henry Clay the chief instigator. But the more he turned it over in his mind, the more he began to suspect Calhoun. Nor did the President discount the information that flowed in from many sources he considered reliable that Calhoun was deliberately differing with his policies to build a platform for himself when he succeeded to the Presidency. Nothing was better devised to anger him than to have the friends of Calhoun blandly taking the succession for granted, and this after the President had not even completed the first year of his term.

Ambitious men who had access to Jackson, and recognizing his prejudices, his gathering suspicions, the willful self-assurance of the old man, played upon them to Calhoun's disadvantage. Major Lewis, Amos Kendall and Isaac Hill, all holding recess appointments, all careful judges of Jackson's volatile temperament, recognized that their future lay in his hands. So finely did they tune themselves to Jackson's moods that they became almost extensions of his personality.

These individuals, soon dubbed the "Kitchen Cabinet," though never a cohesive entity, were careful to stay in the background, never gave advice unless asked and rarely, if ever, intruded ideas at variance with those of the President. Rather they reflected and enlarged upon the President's ideas in both politics and policy. Even Lewis and Donelson, closest of them all in terms of relationship and familiarity, acted as a sounding board for Jackson's diatribes, never as active agents against Calhoun.[9]

Van Buren was now convinced Calhoun's ambitions lay at the bottom of an intrigue that had escalated so slowly over the past year as to become inextricably woven into the very fabric of the party and government. He shied away from slow-moving, clandestine schemes for prudential and moral reasons deeply engrafted in his character. Though Van Buren was ambitious, perhaps as ambitious as Calhoun, he was not responsive to any talk of rivalry between them. He was disturbed when Noah, who had merged his paper with J. Watson Webb's *Morning Courier,* either in an excess of enthusiasm or of mischievous meddling, nominated Van Buren as Jackson's successor. Noah's editorial was the only mention of what instantly became a forbidden topic. While the self-assured Duff Green grew bolder in his espousal of Calhoun, other Administration prints were silent on the subject.[10]

Green's indiscretions, the monumental debate between Webster and Hayne in the Senate, the Peggy Eaton affair, all combined to weaken Calhoun's stature in the White House. An astute political observer, the Virginian John S. Barbour, described four distinct parties, "The slugs and grubs and worms crowded into office and want Jackson's reelection, Calhoun's party, Van Buren's and Clay's." He added that except for Van Buren and Eaton, "the real powers lie behind the throne, the other members of the Cabinet are Calhounites."

Jackson approved heartily of Webster's patriotic sentiments in his reply to Hayne. What really caught his attention as he read Hayne's speech was his argument that a sovereign state had every right to judge for itself federal legislation, and if contrary to its interest, refuse to obey it.[11] Jackson was no scholar of constitutional law and theory, but the federal union was an article of faith with him, the nation a projection of his personality and his iron will. He saw the danger and, remembering Calhoun's supposed authorship of the *Exposition and Protest* to the South Carolina legislature, promptly concluded that a conspiracy existed between Calhoun and Hayne to destroy the Union.

The South Carolinians in Washington, some Virginians and some Georgians, subscribed to the theory of ultimate authority remaining with the states, but they were not engaging in any conspiracy to undermine or destroy the Union. Rather they were determined to work for a revision of the tariff which, if successful, they believed would injure the prestige of Van Buren, its author, and improve economic conditions at home. The President was all-important in the planning. They were impressed with his popularity that had increased dramatically during his first year in office, despite the running sore of the Eaton affair. Success depended entirely, they felt, upon enlisting his support for their anti-tariff stance.[12] Van Buren could then be identified as a sectional tool of northern manufacturers, isolated and destroyed as Adams had been.

No public man in Washington was more alert to the faint outlines of intrigue freshly conceived than Van Buren. He understood at once the meaning behind the announcement in the *Telegraph* of a grand subscription dinner at Jesse Brown's Indian Queen Hotel for the celebration of Jefferson's birthday on April 13, 1830. Quite obviously, the dinner would provide a setting for Calhoun's ideas on the nature of the Union, cloaked in the canonical republican doctrine of 1798, which would popularize and conceal any ulterior motives.

Both Van Buren and the President had caught the implications, which they discussed fully after invitations had gone forth all over Democratic Washington. Benton had been one of the prime movers, hoping for public recognition of a sectional alliance between the West and the South against Websterian New England. Calhoun and his adherents had welcomed the movement, hoping Jackson, a westerner, a southerner, a slaveholder and a planter, would sustain their views; but they gravely misunderstood his character and underestimated his political ability. The President himself had been looking for a pretext to draw Calhoun into a public declaration of what he assumed were his nullification ideas. That the Vice-President should actually provide the means for his own exposé in the presence of the party leaders was an opportunity not to be lost.

During their discussions of the Jefferson dinner, which both decided to attend, Van Buren skillfully built upon Jackson's suspicions of Calhoun's motives, the menace his views posed to the Union. Like Calhoun, he was anxious to enlist the President's power and popularity. But unlike the South Carolinian, he followed Webster's line of reasoning, any sectional combination as Hayne and Benton were espousing substituted regional interests for states' rights. These interests, when expressed in political and ideological terms, as Hayne had done, with the undoubted support of Calhoun, threatened his own section and ultimately the Union.[13]

The question then became, how was the country and the party to be informed? Whatever the President said would be the theme of the Jacksonian press all over the nation. Van Buren and Major Donelson met with the President an hour or so before the dinner to talk about his volunteer toast. All agreed it would be crucial, and in short order they drafted a brief phrase that they felt had just the right tone, emphasis and meaning. Never, thought Van Buren, did so much rest upon so few words. Next they worked out Van Buren's toast which would be made third, directly after the Vice-President's. "Thus armed," said Van Buren, "we repaired to the dinner with feelings on the part of the old Chief akin to those which would have animated his breast if the scene of this preliminary skirmish in defence of the Union had been the field of battle instead of the festive board."[14]

Brown's great dining hall had been set up in a pattern of two side tables facing each other and joined at one end by a head table. Six vice-presidents for arrangements presided at the side tables. Jackson sat at the head table next to an old friend of Jefferson, the septuagenarian Congressman from Virginia, John Roane, who the committee on arrangements had appointed chairman. Calhoun, outwardly composed but tense, was seated on the other side of the chairman. The unventilated hall was crowded to capacity, the air stifling with the fumes of hundreds of pipes and cigars, the aisles liberally spotted with tobacco juice, and the noise level in the hall must have been deafening what with the expansive conversation of the hard-drinking company, the clatter of the cutlery and the tramping of the servants as they rushed mounds of food to the guests and replenished empty glasses.

Eventually, old John Roane judged the time appropriate for the speeches and the din in the hall slowly ebbed away as he called upon Hayne for the address of the evening. The gifted South Carolinian was in good form as he denounced the tariff unsparingly, that "Tariff of Abominations" as it was now called. Hayne's remedy was repeal; Hayne's weapon in case the Congress should not heed his advice was a vague reference to state action which he did not enlarge upon. When the Senator sat down to thunderous applause from those who understood and supported his position, those who misunderstood it but admired his oratory, and those so drunk they would applaud anything, the regular toasts began.

Jackson remained silent through it all, his long, angular face expressionless. At length the regular toasts droned on to their conclusion, and the chairman introduced Jackson for the first volunteer toast. His features still impassive, the President slowly drew himself up to his full height. Many rose, among them the vice-president in charge of the section where Van Buren sat, Felix Grundy, a Senator from Tennessee. Grundy's broad back blocked Van Buren's view, forcing him to stand on his chair, his glass in hand. Jackson turned toward Calhoun and fixing him with his bright blue eyes, cold and measuring, said in tones that carried to the farther reaches of the great chamber, "Our Union: It must be preserved." Then he held up his glass, an indication he wanted everyone to stand and drink to his toast. Calhoun, his face pale, his hand trembling slightly from pent-up emotion, stood with the rest and sipped his wine.

After everyone was seated, all eyes turned to the Vice-President, whose nerves may have been on edge but whose quick wit had not deserted him. "The Union: Next to our liberty most dear," he said in a firm, clear voice, a memorable response had he only ended it there. But he continued so that his point would be unmistakably clear. "May we all remember," he added, "that it can only be preserved by respecting the rights of the states and by distributing equally the benefits and the burdens of the Union."

Calhoun's subsequent remark not only weakened the impact of his toast, but it made Jackson even more suspicious of his intentions, though both he and Van Buren drank to it. Then Van Buren rose, again climbing up on his chair and in his rapid manner gave a toast that complemented the President's, yet in a way softened its dramatic thrust while pointedly responding to Calhoun. "Mutual forbearance and reciprocal concessions," he said, directing his gaze outward over the hall, "thro their agency the Union was established—the patriotic spirit from which they emanated will forever sustain it." Other volunteers proposed toasts, among them Hayne, who prefaced his by asking the President from the floor if he would insert the word "Federal" before Union in his. Jackson was delighted to comply because the term "Federal Union" expressed exactly his intent and in fact was written on the draft in his pocket. He had simply forgotten to include the adjective.[15]

Jackson had carried the day. Scores of Democratic papers carried full column editorials, taking Jackson's seven pungent words as their text for Union, the solemnizing of Daniel Webster's hyper-emotional, grandiloquent conclusion in his second reply to Hayne. The country now knew precisely where Jackson stood. And John C. Calhoun would never again underestimate him. As for Van Buren, he followed in the same well-worn track he had always used to advantage during his political career. Let events shape themselves, be wary and be alert to personal ambitions and sectional cross-currents! React when necessary and react decisively but never on impulse!

Van Buren did not understand at this point the problems his rival was facing, not just in his home state, but in all the seaboard slave states. He was unaware of the moderating influence the Vice-President was exerting over the firebrands of South Carolina, the fact that his political career was at stake at home. What he took for intrigue was, in fact, Calhoun's response to these pressures as well as his burning ambition for the Presidency, which, as far as Van Buren could see, was promoting discord in the cabinet and weakening the Administration. "The question as to the succession of Gen'l. Jackson has engendered some ill blood and as it is supposed produced some intrigue," he wrote William C. Rives. Van Buren's solution was a second term for Jackson during which many of the mistakes, the rivalries, the petty jealousies that so far had tarnished the Administration would be resolved in the fullness of time.

Well before the Jefferson dinner, he and the Kitchen Cabinet had convinced Jackson that he had to make the sacrifice for the sake of the Union.[16] They leaked the second-term decision as cautiously as they could and by the date set for the dinner it was generally known in Washington. Van Buren was confident that the knowledge of a second term for Jackson would quash the intrigues and put an end to any doubts about the confirmation of several important recess appointments. It did not quite work out that way, however.

The most important of these appointments, that of Amos Kendall, won confirmation when Calhoun broke a tie. Noah and Hill were rejected. But after some logrolling in which Calhoun was certainly involved, Noah was confirmed as surveyer and inspector of the port of New York. Hill went back to New Hampshire vowing vengeance on his traducers, the Calhoun Senators who had voted with the National Republicans against him.[17]

Any chagrin the President felt at Hill's defeat, which he regarded as a personal attack upon himself, he kept in abeyance because his attention had become riveted on another issue. Major Lewis had spoken of Calhoun's opposition to his course in the Seminole War in 1827 and again in 1828. Jackson had heard about the South Carolinian's efforts to have him censured for insubordination during the Monroe Administration. His source had been Crawford, who disliked Jackson but detested Calhoun. Jackson, who needed Calhoun's support for his presidential campaign and had no more use for Crawford than Crawford had for him, discounted the evidence. Others in Jackson's entourage, however, ran down the lead through Van Buren's emissary to Nashville, James A. Hamilton. Lewis revealed the information to Hamilton and asked him to verify its authenticity in an interview with Crawford. Hamilton missed Crawford, but he did visit with one of Crawford's closest friends, John Forsyth. And Forsyth promised to seek a statement from Crawford.

Crawford had reconciled himself to the Jackson candidacy but was exerting whatever influence he still retained against Calhoun. Despite the comity that cabinet members were supposed to display about confidential cabinet meetings, Crawford promptly gave Forsyth a statement, part of which he quoted in a letter to Hamilton. Forsyth's version of Crawford's comments portrayed an angry Calhoun demanding nothing so mild as a censure of the tempestuous Jackson, but his arrest and court-martial. Hamilton had filed away Forsyth's letter, not mentioning it to Van Buren, who he knew was strongly backing a Jackson-Calhoun coalition. But in the spring of 1828, while Lewis was visiting New York on a brief holiday, Hamilton showed him the letter. Lewis, too, was unwilling to jeopardize the coalition, though he recognized the importance of Forsyth's letter. He judged the time was not ripe, if it ever would be, to disclose its contents to Jackson.

Now, in the spring of 1830 Lewis thought a revelation would be appropriate considering Jackson's state of mind. One evening not long after the Jefferson dinner when the two men were alone, he related the story to the General, the details of which lost nothing in the telling.

What Jackson had pushed aside two years before, he now viewed in a different light. He demanded, and in due course received, copies of the correspondence. Crawford confirmed Calhoun's adverse posture, though there was no mention of arrest and trial. To Jackson's astonishment, it appeared that Adams had been his strongest supporter in the cabinet. For years Jackson had believed Calhoun his only constant friend in the Monroe Administration, Adams his unrelenting enemy. Obviously, the Vice-President had lied to him and had defamed other cabinet members as well.[18]

When Hamilton learned of the disclosures, he gave the full story to Van Buren who recoiled from having anything to do with what his sixth sense told him would inevitably lead to a rupture between the two men. Jackson sent him Calhoun's reply and relevant correspondence for his perusal and advice; he again declined being drawn into any quarrel between the President and the Vice-President. Any involvement on his part at this time he reminded Jackson would be improper and misinterpreted. Hamilton had been his personal aide; rivals and political enemies would inevitably accuse him of plotting against Calhoun when the two men were working together in friendly accord. Impetuous in so many things, now incensed at Calhoun, the President nevertheless saw the wisdom in Van Buren's frank refusal to be come embroiled in the controversy. "I reckon Van is right," Jackson said to Lewis, "I dare say they will attempt to throw the whole blame upon him."

During this period, several efforts were made to patch up the quarrel, which everyone except the President feared would split the party and make Henry Clay the next President. After several months of charges and countercharges, explanations of this point and that from Calhoun, members of the Monroe cabinet and other interested parties, the dispute died down. But the South Carolinian, brooding over what he considered a blatant misrepresentation of his duty, was biding his time and marshaling his resources for a massive counterattack. He wanted a complete refutation that would expose, he believed, Van Buren's treachery.[19] The coincidence of the President's toast, followed in short order by the Crawford-Forsyth-Hamilton correspondence and finally near the end of the session a veto on both practical and constitutional grounds of an appropriation for building a turnpike between Maysville in eastern Kentucky and Lexington in the center of the state, was conclusive evidence to Calhoun that Van Buren was the marplot who was using the President to destroy him through the manipulation of false issues.[20]

Calhoun had never wavered in his enthusiasm for internal improvements at federal expense. He looked with envy and suspicion on the Erie Canal and the burgeoning prosperity of New York. He feared the population growth of the northern states when compared with those of his own section. The Tariff of Abominations was a first stroke in draining slender southern resources to enhance the wealth and the political power of the North, especially of New York and New England. The supple Van Buren, selfish, narrow and devoted to his own state, was not only bent upon wrecking Calhoun's political career, thwarting his ambition for the Presidency, but with his hold upon Jackson was utilizing the resources of the entire Union for the benefit of New York. Harboring such views, Calhoun found it easy to understand Van Buren's consistent course against internal improvement.

Opposition to the Bank of the United States, opposition to federally sponsored internal improvements, support for a protective tariff, were they not being cast into a set of party principles on which sharp division would occur? On all these questions, Calhoun differed in varying degrees, his harshest criticism always reserved for the tariff. What upset the Vice-President most perhaps was that Van Buren's stand was squarely in the Old Republican tradition of Jefferson, John Taylor, John Randolph and Nathaniel Macon, all southerners whose views on states' rights, strict construction, a wise and frugal central government, had been the traditional, the revered principles of his own party and his own section for thirty years. Calhoun felt himself slowly but inexorably being made a political hybrid in his own land, a public man whose position in the Union, whose very future at this stage in its development, was threatened by an outworn, though hallowed creed. Paradoxically, the old maxims developed for the protection and extension of the agrarian way of life, were now more suitable to the new, bustling commercial, financial and manufacturing states of the North, which were moving away from dependence upon agriculture and growing populous and rich in the process.[21]

Internal improvements had always been popular with westerners and southerners and especially with land speculators. After the land boom collapsed in 1819 there had been a period of relative quiescence, but by 1830 there were distinct signs of renewed speculative activity and with it a growing popularity among members of Congress for internal improvement bills. Although Van Buren knew the President was deeply absorbed with the persistent Eaton affair and with the formulation of his Indian removal policy, he introduced the question of internal improvements during one of their daily discussions.

Well aware of Jackson's past record, and his belief in the military importance of improved transportation, Van Buren was careful to broaden his own views on the subject. Internal improvements of wide national interest that were interstate in nature would not violate constitutional precedent. But the use of federal money for projects like the Erie Canal purely within state boundaries were, he argued, unconstitutional and further, if tolerated would breed a host of schemes that would drain the Treasury, enrich the few at the expense of the many and lead to fraud and corruption.

Jackson had pledged himself to reform, retrenchment and economy in government. Should "the ten thousand schemes" said Van Buren, that arose during the Adams Administration become law, they would make a mockery of Jackson's policies. Exclusion of intrastate projects (scrupulously examined, of course, for their national interest), references to the Adams Administration, to Clay and Calhoun's expansive views, to fraud and corruption and special interests, quickly convinced Jackson that Van Buren's policy was a sound one.

On one of their evening rides over the hills of Georgetown, Van Buren spoke with him about a bill for a federally financed road then before Congress which had all the constitutional objections they had discussed. The road began and ended in Henry Clay's state of Kentucky. It had, he thought, a good chance for passage—the proposed Maysville Road. While the bill was making its way through both houses, Van Buren wrote out in his sprawling, hand a brief, arguing specifically against the project. He handed his manuscript to the President who spent the entire day, "as long as my head and eyes would permit," poring over the thick sheaf of pages. "As far as I have been able to decipher it," he wrote Van Buren, "I think it one of the most lucid expositions of the Constitution and historical accounts of the departure by Congress from its true principles that I have ever met with." When the bill passed in Congress, much to Van Buren's satisfaction, and to Henry Clay's, who saw internal improvements as a popular campaign issue, a rough draft of the veto was ready.[22] At this point, Van Buren made another suggestion that cleverly highlighted Administration policy, yet subtly undercut Calhoun's stand on the tariff. He proposed that Jackson have the secretary of the treasury provide him with a report on the appropriations made by Congress thus far in the session and compare these with the anticipated balance in the Treasury as of January 1, 1831. Jackson eagerly grasped at the idea. Ingham's estimates showed clearly that the appropriations already passed, including funds for the Maysville Road, would exceed the Treasury balance at the year's end. When the bill came before Jackson, he used Van Buren's draft, together with Ingham's estimates for his message, the first major policy pronouncement of his Administration.

As Van Buren anticipated, there were not enough members willing to vote for an override, despite the popularity of internal improvements. He had again succeeded in reemphasizing Jeffersonian principles, thus strengthening the Administration in Virginia, in Georgia and even in South Carolina. He scored heavily against Calhoun's avowed position. In his own state where realistic appraisals rather than abstract constitutional principles were the governing force, the Maysville veto eased fears that Pennsylvania would rapidly complete its east-west canal projects, drawing down the tolls of the Erie and enriching the port of Philadelphia at the expense of Albany and New York City. Even the *New York American* and the *Albany Evening Journal* interrupted their vitriolic opposition to all things Jacksonian with a brief paean of praise for the Maysville veto.[23]

Congress adjourned and Calhoun went back to South Carolina, a bitter man. While he was at home trying to head off a convention that would nullify the tariff, the Kitchen Cabinet was devising additional means to curb his already weakened influence. Amos Kendall and Major Lewis were the principal architects. Both men had long observed Duff Green's independent editorial course; both had noted how the *Telegraph* seemed more an organ for Calhoun rather than for the Administration. And both men, it can be safely said, brought the most telling examples of its inconstant course to the President's attention.

The careless, exuberant Green had been losing ground in Washington even before Jackson's inauguration. But he had prevailed, and the Twenty-first Congress elected him public printer. More important than congressional opposition was the attitude of the President. When the Kitchen Cabinet first broached the idea of another Administration paper to Jackson, he peremptorily rejected it. Green still had a strong claim on his loyalty for his services in the presidential campaign. Nor did the President want two papers in competition, which he was certain would occur and which would simply add to his troubles while weakening the defense of his Administration.

But the campaign for the President's support went on cautiously until Green's editorial course played into their hands. On his way home to Nashville in late June 1830, Jackson wrote Major Lewis from Wheeling that Green's "idol controls him as much as the showman does his puppets, and we must get another organ to announce the policy and defend the Administration." The signal was clear enough and Kendall, after consulting with Barry and Lewis, began negotiations with Francis Preston Blair, a keen editorial writer, and something of a political power in Kentucky.

Van Buren knew nothing, or very little, about the negotiations. He had left Washington on July 5 and was in New York and New England until August. When he returned to Washington, Kendall was making the rounds of the Departments soliciting their printing for the new venture. Van Buren refused to assist on the grounds that if the paper were established "its origin would be attributed to him and he was resolved to be able to say truly that he had nothing to do with it." He had seen through Kendall's professions that the new paper would simply assist the *Telegraph* in supporting the Administration. Sooner or later, he judged, it would come into direct competition with the *Telegraph* and with the Vice-President's ambitions. The President in the meantime had agreed that Blair, whom he did not know, but whose striking prose in defense of the Administration he had seen, should come to Washington and establish an Administration paper.[24]

Blair arrived in Washington about mid-November 1830, virtually penniless, dressed in cast-off clothes, an altogether unprepossessing figure as he presented himself at Major Lewis's office. Almost six feet tall, the Virginia-born Kentucky editor weighed scarcely a hundred pounds. He may have descended from the best of Pennsylvania and Virginia stocks, related to the Madisons and the Prestons, but he looked like a caricature of a man with his extremely broad temples that emphasized sunken cheeks, small sharp pointed chin and a "mouth constantly open, like the spout of a tea kettle, a pair of goggling eyes, a set of teeth [that] would make an admirable harvest rake."

Nonetheless, Jackson took to him at once and eased the way for his advent in Washington. Acting through Kendall and Lewis, the President saw to it that the Departments, except Van Buren's, gave the new press a part of their printing. All those politicians who would curry favor with the White House took out subscriptions and were urged to have their friends do likewise. Within a week sufficient government advances and subscription money had been received to begin publishing. The first issue of the Washington *Globe,* a four-sheet, semiweekly paper, appeared on December 7, 1830. Van Buren took a subscription, but would neither tout the new paper to his friends nor support it in any other way for the time being. He did, however, meet with Blair socially, where he took the measure of the man and was impressed.[25]

An alarmed Green viewed the *Globe* with grave misgivings which prompted him to rash displays of loose talk before individuals whose confidence he could not count on. As for the *Globe,* it received enough revenues during the first month of its existence to begin printing a daily edition. Calhoun, who did not arrive in Washington until late December, instantly saw how much his rival had gained at his expense in the White House, in the Congress and in Pennsylvania, formerly a strong point.

Armed with all his documents and the burning conviction of a man who had been grievously wronged, he began preparing a statement arguing his side in the controversy with the President. Calhoun still hoped for an accommodation with Jackson, though not at the expense of his standing in the party and the country, or, as he suspected, of having to play into the hands of Van Buren. Samuel Swartwout, Calhoun's collector in New York, Senator Richard M. Johnson of Kentucky, one of the most popular public figures in Washington, and Felix Grundy of Tennessee, a close friend of both the President and the Vice-President, were all anxious to close the widening breach.

Jackson himself had finally become more receptive. His decision to run for a second term had just been announced in the *Globe.* Henry Clay was busy putting together a party and building up issues—Indian policy, internal improvements, the rechartering of the Bank of the United States. The Jacksonians were having troubles in Pennsylvania, New York and Ohio with the anti-Masons, and the Workingmen's movement. Nullification threatened to take South Carolina from the Jacksonian column and gave some indications of spreading into neighboring cotton states.[26] Calhoun was the only statesman of sufficient status to keep the South in the fold until tariff revision could be decided upon. So thought many of the worried Jackson politicians in Washington.

Jackson asked Van Buren his opinion of a compromise settlement. "I advised him earnestly and sincerely to consent to any amicable arrangement of the subject that would be consistent with his honor," he wrote, recalling the incident many years later. By now, Calhoun had finished his vindication that throughout implied Van Buren was at the bottom of a plot to destroy his credibility with Jackson. He submitted the draft of his pamphlet to Grundy and Johnson. Both men went over the copy carefully and toned down what they felt were some of the more objectionable phrases. They also suggested that the "Address" be submitted to Jackson for his consideration before publication.

Johnson and Grundy called on Blair with this proposition, but the wily editor resisted their flattering attentions. After reading the "Address" he told them frankly that if it were published it would ruin Calhoun. Blair spoke of the pamphlet with Jackson, and he also told Green that if it appeared there would be "a war between Gen'l Jackson and Mr. Calhoun."

Johnson then gave a draft of the "Address" to Eaton for Jackson's perusal, the understanding between the two men being that if the President had any comment to make about publication it would be accepted. But Eaton, so long the victim of what he regarded as Calhoun's unrelenting ambition, neglected to show it to the President or even make mention of its existence. With the White House silent, Calhoun published "the Address" on February 7, 1831. Jackson read the pamphlet carefully and with mounting anger; Eaton had his revenge at last; the break was now final. The *Globe* opened up the next day with an editorial blast against Calhoun in the slashing manner of Blair at his most caustic.[27]

Calhoun had anticipated an editorial offensive and the word had gone out to his editors that Van Buren must be the object of the counterattack, the President was not to be mentioned. But Kendall, Blair and Lewis knew how to goad Green into indiscretion. And for two weeks, while the *Globe* and the *Telegraph* traded editorial jibes, their respective supporters in the various states did likewise. All of the principals were objects of praise or censure; Calhoun, the Southern statesman or the contemptible liar; Van Buren, the innocent bystander or the base intriguer; Jackson, the noble old hero, or the vengeful tyrant completely in the hands of his self-serving court.

For once Van Buren let the *Telegraph*'s attitude affect his personal relations with a political adversary, albeit within his own party. At social affairs where Calhoun was present, the two men avoided each other. As an excuse for not being drawn into any conversation about their relations, immediately upon arrival Van Buren got up a table of whist. Edward Everett explained that the card table "prevents his falling in with company which he does not like." Van Buren's position was difficult at best, especially prone to embarrassment when he went out in society with the party divided between himself and Calhoun. Green had at least achieved this dubious objective.[28]

The intensity of the newspaper war and the libelous attacks leveled at him prompted Van Buren to publish a notice in the *Telegraph* on February 26, 1831. He flatly denied that he had any connection with the Hamilton-Crawford-Forsyth correspondence, "to prejudice the Vice-President in the good opinion of Gen. Jackson, or at any time. . . ." Although he could produce unimpeachable evidence of his innocence, he never went beyond the statement, still hoping Calhoun would abide by his word and somehow restore harmony in the government. Van Buren's notice failed in achieving that aim, but it did serve to clear his name among many important Jacksonians who wondered after reading the "address" and the newspaper accounts whether he was indeed the treacherous scoundrel he had been portrayed in the *Telegraph* and other Calhoun papers.

Ritchie had been as usual sensitive to the editorial blows. Green's incessant innuendo against Van Buren, his political past, his northern birth, had made a significant impression in the Old Dominion, leading credence to Calhoun's "Address." "It is in vain," wrote Ritchie, "for him to disclaim any 'allusion to one particular individual'—

he does intend you, and so every man who reads the publication will suspect. The prompt declaration of the President has not been sufficient to clear you from the imputation. Many do believe it who were your friends and his." Van Buren's notice satisfied Ritchie and other like-minded conservative southerners, but the public controversy continued unabated despite the lack of new factual evidence on either side.

Van Buren watched the fray with mounting concern about the integrity of the party. He still hoped the editorial brawling could simmer down and that somehow the President and Vice-President would call a halt to the public dispute even if they could not compose their personal differences.[29] But Green would not agree to any cease-fire on terms Blair and Kendall thought reasonable, so the *Globe* resumed the attack, shifting its ground to identifying Calhoun with nullification.

Jackson had warned the nullifiers in his second annual message that he would oppose them with all the powers at his command. Blair saw his editorial policy as simply carrying out orders, but he was striking at the Vice-President's most sensitive point when he accused him of abetting dissension.[30]

Congress had finally adjourned, but the acrimonious dispute continued, pushing would-be neutrals into opposite factions. Van Buren realized that the division was now far too deep for any compromise solution. His own position had changed. He was talked of as the heir apparent, particularly in the southern states, a development at once alarming and disheartening. Inevitably, the opposition press, whether National Republican or Calhounite, would stereotype him as the ultimate conniver, the master intriguer, the consummate manipulator. The only way out of his predicament, he concluded, was to leave the cabinet.

As its most controversial figure, his resignation could restore harmony within the party and stabilize an Administration that was being torn apart by internecine feuds.[31] If he went, Ingham, Berrien and Branch would have to go too. Berrien and Branch had joined Ingham as supporters of the Vice-President and were at odds with Administration policy on internal improvements, central banking and the tariff. A clean sweep of Calhoun influences would give Jackson a free hand in choosing advisers who would support his measures.

There were personal reasons, too, which made withdrawal seem so inviting. As he put it, "the incessant defamation added to the thousand vexations to which official station is otherwise exposed . . ." had so worn him down that he longed for the green hills of Albany and Kinderhook, the easy companionship of his friends and relatives. He worried about his private affairs, whose management he had confided to John, his brother Lawrence and Butler in Albany. He wanted to sell two farms in Kinderhook "unless Uncle Lawrence thinks best for the sheep," he wrote John. "How do they and he come on?" he asked. "We have been kept in such hot water so much here that I have scarcely had time to inquire about him or anything at Kinderhook."

Far from feeling triumphant at Calhoun's reverses, Van Buren was distressed and sorry for the President, who was lonely and sad and frequently ill. The major obstacle was how he would present the idea to Jackson in a form the Old Hero would find acceptable. Van Buren knew he ran the risks of being considered disloyal, of weakness in the face of the enemy, perhaps even of opportunism. In Jackson's current mood he might expect such a reaction and he prepared himself for it.

On their rides during late March, 1831, Van Buren had three opportunities to broach his plan. His nerve failed him on each occasion, much to the amusement of his easygoing son, Abraham. Then in early April, Jackson provided an opening that Van Buren could not resist. The day was warm and beautiful and Jackson, responding to the fine weather, was more cheerful than usual. As the two horsemen turned away from the Potomac toward Tenallytown, a Maryland village, Jackson, who had been speaking about the troubles they had been through, was confident the worst of them were over. "We should soon have peace in Israel," he said. "No! General," Van Buren replied, "there is but one thing can give you peace." "What is that, sir?" "My resignation," Van Buren said. A startled Jackson declared sharply, "Never, Sir," and with increasing vehemence, "even you know little of Andrew Jackson if you suppose him capable of consenting to such a humiliation of his friend by his enemies."

For a moment Jackson's angry tone and his response placed squarely on a personal basis confused Van Buren, but only for a moment. During the remainder of the ride he calmly set forth the reasons why his resignation would not be bowing to the enemy or any kind of retreat on his part, but an affair of state that would restore harmony in the cabinet and permit the Administration to develop clearly and firmly its policies.

When the President asked him what he would do after resignation, Van Buren mentioned the English mission which McLane would soon vacate in accordance with a prior understanding. Nothing was settled, though Jackson had listened patiently and thoughtfully as his secretary of state built his case step by step over a four-hour period. When he had concluded, he enjoined the President not to mention anything about their conversation to anyone, including Lewis and Eaton, so that both of them could be free of any outside advice.

As he had another engagement, he excused himself with the suggestion they meet again for further discussion on the morrow. Jackson agreed.

A worried Van Buren appeared at the White House early the next morning. Although the President seemed to accept the force of the argument, Van Buren noted how loath he had been to concede its validity. His anxiety was confirmed when he saw every evidence of a sleepless night on Jackson's haggard features. The President's opening remarks were cold and formal. "Mr. Van Buren," he said, "I have made it a rule thro' life never to throw obstacles in the way of any man, who for reasons satisfactory to himself, desires to leave me, and I shall not make your case an exception." Van Buren, who was expecting the worst, was ready.

Instantly he affirmed his belief in the strength of his policies, his loyalty to the President. "Come what may," he said earnestly, "I shall not leave your Cabinet until you shall say, of your own notion, and without reference to any supposed interests or feeling of mine, that you are satisfied that it is best for us to part." His final comment was a moving appeal to Jackson's sentimental nature expertly balanced with his sense of duty. "I shall not only stay with you," he said, "but feeling that I have now performed my whole duty in this particular, I shall stay with pleasure and perform . . . whatever it may be proper for me to do." Jackson, his eyes moist with emotion, seized Van Buren's hand. "You must forgive me my friend, I have been too hasty in my conclusion—I know I have—say no more about it now, but come back at one o'clock—we will take another long ride and talk again in a better and calmer mind." The two men again discussed fully the proposal, Jackson probing, Van Buren answering the searching questions that tumbled out, one upon another.

Satisfied now that Van Buren was right, the President asked what he thought of bringing Barry into the picture. Van Buren, sure of having won over Jackson, agreed, suggesting that Lewis and Eaton be also consulted. That evening Jackson presented Van Buren's plan to these gentlemen, all of whom concurred with its wisdom. The President held another discussion the next day at the White House with the same group. Afterward, all except the President rode over to Van Buren's home for dinner. En route, Eaton suddenly asked everyone to rein in their horses for a moment.

Years later, Van Buren remembered his blunt words that immensely strengthened the impact of the move and would pave the way for a thorough change, restoring not just harmony in the cabinet but in the President's personal life. "Gentlemen," said Eaton, "this is all wrong! Here we have a Cabinet so remarkable that it has required all of the General's force of character to carry it along—there is but one man in it who is entirely fit for his place and we are consenting that he should leave it." The others, familiar with Eaton's outbursts, passed his remarks off with a laugh, but when they reached their destination, he asked Van Buren, "Why should you resign? I am the man about whom all the trouble has been made and therefore the one who ought to resign." Van Buren felt a thrill of exultation. "It was a consumation devoutly to be wished but I would have assumed hopeless," he wrote, "and for that reason, I suppose, I have never given it a moment's entertainment." But if Eaton should go out along with him, he would take the disruptive Peggy with him, paving the way for the return of the Donelsons who had gone back to Tennessee and whom the President loved deeply and missed dreadfully.

He concealed his satisfaction, however, and inquired casually "what Mrs. Eaton would think of such a movement." Eaton was certain she would offer no objection and was right. Peggy was not at all happy about the arrangement, but evidently dared not interpose any objection after she learned Van Buren was leaving too. When Jackson was told of Eaton's decision, he grasped at once the perfect solution it offered.

Van Buren wrote out his resignation on April 11, 1831, a few days after Jackson received Eaton's. The President accepted it on the following day, but the news was not made public while the President dealt with Ingham, Berrien and Branch. None of them offered their resignation gracefully and Jackson had to make it explicit before they sent in their letters.[32] He needed one confidential friend in the cabinet near him so Barry stayed on, though much embarrassed at his solitary position. The *Globe* announced the sweeping changes, unprecedented in the public life of the young republic on April 20.

Thinking men everywhere were shaken, though something of the sort had been expected since the publication of Calhoun's "Address" and Jackson's response. On April 16, 1831, Van Buren had informed the Regency of his decision and the next day wrote a fuller letter to Thomas Ritchie stating that he had resigned—"the most important step of my life"—and that he was not hopeful about the consequences to him personally. "But I am willing to do my duty and leave the result with my fellow citizens. That my motives will be . . . misrepresented I am prepared to expect but to that I have been exposed, through my whole life."[33]

While the public furor was at its height, Jackson and Van Buren talked over the make-up of the new cabinet; Van Buren's influence was unmistakable in the new appointments. For the War Department Jackson wanted Hugh Lawson White, a longtime friend and presently a Senator from Tennessee. Woodbury, who had lost his seat in the Senate to Hill, ponderous, pretentious, but capable, was a logical choice for the Navy. His section, New England, had important maritime interests that had not been recognized in the first cabinet. It was also a stronghold of the National Republicans, a section where Van Buren and Calhoun, for that matter, had been trying to build up party strength. Van Buren probably counted on Woodbury and Hill to bring over Calhoun's trainband of officeholders in Boston and elsewhere. Louis McLane, who had won plaudits for his successful negotiations with England on the West Indies trade, was offered the Treasury post, again at Van Buren's recommendation. Everyone, even the Calhoun faction, thought he was eminently qualified. For his own replacement, Van Buren recommended Edward Livingston. To his surprise the President doubted whether Livingston had the executive ability to handle the post. Van Buren finally persuaded him to appoint Livingston.

The primary reason he fought so hard for Livingston was that he would not involve himself in presidential politics. His advanced age and lack of any political base ruled him out of that dangerous game, if he had any such ambition, which Van Buren very much doubted.

Jackson was upset when White declined the War Department, not just because of their friendship and their shared attitudes on Indian policy, but because he had hoped Eaton could take White's seat in the Senate. Next he offered the post to Van Buren's old mess mate, Colonel William Drayton, who, along with Joel Poinsett and Daniel Huger, were leading the fight against nullification in South Carolina. When Drayton declined too, at Van Buren's suggestion he turned to Lewis Cass, governor of the Michigan Territory, and a widely known lesser hero of the War of 1812. A supporter of Jackson's Indian policy, the bejowled, balding Cass concealed his political ambitions behind a phlegmatic demeanor. And he, like McLane and Livingston, had no political base. Though a native of New England, he was thoroughly identified with the West. The President had just as much difficulty finding an attorney general as he had in securing a war secretary. He offered the post to Philip P. Barbour, largely because he was a Virginian and closely associated with Ritchie whose influence needed bolstering after Tazewell and John Floyd had joined the nullifiers. Barbour declined, and the post finally went to Roger B. Taney, a little-known Maryland lawyer and judge, formerly a Federalist like McLane, who had gained local renown for his legal ability. Taney was the most powerful Maryland Jacksonian in a state where the National Republicans were rapidly building up their party.[34]

Calhoun, in seclusion at his plantation, Fort Hill, posed a rhetorical question for Samuel Ingham when he learned the news. "Shall we," he asked, "unite in the reelection of Genl Jackson, knowing as we do, his incompetency, intellectually and morally and the wicked influence under which he acts and must continue to do . . . ?" His answer was a grudging affirmative. "Our true course is at present not to take ground against the Administration, but *at the same time* to take a position independent of them as far as it can be done, short of actual separation."[35]

« CHAPTER 16 »

Secretary of State

The regular packet ship, *President,* from New York to Southhampton, weighed anchor on August 16, 1831. Although the wind was fair for the east and the weather pleasant, the *President* carried only six passengers. One of them was the distinguished, if controversial personage, the new minister to England, Martin Van Buren. His son John accompanied him along with Aaron Vail who would replace Washington Irving as secretary of legation. Vail had been a Regency politician, and a State Department clerk under Van Buren. A man of some ability, he was both a friend and an aide who, like his superior, instinctively made or carried out decisions with reference to their political implications. They saw little of another passenger, the shy, moody son of that sinister chief of Napoleon's secret police. Fouché.[1]

For the first time in his adult life, Van Buren had neither the cares nor the pleasures of politics and public office. Even his summer vacations, when they were not taken up with legal business, had been devoted to political affairs. And for the past two years, the continuous work at the Department of State occupied what time he could spare from other public business, the tempest of politics and patronage in Washington, all of which followed him wherever he went. During these busy years he had, he knew, neglected his family, though for most of his term as secretary, Abraham and Martin, Jr., lived with him. John had made frequent visits whenever Van Buren judged they would not interfere with his study for the bar. And Smith, too, the son who resembled him most in physical appearance, came to Washington during the recesses of his school.

Despite his frequent absences and his heavy burden of work, Van Buren had done his best to watch over his sons. Affectionate and indulgent toward them, he had not spared their feelings if he thought they were not measuring up to his image of what a gentleman should be. Even Abraham, a grown man and an officer in the army, came in for an occasional "scolding." John, the ablest and the handsomest of his sons, caused him the most concern. As any loving parent, he wanted his sons to be a credit to him and to enjoy comfortable, useful lives. His position, however, made the achievement of these goals for his family difficult because their father was always in the public eye. Schoolboy pranks became acts of vandalism, irregular attendance at church evidence of unsound morals. Well chaperoned, innocent excursions with ladies of their own age were frequently made the subject of insinuating notice in the opposition press. Unfortunately stories of John Van Buren's drinking and gambling, his consorting with the *demimonde* of Albany and New York had a semblance of truth, and word of John's excesses reached his father's ears, prompting sharp lectures, reminding his son in forceful terms of the position the family must uphold.[2]

John Van Buren had many of his father's talents, his quick mind, his likeable personality, his ease with people of all stations in life. And in one respect he was superior—his gift as an orator which was displayed during his student days at Yale. He also gave promise of being his father's equal as a lawyer. What this attractive son lacked was balance.

Van Buren enjoyed gambling on elections, on horseraces, at cards, but always with moderation. He had a taste for wines and spirits, but he never permitted himself to become intoxicated, and indeed, never drank so much that it blunted his critical faculties. Unfortunately, John lacked this sense of proportion and more than once his style of living and his escapades gave his father much anxiety. But of late, John had been a credit and a source of satisfaction. He had been admitted to the bar and passed his counselor examination. Van Buren, who always enjoyed his company more than any of his other sons, was delighted to have him along on this mission, which he believed was a turning point in his own life.[3]

As he pondered the tumultuous past two years during the first night at sea, he wondered whether he had made the right decision. He knew some of his closest friends disapproved. But the next morning as he walked the deck, he regained his confidence. The *President* had made a good run while he slept. Familiar landmarks of New York harbor had disappeared, his native coast but a smudge on the western horizon. It was a time for reflection over the abrupt change his fortune had taken. There were no letters and documents demanding his attention which had been, as he recalled, "regularly and plentifully emptied upon my table from the mailbags with the spoken alarms of timid croakings, of complaining, and rarer congratulations of satisfied friends. . . . there were no crowd of petitioners, visitors" and messengers from other Departments or the

White House with notes for his immediate attention, no newspapers to scan.

He realized that he missed the hustle and bustle, the intrigues of designing men, the rumors, the gossip, even the routine that had claimed so much of his life thus far; and as far as he could see, this would now taper off, if not cease entirely. His ambitions for the Presidency he considered at an end as he had told Jackson when he first discussed his resignation. "Reason and experience," he felt, "forbade the expectation that any political party would voluntarily encounter the risk of selecting as its candidate an individual peculiarly obnoxious to its adversaries and of whom strong jealousies were cherished by rival leaders within its own camp, after he had . . . left those rivals in undisputed possession of the field of competition." Yet he was certain he had made a wise decision not just for himself personally, but for the public good.

He left behind him, a record of substantial achievement in his two years as secretary of state, though there had been some mistakes. Still, the verdict his friend and former law partner, Butler, had pronounced when informed of his decision to resign was closer to the mark than Van Buren's retrospective musings. "The moment also," Butler had written, "at which you are to withdraw, is peculiarly propitious. You have held the office just long enough to show that you were fully equal to its duties, but not so long as to be exposed to the prejudices which of late years are so intimately associated with the long possession of this office."

Butler's remarks were particularly apt, not because of their allusion to previous secretaries, but because the major diplomatic problems pending when Van Buren took office were all satisfactorily adjusted or seemed about to be solved. That Van Buren, a man with no prior diplomatic experience, should have turned in a better record than Clay, had settled disputes that defied the best efforts of Adams and Gallatin, was little short of remarkable. He had achieved these results by building on the work of his predecessors, by his choice of singularly capable ministers for the most difficult assignments, by rapidly changing economic and social conditions in England and France that favored the improvement of relations with the United States, and perhaps most importantly, by implementing a policy that differed from his political style.

In politics he played a waiting game, letting the opposition entangle itself in its own mistakes before he struck hard with his disciplined organization. In the foreign policy he developed over almost two years of study, he was direct, open and conciliatory. Van Buren had come to the conclusion that reciprocity would be the goal and that this best accorded with old Republican principles.[4]

But first he understood he must cultivate the diplomatic community in Washington and show them that the new President was not the illiterate, unbalanced frontiersman they had read about during the election campaigns, the impulsive gamester who would endanger European capital invested in the United States on impulse and export his dangerous leveling notions abroad. Van Buren had not been in Washington a week when he arranged for the diplomats to meet the President, and he coached Jackson on what he should say to them.

When they arrived at the White House and were ushered into the East Room, Van Buren, in accordance with strict protocol, introduced each member to the President. Jackson, dressed in well-cut black evening attire that set off his tall, gaunt figure, and ruff of grey hair, greeted them with the proper touch of formality and friendliness. His brief speech after the introductions were made sounded just the right note for the occasion. Here was no excitable, ill-tempered, ill educated demagogue, but a gentleman of natural presence, even if his western accent seemed odd to the cultivated ear. The doors of the state dining room were then thrown open and the old General presided with ease and charm at a sumptuous repast.

The English minister, Charles Vaughan, the French chargé, Count Menou, and the Russian Baron Krudener were impressed; Van Buren followed up this formal occasion with a series of small informal dinners. He was soon on terms of familiarity with all the diplomats, though Vaughan, Krudener and Chevalier Huygens, who represented Holland, became his closest friends in the diplomatic community. Van Buren wanted to dispel the malevolent gossip that had been circulating in Washington about Jackson's incompetence during the first, fumbling weeks of his Administration when no provision had been made for the envoys to be presented at the White House, or to establish a means of communication between them and the new Administration.

The promotion of warm personal relations between himself and the diplomatic community would place an affirmative stamp on their correspondence and their reports. Besides, Van Buren enjoyed the company of these worldly-wise aristocrats, whose manners and dress perhaps made too much of an impression on the tavernkeeper's son. Yet keen observer that he was, he learned much about the personages and the characters of the foreign courts. What he did not pick up in casual conversation he supplemented with his careful reading of the *London Times* and other British newspapers and translations of pertinent articles in the French press.[5]

The President had alluded to the prompt solution of the nation's outstanding diplomatic problems in his first annual message. But well before that Van Buren had added to his already substantial body of information on these issues through the counsel of his clever friend Cambreleng, whose mercantile and his congressional career had made him something of an expert on the West Indies trade that had been more or less closed to direct American shipping since the Revolutionary War. James A. Hamilton and McLane, the new minister to England, also contributed significantly in formulating policy and tactics.

William Pitt Preble, a shrewd member of a prominent Maine family who was designated minister to the Netherlands, was helpful in all aspects of the long-standing dispute with England over the northeastern boundary with New Brunswick. Preble had been one of Gallatin's assistants when, as Adams's minister to England in 1826, he had laid the groundwork for arbitrating the boundary dispute. He had the double advantage of being knowledgeable about politics and public opinion in Maine, a most important consideration to Van Buren. For he would accept no settlement that did not have the approval of Maine itself. And Van Buren was hopeful that Preble would exert enough influence on the king of the Netherlands, who had been accepted as arbitrator for a favorable settlement.[6]

In his approach to the dispute between France and the United States, Van Buren had less information to guide him. During the Napoleonic War, the French government had expropriated American shipping in ports under its control and the French navy had seized or destroyed American shipping on the high seas with a flagrant disregard for neutral rights. American merchants claimed upwards of $13 million in damages for the loss of their property. But the French had emerged from the Napoleonic War with an empty treasury, a crippled economy and a heavy burden of reparations to the victorious allies.

The ministries of Louis XVIII and Charles X, while tacitly admitting the claims, had postponed serious discussion of them on one pretext or another. Senator Samuel Smith and Albert Gallatin were the Americans most familiar with the French question. Smith's advice was to allow French claims of indemnities arising out of the Eighth Article of the Louisiana Purchase Treaty. The Tariff of Abominations had lowered duties on French wines, brandy and silks. These reductions had stimulated trade between the two nations, and Smith urged that indemnity claims and the tariff be used as bargaining counters.

Basically, his instructions to McLane would lift American restrictions against British vessels and cargoes trading from the West Indies on the condition that Great Britain place the United States on the same basis as other nations that traded with the West Indies, the most-favored-nation principle.[7] This concession had been allowed by an Order-in-Council of 1825, but because of a series of delays on the part of the United States it was withdrawn. For years there had been no direct American trade with the islands, forcing merchants to use the ports of the Danish islands, St. Croix and St. Thomas and the Swedish island, St. Bartholomew. The indirect route was inconvenient, lengthened voyage time, affected insurance rates and excluded American vessels from the lucrative interisland trade.

Van Buren urged a spirit of conciliation, rather than an Adamsian insistence on a right when a privilege would serve as well. Reciprocity was the cornerstone, but in case of rejection, Van Buren reserved retaliation—the embargo of trade with Canada and heavier restrictions on British commerce with the United States. As for the tariff, if that should prove an obstacle McLane was to emphasize the burden on American farmers, the Corn Laws imposed. Although he did not include it in his instructions, he was willing to agree upon a restriction in the export of sugar and coffee on American vessels. And he might even consider waiving the drawback then existing on the re-export of West Indian produce from American ports. Van Buren could not resist injecting an implied, but nevertheless critical tone in his historical background and statement of policy that placed the previous Administration in an unfavorable light. Of course, the instructions would not be published until success had been achieved, and he meant to extract from them as much political capital for the Jackson Administration as he possibly could.

Upon completion, he decided he would personally deliver McLane's instructions to him at Wilmington. He wanted to discuss them with McLane and hoped for a bit of relaxation away from the cares of Washington. He also asked Cambreleng and James A. Hamilton to meet with them for further talks and suggested they all foregather at Cape May.[8] But after he had written Cambreleng, he had second thoughts about the personalities involved. Might not the opposition press blow their meeting out of proportion, distorting its purpose and alarm the business and commercial interests unnecessarily? In an offhand manner he mentioned these factors in his invitation to Hamilton that went out the next day. "I wrote Cambreleng, advising him and you to meet me at Delaware and to take a trip to Cape May with me. If you do not apprehend that a meeting with an anti-tariff champion at such place would, in connection with what has already been said, furnish food for newspaper speculation, I should like to have you come. Cambreleng would turn up his nose at this in great contempt but there is more in small matters than he is always aware of."

Cambreleng did not attend nor did Hamilton. Van Buren, accompanied by his sons Abraham and Martin, Mrs. Donelson and her cousin Mary Eastin, arrived at Wilmington on July 25, 1829. The women went off to Philadelphia, leaving Van Buren and McLane free to go over the instructions and canvass the entire manner of approach. McLane would meet with Cambreleng and Hamilton in New York for any last minute conversations that might prove useful before he embarked on the U.S.S. *Constellation* for Europe.

Van Buren was optimistic that they had hit upon the right formula for success. And McLane shared his enthusiasm, though his fretting about the dangers and discomforts of the voyage to England got on Van Buren's nerves. "For a truly great man, he has more littleness about him than usual," he wrote. Cambreleng had indeed scoffed at Van Buren's concern "with small matters," but it was well he had taken them into consideration. For his trip was immediately seized upon as being another trick in the magician's bag. Duff Green had found out that Van Buren was delivering instructions to McLane and broadcast in the *Telegraph* that they would scuttle reciprocity. Fortunately nothing came of Green's pronouncement though Van Buren made haste to have it corrected without disclosing his position. In fact, his prompt action brought an apology from James Hamilton, Jr., now a confirmed nullifier, for Green's being "so quick on the trigger to repudiate the idea that a reciprocity treaty was thought of."[9]

Van Buren felt less hopeful about the French mission than the English, and the preparation of instructions for Rives gave him a good deal more difficulty than those for McLane. The condition and attitude of the French government was not then conducive to a successful adjustment. After he had gone over the mass of dispatches and previous instructions of six American envoys and the evasive notes of eight French ministers of foreign affairs, he despaired at first of having any fresh ideas. After composing a draft he was still dissatisfied and rewrote it completely, finishing the twenty-five-page document on July 12. First, he described briefly the two points at issue between the United States and France, the claims of American citizens against France and the counterclaims of the French government arising from its interpretation of the Louisiana Treaty of 1803. Next, he had his small staff classify all the claims, placing them in five categories. He then reviewed the tedious history of the negotiations for Rives's benefit, after which he analyzed the American and French positions in, for him, a crisp, rather spare style. As a legal brief, it was a clear, logical and persuasive argument that step by step demolished the various approaches the French foreign minis-

tries had taken in the past. In true lawyer fashion, the instructions tried to anticipate new arguments that might be raised together with possible means to answer them.

Despite Senator Smith's advice, Van Buren instructed Rives not to combine the French interpretation of Article Eight of the Louisiana Treaty with the claims. As in the McLane instructions, Van Buren could not resist a criticism of the Adams Administration's handling of the dispute, while following in the main the precedent it had laid down. France's refusal to negotiate on these terms "was called out by the emphatic manner in which Mr. Adams in his note to Count de Menou had asserted them." But again following precedent, Van Buren was willing to arbitrate the Louisiana claims separately. Where he adopted a more flexible position was in a willingness, if it became necessary, to negotiate on all issues outstanding; but American claims were a matter of right and the principle was not to be compromised. In forceful language, he declared that Louisiana was an intrinsic part of the United States, not a colony, and that France could not claim special commercial privileges in New Orleans over other trading nations. Repeatedly he characterized the attitude of the French government as "ungracious" as well as contrary to international law and usage. Should it agree to its responsibilities in principle, the American government was willing to accept a gross sum, which it would distribute to the claimants, and was willing to scale down the amount demanded. Van Buren left the bargaining of these details, which amounted to a concession, up to Rives. Noting that the issues had been so long under discussion between the two countries, Van Buren thought the French government "must long since have made up their mind in regard to them." He was hopeful for a settlement but cautioned Rives against pressing too hard. "Should you find reason to apprehend a bias in the minds of the public functionaries of France so unjust to us, and so adverse to the honor and true interests of France herself," he wrote, "you will not fail to make every effort in your power to produce a better feeling and wiser and worthier views."[10]

After these difficult and burdensome tasks had been completed to his own satisfaction, that of the ministers involved and of the President, he turned to less complex but no less onerous chores with a sense of relief. Diplomatic relations between the United States and Spain and Russia were good. The American minister to Russia, Henry Middleton, an Adams appointee, would eventually be replaced; and Van Buren was even then considering John Randolph as his successor. But this decision would wait. The Van Ness appointment to Spain was meant to reward the Jacksonians of New England for their valiant but unsuccessful stand in the presidential election. And, of course, it was meant to encourage those New England Jacksonians who were supporting Van Buren against Calhoun.[11]

Van Buren had hoped for a respite before tackling what at the moment seemed minor matters, the claims against the Kingdom of Naples and Denmark for Napoleonic spoliations, a commercial treaty with Turkey, the Van Ness instructions, in all of which the Adams Administration had done most of the preliminary work. But it was not to be so because conditions in Mexico suddenly took a threatening turn, and Jackson himself had a personal interest in acquiring the Mexican province of Texas.

Affairs in Mexico were troubled as usual. There had been no less than four revolutions in the past five years. American merchants had had to compete with an exceptionally vigorous British campaign to monopolize the trade of Mexico and, indeed, of all Latin America. The United States was fortunate in having as its minister to Mexico, Joel Poinsett, a man of conspicuous ability, courage, a complete understanding of the bizarre convolutions of Mexican politics and absolute dedication to American interests. Like his British counterpart, Poinsett had involved himself in Mexican politics and for a time had been most successful. By 1829, however, public opinion in Mexico had taken a decidedly anti-American turn, even though members of the faction most friendly to him were in control of the government. At this point, a South Carolina born Mississippian, Anthony Butler, visited Mexico City and returned full of enthusiasm for acquiring Mexican territory west of the Louisiana border. The United States had once held a claim to this territory, but John Quincy Adams relinquished it for the cession of Florida. Since then there had been a good deal of American immigration west of the Sabine and a good deal of lawlessness in the region that Mexico had been unable to curb.

Butler believed the weak Mexican government would welcome a cash settlement of all or part of the territory between the Sabine River and either the Brazos or the Trinity and running northwesterly to the Louisiana Purchase line. Jackson, who had always been deeply interested in the protection of the port of New Orleans, and an early apostle of westward expansion, needed little encouragement from Butler. Van Buren, who had spoken with Butler and who knew of the impression he had made on Jackson, prepared a document suggesting a purchase of Mexican territories west of the Sabine. The President was in full agreement and indicated that the United States might pay Mexico as much as $5 million for the territory. He embodied his views in a set of notes and an outline of instructions for Minister Poinsett. It confirmed their earlier understanding and offered a series of points that Poinsett might use as bait for the purchase. Donelson completed the document, spelling out the President's ideas. These Van Buren incorporated in a dispatch that reflected faithfully Jackson's expansionist views and, if anything, enlarged upon his original arguments that it would be in Mexico's best interest to cede the Texas provinces. The document was entrusted to Butler, who armed with a note from Jackson, set off for Mexico. It all came to naught, however. President Vicente Guerrero, bowing to public pressure, had reluctantly asked Jackson to recall Poinsett. Anthony Butler, who succeeded the worthy envoy, was incapable of making any arrangement with the Guerrero and succeeding Mexican governments. In fact, his seven-year term as minister to Mexico gravely weakened the American position by his persistent and underhanded schemes to detach Texas from Mexico. For Van Buren, the "Texas Project" as it was called, simply weighed him down with an additional burden on a subject that he had given little or no study. Apart from Poinsett's lucid dispatches, his only other source of information had been Butler, whom he had just met.[12]

The "Texas Project" finished and Butler preparing for his departure to Mexico, Rives and McLane, whom he had enjoined to maintain close relations and communication with each other, already on the high seas aboard the U.S.S. *Constellation,* Van Buren looked forward to a more leisurely period. He had already decided on the best method of securing a commercial treaty with Turkey. The Ottoman Empire had suffered a series of grievous defeats, its sea power crushed by the British and French fleets at Navarino, its armies presently under attack from Russia. Van Buren had taken Baron Krudener, the Russian minister, into his confidence and had been assured of his nation's support of an American treaty with Turkey.

A year earlier, Captain William M. Crane, then American naval commander in the Mediterranean and David Offley, the American consul at Smyrna, both commissioners the Adams Administration appointed to negotiate a treaty, had failed in securing for the United States the same privileges enjoyed by other nations in the ports of the Ottoman Empire and especially the Black Sea. Van Buren decided to continue Offley, and since Crane had been relieved by Captain James Biddle, he was appointed a commissioner. But he added a third member, Charles Rhind, a New York merchant of Scottish birth and pugnacious disposition though Van Buren thought him to be "a very sensible, worthy man." Rhind's mission would be secret except to the other commissioners. His "cover" (Van Buren's exact word) was his appointment as American consul at the Russian Black Sea port of Odessa. Van Buren gave Rhind exclusive powers of negotiating with the Sublime Porte because after reading all the correspondence and consulting with Baron Krudener, he decided Offley had worn out his credibility in the Ottoman court. Captain Biddle, like his predecessor Crane was made a member primarily to show the flag.

The instructions were simple and strightforward. Rhind was to secure a treaty that extended the most-favored-nation principle to the United States, but he was not to be a stickler like Offley "upon terms, though not precisely so favorable as those which may be granted to others, yet neither materially affecting the privilege obtained, nor disparaging to our national credit." Van Buren, ever the realist, did not think Rhind would succeed simply through diplomacy. "Everything will depend upon the state of the war when he reaches there. If the Russians are successful, our success is certain."[13] In an almost gleeful mood he finished off the last two bits of pressing business, instructions to Preble and to Van Ness. Randolph had accepted the Russian mission but was granted permission to delay his departure until the next spring.

Van Buren had been ill briefly but made a quick recovery and in the glorious autumn weather of Washington relaxed the pace he had kept up since his arrival at the capital seven months before. The routine of the Department was in the safe hands of Daniel Brent, the experienced chief clerk and his own chosen subordinate, Aaron Vail. Van Buren turned to the more congenial tasks of jotting down ideas for the President's annual message and waiting the first reports from his envoys abroad. These would not be arriving for several months and though Jackson would have liked to make some specific mention of progress in his message, he resigned himself to the general phrases Van Buren supplied him with. Still Van Buren had shaped them in such a way that they breathed a spirit of accommodation, of fair play and of mutual interest. The subject of claims against France for a settlement of Napoleonic spoliations was cast in a firm and righteous though conciliatory tone. The President asserted "the established policy of all enlightened governments as on the known integrity of the French monarch that the injurious delays of the past will find redress in the equity of the future."[14]

Van Buren heard from Rives in early December before he had any word from McLane. Prince Polignac, the French foreign minister, as expected, sought to disavow the Bourbon Monarchy from any acts of the Napoleonic regime, and to see a difference between American claims and reparations already paid to the victorious allies. Rives came out well in the encounter, citing long-established principles of international law and the patent fact that the government of Napoleon had not only the support of the French people but had been recognized by all foreign powers as the legitimate government of France. Rives also seized the opportunity to remind Polignac that the Tariff of 1828 had accorded favorable treatment to French imports.

At a later discussion, again as anticipated, Polignac tried to connect French claims under the Louisiana Purchase Treaty with American claims against France, and had also raised the issue of unpaid debts owed to the heirs of Beaumarchais whose company had supplied arms and munitions to the American government during the revolution. Van Buren was willing to deduct one million francs from American claims to satisfy this long-standing obligation. But his and Jackson's stand on the Louisiana issue remained unchanged—simply a convenient pretext to deny valid claims. "It would, therefore, be as unprofitable, as useless, at this time," he wrote Rives, "to renew the discussions touching those principles." Week after week Rives fenced with Polignac over this issue and finally hit upon a means of solving it. The United States would lower even further its duties on French wines if the French government would withdraw its claims under Article Eight of the Louisiana Treaty. Jackson and Van Buren approved of the tactic, which Polignac accepted. By the end of May 1830 Rives was certain the King would approve a treaty that substantially met all of the American claims.

Unfortunately, revolution occurred in France, the government fell and the King abdicated. Jackson, Van Buren and even a frustrated Rives approved of the liberal government of the new King Louis Phillipe, though they knew the case would have to be made all over again. Rives must accommodate himself to a new set of ministers, a new style of government, quite probably the same obstinacy when it came to paying out cash whether the claims were legitimate or not.

Despite this setback, Van Buren and Jackson were well-satisfied with Rives's performance. His dispatches were clear and full, his presentation of the American case was lucid, and his response to the points the wily French foreign minister kept throwing in the way was quick and generally successful. Rives had, as Van Buren wanted, corresponded freely with McLane; both men profited from their cross-channel discussions. Both men moved easily in society and established good rapport with the governments they were accredited to. When the wives of the two ministers arrived, they too were of notable assistance, not just in a social sense, but in picking up the gossip of the courts and in the case of Mrs. Rives, helping her husband with his dispatches.[15]

Van Buren had adjudged McLane's mission to be less fraught with difficulty than Rives's. His reasoning was based on the sound principle that the planters of the West Indies could get cheaper and better produce from the United States in direct trade than from the indirect trade Britain's policy was imposing. Their influence, he felt, would be made in Parliament, as it had been made in earlier negotiations.

He had some fears that the temperamental McLane might be too aggressive with the British and push too fast. In one of his early notes he said that while the President was anxious for a settlement, McLane was free to proceed at his own pace. McLane was pleased and relieved that he would have all the time he needed without any undue pressure from Washington. And Van Buren need not have worried himself about his minister. However excitable McLane may have been at home, he controlled his high spirits and was all cordiality and business when dealing with the friendly Scotsman, Lord Aberdeen, foreign minister in the Duke of Wellington's government.

McLane had his troubles, too. The British government procrastinated as it tried to balance the complaints of the planters with the insistence of shipping interests from the Canadian maritime provinces that they retain their monopoly of the direct trade. Still, with influential members of Parliament like Robert Peel, William Huskisson, and in the government Aberdeen himself, McLane made some headway in the midst of the welter of issues, domestic and foreign—Catholic emancipation, electoral reform, growing political disturbances in France, the Turko-Russian war—that absorbed the energies of the Wellington cabinet.

Aberdeen's foreign policy of peace and nonintervention made him more susceptible to settling grievances of long-standing provided they were not detrimental to the British economy. In late February 1830 while the Board of Trade was considering the question, McLane decided on his own to make a stronger push and in a note to Aberdeen warned that failure in the trade negotiations would damage friendly relations between the United States and Great Britain, two nations with so many common ties and interests. Aberdeen's reply and his informal conversations were so cordial that McLane felt he would succeed if he could persuade Van Buren and Jackson to have Congress lift the restrictive legislation against British commerce in American ports that preceding administrations had secured from Congress in retaliation for the closing of West Indies ports to American trade.

While Rives's note describing what he deemed was a changing affirmative attitude in Paris was making its way across the Atlantic, Jackson had become restive at McLane's lack of progress. In one of his rare memoranda on foreign affairs, he wanted contingency legislation prepared should the British reject American terms. A considerable volume of trade went on between the United States and Canada. The Maritime Provinces and Lower Canada were, in part, dependent on the import of American foodstuffs because they were cheaper and better than British imports. Van Buren had written an embargo of American-Canadian trade into McLane's instructions, which he could present if negotiations broke down. Jackson now wanted a message to Congress that would recommend an embargo and sufficient revenue cutters to enforce it along the northern and northeastern boundaries.[16]

Before Van Buren had a chance to prepare the necessary documents, McLane's first note arrived, and shortly afterwards several more communications, increasingly optimistic in outlook. And toward the end of May 1830, only four days before Congress adjourned, both Houses received a communication from the President declaring he had sufficient evidence of Britain's removing all restrictions on direct trade with the West Indies and recommending he be given authority to lift American restrictions provided Britain actually reciprocated. Congress passed the necessary legislation with no debate and in record time.

Van Buren had put all of the negotiations on a positive note enlarging on McLane's optimism in the communication to Congress he drafted for the President. But he was careful to point out "no decision had been made" as yet and asked that the President be authorized to act for Congress during the recess by proclamation "in the event of an unfavorable decision." He had, at the same time, written McLane to play the embargo card against Canada if he thought it necessary. McLane never took this drastic action. The Act of Congress satisfied the British government of the United States' good faith in the matter. An American initiative provided the unpopular Wellington ministry with the face-saving device it needed. There were some quibbles about the interpretation of the American statute, but nothing substantial. When Van Buren received the British decision in September, he was as overjoyed as McLane at the final solution of a long-festering problem, and especially pleased that it had come in time for the elections in New York, Maryland, Delaware and Maine.

Jackson was so delighted with the news that he cut short his vacation at the Hermitage so that he could be ready to issue his proclamation when it was thought to have maximum political impact. He issued the proclamation on October 5, 1830. The British government responded declaring that American vessels could now trade directly with the West Indies. Had the receipt of the proclamation in London been delayed another two weeks, McLane probably would have had to contend with the same problem that had halted Rives's negotiations when he was on the brink of success. The Wellington ministry fell and a new Whig ministry under Lord Grey came in with Lord Palmerston as the new foreign secretary. A much more difficult individual to deal with than Aberdeen, Palmerston was no friend of the United States and an aggressive minister whose policies smacked much of mercantilism.

It had been a near thing where Van Buren had acted promptly and decisively in having the United States take the initiative, even though it did so on the basis of opinions rather than facts. McLane had suggested this mode, but Van Buren's was the responsibility. And even with the caveat he insisted on in the President's communication, a collapse of negotiations after such hopes had been raised would have damaged the public trust in the Administration, furnished fuel for the opposition and unquestionably weakened the Jacksonians in the fall elections.[17]

The Jackson press went into action immediately enlarging upon the success and making invidious comparison with the Adams Administration. National Republican newspapers, headed by the *Intelligencer,* claimed that Jackson had bargained away important tariff schedules in the name of reciprocity. Seizing upon the British notice of adjusting its duties in the near future it declared that high discrimination tariffs on American produce and manufactures would place the United States in a worse position than it had been before the negotiations. But the most telling blows of the opposition press were that the West Indies trade was far less important than it had been. The Jackson Administration had wasted time and money in securing privileges that were worth next to nothing.

Actually, the so-called "Reciprocity of 1830" was a compromise. The opposition was in part right when it asserted that Britain would achieve through discriminatory duties what it gave up in its exclusionary Navigation Act. Yet the United States had for the first time as an independent nation gained direct trade with no restrictions to the West Indies on almost all products of both countries. It was a far better arrangement than what the British had offered in 1825. As for discriminatory duties on cargoes, the Grey ministry did enact them, though they were less burdensome than Van Buren had expected. Clearly the arrangement favored American producers over American shippers, but as the National Republican press had said, the American carrying trade was now so diversified that the West Indies was but a small and declining part of its business.[18] The political impact that Jackson and Van Buren had counted on in close states did not materialize, but the triumph was justly theirs, and the energetic diplomacy they inaugurated would have a cumulative effect that would eventually show at the polls.

Van Buren needed the West Indies success to cover up what were two minor blunders the opposition immediately exploited. Randolph's mission to Russia was a complete failure, though the legation's secretary did made a commercial treaty that enhanced American trade in the Baltic. The eccentric Virginian had neither the physical stamina nor the mental stability to take on a diplomatic mission, let alone one with a major power. Van Buren and Jackson had acceded to his every request, and Van Buren himself had traveled to Norfolk personally to present him with his instructions. Randolph remained but twenty-six days in St. Petersburg where he made a bad impression on the Russian court and the remainder of the six months he lived in London.[19]

Van Buren's other mistake was one of much greater magnitude than the unfortunate Randolph episode—the appointment of William Pitt Preble as minister to the Netherlands. At the time Preble seemed an ideal choice. Since Britain and the United States had agreed that the king of the Netherlands be the arbitrator, he would satisfy the interests of his native state and bring his expertise to bear upon the settlement itself.

Preble proved to be a disaster as a diplomat, however. A self-centered, self-righteous individual, he scoffed at precedent if it conflicted with his own preconceptions. Had Van Buren looked into his record, or even bothered to ask Gallatin about Preble, he would never have had him appointed. Decidedly he was not a suitable agent for exerting influence in the Dutch court or in a careful evaluation of any settlement. He seems never to have comprehended that the essence of arbitration was compromise. Nor did he give due weight to national interests when they might conflict with the exaggerated pretensions of Maine.

When he became minister to England, Van Buren tried to put the best possible light on the Dutch king's settlement. And by emphasizing to the United States the benefits of the British acceptance, he advised Jackson and Livingston to support the arbitration in the Senate. Jackson hesitated; Preble's denunciations and the uproar in Maine stayed his hand. Both nations then withdrew from the settlement.[20]

Van Buren's choice of Charles Rhind to negotiate a commercial treaty with Turkey at first glance seemed just as flawed as his appointment of Preble. Rhind shared many of Preble's weaknesses, his vanity, his impulsiveness, his abrasive personality. But where Preble destroyed a favorable settlement, Rhind made a good treaty with the Ottoman Empire. Van Buren had paved the way for him by betting on a Russian victory and by enlisting Russian support for Rhind's mission. Even then, after Rhind had negotiated the treaty with a secret clause that was a mere face-saving device for the Sublime Porte, the mission very nearly foundered when his fellow commmissioners Biddle and Offley refused at first to sign. In this case, Rhind's stubborn egotism stood him in good stead. He would take the responsibility for the secret clause that permitted the Sultan to make private contracts for the purchase of naval stores and ships in the United States, a privilege Rhind declared vehemently would never be exercised. Offley and Biddle signed, only after firing off indignant letters to Van Buren roundly condemning their association as an unscrupulous intriguer.

Van Buren dismissed all the irate complaints cheerfully and seems not to have been concerned when the National Republicans sought to make political capital out of the secret proviso. The Senate ratified the treaty, but it struck out the proviso. The Ottoman government made no effort to cancel the privilege of most-favored nation it had extended to the United States as Rhind had advised Van Buren in his last dispatch. It was hardly in any position to do so after the humiliating Treaty of Adrianople with Russia. The Turkish treaty was a minor affair, but Van Buren had been astute enough to cast his lot with Russia and use the might of the Empire for the commercial advantage of the United States.[21]

When he resigned as secretary of state there was but one major negotiation yet unresolved, the spoliation claims against France. But Van Buren was certain, however distasteful the payment of claims would be, the new French government would settle for nearly all that was asked. Lafayette was now helping Rives treat with the new regime; and just before he stepped down, Van Buren heard from his minister that he was about to reach an agreement of the amount of reparations with Count Molé, the French foreign minister. It appeared likely that the sum would be 25 million francs, about 90 percent of the true value claimed exclusive of interest, which had been waived.[22]

Van Buren remained in Washington carrying out the duties of directing the State Department until Livingston could settle his affairs and be briefed on the policies, pending negotiations and routine business of the office. It was not until May 29, 1831, that Livingston assumed the office. That evening Van Buren dined with Jackson "for once a private citizen" as he scrawled out in a note to three of his four sons. John had come on to help his father and brothers break up the household. Van Buren planned a little family celebration at home after leaving the White House and only wished that Smith who was away at school could have been with them.

The next two weeks Van Buren and his sons supervised the sale of some furniture, the inventory, crating and shipment of the rest to Kinderhook, writing recommendations for servants, sorting and collecting his personal papers, "destroying such as are no longer useful" and removing from the department files what he felt he needed.

Livingston took over his house which by now had acquired almost the status of the official residence for secretaries of state. Van Buren sold his carriage and horses to the new French minister, Louis Serurier. By early June he was all packed up and ready to leave after a final, rather emotional farewell to the President. The fact that he was to be nominated for the British mission was still a well-kept secret, though the letter writers in the press had not overlooked that possibility.

Speculation died down during the summer and was only renewed briefly when the President made the appointment on August 12, 1831. During the interval Van Buren, who was homeless again, spent most of his time in mending political fences on a trip with Butler across the state and in relaxing at Albany where he took rooms at Congress Hall, at Saratoga and at Kinderhook. Here, amid a troop of friends and relatives, Van Buren confessed that though his health was good, he was "at the moment suffering somewhat from the effects of Kinderhook hospitality." On August 1, he left the hearty Dutch fare of his native town for Albany.[23]

He had, however, made some preparations. From Livingston he learned that Washington Irving, for the past two years secretary of the legation at London, wished to leave. With Jackson's permission, he offered the position to Aaron Vail, who was happy to accept. He planned to take the regular commercial packet, departing without fanfare from New York, but Jackson would not hear of it. He had Woodbury order the U.S.S. *Potomac* to New York for his conveyance.

McLane had arrived when Van Buren wrote Hamilton that he was going to Albany, and so it was from that city he hurried with his son John to New York City for the long-awaited conference. McLane, Van Buren, Hamilton, Cambreleng and Aaron Vail spent a day or so discussing politics, the Washington scene, the European situation and the attitude of the British government and court toward the United States. All things were now falling into place. Vail reached New York City on August 3 bearing Livingston's instructions and Van Buren's commission. Van Buren did not want the political attention his departure on a ship of war would certainly arouse. He managed to make his wishes known to Woodbury without the President's knowledge. That accommodating individual found a number of sound reasons why the *Potomac* should not convey the new minister. A relieved Van Buren boarded the regular packet, the *President,* on August 16, 1831.[24]

After an uneventful voyage of twenty-four days, Van Buren's group landed at Cowes on the Isle of Wight where he reported his safe arrival to Livingston before taking the ferry to Southhampton and the post chaise to London. He took quarters at Thomas's Hotel in Berkeley Square, which would be his headquarters until he could find a suitable house for his embassy. McLane had lived here before he moved to a permanent house on Chandos Street and had recommended it to Van Buren. He had supplied him with other useful suggestions: "Adams for carriage and horses, avoid a human named Marsh—Lady Wellesley and Miss Easton will instruct you in the mysteries of fashion and (Joshua) Bates, your banker of Portland Place—whom you will find the kindest man in the world, will instruct you on all matters of business and housekeeping."[25]

Like all his predecessors, Van Buren's pay and allowance were barely adequate to cover his expenses and the modest amount of entertainment expected of the Americn minister. Thus, Van Buren looked for a house near the less-expensive center of London fashionable enough to uphold the dignity of the United States. He found one at No. 7 Stratford Place, a five-story town house that was spacious enough to meet his family needs and social obligations, reasonable enough to meet his stringent budget. "I pay £500 sterling for my home, from which taxes say £50 are to be deducted, and it is considered very cheap. . . ." he wrote Hamilton. "I pay £300 for my carriage and about 2600 dollars for my servants including their board in the house. . . . we go, however, upon notions of strict economy." As embassies went in the 1830s, Van Buren's household staff was very small. He hired a coachman and a footman, a butler-steward, two maids, a valet, and a housekeeper who doubled as cook, seven in all. To save on expenses he imported his wine directly from France. For very special occasions, he had a sufficient quantity of Lynch's best champagne, but he ordered sixteen-dozen bottles of cheaper variety and an equal amount of claret at three francs a bottle, "which will be good enough to give to my friends. I would say," he explained to Rives, who was acting as his wine merchant, "a few degrees above what is usually used with water."[26]

Van Buren's official duties, except for the Maine–New Brunswick arbitration, were not pressing. He had missed the coronation of King William IV and the social activities surrounding that event. Washington Irving greeted him in London and promptly arranged an interview with Lord Palmerston, the foreign minister in the new Grey ministry.

Van Buren and Palmerston seemed to have liked each other from their first meeting. The plainspoken, sensualist, stood half a head higher than the jaunty American minister who presented his credentials with a certain air of candor. The Irish peer was one of the ablest public men in England, but his offhand manners and his lapses into vulgarisms had set the more fastidious of the diplomatic community on edge. Though the two men could not have been more distant in background, education and taste, they both shared a remarkable ability in judging men and women, and events.

The tavernkeeper's son who had risen so far in American politics interested Palmerston, who took his measure and marked him down as an agreeable companion with whom he could do business, not the sort of high-strung fellow McLane had been. As for Van Buren, if he felt any awe at meeting the blustering aristocrat, he concealed it cleverly. For the remainder of his short stay in England, Van Buren and Palmerston enjoyed a friendly relationship that went beyond their official duties.

Van Buren also made a good impression on the new king and his consort, the plain, strong-willed Queen Adelaide. King William was fat, and far from prepossessing in appearance and in ability. Like Palmerston, his manners were most casual. The King, who served for many years in the navy, had imbibed the manners of the quarterdeck, not of the great landed families with whom Palmerston was at home. Indeed, the King's naval service had circumscribed his limited intelligence and outlook, while it enhanced a streak of obstinacy that seemed a family trait. Yet Van Buren, who was used to cultivating inferior minds if they could be turned to his advantage, got along well with King William, or "Silly Billy" as many of his irreverant subjects called him.

Nor was Van Buren at all surprised or offended at the large brood of William's bastards by the actress Mrs. Jordan, which was much in evidence. Basically, his own rural upbringing had taught him to be as casual as Palmerston about such matters unless they posed a danger to himself or his immediate family. He had finally come to terms with his son John's peccadillos, fretting only about the money he spent in social pursuits. "If John were not with us," he complained to Cambreleng, "I think I should be able to keep my wages in balance with expenditures."[27]

Van Buren, himself, had more time on his hands than ever before in his life. The shaky Grey ministry was so deeply involved in domestic politics and social unrest that Anglo-American relations received short shrift. Behind Palmerston's cordiality was a firm policy to avoid any further complications when there were so many at home. The first major political reform since 1689 was being agitated in Parliament and in the country. Times were hard and acts of violence against land owners and factory owners were widespread throughout the kindgom. A revolution had broken out in Holland; there were signs of a revolution in Spain between two opposing claimants to the throne. Van Buren, sensitively attuned to all these distractions facing the British government, patiently put up with Palmerston's long delays to his proposals on outlawing impressment, which he realized was not an urgent issue; or even compensation for slaves who had been liberated in the British West Indian islands on which their transports had been shipwrecked. Most important of all, he found Palmerston willing to declare informally that Britain would accept the Maine boundary arbitration.

Van Buren had more to do with Rives's mission than his own. The young minister reported fully on his negotiations and sought his advice on whether he could exceed the letter of his instructions and lower the duties on wines and brandies in exchange for the settlement of the Louisiana claims. Van Buren backed him up and wrote to both Jackson and Livingston in support of Rives's position.[28]

Other than that, he was almost as active in the glittering London social scene as his son John. He and Aaron Vail, who acted as an interpreter, dined frequently with the aged Talleyrand and his talented daughter, the Duchesse de Dino. Lord Grey, the Prime Minister, and Lord John Russell entertained the Van Burens at country weekends. Lady Wellesley and Lady Holland found the American minister with his easy flattery, his slightly exaggerated manners, his bonhomie a unique relief from the correct and somewhat stiff American envoys they had known. And young John, as urbane and a good deal wittier than his contemporaries among the English "bloods" who lounged around the whist tables at Holland House or Apsley House, the Duke of Wellington's princely London residence, attracted much attention from the doyennes with their marriageable daughters.[29]

In his letters home, when Van Buren was not discussing his diplomatic maneuvers or commenting on British politics and statesmen, he gave descriptions of his travels and the historic personages he met. There was no analysis, not even mention of the desperate social conditions in England—the hideous poverty and squalor of working-class London within walking distance of Buckingham Palace, of Marlborough and Apsley House, of Regent Street and Mayfair. Nor did he report on the plight of the tenant farmers driven by near famine conditions to desperate acts of violence against the property of their indifferent landlords.

Van Buren, like most members of Britain's ruling classes, simply did not comprehend the social upheaval of the times as the industrial revolution in Britain took its heavy human toll, shaking the social order to its foundations. His only reference to poverty was in connection with the cholera epidemic that broke out in Sunderland. The comment, a passing one to Livingston, was the common assumption that poverty, indigence and disease were related and were inflexible aspects of the human condition. Cholera "is not contagious and may be prevented by cleanliness and regular habits," he said.

"The fact that its ravages are principally confined to the very poorest and of course most destitute classes . . . was proof enough that ignorance and vice bred poverty which was as immutable as the seasons."[30]

A political being to the core, Van Buren was no social critic much less a philosopher. Like Lord Melbourne, whom he admired, Van Buren was a kindly man who was unable to detach himself from his own sphere and view the world of the less fortunate with an objective eye. Doubtless his absorption with the political process was one factor in his blindness to reality, and in addition the situation at home seemed to be taking a curious turn.

The President wrote him wondering whether he had changed his mind about not running for Vice-President; he hinted darkly of combinations against Van Buren developing in the Senate. James Hamilton spoke also about opposition in the Senate to his nomination, and James Watson Webb flatly predicted he would not be confirmed, as did Cambreleng. What if a coalition of Calhoun supporters and National Republicans rejected him? On what basis could they do this without exposing themselves to charges of naked partisanship? If they braved criticism which was sure to arise from such a conflict of interest between purely political concerns and the public interest, what effect would it have upon him? Embarrassment surely! The British government and the court would not understand how a minister whose public career had been so illustrious could be recalled at a throw of the political dice. Yet he must prepare himself for the contingency. January was an anxious month at 7 Stratford Place.

Over the Christmas holidays the Van Burens joined Washington Irving in a tour of the English countryside. While at Barlborough Hall, a rambling Tudor vicarage in Derbyshire, Van Buren received a large packet of letters. The one that caught his eye first was an envelope addressed in the bold hand of the President. It was a long letter containing news that the National Republicans had followed the anti-Masonic example and held a national nominating convention in Baltimore. Henry Clay and John Sergeant were nominated for President and Vice-President. But two sentences fairly leaped from the page. "If I am reelected, wrote the President, and you are called to the Vice Presidency, I wish you to return to this country two years from now. The opposition," continued Jackson, "would if they durst try to reject your nomination as minister, but they dare not, they begin to know if they did, that the people in mass would take you up and elect you vice Pres. without a nomination; was it not for this, it is said Clay, Calhoun and Co. would try it."[31] Van Buren gave a good deal of thought to his reply and wrote a very careful letter to the President in which he deftly planted the idea of being Jackson's running mate, but at the same time renewed his opposition to such a course. With what he had heard from Washington, he wanted to be ready for rejection. In late February 1832 the bombshell he was expecting dropped upon him, but not in the way he had imagined.[32]

Aaron Burr, about 1801, when Van Buren first met him in New York City.

De Witt Clinton. Van Buren's great rival for political mastery in New York. (Courtesy of The New-York Historical Society, New York City)

Daniel D. Tompkins. War of 1812 governor of New York State in uniform as commander-in-chief of the state's military and naval forces. (Courtesy of The New-York Historical Society, New York City)

Hannah Hoes Van Buren. (The Library of Congress)

John Van Buren. Second and favorite son of Martin Van Buren. (Courtesy of Mrs. Rufus King Duer)

Mordecai M. Noah as a Van Buren editor in New York City. (Frick Art Reference Library)

William Morgan whose mysterious disappearance led to the formation of the anti-Masonic party in New York.

Azariah Cutting Flagg, Van Buren's right bower in the Regency. (Albany Institute of History and Art)

U.S. Senator Martin Van Buren. (Library of Congress)

William H. Crawford. Van Buren's presidential candidate in 1824. (Library of Congress)

Louis McLane, close friend and political associate of Van Buren in the 1820s. (The University of Delaware)

Thomas Hart Benton, U.S. Senator from Missouri, in the full flush of resounding periods and furious denunciations.

Churchill C. Cambreleng, "Cam" as Van Buren called him, Chairman of the House Ways and Means Committee.

President Andrew Jackson. (The Library of Congress)

Political cartoon satirizing Van Buren's safety fund law. (Courtesy of The New-York Historical Society, New York City)

A view of Washington during the Jackson Administration. (The Library of Congress)

Peggy Eaton in a somber mood. (The Library of Congress)

John C. Calhoun as he appeared at the height of the Eaton affair.

Van Buren as Secretary of State. (Courtesy of The New-York Historical Society, New York City)

William C. Rives, Virginia aristocrat. Minister to France, U.S. Senator, political ally, then enemy of Van Buren.

Opposition cartoon on Van Buren's rejection as Minister to Great Britain. (Courtesy of The New-York Historical Society, New York City)

James K. Polk of Tennessee, capable, industrious Speaker of the House of Representatives.

« CHAPTER 17 »

Minister as Martyr

Van Buren felt out of sorts when he awakened on Monday morning, February 20, 1832. He decided not to attend the Queen's first drawing room of the season and he was still in bed when Patrick Lynch, his valet, appeared with an unusually large bundle of letters. Van Buren saw the familiar handwriting of Cambreleng on the pile and opened it first. He was astonished at what he read. That he would probably not be confirmed he was expecting, but that a tie should have been contrived so that the Vice-President could vote against him, he had never foreseen. Instantly, he recognized Calhoun's fatal mistake and did not need Cambreleng's joyous conclusion to read the situation accurately. "You will be our V.P. in spite of yourself," Cambreleng wrote. "Come back as quick as you can—we have no triumphal arches as in ancient Rome, but will give you as warm a reception as ever a conqueror had."[1]

Absorbed as he was in the news and the accounts, Van Buren forgot about his discomfort and joined Vail, Irving and John at the breakfast table. All of them had read the morning papers and were anxiously awaiting his reaction. Patrick had already told them that Van Buren was unwell and relayed his request that Irving attend the Queen's drawing room in his place. When Van Buren appeared, he handed Irving the Cambreleng letter, which, after he had read it, he passed on to Vail and John. Irving made the first comment. If Van Buren felt well enough he should attend the reception where he could clear up the doubts and the speculations of the London press, reassuring members of the government that the Jackson Administration was not only stable but popular. Irving conceded that it would be difficult for him. He would be the cynosure, an altogether humiliating experience, a minister whom his government had apparently repudiated. But he must do it for his own defense, for that of the President and for the honor of his country, which had been exposed to ridicule before the most powerful government in Europe.

Van Buren agreed and braced himself for an ordeal that turned out not to be much of an ordeal at all.

Lord Palmerston had been on the lookout for him and as soon as he arrived at the palace conveyed him to an anteroom for a private chat. Palmerston was fully informed of the Senate proceedings, having been sent dispatches covering the affair from the British chargé on the same packet that brought Van Buren the information. He understood what had happened, that it had been a partisan affair and he had so informed the King. For himself, the government and the King, he reassured Van Buren that his standing was unaffected and that he should not feel embarrassed in any way. During the days that followed, others in the government and in the opposition, many of whom Van Buren did not know, called and left cards or invited him to teas and dinners.[2]

The next two packets brought bundles of letters and newspapers from home, describing in detail the vote in the Senate, the reaction of the party and the public, and offering suggestions as to Van Buren's future course. Walter Lowrie, one of the Pennsylvania Senators and a longtime political associate, wrote, "At first your friends were rather vexed at this result. No party wishes a defeat, even if that defeat may be supposed to lead to results which they much desire. Now, however, altho but one day has passed, some of your friends, I know and amongst them myself, would not change the result if they had the power."[3]

The President wanted him to return as soon as possible and assist him in giving battle to the enemy. Cambreleng had shifted ground from immediate return to arrival just before the Baltimore convention. Most of the letters assumed that he would be the party's candidate for the Vice-Presidency. But some, including those from McLane, Lewis, John Forsyth of Georgia, and the President himself, thought he should take Dudley's seat in the Senate.

The Regency was divided. Wright and Flagg, worrying about the state election, thought he should run for governor in place of the lackluster Throop, then make a run for the Presidency in 1836. Marcy, Cambreleng and James A. Hamilton considered it unwise for Van Buren to be sidetracked with the governorship when the Vice-Presidency would place him before the nation immediately as the heir-apparent, yet shield him from assuming controversial or divisive stands.[4]

Van Buren mulled over all the suggestions, paying particular attention to comments on the contenders for the vice-presidential nomination from states and from sections. The South and the West especially were insistent on recognition and were skeptical about Van Buren's role in shaping the Tariff of Abominations. Blair offered a pertinent observation that caught Van Buren's eye. The combination of Calhoun, Clay and Webster was a formidable one. "They want the Senate and the Vice-Presidency," said Blair, "to hold the Executive in check, and they have the frenzy of party so perfectly raised in the South that if they can get the Presiding Officer, the President may be said to be check-mated for the next term." Blair had meant this as a spur for Van Buren, but as yet he did not know his man. A political realist, Van Buren was not the person to throw himself needlessly into the Senate as the President's standard-bearer, or prematurely into a vice-presidential race that from all accounts involved a risk to himself and disruption in the party.

Nor did a campaign for the governorship of New York seem either wise or attractive. Although he had canvassed the state before he left for England and found the organization sound, he was worried about an eventual fusion between the anti-Masons and the National Republicans. The situation in New York City, too, where a vigorous Workingmen's party was challenging the Tammany old guard, could be troublesome.[5] Until he had a look for himself, he would not commit himself to the pleas of Wright and Flagg. Private life held out its allures. His financial position was secure, and at times he became disgusted with the darker side of public life. But if he were to give it all up, it must be on his terms, his reputation intact, his character unblemished.

Even as he thought of retirement, Van Buren yearned for public acclaim and the psychic rewards of public service. He wanted to be President fully as much as his rivals; yet he lacked the remoreseless perseverance of Calhoun, the fanatical persistence of Webster, or the supple, cynical management of issues that marked Henry Clay as a political gamester.

Van Buren cared for broad national issues; and though he advanced and defended those that incidentally promoted his native state, he never advocated a policy that was exclusively in its interest or in the interest of its region. Calhoun had some of these qualities and so did Webster; but both, through circumstances and through geography, were eventually forced into sectional positions not of their own choosing. Both men lacked the crucial ability as political leaders to forge an organization that transcended narrow special interests for the greater good. Nor could they discipline their followers and direct their energies, their frustrations, their ambitions, into broader channels.

Jackson on the other hand, through instinct and force of personality, commanded loyalty. And Van Buren, through careful, often plodding examination of all possible contingencies, was usually able to arrive at a plan of action that was persuasive to his followers—

though not always right and not always successful. If, as John Randolph once remarked, "he rowed to his objective with muffled oars," he usually had charted his course with scrupulous care, whether it be in the formulation of public policy or whether it concerned his personal future.

At this dramatic change in his fortunes, he resisted the appeals of his friends to take any overt action. The opposition would splinter, he reasoned, if left alone; it stood a good chance of coalescing should he return with undue haste for vindication. Calhoun might well rally the South on the charge that he had all along connived for just this result, that he had, in Marcy's blunt terms, "seduced . . . not a woman—but the President—made a break between him and our worthy presiding officer—you were a great intriguer—the author of sundry plots . . ."[6]

He would remain in Europe until well after the Baltimore convention. On February 20, 1832, Van Buren wrote Jackson of his intentions and his reasons for delaying his departure. His lease on No. 7 Stratford Place had ten months to run. Custom demanded he relocate his servants in other households. Carriage, horses, wines must be sold. Goods like silver plate, cut glass, fine English cabinet work, which he purchased, must be crated, insured and sent home. All these arrangements would take time, if he were not to sustain a substantial monetary loss. His son John, who had left for a tour of the Continent just before the news of his rejection had arrived, must be contacted and return, a matter of a month at least. These were all minor, though pertinent arguments. The major points were essentially political and couched in ways that Jackson would understand.

In taking up the idea of his speedy return and election to the Senate, Van Buren remarked that there was no vacancy at the moment. Nor was it known whether Dudley, the incumbent, would be willing to resign. "There are many strong reasons," he said, "why I would not for the world suggest or cause it to be suggested to him. I have before had occasion to tax his patriotism by asking him to decline a nomination for the good of the cause; and under circumstances which excited regret on the part of his friends." Then, Van Buren added the telling point. Should he replace Dudley in the Senate, he could not possibly do so until very near the end of the session, a month or so at the most, not enough time for him to make any contribution, but sufficient time for his appearance to seem like "a systematic scheme to hurry home and make the most of the sympathy produced by the violence which has been done me . . . and the false imputation of an original design on my part to run for the Vice-Presidency."

He would put his personal affairs in order, turn over the embassy to Aaron Vail, take a short trip on the Continent with his son, relax and observe conditions in various European countries, returning to the United States in "May or early in June . . . affording another proof of the falsehood of the accusations against me." He would leave his political future to "the unbiased disposal of our political friends with a determination from which I shall not depart to abide by their decision but to be satisfied with it, they always remembering my strong repugnance to the place of Senator."[7]

As far as the governorship, it would appear that if Van Buren were not nominated at Baltimore, he was reserving that option. The next packet brought with it the usual bundle of letters and papers telling of mass meetings in New York, Pennsylvania and other states, and a series of resolutions from the Jacksonian members of the New York legislature roundly condemning the action of the Senate as an insult not just to Van Buren but to the President, the State of New York and to the majesty of the United States. After reading these accounts, Van Buren changed his mind about writing a public letter and decided on the better course of responding to the resolutions of the legislative caucus, which he knew would be forthcoming.

In his reply to the legislature, he mentioned no names but left the action of those in the Senate who voted for rejection to their individual consciences. Where Van Buren struck his sharpest blow was his declaration that the Vice-President and Senators who voted against his confirmation had demeaned liberty and representative government in the eyes of those oppressed Europeans who looked to the United States as their example:

> Whether in their eagerness to strike me down, my adversaries have not reached beyond their aim and given a blow to the character of our institutions in the eyes of foreign nations, is a question not between them and me, merely; nor again between them and their country, alone; but between them and the friends of constitutional liberty throughout the world, who look to the conduct of our legislative bodies as the criterion of that representative system for which they are all so anxiously struggling.[8]

Meanwhile, the mass meetings went on, especially in New York State as the Regency sent out the word, and the *Argus,* the *Evening Post* and the *Courier and Enquirer* engaged in dazzling rhetorical displays that were echoed and reechoed by dozens of Jacksonian dailies and weeklies. In Washington the *Globe* wrought itself into a verbal frenzy as Blair matched Jackson's rage and gored his rival, Duff Green. Ritchie, with more restrained language, assailed Calhoun and Clay as ringleaders of a dastardly plot. "The State of New York is literally in flames," wrote Van Buren to John, urging him to cut short his trip and meet him in Paris. In justifying his Fabian tactics, Van Buren wrote a series of letters to Jackson advising the President to rely upon Marcy and Cambreleng for advice on New York matters. He also set the stage for Throop's replacement when he tactfully hoped there might be a position of dignity for him in Washington or in New York.[9]

The President recognized the wisdom of delay and left the campaign for the vice-presidential nomination (which Van Buren had never sought, at least explicitly) in the capable hands of the Kitchen Cabinet and the Regency in New York. Van Buren himself called the attention of the President to the weaknesses in Clay's and Webster's attacks on Van Buren's instructions. Clay had followed virtually the same course of conciliation in his instructions to Gallatin on September 22, 1826. As a good lawyer, Van Buren had studied all the diplomatic correspondence with Great Britain over the West Indies trade and had taken his notes with him to England. Although 4,000 miles away from the scene of action, the Magician had matters well in hand. Squarely behind him was the great personal popularity and inflexible determination of the President, a cohort of the best politicians and the best editors in the nation, a cause which had now become the cause of liberty, of principle over partisanship. The one area of uncertainty was the attitude of the South, but his reading of Ritchie's *Enquirer* and the southern correspondence that it reprinted, while showing a potential danger over the tariff, indicated that a majority of the politicians and editors in the slave states were looking to a compromise solution. With everything set in motion, Van Buren left for the Rhineland.[10]

The Calhoun papers charged that the convention was another Van Buren contrivance. They had no proof, however, and were unable to halt the proceedings. Jackson was the unanimous choice of the party, so the convention would simply choose the vice-presidential nominee and draft a platform. There were a half-dozen candidates, but only three emerged as real contendors, Philip P. Barbour of Virginia, Calhoun's proxy, Richard M. Johnson of Kentucky, a compromise candidate in case of deadlock, and Van Buren, who everyone conceded had the most strength. There had been opposition in circles close to the President; John Overton, one of Jackson's closest friends and a delegate to the convention, felt the nomination of the Magician would not be acceptable to the South. Eaton, too, feared Van Buren's nomination would split the party. McLane, ever ambitious, encouraged doubts, hopeful that the mantle would fall upon him. Overton's and Eaton's qualms, if they ever reached the President, would have been dismissed out of hand. Had McLane's moves been known to Jackson, he would have further jeopardized his position in the cabinet where his controversial Treasury Report praising the Bank of the United States had already raised doubts about his loyalty to the White House.[11]

Knowledge of the President's wishes and the smooth management of the Regency delegates brushed aside all opposition without any serious division. Van Buren was nominated on the first ballot with 208 votes to 75 for his two opponents, Barbour and Johnson. His supporters, in an excess of misguided enthusiasm, voted a two-thirds rule rather than a simple majority for nomination as a demonstration of party solidarity behind the candidate.[12]

Nathaniel P. Tallmadge, a state senator from Dutchess County, formerly a Clintonian, did yeoman work for Van Buren at the convention. Flagg and Wright were active among delegates from the North. But the most ardent worker for Van Buren was the adjutant general of New York, John A. Dix. Originally one of Calhoun's bright young army officers who was to help him make his political fortune in the North, Dix had swung into the Regency camp after Jackson's election.[13]

News of his nomination reached Van Buren when he was preparing to embark on the packet *New York* from Liverpool for home. The return trip was unusually long, some thirty-five days during which the Regency politicians made extensive plans for a triumphal welcome. They were to be disappointed, however. Van Buren had anticipated them and wanted to maintain a low profile for both political and personal reasons. After the tedium of more than a month on shipboard, he was not in any mood to undergo the fatigue of a public celebration. Ever mindful of the adverse political effects that could come from any sort of extravagant lionizing, he used the cholera epidemic, which had finally reached the United States, as a very real pretext for declining the public dinner Tammany had planned.

His eldest son Abraham was waiting for him when he stepped ashore. Jackson had sent him ahead with a letter briefly stating that he was working on his veto of a bill rechartering the Bank of the United States for another twenty years. He wanted Van Buren in Washington as soon as possible. The President highlighted the urgency of his letter when he said that Hamilton had papers that he wanted Van Buren to examine on his trip and be prepared to give his counsel upon. After a short visit with Smith, who was still living with the Hamiltons, he hurried off with Abraham for Washington.[14]

Van Buren arrived in Washington on the night of July 8, 1832, and went directly to the White House. He found Jackson in bed, exhausted, "a spectre in physical appearance but as always a hero in spirit." The past few days Jackson had spent in pondering, analyzing, and editing a rough draft of his veto message of the Bank recharter bill. Kendall, an opponent of the central bank for years, was its principal author. Other members of the team who toiled at the White House writing and rewriting from early morning far into the night were Attorney General Taney, Secretary of the Navy Woodbury, and Donelson. Taney, in his first major policy work for the Administration, supplied the constitutional history and argument, and Jackson weighed each sentence for its political and economic impact. The tension had been high, because they were dealing with one of the most powerful institutions in the nation and with its leader, a man as indomitable as Jackson himself, Nicholas Biddle.

Congress had passed the recharter bill with substantial majorities. Everyone in the White House knew that their own followers in the House and Senate were split on the Bank, while Clay, Calhoun and Webster and all their adherents were united in favor of it. They also knew that Henry Clay, through Biddle, was counting on a veto; and that he would make it the major issue in his campaign against Jackson. The President could not know whether the veto would be a political success or a disaster; whether it would shake the confidence of the country in the Administration or strengthen it. What he believed was that the Bank represented special privilege and that its rechartering, when it still had four years to operate under its current charter, was an electioneering trick, not a measure in the public interest. He had read Thomas Hart Benton's dramatic speech in the Senate attacking the Bank as a "monster" that was draining the productive West for a small group of eastern aristocrats and foreign parasites. His own experience with the Bank during the lean years of the early twenties had shown him how such a powerful and impersonal institution could affect adversely the fortunes of so many. The entire cabinet, except Taney who was in Maryland on business, counseled a veto which would simply argue that the issue was premature. On the other hand, Kendall and Blair, both of whom had encountered the heavy hand of the Bank in Kentucky, were bitter opponents. Their ideas reinforced those of Jackson, as did those of Taney, who left a long opinion distinctly adverse to the Bank before he left town.

It was a long message, some of it demagogical, more of it sound argument, which emphasized the privileged position and the awesome power of the Bank over the simple economy of the day. Though there were references to benefits the Bank had conferred, these were slight deviations from the theme that no private institution operating behind closed doors should reach into every pocket and purse of every citizen. The thrust was political as the rechartering had been, and if there was exaggeration, occasional lapses of logic, the message was clear, the principles Jeffersonian. But the effort had been so fatiguing to Jackson that it brought on the usual symptoms of nervous exhaustion.

When he saw Van Buren at the door of his bedroom, the old man rose from his bed and passing his long thin fingers through his coarse gray hair, he said calmly, "The bank, Mr. Van Buren, is trying to kill me, *but I will kill it!*" He asked Van Buren to read the message and make any comments or suggest any changes he saw fit. As usual he was prepared, in fact had been thinking about the Bank and its collision course with the Administration for several months.

Before he learned of his rejection by the Senate, he had received the President's message and McLane's Treasury report, which not only spoke of the Bank but urged the renewal of its charter. Throughout McLane's report ran a vein of Federalistic doctrine. Van Buren was chagrined that his hand-picked secretary could not resist his Hamiltonian background; even if Jackson tolerated an honest difference of opinion, McLane's line of policy was unacceptable to him and to all the Old Republicans. "I regret with you the discrepancy between the views expressed by the President and Sec'y of the Treasury," he wrote Rives. "You will perceive that the report is very unpalatable to the great body of the Old Republican Party."[15]

Van Buren had spent much of his political life in dealing with banking institutions, and from the beginning he had been skeptical of their operations and their influence. Until he resigned the governorship and went to Washington as secretary of state, he could not remember a legislative session that did not have applications for bank charters before it. As the economy of the state expanded and especially after the completion of the Erie Canal, the pressure for new banks grew ever more intense and more inextricably woven into the political texture of the state.

There had been several notorious scandals where lobbyists for competing bank groups bought and sold legislators as if they were so many bushels of grain or so many barrels of salt pork. Van Buren himself had been instrumental in securing a charter for the Bank of Hudson, whose cashier, Gorham A. Worth, the witty chronicler of small-town worthies, was one of his personal friends, though they differed politically.[16] He had also helped secure the charter for the Farmers' and Mechanics' Bank of Albany; and he and other Regency men were from time to time directors and officers. This bank, too, despite frank political connections, was conducted with honesty and ability. Its cashier Thomas Olcott, one of the ablest bankers in the state, was a careful and efficient businessman. The Farmers' and Mechanics' Bank did receive some favors from the Canal commission; but its investment policy was, on the whole, most beneficial, particularly in the development of western New York, its trusteeship for a portion of the state funds beyond reproach, its loans made according to sound banking practice, as it was understood at the time. Though it was frequently attacked as a political machine, even Thurlow Weed was never able to uncover any evidence of partisan mischief in its operation.[17]

In essence, Van Buren regarded banks as necessary instruments for economic growth, but he was alert to any abuse of their inherent power over the community, their tendency toward monopolizing credit and making excessive note issues. At the onset of his political career he had tried unsuccessfully to abolish the legislative practice of demanding bonuses for bank charters on the grounds that they stifled competition and stimulated corruption in the state government. He sponsored a measure to raise funds and to restrict banking excesses after the Panic of 1819 through a tax on bank notes. His bill called forth a violent response from De Witt Clinton. "The Van Burens and other would-be great men of the day are in favor of it," he said pompously. "Opulent men must uphold the substantial interests of the country . . . against the heresies of faction and the votaries of Jacobinism."[18] But Van Buren could not, and would not, stem the tide of banking incorporations. What he did was make the banks of the state responsible for the economic health of each other under a board of bank commissioners that was empowered to impose a common standard of probity and prudent ratios of notes to assets.[19]

When Van Buren joined the Administration, he and Amos Kendall were the only members close to Jackson who understood the nature and practice of banking. He soon learned of the President's intense feelings about the Bank of the United States, his firm belief that it used its monopolistic power over credit for private gain over the public interest. In principle Van Buren agreed with his chief, but the Bank was so closely involved with all the great interests of the country that he feared any impulsive action would worsen the depression still clinging like a limpet to the waterlogged economy. Ever wary, he would test the political implications of any move; the best way to achieve that end was to avoid any mention of the Bank in the President's first annual message, thus providing time for a careful polling of opinion in the Congress.

Given Jackson's attitude, if this were not possible, or a general statement not acceptable, then the merest hint of reform in the message would test the reaction of the country and of the Bank, whose policy at the time was bipartisan. Major Lewis fell in with Van Buren's ideas, but they had to move carefully. They knew of Kendall's aversion to the Bank and his influence with Jackson. Van Buren had read the message, and indeed had written a large portion of it, but not the passage dealing critically with the Bank. These strictures disturbed him because of their extravagant language and their obvious bias against the Bank as an undemocratic institution. There were no specific recommendations; Congress would determine proper remedies. Such a voicing of policy, he thought, without any preparation, would have damaging political and economic repercussions.

Uncertain of Jackson's position, Van Buren decided an indirect approach was best. Accordingly, he asked his former deputy, James A. Hamilton, to intercede with the President. Hamilton had ingratiated himself with Jackson and stood the best chance, he felt, of getting any mention of the Bank deleted or at least toned down, without giving offense.

Jackson complied; Hamilton spent the entire night at the White House working over the message, the few pages of copy on the Bank claiming his primary attention. At one time, early in the morning, the General, in his nightshirt, wandered in holding a candle and, ever solicitous, had his servant keep a constant supply of logs on the fire. Hamilton finally produced a draft he believed would satisfy Van Buren and that the President might accept. After a few hours sleep, he went to the President's bedroom at 8:00 a.m. and reported that the work was done. "What have you said about the Bank?" Jackson asked. "Very little," replied Hamilton, who then read him the few lines he had written.

The substance of his remarks was that since the Bank charter did not expire for another seven years, any action at this time might have adverse effects considering the fundamental monetary and fiscal issues involved. "Do you think that is all I ought to say?", asked Jackson. "I think you ought to say nothing," said Hamilton. "Oh! my friend," said Jackson, "I am pledged against the Bank, but if you think that is enough, so let it be." Hamilton breakfasted with the President and then left for Van Buren's. "Well, Hamilton, what is done?" asked the anxious Secretary. "The work is finished. I could not induce him to let me omit everything as to the bank and here is what he agrees to." Van Buren read it over and was satisfied that it would serve the purposes he had planned. The hint of uncertainty about the Bank's future was sufficiently broad for the kind of probe Van Buren had in mind without damaging further an already embattled Administration.[20]

Van Buren watched and listened and read the newspapers gauging the intensity of discussion on the benefits and the disadvantages of the Bank. During the year the Bank became the subject of heated debates in Congress and, as he had anticipated, a controversial topic in the press. Biddle felt so uneasy that he spent large sums for favorable notices in papers whose proprietors saw no conflict between paid editorials and paid advertisements. These encomiums in turn brought forth sharp rejoinders from editors who felt as Jackson did about the Bank, though division as yet was not on party lines.

The able president of the Bank himself journeyed to Washington, where he spoke with Jackson, Van Buren and other cabinet members. He maintained lobbyists in Congress, subsidized Daniel Webster and Henry Clay and dozens of lesser members of both houses. The campaign was going forward, slowly but inexorably, and Van Buren could see a popular issue developing. The second annual message, while it enlarged on the Bank issue, did not disguise the Administration's hostility to the institution as presently constituted. Van Buren softened the language of the draft he received on the Bank. He left untouched the President's suggestion that Congress establish a bank within the Treasury Department for the deposit and disbursement of government funds but with no power to hold property or lend money. The Bank of the United States would continue to regulate credit by accepting state bank notes so long as they were redeemable on demand in specie. The Bank would continue its profitable role in discounting state bank notes, and the people presumably would benefit from its regulatory powers. But it would no longer enjoy its monopoly of the public funds, a major source of its power.[21]

Significantly, Van Buren was not as worried about the Bank as he had been a year earlier, though this time the threat was more direct; and he must have known Biddle would never accept a plan that curtailed so drastically the Bank's sphere of operations. Beyond Biddle, he foresaw the forces gathering that would encourage a great public debate and further clarify the political process along sharply and clearly defined issues. Was not Henry Clay a great champion of the Bank with its powers unimpaired? Was not the Bank the great centralized feature in his so-called American system? What of Calhoun? Had he not drafted the legislation creating the Bank; had he not sponsored the "bonus bill" that would use funds paid to the government for its charter to finance internal improvements?

There was considerable risk in thus challenging the Bank, the certainty of large financial power welding together a coalition of such unlikely political leaders as Calhoun, Clay and Webster. The combination would be formidable and would transcend sectional boundaries. He saw in the Bank, internal improvements and the tariff, a political mixture that reflected the centralizing idea of Hamilton and the Federalists. State banks, locally funded internal improvements whose character was essentially local, tariffs for revenue only after the national debt was paid off—these were as Jeffersonian and equalitarian as the neo-Federalism of Clay was not. Sufficiently idealistic to rest upon these principles, he was sufficiently realistic to count on Jackson's popularity for carrying them through. If a basic conflict was developing the screws were to be tightened slowly, compromise on Jeffersonian terms always being acceptable.[22]

While he was abroad, the coalition formed much along the lines he had anticipated. His rejection as minister he regarded as simply an outward manifestation of these inner tensions. And when he read the veto message on July 9, 1832, he thoroughly approved of the document. The alignment of parties on this issue was as mature as it would ever be; the economy had improved, the time was right. Clay and the National Republicans had been nudged into making the Bank *the* political issue of their campaign. They in turn had played upon Biddle's arrogance, his blindness to Jackson's popularity, and had at length persuaded him that a private corporation was a fit subject for judging a presidential contest.

Jackson was so pleased with Van Buren's praise for the veto message that he felt better at once. There remained a bare possibility that the Senate would override. The bill had passed overwhelmingly in the House but in the Senate by a majority of only eight. Since Biddle, Webster, Clay and other supporters of renewal anticipated a veto, they had worked feverishly on the waverers hoping to gain enough additional votes to override. Jackson wanted Van Buren to visit the Senate chamber the next day when the veto message went in and remain there during debate, using his influence coupled with his powers of persuasion should any party member show signs of voting with the coalition. The President also asked him to lobby for the passage of a new tariff bill modifying or removing most of "the abominations." He hoped the compromise measure would head off the nullifiers, yet not alienate the North and the West.

As a gesture of his firm support and his friendship for his vice-presidential nominee, Jackson insisted on conveying him to the Senate in his own carriage. The next morning just when the members of both houses were entering the north gate of the capitol for the day's session, the familiar carriage with its magnificent team of matched gray geldings swept up to the gate. Senators and Congressmen crowded around the straight, spare figure of the President, appearing a good deal taller in his tall white beaver hat, the old-fashioned badges of mourning for Rachel still attached to its brim, smiling as always on such occasions. His companion, Van Buren, dressed in the latest London fashion, moved easily through the throng, shaking hands and greeting friends. After a few minutes, Jackson signaled his coachmen to move away.

None of the President's opponents had been in the group at the entrance. As Van Buren made his way into the Senate chamber, he was surrounded by friends: burly, long-faced Benton; Felix Grundy, the short, heavy Senator from Tennessee who had given up trying to heal the breach between the President and Calhoun; Mahlon Dickerson of New Jersey, whose smooth features and bland personality masked a captious disposition. Van Buren marked the familiar surroundings where he had spent seven years of his working career, but which for the past three years he had rarely visited.

In the rostrum topped by the large gilt eagle sat Vice-President Calhoun, who studiously avoided any notice of his entrance. Clay and Webster were not in their seats. Van Buren was told they were in the House chamber, and after a short interval he left the Senate and walked across to the House. Just as he raised his hand to reach for the latch on the outer door, it opened and Webster appeared right in front of him. Both men were startled; before Van Buren had a chance to greet him, which he certainly would have done, Webster averted his eyes and hurried off. Van Buren took this gesture as a sign of shame; embarrassment would have been a more accurate reading. In the darkling plain of politics, Webster rarely showed remorse for any action taken against an opponent, particularly one he felt was undermining him in his own region.

After that abrupt encounter, Van Buren, who quickly recovered his poise, entered the chamber. Members of the House crowded around him and while he was shaking hands he saw the lanky figure of Henry Clay making his way toward him, likewise stopping frequently for a greeting or a brief chat. The gradual approach of the two rivals riveted the attention of the entire gathering. Would Van Buren snub the Kentuckian after what he had done in leading the coalition that rejected his confirmation?

Clay, playing to his audience, affected not a care in the world, shaking hands, taking snuff, making small talk, until he reached Van Buren. The courtly Virginian William Archer, an old friend of both, broke the ice by standing on his chair and declaring in solemn tones that he claimed the right of moderating this historic event. His remarks were greeted with laughter in which both Clay and Van Buren joined. As soon as the House quieted down, Clay strode rapidly toward Van Buren. They shook hands and spoke with each other in the most cordial manner, as if nothing had ever come between them. Clay led Van Buren to a sofa where they had a brief conversation on foreign affairs. Van Buren, who liked Clay, was much relieved that their political differences had not ended their personal relations as had been the case with Calhoun and apparently with Webster. Whether Van Buren's presence had been needed or not, it certainly removed any doubt about the veto or the tariff bill. The coalition could not muster enough strength to override, and despite some southern opposition the modified tariff bill passed.[23] His work done, Van Buren left Washington shortly after Congress adjourned to mastermind the campaign in New York, where the political scene was not encouraging.

The anti-Masons and the National Republicans seemed on the verge of consolidation; the Workingmen's party, though faction-ridden, had spread from New York City to Hudson, Albany and the western towns along the Erie Canal. In New York City, it still opposed Tammany, and throughout the state, Regency control. Several leaders of stature had emerged among the parties opposed to the Jacksonians. Most of them were westerners of New England stock from the rapidly developing sections of the state, such as Thurlow Weed, editor of the increasingly popular *Albany Evening Journal;* William H. Seward from Auburn, a slight, young, red-haired lawyer; Albert H. Tracy of Buffalo, who looked like Thomas Jefferson and had many of that great Virginian's personal qualities including his fondness for politics; and Judge Ambrose Spencer's son, John C. Spencer from Canandaigua, who Van Buren considered his most powerful opponent among the anti-Masons.[24]

Despite these political developments, it was an optimistic Van Buren who arrived in Kinderhook on July 25 to spend several days resting among his family and friends. He had taken soundings in Philadelphia and New York City and did not think the Workingmen's party posed a serious threat to the Jacksonians in the national election. Nor did he believe that it could resolve its differences in New York City before the state election, though of this he was less certain. For obviously the anti-Masonic–National Republican combination, if it could be effected, would hold out all sorts of inducements to the Workingmen leaders.

Van Buren had kept in touch with New York politics while he was in Europe. And after his rejection as minister, his interest quickened in response to the pleas from Wright, Flagg and others that he run for governor. Throop, they said, lacked the stature to lead the party in 1832 against a unified opposition. Should he be the nominee, the Jacksonians risked a defeat that would damage the delicately balanced Regency organization.[25] Of course, there were other weaknesses in the organization besides Throop. As Van Buren knew from experience, Wright and Flagg always exaggerated the strength of the opposition. Yet their warnings could not go unheeded, nor could those of the more pragmatic Marcy. Throop had provided a stable caretaker government that leaned on the Regency for advice and support. Now, the situation was quite different, the foe much better organized, the issues more sharply defined. Until careful analyses were made of public opinion outside of the rhetorical displays in the partisan press, neither Van Buren nor any member of the Regency could know how the Bank veto, the tariff, even the now dormant

Eaton affair and other widely publicized episodes of Jackson's colorful Administration would go over with the electorate of New York.

Van Buren owed a debt of gratitude to Throop, who had been loyal to him and who had performed well considering his limitations and the ambitions of the powerful men around him. He must, in the unwritten rule of the Regency, one Van Buren had devised and insisted upon, be rewarded for his services and the further sacrifice he would be asked to make. There would be no difficulty in compensating him with an office appropriate to the dignified station he had occupied. Van Buren had received assurances from Jackson when he broached the subject.

Far more difficult was the choice of a nominee. The Regency must nominate its strongest man. In Van Buren's mind the choice narrowed down to Wright or Marcy, both of whom were content in the posts they were occupying and neither of whom showed the slightest inclination to take the risk. For several reasons, Van Buren decided that Marcy must be the candidate. He was better known than Wright among the people and especially the local agents of the Regency. Marcy had held the post of comptroller for six eventful years with conspicuous ability. During his term of office, the Erie Canal was completed and he was credited with keeping the huge sums for Canal expenditures within estimates. The expenditures and revenues of the state, particularly land sales, amounting annually to millions of dollars, were so well managed that when audited, the books balanced down to the last cent.

During his tenure Marcy had traveled extensively through the state examining for himself projects for which he was responsible as chief fiscal agent. He had also served two years as an associate justice of the New York Supreme Court before being elected to the United States Senate. As a judge, he was conspicuously before the articulate public of the day, the lawyers and their clients. His duties carried him to the regular sessions held at Utica, New York and Albany and to the court of errors that met in New York City to decide appeals in equity from the chancery court. His heavy-set figure, shaggy eyebrows and craggy features were a familiar sight to thousands of political friends and opponents. Blunt and plainspoken, Marcy was a forceful individual who inspired confidence. To those who knew him better, he was a cultivated man, an omnivorous reader with a good sense of humor. Paradoxically, these very qualities that made him a distinguished public man in New York, had not served him well in the United States Senate. He was a poor speaker and an indifferent debater, in a place where a high premium was put upon those qualities. Nor was he a good committee man, because he lacked tact. Gruff and uncompromising, he barely tolerated colleagues who were neither as capable nor as industrious as he.

Wright, the incumbent comptroller, was also managing the duties of his office with distinction and honesty. But the responsibilities were not as burdensome as they had been under Marcy when expenditures for the Erie Canal and the Champlain Canal had reached their peak, receipts from western land sales their greatest volume. Wright spent most of his time in Albany and except in his own district could not command broad popular support.

Van Buren was thoroughly familiar with Wright's career in the state senate. He had heard him speak and debate, not just at the legislative sessions in Albany but in the House at Washington, where he had worked with him on the Tariff of 1828. Although he was not an accomplished public speaker, he was capable enough. In debate he excelled because, like Van Buren himself, he always thoroughly mastered the subject under discussion before he entered the lists. Wright would make an admirable Senator, and the post of comptroller could then be given to Flagg, who was also able and deserving. Flagg's elevation would reward John A. Dix, who would be promoted from his post as adjutant general to succeed Flagg as secretary of state.

Well before he reached New York, Van Buren had worked it out. All that remained was convincing the principals involved. As he expected, Marcy and Wright were reluctant. Marcy was all too familiar with the political imponderables of New York. His seat had four more years to run, and he was not enthusiastic about braving the rough and tumble of a campaign where a defeat damaging to his reputation was a distinct possibility. Emphatically, he agreed with his friend Lot Clark who said, "No man in these times must expect to ride in on a bed of down."[26]

Wright had none of Marcy's pretensions. A modest person, he consistently underrated his abilities. He was happy and comfortable in the neighborly atmosphere of Albany or Canton, and having experienced the discomforts and the pressures of Washington, he did not look forward to a change. Wright knew he must resign himself, as he had before during his term in Congress, to a lonely existence in a boarding house, which would add to his drinking problem. Neither man could refute the force of Van Buren's argument; nor could other members of the Regency who were consulted. For the good of the party, and of the Jackson campaign, they agreed.[27]

These arrangements made, Van Buren set off to make a personal canvass of the state. He spent four days en route to Oswego, meeting with local party leaders and editors on the way. From Oswego he traveled to Governor Throop's home in Cayuga County, carefully sifting opinion in this seed bed of anti-Masonry. His visit to Throop had its painful aspects because he had to persuade the Governor that he must make way for Marcy. Throop by now knew what was coming, in the silence of the *Argus* and the *Evening Post* on the subject of the governorship, and the cruel remarks at his expense which were being bandied about in Albany. Other than the unpleasant chore of telling a man he sincerely liked, and to whom he was under a distinct obligation, that the party had repudiated him, Van Buren enjoyed his trip. His spirits were buoyed up at the favorable reception of Jackson's veto message among the farming communities along his route. The modified tariff, too, which still offered protection to woolen producers and manufacturers, seemed popular. In an optimistic vein, he wrote Cambreleng "the body politic is sound to the core."[28]

By mid-August he was at one of his favorite resorts, Lebanon Springs, where he relaxed and exchanged ideas with his friend, the politically astute druggist, Elam Tilden, and his precocious son Samuel—Sammy to Van Buren. There, in the peaceful surroundings of the Columbia Hotel, he thought over the impressions he had gleaned. The enthusiasm for the veto had surpassed his expectations. "I must again repeat to you," he wrote Jackson, "that the veto message has verily proved to be the most effective document amongst the people, that has ever issued under like circumstances." At the same time he betrayed some nervousness when he admitted that "the union between the different sections of the opposition is, on the face of it, quite imposing." Van Buren planned to make Throop's house his headquarters until the end of the campaign. From there, he explained, he would make "excursions amongst our friends."[29]

The Regency was working hard, not just in New York, but also in the neighboring states of Pennsylvania, Connecticut, Massachusetts and New Hampshire. Butler had written a brief biography of Van Buren which Croswell serialized in the *Argus*. It was then published in pamphlet form and distributed throughout the state. A plain, direct document, the biography explained apparent inconsistencies in Van Buren's political career that the opposition had seized upon as evidence of his untrustworthiness.[30]

If the tariff was not a serious issue in New York, it had become one in other states, particularly in the South. Jackson made light of the nullification movement, but Major Lewis was alarmed; there was no doubt that anti tariff sentiment was spreading rapidly through the seaboard slave states. After much discussion, the Regency concluded the tariff deserved a prominent place in the resolutions the state convention would adopt at Herkimer in the fall. Dix and Butler were the authors. Marcy and Wright looked them over and helped shape the tariff resolution. Van Buren also made extensive editorial changes, particularly in the part dealing with foreign affairs. The section on the tariff praised the modifications Congress had made;

and while it supported a policy that would limit revenue to the needs of the government, it also backed protection "of all the great interests connected with the permanent prosperity and independence of the nation." A resolution arraigned nullification in South Carolina as a dangerous and evil thing foisted on a patriotic majority by a small group of political desperadoes who kept common cause in the "north with eastern monopolists and the interested political agitators."

On the Bank issue and the veto, the resolutions were most extravagant in their language. The question for the decision of the American people was whether they could have Andrew Jackson and no Bank, or the Bank without Andrew Jackson. Van Buren was praised, but in restrained language that was modestly brief. For a document that would presumably direct the campaign in New York, there was not one sentence devoted to state issues. The Regency would chance it on Jackson's popularity and his record. Marcy, the gubernatorial nominee, was given brief favorable notice as was his running mate, John Tracey, from Chenango County. The choice of Tracey for lieutenant governor can be viewed as the only gesture involving an important local issue, the Chenango Canal. Tracey was a fervent supporter of this lateral canal that would link the southern tier of counties with the Erie. Van Buren had supported the canal when he ran for governor because he desperately needed votes from that area of the state. Fortunately, his term as governor was so short he never had to make good on his promises. The counties along the Erie Canal and Champlain Canal had effectively checked Throop's feeble efforts to have the legislature approve the new canal. Now, with the nomination of Tracey, the Regency was signaling the southern tier that the party would back its demands. But the resolutions themselves were conspicuously silent on that point and on every other local issue.[31]

While the convention was in session Van Buren was making his "excursions" from Throop's home in Owasco. Several months before he had received a letter from a group of Calhoun supporters who had passed a series of anti tariff resolutions at a meeting in the town of Shocco Springs, North Carolina. The committee, passing itself off as representative of several political parties, wanted Van Buren's reaction to their resolutions. For some time he was undecided whether he should be drawn out on the subject, at least by this committee whose aim was to lure him into making politically embarrassing statements. But as the campaign developed and nullification ideas became more widespread through the South, he resolved to make the plunge. On the jolting stages over the dusty roads of the western counties, he began covering sheets of paper with his scrawls. Cleverly he did not confine himself to the tariff alone, but to all of the points at issue during Jackson's administration, the national debt, internal improvements, the Bank and nullification, not as a tariff issue but as a political doctrine.

Van Buren conceded that his position on protection made an explanation appropriate; then he shifted adroitly to the power of Congress over commercial regulation. In language whose meaning was unmistakable, he asserted the Constitution gave Congress that power. But he quickly qualified this statement when he said, "It has never been my wish to see the power in question exercised with an oppressive inequality upon any portion of our citizens or for the advantage of one section of the Union at the expense of another." With a flattering reference to Jackson, he reminded the committee of the rapid liquidation of the national debt which would permit a further scaling down of the tariff. This statement presaged a direct assault upon Clay, when Van Buren laid down a stringent policy on federally funded internal improvements. Implicit in his argument was the idea that if lavish appropriations were made for internal improvements, tariff rates must remain at high levels to pay for them. He admitted there were improvements of a national character which Congress could and should finance. Even these projects, however, must be deferred until the debt was paid off, always excepting any that affected the security of the nation, or interstate and foreign commerce. Federal funding of projects that were exclusively within the boundaries of a given state was unconstitutional, wrote Van Buren in a succinct restatement of the Maysville veto.

On the Bank issue he was unusually explicit in his opposition to a renewal of the charter, and just as explicit in his emphatic approval of the veto. That was all he had to say on the Bank, wisely avoiding any comment on its constitutionality. For he had made no effort to conceal his belief that Congress had the power to charter the Bank, thus differing with the lengthy argument Taney had made in the veto message. As he was opposed to recharter, he was just as vigorously opposed to nullification, "as it is called." He did not doubt that the nullifiers were sincere in their convictions; as for himself he thought their doctrine was not only unconstitutional but if unchecked would certainly lead to the destruction of the Union. While the Bank was a threat to liberty and equality of opportunity, nullification was a threat to the integrity of the nation: Van Buren linked the Bank with nullification as a political issue of grave importance, one aimed at subverting the idea of freedom, the other the idea of nation.

Of all the campaign literature that poured forth from the opposing parties, Van Buren's reply to the Shocco Springs committee was the most convincing, and for him, one of the most direct political arguments he would ever make. Years later, in rereading the document, he said, "I have never prepared a paper of that character with which I have been better satisfied." The *Albany Argus* published it in its daily edition and then as a pamphlet which, like the Herkimer Resolutions, was sent all over the state. In Washington, Blair and Kendall immediately appreciated its importance. The *Globe* printed up thousands of pamphlets, which were franked primarily to Jacksonian leaders in the southern states. And Kendall drew on Van Buren's exposition for the special campaign edition, the *Extra Globe,* which he had Barry force on postmasters all over the country.[32]

The campaign was an animated one. The talented but irascible J. Watson Webb and his partner, the equally talented, suave Mordecai M. Noah, had broken with the Regency, ostensibly over the veto, but actually because they could not break into the *Argus*'s share of the state printing. Their defection cost some votes in New York City. Francis Granger was again the opposition nominee for governor, but this time leading a united party of anti-Masons and National Republicans, he had the benefit of organization and the gifted editorial leadership of Weed's *Albany Evening Journal.* Neither Croswell nor Bryant, nor any other Jacksonian editor in the state, could match Weed's instinct for the political jugular.

Somehow Weed unearthed one of Marcy's expense accounts when he was a justice of the state's supreme court on which he had listed an item for fifty cents to repair the seat of his pantaloons torn in a stage accident. Weed's amusingly whimsical account of the broad-beamed Judge calculating the damage had the entire state laughing. Much to Marcy's annoyance, even his closest friends chided him humorously about the incident. The clever jibes from the *Journal,* the ingenuity and the industry of the anti-Masonic and National Republican leaders, could not prevail. Factions within the Workingmen's party neutralized it as a political force.

The depression had finally lifted; the acres of docks along the waterfront of southern Manhattan were busier than any merchant could remember. John Jacob Astor led a growing horde of speculators in city real estate as the center of population moved north on Manhattan Island. The Erie Canal and Champlain Canal, crowded with barges from the West, and the North, were earning the state a net return of well over $1 million a year on its $16 million investment. Villages along the routes just six years after completion of the Erie Canal had become small cities, and Albany had doubled its population. Van Buren saw his lands in Oswego County double in value and then double again. The coalition party could not prevail against this growing prosperity, the discipline of the Regency organization, and the respect for Marcy and for Van Buren.

The contest was a close one, however. In a total of almost 325,000 votes cast, Marcy defeated Granger by less than 10,000 votes, about 4,000 votes less than the vote given the Jackson–Van Buren electors. Since the district system for presidential electors had been abolished, New York's 36 votes were given to the Jackson ticket.[33] An interested observer in Ohio, John McLean, wondered about the election results in the Empire State. McLean, ever restless in his seat on the United States Supreme Court, was again visualizing himself in the White House. "Has Mr. Van Buren's strength been increased by the late contest in New York?" he inquired anxiously of John W. Taylor. "How do you account for the increased numbers of the House friendly to Gen. Jackson, who have been elected? Are these all devoted to Mr. Van Buren? Does not Gen. Jackson's strength in New York greatly exceed Mr. Van Buren's?" Whatever answer he received, McLean's questions were on many other minds.[34]

« CHAPTER 18 »

In Reference to a Possible Future

The sky was overcast and a sharp northwest wind caught up flurries of fresh snow and swirled them around the porticos of the unfinished capitol. For the past several hours carriages and hacks in a steady procession had arrived at the north entrance of the massive building, discharged their occupants and moved off either down Pennsylvania Avenue for more fares, or if they were private vehicles, to either side of the capitol where the coachmen and footmen waited, bundled up against the unseasonable chill. Winter bonnets, long-frocked coats and tall beaver hats marked the steady flow of pedestrians who entered the building. This lowering day was March 4, 1833, and the traffic on Capitol Hill presaged the second inauguration of Andrew Jackson, the swearing in of a new Vice-President of the United States, Martin Van Buren of New York.

Bad weather had forced a cancellation of the customary parade and the holding of the simple ceremony outdoors on one of the porticos. The hall of the House of Representatives, that vast and drafty "cave of the winds" was designated as the site for the inauguration. By noon, the galleries were filled to overflowing and so was the floor, apart from a section in front of the Speaker's rostrum reserved for Senators and Representatives and other dignitaries. The room was stuffy, noisy, overwarm and the jam so great the sergeant at arms and his assistants were constantly on the move clearing aisles and closing doors, even on those with invitation cards. At 12:30 p.m., the clamor died down as Speaker Stevenson left his chair and the President of the United States was announced.

For those who had witnessed the first inauguration, Jackson had aged perceptibly during the past four years. He was gaunter, his face more deeply furrowed, his chin sharper because he had recently lost all of his front teeth and could not bear the discomfort of the ill-fitting false ones made for him. His posture was erect, however, and there was no hint of feebleness in his gait, which was firm and steady as he entered the hall and seated himself at the Speaker's desk.

By his side, the short figure of a balding Van Buren was in decided contrast to the tall President. He, too, was seated at the Speaker's desk, on Jackson's left. Following them came the justices of the Supreme Court in their robes, members of the cabinet, diplomats in their colorful uniforms, and the members of Congress and the mayor and city council of Washington, all dressed almost uniformly in sober black with high white stock.

They took their seats in front of the Speaker's rostrum below the President and Vice-President-elect. When everybody was settled, Jackson rose and waited as the old and enfeebled Chief Justice John Marshall, made his unsteady way, Bible in hand, to his side where he administered the oath. The President then delivered a short speech which most of those present could not hear distinctly. Then the Chief Justice swore in Van Buren. All was over.[1]

While the President was speaking, Van Buren's sharp eye roved over the Congressmen below him. As he expected, Calhoun, Clay and Webster were conspicuously absent. It would have been remarkable if Van Buren had not mentioned this to the President on their way back to the White House. For the short session of the Congress which had just adjourned had been a turbulent one during which the centralists Clay and Webster had been united with the nullifiers in their common antipathy toward the Administration's economic policies. Webster, however, in what Van Buren unfairly chose to believe was strictly self-interest, had supported Jackson's policy on nullification.

Van Buren had only been in Washington for the past week. It had been nearly seven months since he had last visited the city briefly after his return from Europe. And there were those who could not understand why he was not at Jackson's side as soon as it was known he had been elected Vice-President.[2] But Van Buren had two cogent reasons for not complying with the wishes of his friends in Congress. He felt, that his presence would be misconstrued; that it would make it more difficult for the President to cope with tariff reform and with nullification then reaching their critical stages.

Although Vice-President-elect, he had no duties and no responsibilities, even symbolic ones, until his inauguration. Should he be physically on the scene as an advisor, he would simply magnify and focus the attention of the partisan and the troublemaker who would see some political scheme in the President's action on matters of grave public policy. He agreed with his old friend, Senator Kane of Illinois, who said, "Let these sweet souls get at loggerheads before you show yourself here."[3] This possibility Jackson had already grasped with his sharp awareness of the tensions that were building in Congress. There had been enough carping in the public prints and in the lobbies of Congress about the influence of the Kitchen Cabinet to make Jackson sensitive on the subject.

Van Buren's second reason for absenting himself was not so high-minded. The issues involved were politically sensitive. At this stage, he was quite content to steer clear of any direct involvement. But most important to him and to the Jackson Administration itself was the strengthening of the Regency organization, which had weakened during his absence in Europe.

The Regency was a better disciplined group than it had been even five years before, but it still required the skill of Van Buren's fine hand to keep all these able, strong-willed, ambitious men working together in harmony, restraining their own individual opinions for the greater good. He had not been in Albany more than a week after his return from Washington in early July 1832 before he saw the beginnings of a schism in party leadership over the tariff. A modest wool producer himself, he was alarmed at the large-scale speculations Benjamin Knower and others in Albany were making in raw wool and also in shares of woolen manufacturing companies. Knower had been troublesome on the tariff four years earlier; he could well head up a more divisive element now. Where would this place Knower's son-in-law, Marcy, who Van Buren judged the best candidate for the governorship? In New York City, his friend Cambreleng was the spokesman for free trade shipping men, merchants and auctioneers who held together the Tammany organization against the forays of the younger radicals like George Henry Evans and his *Workingmen's Advocate.*[4] The Knower and the Cambreleng factions must be brought into some sort of balance more fundamental than the rhetoric and the organization and the momentum that achieved victory for the Administration and the Regency in November. Van Buren wanted a solid measure of support for the Administration's policies from the legislature and no wrangling over the Senatorship. He had few qualms about the Bank veto which he found popular among the farmers, the state bank interests, manufacturers and merchants.

Tariff and nullification were the potential disrupters, both within the legislature and within the southern strategy he had devised to elect Jackson in 1828. He was acutely sensitive about southern suspicions of his stand on the tariff. Both Jackson and John Forsyth of Georgia had urged him to put himself "right on the subject of the tariff of '28." "You are accused of inducing the southern members to make it odious," wrote Forsyth, "in order to justify its rejection and then voting its adoption contrary to the pledge made or understood." It had not helped Van Buren's image when Dudley in the Senate went right down the line voting for increases on woolens over the 1828 rates in defiance of the Administration, which was seeking a lowering of those very schedules.

Speaking for Virginia, Peter Daniel echoed Forsyth's sentiments and chided Van Buren for not having been on the scene earlier.[5] Duff Green's campaign in the *Telegraph* had made an impression. Van Buren's and Wright's votes and speeches in the Twentieth Congress had been analyzed in such a way that both New Yorkers had devised the tariff as a shameless political trick to gull honorable southern Congressmen. Green, always a master of the quotation out of context when there was a point to be gained, made an ingenious cut and patchwork quilt of Van Buren's speech before the Albany mass meeting in 1827. When he was through, he had omitted all the qualifications, all of the carefully reasoned arguments that Van Buren had labored over to cleanse any tariff he would support of sectional favor or special interest. Southern readers of the *Telegraph* and those Calhoun papers which followed its editorial line, found Van Buren a stauncher advocate of protection than Henry Clay.[6] Of course, the *Argus,* the *Evening Post* and the *Globe* responded furiously to Green's imputations, and Ritchie's *Enquirer* printed many of their rebuttals. Yet Ritchie's editorials, though still cordial, betrayed nervousness on the issue.

Van Buren, who knew Ritchie of old, was not surprised at his vacillating course and was only thankful that he had not, thus far, become a neutral, considering the controversy that was raging around him. Powerful Virginians either in Calhoun's interest, like Governor Floyd, or suspicious of Van Buren's southern strategy, like Senators John Tyler and Littleton Tazewell, were openly and vigorously bent on bringing the Richmond editor to heel.[7]

While Van Buren worked to draw the two wings of his party together on the tariff, he sent his ideas on the Bank and on internal improvements to the President. Looking ahead to the time when the charter of the Bank would expire, he took it for granted that the veto message precluded any banks being established under national auspices in the states. "My advice," he counseled Jackson, "would be to make another fair effort to get along without a bank and if experience should show that one is indispensable to the safe conduct of public affairs," then he would try a scheme Jackson himself had proposed, a government bank of deposit in the District of Columbia. But Van Buren's major contribution to Jackson's fourth annual message was the section on internal improvements. His ear cocked to the complaints of the West and the South, he suggested a middle course that, however, did not infringe upon the distinction he had originally made between interstate and intrastate improvements. Jackson adopted most of his recommendations which drew some criticism from western spokesmen who thought the message was not forthright enough on specific directions the government would take.[8]

Besides these contributions, Van Buren urged successfully that the new Administration should settle itself in office before there was any major reshuffling in the cabinet. After the entire cabinet except Taney had opposed his veto of the Bank bill, Jackson wanted an immediate drastic change; but he took the precaution of sending McLane to New York for a consultation with Van Buren on the subject. Again mindful of the nullification and Bank issues, Van Buren suggested a cautious approach. "The course," he felt, "is not a clear one and we must trust something to time and it is moreover desirable that you should commence this session with a steady Cabinet and without the embarrassments of change."[9]

Absorbed as he was in these matters, he had no idea how dangerous the nullification crisis had become. That the nullifiers were passing flamboyant resolutions and were even said to be drilling troops, he read in the newspapers. Jackson's letters left no doubt where he stood, but they dealt in general terms. Though full of furious denunciation of the "nullyfiers," there were few specific details. Hoffman wrote frequently from Washington, painting a grim picture of despair for the Union. But then Hoffman was excitable and given to outbursts that overly dramatized any tense situation.

Like most politicians removed from the scene, Van Buren considered the problem in South Carolina as mainly a partisan one, the proclamations, the fiery speeches, even the military display, more pressure tactics rather than real threats. Had he been privy to the notes flying between Jackson and Joel Poinsett, the Unionist leader in the State, and the preparations of the war and navy secretaries, he would have been horrified. For they portended not just outright armed defiance of federal authority, nor the President's preparation for meeting force with force, but civil war within South Carolina itself.[10]

It had come as a shock to him when he learned that the South Carolina legislature nullified the tariff to take effect on Feburary 1, 1833. But he expected Jackson to take a bold stand condemning this defiance of federal authority, which he believed was squarely within the strictist construction of the Constitution. It came as a greater shock when he read the President's lengthy proclamation condemning nullification, not only an unconstitutional act but a treasonable doctrine, and clearly asserting he was ready to enforce the law with all the powers at his command. On the same day he issued the proclamation, Jackson wrote Van Buren a long letter detailing for the first time his plans for a military invasion of the state, if it did not back down. Van Buren already knew General Scott was bolstering the fortifications in Charleston harbor. He also knew that armed revenue cutters were in the harbor with their guns trained on the wharves.[11] These steps he thought a painful necessity to uphold the dignity of the government, but the proclamation was something else.

Several days after he received Jackson's letter and a copy of the proclamation, his mail contained a long discussion of the document from Cambreleng. The rotund Congressman applauded Jackson's determination, but was disturbed at the lengthy exposition of constitutional principles, which he found centralist in nature, a loose interpretation that would have done justice to such an inveterate Federalist as Gouverneur Morris, a manifesto calling up bitter memories of the Alien and Sedition Acts. He detected the hand of Edward Livingston in this section of the message and sarcastically referred to him as "the Montesquieu of the Cabinet." Even so, Cambreleng glided over the constitutional justification as not that damaging to Jeffersonian principles. "Happily, the mass of the people sleep over such parts of it and dwell only on those which make them think and feel like men," he said.

Van Buren felt no such easy assurance. Perhaps the masses would pay little attention, but Thomas Ritchie would, as would John Randolph, Calhoun's supporters everywhere and all those southern and northern men who had staked their careers on the fallen star of William H. Crawford, who venerated strict constructionists like Macon and John Taylor of Caroline.

Although Van Buren applauded the President's firm stand, he saw the delicate web so carefully spun between the northern and southern branches of the party breaking strand by strand. In his anxiety, he wrote Jackson, "You will say I am on my old track—caution—caution: but my Dr. Sir, I have always thought that considering our respective temperaments, there was no way perhaps in which I could better render you that service which I owe you as well as from a sense of deep gratitude as public duty." What he particularly regretted was the constitutional discussion which need not have been so elaborate or so frankly Federalist in tone. As set forth, it could cause a "collision with Virginia." If Virginia were lost to the Administration, it would probably take with it all the seaboard slave states, too. He assured the President that the body of public opinion was with him. This was not the time to engage in an abstract discussion on the nature of federal power. "South Carolina has not, and will not, secede." "She will," he predicted accurately, "avail herself of the mediation of Virginia and postpone the operation of her ordinance. Of this there cannot be reasonable doubt."[12]

Van Buren's words seemingly made no impression on Jackson, who went ahead with military preparations. Dismissing his comments, Jackson wrote, "No, my friend, the crisis must now be met with firmness and secession put down forever—for we have yet to learn whether some of the eastern states may not secede or nullify, if the tariff is reduced." He did assure Van Buren that he would "act with forebearance." When Jackson wrote this letter, the Congress was debating a compromise tariff bill that met many of South Carolina's demands. Under presidential pressure, Louis McLane had swallowed his protectionist principles and drafted a measure that Gulian Verplanck introduced in the House. Both Cambreleng and Thomas Hart Benton, good judges of the congressional mood, were certain that whether the Verplanck bill passed in its original form or not, substantial concessions would be made to conciliate the South.

Relieved by these positive signs of returning sanity in Washington, and he hoped in South Carolina, Van Buren turned his attention to the tariff problem at home. He worked with Governor-elect Marcy on his message that, among other things, would affirm New York's support for the President. Marcy was under heavy pressure from his father-in-law, Knower, to come out for no compromise on the tariff. But he thought Knower's "ultra" views were politically impossible. "The South and the West," he said, "feel that they have no interest in sustaining the tariff. They believe it to be a withering curse which has blighted their prosperity and is fast spreading desolation over the country."[13]

Van Buren wanted Marcy to support the Verplanck bill in his message to the legislature. Despite his lofty position, his persuasive talents and his leadership of the party, he did not succeed. The most Marcy would say was that he favored "substantial relief to every real grievance" coupled with adequate protection. The new governor did condemn nullification unsparingly, yet in such a way that he reaffirmed a rather uncompromising view on the tariff.[14]

Van Buren concealed his disappointment in Marcy and began the intricate process of marshaling legislative opinion in support of Jackson's proclamation. While deeply engaged in calming the fears of representatives from manufacturing districts, he was startled to receive a call for a mass meeting on January 24, 1833, at the city hall in Albany where the tariff issue was to be debated and resolutions passed. Heading the list of signers was Benjamin Knower, who, Van Buren knew, was a heavy speculator in the wool market and in woolens shares. Gossip in Albany had it that Knower was in over his head and faced bankruptcy. It did not require much discernment to assume he would be adamantly opposed to any reduction of the schedules on wool and on woolens.

The meeting was billed as nonpartisan and Van Buren noted that though many of his close associates had signed the call, most of them were either manufacturers or woolen dealers. A substantial number of prominent National Republicans were among the sponsors. Without taking the Governor into their confidence, the Regency discussed the meeting and prepared itself for any contingency. If the meeting was open and aboveboard, they would seize the opportunity of passing resolutions calling for a modification of the tariff. If, as they suspected, Knower and his fellow protectionists had engineered the meeting to air their views, Regency members would attempt to seize control. Should this prove unavailing, they would call upon their supporters to withdraw and meet in the capitol where they would secure approval of resolutions supporting the Verplanck bill. John A. Dix would prepare a set of resolutions that would fit any of the circumstances.

Thus prepared, Van Buren's lieutenants attended the meeting and sat silent while one of Knower's associates read a series of resolutions condemning nullification but rejecting any compromise on the tariff issue. Dix was immediately on his feet to propose his substitute resolutions, the only major difference being those which dealt with the tariff. Hard-line protectionists tried to shout him down, and for a brief interval no one could be heard in the hall. Former Lieutenant Governor Edward P. Livingston took advantage of a temporary lull to command attention with his clear penetrating voice: "Friends of Andrew Jackson and of the substitute to the Capitol! Here we have no fair chance of being heard!" As many as two-thirds of those attending followed Livingston to the assembly room of the capitol where they quickly organized themselves and listened to Benjamin Butler give a rousing speech. Dix followed with a burst of eloquence, and again read his substitute resolutions which the meeting adopted with enthusiasm.

Van Buren was quick to exploit the spirit of the capitol meeting.[15] Over the next two days, using Dix's resolutions as a base, he wrote an elaborate preamble in which he reconciled Old Republican doctrine with the nationalist sentiments Livingston had developed for Jackson's proclamation. He argued that nullification was a perversion of states' rights as dangerous to liberty with union as the pronouncements of the ultra-Federalists had been at the Hartford convention. The President had no recourse but to make the comments he had made in the proclamation, for the only alternative was secession and anarchy. Yet he urged conciliation and significant reductions in the tariff, in short, the Verplanck bill.

From the beginning of the controversy Van Buren had pondered on the best means of having the legislature adopt measures of a support for the Administration. The mass meetings had served as just the right catalyst. Although irritated at the position of Marcy, his father-in-law Knower, and other protectionists in Albany, he wanted no break in the party, no weakening of Regency control. The best way, he decided, to hold the line between the manufacturers in the party, the farmers, the free-trading merchants of New York and the banks tied to both of these interests, was to co-opt one of the leading spokesmen for the manufacturers. State senator Nathaniel P. Tallmadge of Poughkeepsie was the person he had in mind. Van Buren had him made chairman of an *ad hoc* joint legislative committee that would issue his resolutions as a "Report" for legislative adoption.

Tallmadge, a former Clintonian, vain and ambitious, readily went along with Van Buren's arguments, especially with the promise of Dudley's Senate seat. It was a master stroke that effectively muffled Knower and his allies. After some critical fumbling in the caucus, Tallmadge was nominated. Silas Wright's nomination for the other seat in the Senate encountered neither opposition nor confusion.[16]

The day of the mass meetings, or "the blow up" as Van Buren called it, an impatient Jackson penned a querulous letter to Albany. "Why is your Legislature silent at the eventful crisis?" the President demanded. "Friendship with candor compels me to say to you that your friends are astonished at the silence of your Legislature and gives rise to dark innuendos of your enemies that you command them, and are waiting the result of the Virginia Legislature."[17] Jackson was, of course, unaware of the complexities and uncertainties Van Buren was facing in New York. Nor had the President, all of whose energy and will was concentrated on Calhoun and the nullification controversy, paid much attention to his political coalition. He simply reflected prevalent opinion that Van Buren was playing politics with a national crisis."[18] Before Van Buren received the President's stern warning, the "Report" had been submitted to the legislature where a combination of National Republicans, anti-Masons and protectionist Democrats delayed its passage until February 24.[19]

Van Buren, however, had wasted no time in sending Jackson a copy of the "Report." Donelson, who screened the President's mail, read his explanatory letter together with newspaper accounts detailing all of the preliminary activities. Jackson listened carefully, then without comment had Donelson file the material. Now the President knew what his deputy had faced, why the resolutions had been delayed and how he managed the affair. Van Buren's actions had substantially assisted the Administration in further isolating South Carolina.

The "Report" supplied a strong constitutional argument against nullification and cut it loose from any possible connection with the Virginia and Kentucky resolutions. If any doubt yet remained, the temper of the protectionist press in the state had dispelled it. The *New York American,* now the chief Clay paper in New York City, made the "Report" the basis for a violent personal attack on Van Buren. And the *Commercial Advertizer* found it "jacobinist beyond endurance." Spirited defense on the other hand came from the *Richmond Enquirer* and the *Globe,* both of which found it a lucid and correct exposition of the Virginia and Kentucky resolutions, which by Madison's own testimony, could in no way be construed as upholding nullification of a federal law by a single state.[20]

Ritchie's praise for the "Report" was the most significant; for as Van Buren had warned the President, the centralist cast Livingston had given the constitutional arguments in the proclamation was hurting the Administration, not just in Virginia but in other neighboring southern states where majority opinion opposed nullification but was extremely touchy on the subject of states' rights.[21]

Van Buren's sensitiveness about southern reaction to the mounting crisis had gone beyond his initial concern for the fate of the Jacksonian coalition to a genuine fear of disunion. The practical-minded, steady Silas Wright had written him an unusually alarming letter on the state of affairs as soon as he reached Washington and called on the President. Wright had only a brief conversation with Jackson and Secretary of War Lewis Cass in a crowded room at the White House. Both were anxious about prompt resolutions of support from the New York legislature, and both expected hostilities with South Carolina. Nor did they feel that any tariff bill acceptable to the nullifiers would pass Congress.

Cass was more specific and more pessimistic than the President. His major concern was whether the Administration could rally the South while it disciplined South Carolina. The only hope he held out against secession of the upper South was the passage of Verplanck's bill, which he very much doubted would survive in Congress. Wright had other unpleasant news to report. The New York delegation he found split on the tariff: one group for standing fast on current schedules, the other for sweeping aside all protection, and none for any compromise.

Before Wright left Albany, Van Buren had heard from sources he considered reliable that the President was considering seriously another even more drastic move than the proclamation. Again he hoped nothing would be done that could play into the hands of the firebrands, nullifiers and, as he put it, "other mischievous persons" before it was absolutely necessary.

He wrote Jackson of his apprehensions in a letter Wright took with him to Washington. But the General took no heed. On January 18, 1833, Van Buren's worst fears were realized when he read in the papers that Jackson two days earlier had requested from Congress the authority for using military force to collect customs. Instantly labeled the "Force bill," the Administration found itself in the uncomfortable position of being supported by its political enemies while its friends were divided.[22] Van Buren had worried about Jackson's impetuous nature and his possible use of military action ever since he issued his proclamation on nullification. Well supplied with conjecture and opinion from the Washington rumor mill, he realized the President would not wait for the South Carolina ordinances to become effective before he took some action. That Van Buren's fears of strenuous executive action were not confined to him alone were evidently shared in the South and among his supporters in the New York congressional delegation.[23]

The Force bill that concerned Van Buren upset the southern wing of the party far more. If the proclamation had strained the bonds of party, the Force bill threatened to snap the cords of the Union. As Cass had feared, congressional representatives of the upper South were decidedly hostile to the measure. William C. Rives, now one of Virginia's Senators, a close friend of Van Buren, and a firm Unionist, told Wright "that the Southern men could not vote for any provisions of force." Thomas Jefferson Randolph, Jefferson's grandson and literary executor, thought that the Force bill would sweep Virginia into the nullification camp. "The condition of Virginia," he wrote Rives, "is very critical. The reprobation of S. Carolina would have been decided; but the proclamation has thrown the nullifiers in principle but not in name from utter prostration into a majority." Ritchie, as expected, was full of lamentation. But the westerners in the party were rallying behind the President, while many northerners hesitated.[24]

After the initial shock, the politicians began to see that Jackson's firm policy had a stronger measure of popular support even in the South than they had supposed. Still the situation was extremely tense in Washington. The northern members of the party, even ardent protectionists like Mahlon Dickerson, a New Jersey Senator who thought the Verplanck Bill and McLane's share in it as stealth beyond the schemes of Symmachus the apostate, was sufficiently chastened to reconsider some of his extremist sentiments.[25] General relief swept over the Congress regardless of party when word arrived that the South Carolina legislature had backed down. The ordinances would not go into effect until the tariff issue was resolved.

Van Buren had made it clear that Wright spoke for him and he urged his supporters in Congress to vote for any compromise, on the tariff acceptable to the South. Through Flagg, he advised Wright that "the deep feeling for the preservation of the Union was paramount to all feelings for the tariff." Van Buren did this, suspecting that Clay and Calhoun would bury their differences and scuttle the Verplanck bill in favor of another they would write. He had been repeatedly warned of a combination between these old political enemies aimed specifically at him.[26] Thus he was prepared when Clay introduced his tariff plan on February 12, and in no way surprised that it received Calhoun's blessing. Clay's bill, which would scale down schedules over a ten-year period, was more favorable to northern manufacturers than Verplanck's bill, even with its many amendments. Obviously, nullification and tariff were moving rapidly in a partisan direction. It remained for Wright to salvage what he could for his chief as the Clay-Calhoun combination took charge.

But about all he was able to do at this point was to unify the Jacksonians in the New York delegation, a difficult task because of the obstinate opposition of the protectionists. Wright, in a rare moment of exasperation, castigated Knower and Marcy for their interference. Van Buren himself had begun to suspect Marcy of misusing his position and his influence in the party for family interest, especially a rescue effort on behalf of his father-in-law Knower. He had several conversations with Marcy, polite and informal; but he left no doubt in the Governor's mind that their old relationship of friendship and trust had undergone a change.[27]

As Van Buren prepared for his journey to Washington, the Clay bill was being rushed through Congress, but so was Jackson's Force bill, a measure that frightened the New Yorkers in Washington almost as much as it dismayed the Virginians. In complete agreement with Wright, who had finally unified the state delegation, Van Buren hoped the tariff bill would take preference and once passed, the Administration would let the Force bill quietly die. But knowing Jackson as he did, he must have realized the General would insist on this symbol of executive authority. Van Buren made no reply to the entreaties of his friends in Washington; nor did he offer any advice to the President. Clay's tariff bill and the Force bill passed in Congress on March 1, 1833, and were approved by Jackson the following day.[28]

One more blow was launched against Van Buren, though its effect simply strengthened Jackson's support for him. By a very close vote, Gales and Seaton were elected printers of the House over Blair and his new business partner, John C. Rives; and in the Senate, Duff Green prevailed. The campaign of 1836 had already begun, and there were many who cast aspersions on Van Buren's motives for remaining in New York while the President grappled alone with nullification, the tariff and the Bank. The ambitious Supreme Court Justice John McLean was angling for an alliance with Calhoun and busy strengthening his position in the Northwest; Richard Mentor Johnson was actively pushing his own availability in the Southwest.[29] Van Buren was alert to all these moves, but remained quite unconcerned. The important thing was a solid party apparatus in New York and this he finally achieved after his mild reprimand of Marcy, and rearrangment of the federal patronage in New York. Throop was made navy agent despite Jackson's thought that the office was not dignified enough for an ex-governor of New York. "Time and the virtue of the people," Van Buren blandly wrote Wright, "will bring all right. More importantly, our Albany difficulties have all blown over and the stability of the party has been greatly increased by the abortive attempt to distract it."[30]

« CHAPTER 19 »

A Perilous Course

For weeks Van Buren had felt rather weak and feverish. He was not, he knew, as alert as he ought to be at a time of the utmost delicacy in his affairs when the least misstep, the smallest miscalculation could result in disaster to him personally. Everything depended on the trust and the favor of the imperious, impulsive man in the White House. Even that could become a liability if public policy went awry, if the Bank he was determined to crush carried down with it the nation's economy.

Each day he spent hours with the President, listening, flattering the old gentleman, but not hesitating to differ if he thought it necessary. After these wearying sessions, Van Buren would return to his stuffy quarters at Gadsby's and the hot coal fire which he believed was the prime cause of his ills. He may well have been right, for the windows were sealed and the draft not what it should have been. The level of carbon monoxide in Van Buren's suite must have been high, probably high enough to affect his health, and the well-being of other guests for that matter. Later he would explain to Jackson, perhaps by way of excusing his diminished perception, "I suffered more than I can well express from that source."[1]

Whether it was Gadsby's hot coal fires and poor ventilation, or a recurrence of Van Buren's old complaint that he labeled "dyspepsia," or simply the consciousness of his new position and the perils surrounding it, he found himself being forced into a defensive position on the Bank. For the first time he sensed the power of Jackson's informal advisors, particularly Blair and Kendall, reinforcing and elaborating the President's ideas. Jackson had taken his triumphant reelection as a mandate from the people against the Bank, and he wished this issue resolved promptly by removal of the government deposits.

On this crucial question Van Buren differed with the President. Not that he was opposed to removal; but ever the careful politician he urged a postponement until public and political opinion could be tested, then prepared, if necessary. Biddle had so involved the Bank with the fate of other banks, especially in the Northeast, with the powerful merchants and shipping interests in New York City and other ports along the Atlantic seaboard; and he was financing so many railroads and other private ventures that any rash move from the Administration could produce an economic collapse with its inevitable political backlash.[2]

In common with other politicians and with many businessmen, too, Van Buren had no way of estimating the power of the Bank beyond the fact that it was the largest single source of liquid capital in the country. Only the President, Kendall, Blair and Taney seemed unconcerned about the economic ramifications of removal. With the exception of Kendall, Jackson and his advisors saw "the monster" in moral and political terms, an aristocratic abomination that bilked the people, corrupted its representatives, purchased the press, and constantly sought to rule the nation for the benefit of a small group of rich American and European stockholders. The popularity of the veto message they took as positive proof that the people understood the dangers the Bank posed, not just to their material well-being, but to their political liberties as well. Van Buren saw this side of the question as clearly as Kendall, Blair or Taney. But that small group of zealots had little or no political bases to cover. After all, it was his future that was at stake, not theirs, not even the President's, who would retire in four years, if he lived that long.

Van Buren had the most to lose if anything went wrong. Yet Blair and Kendall were now major figures in the President's circle of intimate advisors, even more influential than Major Lewis. Any reader of the *Globe,* and Van Buren was a close reader, could measure Blair's influence in Jackson's utterances on removal. But Kendall, whom he suspected, and rightly so, was the major architect of tactics and strategy. This industrious, shrewd Yankee had studied the Bank and banking since he had moved to Kentucky in 1817. Next to Nicholas Biddle himself, and possibly Louis McLane, he knew more about "the monster's" affairs than its own officers. His expertise coupled with a burning hatred of the institution made him ever alert to the Bank's weak points and ever inventive on how to exploit them. Blair, too, was as anti-Bank as Kendall; his experience with banking in Kentucky paralleled Kendall's. The bellicose editor's comments carried weight with the President, though his influence lay more in reflecting accurately the President's thinking and expressing it in caustic editorials. He was, Van Buren thought more the instrument in the operations being planned against the Bank rather than the substantive consultant.[3]

McLane confirmed Van Buren's analysis of Kendall's role. A few days before the inauguration McLane, the treasury secretary, who also suspected Kendall of being the major influence, had a discussion with him that lasted over an hour. Flattered at McLane's interest, and hoping to enlist him in the cause, Kendall was unusually open in his opposition to the Bank and the means he hoped the Administration would use to destroy it. Though McLane opposed removal, he implied he would support the President if he insisted on that course. Confident he could bring McLane into line and sensing an opportunity to make yet another case for the President, Kendall wrote a memorandum to McLane setting forth the measures he thought should be taken against the Bank. If McLane did not show Kendall's memorandum to Van Buren, the President surely did. In any event, Van Buren soon decided that Kendall was Jackson's expert on removals and between bouts of weakness at Gadsby's, made his plans accordingly.[4]

Just before Congress adjourned, the House had overwhelmingly endorsed the Bank as a safe institution for the government deposits. Van Buren was not the sort of person who would dismiss lightly a signal as clear as this one. He was, of course, mindful of Bank pressures on individual Congressmen, but he could not ignore the vote, 109 to 46, as a manifest of political opinion. In fact, he was so disturbed at the President's complete indifference to the action of the House that he invited Kendall for a discussion of the removal issue. No sooner had his visitor arrived than Van Buren put him on the defensive. The Vice-President declared emphatically he was opposed to any removals. Then he sat back and listened to an anxious Kendall deliver a point by point indictment of the Bank, emphasizing its capacity for evil, its manipulation of the public credit for private advantage. As he warmed to his subject, Kendall dropped his customary reserve and warned Van Buren that if the Bank were not shorn of its power, it would control the next presidential election.

That Kendall should have the temerity to lecture the Magician on politics seem to have touched Van Buren's pride. For in a heated rejoinder, he sharply disagreed with his visitor, indulging a temper he almost always controlled, whatever the provocation. Kendall, too, was angry and said if proven wrong he would leave Washington and no longer interest himself in public policy or in politics. Springing up from his chair he said, "I can live under a corrupt despotism as well as any other man by keeping out of the way which I shall certainly do." The two men parted at that point both, according to Kendall, "somewhat excited."[5]

Van Buren may have regretted his outburst with Kendall, for it became increasingly plain that Jackson was determined to have the secretary of the treasury begin removal of deposits no later than the end of September, 1833. These funds would be placed in state banks, selected for their strength and their conservative financial policies. Amos Kendall had outlined the procedure, which Jackson adopted enthusiastically. The President had, however, convened his cabinet and read them a questionnaire he had prepared that covered the entire bank question in broad terms. He wanted written opinions from each member. Taney was the first to respond with a long, cogent brief in support of removal and deposit in state banks. His paper blended elements of political necessity, beneficial public policy and executive prerogative.

By early April, Jackson received an opinion from Woodbury but not from Cass, Livingston and most important of all, from McLane. The President knew that the secretaries of war and state were opposed to removal and was not concerned about their opinions. Livingston was slated to leave his post for the mission to France, while Cass, deeply involved in implementing the Indian removal policy, had pleaded his ignorance of the Bank question. McLane's delay was a source of deep concern to the President. Everyone acknowledged his ability in analyzing problems of public finance. And Jackson was eager for his expert advice on the technical aspects of Kendall's plan. His closeness to Van Buren also could mean that he was speaking for the Vice-President as well. Van Buren had not as yet committed himself except in his conversations with Blair and Kendall. But Van Buren had no hand in the drafting of McLane's response which was handed in on May 20, 1833. The Vice-President was out of town in Pennsylvania mending political fences and sounding out public opinion on the Bank in its own territory.

In only one respect did the secretary of the treasury agree with the President—that the Bank should not be rechartered. Otherwise, with a wealth of detail and reasoned argument, he painted a dark picture of the nation's economy if government deposits were removed. Jackson, though in fundamental disagreement with his secretary, found many strong points in his paper; but he decided that McLane could not remain any longer in the Treasury Department.[6]

Jackson and Van Buren had discussed changes in the cabinet on various occasions going back a year in time. Relations between Taney and McLane were far from cordial, each individual ambitious for himself and jealous of the other. Blair disliked McLane and sought at an early stage to warn Van Buren of his efforts to secure the vice-presidential nomination. Van Buren dismissed Blair's allegation as part of a palace intrigue, his personal fondness for McLane and his admiration of his talents unimpaired.[7]

A few weeks after the election, McLane visited Van Buren in New

York City where they discussed cabinet matters. Livingston was anxious to resign his post and Jackson just as eager to see him go because of his opposition to the Administration's Bank policy. McLane made no secret of his desire to take Livingston's place. When Van Buren mentioned Taney as a good appointment for the Treasury with his own protégé Butler taking over the attorney general's office, McLane raised various objections. Van Buren had not pressed the issue, and both men agreed nothing should be done until the spring. When the volatile McLane returned to Washington, he convinced the President that his successor should come from Pennsylvania; after canvassing various possibilities Jackson, with prompting from McLane, hit upon William J. Duane, then managing the General Bank in Philadelphia. McLane had not passed on Van Buren's suggestion of Taney for the place; nor had he mentioned Butler.[8]

Now was the time for the arrangement to take effect. The secretary of the treasury had a special mission to perform as the agent for the President in removing the deposits, a mission McLane would not undertake. While Van Buren was traveling in Pennsylvania, McLane resigned the Treasury portfolio, and Livingston resigned from the State Department for the French mission. Duane, to the astonishment of many important politicians who knew of him primarily through his father, the fearless editor of the Jeffersonian *Aurora*, took over the Treasury. Jackson promptly named McLane as secretary of state. At a later date, Van Buren regretted the cabinet reshuffle which would bring some unnecessary embarrassment for the President and himself. He admitted the ambitious McLane had manipulated the Administration for his own ends and attributed McLane's success to "the infatuation by which both the President and myself were infected in respect to him, and of the consequent influence he was capable of exercising over us."[9]

Meanwhile, Van Buren's probings in Pennsylvania convinced him that Jackson's popularity far outstripped any power the Bank might exert. He decided that he had been hasty in opposing removal; but still cautious, he would if possible slow down the process until Congress assembled in January. Confident that nine months of preparation would make for a smoother transition, he was not unmindful of the fall elections in New York and many other states where the issue of removal would have a fair test. If, as he believed, the President's popularity would carry the day, congressional sanction would follow as a matter of course. And he had already set the stage for an impressive build-up of Jackson's popularity in the important states of Maryland, New Jersey, Pennsylvania and New York, all of which he himself must carry if he were to succeed Jackson.

Before Van Buren left Washington and the President exposed, as he believed, to the dubious advice of Kendall, Blair and a newcomer he did not trust, Reuben Whitney, a former director of the Bank with a reputation for twisting the truth, he had gotten a commitment out of Jackson for a triumphant tour of the Northeast as soon as the cabinet changes had taken place and the weather improved sufficiently. Besides the Middle Atlantic tier, Van Buren charted a course through New England, a region where the Jacksonians were slowly building up their strength.[10]

There had been precedents for presidential tours of the country. Washington had visited all the sections of the Union during his first Administration. Most recently, James Monroe had followed his example. None of these official visits had been explicitly political in nature; none was designed to elevate a party or benefit an individual. But Van Buren at any rate, was counting on the President's appearance, to improve the party's prospects in the fall elections especially in faction-ridden Pennsylvania and in New England—the stronghold of Daniel Webster, Clay and other political enemies of the Administration. When Van Buren learned of Jackson's serious illness in May, he was so alarmed the trip would be cancelled he could not conceal his anxiety. His letters became lectures on responsibility in a tone few dared use with the masterful old egotist. "Be assured," he wrote, "that nothing short of absolute incapacity will reconcile the people to the disappointment of not seeing you in this quarter. They have been so thoroughly impressed with that expectation that this gratification has become a sort of duty."[11]

Van Buren's peace of mind was restored when Jackson made a quick recovery. Yet he was not completely reassured until he spied the tall figure on the steamboat *York* as it made its way into the small harbor of Perth Amboy, New Jersey. Van Buren was standing on the crowded deck of the New York committee's ship, the *North American,* which would convey Jackson to Castle Garden. From there he would lead the procession up Broadway to City Hall.

New York City had done itself proud in its warm greeting of Jackson. As the *North American* made its slow progress through the narrows, accompanied by the Hudson River day boats, the *Ohio, Hercules* and the *Rufus King,* their spars bedecked with bunting, Forts Hamilton and Diamond and the batteries on Bedloe's, Staten and Governor's Islands roared out salute after salute. In the harbor itself, all of the shipping was clad in holiday attire with the red, white and blue of the national ensign predominating.

Yet the most striking feature of it all was the people, thousands upon thousands of them, clustered like gigantic swarms of bees on the streets, the wharves, the windows and the roofs of the tall warehouses and shops at the foot of Broadway. "There was an ocean of folks," said Seba Smith, a Jacksonian editor, through his character, Major Jack Downing, "cutting up capers as high as a cat's back." Smith, a notable humorist, could not resist retailing to his readers the collapse of the dignitary-laden bridge that connected Castle Garden with the mainland. Jackson and Van Buren, who led the group, escaped unscathed, but not so members of the cabinet who were traveling with the President and many other celebrities. They were all unceremoniously dumped into the shallow muddy water.

Fortunately, no one was hurt. The opposition also had its humorous commentator who, when describing the scene, had Van Buren saving himself by grasping the tail of Jackson's horse and when last seen was "streaming out behind as straight as old Deacon Willaby's cue when he is a little late to meeting. Some folks said it looked like the Flying Dutchman."[12] After the drenched and muddied notables had cleaned themselves up as best they could, the procession started with Jackson on horseback, doffing his tall beaver and bowing left and right while he acknowledged the applause from the crowd lining both sides of Broadway. Van Buren, also on horseback, followed, deeply impressed by this tremendous ovation from his fellow citizens. At the city hall, Governor Marcy was waiting with a short speech of welcome to which Jackson made an appropriate reply.

But as Van Buren had promised Jackson, the New Yorkers expected nothing more than a good look at their President, the general who defeated Wellington's veterans, vanquished nullification, was hacking away at "the monster," the mysterious Bank of the United States, and protecting the nation from all other enemies at home and abroad. "They are prepared to have you decline a dinner, to omit addresses [and] not put you to any fatigue or inconvenience beyond what will be unavoidable," Van Buren had said before the trip and he was as good as his word. Beyond the procession, the greeting at city hall, the only official function Jackson performed was a review of a militia regiment at the Battery and a brief inspection of the Navy Yard and the old Revolutionary battlefields in Brooklyn and Manhattan.

He did attend several receptions and shook thousands of hands; yet far from taxing the energies of the old man, Jackson seemed to draw new strength and exuberance from this outpouring of good will. In high spirits, his only complaint was the severe sunburn he received on his head and face from riding "hatless too long in sultry Philadelphia." "I have witnessed enthusiasm before but never before have I witnessed such a scene of personal regard as I have today and ever since I left Washington," he wrote his adopted son. Ex-Mayor Philip Hone, politically opposed to the Jacksonians and no admirer of Old Hickory, had to admit that he "is certainly the most popular man we have ever known. Washington was not so much so."[13]

Jackson had been struck by what he took to be the absence of partisan spirit in the New York City celebration. Despite his optimistic comment on the political scene, the astute veteran knew the calm was a transient phase; that the Bank issue was opening a deep rift in his second Administration before it had fairly started. After the plaudits, after the cheers, after the thousands upon thousands of ladies waving handkerchiefs, the partisan fires would flare up again. Certain as he was of his course, and inflexible his purpose, he was not as indifferent to the consequences of his policy as many thought. He was especially anxious about Van Buren's posture and the information he had gleaned in Pennsylvania and New York.

One other was more concerned, Louis McLane, who had accompanied the President from Washington but was now parting company. The new secretary of state desperately wanted a confidential talk with the Vice-President on the removal issue. He anticipated, however, that there might be neither the time nor the occasion for one; so before the presidential party left Washington, he wrote a note to Van Buren. When it turned out he could not speak privately with Van Buren before he left, he had Major Donelson deliver it. Almost as a suppliant, McLane begged Van Buren to consider carefully the lengthy brief against removal which he had furnished the President on May 20. "God knows I have no love for the present Bank," he said, adding, "something undoubtedly is due to the new Secretary who should not be driven to the step at least before the Senate have acted upon his nomination."[14]

Van Buren had known well in advance that Jackson wanted his advice on the removal question and on Kendall's plan for using selected banks in coastal cities. Before Jackson's arrival he had received a long and able letter from Kendall that answered many of McLane's arguments against removal and eased many of Van Buren's doubts. Any bank chosen for deposit must make monthly reports on its financial condition to the secretary of the treasury, and more frequently if deemed necessary. Deposit banks must furnish all the services the government might require at no cost, including the expenses of auditors who would examine their books at the discretion of the secretary. Most important, the Kendall plan called for phased removal. No new deposits would be made after a certain date, but funds would be withdrawn gradually, the rate determined by the current expenses of the government. As Kendall explained it, "This has the double recommendation of shielding the government from every charge of harshness towards the U.S. Bank and enabling the selected state banks to make terms with their rivals and neighbors before they may be called on for payments by the Treasury."[15]

During the three nights he remained in New York City and at various times during the trip through New England, the President and Van Buren went over Kendall's plan and various notes that had been prepared before the trip including McLane's brief. Their discussions were completely private. Secretary Woodbury, who joined the party in New York, and Cass were not present, nor were Donelson and Ralph Earle, the painter who was acting as an aide to the President. Impressed by the President's determination and the various safeguards that Kendall had broadly sketched in his plan, Van Buren agreed that the government discontinue making deposits in the Bank of the United States and did not object when the President set an early date between September 1 and 15 to begin the shifting of funds. They had reached agreement and Donelson was copying two documents for Duane's guidance when the party reached Hartford on June 17. The first document was a letter of instructions based almost exclusively on the Kendall plan. Five state banks in Baltimore, Philadelphia, New York and Boston would be chosen. Nothing was said about the political character of any particular bank. The choices were left up to the secretary but, said Jackson, "it is nevertheless extremely desirable to secure their good will and friendly cooperation." The President took it for granted that Duane would have to discuss arrangements with the bankers and suggested Amos Kendall be the negotiator with the banks.

The second document appears to have been drafted at Van Buren's suggestion and with his help. In effect a position paper, it gave a history of the Administration policy toward the Bank primarily for Duane's guidance and benefit, but it also answered various points McLane had raised. In this lengthy document Jackson made a vital concession which it was hoped would ease any qualms Duane might have about carrying out the President's instructions. Jackson accepted full responsibility for the transfer. Neither Van Buren nor Jackson wanted another cabinet blow-up; and Van Buren knew by now that Duane was McLane's man. If Duane followed the lead of the secretary of state, he would balk at the timetable; if he resigned, Cass and McLane would in all likelihood follow.[16]

When the presidential party reached Concord, New Hampshire, Jackson, who had been indisposed in Boston, became seriously ill. Until then, his always precarious health had held up admirably under the strains and fatigues of countless public appearances, the weariness and discomfort of cramped steamboat quarters, short jolting rides on the primitive, new trains, longer ones in swaying carriages frequently blanketed over with clouds of dust from stretches of dry turnpikes and country roads. At Concord, however, the inevitable occurred. Jackson began coughing blood; he developed a fever and a resurgence of pain in his side and chest made it difficult for him to breathe. Without gratifying the Jacksonians of New

Hampshire, the presidential party turned south and by the fastest means possible returned to Washington.[17]

Van Buren remained with the President when they reached the White House. While he and other anxious members of the President's personal and official family watched at the bedside, McLane arrived. In Jackson's private drawing room, Van Buren explained that he had changed his mind on the deposit question. "You now advocate the removal in obedience to the wishes of the President," the crestfallen secretary of state said. "I found the President so determined that I could not oppose him," was Van Buren's somewhat cryptic reply.[18]

The President had reached the White House on July 4 and within four days had recovered sufficiently to be up and around, though still remaining in his bedroom. On July 10, he received a message from Duane that he had not expected. The secretary of the treasury took issue with Jackson's instructions on the deposits. Their only point of agreement was that they both opposed rechartering the Bank. On deposits he argued that Congress, not the President, must decide, and he questioned whether selected state banks were the proper facilities for receiving government funds. Duane based his case on the resolution of the House declaring the Bank a safe institution for performing that function. Then, without offering any concrete evidence, he roundly condemned state banks as "unfit agents." Considering the direct contradiction of his instructions, Jackson's reply was remarkably mild in tone. But his advisors, Van Buren among them, were even then working on an extensive refutation of Duane's position. The President was hopeful the weight of their argument would bring his secretary to his senses and avoid any cabinet crisis.[19]

During the next five days, Jackson and Duane exchanged views, the stubborn Secretary of the Treasury clinging to his opinions. One of his main arguments had been that sound state banks would not accept the deposits, fearing the power of the Bank of the United States to which most of them were under financial obligation. Perhaps this assertion of Duane's eased Jackson's wrath, which could be mighty if crossed, or stayed his impatience; or perhaps even Van Buren, always cautious, exercised restraint over his impetuous advisors. Whatever the reasons, the President was unusually temperate with the recalcitrant Secretary.

Duane eventually agreed that Kendall should make a survey of the state banks and find out whether they would accept deposits after a certain date. His instructions, however, were so prejudiced against removal that Kendall's effectiveness with the bankers would have been seriously jeopardized. After reading them, Jackson gave his agent authority to edit them as he wished and Duane had no recourse but to accept the new version. In late July, Kendall left Washington as the Administration's agent and negotiator. Van Buren departed for New York City about the same time; and the President a few days later went to his favorite summer retreat near Washington, the Rip Raps, a small island in Chesapeake Bay near Hampton Roads, Virginia. Besides his extensive household, Van Buren's son, Abraham, accompanied him and later Blair joined the party.[20]

Jackson's patience was running out. Before he left Washington he wrote a candid letter to Duane in which he said that if he would not execute the Administration's policy he must resign. When Jackson related this information to Van Buren he asked for his opinion on whether the government should begin removing deposits at once or defer the action until January 1, explaining at the same time that a majority of the cabinet favored the later date. Jackson was troubled about the cabinet's attitude. He wanted Van Buren's advice, but he also needed his support.

Van Buren had left Saratoga, where he had gone for a rest, and had hurried to New York to see his youngest son, Smith, off for Europe with the Livingstons. He was also expecting a visit from McLane, whose views on the Bank issue he was anxious to hear.[21] He was entirely unprepared for any commitment beyond stating he would reply "in season" after he had discussed the timing with Wright. From what he knew about Duane's appointment and his position, Van Buren wondered whether he was speaking for the secretary of state and, whether, as he had heard, McLane was controling all of the Cabinet except Taney.

There was another important reason for delay in answering Jackson's query about the timing of the process. Van Buren wanted Kendall's views on how the major banks of the east coast viewed the deposit question. Would they accept government funds even though most of their officers and stockholders were political opponents of the Administration? What of the Bank itself, how would it react? If Biddle, who had been extending loans and discounting the paper of state banks at an increased rate since the veto, reversed himself and began demanding redemption in specie or the Bank's own notes, were the assets of the major eastern banks strong enough to withstand what would inevitably be a tight money market? Could the Treasury lend support, could it do it lawfully, and if so, how much and to what extent should the major state banks be incapable of standing alone against Biddle's reprisals? He had talked with James A. Hamilton on the subject, and though he had come to recognize Hamilton's limitations, he could not dismiss lightly his views opposing removal. All thought such a step, according to Hamilton, "would be difficult of accomplishment and disastrous in effect on the business of the country."[22]

Despite these distress signals from McLane, from Hamilton, and presumably from the New York banking community, Van Buren had gone along with the President. Yet, apart from the views of Hamilton and his friends, the only other outside opinions he or anyone else in the President's circle had about possible economic effects were an elaborate paper of Reuben Whitney and various detailed letters of Thomas Ellicott, a Baltimore banker, close friend of Taney and opponent of the Bank.[23]

Whitney, Ellicott and Taney may have persuaded Van Buren with their arguments; still he recognized special pleading when he saw it whether from Hamilton or from Reuben Whitney. Once committed to the removal, he wanted hard evidence, specific facts that Kendall would supply, and fresh information on politics and policy from McLane. For this kind of in depth evaluation, Van Buren had arranged for Olcott to talk with Kendall. He himself would quiz McLane, and then he and Wright would consider the information.

Olcott was an impartial judge of the New York City banking community. As cashier of the Farmers' and Mechanics' Bank of Albany, he was thoroughly familiar with the policies of the city banks but had little or no business ties with them. Wright's experience was broader. During his term as comptroller of the state he had dealt with all of the major banks, while his congressional experience had made him something of an expert on the wider implications of politics and public policy. It was Wright's deliberative mind, however, his capacity for sorting through quantities of material and selecting out just the right data for a reasoned, balanced conclusion that made him such a valued counselor. If there were weaknesses on either side of what Van Buren knew was fast becoming the preeminent political and economic issue, Wright would find them and weigh them judiciously.[24]

Van Buren's suite in the City Hotel with the inside blinds drawn provided scant protection from an early August heat wave. Yet here in the stifling parlor, Van Buren and McLane talked for hours. The Vice-President soon discovered what he had expected: Duane shared McLane's views on the deposits. As always he was fascinated at the agility and the scope of the Secretary's mind, attracted to his magnetic qualities of speech and personality. He was even impressed with McLane's argument that nothing be done until Congress convened in December. At that time McLane said, the secretary of the treasury would ask Congress for the authority to discontinue deposits in the Bank and ultimately draw them out as the needs of the government required. Should Congress refuse, he and Duane would support immediate removal on executive authority. But Van Buren saw the weaknesses in McLane's proposal. What could Congress do before the secretary of the treasury had acted except pass advisory resolutions with no force of law? Of much more serious import was the matter of responsibility. The Administration was charged with executing the law. For the President to ask the legislative branch how it should be done would seriously weaken the President's prerogative and set at defiance the distribution of powers within the government.[25] He did not question McLane on these points, however.

Kendall arrived in New York on August 10 and put up at the City Hotel. The next morning he was surprised to see Van Buren and McLane, both of whom were anxious to hear about his reception. After he relayed to them his experience in Baltimore and Philadelphia, they realized his mission had been more successful than they had believed possible. The private banks of Baltimore which had a long history of resisting the Bank had been expected to be friendly. But Kendall's success with the banks in Philadelphia, directly under the eye of Nicholas Biddle, had been little short of astonishing. Of the nine banks in the city that met the standards Kendall had set forth in his circular he left with them, four were quite willing to accept deposits, two declined and three had not responded when Kendall reached New York.

After weighing this information, Van Buren sat back and let McLane explain again his plan of involving Congress in the process. Kendall was impressed when McLane said that he and Duane would use all of their influence in Congress for a favorable resolution. At least he reported to the President that "as one individual I should be delighted with this *provided* Mr. McLane and Mr. Duane will exert their influence in Congress to sustain the measure." But Kendall skillfully covered himself. In presenting the other side he asked, what if the two secretaries reneged? The Administration would in all probability be defeated by a combination of the forces against them, Nicholas Biddle, every member of the cabinet save Taney and Barry, the National Republicans and those would-be Jacksonians who were under financial obligation to the Bank. Kendall was a shrewd operator. He was careful not to involve Van Buren with McLane and Duane, and he knew enough of the President's intentions to give a personal opinion for prompt removal on executive authority; yet he left the door open for congressional participation should Jackson change his mind.[26] Kendall was finely attuned to the President's mercurial disposition and he was uncertain about Van Buren's influence or his intentions, as the Magician had meant him to be.

After his impromptu discussion at the City Hotel, Kendall met with Thomas Olcott who introduced him to the various bankers in New York City and sat in on their talks. Afterward, Olcott and Kendall went over the bankers' responses, noted their political affiliations (primarily opponents of the Administration), evaluated their assets and their liabilities. In the process, Olcott, through adroit questions and carefully phrased comments, formed an estimate of Kendall's abilities, the feasibility of his plan, attitudes and the reaction of the New York City banks. On the whole, Olcott's summary was favorable, but he worried about possible speculation with government funds and suggested various restraints that might be imposed on all participating banks. He also opted for a phased removal over the next six months to ease the impact of any dislocation in the money and credit supply or in any impairment of business confidence.[27]

As he had done in Baltimore and in Philadelphia, Kendall left his circulars with the principal New York banks before he set out for Boston. There he encountered a wall of resistance. Of the thirty banks doing business in Boston only two were willing to be considered depositories. Nor were the New York City banks at first enthusiastic about the plan as Kendall learned on his return. The three largest banks, each of which had a capital of $2 million or more, balked at the requirements listed in Kendall's circular. For years they had cooperated with the Bank of the United States and had prospered. They had, in addition, a healthy respect for Biddle's power and would not willingly court reprisals. Again Olcott helped out and eventually George Newbold, president of the largest New York bank, agreed to accept the funds under the conditions Kendall set down. The other two, including the Manhattan Bank, which was not a member of the Safety Fund system, followed Newbold's lead. But the task had not been an easy one for Kendall who had been subjected to a merciless examination of Jackson's policies.[28]

While Kendall was explaining the government's plan to skeptical New York and Boston bankers, Van Buren fled from New York's heat, noise and squalor. Earlier he had planned his summer with his accustomed care. He would spend a month with his son, Cambreleng and other close friends in Saratoga, take a brief tour in western New York for his usual political soundings, and then what he looked forward to the most, a ramble through the old Dutch villages of the Hudson River Valley with that most delightful of companions, Washington Irving. As he sweltered in the City waiting for McLane and the Livingston family, he asked Cambreleng to engage quarters for himself and family at either the United States Hotel or Congress Hall in Saratoga. Careful about money affairs (though he was well enough off to lend his friend David Evans $15,000 in cash) he insisted he not be charged double simply because he wanted two rooms. By way of explanation he referred to the expenses of his sons even though John, the most extravagant, was practically self-supporting. "These boys," he said humorously, "run me so hard that I must look out for number one."[29]

He had no sooner reached Saratoga, however, than he was confronted with a substantial manuscript in the hand of Major Donelson and a letter from the President. Jackson had dictated his views on removal, which he was now determined must take place as soon as possible and well before Congress met in December. He wanted Van Buren's opinion on his paper that catalogued the Bank's corrupt practices and misdeeds as he understood them. Fortunately, Wright, to whom Van Buren had written requesting a meeting, was waiting in Saratoga. The two men went over the material carefully, Van Buren making small corrections here and there but not challenging the President's case or "exposé" as he called it, nor his decision for early removal.

Wright begged for more time and further consultations with the bankers and the Regency men in Albany. Van Buren agreed and in response to Wright's warnings about speculators with inside information, he wrote the President explaining the reasons for delaying his opinion. Reminding Jackson that there are always two sides to any question, he wanted further discussion with Wright, whose judgment he praised warmly. He was leaving, he said, on the morrow for western New York and had arranged for Wright to meet him at Oswego. In two weeks time he would be back in Albany and would then send in his opinion. He closed with the barest hint of a difference between them. "I hope we shall in the end see the matter in precisely the same light."[30]

At the Rip Raps, Jackson was receiving not only favorable reports from Kendall, but evidence of Biddle's partisan activities paid for out of a special fund. From these and other materials he was adding to his case against the Bank further justifying his decision to cut off deposits at an early date. Kendall had also passed on some information from an admittedly biased source that reflected on McLane. Jackson reacted sharply to Kendall's gossip-mongering. Once he had made a decision he demanded it be carried out promptly.[31]

He was now suspicious that McLane was behind a movement to thwart his will; and he was not altogether satisfied with Van Buren's tardiness, especially after he received Kendall's account of his meeting at the City Hotel in New York. Personally, he exculpated McLane but then declared that the public believed him guilty; and he wrote Van Buren "that Mr. Duane's course has been such to confirm it in public opinion." He would now act more vigorously to protect McLane's reputation, Jackson continued, and then a dangerous note crept in, "and particularly your popularity, it being well understood the confidence and friendship that exists between you and him. it [sic] is already hinted that you are opposed to the removal of the deposits, and of course privately a friend to the Bank. this must be removed, or it will do us both much harm." On August 17, 1833,

Blair, writing Van Buren from the Rip Raps, censured McLane severely. Considering the source and the relationship Blair bore to the President, Van Buren must consider the editor spoke for the President as well as himself. Van Buren was traveling when this letter arrived in Albany. By August 30, he had restudied Jackson's "exposé" and promised the President that he would have his views in a few days.[32]

If Van Buren was disturbed at Jackson's menacing letter of August 16, he gave no sign. In fact, he would not answer the President until he had Wright's opinion, which he did not receive until September 3, three days after he reached Albany. Now he disassociated himself from McLane's, and presumably Duane's, plan, which he described briefly as one of three courses that might be taken. The second was the President's plan of removal "some time," as he phrased it, before Congress met; the third suggestion, which Silas Wright had recommended, was in effect a compromise between the other two.

Wright had spoken with the Regency men, whom he found divided on when and how the deposits should be removed. His version, which Van Buren preferred but was "not so strenuous about it, as to feel that I should be unable to sustain the second one," would make all the preparations for transfer; compile all the evidence of alleged mismanagement, corruption and partisanship of the Bank; exhibit the condition of those banks chosen for deposit and the regulations under which they would handle the public funds; and give the order for removal to take effect on January 1 (which Van Buren and Wright assumed mistakenly was the beginning of the new fiscal year). Wright thought, and Van Buren agreed with him, that this short delay would head off any collision with Congress while carrying out the President's objectives. The case against the Bank was so strong he anticipated no threat from Congress despite its being relegated to an advisory role.

The Wright–Van Buren plan was simply a political device that would ease congressional resentment, but in a way it had the same weakness of the McLane–Duane proposal, a weakness Van Buren recognized. As he put it, "by leaving the matter open you (Jackson) seem in some sort [of way] to invite the action of Congress on the subject." Still he was convinced Congress would pass a resolution on the subject anyway; and if the Administration acted while Congress was in session after giving due notice, it would receive a strong vote of confidence that would maintain good relations between the executive and legislative branches, and calm any fears the business community might have about the transfer. Implicit in his presentation was an earnest desire to stay the powerful hand of Nicholas Biddle from creating a damaging credit shortage in the commercial centers of the Northeast. Wright's worries about inside speculation may have had some bearing too. James Gordon Bennett, now editing the *Pennsylvanian* in Philadelphia, had written him a six-page letter accusing Kendall and Blair of planning to speculate in Bank of the United States stock, once they were sure removal would take place. Though Van Buren must have discounted most of what the vindictive editor said, he could not dismiss it completely. For it might just be true.

Van Buren was very worried about the banks in New York City. "Great care," he advised the President, "must be taken in the selection. . . . certainly not less than three should be taken, and if possible *four*. Those engaged in them, like the rest of their Fellow Creatures, are very much governed by their own interests, and it would be well to consult someone out of the city as to their proper collection."[33]

Jackson was not at all happy with Van Buren's opinion and seems to have misconstrued his language on one very important point, the McLane plan. Although the meaning of the sentence in question is not clear, the President's conclusion is unmistakably so. He wrote to Van Buren that he regretted "the division of my Cabinet . . . and I counted on your support firmly in taking the stand with me to remove the deposits on the first of October next if Mr. Kendall's report was favorable to the safety of the State B.'s." This sentence, stressing the words "counted" and "firmly" taken with the earlier reference to McLane's position and to the letter of warning he had sent Van Buren on his relationship with McLane, seemed to place Van Buren among his opponents.

At least this was Van Buren's impression and he struck out "vigorously and clearly." He denied that either Wright or himself approved of McLane's plan. He had been asked his opinion on when the removals should take place. He had given as his preference January 1 and his reasons for that date. He had not changed his mind, though he admitted the report of the government directors, which Jackson sent him, that charged Biddle with diverting funds for publishing partisan speeches and widely distributing articles and editorials opposing the Administration had materially weakened his case. "Upon the whole, do as you think best for the honor and interest of the country," he said, "and count with the utmost confidence on being sustained with immovable constancy, by myself, Mr. Wright and all our true friends in this state."

On rereading the President's letter he noted a postscript asking him to be in Washington during the first week in October, which in his haste to answer he had overlooked. During that time as he had told the President, he would be traveling with Washington Irving.

He wrote to Jackson immediately [to repair his omission] and fell back on a familiar argument why he should not be in Washington at the date appointed for the removal.

> You know that the game of the opposition is to relieve the question, as far as they can from the influence of your well deserved popularity with the people, by attributing the removal of the deposits to the solicitat[i]ons of myself and a monied junto in N.York, and as it is not your habit to play into the enemies' hands you will not I know request me to come down unless there is some adequate inducement for my so doing.

As he had when Van Buren remained in New York during the nullification crisis, Jackson agreed with the Magician's rationale or, as some would have said, sleight of hand. Van Buren took the opportunity in the same letter of again divorcing himself from McLane's views. "He, and I differ, *toto coelo* about the Bank and I regret to find that upon almost all public questions the bias of our early feelings is apt to lead us in different directions." Professing the strongest personal friendship for the secretary of state and alluding to the many times he had interceded himself in his behalf, he would not "desert him, as long as I have it in my power to serve him and his."[34]

McLane was a widely respected public officer with no political base worth mentioning; his continued presence in the cabinet shielded the Administration from the attacks of excessive partisanship in its appointments. That a former Federalist from a tiny state whose government was in the hands of the opposition should occupy the most prestigious office in the cabinet gave the lie to those who were accusing the Administration of practicing a rampant spoils system, a charge that concerned the Virginians especially. Looked at pragmatically, and Van Buren was nothing if not pragmatic, McLane was an important factor in maintaining the respectability and the solidity of the Administration.

Van Buren's forthright letter mollified the President, who praised its spirit and was now ready for specific action, setting the date for the public announcement of removal as soon as the details with the state banks that had been chosen were worked out. Three days later Jackson read his "exposé," which Taney had rewritten, to the cabinet and gave them notice that the *Globe* the next day would notify the public that the deposits after October 1 would be placed in state banks.[35]

Even those doubters in the cabinet were impressed with the evidence against Biddle's management and the President's forceful language. But Duane still clung to his original position and after an ultimatum from Jackson told the President that he would neither resign nor carry out the removal. Jackson managed somehow to control his temper and wrote out a brief letter dismissing him. A second note appointed Attorney General Taney in his place. Taney immediately resigned his post and was sworn in as secretary of the treasury.

The excitement of the past few days improved Jackson's health. As he explained to Van Buren, "I have been a good deal indisposed, would no doubt have been much worse, if it had not been for the *exciting* pills administered by Doctor Duane." The President wanted Van Buren's advice on the appointment to the vacant attorney general's post, adding that Woodbury thought the appointment should come from the South. Jackson wavered between Pennsylvania and Virginia, suggesting Judge Richard E. Parker of Virginia, a member of the Richmond Junto and Ritchie's cousin, or George M. Dallas, head of the Family party in Pennsylvania, a staunch Jacksonian but an equally staunch rival of James Buchanan's Amalgamators, the western faction of the party.[36]

During these exciting moments in Washington which Jackson relished and which took his mind off his physical infirmities, Van Buren and Washington Irving were in the midst of their trip. In the beautiful weather of early autumn with just a touch of crispness in the air, the two men explored the romantic scenery of the Highlands with its steep crags and rushing streams. From their open carriage they viewed the variety and the beauty of the placid Hudson filled with shipping of every description heading north and south—small sloops and larger schooners tacking their way upstream against the current or catching the predominant northwest wind gliding wing and wing downstream. Where ten years before steamboats were few, now the plumes of smoke from their lofty stacks were found everywhere. They saw blunt-nosed passenger vessels with their sidewheels slapping rythmically through the stream as their captains swung vessels around sandbars following the channel markers.

At the base of the Catskills, the carriage moved along at a leisurely rate crossing and recrossing the Hudson, through Kingston where the blackened foundations of houses the British had burned in the Revolution could still be seen from the road, to Goshen, Haverstraw, Tappan and Hackensack on the west side of the river and then across to Brooklyn.[37] Here reluctantly Van Buren had to cut short his trip. Letters from Washington bespoke another cabinet crisis. As he saw it, McLane was the focal center of the problem and gave every indication of resigning. If he went, Van Buren felt sure Cass would follow.

There were two letters from Jackson. One postmarked the twenty-third, detailed the final episode with Duane and asked again for suggestions on filling the attorney general's post Taney had vacated. The other was written the next day, just after an interview with Cass and McLane. Jackson hoped they would remain, "but if it so happens that they do not the question remains—whom shall I select for the State, War and Attorney General?" The General was defensive and contradictory, obviously in need of guidance. For the past two months he had been bombarded with letters from close friends, political associates of long-standing, like Ritchie, urging him to take no action against the Bank. Even the venerable, much respected Nathaniel Macon spoke up for the Bank from his plantation in North Carolina. On the other side, Blair was constantly at his elbow arguing that he must stand fast whatever the repercussions in the cabinet and in Congress when it convened. Taney could see no harm if Cass and McLane left the cabinet, and in fact welcomed the possibility of their departure.[38]

Aware of these tensions, Van Buren acted promptly and decisively. McLane was all important. He must be persuaded to stay and what better means of persuasion than Washington Irving, his traveling companion. The kindly, perceptive and completely apolitical Irving was one of McLane's closest friends, practically a member of his family. If anyone could calm the tempestuous, sensitive Secretary, have him see why his presence was so desperately needed in the Administration, Irving could do it. Van Buren easily persuaded the writer of the necessity and then briefed him carefully on how to present the case to McLane. When Irving arrived in Washington, McLane and Cass had both decided to remain in the Administration, but McLane was in a dangerous mood, hurt at what he considered underhanded attacks upon him from Taney, Blair and Kendall.[39]

Not knowing at the time whether McLane and Cass would stay or not, Van Buren responded to Jackson's queries about the three posts. He agreed with Woodbury that a southerner should fill the attorney general's post. His preference was John Forsyth of Georgia, who had done much to stay the nullification current in his state. Van Buren had been a close friend of Forsyth's since they both campaigned for Crawford. If not Forsyth, he leaned toward Parker, whom Jackson had mentioned in an earlier note, and he spoke favorably of Thomas Griffin, a North Carolinian and another Crawford veteran. But above all, he urged reflection and care in any appointments and he delivered a brief homily to the President on the subject:

> There is one point you may depend upon, my dear sir, and that is that there is an extreme anxiety on the part of the Democracy of the country—your stay and support—that you should infuse a little more

> of their good spirit into your Cabinet than it now possesses. . . . our quondam friend, Duane, was either beyond or behind the age. Do not be in haste.[40]

Although he knew Jackson wanted him and needed him in Washington, Van Buren felt his presence there during the cabinet crisis would be misconstrued. The opposition press would roundly attack him for interference and would belittle Jackson by insisting he was a puppet in the hands of the Magician. He was, of course, right; yet at the same time, his motives, appropriate for the occasion, were also playing it safe for himself. But he made sure that he could be reached easily and quickly. He cut short his trip, remaining in New York City where, as he explained to Jackson, he "would be within striking distance . . . and we can communicate by letter every 24 hours," a measure of improving transportation along the Atlantic seaboard where the mails now went by rail most of the way between the two cities.

Van Buren did not expect his absence from Washington would go unnoticed, nor that he would escape criticism. No sooner had the *New York American,* the *Courier and Enquirer,* and the *Albany Evening Journal* learned about Duane's dismissal from the *Globe* on the twenty-fourth than they saw the deft hand of Van Buren manipulating Jackson's cabinet for his own benefit, hiding behind the shield of the General's popularity. It was strongly hinted that Van Buren was in league with the big New York banks. Thurlow Weed thoroughly raked over John Van Buren's recent election to the board of directors of the Farmers' and Mechanics' Bank in Albany. Olcott's visits to New York City, his being seen with Kendall there, provided sufficient evidence for unsubstantiated charges.[41]

As Van Buren had been through it so many times before, he paid scant attention to the distortions of the opposition press beyond sending Jackson a clipping that proved his point. "You will see by the enclosed," he wrote, "that the opposition have commenced the game I anticipated. . . . They have found by experience that their abuse of you is labor lost, and they conclude wisely, that if they succeed in shifting the Bank question from your shoulders to mine, they would be better able to serve the Mammon than they are at present." The President agreed completely. "We act solely on the defensive and I am ready with the screws to draw every tooth and then the stumpts. Fear not therefore, for the abuse, the more they heap upon you, the more will the virtuous portion of the country appreciate you."[42]

So confident was Van Buren of victory at the polls in the state that he left for Washington well before the election took place in November 1833. He had rented one of the houses on the corner of Pennsylvania Avenue and 19th Street that were known as the Seven Buildings. His son Abraham had had the rooms repapered and other interior arrangements made so Van Buren could convert either a lower or an upper chamber into a dining room. But he had not as yet moved in his furniture or made necessary additional purchases. Painting and carpeting were incomplete and he wanted to supervise the work personally. Either Van Buren or his son mentioned these domestic matters to Jackson, who promptly invited him to stay with him at the White House. "I now invite you here. I have dear Sarah's room prepared for you until your own house can be put in order," he wrote.

The acute stage of the crisis had passed but many, many problems remained. Jackson needed Van Buren—his even temper, his companionship, for he had come to look upon him as his closest personal friend in Washington. His charm, his convincing arguments, his perception of politicians and policy were indispensable to the old man who, now the excitement had passed, was beginning to feel unwell again. Taney was "a host," energetic and of great ability in Jackson's eyes; but the dour Marylander who was efficiently working out the means for removal of the deposits was no Van Buren. The taciturn Taney, who was usually enveloped in clouds of cigar smoke, lacked the ready wit, the balanced judgment that was second nature to Van Buren, those very qualities Jackson prized the most. Barry, the only other cabinet member, whose loyalty and friendship were beyond question, was a sick man. It was common knowledge in Washington that the Post Office was being badly mismanaged. Charges of corruption and of inefficiency had already come to the President's ears and distressed him. The removals could be safely entrusted to Taney and for the time being he could direct the attorney general's office. But the post had to be filled and there was the annual message to be drafted, a strategy developed in dealing with the opposition in Congress to the Administration's removal policy.

Jackson had become wary of plausible men with ability. Eventually, he offered the post to Peter V. Daniel, like Parker a member of the Richmond Junto, probably at Van Buren's behest. Daniel, a man of talent, a good speaker, capable lawyer and firm supporter of the Administration, was an excellent choice. Unfortunately after he came to Washington and discussed the duties with Taney, Daniel decided that the work would interfere seriously with his personal life. He declined the offer, but not before Blair carelessly published his acceptance in the *Globe*.

Van Buren now proposed Butler. The office had been offered to the South and to Virginia, and declined. The Administration was thus relieved of any further deference to sectional feeling. Professionally, Butler was an ideal candidate. He was a fine speaker in the flamboyant style so admired of public men. He stood at the head of the New York bar. He was abstemious in his personal habits, a devoted family man, absolutely loyal to Van Buren and Jackson. Above all, he was an incredibly industrious person whose ability and learning had never been questioned in the hottest of partisan combats.

The only problem, and the most difficult to surmount, was persuading Butler to accept. Van Buren wrote him a long, warm, carefully phrased letter. He explained that the publication of Daniel's acceptance had been a mistake, but a fortunate one because it removed any political liability that stood in the way of Butler's appointment. "Before this had occurred, I would not myself have proposed it to you, had the matter been at my disposal. Now I think it free from difficulty and objection," he said. Anticipating Butler's reluctance, Van Buren argued how beneficial the office would be for Butler's future and the future of his family. "Although you are not a slave of mad ambition," he wrote, "you are, as you ought to be, tenacious of your professional standing. This cannot be increased at home, and can only be made *national* by being identified with national concerns." Who would ever have heard of Wirt, Webster, Pinckney or Taney, none of whom Van Buren thought better qualified than Butler had they not made their reputations in Washington?

Speaking further of the rewards of the post, Van Buren exaggerated them while he minimized the burdens. "You can enter upon the business of the Supreme Court of the U.S. with advantages which, if not immediately equal to those of Webster (who makes his thousands not to say tens of thousands by it), very soon would be." The President, he went on, would not object to Butler's carrying on his private practice in New York. In fifteen hours Butler could reach New York, another day be in Albany. Having spoken of the office in most glowing terms, Van Buren shifted his argument to Mrs. Butler and described Washington society as most suitable for the family, if she wished to accompany her husband. "Mr. McLane, Mr. Taney, Mr. Woodbury and Gov. Cass have each houseful of little girls of the very finest character, and I am quite sure that society for Mrs. B. and the children, would be at least as good as in N. York, and if she cannot possibly do without hearing something on the subject of temperance, she can count on Gov. Cass as a never failing source." The letter was couched in Van Buren's best style, solid argument, touches of wit, appeals to patriotism and to the profession Butler revered. How could he refuse, how could Harriet resist? They could not. Butler accepted, much to Jackson's and especially Van Buren's satisfaction. Butler would be a firm anchor for the Administration and for Van Buren's future. There was little doubt in his mind that McLane would soon leave the cabinet. The presence of Butler, whose character and ability would soon be recognized in Washington, should do much to make up for the loss of McLane's reputation. In the isolated, disorderly little capital, the cabinet set the character of the Administration. The opinions of the Washington press and those expressed in the constant stream of letters from Congressmen to their allies and constituents had a powerful impact on what the nation thought of its leaders.[43]

« CHAPTER 20 »

A Pinch of Snuff

Daniel Webster rose from his desk and solemnly read a petition for relief from financial distress. It was the hour the Senate set aside each morning to hear petitions; and when the clerk called Massachusetts, Webster, as he had been for the past three months, was ready with a memorial. On this morning in March of 1834, the petition he chose to read was from the master builders of Philadelphia. The subject was always the same: a tyrannical President was destroying the economy, imperiling the livelihood of thousands of workers. Other opposition Senators, or Whigs as they were now calling themselves, had each in turn presented long and dolorous accounts of empty factories, deserted farms, silent forges. Protests against the Administration's financial policy far outnumbered those petitions defending it.[1]

Vice-President Van Buren seemed impervious to the declamations on the floor; he barely listened to the deep, measured voice of Webster. Petition day was one he dreaded, for the tedium and repetition of the memorials caused his attention to wander, though he knew he must be ever alert with Clay, Webster or Calhoun ready to use any means that might embarrass him, especially if it meant some subtle circumvention of the rules that he had failed to pick up.

Webster finished reading his petition and had no sooner taken his seat than Henry Clay was on his feet, seconding Webster's petition and making a motion for reference to committee, all routine; but then Clay suddenly advanced to the front of the Senate chamber. He looked directly up at Van Buren and addressed him. Although his mellifluous voice was low and conversational, it reached everywhere in the small chamber with its surrounding galleries. Clay had not uttered two sentences before Van Buren realized that he was making a personal plea. The Vice-President was the last hope of a suffering nation, the only person with enough influence to reason with Jackson and restrain him from his reckless course. "Go to him and tell him, without exaggeration, but in the language of truth and sincerity, the actual condition of his bleeding country," said Clay earnestly, his mobile face reflecting the sorrow, even the anguish in his voice.

Van Buren, though taken aback by this sudden move and what he recognized as yet another bit of theatricals that Clay had concocted to belittle him publicly, maintained his accustomed calmness, listening intently and respectfully. When Clay had finished, Van Buren stood up and called Senator Hugh Lawson White—the president *pro tempore*—to the chair. The floor and the galleries were hushed as the Vice-President walked over to Clay's desk and with a slight smile asked "for a pinch of [your] fine maccoboy snuff." Clay, of course, complied and Van Buren, after a courteous bow, returned to his chair. For once, he had deflated the Kentuckian and, in fact, held up to ridicule all the memorials presented that day, if the applause from the galleries was any measure of popular opinion.[2]

Clay's histrionic gesture was simply one of the many maneuvers—rhetorical, parliamentary, partisan, obstructive, clever and crass—he had used to gain advantage for himself, the Bank and the Whigs over the Administration during the first session of the Twenty-third Congress, or in the common parlance of the day, "the panic session." His nimble mind, ready wit and ingenious tactics had made him the darling of the galleries and the envy of his colleagues, friends and enemies alike. Calhoun, with all his brilliance, felt he had to follow Clay's lead if he were to maintain himself and his small following against the hated Jackson and the despised Van Buren. Clay found Webster more difficult to manage, and for a time it appeared that he would move away from the coalition.[3]

Webster had watched Jackson's stock rise in New England. The President's vigorous handling of the nullification crisis appealed to the nationalism of the Northeast. And Webster, despite his lofty pose, his grandiose eloquence, his strident opposition to Administration policy, saw political advantage to be gained if he could preserve his identity and that of his faction, yet throw his weight behind the Administration on critical issues. Webster had been playing this independent role for the past year ever since Clay had assumed leadership of the opposition and with Calhoun's assistance pushed through the Compromise Tariff. Van Buren had watched the Webster strategy unfold, but said nothing. Jackson, grateful for any support that would weaken the Clay-Calhoun coalition, accepted Webster's position in good faith. Though both men continued to keep their distance, there was a growing rapport between them. Other astute observers may have been mystified at Webster's seemingly changed position but not Van Buren. He appraised it for what it was, Webster's bid for the succession and the creation of a new party out of an amalgamation of Democrats who distrusted Van Buren's motives and Whigs who disliked Clay's pretensions.[4]

When the Congress convened on December 2, 1833, Van Buren was back in Albany engrossed in local politics. He had rented the ladies' parlor at Bement's and thither streamed a constant procession of politicians, would-be office holders, members of the legislature and of the Regency.[5] But his attention was centered upon Washington. Every day he received a bundle of newspapers and letters from the capital keeping him abreast of the news. Van Buren, following the precedent set since 1823, would not appear and take his seat as presiding officer of the Senate until that body had organized itself under the president *pro tempore.* It was especially important for this session because the president *pro tempore* was Senator Hugh Lawson White of Tennessee, ostensibly a friend and political supporter of Jackson, but in reality moving away from Administration policies. Under the Senate rules, the Vice-President had been delegated the authority to appoint committees, Van Buren's predecessors had remained absent until the president *pro tempore* made the appointments, thus preserving a completely neutral stance. White, however, was in a politically awkward position. Not completely committed to the opposition, he was loath to exercise the appointive power. Nor was Clay at all anxious that he should.

Under the circumstances, White followed the only course open to him. He asked to be excused from voting. His associate from Tennessee, the quick-witted Felix Grundy, at once recognized what was being attempted and was determined if at all possible to make White take a stand. He failed in his attempt, however, and with White abstaining, the Senate repealed its rule.

Meanwhile, Webster had approached Grundy and offered his support in selecting committees whose majorities would be favorable to the Administration. The proposal astounded the Tennessee Senator. Why should Webster, great and good friend of the Bank, lend such direct aid and comfort to an Administration he had attacked so mercilessly for the past four years? Was he sincere, and if so, what price would he demand? With these unanswered questions in his mind, Grundy hurried to the White House for discussion and direction. Jackson, not unaware of Webster's flexible attitude toward his Administration, was tempted to accept the offer and with it control of the Senate. Yet at the same time he was suspicious of such a valuable gift from a longtime political foe. Why not test Webster's word and the weight of his influence. He suggested that Grundy offer a resolution to postpone selection of the committees until December 16. By then, Van Buren would be on hand for his valuable counsel.

With the presidential blessing, Grundy made a deal with Webster; and now it was Clay's turn to be astonished. When the resolution came up on December 12, Grundy moved postponement until December 16. His motion carried 28 to 13, Webster along with seven New England Whigs voting with the Jacksonians in the affirmative.[6]

Late at night, December 14, 1833, Van Buren reached the White House where he would be a guest of the President's until his furniture arrived at the Seven Buildings. He found in his room a message from the President. Would the Vice-President meet with him and with Senator Grundy as soon after breakfast as convenient? Early the next morning, he went to the President's office where he found the two men standing before the fireplace in deep conversation.

After a warm greeting, the President briefly outlined the situation in the Senate and turned to Grundy for a fuller explanation. Van Buren listened as the Senator enthusiastically presented a case for cooperation with Webster. The Democrats already controlled the House and if they were associated with Webster in the Senate they would force the Clay-Calhoun coalition into a helpless minority. Grundy argued convincingly for the deal, and though Jackson said nothing while he spoke, Van Buren could sense his tacit approval. After all, if Congress was secure, Biddle and his Bank would be smashed with ease and other troublesome problems that had beset his first Administration would not disturb his second. Clay and Calhoun, whom he blamed for most of his difficulties with Congress, would go the same way as Biddle. The next four years would be more tranquil, permitting his political principles, which he passionately believed to be the only true ones, to mature without distracting divisions.

Van Buren thought otherwise. Any arrangement with Webster violated all the political precepts he had been struggling to maintain for more than ten years. Amalgamation with Webster's faction, for that was what Grundy was proposing, would return to the shopworn, disorderly, potentially threatening personal politics of the Monroe years. After spending a decade promoting a political system that was based squarely on measures, not men, on nation, not section (a federal balance where the states had their rights within the Union) where politics reflected clear divisions on public policy, not the aspirations of individuals or of disparate special interest groups, Van Buren would not compromise for any fleeting advantage the Administration might gain in Congress. Every instinct of his being recoiled at the thought of an alliance with Webster, of all politicians, a man he believed to be personally corrupt, a mere creature of the Bank and of all Federalists who always placed their personal interests over their country's. Webster's moves toward cooperation with the administration was, in Van Buren's opinion, mere bombastic claptrap to cloak the nefarious doings of Massachusetts millionaires behind Jackson's popularity.

Nor was it lost on Van Buren that such a move was aimed squarely at him. His succession to the Presidency, now so tantalizingly within his grasp, would only be snatched away if this bargain were sealed. The opposition would surely find in it another flagrant example of the Magician's sinister management, another proof that he was utterly unfit for the Presidency. His public image among the movers and shakers had been distorted enough in the Whig press. Opposition editors had too often echoed John Randolph's unflattering description. "Too great an intriguer," this indigenous curiosity had written Jackson in an excess of malice, "And besides wants personal dignity and weight of character . . . an adroit, dapper, little managing man but he can't inspire respect much less veneration."

Van Buren's response to Grundy was prompt and incisive. He would not dignify the proposed arrangement by asking if it were practical, but asserted that Grundy "must be as little favorable as I could be to a political coalition with Mr. Webster." Webster, he continued "had always occupied in time of peace and in time of war, and especially in relation to the bank which we all knew would be the principal subject of the session" a position diametrically opposed to the policies and the fundamental principles of the Jacksonians. Surely there were troubles ahead; the Bank was powerful, Biddle an able and stubborn opponent. The Whigs, though not as well organized as the Democrats, had clever leaders like Clay and a well-financed press. But Van Buren was certain the Administration would prevail provided it kept its independence and entered into no alliances with former enemies. Should Webter's offer be accepted, it would confuse the Democracy. At the same time it would furnish Clay with the means for besmirching the President's high position of disinterested leadership with the charge that he was simply a partisan adventurer who would stoop to any political contrivance.

The entire conversation had been held with the three men standing, Jackson listening intently as he rested one hand on the mantle of the fireplace. After Van Buren finished, he looked at Grundy, giving him the opportunity for a rebuttal. But the Senator remained silent and Jackson, after a moment or two, thought the deal should be dropped. Grundy promptly agreed and as promptly left. The three men never mentioned the subject again.[7]

Although he diverted a potentially disastrous move, Van Buren realized what an exposed position he was in. Next to the President, he was the most influential man in the party, yet as Vice-President, he had no power, beyond the politically perilous one of breaking a tie. Whig Senators could, and he was certain they would in the course of speech and debate, seek to embarrass him, seek to prove to the country that he was unfit for the Presidency. In the face of unrelenting attacks he had no recourse save holding his temper and his tongue under what he knew would be extreme provocation.

When he appeared in the Senate chamber on December 16, he made a little speech admitting he had not reacquainted himself with the rules and asking the Senators' forebearance should he make mistakes before he had thoroughly mastered the procedures.[8] Van Buren made no mistakes worthy of notice as presiding officer, but he had not been in the chair more than a week or so when the Whigs began their assault. With the failure of his approach to the Democrats, Webster and his New England contingent returned to their original stance of opposition. From then until the end of the session the Whig Senators held together, manipulating the rules and the committees to their advantage. Van Buren had to remain silent as the Whig Senators subjected the Administration to a withering attack over the removal of the deposits. A model of composure, he bore with his usual equanimity their remarks, freighted as many of them were with critical references to his own political behavior.[9] Even the most pointed jibes from the vain and vulgar George Poindexter, a Mississippi Senator, failed to disconcert him. Invariably civil to Poindexter in the lobbies or on the Senate floor, he maintained his urbane pose on those few occasions when the two men met socially.

Clay, Webster and Calhoun's spokesman, the northern-born Senator from South Carolina, William B. Preston, while not above badgering the Vice-President, earned his respect for their ability. Calhoun remained for the most part silent during the raging debates. Van Buren, courteous to them all, was warm and friendly with his most articulate detractor, Henry Clay. He could not resist the Kentuckian's charm even when he knew he was the butt of rhetorical obloquy that would look well in the papers.

With Webster he was never on more than formal terms. He distrusted his public morals; and since the Grundy exposé, he was vigilant than ever to Webster's stealing a march on him in the ever-shifting power balances that marked the panic session. While the debate went on, Van Buren concluded that Webster was little more than the well-paid tool of Nicholas Biddle. Not even the State Street bankers and business interests of the southern New England states could command such loyalty from their spokesmen as did the Lord of Chestnut Street. That he should represent the Bank of the United States rather than the State of Massachusetts on the floor of the Senate was in its own way more repugnant than the scurrilous posture of Poindexter. Van Buren had known Webster for years. He had been associated with him as counsel on several important cases before the supreme court of New York. Webster's ability as a lawyer, an orator, a debater, went unquestioned; his character as a public man was something else.[10]

The aloof Calhoun remained a tragic figure in Van Buren's eyes. Like Clinton he seemed the victim of his own brilliance, but he lacked Clinton's supreme finesse. And it was this facet of Calhoun's character that could make him a dangerous, unforgiving opponent. There was also a stiff-necked pride about Calhoun, who coldly rejected Van Buren's advances despite their daily meetings in the Senate. The little New Yorker of obscure origins and little education had, he believed, been the principle agent in thwarting his ambition.[11]

In striking his own political balance of account, Van Buren saw clearly how Clay's leadership on the Bank issue and on the Compromise Tariff would weaken Calhoun's position among the Old Republicans of the South. He was letting his personal grievances dictate his political course. The strange combination of the nullifiers supporting such ardent centralists as Webster and Clay was an anomaly so gross after all his public utterances in defense of state power that even disciples like John Floyd of Virginia and Duff Green found his stand in the Senate not a little puzzling.

The old Crawford men who made up the leadership of the Democratic party in Virginia, North Carolina, Georgia and even in Calhoun's home state, were lining up behind Van Buren's candidacy if for no other reason, the lack of any alternative in the South. Cambreleng unerringly put his finger on Calhoun's bewildering course. "The bank system," he wrote Ritchie's conscience, Claibourne Gooch, "is destined to be your regeneration—the deposit question—the contest with the Treasury and the quarrel with the Executive will soon be merged in the greater question of the bank or no bank and the bank against the people. . . . you will find the nullifiers going for the U.S. Bank, without, as I think a considerable portion of the rank and file."[12]

Still the coalition might just bring it off. While the Whig Senators fulminated and the Whig presses denounced Jackson's tyranny and Van Buren's obsequious sycophancy, Biddle began a systematic calling in of the Bank's loans. Through the winter and spring of 1834, the pressure mounted on the mercantile and financial communities of the coastal cities, small factories, fledgling railroad lines and speculators everywhere.[13]

One of the first to go down was Governor Marcy's father-in-law, and onetime Regency stalwart, Benjamin Knower. For some time Knower's family and personal friends had been aware of his far-flung speculations, but they had believed his resources sufficient to cover his indebtedness. When Biddle began demanding redemption from the state banks, Knower could not renew his notes. The Farmers' and Mechanics' Bank, hard pressed at the sudden scarcity of money, was unable to sustain him. Knower liquidated all of his assets in a falling market, even his home in Albany; yet he fell far short of covering his indebtedness and was forced into bankruptcy, which shook the economy of Albany and was felt in the bustling new cities of western New York. Marcy's unsuccessful efforts to rescue Knower did not endear him to his Regency associates who were grumbling about his pretensions and his tendency to mix his personal affairs with his public duties.[14]

No other failures as spectacular as Knower's occurred, but the stringency of credit hampered business operations and resulted in memorials to Congress, mass-meetings in New York City, Albany, and other urban centers throughout the State. General delegations of merchants visited Washington with their complaints and found a ready audience in Congress and among the editors of the anti-Administration press. It was certainly true, to the discomfort of the New York Democrats, that the scarcity of credit was not just affecting merchants and speculators in New York City and Albany. All along the line of the canal system trade had come virtually to a standstill. Wholesalers and jobbers had in the past bought up the late harvests and warehoused them while the canal system was frozen during the winter for spring sale when navigation resumed. Silas Wright was forced, under the probing analysis of Henry Clay, to admit that the local banks would not furnish credit to the middlemen while the threat of further curtailment hung over their heads.[15] The harvested crops as far away as Michigan territory were not being moved, creating hardships for the small farmer and small businessmen, mainstays of the Jacksonian electorate.

Wright and Tallmadge did the best they could in trying to fix ultimate responsibility upon the Bank. But Clay and Webster drove them into damaging contradictions where they admitted that the state banks of New York were solvent; yet they had uniformly refused loans. In view of Secretary Taney's report, which had praised the public service and the intrinsic strength of the state banks, the admissions of the New York Senators were embarrassing, indeed. New York banks and New York interests were the focal points of Whig tactics.

Clay, in a series of attacks, claimed that Van Buren was the advance agent for the New York banks which were determined for their own special advantage to destroy the Bank of the United States so they could monopolize the profitable loan business of the country. As Van Buren wrote to his son John, "The search after something in relation to the disposition of the deposits that would give countenance to the charges that they [the deposits] . . . promote the views and sustain the interests of New York, has been most scrutinized but in vain. They would give the world, if they had it to give for such a fact." Clay hinted darkly that as the Safety Fund gave political advantage to the Albany Regency, so removal of deposits would give a similar advantage to the Administration.[16]

There was every reason that the Bank could not continue on its present course without ruining itself in the wreck of the nation's economy. The quick-witted Cambreleng made this point in a speech that revealed the Bank's precarious position with respect to its obligations, not the least of which was paying the interest in specie on its government-owned shares. Yet, even so optimistic and seasoned a veteran as Van Buren was at times overwhelmed by the fury of the opposition in the Senate, the inroads the Bank was making among the Democratic majority in the House and the legislatures of crucially important states like Virginia and Pennsylvania. The mails were flooded with pamphlets denouncing him and the President and the secretary of the treasury for using reckless means to achieve corrupt or dubious ends. Biddle and the coalition leaders in the Senate, Van Buren had to recognize, united energy and talent with money, a powerful combination.

In mid-February, Van Buren thought "the fate of the Administration if not the character of that which is to come after it" was in danger. "Situated as you are, you have it not in your power," he wrote his son John, "to form the slightest idea of the force and intent of the opposition which the President has to encounter. . . . Never could it have been said with as much truth that heaven and earth and the other place too are raised to defeat him."[17] In the cabinet there were still misgivings about Taney's report justifying removal. McLane, Woodbury and Cass were unsympathetic.

When word of "noncommitalism" in the cabinet leaked out, Van Buren heatedly denied it; yet Jackson himself thought it necessary for Butler to prepare an opinion on the constitutionality of the removal. In one of his most powerful papers the Attorney General defended Taney's actions only to bring down on his own head a plausible though spurious challenge from Clay.[18] At the height of the tumult, New York City held its charter elections. The regular Democratic party and Tammany for once acted as a unit against an opposition that was not only as well organized as they, but had substantial support from a divided mercantile community and the irresponsible but effective pen of J. Watson Webb, now sole editor of the *Courier and Enquirer,* which had the widest circulation of any paper in the City.

The City banks were not pleased at the action of the bank commissioners which drew off $300,000 in their Safety Fund accounts to assist the Albany financial community. Not one of them would discount new paper, but would only renew their loans at significant advances in interest rates. Though beneficiaries of government deposits, the banks had not as yet received enough funds to make any substantial impact on their cash reserves. As hard-pressed merchants reduced their imports, customs receipts declined. Worried Democrats called for state aid in a variety of proposals—a state bank with a capital of at least $10 million, issuance of new canal stock, direct loans from the state treasury—were the most talked of.[19]

The Whigs meant to capitalize on the financial uncertainty in the City. They nominated a recent convert, Gulian Verplanck, formerly one of Van Buren's "high-minded" Federalists, and a three-term Jacksonian Congressman, for mayor. The Democrats put up Cornelius Lawrence, a Tammany sachem, Congressman and well-known merchant who, unlike Verplanck, was a steadfast opponent of the Bank. On election day, riots broke out at the polls in many of the wards. Urged on by the *Courier and Enquirer* and the *New York American,* bankers, merchants and their clerks from Pearl and Wall Streets, invaded traditional working-class wards. Aroused Democrats defended themselves and retaliated in kind. Mayor Gideon Lee and the city watch finally managed to restore order, though not before a good many citizens had been injured in the scuffle. Lawrence eked out a victory of about 200 votes in an election where at least a thousand fraudulent votes were cast on both sides.[20]

Though serious enough, the New York election riots occurred after the crisis had passed. The Democrats had moved rapidly to restore confidence among vacillating Congressmen, frightened businessmen and farmers. Early in the session, Van Buren drafted, and Wright forwarded to Albany, a series of resolutions supporting the Administration for the New York legislature. And Van Buren himself was active in persuading members of Congress from other states to have similar resolutions passed if their legislatures were in session and had secure Democratic majorities, or if not, to encourage mass meetings. When opposing delegations appeared in Washington, Van Buren and others saw to it that they were followed by friendly groups of citizens from the same locale. If the Democrats could not match the Whig memorials in quantity and geographical distribution, they did their best to counter with their own, accusing Nicholas Biddle of creating the panic to advance his own power and destroy democratic constitutions.

In the newspaper duels, Blair's *Globe* far surpassed the stodgy *Intelligencer* or the bombastic *United States Telegraph* in its swinging, cutting style, its spare yet biting arguments. But in New York, the Administration did not fare so well. Webb's *Courier and Enquirer* had few equals in the country for unrestrained abuse. The *Evening Post* under Bryant and Leggett, when not treading on the forbidden ground of abolition, was feuding with the other Administration paper in the City, the *Times,* or responding to Webb's diatribes with indignant denials. Nor in Albany was Croswell's *Argus* any match for the venomous wit Thurlow Weed injected in the columns of his *Evening Journal.*[21]

The results of all these efforts, though in the main checking the flood tide of Bank propaganda, were not entirely successful. Calhoun's influence was strong enough in the Virginia legislature for it to adopt a series of pro-Bank resolutions, highly critical of the Administration's policies. Thomas Ritchie was unable to reverse the pro-Bank momentum in the legislature. The Whig majority, after castigating Jackson, took the bit in its teeth and instructed its Senators to vote for a return of the deposits to the Bank. Tyler was more than willing to accede, but his colleague Rives, who had just delivered a ringing speech in favor of the Administration's policies demurred. Van Buren sought to restrain him from taking any impulsive action. Rives, however, had too much of the special Virginian's concern for states' rights in his make-up to accept Van Buren's advice.

A saddened Van Buren watched as Rives made his final speech in the Senate, ending with his resignation because he could not in good conscience obey the instructions of the Virginia legislature. That body promptly replaced him with Benjamin W. Leigh, a Calhoun supporter and sworn enemy of Van Buren. The Jacksonians, a minority in the Senate, had lost one of their strongest voices, and Van Buren's stock throughout the country suffered at this rebuff from the Old Dominion.[22] The New York resolutions followed by those of Pennsylvania, which the Jacksonian's Governor George Wolf had engineered, took some of the sting out of Rives's resignation. Van Buren's colorful balancing act among the various Pennsylvania factions the previous spring along with a timely assist from James Buchanan, who had just returned from Russia, had for the time being lined up a Democratic majority in Pennsylvania against the Bank.[23]

A third blow at the Bank and perhaps the most telling of all was the decision of Governor Marcy to ask the legislature for a state loan of from $5 to $6 million to the banks so that they would not be forced to suspend specie payments. When he learned of the Governor's message in late March, Silas Wright had one of his rare disagreements with Van Buren. Imbued as he was with a deep-seated suspicion of the New York banks, Wright felt the loan was not only unnecessary but would simply substitute Wall Street and Pearl Street in New York City for Chestnut Street in Philadelphia. The banks, he feared, would dictate state policy whenever it suited their purposes and their purses by creating local credit shortages and demanding more aid.

But Van Buren saw the broader picture in its full range of economic and political implications. The Bank, he was certain, was on the verge of reversing its tight money policy. With the resources of the Empire State squarely behind its banks, Biddle had no leverage in the nation's greatest money market. He doubted whether the fund would ever be needed, and if it were only a small amount might be necessary. The gesture and the threat it contained would put the final pressure on Biddle. Wright urged Van Buren to write his views "fully and freely" to the Regency politicians at home. But the Vice-President would not commit himself on paper, especally since Marcy had already taken the initiative. He remembered only too well the charges of Clay and other Whig Senators that he was using his public position and his relationship with the President for the advancement of New York's special interests. "When I find myself differing from this sagacious man and from my best friends at home," wrote an apologetic Wright to Flagg, "I doubt myself, and . . . will cheerfully submit to what they think for the best."[24]

Almost simultaneously with the passage of Marcy's relief legislation, the admired old financier and public servant, Albert Gallatin, as a member of the Union committee of the New York banks, warned Biddle to ease the pressure on the financial institutions of the City. Public opinion against the Bank had now risen to such heights that Biddle reluctantly reversed policy. Within a month the economy had stabilized and evidences of renewed prosperity could be seen at every hand.

The "Panic Session" was a trying affair for Van Buren, as trying as any period in his tumultuous public life. He found himself in the exposed bastion of the Administration, surrounded on all sides by the very center of the opposition, the constant target of its heaviest guns. Of course, Jackson came in for his share of partisan thrusts when the Senate rejected his nominees for government directors of the Bank or more directly when it refused to confirm Taney as secretary of the treasury and Andrew Stevenson for the mission to England.[25]

On March 28, 1834, when it was clear to the most hopeful champion of the Bank that the deposits would stay where they were, Clay resorted to his last desperate expedient. He resurrected a motion of censure on the President he had presented early in the session but let remain on the table. Applying extreme pressure on the Whig majority he obtained the Senate's censure of the President for an unconstitutional violation of the public trust in removing the deposits without securing the approval of Congress.

What might have looked humiliating in the opposition press, a setback for the Administration, Van Buren judged would react upon the perpetrators themselves. The close vote had scarcely been recorded when the *Globe* was in full cry against Clay for dragging the Senate into such a mean, unwarranted move. Thomas Hart Benton made the Senate hall ring with his melodramatic condemnation of the move. Van Buren helped in drafting a protest from Jackson. Again he felt a sense of satisfaction when Clay forced his colleages to reject it, not even permitting the protest to be entered upon the official record.[26]

Van Buren wondered how such a skillful politician as Clay, such a master of parliamentary maneuver, a public man of outstanding talent, could not learn from experience. How could Clay forget that he helped make Van Buren Vice-President and Jackson's political heir? Yet, apparently he learned nothing from that spiteful episode. For in censuring Jackson, he had unnecessarily used his parliamentary skills to rebuke on the most dubious of grounds the most popular President since Washington, and had done so not once but twice.

If the Senate action was aimed at weakening Van Buren's chances for the succession, as seemed likely, it miscarried completely. Even the opposition press had difficulty in understanding and communicating to its readers the precise constitutional grounds for Clay's resolutions. Clay had let his vanity, his penchant for gambling, his sense of frustration, his fatal infatuation with brilliant effects, control him in wagering all on the throw. He would now take his place among those whose rampant ambitions had blinded them to reality, who, instead of accepting defeat and waiting for the wheel to come round again, plunged and lost. Clinton, Calhoun, Webster and now Clay were all men of broader gauge than Van Buren, but their fatal error was that they were too conscious of their superiority and in acting accordingly, acted wrongly.[27]

The daily attendance in the Senate was so distasteful and so wearing that Van Buren had more than his share of colds and debilitating upsets that confined him to bed for short intervals. But he had never been more content with his domestic arrangements in Washington. His home in the "Seven Buildings," on Nineteenth Street, which once housed the State Department and which he rented from a new friend, the litterateur and wit, James Kirke Paulding, was comfortable, and within walking distance of the White House. His sons Abraham and Martin, Jr., made up his household. Abraham, a rather staid young officer, was his father's confidant. He was a good listener who now and then offered sound advice and who was always a source of steady reassurance. Martin, Jr., acted as secretary, making legible copies of Van Buren's sprawling hand and unraveling the worst of his rambling sentences. John visited now and then, delighting the family with his lively accounts of Albany society and his verbal caricatures of the local politicians.

Van Buren was now well satisfied with his second son's progress and had given up worrying about his casual style of living. John was fast becoming one of the leading lawyers of Albany, in partnership with the talented, ambitious John McKeon, who was serving his first term in the lower house of the New York legislature. The two young lawyers had as much business as they could handle.[28] Van Buren was heartened to learn that his son was frequently in the company of the lovely daughter of his boyhood friend James Vanderpoel, also a prosperous lawyer in Albany and an extensive land owner in Kinderhook. Father and son kept up a steady correspondence. Van Buren relied on John for political news and for the management of his business affairs at home.[29]

That John Van Buren was prospering is evident in his stock speculations. Despite tight money in New York, he had enough cash to purchase about $10,000 worth of stock through Jesse Hoyt. Young Van Buren had inside information about Marcy's message calling for a state loan to support the banks. Anticipating the end of the panic, he had Hoyt purchase stocks in the Mohawk and Hudson and the Boston and Providence railroads. In a profane choice of words, an impatient John wrote, "Why Goddamn you Jesse! buy my stock and draw upon me at sight. You must be poor bitches down there if you cannot raise their two-penny sum. If the stock has gone up let it go to Hell. The Bank will come up against the Safety Fund Banks and depress stocks—the Governor's message will eventually relieve the country."[30] While the elder Van Buren was no stranger to speculation and shared with his son a passion for betting on elections, he would have been disturbed at this correspondence which, if it had gotten into the wrong hands at this time, would have caused an uproar in the papers.

In late January, Abraham was ordered north at his own request and Van Buren invited Butler to take his rooms. The Attorney General was happy to accept for he had found Fuller's boarding house a rather lonely place. Van Buren set a good table and though Butler did not drink, and paid little attention to his fare, he enjoyed the lively conversation among close friends at the Seven Buildings. Wright, Tallmadge and other Democratic Congressmen from New York were frequent visitors, as were cabinet members and occasionally the President himself. Despite his abstemious habits, Butler was no prig; he could not have been and remained a guest of Van Buren.

Wright was a heavy drinker who indulged himself when he paid a visit, as did most of the other politicians who called or dined at the Vice-President's hospitable table. Nor were the dinners exclusively male. Van Buren enjoyed the company of women, whom he flattered outrageously; and they, in their turn, were fond of him, whatever their political preferences might be. Of course, Van Buren had special favorites like Mrs. William C. Rives or Mrs. Stephen Decatur, Jefferson's two daughters, and always Jackson's bevy of lovely nieces and in-laws.[31]

And he needed all the relaxation he could get because no sooner had the Bank issue seemed to be resolving itself than trouble with France suddenly developed. At the height of the Bank controversy, and possibly affected by the apparent divisions in the American government, the French Chamber of Deputies, by the close vote of 176 to 168, rejected the indemnity Louis Phillipe's ministry had negotiated with the United States. Jackson, his nerves on edge as he battled with the Senate and the Bank, was outraged. Even his bitterest opponents were shocked at what they regarded as French duplicity, a contemptuous slur on the nation's honor. James A. Hamilton, who differed with the President over the Bank and deplored the political excesses in Washington, urged the government to take a firm stand against France. McLane had even stronger views and, as secretary of state, he was in a position to influence directly the tempestuous President.[32]

As soon as Van Buren learned of the news from France, he noted four salient points: first the treaty was a government measure and, second, though irritatingly delayed as the ministry maneuvered it through the maze of French politics and public opinion, it had failed by only six votes. The third point was that Louis Phillipe's foreign minister, the Duc de Broglie, resigned, indicating that the King and ministry had acted in good faith. Finally, dispatches from Livingston, which he had seen, convinced him that the King and the new French foreign minister, Henri Gautier, were acting honorably when they gave "the strongest personal assurances" the ministry would resubmit the treaty. Elections were to be held for a new Chamber and the King was certain a majority favorable to the treaty would be secured. Unfortunately, the new Chamber would not meet except briefly to organize itself until December, eight months away, a lengthy period of time during which it would be difficult to restrain the impatient President and his belligerent secretary of state.[33]

Van Buren immediately turned to Rives for advice. No one in the country was in a better position to supply it; he had negotiated the treaty, and was familiar with French politics and leading French political figures. A seasoned American politician, he was likewise knowledgeable about public opinion, especially southern opinion.

Rives was decidedly opposed to reprisals, any drastic action like seizure or impounding of French property at this time. "What I would suggest," he wrote, "would be the imposition of heavy duties on the principal productions of their industry, particularly their silks, which are now duty free." Rives estimated a stiff tariff would cost the silk industry of Lyons 50 million francs a year. Tariff schedules on other importations such as French wines, should be raised also. The President could ask Congress for the necessary legislation which automatically would go into effect if the French Chamber in December again rejected the treaty. Rives's recommendations were similar to those Livingston suggested, though not as harsh or as peremptory.[34]

Although Van Buren had not been consulted and made it a practice never to obtrude his advice when it concerned an executive department unless the President asked him, he was always prepared should the summons come. He was certainly aware of strong differences of opinion within Jackson's cabinet, and knowing McLane as he did, he assumed the fiery ex-Federalist was pressing for strong measures. How strong, he did not know until a distraught Taney came to his house and pleaded for help.

Taney had visited the President just after the news arrived at the White House from France. He found Jackson angry and excited at what he felt was the French ministry's playing politics with a solemn obligation. Despite his intense irritation Jackson remained aware of the structural division in the French government where the treaty-making powers were lodged in the ministry and the appropriation of funds in the legislative branch. He was content for the moment to accept the explanations and the assurances of the King and his ministers, but he insisted on instructing Livingston that he would hold them strictly accountable.

Summoned to a cabinet meeting a few hours later, Taney was shocked when the President, in firm, clear tones, announced he was preparing a special message to Congress, elaborating on the perfidy of the French government and asking for legislation that would authorize letters of marque and reprisal. Taney's long, saturnine face took on a more melancholy cast as he listened while Jackson, warming to his topic, began showing all the familiar signs of coming to a firm resolution of policy. Taney was not as much worried about the risk of war that might follow if these draconian measures were adopted as he was about their political implications. Whether the legislation was secured or not, he saw a grave loss of prestige for the President.

Confidence in the Administration was gaining steadily, but the Bank battle was by no means over. Attempting to involve the United States in a war with France at this stage would, he thought, injure Jackson's credibility, prove what the Whig politicians and press were proclaiming to the nation: the President was a reckless, autocratic despot. He had already injured the economy simply to satisfy a personal score against Nicholas Biddle; he would now risk a war with a traditional ally over a fancied insult. The Bank, though crippled, was still a powerful institution. Taney could see it ride back to popular favor on a surge of revulsion against what would be deemed Jackson's irrational policies.

Of course, personal reasons colored Taney's somber assessment of the dire consequences the President faced in adopting such a policy. McLane, whom he detested and who returned the animosity in full measure, must have changed the President's mind. This inference was borne out when Taney argued against the precipitous move. For McLane promptly answered the points he raised, closing his remarks as Taney remembered them "in earnest and decided terms advising that the message should be sent." Cass supported him but Woodbury remained silent, and Butler and Barry were not present. But Taney was a stubborn man. When he thought he was right he could be as determined as the President. After McLane had finished, Taney emphatically reminded Jackson how the Bank had succeeded in deranging the currency and alarming the public mind. Warning that war might result, he stated flatly that the country was in no condition for hostilities with a major power. And then, he enumerated specifically the grave consequences to the nation should a war be precipitated. McLane replied; and so the argument went, with the President occasionally interrupting. When the conflicting viewpoints had been completely aired, Jackson ended the discussion. Taney, whose judgment he valued, had made some impression. The President reserved his opinion for another discussion, but Taney was certain from what he had said that he was leaning toward McLane's view.

"I left that Cabinet meeting," recalled Taney, "in a state of greater anxiety and alarm than I have ever felt at any other moment of my public life—I saw that I had failed to convince the President." Taney was certain Congress would never approve the issuance of letters of marque against France, but he quailed at the political defeat the Administration would suffer. Only Van Buren, he thought, had the stature and the calm and reasoned argument to derail the politically ruinous course McLane had set the President on. Thus, the unexpected arrival of a worried Taney at the Seven Buildings.

Van Buren listened as Taney described the cabinet meeting, McLane's influence and the President's probable course of action. As Taney had anticipated, Van Buren agreed completely with his position. At all costs, Jackson must be restrained from taking a step that would play directly into the hands of the opposition; that would foreclose any payment of the indemnity and might even result in war.[35] Not a moment was to be lost. Van Buren went to the White House, possibly right after Taney left, more probably the next morning as the interval would have given Jackson time to reflect on the points raised in the cabinet meeting and put him in a more conciliatory mood. Armed with Rives's practical suggestions, fully apprised of Taney's and McLane's arguments, Van Buren calmly laid out for Jackson all of the points at issue, all of the consequences, and proposed essentially Rives's plan of economic pressure to take effect if the Chamber again refused the appropriation in December.

The two men had several discussions before Jackson came round completely. As always, when Van Buren called up his logical and persuasive powers to their utmost, he was not only able to sway the President but also to gratify his egotism in such a way that Jackson concluded he had always held this position; that Van Buren had merely clarified it for him. Still the talks were "embarrassing and painful," the decision to scuttle McLane's policy being reached with great reluctance. Van Buren expected his relations with the Secretary would be strained.[36]

Since his return from England in 1832, Van Buren had received information from time to time of McLane's alleged intrigues against him. It was no secret that Blair, Kendall and Taney disliked McLane personally and opposed him politically.

With the end of the Bank war in sight and now this irresponsible, potentially disastrous policy McLane was urging upon a susceptible President, Van Buren began viewing him in a different light. No longer was it a matter of services McLane could perform for the Vice-President in the shifting power plays at the White House. Van Buren was in line for the Presidency; he could not afford the luxury of maintaining McLane in his place against the Kitchen Cabinet after two differences of opinion over crucial issues of public policy, one of which risked alienation of Jackson, the other the wreck of the party and the end of his pretensions to the Presidency. Yet other than persuading the President to override his secretary of state, he said nothing against McLane, nor made any move to undermine his position; rather he would let matters take their course.

One or two days after the meetings with Jackson which resulted in the dropping of McLane's proposal, Van Buren attended a party given by Mrs. Benjamin Ogle Tayloe, a famous Washington hostess. While moving through the crowd, he saw Mrs. McLane, whose company he had always enjoyed. Just as he was about to present himself, Mrs. McLane raised her voice and said to one of the guests, "Well, thank Heaven! It is over at last." What, Van Buren wanted to know, had happened? "Why," she answered, "I referred, of course, to Mr. McLane's resignation!" When Van Buren expressed surprise at the news, Mrs. McLane was incredulous despite his professions of complete ignorance. And, in fact, Van Buren was taken aback. He knew McLane would be deeply hurt, but he did not think he would act so precipitously. McLane had no political prospects, very little estate, a large and expensive family and no law practice to speak of. Before leaving Mrs. Tayloe's party Van Buren sought out Major Donelson who confirmed that McLane had resigned, and he, too, expressed surprise that Van Buren knew nothing about it.

Here was cause for alarm. Might not McLane, in resigning, imply to the President that Van Buren had forced his hand? If indeed he had done so, how was Jackson to regard such an abrupt change in the relations of the two men as anything but disloyalty on Van Buren's part toward the Administration? He decided on the spot to visit the President and asked Donelson to accompany him. They found Jackson stretched out on a sofa in his office looking tired and sad. Van Buren immediately explained his complete surprise at McLane's resignation, demonstrating to the President's satisfaction that he had had nothing to do with it. Jackson, who had been harboring the very thought Van Buren had intuitively ascribed to him, was relieved. Now he saw McLane's resignation as a desperate man's attempt to foment suspicion of Van Buren's motives and promote discord within the cabinet. Jackson had Donelson read McLane's letter of resignation and the reply he had drafted for his signature. What was Van Buren's opinion? While sympathizing with the President's desire to maintain his friendship with a longtime valued public servant, Van Buren thought some of the phrases in Donelson's draft had the appearance of conceding errors. Instantly Jackson handed Van Buren his pen and inkwell and asked him to make what changes he thought necessary. This being done, without even reading the edited draft, Jackson directed Donelson to copy it.

While Donelson was engaged in this task, the President wanted Van Buren's advice on McLane's successor. Evidently Van Buren had been considering this question; he was ready with a nomination—John Forsyth of Georgia, who he knew would accept the post.[37] Jackson agreed and had Donelson make out the nomination at once. The Senate confirmed Forsyth in a universal display of harmony on June 27, 1834, three days before it adjourned.

The next day Van Buren visited McLane's house where Mrs. McLane graciously received him. She had a servant send word of his arrival to her husband who, she said, was in his study. A period of time elapsed during which she and Van Buren engaged in small talk, but McLane did not appear. When his continued absence became embarrassing, Van Buren took his departure. The two men never spoke again.[38]

« CHAPTER 21 »

Mediator

Senator Willie P. Mangum of North Carolina was a troubled man as the year 1834 dwindled away. He had journeyed through the North and spent several weeks in New England seeking to discover the identity of his party, the Whig party. How could it be that with such great principles at stake, indeed the very fate of the legislative power against the encroachments of executive tyranny, the Whigs could not mount a cohesive offensive? Though as able and energetic as the Democrats, why could they not discipline themselves and act together? It was all a matter of interest over principle, thought Mangum, but then what of Calhoun? His followers constantly sought to emphasize that the South Carolinian was the only public man who stood squarely on principle. Mangum was far too practical to cast his lot with such visionaries. If the southern Whigs went for Calhoun, they would simply be handing his own state and Virginia to Van Buren. "Mr. C. cannot get the vote of either state," he wrote his friend John Beard. To his surprise he found much political idealism among New England Whigs, but of a kind that he feared was incompatible with southern institutions. "And deeply as I abhor the treachery and the usurpation of the present administration," he said, "I fear their weak and rash excesses much less than I do the settled steady and persevering policy of New England." Yet he liked his New England associates—"industrious and moral and patriotic . . . the most civil people to a stranger in the world." It was only their purposes he mistrusted because he felt they would never understand what a southern Whig stood for. Perplexed as he never had been before in his political life, he saw that geography and custom were stronger than ideology; that the Whigs were sorely beset in search of their own destiny. If only they had more of the old Revolutionary Whig spirit about them. Modern Whigs—that was another thing, "tho I quarrel with no man for calling me a Whig—yet I feel it no compliment."[1]

While Mangum was lamenting the lack of direction among the Whigs from his hotel room in Philadelphia, two hundred miles to the north in his home town of Kinderhook, Vice-President Van Buren was being subjected to the full weight of an organized, well-financed Whig party in New York. In fact, the reason he was in Kinderhook at the time was the extraordinary effort being made to defeat him in his own county. He, as well as his opponents, recognized the value of carrying the county in the November elections. For Van Buren, victory there would give a further boost to his candidacy. Kinderhook was in Columbia County, long a Federalist stronghold, yet also in the district which he had twice carried as a Republican at the beginning of his political career. Disturbed as he was with bad news from Virginia, the New England states, Ohio and Tennessee, Van Buren would have disputed Mangum's claim that the Whigs were a discordant, irresolute mass, scarcely a party. In New York itself, the disciplined ranks of the Democracy were being assailed with all the ingenious diatribes the fertile minds of J. Watson Webb and Thurlow Weed could conceive in the editorial columns of their papers.

A new element had entered the campaign; the state banks, especially those in New York City, were underwriting a campaign to discredit the Safety Fund. They had always opposed Van Buren and the Regency, but now, through pamphlet, broadside, the columns of the Whig press, they were charging the Regency with using the Safety Fund for political advantage. Prodded on by Webster and Clay, the *Evening Journal,* the *Advertizer* and the *Courier and Enquirer* questioned the stability of the system and roundly abused Van Buren for foisting it on the state.

If interest rather than principle was the fundamental tenet of the Whig party in New York as Mangum had said, it was not as durable as he supposed. Despite their well-coordinated campaign, Columbia County and the rest of the state went solidly Democratic, reelecting Marcy and Tracy by a majority of 11,000, carrying seven of the eight senatorial districts that were in contention, and sweeping the lower house by over three to one. The landslide victory was a personal triumph for Van Buren and virtually confirmed that his party would nominate him for the Presidency.[2] He had, of course, been a candidate since his election to the Vice-Presidency two years earlier, but many obstacles had to be overcome (the recent New York election being the most serious) before he could be sure of the party's endorsement.

At the onset there had been two rivals in the field, two westerners, John McLean of Ohio and Richard Mentor Johnson of Kentucky. Van Buren knew that Lewis Cass and Levi Woodbury also had ambitions, though he did not take them seriously. McLean and Johnson were a different matter. The clever, intensely ambitious McLean had maintained the legion of political contacts he had carefully nurtured as postmaster general during the Adams Administration. He was also a prominent Methodist and never missed an opportunity to promote himself through church assemblies, camp meetings and the national hierarchy. As a Justice of the Supreme Court, he kept himself in the public eye of the Washington community.[3]

Johnson, a colorful character, an old Indian fighter and a gregarious politician with a good memory for names, was appealing to the western farmers and restless frontiersmen. He had some of Jackson's qualities, his open democratic manner and his capacity for maintaining deep personal attachments. More important than any of these personal attributes was his control of the Democratic organization in Clay's home state. In the easy-going society of the West, Johnson's black mistress, Julia Chinn, to whom he was devoted, and his two daughters by her, had not hurt his political career. But to many in the slave-holding South, particularly along the seaboard, his unorthodox domestic arrangements were looked upon with disgust that bordered on loathing. The *United States Telegraph* constantly played on this theme with racial slurs on Johnson's domestic affairs.[4] Jackson, Blair and Kendall saw nothing socially wrong with Johnson's private life. The President and Johnson had been close friends for years. Like all wounded veterans of the War, Johnson held a special place in Old Hickory's estimation. He had proven a successful and courageous commander of militia troops at the Battle of the Thames, a victory over the British and the Indians. In the Battle of the Thames, Johnson killed an Indian chief in a hand-to-hand encounter. Whether that chief was Tecumseh, then the scourge of the northwestern frontier or not, Johnson promptly claimed his adversary was the great Indian leader and his claim went undisputed among the credulous western folk, adding immensely to his popularity.

Johnson's friendship with Jackson, his western roots, his undoubted loyalty and his inherent simplicity appealed to Blair and Kendall. With Johnson as Jackson's successor they saw a continuation, even a strengthening, of their role as intimate advisors to the President. They had to tread softly, however. The President may have been easily susceptible to flattery; to notions of western nationalism Kendall and Blair kept before him; to timely reminders that Johnson was the only check against the hated Henry Clay in Kentucky; but they were careful not to overstep bounds, sensitive to his moods and respectful of his political acumen.[5] Not as clever as Blair and Kendall, the hulking William B. Lewis was no friend of theirs when it came to jockeying for influence.

Once a Van Buren man, Lewis had gradually come round to covert opposition. In the shifting politics at the White House, Donelson, who had once opposed Van Buren, was now using what influence he had with the President to further the New Yorker's cause.[6] Lewis and Donelson disliked and distrusted each other and they both resented the influence of Blair and Kendall.

Van Buren, quite at home in palace intrigue, understood the motives of the two Kentuckians and never underestimated their abilities. As he watched McLean and Johnson, he played his passive role, content that able men like Taney and his devoted protégé, Benjamin Butler, both of whom had earned Jackson's respect, as members of his cabinet, would protect and enhance his position.[7] In the Senate, no man defended the Administration's policies more calmly and cogently than Silas Wright. In the House, Cambreleng guarded with eloquence and experience Jackson's fiscal policies. And no two men in Washington were more devoted to Van Buren. With such a phalanx behind him and with his own consummate powers of flattery and persuasion, Kendall and Blair made little headway with the Johnson candidacy. But they were sufficiently influential to begin maneuvering Jackson behind the Kentuckian for Vice-President, a movement Van Buren deplored, but considered too risky for any overt interference.[8]

Despite his firm political base in New York, Van Buren felt he could not gain the nomination or win the election without Jackson's wholehearted backing. Personally, he preferred William C. Rives for the vice-presidential nomination. Ostensibly in retirement at his plantation, Castle Hill, Rives was leading the Democratic forces in his crucial state. Rives had been early in the field lining up support for his own nomination. A Van Buren–Rives ticket would follow in the tradition of alliance between the Old Dominion and the Empire State.[9] Unfortunately, Rives had made a bad impression on the President when he paused in Richmond on his return to Washington during late September, 1834. Acutely sensitive to the French indemnity problem, Jackson was in no mood to hear Rives explain the causes of the recent delay.[10]

When the President reached Washington and read the dispatches from Livingston to Forsyth he was outraged at what he regarded as deliberate delay on the part of the French government. Livingston had urged Louis Philippe not to prorogue the Chambers until the indemnity bill had passed. Livingston had wanted action and he hoped positive action on the part of the French in time for Jackson's annual message to Congress. In his unofficial correspondence, Livingston implied that the King had misled him and the American government. Jackson jumped to the conclusion that Louis Philippe was the worst of hypocrites.

A recipient of many of Livingston's letters, Van Buren had formed a roughly accurate picture of the negotiations and correctly assessed their impact on the President. Then he received a letter from Rives describing his meeting with the President and adding that Livingston was a sick man who had been absent from Paris when the King dissolved the Chambers. Rives again counseled patience and diplomacy, which he was certain would result in the payment of the indemnity. Van Buren was disturbed. It was no secret in the group around the President that he was close to Rives. Anyone who took the trouble to examine the history of the Virginia–New York combination Jefferson and Madison had put together could infer that Rives was Van Buren's candidate for the vice-presidential nomination. Or if he needed further verification, he might weigh the evidence of the more recent coalition Van Buren had devised that helped elect Jackson in 1828. It would appear Rives had damaged himself with the temperamental Jackson and provided Blair and Kendall with a lever that might be used to undermine Van Buren's own standing with the President.

For the time being he must cover himself even if it were at Rives's expense. He wrote Jackson, enclosing a copy of Rives's letter to him. "By the enclosed," he said, "you will perceive, as well, that our friend Rives is getting to be quite confident of his own success and that he has fallen into the prevailing error of the Virginia politicians, who appear to be always on the lookout for some rash act on your part—a fault which it seems that no success of your measures can entirely cure them."[11]

After sealing his letter to the President, Van Buren wrote a very long letter to Rives assuring him that he would be in Washington the third week in November and would present his views directly to Jackson.

> You may take it for granted that a strong (but I trust prudent) statement of the injuries we have rec'd from France will be made. . . . you in Virginia or I might perhaps rather say they are prone to apprehend rash acts from the Gen'l., but they must admit that after all he has been oftener in the right than others. . . . I have been a strenuous advocate for forebearance heretofore (more than you are aware of) and will continue to be.[12]

Jackson's response to Van Buren's letter and enclosure was explosive. "Surely Mr. Rives has not weighed the subject well, he is badly advised from Paris as to Mr. Livingston's absence," he wrote. Indignantly, Jackson posed a rhetorical question. "Can the Executive under the circumstances be longer silent? If he speaks to Congress, it must be in the language of truth, and he cannot refrain from recommending to them to legislate provisionally on the subject."[13]

Van Buren may have diverted some of Jackson's anger to Rives, whom he felt would not suffer any permanent setback. When the indemnity was paid and friendly relations with France restored, when Rives was reelected to the Senate, Jackson would have long since forgotten the episode and have restored him to favor. Meanwhile, he had to improve his own situation as Jackson's principal advisor on foreign affairs. He and Taney must bear the major responsibility for the temperate policy toward France that Jackson had adopted reluctantly. Van Buren was certain, despite setbacks, that the policy had been right. Caution, prudence, time and patient diplomacy would eventually settle the differences between the two nations. After all, the indemnity was paltry—$5 million—a fraction of the annual trade between the two countries. The worst alternative and one that could not be discounted, was war with France. Even if purely naval, such a war would cost far more than that. Merchants of his own state, of New England, of Philadelphia, Baltimore, Charleston and New Orleans would be damaged severely, perhaps ruined. Political implications were as bad as they had ever been. The Whigs, who were spoiling for an issue, would come down hard on Jackson's rash behavior and would exploit it for partisan gains.

A prudent or a logical course, however, was not always possible with Jackson. Van Buren, nevertheless believed in being prepared. Somehow he hoped he would be able to tone down the President's message, which he feared would overstep the line and at one blow reverse all that had been accomplished over the past four years of negotiations. Its impact on his own political career at this critical juncture could be disastrous. Yet he could only go so far or he might lose the President's confidence and with it his support. His enemies within and without the party would quickly exploit the advantage; nor could he be certain that Blair and Kendall and Donelson would remain friendly spectators.

Whatever Rives's indiscretions, he was a highly intelligent man who knew more than any other immediately available citizen about the current state of French politics and the character of France's leading men. Van Buren continued to ask his advice. He also consulted with his life-long friend and political confidant, the Albany merchant Lucas Elmendorf. Both individuals supplied him with telling arguments against any precipitous action, but they asked for a firm statement from the President.

Rives thought Jackson should make a bid for world opinion by emphasizing American forebearance in the face of flagrant procrastination on settling a just claim. Any forced collection of damages through reprisals or the extreme measure of requesting Congress for the power to unlease privateers on French shipping, Rives opposed vehemently. He noted that the French King at the recent organization session of the Chambers gave notice his government would ask for the appropriation to put the indemnity treaty into effect. Until the Chambers rejected this announcement from the throne, it would not be productive for Jackson to issue any bellicose statements that would inflame French public opinion.[14]

Van Buren arrived in Washington to find it buzzing with war talk. As he had the previous year, he spent a few days in the White House, while his sons, Smith and Martin, Jr., renovated the Paulding house in the Seven Buildings. He found Jackson adamant on a tough message of warning to the French. Van Buren agreed with his request to look over all the Department reports and make whatever editorial suggestions he thought necessary. When he moved into his home, he had a sudden flareup of what he called influenza, which had laid him low for several weeks the past summer. Feverish and confined to his bed for some days, he diligently pushed on with the task the President had assigned him. He polished up his own draft on internal improvements for Donelson to copy and had Woodbury rewrite the section in his report dealing with the deposit banks to make it more readable for the general public. On the most sensitive subject of all, relations with France, Van Buren himself wrote an able history of the indemnity issue which, while he put the burden of proof on France, he softened here and there with passages on American forebearance and good will.

The President rejected any suggestion of commercial restriction and insisted on warning the French government he would recommend reprisals (a harsh word in diplomatic usage) if the new Chamber did not appropriate the funds.[15] Van Buren bowed to the imperious will of the President and included the language he wanted and that which Kendall had written at the end of his draft. Although he felt Jackson's forceful declaration would inflame public opinion in France and delay, perhaps preclude, any settlement, he was hopeful it would appeal to the patriotic instincts of the people. As he wrote Rives, who was involved in a difficult campaign for election to the Senate, "taken as a whole it can not hurt but must help you at Richmond to which every eye is now directed."[16]

When Congress convened on December 1, Van Buren was well enough to appear in the Senate Chamber, but he could scarcely hold up his head. For the next week as he slowly regained his health, he realized Jackson's demand for reprisals would suffer short shrift in the Senate. The Whigs controlled all committees, and Henry Clay had minced no words in condemning the President's course on the French question. But it was embarrassing to him personally when the entire Senate, including the members from his own state, refused to support Jackson's policy. Van Buren in a way was relieved at the Senate's action which he thought would probably remove any harm Jackson's strident language had caused. All he would ever say on the matter was a playful remark to Rives before the Senate had taken any action. "The Gen'l is in fine health and rampant spirits," he said. "If Louis Philippe were in striking distance he would hold his throne by a brittle tenure."[17]

Of course, Jackson could be right in making such a vigorous display of firmness. In any event, he assumed the French minister, Louis Serurrier, would provide a clue to the reception of the message in France. The President had sent out invitations for the first state dinner of the season. Would the Serurriers attend? Alas, Monsieur Serurrier had suddenly become ill and sent in his regrets. The absence of the Serurriers foretold an ominous reaction in France, which was not long in coming. Almost every member of the Chamber now in session regarded the message as a slur upon the national honor and a threat that could not be ignored. The French press of all political persuasions rang with belligerent pronouncements condemning Jackson and the United States unsparingly.[18]

Louis Philippe's government had no recourse but to order home its minister. At the same time, it informed Livingston that his passport was ready, but significantly did not request his departure. In the absence of any instructions, Livingston decided to remain at his post and though to no small degree responsible for Jackson's message, through his forceful dispatches put all of his formidable legal talents into softening its offensive passages while emphasizing those which underscored the peaceful intent of the United States. He took pains to point out that Jackson's recommendation of reprisals reflected the policy of one branch of the government and was provisional only. Congress alone had the power to give them the sanction of law.

Livingston's note had a calming effect, but of much more importance in soothing French opinion as Van Buren suspected, was the Senate's unanimous refusal to pass the legislation Jackson had requested.[19] On March 28, the ministry formally proposed the appropriation act. The new foreign minister, alluding to the action of the United States Senate, declared the offensive passage in Jackson's message did not represent the position of the American government. After six weeks of heated debate, the Deputies finally approved the legislation on April 18, 1834, though not without making payment contingent upon an apology from Jackson for his language which it said impugned the honor of France.

By now Livingston had received instructions from Forsyth that if the French government did not comply with all the provisions of the treaty, he was to consider himself recalled. The demand for an apology, he decided, was a proper reason for requesting his passport and returning home. He left his son-in-law Thomas Barton, secretary of the legation, in charge.[20] When the news of the turn taken by the Chamber of Deputies reached Washington, Jackson, as expected, utterly rejected any apology.

Swept along by events, Van Buren was for a time unable to reason with the determined Jackson; nor was he able to come up with a formula which would defuse an increasingly dangerous situation. It was intensely frustrating to be powerless in untangling what he believed to be simply the result of a misunderstanding. Slow communication with Europe gave him some breathing time. Congress had adjourned by the time Livingston arrived and gave his personal account to the President. However incensed Jackson might be, he was far too practical a politician to call Congress into special session over a matter of injured pride. All he could do was direct Forsyth to have Barton demand the money and if it was not forthcoming, return home. He refrained from the far more serious step of breaking off diplomatic relations.[21]

The President had a majority of his party behind him. Van Buren, while still deprecating that matters had reached such an impasse, approved of this action. The next round in the exchange of international bombast occurred after Van Buren received the Democratic nomination. His position was now considerably strengthened, his advice carried more weight with the President. Yet oddly, he did not bring his enhanced influence to bear on a particularly sensitive episode in the relations between the two countries.

Had he been on the spot rather than electioneering in New York, he might have saved a good deal of embarrassment for Jackson and for the French government. In September, the French chargé, Alphonse J. Y. Pageot, son-in-law of Major Lewis, received an unofficial note from the French foreign minister, Broglie. In the same envelope were instructions to read the Appropriation Act of the French Chambers and his note to Secretary Forsyth. The law, when placed beside Broglie's interpretation of it, did not demand an apology as such. All the French government needed to put it into effect was an explanation by the President that he had not meant to threaten France. It was a most conciliatory note and, in fact, skillfully picked up on several phrases Jackson had used in his message. Acting on his own responsibility (the President was still in Tennessee preparing to leave for Washington), Forsyth refused to receive the note or even a copy of it for transmission to Jackson. The result of Forsyth's thoughtless intransigence was a further deterioration of relations with France.

Van Buren did not wait in Albany for the New York election but hurried off in late October for Washington. En route he stopped over in New York where he kept himself clear of the ruckus in Tammany Hall that gave the political world a new name for the radical faction, "Loco-Focos." Though obviously troubled by the split among the New York City Democrats, he was more disturbed about affairs in Washington where Jackson and Forsyth seemed to be going out of their way to block a reasonable solution to the indemnity question. Wars had been fought for less and, in Van Buren's opinion as he thought about 1812, rarely settled anything of consequence. Yet those who saw him in New York came away believing he had not a care in the world. "He looks very well," said Philip Hone, "and from ease of manner and imperturbable good temper, it might be supposed that he had less to occupy and trouble his mind than any man in New York."[22]

But he had scarcely settled himself in Washington when Pageot again brought Broglie's letter to the attention of Forsyth. And again the Secretary refused to receive it because it was not an official communication from one government to another. This time, however, he had Jackson's backing; if Van Buren argued against Forsyth's technically correct but provocative action, there is no record of it. Perhaps his reason for not opposing Forsyth was that he thought he had found a formula which would satisfy both sides. Like Broglie, he drew his inspiration from the last annual message where Jackson had repeatedly emphasized, while presenting the American case, his hope for maintaining peaceful relations with France. He argued that the President could not offer an apology when no injury to the pride of France had been meant. In fact, he maintained that Jackson had made special efforts not to offend. The French grievance was characterized as a "supposed insult." Then, with no apology, he would have Jackson make a statement that it had never been his intention "to menace or insult the Government of France." Further, as President, Jackson had the duty of reporting fully and freely to Congress on foreign affairs.

Following the lead that Livingston opened up, Van Buren underscored the fact that communication between two branches of the American government was purely a "domestic affair." Although it might be a further embarrassment to the French, Van Buren, with an eye to domestic politics, prevailed upon Jackson to include publication of selected correspondence along with the message.[23] The conciliatory tone Van Buren had interjected met with enthusiasm in some quarters, relief in others. The *Globe* took care to emphasize the pacific gestures in the message and even the *Intelligencer* saw merit in the President's policy. Van Buren congratulated himself on being able to convince Jackson, Forsyth and the Kitchen Cabinet that the kind of answer Broglie could accept should be included in the message. It was a gamble, of course, but from what he knew of Broglie's letter, he assumed the minister would find the substance if not the precise form of an apology. In sending a copy of the message to Elmendorf, he asked that frequent critic of Jackson's diplomacy, "Will this do?"[24]

In France, Barton, under direct orders of the President through Forsyth, asked Broglie if the French government's policy on an apology had changed. Upon receiving an emphatic negative, he asked for his passport and sailed for home in December of 1835 before Jackson's more temperate message had been received in the French capital.

News of Barton's departure and of warlike preparations in France were received in New York well before the chargé arrived. His ship had made a slow voyage, and it was not until January 10, 1836, that he arrived in Washington where he went directly to his father-in-law's residence. He and Livingston then made for the White House, picking up Van Buren and Forsyth on the way. The imperturbable Vice-President for once could not conceal his anxiety, though ever courteous, he deferred to the Secretary of State. "Well, Sir, what are you going to tell the President?" asked Forsyth, expecting some extended explanation. All he got from Barton was the laconic response: "I am going to tell him the whole truth as I understand it." But near the steps of the White House, the tension in the group had become so pronounced, Barton finally decided he ought at least find where they stood on the issue before he reported to Jackson. "Gentlemen," he asked self-consciously, "do you want oil poured upon the flames or water?" "Oh, water, by all means," said all three. "That will be the effect of the little I have to say." They found Jackson in a frenzy of impatience, striding back and forth in his office. Brushing aside Barton's explanation of his delay in reporting, Jackson asked peremptorily, "Tell me, sir, do the French mean to pay the money?" Barton replied that they would not.

After being given permission to describe the situation in his own way, he said that French public opinion had been so worked up over the affair that the government dared not carry out the provisions of the treaty without some concession from the United States in the matter of language. The throne itself would be in jeopardy. Powerful interests in France wanted a settlement. The King and his ministers were deeply committed to one, but the politicians and the press had so aroused the masses that they could do nothing at the present time.[25]

After Barton left, Van Buren, Livingston, Forsyth and Jackson discussed the matter. Jackson now realized what Louis Philippe and his government were facing, that they were not deliberately procrastinating but had been acting honorably in their sincere desire for a peaceful settlement. Livingston prepared a special message for Congress, giving the two houses the substance of Barton's report. Considering the circumstances, he made it as conciliatory as possible.[26] While the committees on foreign affairs were considering the message, Charles Bankhead, the British chargé, received a dispatch from home that the French government had come to terms. It had been the phrases Van Buren inserted in Jackson's annual message that satisfied Broglie and the King; they had what amounted to an apology. Louis Philippe proclaimed the treaty to be in effect, and the first installment of the indemnity was paid to the Rothschilds, who were acting as the American agent.

The long-standing dispute that had hung over three administrations and reached a crisis in a fourth, was settled without loss of face to the United States. The country rang with praise for Old Hickory's bold stand. But those who studied the language of the messages and the dispatches that were promptly displayed before Congress and in the press could not fail to miss Van Buren's hand in the affair. Washington Irving was one of those careful observers. "I am inclined to give you great credit for the happy management of this matter," he wrote Van Buren on February 24. "And for the able manner in which the collisions between the two countries have been prevented from striking fire."[27]

Van Buren could not afford to bask in the luxury of praise from close friends who knew about his part in the solution of the French crisis. While devoting long hours to diplomatic matters, he was working out his response to another problem, more serious to him politically than the French question—abolition.

Since 1830, William Lloyd Garrison, James G. Birney, Samuel May, Wendell Philips and Gerrit Smith had been agitating for the emancipation of the slaves. A tiny group, they were deemed simply a nuisance during local political contests. But after the Nat Turner slave revolt in 1831, abolitionism began to be taken seriously in the South. Garrison's newspaper the *Liberator,* which had staked out an uncompromising position against slavery from its initial publication, may have gained few adherents in the North, but its fiery language ground on southern sensibilities; for the first time in more than twelve years, debate over slavery in its economic, political and especially its moral aspects slowly intruded itself into the public mind.

There were still slaves in New York when abolition became an issue. Though a free state for over thirty years, the legislature had decreed gradual emancipation. Van Buren had grown up in a slaveholding household, though the kindly Abraham had treated his bondsmen as members of the family. Van Buren himself had briefly owned a slave named Tom, who ran away and whom he never sought to reclaim. After ten years had elapsed a man in a neighboring town located Tom in Worcester, Massachusetts. He offered to buy him, and Van Buren was quite willing to sell him for fifty dollars provided he was "taken without violence."[28]

Sometime after this episode in 1824, he decided that slavery was unjust, a moral blemish on American society that professed freedom and equality. As a lawyer, however, and as a constitutional scholar of the Jeffersonian school, he just as firmly believed that slavery was a local matter, a question for the states and their citizens to determine for themselves. Like other perceptive politicians of his day, both northern and southern, Van Buren saw in slavery the one deeply divisive issue that threatened the foundations of the Union. In the face of it, the hotly contested issues of the tariff, the Bank and internal improvements, were transient flurries around which passions, rhetoric, political intrigue and organizations whirled with seeming fury. Politicians would deny this assertion vehemently; but at bottom they knew John Randolph had been right when he pointed his skeletal finger at some imaginary point in the Congress and in public meetings and warned his fellow citizens that the specter of slavery was stalking the land.

Van Buren's public record on slavery had been legalistic rather than humanitarian, a factor that countered his negative image in a South increasingly suspicious of all northern politicians. On the matter of race, which as yet had not become much of a political issue anywhere, he was relatively free of prejudice. As a United States Senator, Van Buren had joined his colleague Rufus King in opposing the organization of Florida as a slave territory, following the guidelines Jefferson had staked out thirty-five years before and the recent resolutions of the New York legislature on Missouri.[29] That fall Van Buren had led the moderate faction in the New York constitutional convention that abolished property qualifications for suffrage, but retained them for free blacks. His position on this question, which never changed, reflected his concern about their economic and social status.

« CHAPTER 22 »

Partisan Leader

Van Buren's problems with abolition began well before his nomination for the Presidency at Baltimore in May, 1835. The issue was raised in that important but deeply conservative slave state Virginia by that influential editor Thomas Ritchie.[1] Van Buren was not only a northerner, a citizen of a free state, but in his public career he had backed antislavery proposals. As the abolitionist campaign attracted more attention in the South, so Duff Green saw the shape of a winning issue. Along with the catalogue of Jacksonian sins with which Van Buren was identified, abolition in Green's editorials became a much more important issue than his stand on the tariff, internal improvements, the Bank. The abolition question skillfully exploited in 1834 was a major factor in uniting the Whigs and defeating the Junto—for it was generally known that Ritchie and his group were backing Van Buren.[2]

Recognizing the need for a presidential candidate to help consolidate their disparate factions, the Whigs backed Hugh Lawson White, who by now had severed completely his political ties with Jackson. Though lacking the organization of the Democrats, the Whigs in Virginia were the majority party and their candidate was both a westerner and a slaveholder. Duff Green now added a practical dimension to the abolition issue. Anticipating Van Buren's constitutional argument that slavery was a state matter, Green asked in an editorial without naming him or any other candidate, where the South stood on the abolition of slavery in the District of Columbia. There was, of course, no doubt about Congress's power over the District. Green's question was of immediate moment in the Virginia election.[3]

Ritchie and his lieutenants were slowly overcoming the Whig onslaught when the *Telegraph* stepped up its efforts to link Van Buren with antislavery in the North. Although completely satisfied that the Vice-President's political principles were, if anything, more orthodox than Jackson's, Ritchie was disturbed. Like Mangum, he was never entirely comfortable with a northerner, even as charming and cheerful a northerner as Martin Van Buren. The Junto was facing heavy weather with bitter opponents like Governor Littleton Tazewell, Randolph's political heir, ex-Governor John Floyd, Calhoun's representative, and the Clay-leaning Senators, John Tyler, cultivated, acerbic, and Benjamin Leigh, merely acerbic. Only dire necessity would have made Ritchie sound out Van Buren on abolition. Custom, precedent, courtesy, all hallowed in the Old Dominion, dictated that presidential candidates be not directly importuned for their opinions on current issues. Ritchie was nothing if not a respecter of tradition. Yet after dueling with the *Telegraph* over the issue for almost a year and making his probes of local public opinion, he broke with custom and asked Van Buren for his views on what he judged to be an extremely sensitive political subject.[4]

Van Buren too had been a careful reader of the *Telegraph,* and knew that sooner or later the Whigs would seize upon slavery as a means of dividing the Democrats in swing states—Virginia, Tennessee, Kentucky, Maryland. A month earlier, Colonel Joseph Watkins of Goochland County, Virginia, a firm Jacksonian, had asked Silas Wright four questions about Van Buren's political career. Why had he supported Clinton rather than Madison for President in 1812? Why had he backed Rufus King for the Senate in 1820? What was his course on the Missouri Question? Why did he vote for the Tariff of 1828? From a political viewpoint, three of the questions posed no difficulty, but the Missouri controversy required more careful language. The account must be a truthful one; and Wright, with Van Buren's help, made sure there were no inadvertent phrases that might be twisted out of context.

It was important that Rufus King's espousal of free soil be disconnected from the political reasons that made him Van Buren's candidate for the Senate. This done, Wright pointed out that Van Buren was not a member of Congress during the debates and was, therefore, not responsible for the views expressed on either side or the results, whether they were beneficial or not. His connection with the debates was thus reduced to two acts on his part: the resolutions of the New York legislature in favor of restricting slavery and his signing the call for a mass meeting in Albany to consider the question. Wright freely admitted that Van Buren went "with his state in approving the resolutions." But in explanation he pointed out that Van Buren had not introduced the resolutions in the state senate nor had he spoken for or against them. On the tariff, Wright enclosed a copy of Benton's letter which, he said, gave the facts and reasons for Van Buren's vote.

Seven weeks earlier, Thomas Hart Benton had written the first major campaign document for Van Buren. A committee of Mississippi Democrats had nominated him for Vice-President; and in declining, he wrote a lengthy letter in which he endorsed Van Buren for President. On the tariff of 1828, Benton declared Van Buren had opposed the measure and only voted for it because he had been instructed to do so by the New York legislature. Had Van Buren arranged these instructions so that he could hide behind them as so many of his critics charged? Benton categorically denied this imputation and so did Wright, both of whom would not willingly lend themselves to distortion or falsehood. Wright, a state senator at the time, was certainly in a position to know the truth of the matter.[5]

Van Buren was surprised when he received Ritchie's letter. Though he had long since taken account of the editor's touchy personality, he judged the situation in Virginia to be very serious for Ritchie to take such a step after Wright's letter had already appeared in the *Enquirer.* What must have moved Ritchie was that Wright's letter to Watkins had expressed no opinion on slavery in the District of Columbia. Now, at Van Buren's request, Wright addressed himself to that question. He conceded to Ritchie that Congress' power over the District of Columbia was "exclusive" but the good sense of the people through their representatives would never vote to alter its social institutions without the consent of Virginia and Maryland.[6]

Neither of the Wright letters presented a cohesive account of Van Buren's career, nor did they explore fully controversial political points. Letters from all over the country, but primarily from Virginia, kept piling up on his desk asking for more specific, more detailed information. "Never was a poor devil subjected so to bear a cross-examination as I have been by the Old Dominion" he said.[7] Van Buren turned to Butler for assistance.

Although overworked with an unusually heavy schedule of appearances before the Supreme Court and the usual stack of requests for opinions from the President and other executive officers, Butler agreed to rewrite a campaign sketch that he had done in 1832. Using the profile as a base, drawing on Wright's letters, back issues of the *Argus,* Van Buren's private letters and recollections, the proceedings of the New York legislature, Butler wrote a comprehensive treatment of the Vice-President's public career. Biased of course, the Butler narrative gave the appearance of objectivity, leaving it for the reader to decide whether Van Buren was wrong on any given issue. Such a conclusion was invariably preceded by a subtle argument and an arrangement of facts that prompted the reader to conclude no mistake had been made.

Van Buren had thought long and hard on the matter of slavery and had decided that a bold move must be made. The charge of noncommittalism that had always rankled him must be quashed once and for all on what he judged would be the most important issue of the campaign. Sensitive to his northern background, he had Butler draw on that epitome of northernism, Daniel Webster, to explain his position on slavery and abolition. The quotation from a letter the Massachusetts Senator had written in the spring of 1830 was explicit: "In my opinion, the domestic slavery of the Southern States is a subject within the exclusive control of the States themselves and this I am sure is the opinion of the whole North. Congress has no authority to interfere in the emancipation of slaves or in the treatment of them in any of the states."

With respect to abolition of slavery in the District of Columbia, again relying on Webster, Butler noted that petitions in favor of it had always been tabled. Butler let Webster's remarks speak for themselves. He did not enlarge upon the topic by discussing whether Congress had the power to legislate on slavery in the District; nor did he comment on Wright's allusion to this abolitionist goal. On the tariff issue, he inserted the section dealing with that matter in Van Buren's letter of October 1832 to the Shocco Springs, North Carolina Committee.

Butler argued more persuasively than ever before that Van Buren had not engineered the instructions of the New York legislature under which he felt compelled to vote for the tariff. What he omitted was that Van Buren had favored the tariff; and even if he had nothing to do with the instructions from New York, he grasped them as an opportunity to shield himself from any adverse criticism.

Again quoting liberally from the Shocco Springs letter, Butler declared unequivocally that Van Buren shared Jackson's opinion on internal improvements. The federal government did not have the power to build roads and canals within states or to subsidize such projects. Passing on to Van Buren's role in the Bank controversy, Butler effectively squelched the frequent charge that he had sought the ruin of the Bank for the benefit of state-chartered banks in New York City.[8] The remainder of Butler's article was a rewrite of Van Buren's speeches and letters that expressed his views on the structure of the American government. The argument was historic in nature and developed fully Van Buren's notions that free governments could only exist on an adversarial basis where clearly defined opposition points of view on public issues created the dynamics of democracy. His views were, of course, opposed to what he called the "Old Tory centralist ideas" on government; but these ideas, though wrong, had always existed and would continue to exist. Citizens of a free society should welcome their persistence because they offered such a sharp contrast with the true principles of Jefferson. Public men of right mind could articulate them clearly for the people and act upon them properly in the governance of a nation. These generalizations rationalized a party system where there would always be at least two conflicting opinions on important public issues.[9]

Butler sent the expanded profile to Hugh Garland, editor of one of Calhoun's newspapers in Virginia. When Garland published it in April 1835, the hesitant Ritchie had no more questions to ask, no qualms about Van Buren's northernism. The *Enquirer* carried the Garland letter in installments with highly favorable editorial comments, as did the *Globe.* Both papers ran off thousands of copies in pamphlet form that were distributed under various congressional franks throughout the country. But knowing Ritchie as he did, Van Buren had paved the way for the Garland letter with one he wrote for Senator Elias Kane's signature and another for Silas Wright's. In both letters he condemned abolition of slavery in the District of Columbia as an effort "to distract Congress and the country . . . particularly in the midst of a Presidential canvass when all parties are so ready to seize upon everything that can be turned to advantage."[10] These explanations made, Van Buren felt he had done everything possible to satisfy Virginia and the slave-holding states that he was a northerner who could be trusted to protect their rights.

It was not to be so. Virginia politicians in particular, but other southerners as well, catechized him on points he thought he had answered fully. "I see you have to suffer for my sins," he wrote to Mrs. Rives in a bantering way that could not quite conceal his annoyance at the deep-seated suspicion of his motives that existed throughout the southern states. "God knows I have suffered enough for my southern partialities as the apologist of southern institutions and now forsooth you good people will have it—*nolens volens* that I am an abolitionist and general other 'ists that do not deserve to be mentioned."[11]

Virginia may have been a prime source of worry for Van Buren, but he could not ignore threatening developments in Pennsylvania.[12] The situation there was complex with the Democratic party split in two factions, the Amalgamators headed by James Buchanan and the Family party of George M. Dallas, an aristocratic Philadelphia lawyer with a taste for politics. The Amalgamators drew their strength from central and western Pennsylvania, while Dallas's Family party represented the more populous, more prosperous east. The Bank issue had affected both factions; deep personal feuds rankled, which, lacking the touch of a Van Buren, constantly challenged party leadership. Opposing the divided Democrats, was a powerful anti-Masonic party allied with the Whigs.

After much soul-searching, Dallas threw in his lot with the Administration over the Bank question. Biddle had made extraordinary efforts within all parties and all factions to build a solid front against the Administration. But Dallas's defection, and with him the leadership of the Family party, put a brake on the impetus of Biddle's drive. The Amalgamators had also turned against the Bank, though important elements in both factions continued to support Biddle. These splinter groups with backing from the anti-Masons and Whigs had given the Bank a strong position in Pennsylvania.

The fragility of the Democratic organization in Pennsylvania became obvious when the shaky alliance between the two factions began to come apart just as the political cauldron in Virginia reached a boiling point. Pennsylvania Governor George Wolf, nominally a Family party man, insisted on renomination over the objections of Dallas. Wolf's administration had been tarnished with scandal. His refusal to back an ambitious program of state-supported canals and railroads had earned him many enemies. Of course, the Bank did everything it could to prevent his reelection after his turn-about on the removal issue.[13] Nevertheless Wolf was renominated at the state convention in March 1835 after a substantial number of anti-Wolf delegates seceded. Subsequently they held another convention at Lewiston where the bolters nominated the highly independent Henry Muhlenburg.

Van Buren's personal preference was for Muhlenburg. His friends in both factions, the rich bibliophile from Philadelphia, Henry Gilpin, Dallas and David Petrikin, a politically potent member of the Pennsylvania Legislature, if they did not support Muhlenburg, were decidedly opposed to Wolf. Buchanan first backed one candidate, then the other, eventually settling on Wolf. Duff Green, eyeing the divisive politics of Pennsylvania, tried his hand at further widening the rupture.[14] Well aware that neither he nor New York were popular in Pennsylvania, Van Buren picked his way through this political quicksand with infinite care. He deflected the frantic letters from Petrikin, the somber warnings of Buchanan. But he permitted some close associates in New York to float Muhlenburg in the columns of the *Argus* more to test his strength in Pennsylvania than for any reason of partiality. When the Philadelphia politicians he did trust, Roberts Vaux, a Philadelphia editor and sometime ally of Dallas, and Henry Gilpin warned him that Wolf had the edge over Muhlenburg, Van Buren decided to remain entirely neutral, reasoning that if the factions were irreconcilably divided in the state, they would support the candidate of the party for the Presidency.[15]

As if Virginia and Pennsylvania did not present enough cause for anxiety, New York City politics were in such a state of chaos they seemed insoluble. The old Workingmen's party, which had counted as the radical opposition to the entrenched party organization of Tammany, had broadened its appeal, changed its leadership, and bore little resemblance to the original except opposition to monopoly in any form, economic or political. Under the slogan of "Equal Rights," it voiced its opinions through the *Evening Post.* William Leggett, who was editing the *Post* during Bryant's absence in Europe, was a highly talented but tempestuous editor. Formerly a naval officer, until court-martialed and cashiered for dueling, Leggett brought the combative spirit of the sailor, the bluff manners of the gun deck and the soul of a dedicated humanitarian to his partnership with Bryant.

Bryant exercised constraint over Leggett and indeed brought out his peculiar genius for eloquent journalism. But when Bryant left for Europe, placing Leggett in charge, the young editor immediately plunged into what is best described as an idealistic frenzy. The *Post* eagerly embraced equal rights, denounced the Regency as a political monopoly and the *Argus* as its corrupt mouthpiece becoming rich on state printing contracts.

Leggett did not confine himself to Regency publications. He taunted Noah, now editing the *Evening Star,* and the choleric J. Watson Webb of the *Courier and Enquirer* as being subservient paid lackeys of Nicholas Biddle. Leggett's vituperative feud with the *New York Times,* Tammany's advocate in New York City, however, went beyond the bounds of propriety, as loose as it was in those days of billingsgate journalism. The Regency organization was enraged. "Is Leggett wicked or crazy?' asked young John Van Buren of his crony, Jesse Hoyt. Van Buren, who secretly admited Leggett's style and his passion for equal rights, had to respond. When the *Post* began drifting closer to abolition, Van Buren saw to it that federal patronage was cut off.

Other than that, he kept clear of the contest, realizing that tempers, interests and positions had so solidified in the City that any interference on his part or on the part of any of his lieutenants stood a chance of bringing down upon his own head the ire of both factions. Young Samuel Jones Tilden was of like mind, giving an early foretaste of his political acumen. "I wish you would suggest to Croswell," he wrote his father Elam, one of Van Buren's trusted counselors, "to let the *Post* alone. Enough has been said. The constant attacks of the *Argus* not only enable but force Leggett to keep up the warfare."[16] It was nerve-wracking all the same, this waiting game; for Van Buren was importuned by both sides and his neutral posture had to be explained again and again in terms acceptable to warring groups without being charged by both with that dreaded word "noncommitalism."

Nor were Van Buren's worries purely political and diplomatic during those spring days of 1835. Major Barry, the last member of Jackson's original cabinet, had been ill for some time and unable to function as postmaster general. Even in the best of health, Barry had been an incompetent administrator. Unscrupulous contractors had bilked the department of thousands of dollars; postmasters and deputy postmasters, some 13,000 of them all over the country, if they did not engage in petty graft, had no well defined system of remittance. Additional thousands of dollars were siphoned off or locked up in their offices for long periods of time and returned to Washington on no specific schedule. The central office was not free of embezzlement. Early in 1835, Jackson finally learned about the scandals in the Post Office. There was not enough cash on hand to honor its contracts, and the President was mortified and furious by turns when creditors threatened to sue the government. Barry must go! Amos Kendall was the only man in the country who could cope with this problem.[17]

No doubt Van Buren agreed with Jackson that Kendall was the best man available to clean up the mess in the Post Office. From a political point of view, however, Kendall's appointment was troublesome. The Whig press had securely fixed his image in the public mind as a wire-puller of the worst sort, the sinister influence around Jackson, chief of the Kitchen Cabinet, manipulator, intriguer. Kendall lent himself to the stereotypes. Of average height, thin, with a long face, close-set eyes, sharp nose, prematurely white hair, a man of few words, though when he spoke it was with a pronounced Yankee twang that his years in Kentucky had not softened. Not often seen in public, he shunned the social whirls of the little capital, while he labored long hours in his office or at the White House.

Van Buren had always regarded Kendall as highly intelligent, but he had also taken his measure as someone who knew how to insinuate himself into Old Hickory's confidence. He had much the same feelings about Blair, but there was a warmth to Blair, a touch of southern charm that mellowed his otherwise acidulous disposition. Kendall was careful, cold, competent and potentially dangerous. More than once he had used his influence with the President to reinforce his prejudices and thwart what Van Buren believed a correct policy. The threat of reprisals that had very nearly wrecked negotiation with France had been Jackson's idea but Kendall's handiwork. Taney, who, like Van Buren, had found himself undercut by Kendall, thought Kendall's was a bad appointment that would be damaging politically. From Virginia came the usual cry of distress.[18]

Whatever Van Buren's thoughts may have been about Kendall as a person and a power at court, he must have been relieved to have him buried for the time being in Post Office problems. Besides, it was to his advantage as much as Jackson's that the Post Office be cleaned up as rapidly as possible. He would have to agree with Old Hickory that Kendall was the best choice for this difficult task. Within a month, Kendall had identified the problem, devised a management policy that solved it and weeded out incompetent and corrupt employees. By the end of the year, the Post Office was completely solvent and service had improved dramatically. Criticism of the appointment died away, and even Kendall's most captious critic had to admit the Post Office had never been so well administered.[19]

While these reforms were pending, the annual campaign in Virginia reached a climax. After an even more desperate struggle than the previous year, the Democrats achieved a complete triumph carrying both houses of the legislature. Since the issue had turned largely on presidential politics, Virginia seemed relatively safe. It was a time for rejoicing at last in Richmond. With the legislature safely in its hands, the Junto planned "to instruct" the two Whig Senators, Tyler and Leigh, out of the Senate. Rives was slated to succeed one of them. But Hugh Lawson White's defeat in Virginia was counterbalanced by his sweeping victory over Van Buren in Tennessee.[20]

The political climate for Van Buren in Tennessee had been threatening for the past two years, ever since White had broken with Jackson over the removal question. White was as popular as Old Hickory in their home state, though they drew their political strength from different areas and social groups. While Jackson was the idol of the small farmers, White had the backing of the planters, the business and professional men, and the mountaineers of his home section in eastern Tennessee. The Administration's restrictive policy on internal improvements, its insistence on hard money, had cost much support from the speculators and all those who favored rapid development.[21]

Jackson fulminated at White and Bell, whom he held responsible for defections from the Administration in Tennessee. But Jackson did not suspect that one of his closest friends in Washington, William B. Lewis, was their secret ally. Van Buren had never been popular in Tennessee or in any of the southwestern frontier states for that matter. Davy Crockett, the Whigs' answer to Jackson's frontier image, had lent his name to numerous ghostwritten books and pamphlets that caricatured Van Buren unmercifully. To the average uncouth, ill-lettered westerner, Van Buren was held up as the very epitome of the effete, blood-sucking East, a pliant, little intriguer, who always ran away from a fight after making sure he had the cash.[22]

Against the White-Bell-Lewis combination, Van Buren had the loyal support of James K. Polk, Felix Grundy, Tennessee's other Senator,

Governor Carroll, an astute politician of the old school, and Donelson. Of course, towering above all these workers in the field was the President. He recognized the damage to his own reputation if Van Buren lost in his own state and was vain enough to exert all of his formidable influence. Unfortunately, the White organization represented the wealthy and the articulate. Sober John Bell may have lacked the flair of the Jacksonians, but he quietly went about his way making the right sort of deals with the right sort of editors and politicians. By May 1835, four months before the state election, even Van Buren's most outspoken supporters, Governor Carroll, Alfred Balch and Judge Catron, had given up.

As soon as Jackson learned of their surrender he took it as a personal insult and penned one of his furious missives to Donelson, who was then in Tennessee. Angry letters, even from the President, were of no avail.[23] When the convention assembled at Baltimore on May 20, 1835, there were 615 delegates present, but no one from Tennessee and South Carolina answered the roll.

The convention idea had become firmly entrenched. Over three times as many delegates were present than in 1832; yet well over half of them came from New Jersey, Maryland and Virginia. In fact, Maryland, with 171 delegates, accounted for more than one-fourth of the total.[24]

Though not so widely representative of the party and the nation as Democratic papers would have their readers believe, the convention in numbers was the largest assemblage held up to that time. Blair, who was attending as an editor and as Jackson's personal representative, "was amazed at the prodigious turnout." Baltimore, which was abundantly supplied with hotels, taverns, and boarding houses, was hard pressed to provide accommodations.

Once Andrew Stevenson, ex-Speaker of the House, was elected chairman and the committees were appointed, the convention came to grips with two serious issues: the nomination of a vice-president and the search for some means to have Tennessee represented. Silas Wright, a delegate and Van Buren's manager, had to bear the responsibility of handling both matters, a task he deplored. For he knew that the vice-presidential question had been settled beforehand. Virginia and the seaboard slave states were committed to Rives. Pressed by Jackson, by Blair and by westerners like Benton and Grundy, Van Buren found expediency the only route at the moment. No doubt, too, he was alarmed at the impetus of the White candidacy and agreed that Richard Mentor Johnson would provide balance on the ticket. Whatever the reasons, he was persuaded to drop Rives, though it cost him anguish, and disturbed Silas Wright even more.

Wright had encouraged Rives candidacy and had implied that New York would back him. It was now his unpleasant duty to inform the Virginians that the signals had changed without alienating them completely and perhaps throwing the South into the White column. One of Rives's friends, on his arrival, introduced himself to the harried, red-faced Wright and asked for his opinion. Wright was frank as he could be considering the circumstances. Personally, he and the New York delegation preferred Rives "but they could take no part being delicately situated . . . if compelled to vote they must take a man who would bring the greatest strength to the party." Again and again, angry Virginians, some Georgians and North Carolinians, believing they had been betrayed, that expediency had triumphed over principle, badgered poor Wright, who sympathized with them but felt bound by Van Buren's decision. As if this were not enough, he acquiesced to a shabby scheme in which Tennessee's fifteen votes were to be counted though the state had sent no delegates.

A petty politician from Tennessee, Edward Rucker, was recognized in one of the taverns. After some persuasion, Rucker agreed to become a delegate and cast the vote of the entire state for Van Buren and Johnson. Had Wright let matters take their regular course, Johnson would not have been nominated. Tennessee's vote carried him over on the first ballot. An honorable man, Wright was disgusted with the role he had played. As he wrote about it to Van Buren, he said, "I have never in my life had such a seige and I would not undertake another for any less consideration than that which induced me to undertake this."

The Virginians were furious and proclaimed their state would not support the ticket. Dr. R. C. Mason, who had argued heatedly against Johnson's nomination, expressed their sentiments exactly. "How could it happen," he asked Rives, "that regardless of character and personal standing as well as of qualification and political principle, the convention's nominee has prevailed . . . by taking 'Old Tecumseh' on the grounds that he was kind to his heighbors, to widows and old soldiers, and that he wrote the Sunday Mail Report?"[25] Van Buren felt it necessary to caution Rives about any separatist movements. Rives pledged himself to support the ticket; but it was clear from his reply that the warm relations that had existed for so many years between the two men had taken an ugly turn.[26]

Nor had the situation in Pennsylvania improved. The Wolf and Muhlenburg factions had sent delegates to Baltimore, and there was no apparent division in their vote for Van Buren and Johnson. But the bad blood that existed between the two groups did not bode well for Van Buren's election. "The condition of Pennsylvania is very bad," wrote Wright from Baltimore, "and I do not know that it is to be mended." Buchanan, who kept up a stream of pessimistic letters, thought it would be easier to move the Allegheny Mountains than to bring the Wolf and Muhlenburg factions together. The Whigs, he felt, might carry the state and White gain its electoral votes.[27]

Fatalistic about politics and recognizing that much could happen between now and election day eighteen months hence, Van Buren preserved his usual equanimity. But he did decide that another public letter cast on a high plane, might calm the immediate excitement and, of course, be useful as a campaign document. The letter of the committee informing him of his nomination furnished the pretext for what was a communication to the American people.

Van Buren spoke the truth when he declared he had never in word or deed pursued the Democratic nomination, but then he surely had not discouraged his supporters either. His definition of noninvolvement had to be a narrow one, indeed. Recognizing the ill-will the convention had stirred up, Van Buren stressed the themes of conciliation and continuity. He declared that

> until the wit of man shall be able to devise some plan of representation by which all who think themselves qualified may be at the same time admitted to a participation in the Administration of its affairs, we must not expect to be relieved from the spirit of complaining, nor even surprised to find it most vehement at a period of the greatest prosperity.

Were he elected, he would "endeavor to tread generally in the footsteps of President Jackson—happy if I shall be able to perfect the work he has so gloriously begun."[28]

Never before in any campaign had a candidate spoken so often on public issues; never had a presidential candidate, excepting Jefferson and Madison, published his own ideas on government or so highly praised the federal system. Taken all together, published in a pamphlet, and entitled *Mr. Van Buren's Opinions,* the *Globe,* the *Richmond Enquirer* and the *Argus* printed and broadcast them throughout the country providing partisan editors with political ideas, stump speakers with material for their rural audiences and answers for hecklers in the crowds at militia meetings, festivals and debates.

As Van Buren assumed, the Virginians would eventually calm down, perhaps not vote for Johnson, but would not split away from the Democratic party. Ritchie and Rives and others of the Junto could not go with nullifiers or Bank men and surely not with the Whigs. Pennsylvania and Tennessee remained the weak spots, South Carolina hopeless. To some extent, these weaknesses were eased by significant improvements in New England. Gideon Welles and John M. Niles had finally brought Connecticut into the Democratic ranks. Of all the New England states, only Massachusetts seemed doubtful. But George Bancroft, who was close to the Regency, aided by the Federal officeholders in the state, had created a smooth-working organization of farmers, anti-Masons and merchants. The President lent what support he could, the most important being his control over Blair, whom he stimulated to greater editoral endeavors.[29]

While the summer of 1835 faded away, Van Buren relaxed at his old haunts in Lebanon Springs, Saratoga and Oswego, receiving hopeful reports from the Northwest and the Northeast. The party of the South was showing few signs of division, but the activities of abolitionists continued unabated; then suddenly in early September, the Administration made abolition again a national issue. Kendall gave southern postmasters authority to open packages suspected of containing abolitionist literature and if this material were found to destroy it. A great uproar followed this action, which was obviously an infringement of privacy and free speech. The *Evening Post* excoriated Kendall's order, and many Democratic papers in the North and West joined the Whig press in condemning the action. For once Ritchie and his associates did not worry about constitutional inhibitions. From the *Enquirer* came a "calm appeal" to the North to curb the abolitionists.

Van Buren's campaign organization quickly went into action. At a mass meeting held in Albany on September 4, speakers denounced abolitionists, comparing them with rioters and subverters of the Union. Although the Regency was out in full force, many influential Whigs signed the call and spoke at the meeting. John A. Dix presented the resolutions. On the touchy constitutional question of free speech, he avoided any specific interpretation or recommendation. Rather he asked the citizens to stigmatize any effort that would disrupt the Union or precipitate a slave rebellion in the South.

Van Buren was absent from Albany at the time, but as he explained in another public letter to William Schley, a citizen of Georgia, he took responsibility for calling the meeting and agreed completely with its resolutions, which he enclosed together with the call signed by all the executive officers of the state.[30] This public explanation and mass meetings held all over the North and Northwest stilled for a brief period the qualms about Van Buren that existed within the Democratic party of the South and Southwest. Ritchie labored mightily in the *Enquirer* and Blair in the *Globe* to expunge these latent prejudices, only to have them crop up again under the persistent allegations of the Whigs and the Bank Democrats.

When the Twenty-fourth Congress convened on December 7, 1835, John Quincy Adams opened up the issue anew and with greater vehemence than ever before. For several years now he had been presenting petitions for the abolition of slavery in the District of Columbia, but they had been few in number and had been tabled without discussion. Now, the abolitionists had devised a new tactic. The petitions would, within the prescribed language, indict the institution of slavery. Adams, a staunch defender of the right of petition, was very much available to present whatever petition he might receive, whether it be for abolition or for internal improvements, as he had done in the past. He claimed the right to read any and all petitions he would present. Southern members fulminated against Adams reading what they insisted was abolitionist propaganda under the semblance of a petition. But "old man eloquent," as Adams had come to be called, refused to be silenced. It finally took a rule of the House that was not passed without hot debate to refuse any reading of abolition petitions. A majority in the House voted away an ancient Anglo-American right. Van Buren, to his discredit, avoided the subject, though he knew the action of the House was subverting basic principles of freedom he had espoused his entire political life.

By then Van Buren had rather effectively squelched the rumors and innuendos that he was an abolitionist. Several months before, when the movement to discredit him in the South had reached a peak, he grasped at a life buoy Governor Marcy threw out to him. Marcy was preparing his annual message and wondered whether he should say anything on the subject of slavery. "Your position and that of Mr. Butler enables you both," wrote Marcy, "to give advice to satisfy the expectations of the reasonable men at the south." Van Buren replied at once in the affirmative and asked Butler to give Marcy in general terms his ideas of what ought to be said. Before anything more was done, Marcy wrote again. He had already decided he must say something on the subject, but should he recommend any legislation? "If we attempt to do anything in that way is there not danger that we shall be able to do so little and fall so far short of what will be expected in the south?"

Much as he desired to give the South the strongest possible assurances, Van Buren balked at punitive legislation. There were state constitutional restrictions, and the laws of New York were not to be written in southern legislatures. As communicated by Wright, Van Buren suggested "a distinct declaration that no legislation was at present considered necessary."

For the next three weeks tension mounted at the Seven Buildings as Van Buren awaited Marcy's message. When it finally arrived on January 11, 1836, everyone was relieved. The moderate southern Democrats thought it would check much of the anti-northern feeling that electioneering had stirred up. Rives, to whom Van Buren sent a copy, declared it was "by far the most complete demolition of abolition that controversy has brought forth." Even some of the "nullies," Wright reported, had words of praise for Marcy's thoughtful approach.

The Governor had not flinched from declaring that slavery was considered morally evil in New York; but then he spoke of the "formidable difficulties" of abolishing it in the South. After describing the abolitionists as a tiny group of misguided people, he could not believe they were directly inciting a slave revolt as the more inflammatory southern press and politicians were repeatedly charging. What he thought they were attempting was the cultivation of public opinion in the South that would support emancipation. If he were mistaken, Marcy had words of warning to abolitionists and southern fanatics alike. "It must then be determined," he said, "how far the several states can provide . . . their own laws for the trial and punishment by their own judiciaries of residents . . . intended to excite insurrection and rebellion in a sister state."[31]

Although the slavery question was the most threatening aspect of the campaign, the twenty-fourth Congress concerned itself with a variety of public issues which became electioneering topics—the distribution of the Treasury surplus, the allocation of public lands, the perpetual argument over internal improvements and banks, and the powers of a given Congress to alter actions taken by a previous Congress and entered in its journal. While these debates were going on, a Kentucky Congressman, Sherrod Williams, tried to lure Van Buren into a premature disclosure of his views. Van Buren, of course, saw through the clumsy subterfuge but he filed it away for a public answer after Congress had adjourned.

On June 14, 1836, he answered all of Williams' questions in detail covering much of the same ground he had gone over before, adding only three additional comments. He opposed distribution of the surplus as tending toward corruption and subverting states' rights; he was opposed to federally supported internal improvements on navigable waterways above ports of entry; and he stood by Benton's efforts to expunge the Senate's censure of Jackson in 1834.[32]

When Congress adjourned in mid-July after an unusually long session during which Calhoun, Preston, Clay and Webster had tried to make him the object of calumny Van Buren was glad to get away from Washington. His *Opinions,* now swelled with the acceptance letter, the Schley letter and his reply to Williams, was a bulky pamphlet of forty pages in fine type. For those who were willing to wade through the material, and there were many who did, Van Buren had commented fully on all the controversial topics of the day. It was virtually impossible for the Whigs and the nullifiers and the Bank Democrats to fault him for lack of candor. The burden of the campaign was now on other shoulders.

Except for a few days visiting in Kinderhook and Saratoga, a brief excursion to the western counties of the state with his "privy councillors," John, Martin, Jr., and Smith, Van Buren remained in Albany. The usual rallies, ribald songs, doggerel poems, scurrilous cartoons, flamboyant oratory and ardent spirits marked the progress of the campaign.[33]

Had the Whigs less "great men," less Websters, Clays and Calhouns, each with his own interests and ambitions, they would have avoided the pitfalls Mangum had so graphically described on his visit to the North two years before; and they would have won the election. Not one of their leaders would defer to any other. Typical of the divisiveness among the Whigs was Webster's refusal to take himself out of the contest when he knew by the fall of 1835 that he did not have a chance.[34] Nor could the Whigs provide a coherent program like that set forth by the *Address* of the Democrats, which a convention committee distilled from *Van Buren's Opinions.*

Thus, they made Van Buren the issue, not what he stood for, but the man himself, his appearance, his political style, his alleged lack of scruples, even his parentage was called into question. "They have been driven to the miserable devise of endeavoring to prejudice the public mind by calling him an intriguer—a Magician," said Claiborne Gooch. "If a quick penetrating mind, a cool judgement—quiet and practical wisdom and honorable success be indices of a Magician then the name has been properly bestowed." The Whigs also backed sectional candidates, hoping they could deprive Van Buren of a majority and throw the election into the House. They very nearly accomplished their objective. As Buchanan had prophesied, the split among the Democrats of Pennsylvania gave the governorship to the Whigs and almost cost Van Buren the state's electoral votes.[35]

Van Buren was in Albany when the returns started coming in by piecemeal as far distant districts began the slow process of tabulating votes, certifying them for publication in local newspapers. By mid-November, with Ohio, Kentucky (despite Johnson's place on the ticket), Indiana, Illinois and Pennsylvania all uncertain and his majority in Virginia so slim he could lose the state, victory seemed doubtful.[36] Georgia, Tennessee and New Jersey had gone for Harrison or White. Louisiana and Mississippi seemed to have been lost. The only cheering news that came to him at Cruttenden's, his Albany headquarters, was from New England and New York, though here his support was more in rural districts than in the cities. Van Buren barely carried New York City, where the Loco-Focos and Tammany battled it out to the very end. Bryant's return and the departure of Leggett to establish his nonpartisan *Plain Dealer* helped Van Buren in the closing days of the campaign.[37]

All Van Buren could expect at this point was an election determined by the House. If this should come to pass, he would lose to either of the western candidates, Harrison or White. By December 1, he had reliable information that he had carried Pennsylvania, Alabama, Virginia, Illinois, New York, North Carolina, Missouri, Arkansas and Michigan. The margin at most was 4,000 in Pennsylvania, dwindling to about 1,000 in Illinois. Indiana and Ohio cast their votes for Harrison, and the outcome in Louisiana and Mississippi was still undecided. Even if they went for White, he would still have a majority in the electoral college.

The news that he carried both of those states by about 300 votes each came to him in Washington where he had arrived in time to preside over the Senate when Congress convened on December 5, 1836.[38] He knew by now that he was President-elect, though it had been a close thing. When all votes were in and tallied and certified, he had a popular majority of about 28,000 votes over the combined vote of his three competitors. He could take a measure of pride in the course of his candidacy which had dealt fairly, as he saw it, with the issues. Whether one believed in his interpretation of them or not, they had been more extensively and more cogently presented than any other public questions would be for the next two decades.

Most of the documents he had written himself, while presiding over a tumultuous Senate, advising members of the cabinet, ghosting significant parts of Jackson's messages, acting as the President's chief advisor on patronage and on foreign affairs, feeling his way around the Kitchen Cabinet, while always keeping an eye on New York. The short, balding man had carried a heavy load for the past eight years, a burden that would have crushed lesser men.[39] Quicker, more perceptive, more intelligent than most of his associates, he was the object of jealousy, of scorn, even of hate. The "Red Fox of Kinderhook" they called him in their milder moments, reserving more explicit epithets for the tavern talk of Washington and elsewhere. The nickname had a ring of truth, but those men of ability, of wit, of honest intentions, Washington Irving, Gouverneur Kemble, James Kirke Paulding, Cambreleng, the Regency which governed New York

with distinction for twenty years, would have vehemently denied it. Andrew Jackson, whom he had served well, sensed Van Buren's innate strength. "Your views is [*sic*] always grateful to me," he wrote Van Buren. "They are like my own, always based upon the just grounds of the prosperity for our country and the general good."[40]

« CHAPTER 23 »

"Be Careful of Cataline"

The hall of the House was crowded on the chill morning of February 8, 1837. Promptly at 10 o'clock the Senators, headed by President *pro tempore,* William R. King of Alabama, filed into the chamber for the formality of opening the sealed ballots that would officially declare Martin Van Buren to have been elected eighth President of the United States. While King was announcing the vote of each state, Van Buren and Clay met at the rear of the chamber. "It is a cloudy day, sir!" said the Kentuckian. Van Buren could not resist a playful remark. "The sun will shine on the fourth of March," he replied, with a show of confidence that masked a grave anxiety about the state of the Union.[1]

When the tally came to an end, King announced that Martin Van Buren had 170 votes, a majority, and would become President of the United States on March 4. Johnson, with only 147 votes, had no majority, and as the Constitution prescribed the Senate would elect the Vice-President. When the Senators reassembled in their own chamber, the junior Senator from Tennessee, Felix Grundy, gained the floor. He proposed that the Vice-President be elected by voice vote, each Senator stating his preference from his desk. Grundy's motion prevailed. The secretary of the Senate, Asbury Dickens, who had made a comfortable career for himself in the tiny Washington bureaucracy, read the roll beginning with Judah Dana of Maine who voted for Johnson, followed by thirty-two others, six more than were necessary to elect.[2]

But the President-elect had far more pressing problems than a socially embarrassing Vice-President who had cost him southern votes. For the past year he had become more and more concerned about the economy. He had agreed completely with Jackson's order to the land offices that they received only coin for purchases, the Specie Circular as it was called. While traveling through New York City in the summer of 1835, he had been astonished at the rise in the price of city lots, the speculation in western lands and the spectacular rise in the stocks and bonds of banks and of new railroad enterprises.[3] What had worried him in 1835 deeply concerned him a year later, despite the Specie Circular. The third installment of the surplus, some $9 million, had vanished like a wisp of morning fog in the heat of the money mart.

New York, which controlled two-thirds of the nation's foreign trade, was indulging in an orgy of overtrading. Imports that stood at $37 million in 1829 had swollen to $150 million in 1836.[4] Van Buren may have been alarmed at the frenzied speculation, but to the average workingman in New York City, the rise in the cost of living was catastrophic. The winter of 1835 had been unusually severe throughout the Northeast and the West. Spring had come late and frost early. New York's wheat crop had been severely damaged as had the corn and wheat crops of the West. Food was scarce, even staples like potatoes and flour tripled in price, while tenement rents soared as land and buildings constantly changed hands. In December 1835 a great fire destroyed most of New York's commercial district, bankrupting insurance companies and hundreds of businessmen.[5]

To such an observant man as Van Buren, the evidence of impending panic was everywhere. But when would it occur and what could be done that had not already been done to cushion its impact? Among his friends, Van Buren was the same cheerful, thoughtful person he had always been. Silas Wright noted he was not "delirious" as most men would be who had just been elected President. Yet he was troubled more with dyspepsia than he had ever been, a sure sign of stress; and he was taking a concoction of water, soot and powdered charcoal, a friend had suggested that actually gave him some relief.[6]

A lesser worry, but bothersome for a man of Van Buren's temperament, was the make-up of his cabinet and the whole gamut of executive patronage. He remembered vividly the cabinet problems that plagued Jackson until recently. He valued harmony among his advisors and party lieutenants almost as much as he counted on their ability and honesty. From long experience in the tangled byways of New York state politics he had learned that once a party had established itself, removals for favor courted trouble; they were the breeding ground of discontent. Patronage was a two-headed thing.

He wrote in a memorandum for his own guidance:

> Whilst I would not think it justifiable to remove a capable and faithful public officer, either because his opinions were different from my own or on the ground of personal dislike or for any reason personal to myself, I should feel it my duty to do so whenever I had good

> reason to believe either a change in the office was necessary to carry out the general policy in the administration of the gov't. in favor of which the people decided and which it was then desired to be executed.[7]

Jackson's present cabinet, which Van Buren had been instrumental in assembling, met most of the qualifications. Lewis Cass had resigned a few months earlier. His trusted advisor and close friend, Benjamin Butler, was temporarily acting as secretary of war while he performed his own duties as attorney general. Butler, who had never wanted any cabinet post, was anxious to quit Washington and devote himself to his family and law practice at home. Van Buren was so dependent on Butler he could not bear the thought of losing him. On December 30, 1836, he wrote his son John that nothing had been decided about the cabinet except he wanted the reluctant Butler to remain as attorney general. Would John have some member of the Regency write Butler urging him for the sake of the nation and the party to remain where he was. "It is of vital importance to me that he should. Had you any idea of his great capacity you can have no conception. Whilst conflicting interests will immediately present trouble the moment a place is opened."[8] Pending Butler's decision, which was not forthcoming despite pressure from Albany and subtle though unmistakable hints from Van Buren himself, he eventually decided to retain the old cabinet for the time being.

None except Butler were outstanding men, either as politicians or counselors or administrators. Woodbury was the most capable and had more experience than any of the others. Formerly a Senator from New Hampshire, he had served as secretary of the navy before taking over the Treasury Department when the Senate refused to confirm Taney. A balding, heavy-set man, he had literary pretensions and burning, though he thought well-concealed, presidential ambitions. Industrious, conventional in his thinking on fiscal matters, he represented New England in the cabinet. Secretary of State John Forsyth had less ability. A sensitive, lean southerner, he looked the aloof aristocrat with his acquiline nose and firm chin. In reality, he was a genial, pleasant person, a good man for carrying out established policy, always dependable, but lacking in creative flair. Born a Virginian, long a citizen of Georgia, he had represented that state in the House as well as in the Senate. Unable to keep his state in the Van Buren column, Forsyth added little political weight to a Van Buren cabinet, but he was the only public figure available from the Deep South.[9] Van Buren himself would supply the creative expertise should an unforeseen change in foreign policy be demanded.

The weakest member of the cabinet, and also the oldest and richest, was Mahlon Dickerson. Absentee proprietor of a celebrated iron works near Trenton, New Jersey, Dickerson was a fussy, querulous man who, when he was not ill, which was frequent, devoted more time to his hobby of gardening than he did to the Navy Department. If anything, he was a political liability whose unyielding high tariff views were looked upon with deep suspicion in the South and over much of rural America. Van Buren had early taken his measure and in an indirect way tried persuading Dickerson to resign by offering him the mission to Austria. But Dickerson rather indignantly refused, leaving Van Buren with the decision of requesting his resignation or letting him stay on. Van Buren wanted no public dispute at the commencement of his Administration, so he decided to make the best of Dickerson's intransigence and wait.[10]

Next to Woodbury, Postmaster General Amos Kendall was the keenest mind and excelled his fellow New Englander in administrative ability. Unfortunately, Kendall's health was as fragile as Dickerson's. He tended to be outspokenly and rigidly radical, not just on the expiring Bank of the United States, but on all banks. A gifted political writer, like Blair, he saw government as an opportunity to make his fortune through means that were acceptable by the standards of the day. Kendall earned large profits on the *Extra Globe*s that he edited and Blair and his partner Rives printed. These campaign documents were franked off by the thousands to the postmasters with the not so gentle hint that they be disposd of at a price to the faithful. He also used inside information to make prudent investments and did his share of speculating in western lands, as did most of his associates in Washington, whether in the executive branch or on Capitol Hill. Blair, too, took advantage of the opportunities thus afforded, but both men were reckoned to be honest, immune to lobbyists; and Kendall's stewardship of the public funds was beyond reproach.[11]

Having worked out his cabinet according to the principles he had set down, he had but one post to fill, the War Department, and probably after the closing session of the Supreme Court in April, the attorney general's office. The realities of sectionalism dictated a southern appointment. An obvious choice was William C. Rives, whose disappointment over losing the vice-presidential nomination was common knowledge in Washington. On February 1, 1837, Van Buren offered Rives the War Department, fully expecting the ambitious Virginian to accept. But to his dismay, Rives refused; and though he couched his rejection in polite terms, he indicated he thought the position was not of sufficient stature. In a subtle bid for the State Department, Rives noted that his background and experience had been in foreign affairs. Van Buren begged him to reconsider, holding out the possibility of a change, but emphasizing that Forsyth and Woodbury would remain in the new cabinet for the present. The Virginian again refused, and bowing to the inevitable, he offered the post to Joel Poinsett of South Carolina, who accepted.

Van Buren had known Poinsett for fifteen years and had always been impressed by the versatile talents of the South Carolinian. He had made a few speeches as a member of the House in the early twenties that Van Buren thought were excellent, but his diplomatic career had not lived up to glowing expectations. Yet if any of the youthful dilettantism clung to him—his brief study of medicine at Edinburgh, the briefer study of law at home, the incessant travel to romantic and remote places—he completely dispelled it by his courageous and inspired stand in his native state against the forces of disunion during the nullification crisis. Poinsett matched the intelligence of Butler, and in sheer intellectual capacity and sophistication he was far superior to anyone in the cabinet.[12]

The only disturbing factor yet remaining was how to acknowledge Pennsylvania, whose political complexities were always fraught with danger to the Democracy. Buchanan and his Amalgamators were still in deadly competition with George M. Dallas and his Family party. Both factions were further split on local issues between the adherents of Henry A. Muhlenberg and former Governor George Wolf. If Van Buren appointed either Buchanan or Dallas to the cabinet, he would gain the enmity of the one who was passed over, a risk he could not afford to take in a state he barely managed to carry.

His solution was to offer Dallas the Russian mission, which Buchanan had recently vacated. This diplomatic post, a senior one to be sure, was scarcely the recognition Pennsylvania deserved; but Van Buren gambled that Dallas would be content with a mission abroad and Buchanan would be so pleased to have his rival out of the country he would be pacified. Taking due care to explain the entire cabinet situation in his letter to Dallas, he implied that higher things might be in store. Other Pennsylvanians important to the party, like the wily Simon Cameron, Muhlenberg and Wolf, whom he made Comptroller of the Currency, were disgruntled but Van Buren could afford to disregard their complaints if the major powers were content. And they were; his gamble paid off handsomely.[13]

By mid-February, Van Buren had a balanced cabinet, but beyond Butler, whom he thought he would lose, and Poinsett, surely not a brilliant one. Had he planned it this way? Had he sought harmony and calm over the tempest and the tumult of the previous eight years? Possibly so, given his native caution. Yet, always excepting the Eaton affair, Van Buren knew that outside forces and developments not connected with the caliber of individual members and the temperament of Andrew Jackson had been the principal agents. Although Van Buren wanted a unified cabinet, he was also quite aware that he must avoid the imputation he was not his own man, simply a creature of the impetuous Old Hero. He could not afford to offend Jackson who would still be powerful and popular in his own right after leaving the Presidency. He was genuinely attached to Jackson, knew his sensitive nature, his vanity, his capacity for great warmth and affection. Personal and political reasons, the emotional and the practical, as usual guided his hand.[14]

While dealing with matters great and small, coping with nervous indigestion, Van Buren found the time, and apparently the physical stamina, to involve himself in Washington's social whirl. Jackson became very ill after his return from the Hermitage, his home in Tennessee, and for a time the White House was closed to most visitors, until he recovered near the end of December. Van Buren took it upon himself to fill the void by at least attending the soirees the cabinet members gave in turn. And they were crowded, elaborate affairs where Henry Orr, a free black of great taste, was the caterer. The Woodburys and the Forsyths entertained from one to three hundred at evening affairs. Guests dined on ice cream molded into the shape of doves, castles of quartered oranges glazed with yellow sugar, cakes and other dessert dishes washed down with apple toddy or a variety of French wines.

But dwarfing all the parties, indeed the event of the season, was another evening affair at Reuben Whitney's, whose extravagance proved that the special agent in Washington of twenty or so deposit banks was not underpaid for his efforts. A special committee of the House was investigating Whitney's activities at the time, but this did not deter him from going ahead with his extravagant affair.

The Whitneys, who lived near Capitol Hill, had thrown open most of their spacious house for dancing and dining. They were a large, imposing couple who greeted their guests at the door, assisted by their similarly large and imposing daughter. One guest found such a jam of people in the front parlor that it took him ten minutes before he spotted an opening to the second parlor where a cotillion was in progress to the music of a string orchestra. To his dazzled eyes, the scene was so colorful and rich that it excited some Calvinist forebodings of the wages of frivolity; yet he did not fail to record the party in rather rapturous detail. "No one sits or can sit at these crowded assemblies," he observed, "everyone wears gloves variously colored, some white, others light yellow or slate. . . . Head ornaments of ladies are artificial flowers, plumes, birds of paradise, feathers, crescents of gems." Shoulders and breasts were bared in some instances, so low that they startled, if they did not shock him.[15] Dining was upstairs where three tables were crammed with sugar plums, sugar almonds, creams of all colors, rich jellies, some eighteen varieties of cold and hot meats, fish, game and huge chicken salads. There were the usual castles of oranges, ice cream pyramids and mounds of fresh grapes from Portugal. Champagne, lemonade, Madeira and Sherry were constantly being passed by a score of black waiters in livery. Republican simplicity, it would seem, had vanished in some quarters of the Democratic party.

Van Buren was much in evidence, though he had cast aside his customary brilliant attire and was dressed in black frockcoat and white stock like all the other gentlemen present. These included the entire cabinet, along with Blair, his wife and daughter. The Whitney party seems to have been largely, if not exclusively, a Democratic affair; and if Congressman John Fairfield of Maine was an accurate judge of expense, it cost the bank lobbyist from $1,500 to $2,000, three years' wages for a government clerk, or as many as eight for a skilled worker in one of New York City's smoky boiler works.[16]

The tumblers of charcoal, soot and water Van Buren was drinking every day must have neutralized the rich food and drink; the annoyance of being pestered by would-be officeholders who descended on the Seven Buildings; the worry as he read the letters, one more dismal than the other from New York, describing signs of a growing panic among the merchants and the bankers. For all who saw him and many who conferred with him every day had never seen him in better health or spirits. He was looking forward to the appearance of Smith and John who were en route from Albany for the inauguration.

On the third of March, Van Buren went over to the White House where he and Chief Justice Taney spent the night with the President. True to the flippant prediction he made a month earlier, March 4, 1837, was to a beautiful day. The handsome phaeton, gleaming in the bright sun of the morning stood ready with Jackson's team of splendid matched gray horses, his coachmen and his footmen. For several days visitors from all over the country had thronged the city to witness the Republic's most impressive public spectacle. Delegations of Cherokee and Pottowatamie chiefs had arrived, some in native costume, feathers, beads and paint; others in sober conventional garb, but attracting attention wherever they went.

Around the Capitol a dense throng had gathered well before the President and President-elect entered the carriage and, preceded by a small detachment of volunteer mounted dragoons, trotted through the crowds along Pennsylvania Avenue to the Capitol. Both men were bare-headed, Jackson with his shock of gray hair, a half head taller than his balding companion. Sweeping through the archway beneath the east front of the Capitol the phaeton came to a halt. The tall, thin President descended first and stood for a moment while awaiting Van Buren and bowed to the crowd which stood silent. From the foot of the stairs, the gray hair of the President's head marked his progress up the wide steps; Van Buren walking beside him was lost to view. They entered the House chamber where they took their places in the front row followed by the cabinet, the Supreme Court, army and navy officers, and the diplomatic corps in their colorful uniforms heavy with gold lace. The new Chief Justice administered the oath to the new Vice-President, Richard Mentor Johnson, who for once was not wearing his red vest, having paid some attention to his dress and personal appearance. After this brief ceremony, Van Buren rose and, accompanied by Jackson and Taney, walked out to the east portico where a wooden platform had been erected. When the multitude below saw the three figures appear, the erect Old General briefly dominated the scene. "A murmur of feeling" swept over the throng as N.P. Willis described it. Jackson bowed, and then he and Taney took their seats on the portico. When all was settled, Van Buren advanced to the edge of the platform and deliberately curbing his habit of rapid speech, read his address clearly and with as much dignity as he could muster.

"The air was elastic and the day still; and it is supposed that near twenty thousand persons heard him from his elevated position distinctly," wrote Willis. "I stood myself on the outer limit of the crowd and though I lost occasionally a sentence . . . his words came clearly to my ear." Van Buren's address, which lasted about half an hour, dwelt upon familiar themes—the great experiment of democratic government, its example to the world, the current state of domestic tranquillity, the federal balance between state and national power—with notable exceptions. Speaking of the founding fathers, he said, "I feel that I belong to a later age and that I may not expect my countrymen to weigh my actions with the same kind and partial hand." He would do his best to follow tradition; but he would respond to the new dynamics of a changing society, and he would expect his actions to be tested in new and untried ways.

For the future, though bright, was not without perils for the Union, and the greatest of these was the contention that existed over slavery. He deemed it his "solemn duty" to announce his policy on this "delicate subject." He reaffirmed what he said in his *Opinions* during the campaign; that he was "the inflexible and uncompromising opponent of every attempt on the part of Congress to abolish slavery in the District of Columbia against the wishes of the slaveholding states, and also with a determination equally decided to resist the slightest interference with it in the states where it exists." If Congress should pass such a bill he announced he would veto it on grounds that were political and sectional rather than constitutional. Without mentioning them by name, Van Buren condemned the abolitionists and then seemed to contradict himself by pledging "a strict adherence to the letter and spirit of the Constitution as it was designed by those who framed it." As he approached his conclusion, Van Buren paid a warm tribute to his predecessor. He could not expect to conduct his administration as successfully or ably as Jackson had done, "but united as I have been in his counsels . . . I may hope that somewhat of the same cheering approbation will be found to attend upon my path." A prayer for guidance and he was done.

Taney rose, moved to his side, and administered the oath. Van Buren kissed the family Bible, and the crowd burst into cheers as the four sons of the new President, ex-President Jackson, Chief Justice Taney and a mass of dignitaries closed around him to offer their congratulations. The cheering stopped as Jackson and Van Buren descended from the Capitol to the phaeton awaiting below. But half-way down the broad granite stairs, the Old General hesitated for a moment and spontaneously a mighty cheer burst from the people gathered below. Jackson bowed and then strode down the remaining steps. A series of orders from the captain of the dragoons and the carriage wheeled slowly south through the multitude to the White House. There, Van Buren officiated at his first reception, which was so crowded that even such an enterprising journalist as Willis could not gain entrance.[17]

The opening weeks of his Presidency Van Buren hoped would give him a brief respite from political and social cares, a short period for reflection during which he could analyze the problems before him comparatively free from interruption. Congress had adjourned, and Jackson left for Tennessee on March 8.[18]

For the first time in a number of years, his immediate family were altogether under one roof. It had been settled that his eldest son, Abraham, would resign his commission from the regular army on March 3. He would take Major Lewis's place as second auditor in the Treasury Department, so that he could act as his father's secretary. Congress did not provide for such services; and Van Buren, as his predecessors had done, used this office for his personal assistant. Martin, Jr., would also stay on helping his brother and copying much of the President's private and official correspondence in his clear hand. At first Van Buren paid for his services out of his own pocket, but later he would receive his wages from the government as a clerk in the land office.

Smith, who disliked Washington, would return with his brother John to Albany. A shy, self-conscious young man, he had not as yet found his way, a cause of considerable concern to his father.[19] John, fast becoming one of the leading lawyers of the state and Van Buren's great hope for the family, would look after Smith who was thinking about studying law. By now, Van Buren was entrusting much of his political and personal affairs at home to this handsome, energetic second son whose political and legal acumen most resembled his own.

Van Buren was still concerned about John's personal life. He drank too much and gambled too much in stocks, horses and elections, though he seems to have avoided that great vice of speculating in lands. And he had not been as discreet as his father would have hoped with his love affairs. Opposition editors had discovered this particular vulnerability; and while John managed to avoid any scandal, reports of his escapades with the frivolous beauties of New York's upper classes emerged briefly from time to time in the Whig papers. Again, like his father, John had developed an enormous capacity for work despite his extensive social life. Van Buren would miss his high-spirited son when he left Washington, and tried to make the most of John's brief stay. He found that he was able to play the family man for only a few weeks.

"The state of the economy," said Van Buren's friend Gorham Worth, was "damnable." In a letter to Butler he described a nation of contrast, "blessed by providence, yet restless and unhappy, possessed of a government of law, yet lawless—free, yet slave to prejudice and party, rich and prosperous and yet on the eve of a general bankruptcy."[20]

Five days after Worth's grim comments, on March 17, 1837, the panic so long threatened, began in the commercial community of New York City with the failure of I. and L. Joseph, one of the nation's largest dealers in domestic exchanges. The immediate cause had been the collapse of the cotton market in New Orleans, but the Josephs had been so extensively involved with the banks, merchants and jobbers, large and small, that their failure had the effect of a seismic incident of such magnitude that aftershocks were felt all along the commercial centers of the seaboard.

Within a month every incoming packet from Europe brought sell orders on bills of exchange and, indeed, on all commercial and financial paper. Specie, which had been commanding a premium for several months after the Bank of England twice raised its discount rate, rose sharply as British and French merchants and bankers liquidated their American holdings. Worried New York City bankers, much to the disgust of the *Evening Post,* called upon Nicholas Biddle, now president of the Pennsylvania-chartered Bank of the United States, to assist them with his still mighty resources. The money man from Philadelphia visited the City, but all he offered at the time were words of confidence, no cash. From then until early May, failures in New York City and elsewhere were almost a daily occurence.[21] The only consolation Van Buren received during this gloomy period was Butler's willingness to stay on indefinitely to help his chief prepare for what both of them knew was an impending crisis.

Fearful bankers and speculators, looking for any means out of their difficulties, saw the Specie Circular as the prime cause for their misfortunes; its repeal might buy them enough time to cover this obligation. Since his inauguration, they had been pressing Van Buren to rescind Jackson's executive order requiring that only specie be accepted for the purchase of Western lands. They argued that it was draining specie out of the East for western land sales and thus placing a prohibitive burden on credit where it was most needed. As the money market grew tighter, banks, especially deposit banks, looked to the Treasury for assistance, which it had little to give.[22]

Conflicting reports were coming to the President. Cambreleng, whose financial ability he respected, was certain that the New York City banks were sound. The respected Flagg, who disagreed with Governor Marcy over his soft money policy, would not retreat one inch. "All classes," he wrote Van Buren, "can only be cured by cupping and bleeding."[23]

It was one thing for an Albany or a New York City or a Baltimore bank to press for canceling the Specie Circular; it was quite another when Senators Rives and Tallmadge added their weight to the pleas.[24] Van Buren had already been advised from various quarters that it would make no difference now if the Circular were canceled. He had, on the other hand, received word from Jackson that the Circular must stand or a new and greater orgy of speculation would follow and with it a deeper, more dangerous collapse of the economy. But after reading particularly dire accounts of the situation in New York City from men whose financial judgments he valued, like Gorham A. Worth, James Lee, a banker and merchant he had known for years, and Campbell White, a Democratic Congressman from the City whom he considered one of the country's soundest businessmen, he began to have his doubts about retaining the Circular.

The action of Rives and Tallmadge, the information he had about Governor Marcy's stand and the split over fiscal policy within the Regency, presaged a political division that could injure the party.[25] Van Buren discussed the matter informally with the cabinet, but found no consensus. After reflecting deeply on the question, he decided he would not submit the question for written opinion. He would shoulder "the entire responsibility" himself.[26] The Specie Circular would remain. He had not come to this conclusion as Mahlon Dickerson thought, simply because he was placating Jackson and Benton, but because he agreed essentially with Flagg that the banks would use its repeal to extinguish their debts in their depreciated paper. Of major concern to the President was the fate of the government deposits. Without this restraint in the deposit system, banks would continue to absorb the government's specie and use it to expand note issue in an effort to pay off their grossly inflated obligations in public lands, other domestic loans and foreign exchange. Already a dozen or so deposit banks had closed their doors defaulting on several million dollars of government funds.[27]

Van Buren's decision was met with consternation among New York's merchants and bankers, and they sought to have him reverse himself. In late April they held a meeting which adopted three resolutions: that the Specie Circular be suspended; that payment on bonds merchants gave for customs duties be postponed; and that a special session of Congress be called for the adoption of emergency measures. The meeting appointed a committee to present these resolutions to the President.

Van Buren suspected the move as a political gesture rather than an economic one and prepared himself accordingly.[28] When Isaac Hone, a relative of Philip Hone and prominent in Whig circles of the City, asked for an interview on behalf of the committee, Van Buren agreed but stipulated that any questions the committee might have be made in writing to avoid any misrepresentation. He would respond in the same fashion. Promptly at 2:00 p.m. on May 4, 1837, Van Buren greeted the committee in the White House. Hone handed him the questions, and after a moment or two of small talk, the committee took its leave.

After glancing over the petition Van Buren called in Butler and Woodbury. The three men went over the document line by line. Its preamble, they immediately noted, was purely political and should be rejected as such. Postponement of payment of bonds for duties, a measure of relief Cambreleng had urgently recommended, Butler thought might be useful. The laws governing customs collections were sufficiently broad to permit the President and the secretary of the treasury some discretion in this area. No one believed the panic warranted the repeal of the Specie Circular or the calling of an extra session of Congress, though Van Buren decided to keep both options open for the time being. "I have not been able to satisfy myself," he wrote the committee, "that I ought under existing circumstances, interfere with the order or call a special session of Congress." On the duty bonds he replied that he was investigating whether he had the power to alter the mode of payment.[29]

No one in the White House or even on the New York committee

was aware just how grave the financial situation had become. Four days after their meeting, however, the Dry Dock Bank, a deposit bank in New York City that held $280,000 of government funds and $90,000 of state funds, closed its doors, precipitating a run on all the City banks. In one day, panicked depositors withdrew well over $2 million in coin. On May 10, 1837, the banks in the City suspended all specie payment, including the savings banks which held the life savings of over 50,000 small depositors.

Immediately, yet another committee was formed. Headed by Senator Tallmadge, it left for Albany on May 11 to have the legislature amend the section of the Safety Fund law that revoked the charter of any bank that did not redeem its paper in specie on demand. A panicked legislature faced with collapse of the banking system, not just in New York City but in Albany and in all the country banks west of the Hudson, passed the necessary legislation with scarcely any debate. Governor Marcy signed the bill into law, though not without some misgivings after Wright, Flagg and Dix had urged him to stand firm. Wright saw the whole thing as a power play from special interest groups, among which he classed the speculators and overtraders. He still thought the economy of the state was sound, the farming community prosperous and largely free of debt.[30]

Following the lead of the New York banks, all the banks in the nation stopped paying specie. The suspension immediately placed the Treasury in a perilous position. That "most unfortunate deposit act of 1836," as Secretary Woodbury called it, prohibited the government from depositing any of its funds in banks which had suspended.[31] Before suspension, Woodbury had transferred government funds from bank to bank in an effort to bolster the faltering deposit system. Now he had no option but to stop all payments, making the Treasury its own depository. In addition, the Department required deposit banks to pay the installments of the government surplus in specie. A sum of over $9 million was due for distribution to the states in October from banks that were practically stripped of their cash. Woodbury and Van Buren, anticipating the harmful impact of this payment, began searching for ways of providing some relief. It was obvious to both gentlemen that distribution had to go; the Deposit Banking Act, now a dead letter, must be repealed.

Since the meeting with the New York committee, Woodbury, the solicitor of the treasury, Virgil Maxcy, and Butler had been examining the laws governing customs collections. Maxcy reported that the secretary of the treasury had some discretion in the matter if an emergency like the recent New York fire forestalled prompt collection of duties or redemption of merchant's bonds. There had also been discussion about whether a special session of Congress should be called. But it was not until the banks suspended that Van Buren and his counselors decided a special session was imperative. They all agreed that it would be called for the first Monday in September.[32]

On May 16, Van Buren issued the call and at various times during the summer sent out letters to those whom he regarded as the best financial minds among the Democrats, posing various kinds of action Congress might take and asking for advice. Enclosed with these letters was a proposal drafted by Dr. John Brockenbrough, president of the Bank of Virginia, calling for a separation of the government from all banks. Brockenbrough's plan, or varieties of it, had been discussed for years. Van Buren saw it as a feasible alternative to the emergency and included a copy of Brockenbrough's plan to those whose advice he sought.[33]

Most recipients agreed with the decision for a special session and accepted the idea of an independent treasury. First thought of as an emergency measure, Van Buren had now come to the conclusion that this method was the only way out of the financial morass that was engulfing the economy and imperiling the public credit. He was well aware of the political risks he was taking and was annoyed when Silas Wright criticized the special session, declaring that it would provide a national forum for the Whigs, who would use it with the aid of Democrats like Tallmadge, to reestablish a national bank. Wright also objected to the independent treasury idea because he feared it would break up the party and defeat Van Buren's reelection. By mid-June, Wright had changed his mind. Much to Van Buren's relief, Wright now supported the Administration's program, for Van Buren counted on him to act as the floor leader during the special session.[34]

Van Buren had foreseen the charges of inconsistency between a new policy of separation and the former policy of involvement with state banks. In his opinion, the Administration had no choice. A national bank was unthinkable; the deposit system had proven unworkable; an independent system had yet to be tried, and while it satisfied bullionist ideologues in the party, like Jackson and Benton, it was not as radical as it appeared. Once the Treasury had righted itself, public expenditures would supply the contracted economy with specie that would be supplemented with a renewal of imported coin as trade picked up. Meanwhile, state governments would assume their traditional role in regulating their own banks, much as the Safety Fund system operated in New York.

Van Buren was counting on the banks and the state legislatures to be more responsible than they had been before the panic; that experience had taught some lessons. When the banks resumed specie payment, customs duties and land sales could be made in their own notes which, according to conservative banking practice, would be issued in quantities of four times the specie in their vaults. Whether Van Buren's acceptance of the independent treasury was appropriate to the crisis, whether it was, given the state of economic thought at the time, sound fiscal policy, it was a courageous decision made in the public interest not for political benefit. The charter election in New York, where the Whigs swamped the Democrats, was a sufficient reminder that an independent treasury would scarcely be a popular measure in the cities where banking influence was concentrated. A safer route politically would be the course Ritchie outlined in the *Enquirer* and Croswell in the *Argus:* reform of the deposit system where public money would be more secure but would still be available to support private credit under treasury supervision. Even the *Globe* for a time thought well of this proposal.[35]

All through the hot, humid summer of 1837, Van Buren, Butler and Woodbury labored in Washington. They were assisted by Van Buren's friend, Henry Gilpin, who succeeded Maxcy as solicitor of the treasury and by the Department's chief clerk, William Gouge, a mild-mannered, hard money enthusiast whose *Short History of Paper Money and Banking* was literally a bible for American bullionists. The small group of intelligent, overworked men prepared five recommendations, all of which were designed to meet the emergency. Only the independent treasury measure could be considered as a policy measure.

By July it was evident that the deposit banks could not make a fourth distribution to the states; and if the government demanded the public funds they held, the entire banking system would collapse. Despite Van Buren's initial misgivings, the deposit banks still remained the soundest financial institutions in the nation, holders of a major portion of the nation's specie and notes and loans of its banks.[36]

Highest priority was assigned to securing repeal of the Deposit Act along with permission for the secretary of the treasury to withdraw government funds over a period of time and in such amounts as the individual banks could sustain. The White House group recognized funds would only trickle in. Likewise, revenue from land sales and from customs was declining rapidly as a depression took hold. The government would fall far short of meeting its obligations, including appropriations the Twenty-third Congress had made to cover the deficit. There were two options open; issuance of bonds on a long-term basis or short-term treasury notes bearing interest but redeemable in one or two years. Van Buren decided upon treasury notes, even though such a trusted advisor as Cambreleng advised against their issuance. To the President, the practical advantages of the notes far outweighed the hard money arguments based on the assumption that treasury warrants would perform the same service without relaxing a needful corrective policy. Circulation had fallen so far so fast that all areas of the economy needed temporary relief. Van Buren reasoned that the notes would increase the money supply while at the same time discharging the government's commitments. Another closely related measure of relief was congressional sanction for a temporary moratorium on the payment of duty bonds by hard-pressed merchants.

As the summer wore on, it became clear to Van Buren that his political power bases, Virginia and New York, would be the prime sources for opposition to his program and to his Administration. Rives and Tallmadge were openly opposed to the independent treasury when it was first floated in the *Globe* during July. Governor Marcy, who had favored a rather mild, as he called it, "fixed system of paper and specie" was cautiously moving away from Van Buren's program.[37]

Politically, Marcy could see no difference on this point between Van Buren and the radical Democrats, the Loco-Focos, in New York. These deadly foes of Tammany, of all banks which they saw as monopolistic, and of the regular Democratic organization in the state and New York City, had increased their following dramatically since the panic. With their slogan of equal rights, their editorial champions, William Cullen Bryant in the *Evening Post* and William Leggett in the *Plain Dealer,* their ideas and their organization were spreading far too rapidly for the governor's orthodox views on party management and business enterprise. Marcy had become so deeply involved with the operations of the Farmers' and Mechanics' Bank that he tended to view the economy as if he occupied a bank president's chair rather than that of the governor.[38]

Van Buren appraised Marcy's dilemma accurately and was certain he would support the Administration, however grudgingly. As for Tallmadge, Flagg gave the best estimate of his course in a letter to Butler where he recalled the sinuous maneuvering of the Senator's cousin, James Tallmadge, Jr., when he betrayed both Van Buren and Clinton. "Tallmadge is Tallmadge," Flagg wrote, "and if he is faithless to honest principle, he is true to his blood."[39]

There was still a possibility, Van Buren thought, of converting Rives and Ritchie even after they had rejected the independent treasury.[40] But the President had to discipline the two most important Administration journals, and for precisely opposite reasons, if he were to head off the threatened divisions in New York and Virginia. The *Globe* had been pursuing its own course since Van Buren's inauguration, and this course was a baffling one. While Reuben Whitney and the deposit banks seemed to be on trial before the panic, the *Globe* spent issue after issue defending them. For several weeks a subscriber to the *Globe* might well imagine that it was Whitney's official paper.

After the crash, Blair seemed bewildered and then recovering, made a complete reversal. The enemy now was not just Biddle, but all banks and particularly the deposit banks. Reuben Whitney's name all but disappeared from its columns. Blair, in his usual biting, slashing manner, raged at banks as a last refuge for scoundrels who would fleece the public with spurious extensions of credit and worthless notes.[41]

If Blair had taken the radical line with such vehemence that he disturbed Van Buren's closest supporters, Croswell's editorials and articles bordered on the soft money views of Tallmadge. Van Buren moved indirectly but positively to bring the *Argus* behind the Administration. Although Croswell was not enthusiastic about the independent treasury, he did not fault it and printed favorable notices of the plan from other papers.[42] Van Buren's personal intervention had the desired effect on the *Globe* and the *Argus;* but he was unable to bring Ritchie over, and he had given up completely on Rives, whose thwarted ambition he felt had twisted his political principles and warped his judgment.

In an effort to escape the torrid heat of Washington, Van Buren and Poinsett rented a pleasant country house from Major Andrews, about three miles from the city.[43] Both men took their work with them, of course; twice a week Van Buren rode on horseback to the White House accompanied sometimes by Poinsett and always by his two sons, Abraham and Martin, who looked after the books, letters, papers and reports loaded into packsaddles. Woodbury and Butler were frequent guests for working sessions, though Van Buren saw to it that there was relaxation too.[44]

Among the visitors at his summer home on August 2 was an engaging man from New York who introduced himself as Thomas Allen, editor of a new Democratic paper about to be established in Washington, the *Madisonian.* Van Buren had read the prospectus of the *Madisonian* in the Washington papers. It supported the Administration in general terms, but the President had learned from various sources, Blair being one, that state banking interests had put up the money to establish it.

Van Buren knew a good deal about Allen and members of his influential Massachusetts family. His father, Jonathan, lived in Pittsfield where Thomas was born. This little manufacturing and farming town was only about twenty miles east of Albany; many of the Yankee families who had moved into the river counties of New York came from the Pittsfield area. Allen's father Jonathan, a friend of Elam Tilden, was a force in Massachusetts' Democratic circles.

Thomas Allen had had some editorial experience writing for the *Family Magazine* in New York. He was also a lawyer who had never practiced, preferring a literary career to the drudgery of the law. He had, in fact, done a good deal of drifting after his graduation from Union College in 1833, but he was a man of considerable talent. Allen held soft money views, though he concealed them in the prospectus for the *Madisonian.* His editorial mission, as he saw it, was to rid the Democratic party of its hard money policy which, he agreed with family friend Tallmadge, would ruin the economy and break down those social standards he had known and cherished.

Van Buren was on guard when Allen complained about Blair's attack on his project. The President said he could take no part in such matters, but he should be "sorry to see another opposition paper started here, or anything to produce divisions in the party." When Allen assured him his "sole purpose was to support the Administration and curb, if possible, editorial ultraism," a direct reference to Blair, Van Buren blandly remarked that "the excesses would cure themselves."[45]

The President had been correct in his suspicions. After a few issues it was clear that the *Madisonian* would not support the independent treasury, which had now become the foundation of the Administration's policies.

The *Madisonian* pronounced the independent treasury to be undemocratic because it would promote two kinds of money, what Rives had called the public money and the people's money. It was simply a scheme to create a treasury bank which would crush state banking and convert the public money into a vast slush fund for executive patronage. Despite his slim journalistic experience, Allen was proving to be a good editor, with a calm, thoughtful style. It was in decided contrast to Gales's and Seaton's lumbering rectitude when dealing with public matters, to Blair's furiously cutting editorials or to the fact-crowded columns of Hezekiah Niles' *Register.* Allen's deft pen was helping significantly to raise Rives's and Tallmadge's faction to third-party status when the special session convened on September 4, 1837.[46]

At Major Andrews's house and at the White House [after the painters and plasterers who were refurbishing the mansion had finished their work] Van Buren and his working group drafted all of the measures that would be introduced by Wright's Committee on Finance in the Senate and Cambreleng's Committee on Ways and Means in the House. By August 26, Van Buren and Butler, with some help from Poinsett and Woodbury, had written the message to the special session. Woodbury was compiling his lengthy, heavily statistical report on the state of the Treasury.[47] Anticipating the argument that the independent treasury would be a source of corruption, Van Buren had Kendall prepare a report on the Post Office that demonstrated how effective his system operated to prevent fraud and insure the safe, economic transfer of public funds.

Criticism of the patronage aspects of the plan was not merely confined to the *Madisonian* or the *Daily National Intelligencer,* which spoke for the opposition, but to sterling Democrats like Taney. In a sharply worded letter to Van Buren, the Chief Justice opposed the entire Administration's program. Obviously unaware of the state of the Treasury or the depth of the depression, Taney would offer little or no relief. And like Rives, he shuddered at the thought of entrusting the public funds to a horde of hungry political appointees. Just why he drew a distinction between state banks, in whose vaults he proposed the deposits would be stored but controlled by public officers, and post offices or customs houses, he never made clear.[48] Fortunately, the President realized the tottering economy could never withstand Taney's harsh remedies any more than it could tolerate a strict adherence to Gouge's bullionist theories or Rives's inflationist ideas.

Polk and Wright arrived several days before the session and worked out their plans with the President for pushing his program through Congress. Neither individual was too hopeful. Wright was very pessimistic about whether the Senate would accept the emergency legislation, let alone the independent treasury bill. While conferring with the President on parliamentary tactics, Polk had to conduct some very delicate maneuvers with the soft money Democrats or "Conservatives" as they were calling themselves. It was indispensable to the Administration that he be elected Speaker because a combination of Conservatives and Whigs made up a majority of the House. Should the Speakership go to John Bell, Polk's perennial opponent, the combined opposition would control all of the committees and destroy the Administration program.

Before Wright reached Washington he had spent several days in New York City countering his colleague's efforts. For the most part, he was successful in keeping the City delegation behind Polk.[49] But the key to holding the Conservatives in line was an offer to sacrifice Blair as printer of the House in favor of Allen and the *Madisonian.* As much as Van Buren distrusted Allen, he was playing for larger stakes than making profits for the *Globe.*

With the deal consummated, the Administration on September 4 won its first major victory. The Conservatives joined with the regular Democrats in the House to elect Polk Speaker by thirteen votes. The

Speaker than proceeded ruthlessly to exclude Conservatives from the important committees, an act that they had not anticipated from a politician who held a justifiable reputation for compromise. The Administration was now prepared to follow through with its bargain. After several ballots, Allen was elected printer for the House. Van Buren had outgeneraled the opposition at what he felt was a minor sacrifice. Neither Blair nor Jackson seem to have understood the tactics or the President's part in it, which, in fact, was known only to Polk, Wright and Cambreleng. Both men attributed the victory of the *Madisonian* to the machinations of their *bête noire,* John Bell.[50]

That durable bureaucrat Asbury Dickens, secretary of the Senate, was conscious of the unique gravity of the occasion when he rose from his chair next to Vice-President Johnson's and began reading the President's message to an attentive Senate. In the nation's history there had been three previous special sessions called, and all had dealt with wartime problems of crucial concern. Only two of the Senators, Webster and Calhoun, who listened to Dickens could recall hearing the last presidential message to a special session.[51] It had been almost twenty-four years since then.

Van Buren's message was a plain and concise description of the financial crisis and its causes which, as he saw them, could be attributed to the glut of paper money and of credit in loans for speculation. There was nothing in it that had not already been anticipated, though few Senators had realized until that moment the serious condition of the Treasury; how emergency measures must be taken or the government could not meet its obligations. Nor would some of them be convinced until they studied Woodbury's bulky report issued at the same time as the President's message. The President's comments were somber ones, as they were meant to be, but they were not without an optimistic strain. He made it clear that the agricultural interests of the country were comparatively unaffected. And in an obvious allusion to the cotton crop he expressed confidence that world demand for it would at an early date reverse the downward spiral in prices. All that was needed, he thought, was extension of temporary relief that he had already provided on his own authority, an issue of treasury notes so that the public credit would not be impaired, and the independent treasury.

It was a long message outlining a distinct program for economic recovery. Dickens required the better part of two hours to read it while the Senators were making notes on their printed versions, and glancing at portions of Woodbury's report. Reaction predictably split along party lines, the only surprise being Calhoun's favorable response when he said, "Nobody can deny its boldness." Before the Secretary had finished his reading, Silas Wright had scribbled a letter to Woodbury asking for a meeting, that day if possible. He expected to be elected to his old post as chairman of the Finance Committee and needed additional information on precise points that were sure to be raised, if not in committee, on the floor. For unlike the House, the Administration had a majority and the committee slate had been worked out beforehand.[52]

It had been clear to Van Buren since his correspondence with Rives, his early knowledge of Tallmadge's activities and the equivocating course of the *Richmond Enquirer,* that he must devise a middle course that would not alienate extreme hard money men like Benton, yet would draw enough Conservative votes to pass the independent treasury bill in the House. The bills Wright brought to the Democratic caucuses reflected Van Buren's concern for the fate of the independent treasury and for an increase in circulation of the money supply, which had fallen by more than 50 percent since the panic.

This crucial bill was silent about whether the government would accept the notes of specie-paying banks for the purchase of public lands, excises and imports. Van Buren had calculated that the omission would hasten specie resumption and would afford a modest relief in discharging the debts owed the Treasury from the deposit banks.[53] Van Buren's parliamentary approach was sound and no doubt would have succeeded had not Calhoun in one of his rare moments of political insight seen through the Administration's strategy and resolved to expose it.

On September 11 Wright, reporting from the Finance Committee of the Senate, introduced the first of the bills that made up Van Buren's program. During the next three days all except a bankruptcy bill went before the Senate. Debate on postponement of the October distribution was made the order of the day; and though Rives attempted to table it temporarily, to Wright's surprise Calhoun came to his rescue, helped defeat the motion and cut short debate. The bill passed on the next day, Calhoun voting with the Administration. Four days later Wright was jolted out of his brief feeling of optimism. The bill for the issuance of treasury notes came up for debate. Calhoun secured the floor and made a short speech in which he declared treasury notes by their very nature could not be debated without reference to banking and treasury operations. He moved postponement until September 18 so that he could have sufficient time to prepare an amendment. Wright and Benton objected but they were unable to head it off.[54]

On Monday Calhoun made a long speech that gave the Administration his terms for support, closing his remarks with his amendment that scuttled both Van Buren's carefully laid plans and the independent treasury. Calhoun's amendment would have the proposed independent treasury accept three-fourths of the government's obligations in specie for the year 1838, one-half in 1839, one-quarter in 1840 and specie only from 1841 on. The President had no choice but to accept Calhoun's plan or alienate the ultra hard money Democrats and Jackson. In accepting he realized, he had lost his bargaining power; and while the independent treasury would pass in the Senate, it would, he was certain, lose in the House where a dozen Conservatives held the balance of power between the Whigs and the Democrats. Having accomplished his purpose, Calhoun agreed to have his amendment removed from the treasury note bill and made a part of the independent treasury bill. Then with a speed that amazed Wright, the Senate proceeded to pass four of the remaining bills in a week's time. The independent treasury bill which, in the meantime, carried not only Calhoun's amendment but an amendment of Benton's defining legal currency as gold and silver only was, as expected, the sticking point. The bill finally passed on October 3, 1837, only to have the House reject it by 13 votes eleven days later, the South Carolina delegation dividing equally.[55]

For once Calhoun had worsted Van Buren in a carefully planned maneuver on what he regarded as one of the most important public issues of his career. A deeply distressed Van Buren could only agree with Jackson, who had viewed Calhoun's support with grave misgivings. "Be careful of Cataline, he may be useful but don't trust him."[56]

« CHAPTER 24 »

Beset by Difficulties

The mood in the White House during the fall of 1837 was gloomy despite the President's never-failing good spirits, his careful attention to the wants of his family, his guests and his colleagues. Abraham Van Buren was restless and dispirited at the sheer volume of letters to be answered, people to see the President, people to be turned away until a meeting could be arranged. Martin, Jr., too, fretted about the flood of private letters and documents he was expected to copy. Van Buren made every effort to lighten their load and have them feel their presence was important to him. He treated them as advisors, shared his plans and his problems with them, sought their counsel over the breakfast table and on those rare occasions when there were no visitors for dinner.[1]

The election news from New York had been shocking. For a time even Van Buren's calm was shaken. The Democrats were completely overwhelmed at the polls. When final results were tabulated in mid-November, only 27 Democrats retained their seats in the lower house of the legislature, while the Whigs held 121. Out of the 8 senate seats up for election the Whigs captured 6.[2] Cambreleng denounced Marcy and Croswell for the catastrophe. Both men had been critical of the independent treasury, but they had kept their remarks private. The roots went much more deeply than Cambreleng realized at the time and spread much more broadly than the lackluster posture of the *Argus* or the arguments within the Regency.

Unemployment in the factory towns and villages, acute commercial and financial distress in the cities, confusion over the Administration's policies, were carefully and skillfully compounded in harsh indictments of the Democracy. Thurlow Weed, William H. Seward, and Francis Granger had finally built an organization which penetrated into every hamlet. They profited also from the Conservatives, whom Tallmadge pulled together in support of the Whig ticket. These were the compelling reasons for "the tornado," as Van Buren

called it, that swept away the strongest, most ably conducted political organization in the country.

After his initial shock, Van Buren analyzed the reasons accurately and regained his accustomed poise. "We have seen such things before and will get over this as we have those which preceded," he wrote a concerned Jackson.[3] If anything, the rebuke in New York strengthened Van Buren's insistence on the government's separation from the monied interests. But politics and financial policy were not the only causes for the pessimism that had settled over the White House. Along the nation's northeastern and southwestern borders, disturbances threatened serious complications in foreign affairs. Two revolutions had broken out in Canada, both of which enlisted the sympathies of the border counties in New York and Vermont. Texas had seceded from Mexico and established an independent republic, which the United States had recognized on the last day of the Jackson Administration.

Apart from Texas, relations between the United States and Mexico were at a low ebb. Over the years since independence the unstable Mexican regimes had seized American ships and confiscated American property. American citizens had several million dollars in claims that preceding administrations had been unable to settle. With feeling in the southwestern states already running high in favor of Texas, its annexation to the United States, and the unresolved claims question, anti-Mexican sentiment could embroil the United States in a war it could ill afford—its economy sunk in depression and its tiny army entangled in the swamps of Florida battling the Seminole Indians.

Nor were the Seminoles the only potentially explosive problem with the Indian policy bequeathed by the former administration. Partially integrated Cherokees of Georgia and Alabama had been slated for removal from their ancestral homes to new lands west of the Mississippi. The government had concluded a treaty with a fraction of the Cherokee chiefs who had been bribed at New Echota, a village in what is now western Georgia, providing for gradual removal to be completed by May 1838. Under the leadership of the able, tough-minded John Ross, a majority of the Cherokees managed on one pretext or another to halt removal, angering Jackson and infuriating Georgia and Alabama whites, many of whom were speculators who coveted their fertile, agricultural lands. But Ross had done more than rally the Cherokee Nation against removal; he had traveled widely in the North, cultivating popular sentiment for the Cherokee.

A successful merchant and planter, the magnetic Ross, with only a trace of Cherokee blood, had managed to unite the Cherokee chiefs in denouncing the New Echota treaty and to convince the Whigs in

Congress that a great injustice was being done to these peaceful, hard-working people to satisfy the land hunger of greedy, shiftless whites and corrupt, speculating land companies. Apart from Ross's public relations and his Fabian tactics, numerous other perplexities beset Van Buren with the Cherokee removal policy.

The Senate had approved the New Echota treaty with only one vote to spare, and it had never been ratified by the masses of the Cherokee people or a majority of their chiefs. Strong religious groups opposed the treaty and the removal of the Cherokees, who were primarily Christians, Methodists and Quakers. Humanitarians everywhere, some even in Georgia and Alabama, protested such treatment of a tribe that had fought loyally against the British and hostile Indians during the War of 1812. Ross, himself, had been an officer with Jackson at the Battle of Horseshoe Bend where the warlike Creeks were completely smashed.

Since the publication of Fenimore Cooper's early *Leather-Stocking Tales,* the Indian as a romantic symbol endowed with unique qualities many Americans cherished had become a nostalgic figure of the past. The prospect of his mistreatment at the hands of unscrupulous whites aroused a sense of guilt in the minds of many.

Van Buren was not unmindful of the image of the heroic Indian brave. Nor was he, as a northerner, a New Yorker and a politician, unaware of the feeling against Indian removal the ministers were stirring up among their congregations in his native state. A member of the Senate during the Adams Administration, he had followed the debate over the government's policy toward the Indians and was familiar in a general way with the various treaties negotiated since the beginning of the Republic. And just now speculators of any type were in bad repute. Many Democrats as well as most Whigs were ready to believe that a horde of unprincipled traders were standing ready to plunder the government and the Indians in the removal process, a message John Ross earnestly propagated.

There was also the matter of money. The treaty of New Echota, fraudulent or otherwise, obligated the government to pay $5 million for their lands. Additional millions would be required for the removal itself, and the Treasury was bare. The treaty had decreed that the Cherokees begin removal by the spring of 1838, now only six months off. Politically it seemed a quagmire for the distracted President. Daniel Webster and Henry Clay were thundering in Congress against removal; Horace Everett, a sturdy Vermonter and even Thomas Hart Benton pointed out that the expulsion of the Cherokees would add more slave territory to the Union. But the policy had been set and Van Buren personally thought compensated separation and resettlement in the far West away from the temptations and the tricks of the white adventurers was not just a practical but a humane solution.[4]

With all of these problems, great and small, Van Buren gave first priority to the finances of the government. Mindful of the criticism hurled at his message to the special session and the frequent intemperate outbursts of the *Globe,* he went out of his way to reassure the financial community that neither his Administration nor the independent treasury plan should be considered hostile to banks.

Banks were creatures of the states that had chartered them, and they served local interests which varied throughout the Union, he declared in his first annual message. It was not merely a state right but a state obligation to regulate the banks even as it was a national right and a national obligation to collect and protect the public funds. Briefly he noted how most of the nation's banking capital was dependent upon the action of state legislatures and would therefore exercise political pressure for private purposes. It would not be in the general welfare again for the public funds to be at the mercy of the political process in the several states where deposits that belonged to all would be used for the private ventures of the few. He favored a complete separation, but in a softening of his former position, he indicated he would go along with a special deposit system where government funds would be entrusted under appropriate safeguards in state banks. As if to emphasize this point, he noted that government funds in the deposit banks and the surpluses distributed to the states amounted to $34 million without considering any appropriations that the Congress must make to carry on severely curtailed operations of the government. Hopeful that the deposit banks which had not gone under would pay off a substantial amount of the debt during the year, Van Buren estimated another issue of treasury notes would have to be made. As it had been doing since the panic, the Administration would continue deficit financing during 1838.[5] For those who were interested in the precise condition of the government's finances, he recommended a careful reading of Woodbury's report, which showed a deficit of over $12 million for the year 1837.

Van Buren's more moderate stance and especially the tone of his message on fiscal affairs did much to soothe the fears of Governor Marcy, though he still thought Van Buren's message to the special session had done infinite mischief to the party in strengthening the Loco-Focos. Again Wright and Cambreleng must carry the independent treasury through Congress. But this time Van Buren decided to make a further concession to moderate Democrats in striking out Calhoun's specie clause from the bill. He would risk the desertion of the Nullifier, or States' Right's party as they were styling themselves, in exchange for what he hoped would buy Conservative votes in the

House and heal the schism in the party. The vote in the Senate was a near thing, 27 to 25, with Calhoun voting against it along with Administration stalwarts Grundy of Tennessee and Buchanan of Pennsylvania, under instructions from their legislatures. After an acrimonious debate, Wright made one of the longest, most carefully argued speeches of his career to crowded galleries; the ruddy-faced Senator, with his husky voice, systematically laid out the reasons why the bill should be passed. Interrupted as he was several times, Wright displayed his usual disarming courtesy and his patience in dealing with questions and comments that were repetitive in nature, or irrelevant to the matter at hand.[6]

The effort of piloting the independent treasury through the Senate for a second time left Wright exhausted and in what he called a "hypochondriacal" state of mind, assuredly not helped by his steady drinking. But the stars in the courses seemed to be striving against Van Buren. Before the bill became the order of the day in the House, the Administration lost five supporters through death and a disputed election in Mississippi. With the specie clause stricken out, Calhoun's contingent would follow their leader. Despite the efforts of Polk and Cambreleng, the bill failed again late in the session.[7]

With the Deposit-Distribution Act inoperative, the independent treasury idea had by now descended to the first circle of political rhetoric. Conceived of originally as a policy that would restrain the excesses of private banks by withholding from them a major portion of the nation's hard currency, it had at least presented an alternative to a third national bank or another round of inflation. Like the Specie Circular, which Congress repealed at the end of the session, neither policy would have any effect on the steeply contracted economy. For an unofficial independent treasury had been operating since the banks suspended; and with public lands a drug on the market, public income from that quarter was but a fraction of what it had been. Yet Van Buren had not given up. With him the independent treasury had become central to his political and economic beliefs, an expression of constitutional purity, of states' rights and wholesome fiscal restraint. He was not an inveterate bullionist like Jackson or Benton; he was a fiscal conservative who believed government control of the specie reserve would eventually make the banker and the state more responsible to their creditors, particularly the farmers of the North and the West and the planters of the South.[8]

As Van Buren watched the varying fortunes of his recommendations to the second session of the Twenty-fifth Congress, he was moving to strengthen his Administration through changes in the executive patronage. The customs collectors in Philadelphia, New York City and Boston were all about to complete their second four-year terms. All had used their offices to build up personal party machines and at least one, Samuel Swartwout, the collector in New York, Van Buren suspected was engaging in corrupt practices, as he had predicted long ago to Jackson. Van Buren kept his own counsel about Swartwout's activities during Jackson's Administration. But he planned to replace the collector when his second four-year term expired in January of the new year.[9]

Boston presented a different series of problems. There, Collector David Henshaw, had made himself the party chief, using customs house funds and patronage to ensure his control. Henshaw, a cunning organizer and a successful businessman, ruled his political domain with an iron hand, ruthlessly stamping out incipient revolts and imposing a close censorship on his operation. A wealthy wholesale druggist when Jackson rewarded him with the collectorship, he had struck a covert alliance with Woodbury and Isaac Hill and developed imperial ambitions. Over the years newspapers, banks, insurance companies and railroads appeared when Henshaw, like Prospero, invoked his magic spells. Otherwise, there was no resemblance to the deposed King of Mantua in the hard-featured, small-mouthed Yankee, whose large lusterless dark eyes missed little in his constant pursuit of power and profit.

Henshaw may have been supreme in Boston, but he had challengers in the countryside. Marcus Morton, heir to the Jeffersonian tradition in the Bay State, the anti-Federalist creed of Samuel Adams and of Elbridge Gerry, represented a constant threat to Henshaw's supremacy. Morton, an educated banker and judge, despised the grasping collector, his blatant opportunism and his hard-fisted contempt for anything that smacked of agrarian idealism. On occasion the rivals joined forces for mutual interest against the Whig majority; though they were nothing more than temporary alliances between Morton's rural following and Henshaw's city machine. As in New York, the anti-Masonic fury swept through Massachusetts and a smaller but articulate Workingmen's party arose in manufacturing towns like Worcester outside of Henshaw's direct influence.[10]

George Bancroft, one of the founders of the famed Round Hill School who had interjected himself into Democratic politics, managed to coordinate both the anti-Masons and the Workingmen, no mean feat for a Harvard-educated teacher and scholar. He had established his credentials as an ardent Jacksonian with an article attacking the Bank of the United States that appeared in the *North American Review*.

Bancroft saw to it that Van Buren read this article, other articles, speeches and the first volume of his *History of the United States,* all of which fused a kind of religious zeal with the secular elements of

Jeffersonian democracy. His rhetoric may have been extravagant, but there was no doubt of its sincerity. Van Buren marked him as a potential leader in his plan to expand the Jacksonian organization throughout New England. Bancroft's ancestry and scholarly eminence had also impressed Van Buren, who never forgot his own humble origins and the gaps in his own education. In a practical sense, the eloquent schoolmaster would help refurbish the intellectual luster of the party Henshaw and his cronies had dimmed with their crass machinations.[11] Not the least of Van Buren's interest in Morton and Bancroft was that they represented a challenge to Woodbury and Hill's growing strength in the region.

His first offer of collector went to Morton, who declined on personal grounds, recommending Bancroft in his place. Van Buren promptly acted on the suggestion and appointed Bancroft to the post. But when the Commonwealth Bank failed and a committee of the Massachusetts legislature uncovered an embarrassing involvement of Henshaw and his ring in its affairs, he advised the new collector to be discreet in his dealings with his predecessor. Henshaw was not to be driven out of the party, and the new collector was not to make wholesale removals.

Van Buren had enough factionalism on his hands without incurring more from a well-meaning but untried intellectual.[12] Bancroft soon proved to be as pragmatic a politician as he was an educated idealist. Forging closer links with Morton, he maintained a relationship with Morton's rival, Robert Rantoul, a Gloucester merchant who was also building up a country following. And he courted the shrewd Benjamin Hallett, the ambitious, not too scrupulous editor of the *Boston Advocate.* As for the customs house, he kept most of Henshaw's old retainers, who, in a rapid about-face, gave him their complete loyalty.[13] A relieved Van Buren turned to the Pennsylvania patronage; here he was not so fortunate in trying to meld the factions into a cohesive organization

After Governor George Wolf had been defeated for reelection, Van Buren appointed him Comptroller of the Currency, a worthy, though humdrum job. Buchanan had protested vigorously that he had not appointed a Pennsylvanian to the cabinet and that, in fact, Wolf was the only representative of the state holding a post in Washington. As a recognition of his claim and an effort to balance the contending factions, Van Buren appointed Henry Muhlenberg minister to Austria. But this arrangement merely angered his rival, George Wolf, who promptly resigned his federal job. The President found a way out of this awkward dilemma. He made Wolf collector of customs for Philadelphia. In exchange, he appointed the present collector, James N. Barker, to the job Wolf had just vacated.[14] Barker was a talented dramatist with a ferocious temper and a sardonic sense of humor, who had alienated many Philadelphia merchants and even some of his closest associates in the Family party. But far from bringing harmony, the Wolf appointment aroused fears of another outbreak between the Dallas and Buchanan factions.[15]

Upsetting as his decision on the collectorship was to the stability of Pennsylvania politics, Van Buren brought down on his head the wrath of both factions with some unfortunate appointments in Pittsburgh. Never popular in Pennsylvania, he had at first managed to calm factionalism with a judicious dispensing of jobs. But now the party seemed caught up again with internecine quarrels.

As he turned his attention to New York patronage, he wanted very much to avoid what he termed "the Pennsylvania difficulties." Uppermost in his mind was preserving the tenuous truce between the Loco-Focos and Tammany that had been in effect since the fall of 1837. His means, of course, was the patronage of the customs house, wielded expeditiously. There would be no criticism from the Hermitage when he did not renew Swartwout's term because that worthy had openly upheld the Conservatives in the New York election, unforgiveable treason in Jackson's lexicon.[16]

Much to the disgust of Wright and Marcy, Van Buren appointed Jesse Hoyt to replace Swartwout. Hoyt had once been on good terms with the ex-Collector, and while he had maintained an honest reputation, he, too, had a taste for the good life. A close friend of John Van Buren, he was a successful lawyer and the only politician of any stature in New York City who had friends among the orthodox of Tammany and the radical Loco-Focos. If anyone could use the customs house patronage to heal the breach between the two factions in the City, Hoyt was the logical man.[17] Van Buren was only too familiar with Hoyt's impetuous nature, and his arrogance. What he counted on most to keep him in check was the presence of Butler in New York and his son John in Albany. An integral part of his policy toward the City patronage was the appointment of Butler as federal district attorney for the Southern District of New York at the expiration of the incumbent's term.[18]

Van Buren harvested immediate political benefits from the Hoyt appointment in the charter election. The Whigs still managed a precarious hold over the common council by one vote and reelected Aaron Clark mayor by 99 votes. Recovery from the previous year was dramatic and heartening; the party in the City for the first time in a half-dozen years was united and well organized. Democrats everywhere were cheered at the results which, in Van Buren's and Butler's opinion, was a sure sign they would carry the state in the fall.[19]

The New York charter election a moral victory, specie resumption among the New York banks imminent, Butler now reminded the President of his promise to accept his resignation when the acute tensions of the panic year had eased off. Reluctantly Van Buren agreed. Dickerson was leaving too, when Congress adjourned, so the President had the unenviable duty of selecting replacements.

For a politician so experienced, so deft and quick in his analyses of men and events, Van Buren was singularly inept at his own cabinet making. Forsyth brought his southern birth and the charm of his personality, but little else to the State Department. Buchanan, who yearned for the post and was now consolidating his position as master of the long plague-ridden party in Pennsylvania, would have lent political and diplomatic strength Forsyth could never match. Poinsett was an able administrator, but he came from a state that had not given Van Buren one vote. Nor would he ever be in a position to challenge Calhoun, either in South Carolina or anywhere else. Woodbury was also a hard-working, competent secretary but it seemed the height of folly to reserve the second position in his cabinet for New England, a region that was considered the very den of Whiggery. Had Woodbury been absolutely loyal to his chief there might have been reason to keep him on; but he was not and Van Buren was well aware of his machinations.[20]

With Butler and Dickerson departing, and Kendall, his health ruined, seeking a diplomatic appointment, a cabinet reshuffle would have been timely as Richard E. Parker, his steadfast supporter in Virginia, reminded Van Buren. The President let his friendship for Poinsett stand in the way of a much stronger political appointment from the South, a Virginian like William H. Roane, Ritchie's cousin, or even George Dromgoole, a leading opponent of the Conservatives in the House.

Van Buren's policy on cabinet posts may be summed up as resistance to any change except what was forced on him by resignation or retirement. Apparently ignoring the fact that he was also representing New York, he decided his native state must have the Navy Department. First he offered the appointment to Jacob Sutherland, a figure from his distant past, little known in New York itself, and a complete political nonentity outside of the state.[21]

It is hard to say why Van Buren offered a cabinet post to such an obscure individual, perhaps on the grounds of personal friendship, more probably for some reason connected with New York politics. Sutherland at least had more sense than the President when he declined, explaining he was unfamiliar with naval affairs, little known to the country, and in no position to offer counsel on affairs of state.

Next Van Buren turned to Washington Irving, who had had some

diplomatic experience and was familiar with Washington society. Irving, of course, was well-known as one of the nation's leading writers, whose wit and charm and patriotic biographies had amused and educated a generation of Americans. But this engaging middle-aged bachelor, an ornament to any dinner party, would have been completely out of place administering the navy, small as it was. Anticipating difficulty in persuading Irving, Van Buren had his son John and Butler both try to convince him. Neither was successful. Irving preferred the comfort of his home at Sunnyside, the freedom and the leisure to continue his literary work.[22] He could have still made a strong appointment from New York, John A. Dix, A. C. Flagg or Michael Hoffman, all of whom were experienced politicians, well-known and well connected in Washington. Such a selection may have briefly upset the balance between radical and moderate factions within the Regency, but there were numerous younger men, like Horatio

Seymour and Samuel Tilden, who would restore harmony while providing new ideas and new approaches.

Yet Van Buren turned to another literary figure, James K. Paulding, who, after hesitating briefly, accepted. Although not as prolific or as creative an author as Irving, Paulding enjoyed a solid reputation as a writer of satire and a novelist. He at least had professional experience with naval affairs, having served as secretary of the Board of Naval Commissioners in Washington and as navy agent in New York for many years. He had also kept himself familiar, as Irving had not, with the political and economic issues of the day.[23]

Van Buren's search for an attorney general was less trying despite the fact that apart from the Treasury, the post was the most demanding in the cabinet. "Butler has made a grand sacrifice to enable him to retain his personal situation," said Richard E. Parker in declining Van Buren's offer, "and I could not expect at my time of life to do as well as he has done at the Bar, nor to enter on a new and unprofitable career." Failing to secure Parker Van Buren profited from the contest raging between Senator Felix Grundy and the Tennessee legislature over its instructions, which Grundy obeyed, to vote against the independent treasury bill in the special session. Grundy was debating with himself whether he should resign after the current session of Congress when he received Van Buren's offer of the attorney general's post. He readily accepted it as the best solution to his present dilemma. A seasoned lawyer and a shrewd politician, the corpulent Grundy enjoyed public life but he could not continue in the Senate, as he explained to a friend in North Carolina, "fettered and bound to vote against my own judgement and inclination" by the Whig-dominated legislature of Tennessee.

Grundy brought real strength to the Administration. An excellent debater and organizer, longtime friend of Jackson and Polk, he led the opposition to the Whigs in Tennessee. A resident of a border slave state and a westerner, his presence helped maintain sectional balance. Tennessee would now have two citizens in high national office. Shortly after his inauguration, Van Buren had appointed John Catron, a veteran member of Jackson's Nashville Committee to the Supreme Court, along with the industrious John McKinley of Alabama, who maintained his connections in the border slave states while representing the Deep South on the Court.[24]

Van Buren had an opportunity to replace Woodbury as he was casting about for an attorney general. The politically treacherous New Englander was offered the chief justiceship of the New Hampshire Supreme Court in late May, 1838. Worn down by his constant labors at the Treasury and with the economy seemingly stabilized, Woodbury longed for the comparative ease the judgeship at home would provide. He would also have the time to work with Hill and Henshaw and other anti-Van Buren Democratic leaders in New England and in New York. Van Buren, horrified at the loss of his experienced secretary and perhaps not less horrified at the thought of Woodbury on the loose in the Northeast, persuaded him to stay on.[25]

Cabinet and patronage affairs settled, Van Buren arranged with Poinsett to make a trip to the Virginia Springs. Although he assured Jackson his health was excellent, he had lost weight and for some time looked worried and pale.[26] Van Buren planned to spend from six to eight weeks, giving as his reason to Jackson that "having been here twenty months without interruption, I feel that I need some change of scene." It would have to be a working vacation, he realized; and accordingly Abraham and Smith were to accompany them while Martin, Jr., remained at the White House as his acting private secretary. He was careful to leave behind him his itinerary, after assuring Martin, Jr., he would communicate promptly any changes which might interfere with his prompt receipt of the mails.

Van Buren's comment that he needed a change was an understatement. Not one of his predecessors had been faced with so many crises, so many problems, in such a short space of time. He had concentrated his energies on domestic problems, but as he started west into Virginia, the carriage that followed him with his sons was crammed with diplomatic and political correspondence. At the moment there was a lull in diplomatic activity, which Van Buren knew was only temporary. Nothing had really been settled along the northeast border with Canada. Diplomatic relations had lapsed between the United States and Mexico. The issue of the annexation of Texas not only threatened a split in the party along sectional lines but also to bring on a war with Mexico.

In the midst of the financial panic eighteen months earlier, news reached Washington of uprisings in both Upper and Lower Canada. The most serious challenge to British authority was led by an intelligent, well-educated, but erratic Scots emigrant, William Lyon MacKenzie. A journalist and a member of the colonial assembly for Upper Canada, MacKenzie was one of those passionate lovers of liberty the romantic spirit of the age so admired. Stubborn, impractical and highly opinionated, he was a poor choice for leading a rebellion against established authority. Yet at first his movement, aided by American sympathizers, scored some slight local successes. In an instant, it seemed the whole frontier was aflame at least with gasconade on both sides of the wilderness borders of New York and northern Vermont.[27] Van Buren quickly recognized the inherent seriousness of the rebellion and in early December 1837 wrote letters to the governors of New York and Vermont requesting they preserve a strict neutrality along the frontiers of their states. Instructions also went out to federal marshals and district attorneys in the area to investigate and take appropriate action against any overt action of American citizens. Both governors issued proclamations ordering a strict neutrality but did little more. Marcy, ill and dispirited at the time, thought the excitement would wear itself out. He took no extraordinary measures to focus on the real danger point where the Niagara River separated Upper Canada from the United States near Buffalo.[28] There MacKenzie and his followers, after being defeated by British regulars and Canadian loyalist militia, had fled to regroup on the American side of the border. American sympathizers swelled MacKenzie's ranks, headed by a young popinjay from a powerful New York family, Rensselaer Van Rensselaer.[29] Deeply involved at the time in devising a program for rescuing and eventually restoring the economy, Van Buren left the problem of dealing with the Canadian disturbances to the state authorities.

Late in the afternoon of January 4, 1838, Van Buren was jolted out of his complacency. Reports had reached him that hostilities had broken out along the Niagara River. Rapidly scanning the New York newspapers and other reports, he learned that a British force had boarded a small rebel steamer, the *Caroline,* which was berthed on the American side of the river. In an exchange of fire, Americans had been killed and wounded. After the *Caroline* was taken, the raiding party set it on fire and sent it drifting ablaze over the Falls. The newspaper accounts, most of which grossly exaggerated the casualties, still left no doubt in the President's mind that a serious incident had occurred. This raid betokened an international episode beyond the capacity or the jurisdiction of Governor Marcy.

Van Buren had planned a dinner party that evening, a rather large affair, with twenty-four guests including the Whig leadership in Congress headed by Henry Clay and some Democrats. They would have to wait; for he decided he must call an emergency cabinet meeting. There is no record of the discussion but certain facts are obvious. At all costs, the President wanted to avoid any escalation that would lead to war between England and the United States. The country was completely unprepared for any major conflict, its 8,000-man regular army thousands of miles away in Florida, on the Texas frontier and scattered in various posts along the western frontier from Louisiana to northern Michigan, guarding settlers from Indian forays and the tribes from white squatters. The army's organization was antiquated, artillery and munitions woefully scant, the result of almost twenty-five years of congressional parsimony and indifference. Similarly the navy was tiny, its vessels for the most part obsolete, its officer corps divided up into quarreling cliques.[30]

From the information the cabinet had, the local militia was undependable, most of them sympathetic with the rebels. If they should be called out, there was every possibility they would not respond to military discipline. A strained situation might well become uncontrollable. The cabinet decided that some reliable armed forces from other more distant counties must be called up and posted near Navy Island in the Niagara River where young Van Rensselaer was commanding a motley array of about 300 armed men. The cabinet also agreed with Van Buren's recommendation that he issue a proclamation of neutrality announcing to the nation and to Great Britain the policy of the government.

Beyond powers to call out the militia, to commit the Administration to strict neutrality, to emphasize the nation's peaceful intent and arrest any armed citizens who attempted crossing the border, Van Buren had no additional powers with which he might enforce neutrality. Another decision, made at that tense cabinet meeting was a request that Congress give the President more latitude in dealing with this or any subsequent crisis in foreign affairs, and an emergency fund if needed.

After the meeting adjourned, Van Buren decided that the federal authority on the frontier must be made more conspicuous in the eyes of the would-be invaders, Canadian refugee insurgents and the British forces facing them. The logical man for the job was General Winfield Scott, who happened to be one of the dinner guests waiting below for his appearance. Van Buren had known Scott since 1814 and was familiar with his strengths and weaknesses. He judged the General to be the best high-ranking officer in the army, with a firm grasp of tactics, strategy and leadership; and he had other useful qualities for dealing with just such a crisis. Trained as a lawyer, Scott had sharpened his powers of natural eloquence throughout his military career. A politician, albeit of the opposition, he appreciated the art of persuasion and compromise, precisely those qualities Van Buren thought were needed in mobilizing the peaceful elements of the community to help in restoring order.

When Van Buren finally appeared at his dinner party, he was his usual calm, genial self as he shook hands with his guests. But before dinner was served he drew Scott and Clay aside and in a low tone recounted what he knew of events on the Canadian border. "Blood has been shed," he said. Then turning to the massive Scott, he urged, "You must go with all speed to the Niagara frontier. The Secretary of War is now engaged in writing your instructions." Both Van Buren and Poinsett knew Scott was vain and difficult to handle, and they drew up his instructions accordingly, giving him, as he put it, *"carte blanche,"* in settling the affair, the President emphasizing only that he must preserve the peace.

Scott left the next day and when he reached Albany induced Governor Marcy and the state's adjutant general to follow him as soon as they were able.[31] En route he collected what regular troops he could find, a company or two of recruits, and ordered them to Buffalo. A whirlwind of activity, Scott directed the frontier commanders under the President's authority to use every means at their disposal to prevent raids into Canadian territory, seek the cooperation of community leaders in dampening the public excitement and respond to any calls for posses from federal marshals, attorneys and collectors. Not the least of his efforts was the prompt opening of communications with the British.

In full dress uniform and cocked hat draped with yellow plumes, Scott enlisted leading citizens of Buffalo in a concerted effort to preserve order. Nor did he neglect frontier villages where he counseled restraint in barrooms, in stump speeches and in churches. Exerting all the force of his impressive personality, Scott lectured young Van Rensselaer on the hopelessness of his position on Navy Island. The General's reputation, his appearance and the authority of his arguments convinced Van Rensselaer to withdraw until he had gathered more men and supplies. Van Rensselaer and MacKenzie were soon arrested; volunteer posses collected most of the arms the "Patriots" had confiscated from state arsenals. The situation rapidly returned to normal.[32]

Scott was preparing to leave Buffalo when he learned that another "Patriot" incursion was planned from Ohio. The doughty Scott rushed to the Michigan-Ohio frontier and with a detachment of regulars stationed at Detroit pacified the region. "Hunter's Lodges" and "Patriot" societies would continue plotting and planning; but Van

Buren's forthright policy of neutrality, the energetic and forceful role of Scott, the assistance the federal government received from the governors of New York and Vermont, and the obvious loyalty to the Crown of an overwhelming majority in Upper Canada brought a temporary peace to the frontier.[33]

While Scott was pacifying the American side of the Niagara River, Van Buren was directing a political and diplomatic initiative to maintain good relations with Great Britain but not make it appear his Administration was indifferent to the cause of liberty and independence. In a special message to Congress issued three days after his Proclamation of Neutrality, the President, briefly commenting on the *Caroline* incident, sent in some of the documents he had received from federal officers that described the event in graphic detail. He included a note from Forsyth to British Minister Henry Fox informing him that redress would be demanded of his government for the raid.[34]

Despite this show of vigor, tempered with the statement of neutrality, Congress paid little attention to the frontier disturbance. Enmeshed in the debate over the independent treasury bill, distracted by Calhoun's proslavery resolutions and their sectional implications, border forays in the wilderness of the northern frontier could not command much attention.[35]

Yet Van Buren had to lodge an official protest, and for domestic consumption he must make it as strong as possible. He had Forsyth instruct Stevenson in London to demand indemnity for the *Caroline* incident and regrets at the breach of American sovereignty. Lest the belligerent tone of these representations be misunderstood at Whitehall, Van Buren intrusted his son John with a private conciliatory message to Britain's foreign secretary. In his note, Van Buren took pains to invest John with unofficial diplomatic powers. "My son is well advised of public feeling and the views of public members upon most questions of interest between us," he wrote, "and disposed to speak of what he knows and believes with candor and freedom." The younger Van Buren was planning a trip to London on legal business, so his departure at this time would not be the subject of political speculation in the Whig press.[36]

John Van Buren delivered his father's note a little over a month after Stevenson had lodged his formal protest and at a series of social affairs spoke with Palmerston, Lord Melbourne, the Prime Minister and the Duke of Wellington, head of the Tory opposition. The younger Van Buren, more at home than his father had been in the glittering, gossipy atmosphere of Holland House or the balls and dinners of the London social season, made a fine impression everywhere. The continuous entertainment, where the British upper class went out of its way to entertain the President's son, eventually wore down his iron constitution and was, of course, the subject of all sorts of adverse publicity in the Whig press at home. But the President's note and John Van Buren's personal diplomacy had the desired effect.[37]

At Palmerston's urging, the British government made no formal reply; nor did Stevenson press for one because another series of border forays erupted in the fall of 1838. Minister Fox, exaggerated the situation when he wrote his government that a conspiracy existed involving 40,000 Americans poised to invade Canada from Maine as far west as Michigan.[38] While Van Buren was preparing to react with his expanded powers, which Congress had reluctantly given him under a revised Neutrality Act, news reached Washington that Canadians from the Province of New Brunswick had moved into territory the State of Maine claimed and were removing valuable timber. The governor of Maine, John Fairfield, and the lieutenant governor of New Brunswick, Sir John Harvey, both adopted a hostile stance.[39]

Van Buren had repeatedly urged Maine authorities to have this long-standing border dispute settled by convention between Britain and the United States. Former Governor Kent, a Whig, and a Whig-dominated legislature had obstinately refused. In the fall of 1838 the ardent Jacksonian, John Fairfield, had been elected governor. Van Buren was fond of Fairfield, a plain, unassuming lawyer and farmer with a good sense of humor. Surely, under Fairfield's leadership, the legislature would not mix Whig politics with foreign policy. A convention was now a real possibility.

But the President misjudged the temper of the state, the attitude of the new governor and the extent to which the boundary dispute had become *the* political issue between the two parties in Maine. John Fairfield may have been an unassuming man who darned his own pantaloons and took a room on the top floor of his boarding house in Washington to save a few dollars a week in rent, while prudently dangling a stout rope from his window as a home-made fire escape; but his rustic manners and his seemingly pacific disposition concealed the temper of a romantic patriot in the Jackson mold, a politician who would, if possible, be a more forceful defender of Maine's interests than the opposition.[40] One of Fairfield's first acts after his inauguration was the dispatch of an agent accompanied by a small group of Maine men familiar with the region to search for and expel any trespassers from New Brunswick. His agent and guides were arrested and imprisoned under the direct orders of Sir John Harvey.

The party was released, but Harvey, pressed by New Brunswick lumbering interests, decided to clear the Americans out of the entire area. To that end he issued a warlike proclamation ordering sufficient military force to repel what he called "a foreign invasion."[41] Fairfield received the proclamation and a note from Harvey justifying his stand on February 15, 1839. The documents aroused all of the Governor's latent nationalism and his devotion to his native state. He had copies made, sent one to the legislature together with a message asking for authority to call up 4,000 militia and an appropriation for the defense of Maine. The other copy together with a note he forwarded to Van Buren with a request for assistance. Swept by a storm of patriotic ebullience, the Maine legislature appropriated $800,000 for military expenditures and approved Fairfield's call-up of the militia. With thousands of armed men mobilizing, this was no border frolic of adventurers, no intrusion of bottle-brandishing drifters, but a grave confrontation of two sovereignties. War was in the air.

When Van Buren received Fairfield's request for aid and the Harvey proclamation, he was deeply concerned at the belligerent pronouncements of the two governors. After reading Fairfield's reply to Harvey he was surprised at the fighting spirit of his seemingly mild-mannered friend who had shown so little concern over the border incidents of previous years. The Congress and the President were bound upon application of a state to protect it against invasion, insurrection or domestic violence. Now that the call had been made, he had no choice but to ask Congress for the means to help Governor Fairfield. Yet he was determined as far as it was humanly possible to avoid war. He had every reason to believe that the Melbourne ministry felt similarly. They had recalled the feckless governor of Upper Canada, Sir Francis Bond Head, and had replaced him with Lord Durham.[42] Forceful but conciliatory in appealing for support, Van Buren left the door open for a peaceful solution to the crisis.

Concurrently, he ordered Scott to Washington. The General received the President's message after just completing in winter weather a thousand-mile trip from Detroit east along the northern frontier, again enlisting community leaders to head off the irresponsible "Hunters" and "Patriots." Evidently indifferent to hardship, Scott set off at once.[43]

Van Buren was anxious that the Administration's program for strengthening the military not be caught up and delayed by partisan debate. Appealing to Scott's self-importance and his sense of duty, he enlisted him as a sort of an Administration lobbyist with the Whig members of the congressional Committees on Military and Foreign affairs. The President's adroit move was eminently successful. Bills that extended the militia service from three to six months gave the President authority to call out up to 50,000 men and appropriate

$10 million for war expenditures if these became necessary passed a politically hostile Congress in record time.[44] No sooner had the necessary legislation reached the President's desk than Scott was off again, this time to Augusta, the capital of Maine. Again Van Buren counted heavily on Scott's reputation and his political affiliation for negotiation with both sides. As a Whig with Democratic credentials, he was in the best possible position to assure Governor Fairfield that the opposition would not make political capital out of what might appear a sacrifice of Maine's interests. Scott would, of course, forestall any efforts on the part of his own party in Maine to goad the Democrats into a possible war for the benefit of the next election.

After the General left for the north, Van Buren, acting through Forsyth, opened a negotiation with the proverbially nervous British Minister Fox. The President's strategy now became apparent. If he could enlist Fox in a joint declaration where neither side made any concessions, but the British forces would withdraw from the Aroostook River where they were facing the Maine militia, then the Maine militia could also pull back without giving up any of its claims to the disputed territory. Although Fox and Forsyth disclaimed any "specific" authority in what amounted to a recommendation, they both "earnestly" urged Maine and New Brunswick to withdraw and await a territorial settlement between the United States and Great Britain. Fairfield and the Maine legislature greeted the memorandum with scorn, as did General Scott when he read it after reaching Augusta.[45]

Neither Scott nor the Maine authorities understood that the memorandum, coupled with warlike preparations in Washington and Augusta, gave Sir John Harvey a clear view of American policy, which combined force with persuasion and peaceful intent. It also permitted him to shift a portion of the responsibility to the shoulders of Fox.

When Scott opened secret communications with Harvey and found him receptive to a withdrawal of British troops, he attributed the conciliatory attitude to a friendship he had struck up with the Royal Governor during the War of 1812, never dreaming that the memorandum he had spurned provided just the right diplomatic climate for an easing of tensions, at least on the New Brunswick side. With Harvey's willingness to withdraw if Maine did likewise, Scott was able to convince Fairfield he would not lose face if he demobilized after the New Brunswick forces left the disputed territory. Scott had kept Fairfield's name out of the negotiations with Harvey; and now he worked skillfully to bring the Whig opposition behind a temporary settlement where neither side gave up any claim, but both pledged to keep the peace until a final disposition could be made between the United States and Great Britain.[46]

Van Buren had been at his best in unraveling a complicated and serious situation without disturbing the political equipose in Maine or involving an unprepared nation in the throes of a deep depression with a war which, like that of 1812, would command only sectional support. He had done what Edward Everett had thought insoluble: set the stage for an eventual compromise solution. Scott must be praised as the man on the spot who handled the diplomatic and political aspects of the affair with great skill. But it was the President, after all, who conceived the overall plan and who selected Scott for the détente mission which he performed so well.[47] Nor did Van Buren cease his efforts to obtain a permanent settlement, not just of the Maine boundary but of the entire northern boundary extending to Michigan. And he personally directed efforts to obtain reparations and an apology for the seizure of the *Caroline* while in American waters and pardons for American prisoners convicted or awaiting trial for their participation in the border disturbances.[48]

The disturbances on the northern border had been worrisome and at times, as in the Aroostook confrontation, decidedly dangerous to the maintenance of peace. But from the beginning, Van Buren had counted on his personal rapport with Palmerston and careful cultivation of Henry Fox to keep relations between the United States and Great Britain from serious deterioration. His official policy insofar as he had the authority to act had been one of conciliation, yet at the same time one of prodding Congress into giving the executive more latitude and more resources in case the situation got out of hand. Dispatches from Minister Stevenson, private notes from Palmerston and letters from his son John confirmed his belief that the British government understood this policy and would cooperate as far as it was able, paying due regard to political constraints at home and in Canada.

But the difficulties along the southwestern borders of the Union, involving as they did a successful revolt of American emigrants against Mexico, threatened graver political consequences within the nation. The emergence of the Republic of Texas complicated relations between the United States and Mexico which were already in a perilous state over long-standing claims of American citizens against a series of unstable Mexican governments.

Van Buren would capitalize on Jackson's reputation and indeed the reputation of the country for strong measures. Fortunately, the new Secretary of War, Joel Poinsett was the best-informed man in the country on Mexican affairs. Together, the President and his Secretary formulated a policy that coupled peace and harmony on the one hand with a firm demand that the American claims be settled on the other.[49]

Emphasizing the Administration's firmness, Van Buren selected Robert Greenhow, a State Department clerk, as a special envoy with instructions to deliver a formal note to the Mexican minister of foreign affairs, vigorously setting forth the wrongs done American citizens since Mexico gained its independence. The dispatch cited some of the more flagrant incidents and complained of "unreasonable delay" in settling these claims which the Mexican government had acknowledged to be negotiable. Fifty-seven additional claims were attached to the note, but in conclusion Van Buren and Forsyth merely asked for "an earnest" of the Mexican government's intentions to make reparations; and Greenhow was explicitly instructed to tell the Mexican foreign minister "that the U.S. required an examination and liquidation of claims, but did not demand their immediate payment."

On the recognition of Texas, which Mexico had protested before breaking diplomatic relations, Greenhow was to assure the Bustamente regime that the United States would follow a course of strict neutrality. Though the dispatch was couched in terms of opening negotiations for settlement, the American government declared how seriously it viewed the matter and how determined it was not to accept further procrastination. Greenhow was instructed to present this note to the foreign minister, wait one week and depart whether he received a reply or not. He was also the bearer of a friendly private note from Poinsett to Mexico's President Anastasio Bustamente.[50]

The Mexican government again accepted responsibility for the claims but asked for the usual delay. Van Buren, who anticipated this action, rejoiced at Bustamente's affirmative posture and the resumption of diplomatic relations on the part of Mexico. Prepared as he was for a long negotiation, he was not too disconcerted when the new Mexican minister to the United States, Francisco Martinez, not only arrived in Washington several months after he had been expected, but then waited until just two weeks before the convening of Congress to present the Mexican position on the claims.

Van Buren must have been nettled however when Martinez's note alluded to eight claims, not fifty-seven, only three of which were on the State Department's list. Taking into account the chaotic conditions in Mexico at the time, the President did not wish to weaken the shakey Bustamente regime. Instead of sending a special message with documents to Congress, Van Buren simply referred to the claims in his second annual message, although he was forthright in his denunciation of Mexico's delaying tactics.[51]

In those precious hours the President could spare on the issue he mulled over the question of annexation. He saw clearly how incorporation of Texas would strain a Union distracted with severe financial problems and with sectional divisions from mounting agitation over slavery. Though most northern Congressmen deplored abolition, they were not impressed at the arguments some of their southern colleagues used in picturing slavery as a blessing for the benighted blacks.[52] When President-elect Van Buren had used his influence with Jackson to shift the entire question of recognition and annexation to Congress. His lieutenants succeeded in delaying it until near the end of the session despite intense lobbying of the Texas agents and speculators in Texas lands like Samuel Swartwout, the corrupt collector in New York.[53]

Jackson had taken this opportunity to declare that the sense of Congress favored recognition.[54] On March 3, 1837, he nominated Alceé La Branche of Louisiana as chargé d'affaires to the Republic of Texas. The first and most crucial step had been taken. The United States was now involved in Texas affairs, and annexation would be but a matter of time, or so thought many contemporary observers, men like James Kirke Paulding, Calhoun and Webster.[55] Van Buren began his administration with a policy on Texas which, if he had reluctantly acquiesced in, he certainly did not mean to implement further with annexation.

The Mexican government promptly denounced recognition and explicitly declared it would fight if the United States absorbed what it regarded as still a province of Mexico. However Van Buren might have felt initially about recognition, he was not opposed to territorial expansion in the Southwest. In his cabinet, his trusted advisor on Latin American affairs, Joel Poinsett, was an out-and-out advocate of annexation. Kendall, too, added his weight to the pro-Texas faction. Forsyth, though a southerner, had no fixed ideas on the subject and simply followed Van Buren's lead. But Woodbury, and indeed the entire Administration for that matter, soon became so overwhelmed by the panic and planning for the emergency session of Congress that the Texas issue was thrust aside. Because Congress was not in session and the new Texan minister, Memucan Hunt, was absent in Mississippi, there was a period of calm. By mid-July, however, the calm was shattered when Hunt, brimming with energy, arrived in Washington.

A North Carolinian by birth, a Mississippi planter by occupation, and a Texan by opportunity, Hunt had been one of the horde of adventurers who poured into Texas after Houston's victory at San Jacinto. Typical of the speculators who struck it rich in the fertile delta bottom lands of the Mississippi, he was hopeful of gaining fame and adding to his fortune now that Texas was independent. Unlike his biblical namesake, he was neither wise nor prudent. Rather, he was an arrogant, pretentious man, impatient with the ways of diplomacy, counting on a quick resolution in favor of Texas which would bring him personal acclaim and foster his political ambitions.

On July 6, 1837, precisely at 12 noon, Minister Hunt accompanied by Aaron Dayton, acting secretary of state during Forsyth's absence, arrived at Van Buren's office in the White House. Dayton presented him to the President, and they shook hands. The new envoy then handed him a letter from President Houston and made a short speech in which he hoped "nearer relations than those of harmonious diplomatic intercourse will ere long exist." Van Buren replied with his customary urbanity and, while seeming to encourage Hunt, really said nothing of substance.[56]

For the next month and a half Hunt cooled his heels at the State Department. Forsyth had returned from Georgia but was unresponsive to Hunt's queries about annexation. The Texas minister played every card he had with the soft-spoken Georgian, warning him that if the United States did not move for annexation, Great Britain or France would seize the initiative. Texas agents were already in touch with both governments. Neither threats like this nor the advantages Hunt pointed out made much of an impression on Van Buren. But when the Texas minister filed a twenty-page formal note with the State Department, the President did call a cabinet meeting to consider it. Hunt's note, like his unofficial talks with Forsyth, was a mixture of fact, chronicle, bravado, misinformation and vague warnings that Texas would seek treaties with foreign powers if annexation was not promptly recommended to Congress.

Van Buren may have been partial to eventual annexation, paying close heed to possible constitutional inhibitions and Mexican reaction, but he was preoccupied with efforts to stave off what he feared might be a collapse of the economy. He had developed a program he thought would best preserve the public credit and check the rise of another inflationary spiral. It would be difficult enough to carry out this program against the opposition of the Whigs and anticipated defections in his own party without raising the divisive question of Texas and its dependent issues of slavery and abolition. The reply to Hunt's note was purposely made as harsh and uncompromising as possible, a forcible reminder that the nation faced higher priorities than stirring up a controversy over a remote frontier territory.

Memucan Hunt could not have acted at a worse time and in a manner least calculated to influence such a deliberative mind as Van Buren's. When Calhoun joined forces with the Administration in backing its financial policies, Texas' strongest advocate in Congress remained silent. Talk of annexation was muted among the politicians in Washington and in the partisan press of the country, but

Van Buren and Forsyth were careful not to dash Texan hopes. "If matters were properly conducted at home" Forsyth told Hunt in mid-November, a month after the special session had adjourned, he had no doubt the annexation would be accomplished.[57]

When the second session of the Twenty-fifth Congress assembled in December, both Hunt and another agent, P. W. Grayson, taking note of petitions pouring into Congress against annexation, reported that there was, in Grayson's words, "no solid foundation on which to build a hope that the measure can be carried," though both men believed the Administration to be favorable. Hunt neatly summed up the Government's position when he wrote Dr. R. A. Irion, Texas's Secretary of State:

> Hampered as they are by their party trammels in the one hand and their treaty obligations with Mexico on the other, by furious opposition of all the free states, by the fear of false dealings and injustice, and of involving this country in a war, in which they are now doubtful whether they would be supported by a majority of their citizens, they dare not and will not come out openly for the measure.[58]

And there the matter stood and would stand for the next four years.

« CHAPTER 25 »

"Little Van Is a Used-up Man"

The monumental debate over the independent treasury system, now in its second year, was hurting the President politically. Were the presumed powers of the Magician failing him? This was a question most frequently asked in the lobbies of Congress, the state house in Albany and in other centers where politicians gathered. It would seem so after the Conservative-Whig coalition in the second session of the Twenty-fifth Congress defeated for the third time the independent treasury bill.[1] The Whig party was the prime beneficiary of Van Buren's stubborn insistence on the measure, a fact not lost on the White House. Yet the President did not despair. He was satisfied that the system was essential for the public good, and further that it was an economic expression of Jeffersonian ideals as opposed to the Hamiltonian concept of using the power of the national government for enhancing private means. If the debate had been costly, it had made the political issue clear-cut, expressing as it did Van Buren's definition of the democratic process. As President and party leader he would make the independent treasury or subtreasury a canon of the Democracy, an expression of its cohesiveness, that would force the opposition to develop and clarify its own position.

It would, he hoped, consolidate his own party. Should his undeviating course cement together a yet fragmented opposition into a national party, this was a risk he was willing to take. The ablest political strategist of his generation, he was willing to place himself and his party in jeopardy for an idea that he believed with all his mind and heart expressed the genius of democratic political institutions.

But he was optimistic about the future. In May 1838 the New York banks, aided by the Bank of England, resumed specie payments, shortly followed by those of Philadelphia and other business centers

William Leggett, crusading co-editor of the New York *Evening Post.* Van Buren secretly admired him.

Thurlow Weed, chief mouthpiece and political organizer of the opposition to Van Buren in New York.

Felix Grundy, U.S. Senator from Tennessee, one of Van Buren's attorneys general. Fat, self-indulgent, politically astute.

Angelica Singleton Van Buren, the President's daughter-in-law and White House hostess. (The Library of Congress)

Van Buren runs for President. (Courtesy of The New-York Historical Society, New York City)

President Van Buren. (The Library of Congress)

Richard Mentor Johnson, controversial Vice President of the United States. (The Library of Congress)

Mrs. Henry D. Gilpin, in whose hospitable Philadelphia home Van Buren spent many pleasant hours. (The Historical Society of Pennsylvania)

Henry D. Gilpin, Philadelphia bibliophile and Attorney General in the Van Buren Administration. (The Historical Society of Pennsylvania)

U.S. Senator Silas Wright, Jr.

John A. Dix, the eloquent Senator from New York.

Henry Clay, toward the end of his public career. (The Library of Congress)

Roger B. Taney, a staunch Van Buren supporter in the Jackson Cabinet. (The Library of Congress)

Daniel Webster, a political opponent whose character Van Buren found wanting.

John C. Calhoun, unforgiving defender of Southern interests. (The Library of Congress)

Thomas Hart Benton, bitter enemy of Calhoun. (The Library of Congress)

Benjamin F. Butler, Van Buren's law partner, Attorney General under Jackson and Van Buren. (Print Collection, National Portrait Gallery, Smithsonian Institution)

James K. Paulding, New York wit and novelist, Van Buren's Secretary of the Navy. (The Library of Congress)

Levi Woodbury, Secretary of the Treasury under Jackson and Van Buren. Capable, secretive, ambitious. (The Library of Congress)

POLITICAL JUGGLERS LOSING THEIR BALANCE

Whig cartoon (1840 campaign) attacking the Van Buren Administration. (Courtesy of The New-York Historical Society)

John Forsyth, a charming, loyal but mediocre Secretary of State. (The Library of Congress)

Joel R. Poinsett, Van Buren's able Secretary of War. (The Library of Congress)

John Van Buren as a leader of the New York bar. (The Library of Congress)

James K. Polk, successful Democratic nominee in 1844 and President of the United States from 1845 to 1849. (The Library of Congress)

Whig cartoon assailing Van Buren's stand on public policy before the Democratic nomination of 1844. (Courtesy of The New-York Historical Society, New York City)

Cave Johnson, Polk's friend and confidant from Tennessee; Postmaster General in his Administration.

William L. Marcy, friend and foe of Van Buren in New York politics. (The Library of Congress)

George Bancroft. The Barnburners considered him disloyal. (Naval Photographic Center, Naval Station, Washington, D.C.)

Edwin Croswell, not so steadfast editor of the Albany *Argus*. (The Library of Congress)

Democratic cartoon attacking Van Buren and the Free-Soil party in the campaign of 1848. (Courtesy of The New-York Historical Society, New York City)

Lewis Cass, Democratic nominee for President in 1848.

John Van Buren, as he looked during the Free Soil campaign. (The Library of Congress)

Governor William C. Bouck, New York Hunker leader.

Daniel S. Dickinson, U.S. Senator, violent party opponent of Van Buren. (Print Collection, National Portrait Gallery, Smithsonian Institution)

Last photograph of Andrew Jackson. He died five weeks later.

Van Buren in old age. (The Library of Congress)

throughout the country. Renewed confidence had its effect on the state elections where the Democrats experienced a resurgence of strength except in Van Buren's home state of New York. Despite improving economic conditions, bankers, merchants and manufacturers gave money freely to Whig and Conservative leaders. The major issue was Van Buren himself and his subtreasury system, even though it was obvious that economic conditions were improving without the private use of government deposits.

Governor Marcy, a candidate for reelection, had considerable difficulty reconciling his previous easy money policies with the hard money line of the Administration. His close connection with the Farmers' and Mechanics' Bank in Albany and its credit relationship with the banks of western New York made him especially vulnerable to Whig attacks of inconsistency. The *Argus* defended the Governor and the Van Buren Administration, highlighting their agreement on basic political principles and denying any differences over fiscal policy. Croswell's editorials, however, lacked their old vehemence, their sense of indisputable purpose. And too often they dodged Weed's penetrating thrusts in the *Evening Journal.* In New York City, Bryant's *Evening Post* could find little to praise in Marcy and much to commend in Van Buren. The Democratic press reflected the tensions between Washington and Albany.[2] The Whigs themselves, with a confidence springing from their landslide victory the previous year and well supplied with funds, had finally developed an organization that rivaled the Regency's in its heyday. They nominated a loquacious young lawyer and veteran state senator, William H. Seward, as their candidate and could count on the support of the Conservatives, whom N. P. Tallmadge had effectively consolidated.

Preoccupied with diplomatic problems, the possibility of war with Great Britain over the northern boundary, the continuing difficulties with Mexico, saddened at the sudden deaths of his sister, Jane, and his brother, Abraham,[3] worried over the internecine political war raging in Pennsylvania, Van Buren did not keep as close an eye on developments in New York as he had in the past. He recognized that the contest would be close, but he counted on the Regency organization and the improved economy to reverse the Whig triumph of the previous year. Though he traveled through the State on his return trip to Washington from the Virginia Springs, consulting with the Regency and with Hoyt and others in New York City, he was shaken at Marcy's defeat by about 10,000 votes. Worst of all, Van Buren's old friend and confidant, his personal leader in the House of Representatives, Churchill C. Cambreleng, was beaten.[4]

Yet more troubles of a political and administrative nature had to be faced before he was fairly settled in the White House. In late

November 1838, Jesse Hoyt discovered irregularities while comparing weekly and quarterly reports with their corresponding accounts in the books of his predecessor, Samuel Swartwout. When he called these to the attention of the auditor of the customs house, he was presented with evidence of widespread fraud. Hoyt promptly communicated his preliminary findings to Woodbury, who, after consulting Van Buren, sent them to the solicitor of the treasury, Henry Gilpin, and comptroller of the currency, J. N. Barker, to make a thorough investigation of Swartwout's activities.

After a few days of intense work, they reported back to a shocked Van Buren that the former collector's accounts were short by a million and a quarter dollars over and above all of his assets. Swartwout had been using substantial amounts of the public money he collected for his own private account—speculation in securities, land and various business enterprises. What was more flagrant, he had begun helping himself to public funds almost as soon as his appointment was confirmed in 1829. And Federal District Attorney William M. Price had not only helped him cover up his thefts, but had shared in some of the plunder.

Van Buren and Woodbury were upset and embarrassed. It mattered little at this late date that Jackson had insisted on Swartwout's appointment over Van Buren's objection or that Congress had been woefully lax in establishing a proper system of accounting for customs collections. Secretary Woodbury and his auditors, nominally in charge of all Customs houses had received reports that Swartwout had made up without any means of checking their accuracy. With the collaboration of the district attorney, the only public official on the spot who by law was charged with reconciling the collector's accounts, it was impossible for Washington to detect any irregularities until there was a change of collectors or district attorney. Still, Van Buren and Woodbury realized it was fruitless to make technical explanations and counterproductive to lay the blame on Congress or on the previous Administration. What he proposed to do was recommend legislation that would replace the antique reporting procedure with a system of independent audits. And with rare inspiration he would turn this monumental fraud into a telling argument for the subtreasury, where the collection and payment of all public funds would come under the direct supervision of the Treasury Department.[5]

He knew the Whigs would make political capital of the incident even though Swartwout, had become openly affiliated with them, before he fled the country. Van Buren was not surprised when the House appointed a special committee consisting of six Whigs and three Democrats. That it would be a political *auto-da-fé* was guaranteed after Speaker Polk failed to prevent the Whig-Conservative coalition from selecting as a member Henry A. Wise of Virginia. This splenetic, arbitrary egotist, whom Administration Democrats despised as much for his vicious rhetoric as for his desertion to the Whig party, was certain to intimidate witnesses and create a partisan free-for-all.

As expected, Wise and his colleagues subjected Woodbury to a merciless grilling but were unable to make much headway with the well-informed, articulate Secretary who never lost his composure under the most trying of circumstances. When the committee moved on to New York, they were equally unsuccessful with Hoyt, though he was not as even-tempered as Woodbury. What they did discover was not as useful as they had hoped since it reflected on Congress and on Whigs as well as Democrats. In the absence of law and usage, all of the clerks and auditors in the customs house regarded themselves not as employees of the government but of the collector, an apostate Democrat. Swartwout had ruled supreme, and if his subordinates knew he was dishonest, as most of them did, they closed their eyes and their mouths for fear of losing their posts.[6]

Congress did pass additional legislation which provided some but not all the safeguards the Treasury demanded. And while the Whigs continued to accuse the Van Buren Administration of culpable neglect, they were unable to pinpoint any swindle under the existing laws when the frauds were taking place. Yet politically they managed to tarnish the character of the Administration.[7] Van Buren's inspired effort to make the frauds a weapon in favor of the independent treasury furnished editorial copy for the *Globe* and other Administration spokesmen but was unable to crack the Whig-Conservative coalition on the subtreasury.

Though again disappointed at the defeat of his fiscal system, Van Buren was relieved when Congress adjourned. At least this would ease some of the pressure on him personally, giving him and his advisors a respite during which they could consider the whole range of relationships between the federal government and the economy. Van Buren realized that a fresh approach had to be made if his policies were ever to assist in stabilizing the price structure. Almost all of the deposits that had been frozen when the banks suspended in 1837 had been returned to the Treasury; but in the absence of any public depository, they were still being kept as special accounts in the nation's banks, an embarrassing by-product of the Administration's continued defeat of its subtreasury system.

Van Buren had pared the expenditures of the executive branch as far as he dared; yet a persistent deficit continued to plague his conscience. While he was traveling through Virginia with the Poinsetts the past summer, he had received letter after letter from Woodbury lamenting the state of the Treasury constantly being stripped of its diminishing resources by the demands of the War Department, whose expenses for Indian removal were far exceeding what had been estimated.[8]

The expenses of the Navy Department, too, had climbed as preparations were being made for the exploring expedition of the South polar regions and the Pacific coast of North America. Border troubles had required increased outlays for both branches of the armed forces. Receipts from customs and land sales were not increasing as the economic horizon grew brighter. For the Compromise Tariff annually reduced rates on imports. To meet these extraordinary costs, the Administration was faced with having to issue additional treasury notes or float a loan. Either course was distasteful to Van Buren, but he preferred another issue of treasury notes because he believed they acted as a wholesome check on the banks who were pledged not to circulate them. And the secretary of the treasury could redeem them as the state of the Treasury allowed. They could also act as a means of pumping hard money into cash-short regions, thus helping the banks maintain specie payment.

Additional treasury notes would, of course, bring criticism from Benton and other bullionists in the party, but their complaints were overbalanced by the benefits of flexibility. Woodbury had warned Van Buren that dire consequences might occur if there was another suspension. This possibility had also reinforced his decision to make a second issue of notes as did the fact that he would not have to seek authority from Congress as he would have if he had resorted to a loan.[9]

Actually the costs for Indian removal, frontier defense and the Seminole War reversed to some extent the severe deflationary trends of suspension and government policy. Expenditures of some $40 million above estimate for these unanticipated costs and the issuance of $10 million in additional treasury notes cushioned the economy and contributed to economic recovery. Neither Van Buren nor Woodbury saw it that way; but both men, particularly the President, were determined that federal policy would act as a constraint on the banks. The oft-repeated charges of Rives and Tallmadge that the Administration would establish a gigantic treasury bank if the independent treasury passed had an element of truth, which became apparent when improvement in the economy began to falter in the fall of 1839.

But recovery had seemed certain during the spring of 1839. As Van Buren made his summer plans where he always combined politics with relaxation, he decided he would make this trip a state visit. There was nothing novel about his decision; Washington, Monroe and Jackson had all made ceremonial visits to the North. Where Van Buren broke with tradition was in combining the trappings of a state visit, which would bring out large crowds, with speeches that were political in nature.

Instead of the purely formal response to the chairmen of the local reception committees, he would explain and defend his policies. It would, of course, all be dignified and decorous—no comments on the opposition, no overt partisanship, no recriminations. But the message would be there, and it would be a message of peace and prosperity that lacked only a subtreasury law to strengthen the economy and restrain speculation. In western New York where feeling still ran high against his refusal to support the Canadian rebellions, he would appeal to the stable elements of the border communities, the leading men who would have everything to lose if the border disturbances escalated into open warfare. For the first time in the nation's history an incumbent President was campaigning for reelection. This would not happen again for more than half a century.

Several months before he set out from Washington in mid-June, Democratic organizations along the route had been preparing to dress up their towns, spruce up the local militia for a parade. Secretaries Poinsett and Paulding had soldiers, sailors, forts and ships ready to do the President honor and add color to the occasion. Arrangements in many places were haphazard, with the militia living up to its ramshackle reputation, still citizens came out in great numbers to join in the excitement. Van Buren was heartened by the reception he got in Pennsylvania and New Jersey, where he was amused when his son Smith, after being lowered into one of Mahlon Dickerson's mineshafts, advised his father it would make a first-rate subtreasury. But nothing would compare with the turnout and the reception Tammany organized in New York.[10]

As Van Buren crossed the Hudson from New Jersey, the U.S.S. *North Carolina,* one of the few seventy-five gun ships of the line that was not laid up for the sake of economy, boomed out a twenty-one gun salute. It was a splendid sight, rigging and spars manned by its crew in white, its officers on the quarterdeck in full dress at rigid attention, all coming to salute at the command of their skipper. The *North Carolina,* and indeed all the ships in harbor, were dressed up in colorful signal flags and bunting. At Castle Garden, Tammany Sachem John W. Edmonds gave a short speech of welcome. Van Buren replied in his high, clear voice with a short speech describing and defending the subtreasury system, yet not failing to remind his listeners the debt they owed Jefferson and the party he founded for their individual liberties.[11]

Mounting a black horse that had been provided for him, he reviewed the militia assembled there, then rode slowly as Jackson had done at the head of a mile-long procession preceded by a company of mounted dragoons, and followed by sailors from the *North Carolina,* Tammany braves in Indian garb, artisans in their working clothes, members of young men's Democratic organizations from the various wards and, in fact, anyone who wanted to follow the president. Whig Mayor Isaac Varian and the common council awaited him at the city hall where the greetings and responses were strictly formal. Conspicuously absent were Governor Seward and Lt. Governor Luther Bradish who had been invited but had publicly declined. Nor did these officials greet Van Buren when he arrived at Albany some days later. Elsewhere, Whigs and Democrats and even Conservatives joined their partisan rivals and made the presidential tour appear a popular triumph, especially in the river counties. At each stop Van Buren described and defended his fiscal policies in response to the welcome remarks of the local committee chairmen. When he turned West and neared the border regions, he boldly spoke of the recent disturbances with Britain and expressed his sense of accomplishment in preserving the peace.[12]

Any politician less astute than Van Buren would have seen in the crowds, the excitement, the enthusiastic reception to his campaign speeches, as a sign of solid support for his Administration and a manifestation of his personal popularity. But Van Buren was no ordinary politician.[13] In New York City he noted the high premium on specie, the rising interest rates; behind the cheering crowds, he could see boarded-up shops and warehouses. Still these sobering aspects of a faltering economy were more than balanced by positive indicators. Weekly tolls of the New York Canal system he observed were increasing. Existing railroad lines were doing well and along his route he saw new construction of lines that would provide faster year-round transportation for travelers and the mail.

After six weeks touring the state he felt reasonably confident about the fall election, but he respected the Whig organization and there were too many imponderables that could swing the state either way. Before he reached Washington all signs pointed to an Administration victory. The Democrats carried Ohio, Maryland, Georgia, New Jersey and Mississippi. Amazingly, Marcus Morton wrested the governorship of Massachusetts from Edward Everett by one vote, a tribute to his own and George Bancroft's talents for political organization. In New York the Whig majorities of the preceding year were reduced, but they managed to retain control of the lower house. The Mississippi election augured improving prospects in the Deep South, a political dividend that had resulted in part from Calhoun's alliance with the Administration. That bellwether of southern politics had seen in the Administration's Treasury proposal benefits for the plantation South and particularly the producers of cotton. Northern bankers, like their fellow manufacturers with their protective tariffs, had been taxing southern producers through control of credit, aided by the government. In this process a vicious circle was evident to him: the revenue the tariff produced found its way into the banks, especially the large northern banks, which in turn charged the planters high interest on their loans, or worse, speculated with the people's funds, creating erratic pricing for commodities.[14]

He also shared Van Buren's constitutional philosophy on money and credit, which he now thought were properly a state concern. But Calhoun realized that his alliance of mutual interest with the Administration had to be a gradual one or his support in his own state would be endangered, and his allies in other states like Dixon Lewis of Alabama or R. M. T. Hunter of Virginia would desert him. For the past seven years, the names of Jackson and Van Buren had been anathema among the extreme states' rights men in the Deep South. Tact, reasoning and time for careful explanation were essential if he were to keep his faction together and eventually, he hoped, have it take over the Democracy. But there had to be a beginning.

Calhoun was very much in view at Van Buren's public reception on New Year's Day 1840, chatting amiably with his ancient enemies Frank Blair, Amos Kendall and Isaac Hill, greeting Van Buren with as much warmth as his austere demeanor would permit. One of Rives's friends who witnessed the affair was shocked. Though no friend of Calhoun, he had thought the South Carolinian "had too much of lofty high toned southern feeling to stoop to so degrading an act."[15]

Moving ever so cautiously and seeking reasoned arguments, Calhoun convinced his lieutenants that the realignment was both appropriate and honorable. They in turn explained his changed position satisfactorily to their constituents. The States' Rights party was to phase out, and all were to return to their original allegiance. Van Buren, well aware of Calhoun's ambition, was quite certain he could count on the loyalty of his faction, at least while it was expedient, and that meant support for his reelection. His party in the South, he thought, was in better condition than it had been for some years.[16]

Meanwhile events in Europe provided some short-term political benefits that Van Buren eagerly grasped. The British economy which had been almost as volatile as the American, was seriously threatened after two years of crop failures. Just as recovery seemed to be taking hold in the United States, the British government spent the huge sum of £10 million on the Continent for the purchase of wheat. Had it not been for French and Dutch loans in the spring of 1839, a financial panic of disastrous proportions would have struck Threadneedle Street.[17] The Bank of England sharply increased its discount rate to stop the flood of specie pouring out of the country. But for the time being its restriction on credit had only an indirect impact on American banks, at least those in New York City, Philadelphia and Boston. Specie did not leave the United States in any appreciable amounts until Biddle's bank, still the largest and strongest financial institution in the country, was caught in a vise of credit contraction.

Before his retirement in early 1839, the Napoleon of Chestnut Street had anticipated higher prices for cotton in the world market as the economy improved during 1838. He purchased a substantial amount of the 1838 crop for export with loans from European bankers. By the time this cotton reached the market in the fall of 1839, prices had broken sharply. Most of Biddle's cotton could not find customers among the hard-pressed merchants. The Bank of the United States was unable to supply specie to cover its paper as it came due, despite its frantic efforts to raise cash by selling post (promissory) notes in New York, Boston and Philadephia at ruinous discounts. After liquidating most of its assets, the directors were forced to suspend. For another year and a half, the great Bank lingered on, becoming bankrupt in April 1841, its charter revoked, its tumultuous course in the economic and political development of the nation at an end.[18]

The Bank's operation had put a heavy drain on the specie of the banks in New York, Albany and Boston. But after the Panic of 1837, the New York Safety Fund and the Massachusetts Suffolk Bank system had pursued careful policies in making credit available and in issuing notes. When they resumed payment they had ample reserves and were able to withstand the pressure from Biddle. After the Bank of the United States suspended, carrying with it the banks of the South and the West, New York and New England banks stood firm, though loans were hard to come by.[19]

Van Buren recognized the anomaly and the opportunity at once. The banking community in New York and in New England, which had abused him so roundly as a political demogogue who would destroy the nation's economy with his ignorant shopworn slogans for hard money, were themselves following the same line in their credit policies. At the same time the arguments of Conservative leaders like Tallmadge and Rives so vigorously asserted as eternal verities were shown by this second suspension to have had no foundation at all. Their position that state banks and soft money were the only sure means out of the depression was swept away with the suspension of the Bank of the United States and the partial suspension that followed. Already many of Tallmadge's and Rives's strongest backers in the business community were demanding a review of state banks. Even the Whigs were conspicuously silent on the suspensions. The *Madisonian,* advocate of the conservative cause in Washington, experienced a flood of cancellations. Thomas Allen, its gifted editor, worried about the continuance of his tri-weekly thorn in the side of his rival, the *Globe,* began casting about for another position.[20]

Economic hardship in all classes and sections helped the Whigs carry New York in the fall elections of 1839. This time they gained control of the legislature. Tallmadge, who had been denied reelection to the United States Senate a year earlier, was returned, but not as a Conservative. He had made a complete transition to the Whig party. In Virginia, Rives, unable to secure a majority, would not be a member of the Senate during the first session of the Twenty-sixth Congress. Although many Conservatives joined the Whig party, some returned to the Democracy. And Calhoun's States' Rights party was now Democratic in name as well as practice.

Were Van Buren to make a new departure, this was the time, this the political opportunity he must seize if he were to eliminate what was left of the Conservatives and draw a sharp line between the two parties. But most of all he would suggest a cogent and timely reason for the passage of the subtreasury plan, a reason he had always kept at the back of his mind but had never precisely articulated. The withholding of specie from the market place in government depositories under strict regulation would act as a restraining influence on banks.

Throughout November 1839, working from notes and materials Woodbury supplied him, he addressed himself to this argument for the subtreasury. After rehearsing the past experience of panic and suspensions, he built up his argument step by step, buttressed by an elaborate, knowledgeable assessment of credit and banking both here and abroad that, while recognizing the close relationship between the national government and the business community, effectively dismissed a national bank as incompatible with a democracy and state banks as imprudent guardians of the public funds. Then boldly for a Jeffersonian, he said, "The direct supervision of the banks belong, from the nature of our government, to the states who authorize them . . . but as the conduct of the Federal Government in the management of its revenue has also a powerful, less immediate, influence upon them, it becomes our duty to see that a proper direction is given to it."[21]

Van Buren's third message, his longest, was also his most thoughtful analysis of the state of the Union and his most articulate and forceful declaration of his political philosophy. As a state paper it ranks high in its grasp of the problems besetting the nation and in its bold, even radical, presentation of what his Administration meant when it spoke of public policy in the economic realm. The message drew unqualified praise from Marcy and from the *Evening Post,* signaling an end to the feud between the organization Democrats and the Loco-Focos. With the party now in general agreement on public policy, at least as the President expressed it, and with the Conservatives in disarray after the partial suspension, the outlook for 1840 seemed better than it had been for the past two years.[22]

Although party leaders in the House had firm direction from the President on matters of state, they sorely missed the guiding hands of Cambreleng and Polk. Calhoun sensed their predicament and was ready. He realized that if he could keep his faction together it would hold a balance of power between the Whigs and the Administration Democrats. The object was to capture the Speakership as Franklin Elmore, one of Calhoun's most devoted proselytes explained it to a friend in September 1839. Elmore's candidates were the obese Dixon Lewis of Alabama, so fat he required a special chair in the House which became a tourist attraction, Francis Pickens, a leading nullifier during the crisis of 1832–33, or R. M. T. Hunter of Virginia, who he thought had the best chance since his political stance remained an uncertain quality.

Elected to the Twenty-fifth Congress as a Whig, Hunter had nevertheless supported the subtreasury plan. An engaging young lawyer, a good speaker, he had made a mark almost immediately in the House. And though he had deserted his party on the subtreasury, he had backed all other measures the Whigs had introduced, voting as their caucuses determined. It was not generally known that Hunter was a disciple of Calhoun. Whether it be Hunter or not, Elmore reasoned that with any of the Calhoun clique as Speaker, he would pack the important committees with men partial to southern measures, who would "shape our destinies and give us a southern policy." In an obvious reference to Van Buren, Elmore added "give me a weak northern President whom we can control; who professes our principles and can be made to act up to them, far very far in preference to a strong southern President, all of whose principles and policy are against us."[23]

Elmore underestimated Van Buren's strength of purpose, taking his Jeffersonian posture and his Unionism for weakness. If the President seemed to favor southern political objectives, he did so because he was seeking through compromise a formula that would preserve the Union from the contending forces he could already see were pulling it apart. He was no more a tool of the slaveholding planters than he was of northern bankers, industrialists and merchants. Where he was consistent in upholding a special interest, it was squarely in the Jeffersonian tradition of small, independent farmers like those of his own county whose rights he had defended in his youth against the great manorial lords.

When the Twenty-sixth Congress finally organized in late December 1839, the Democrats, including their recent nullifier converts, had held several caucuses for Speaker. The Administration backed John Winston Jones, a mild-mannered lawyer who was close to Ritchie and others of the Richmond Junto. The Whigs put up John Bell. So closely matched were the two parties that Jones failed to be elected on three ballots, during which Lewis gained steadily. There was a final effort to stampede the Democrats for Lewis which failed when a solid phalanx of Whigs and a number of Administration Democrats kept him from gaining a majority. After ten ballots, Elmore's last candidate, R. M. T. Hunter, whose vote was increasing slowly, grudgingly was given the entire Whig vote and enough Administration support for election.[24]

Van Buren had a nullifier in Whig clothing as a Speaker. But as he made a careful survey of the situation, he decided Calhoun had gained at most a hollow victory. Hunter was young and inexperienced in controlling the rough and tumble of the House. If anything, the maneuvering of Calhoun's lieutenants had exposed them as unreliable allies rather than true party men.[25] Nor did Hunter in the Speaker's chair especially concern the Administration. He had twice supported the subtreasury bill, and there was no reason to doubt he would reverse himself. Hunter quickly confirmed Van Buren's judgment when he appointed a majority of Democrats, all proponents of the subtreasury, to the Ways and Means Committee, naming his competitor John W. Jones chairman.[26] There were now sufficient Administration and Calhoun Democrats to carry the subtreasury bill in the House. The Senate, as it had always been, was secure.

Calhoun Democrats gave another proof of their cooperative spirit as soon as the Speakership contest was concluded. They spurned Duff Green, who had solicited their support for the House printing contract, and then they voted with the Administration in electing Blair and Rives. The Administration looked on with satisfaction as much for the subsidy Blair and Rives had gained as for the split between Calhoun and his veteran supporter. Green, for all his faults, was a bold politician and an experienced editor.[27] His capacity to do mischief, never underrated by Van Buren, was now virtually extinguished, at least within the Democracy. It remained only for Calhoun to make public his reconciliation and formal adherence to the policy of the Administration. And this is just what his colleague Senator William H. Roane of Virginia suggested to him. Speaking for the Van Buren wing of the party in the Old Dominion, Roane had long deplored Ritchie's erratic course, his attachment to the Conservatives and his close relations with Rives, whom Roane thought little different from a Whig in politics. A formal merger between Calhoun's adherents, a small but influential group in Virginia, and the Democracy might well jolt Ritchie out of his indecision, detach him from Rives and again bring his influence behind the Administration. Calhoun and Roane, carrying on their conversation in a carriage of the train from Richmond to Washington, came to no conclusion. Roane mentioned Van Buren had said that he wanted to make the first overtures. Would Calhoun leave his card at the White House? Calhoun declined, but in such a way that Roane believed he was simply postponing the event until he was certain of his political position at home. "He had always intended to restore [their] personal relations, if he [Van Buren] persisted in the course he had taken," Roane wrote, "but that [he] must be the judge of the time and the mode."

The cooperation of his faction with the Administration in the House showed Calhoun that he had nothing to fear politically from mingling his fortunes with those of Van Buren. Still he hesitated. He wanted to be certain that Henry Clay would support William Henry Harrison who had been nominated in Harrisburg on December 8, 1839. If he did not and the Whig party failed to mobilize itself behind one candidate, he might have a chance against Van Buren by drawing off the Southern Whigs on the abolition and tariff issues, while retaining his old power bases in the North.[28] By December 24, with both Houses organized, Calhoun realized that Clay would not split the Whig party. It was either Harrison or Van Buren; and as he listened to the President's tightly constructed message voicing Jeffersonian values yet artfully giving them a new dimension in his exposition of the subtreasury, he gave Van Buren his qualified approval.

However painful it was for the proud South Carolinian, so conscious of his standing, to make the first move, he arranged through Roane to call upon the President. At the appointed hour Calhoun was ushered into Van Buren's office. Sensitive to his discomfiture, the President sought to put him at ease. After all, they had not recognized each other socially for eight years. Van Buren knew Calhoun envied him his position and under it all despised him as the architect of all his political misfortunes. Nor did Calhoun attempt to dispel his feelings about the President. Chillingly, he said that he had not changed his opinion of Van Buren "as to his course towards me, out of which our former personal alienation grew; nor can I."

But as Calhoun agreed with the Administration's policies, he said he intended "to remove the awkwardness of defending the political measures and course of one, with whom I was not on speaking terms and the weakening effects resulting from such a state." Van Buren said he was "gratified, approved the step he had taken," and the short interview was over.[29]

There was now no doubt that the subtreasury would become law. The Administration with Calhoun's support held a majority in both houses. Under the wise direction of Wright and Benton, the Senate approved the bill in a record three weeks time. Speaker Hunter did as he was bid, but because he lacked the respect of the House and proved clumsy in wielding his formidable powers, the bill was delayed through parliamentary maneuvers, finally passing at the end of June. Van Buren waited a few days before signing it so that it would become law on the Fourth of July 1840.

The great struggle over, Van Buren turned his attention again to the campaign. What had looked politically promising as he stepped ashore at the Battery in New York City amid the cheers of his fellow citizens a year before, had suddenly taken a turn for the worse. The major cause was a steep decline in the economy as the depression in Europe finally reached the United States. The Whigs, thoroughly organized for the first time throughout the nation, were mounting a campaign modeled on the Jacksonian—an emotional appeal to the masses—but far exceeding their opponent's approach in effort, novelty and style.

Tactics first employed in several of the state elections since 1836 had been improved and expanded with ample funds from manufacturing and banking interests. There were of course the ephemeral campaign papers and badges. Broadsides were far more spectacular than those the Jacksonians had pioneered. The Whigs supplemented these stock devices with songbooks, mottoes, jingles and always plenty of liquor to stimulate enthusiasm. When Harrison was nominated, appropriately at Harrisburg, in December 1839, Whig leaders determined to exploit his record in the War of 1812 as a military hero.

Mainly at the insistence of Thurlow Weed, they presented Harrison as a simple farmer who wore a coonskin cap and lived in a log cabin on the frontier. The homespun image, though errant nonsense, was believed to be true by hundreds of thousands. Actually, Harrison came from an aristocratic Virginia family; he did not live in a log cabin but in a pretentious home at Northbend, Indiana. Though the commanding general at the Battle of the Thames, an American victory over the British and Indians, he owed his success largely to his subordinates, Richard Mentor Johnson, Van Buren's Vice-President, and Johnson's brother James.

These artifices Weed and other Whig leaders earnestly propogated caught on, and soon thousands of log cabins dotted a land where liquor seemed to spring eternal. Never before had innuendo, slander and gross falsehood erupted in such volume and such intensity. Van Buren, who actually came from the masses, was pictured as an effeminate, overindulged fop who led a life of Babylonian splendor. When he was not being portrayed in gross cartoons or in uncouth stump speeches as a cowardly little fellow with perfumed whiskers, he was made into a high-toned aristocrat who drank French wines and abhorred plain American living. In reality, Van Buren lived as modestly as a head of state could, purchasing almost no new furniture or plate. His predecessors had bought most of the carpets, drapes, chairs and cutlery in use at the White House. The gold table service Charles Ogle, an obscure Whig Congressman from Pennsylvania, claimed Van Buren dined upon was silver gilt. President Monroe had purchased it in the early '20s, and it was only used for State dinners. But many of the credulous public believed in Ogle's blatant distortion of the President's living style. Published as a pamphlet, this piece of demagoguery was an effective campaign document.[30]

The Democrats responded with the *Extra Globe,* ably edited by Amos Kendall, who saw to it that the more than 30,000 postmasters throughout the nation got subscribers. That they were franked out at government expense did not concern him; nor did he see any conflict of interest when he pocketed over $60,000 for his work. The Administration, to the infinite chagrin of the Whig high command, was better able to get their message across through their control of the nationwide network of federal officeholders. Still it was apparent that even without the Whig extravaganzas, Van Buren was vulnerable on many points.

Through his own indecision, he found himself saddled again with Richard Mentor Johnson as his running mate. Johnson, certainly a kindhearted and in many ways a progressive public figure, was a bit too colorful in his personal life for Washington society. Amos Kendall, who had been influential in placing Johnson on the ticket in 1836, visited Jackson at the Hermitage during the summer of 1839. On his return he stopped at Johnson's tavern and was so disgusted at the Colonel's way of life that he denounced him unsparingly to Van Buren.[31] Jackson, too, who had been Johnson's principal backer for the vice-presidential nomination, warned Van Buren repeatedly that he must be dropped or the Administration would certainly lose Kentucky and Tennessee.

As he listened to the growing complaints about the Vice-President's behavior and the drag he would be on the ticket, he could not help but remember how Johnson's native region had rejected him in 1836. The reasons now were more compelling than ever. Yet Johnson, Van Buren was aware, had considerable strength among the radical Loco-Focos in the Northeast; and as a westerner, a colorful veteran and a man of the people, he might help strip away the heroic mythology the Whigs were building up around Harrison. Then there was the problem that Johnson would not bow out. Any overt attempt from the White House to drop him would probably stir up bad feelings, which Van Buren wanted to avoid at all costs. Van Buren finally decided that the convention should make no nomination, a typical compromise arrangement that seemed at the time the most expedient course. As he explained to Jackson, "From the number of friends whose names were brought forward and one of them a member of my own Cabinet, a rigid neutrality on my part became like an act of duty and justice."[32]

The Democrats met at Baltimore on May 4, 1840, and though many more states were represented than four years earlier, South Carolina and Virginia did not send delegates. Again New York and Pennsylvania dominated the proceedings. Grundy, who had resigned as attorney general and had just been elected to the Senate from Tennessee, gave the major address; Isaac Hill presided. Everything went smoothly, even sedately. The convention renominated Van Buren by a unanimous vote and left the vice-presidential nomination to the states. Surprisingly Johnson showed unusual strength in Pennsylvania and New York, both of which Van Buren must win if he were to be elected. Party leaders, more worried about the presidential contest four years hence than they were about Johnson's personal life, swung behind the Colonel.[33]

If the vice-presidential issue were not damaging enough, a recommendation of Joel Poinsett urging a thorough reform of the useless militia establishment became in the sure hands of Whig leaders a political liability of major importance. Responding to the border troubles and the long, drawn out Seminole War, Congress asked Poinsett to see what could be done about the militia in case of a national emergency.

Overburdened even more than usual with the cares and responsibilities of his office, Van Buren seems not to have read over Poinsett's report or, if he did, failed to recognize its political implications. The Secretary had addressed himself to this task with vigor and imagination. After thoroughly investigating the state of the militia, which he found to be a laughingstock, Poinsett developed a bold concept that in essence would create a reserve force subject to basic training under regular army officers.

At first, the program met with considerable approval even from

Whig journals like the New York *Courier and Enquirer,* which called the report "a plain, unambitious sensible document".[34] The *Globe* and the *Argus* also strongly advocated Poinsett's plan; but Thomas Ritchie's *Enquirer* held back while its editor warned Poinsett in a private letter that Virginia Whigs were calling it "tyrannical and oppressive"; they were charging, said Ritchie, that the Van Buren Administration was bent on creating a standing army to convert the Republic into a military despotism.[35] Poinsett replied with a lengthy explanation, denying vehemently that the plan endangered American liberties; but the correspondence rather than dampening an incipient issue merely gave it wider currency, especially in the West where the plan was made to appear a shackling of personal freedom.

Whig politicians caught at the theme of a "standing army," which they claimed the Administration would impose on the country if it could. Whig orators headed by Henry Clay and John Jordan Crittenden, exploiting the spurious premise of a standing army and citing historical precedent, accused the Van Buren Administration of saddling a ruinously expensive, totally subversive military establishment on the country.[36] Realizing that the furor had taken a dangerous turn, Van Buren disavowed Poinsett's plan as unconstitutional because the organizing of the militia was reserved to the states. In an effort to combat the Whigs' reckless charges of military conspiracy, he gave a concise history of militia reform plans beginning with the Washington Administration in a public response to a letter from a group of Virginians.[37] Once Poinsett had become a promising political target with his militia plan, the Whigs saw the prospect of additional gains in criticizing the administration's Indian policy. Through the Secretary, who had direct responsibility, they could challenge Van Buren's record on the removal of 15,000 Cherokees from their tribal lands and the seemingly endless Seminole War. From a partisan view there was ample ground to focus attacks on the Administration.

Van Buren had inherited the removal policy, the War and a corrupt commissioner of Indian Affairs, Carey Harris, from the Jackson Administration. When Harris's speculations in Indian lands came to light, Van Buren replaced him with the competent, honest Commissioner Thomas H. Crawford, who had had some experience in Indian affairs. But by then the first phase of removal had taken place under conditions that reflected economy and dispatch rather than humanity and preparedness. To supervise the actual removal Van Buren relied upon General Scott. As a Whig, Scott would bear at least some of the political opprobrium that would accompany the act. As a capable administrator and organizer he would accomplish the unpleasant task with understanding, efficiency and, Van Buren hoped, economy. Though he differed with the President in politics, Scott had proven time and again that he would execute his orders; insubordination was unthinkable to the General whose obsession on this point had involved him in much previous difficulty.

Without the assistance of Poinsett, who was desperately ill at the time, Van Buren let Scott write an outline of his instructions, which gave Scott the widest possible latitude. The only point the President insisted upon was that removal be prompt and humane.[38] Scott was the best choice he could have made; but neither he nor Crawford had any idea of the problems involved in rounding up, provisioning, and transporting 15,000 unwilling people. When only a handful of Cherokees responded to his summons, Scott ordered his detachment of regulars to scour the countryside and bring them in to a stockade where they would wait until he had accumulated enough clothing, provisions, wagons and farm tools to begin the trek. Scott's soldiers, many of whom came from the dregs of society, committed atrocities on the poor defenseless Indian farmers. And matters would have been worse had it not been for Scott's energy, his quick intelligence, his ability to exert iron discipline and to fend off white traders who had hoped to make a killing on government contracts.[39]

Operating under his broad instructions, Scott gave all the contracts to John Ross, even though the Cherokee leader estimated double the amount the government had appropriated for the removal of each Indian. Bad weather caused delay and hardship. Hundreds of Indians died on the long trip. When they finally arrived at their alloted lands in the dead of winter in 1839, they were confronted with a wilderness and inadequate supplies of tools, draft animals, forage and other essentials needed to erect shelters before they could begin the heavy labor of preparing the virgin forest for a crop.[40]

Political criticism of the Administration's handling of the affair was so widespread that Van Buren was associated in the minds of northern and many southern voters for initiating a policy that actually went back to the Jefferson Administration. He was also accused of waste and extravagance for the contracts that Scott had made, and he accepted full responsibility. Scott escaped all blame for the episode, as perhaps he should have, though if he had acted less precipitiously he would have saved lives.

The conduct of the military during the Seminole War also reflected seriously on the Administration. Every session of Congress since the beginning of Van Buren's Administration featured heated debate over the army's inability to defeat a wretched little band of fugitive slaves and desperate Indians who had refused to emigrate. Secretary Poinsett was undoubtedly the best professionally qualified secretary of war since the founding of the republic. And therein lay his primary fault. Poinsett and his officers tried to fight a guerrilla war under terrible conditions of climate and terrain with traditional European military techniques.[41]

The Seminoles, under the inspired leadership of a young half-breed, Osceola, combined their knowledge of the land, with courage and discipline and were able to defeat or fight off the professionals and the militia men who marched against them year after year. Reports from disgruntled army officers created considerable sympathy for the Seminoles and especially for their leader Osceola, who became a symbol of liberty, bravery and ability, even in South Carolina and Georgia where the white population not only dreaded the Seminoles but were ardent exponents of Indian removal.

When the army flagrantly and deceitfully violated a flag of truce and captured Osceola, the outcry in Congress was strenuous. In the Senate, Clay and Webster scored General Jesup for betraying his honor. John Bell and Henry Wise of Virginia were equally outspoken after Osceola died in prison in 1838. Whig orators held up his fate as that of a noble Indian betrayed. The army, and particularly Poinsett, was charged with covering up ineptitude by guile when they found they could not achieve success.[42] Nor were these accusations completely without foundation. Poinsett had made use of several Cherokee chiefs to treat with Osceola. When it appeared that their efforts could have been a ruse to capture him, John Ross entered into a spirited exchange with Poinsett, charging him with bad faith. All of their correspondence was laid before Congress at the demand of the Whig leaders and distorted all out of proportion in the Whig press. Poinsett and Van Buren were now pictured as heartless cynics who would bring dishonor on the nation.[43]

Pertinent to these partisan blasts were the mounting costs of both removal and the War which by 1840 were reaching the incredible sum of $50 million while the country had fallen back into a steep business depression. The overruns for removal upset Woodbury, who never could find sufficient money after repeated juggling of accounts. He had to resort to the dubious expedient of issuing warrants to the War Department that were not covered by appropriations.[44] As it was, Van Buren alienated many voters in Ohio, New York, Pennsylvania and New England because of the Cherokee and other Indian removals. In the midst of furious indictments on Cherokee removal from the *Albany Evening Journal,* the *Courier and Enquirer* and the *National Intelligencer,* another crisis arose that created a brief political uproar and further increased tensions over slavery.[45]

While Van Buren and Poinsett were traveling in western New York during August 1839, a Coast Guard brig boarded a suspicious schooner, the *Amistad* in Long Island Sound. They found aboard forty-nine Africans, two Spaniards and five black children, one of whom was acting as a cabin boy. The Spaniards had a strange and wild story to tell. The Africans, who had been sold into slavery at Havana, managed to free themselves on the high seas, murder the captain and the cook and by sign language force the remaining whites, who knew something of navigation, to return them to Africa. The whites, however, took advantage of their ignorance and that of their leader who was called *Cinque,* to sail northwest instead of east, which accounted for the ship's appearance in Long Island Sound.

The Coast Guard disarmed the Africans and brought the *Amistad* into New London Harbor as salvage. An overnight sensation occurred which curiously mingled with politics as Whigs and abolitionists whipped up public opinion about the plight of the Africans while the Democrats charged them with mutiny and murder. The federal district attorney promptly asked for guidance from Washington where Forsyth was already in consultation with the Spanish minister on the matter. John Quincy Adams and the best legal talent in Connecticut volunteered to defend the Africans. Through various legal devices they managed to restrain the district court judge from handing them over to Spanish authorities. In Washington, Forsyth conducted negotiations without Van Buren's guiding hand and clear understanding of the points involved—property rights, international law and treaties dealing with the international slave trade.

When the President returned to Washington, he was more impressed with the research and the arguments of the defense counsel than with the presentation of the federal district attorney at a series of court proceedings. Forsyth was so deeply involved in the case that Van Buren did not intervene, beyond signing a warrant for the transfer of the Africans to federal authority in the event the court in Connecticut determined they were to be remanded to Spanish authorities. He did, however, after looking over a summary of developments in the case Aaron Vail prepared for him, direct that the State Department make available all relevant documents to the defense and that the district attorney not object to them being used as evidence for the Africans.[46] The State Department arranged with Paulding to have a naval vessel stand by to transport the Africans to Cuba if, as Forsyth anticipated, judgment favored the prosecution. With a total and reprehensible disregard for the Africans' rights, he wanted them bundled on the naval vessel and on their way to Cuba before the defense could devise an appeal.

Strict secrecy had been enjoined on the captain of the vessel, the U.S.S. *Grampus,* and the district attorney. But the appearance of the ship in New London Harbor caused speculation and after the judge found the Africans not to be slaves and directed they be returned to their homeland, the true mission of the *Grampus* became general knowledge. The Abolitionist and Whig press in the North accused Van Buren of aiding and abetting kidnappers and slave catchers, of prostituting the federal government and his position as chief magistrate to that end.[47]

Van Buren's apparent indifference to the *Amistad* case may be explained by his need to make two cabinet appointments. One of the first letters he picked up from the pile of correspondence awaiting him in the White House was a letter from James K. Polk informing him that the Tennessee legislature now in control of the Democrats wanted Grundy to replace Whig Senator Foster who had just resigned rather than obey its instructions. Van Buren would sorely miss Grundy as an advisor, and as usual he dreaded the intrigues that would immediately spring up among claimants for his position. After thinking about a suitable candidate, he offered the post to Buchanan who preferred to remain in the Senate. Van Buren now tendered the office to Henry D. Gilpin, the solicitor of the Treasury Department and loosely connected with Dallas's faction. Gilpin wanted the post; he was a capable lawyer, though no friend of Buchanan, who was not to be fobbed off. "Every avenue to a Cabinet office during Van Buren's Administration," the wrathful Buchanan wrote Governor Porter of Pennsylvania, "is closed against any Pennsylvania man and the President's disposition towards myself is proclaimed from the house-top." A bemused Calhoun commenting on the appointment, said, "Gilpin is Attorney-General. No one seems to know by whose interest. His appointment gave much dissatisfaction to Buchanan."[48]

The other appointment caused little trouble; nor did it bring any political benefit. Kendall resigned to edit the *Extra Globe.* As his successor, Van Buren chose John M. Niles of Connecticut, long a steady defender of the Jackson and Van Buren Administrations on the Senate floor. A practical man of business, a good editor and a tireless worker, Niles would be available to help Kendall with the *Extra Globe.* Father Niles, as he was called, was happy to receive this mark of Van Buren's regard for his past services and was soon at his post in Washington.[49]

By now the campaign was moving into its final stages. Niles's own state of Connecticut had gone Whig despite the vigorous efforts of a well-developed Democratic organization. Yet the Democrats were beginning to make inroads on Whig gains. Van Buren's plan for action in New York had been simplified and Flagg's central committee was functioning efficiently, getting out pamphlets and handbill materials, organizing Democratic processions, erecting hickory poles in villages and towns.

John Van Buren, who kept his father informed on local developments, had early in the campaign joined Dix and Elam Tilden in urging Silas Wright to lend his popularity to the ticket by running for governor against Seward.[50] A horrified Wright, who had no taste for the problems a governor of New York faced and by no means certain he would win, managed to extricate himself from a draft. He did agree to stump the state, something he had never done before. Though Wright was a poor speaker, no man in the country, except the President, was as well versed as he on the issues of the day. And Van Buren from the beginning had advised a strategy based squarely on the issues. He spent hours responding to letters from all over the Union. His answers were specific and covered not only current questions of policies, they spanned the entire range of his political career. In his letter to John B. Carey and others of Elizabeth City he wrote what amounted to a pamphlet of 8,000 words. This public letter, along with letters of previous years, was printed by the thousands and franked through federal offices as was the *Extra Globe.*[51] In practically every state, the Democrats established campaign papers like the *Rough Hewer* in Albany and the *Crisis* in Richmond.

Whatever differences Buchanan may have had with the Administration, he stumped the state of Pennsylvania and like Wright in New York spoke to huge crowds. In Tennessee, Grundy and Polk fought strenuously to reverse the tide as did Jackson himself. The Old Hero, though seriously ill, had in fact risked his life to visit New Orleans on the anniversary of his victory over the British. During the summer of 1840 he attended several barbecues where he urged his friends to get the vote out for Van Buren.[52]

John L. O'Sullivan, the young Irish emigrant whose idea of publishing a monthly review on subjects of interest to the party Butler and Van Buren had enthusiastically approved, put his editorial talents behind the campaign. The *Democratic Review,* now two years old, had a healthy subscription list. Its feature articles, unlike the political pamphlets and editorials of the day, catered to the human interests of its readers, stylishly embroidered with Democratic dogma. Every issue had a profile of some Democratic leader and carried his picture, often a crude likeness, which made party notables recognizable to the average subscriber.[53] O'Sullivan's *Review* did much to personalize the party and its policies. It also gave an aura of respectability, of literary and intellectual pretension to what was essentially a partisan journal. The well-educated O'Sullivan eagerly sought and published pieces from the very best writers of the time.

Everyone who visited Van Buren's summer home outside Washington in 1840 marveled at the volume of his personal correspondence. William L. Marcy, who was in Washington that summer and spent a good deal of time with the President, wrote his friend Wetmore that Van Buren is "greatly engaged in answering interrogations sent him from all quarters. These replies require to be well considered and . . . answers cost him some labor." Van Buren had enticed Marcy to Washington with a sinecure as one of the commissioners to settle the Mexican claims. His motives were typical of his style. Marcy was emerging as the leader of the conservative faction of the party in the Empire State.[54] He had been out of public life since his defeat for reelection to the governorship in 1838. Van Buren wanted to smooth over the differences that had grown up between them and head off, if possible, what he felt might eventually create a split in the party. He also enjoyed Marcy's company, his dry wit and direct manner. Not the least consideration was access to Marcy's practical advice and his skill as a political writer.

Though doubtful of success, Van Buren kept his thoughts to himself, "firmly persuaded" as Marcy observed, "that his course of policy is honest and right. He is quiet and calm but I will not say an unconcerned observer of the agitations and wild uproar of the political elements around him."[55] Increasingly his attention was fixed on New York where the party was operating under great handicaps. The Whigs had access to much more money than the Democrats. Flagg, Van Buren's manager, could and did use the federal network of offices and made liberal use of the franking privilege, postmasters and marshals. But his corps of activists could not compare with the political importance of thousands of sheriffs and justices of the peace, all of whom were active Whig partisans.

Van Buren was well aware of the advantage these local officials gave the Whigs, as he himself had relied upon them in the past for successful campaigns. But the most burdensome problem confronting the Democrats was not the circus atmosphere of the Whig campaign, nor the size of its war chest, nor even its well-knit organization and skillful leadership; it was the abrupt end of economic recovery as the depression in Europe sharply trimmed American exports and drove the economy back almost to the panic levels of 1837. The stubborn stand of the New York and Massachusetts banks in maintaining the specie standard caused great hardship for workingmen, small businessmen everywhere, as well as the great mercantile houses, not just in New York City, but in all the commercial centers of the Union.[56]

The Whigs took advantage of the depression, blaming the tariff policy and subtreasury system for unemployment in the cities, silent mills in the countryside. Careful to keep their tariff arguments out of the South, Whig stump speakers and Whig politicians everywhere blamed the hard money policy for the depression. Their solution was simple; scrap the subtreasury, loosen up credit and currency, and more dollars would find their way into everyone's pockets. The Whigs were not specific on how this would be accomplished, not daring to risk the loss of state bank partisans through an open espousal of another central bank.[57] Although the actual results would not be tabulated and certified for some weeks, Van Buren acknowledged defeat after he learned of the New York election on which he had placed his last hope. The Whigs carried the state for Harrison and Tyler by about 13,000 votes, almost 3,000 votes being cast for the abolitionist Liberty party's nominee, James G. Birney. Cherokee removal, the Seminole War, Poinsett's militia system and the *Amistad* affair may have made the difference.

In the electoral college, Harrison scored a landslide victory of 234 votes to 60 for Van Buren. But, in fact, the Democrats had done remarkably well considering the distracted state of the economy. They lost Pennsylvania by only 350 votes out of 92,000 votes cast. In several other states that went to Harrison, the difference between the two candidates was not more than 2,000 or 3,000; and the Whigs engaged in extensive frauds, particularly in New York where they imported voters from other states on a grand scale. Wherever they controlled the inspection teams, they were not overscrupulous about repeaters; nor were the Democrats for that matter.[58]

Still they blamed their defeat on widespread election fraud and the triumph of demogoguery over reason, mummery and cant over common sense. Many of them, like Jackson, Blair, and the President himself, wondered whether their faith in the better judgment of the common man had been misplaced. Calhoun thought Van Buren himself was responsible for the defeat. "MVB lacks powers of a high order intellectually and mentally," he wrote, "and excited no enthusiasm and like most men who lack in these particulars, relied too exclusively on address and management." These carping remarks were, of course, self-serving. Calhoun could scarcely have considered Harrison an intellectual heavyweight. Anyone who read Van Buren's public letter to the Democrats of Elizabeth City, Virginia, as Calhoun assuredly had, could not have meant what he wrote had he no ulterior purposes in mind. He was deliberately seeking at an early date to belittle his great rival—to set the stage for his nomination in 1844.[59]

A more sensible explanation than election fraud and Whig theatricals, or Van Buren's alleged lack of intellectual prowess, came from Cambreleng who was in Europe when the Bank of England raised its discount rate and the notes of the Bank of the United States were protested. He wrote Van Buren,

> I feared another crisis in America at a time most unpropitious for our cause and at an admirable moment to serve the intent and purposes of the opposition. This second revulsion and no other cause whatever

(Abolitionism was contained and lost us only one vote) has elected your opponent and would have elected any other man. It depressed every description of property below its extreme point even in the revulsion of 1837 and produced a spirit of general discontent which the opposition easily and readily availed themselves of.[60]

« CHAPTER 26 »

Search for Vindication

Feelings of personal humiliation, of rejection, were mingled with those of relief that the burdens of demanding office would soon be lifted. Had Silas Wright not spoken plainly about duty to the party, Van Buren would have made a public declaration that he would never again be a candidate.[1] But, as Wright reminded him, he was still head of the party. He must rally the crestfallen, inspire the faithful and renew their confidence in the abiding principles of the Democracy. His first task was the message, that would sum up four stormy years of financial panic and depression, of serious border disputes, of Indian removal and abolitionist agitation.

He could have drawn on any number of able men for style and content—Paulding, Kendall, Woodbury, Benton—but he knew what he wanted to say and he wrote the message himself without any consultation and with less than his usual quota of strikeouts, rewrites and rewordings. As befit a lame duck President, the message was more declaratory and descriptive than specific in nature. In three areas Van Buren emphasized a pride of accomplishment: his policies had maintained the peace in the face of grave provocation; he would leave to his successor a Treasury that was lean but solvent, a public credit unimpaired despite unusual drains upon it; and a system of managing the public funds for the benefit of the public not the private interest—or at least that was the way he saw it. Much of the message was a sober yet sharply etched definition of what he considered to be the role of the federal government over the lives of the people. Without mentioning Jefferson, he proclaimed his adherence to strict construction, opposition to a national debt and a national bank, both of which he regarded as unconstitutional, undemocratic and unsafe. His only reference to state banks was an oblique one in the context of the subtreasury system which he assured the nation would be the basis for eventual prosperity. "Under judicious state legislation," the economy would not be damaged by suspension of specie payment or "the explosion of a bloated credit system." There was nothing new or original in the message that seemed in a way as tired as its author.[2]

While he was working on his message a Canadian deputy sheriff, one Alexander McLeod, took a glass too many in a tavern on the American side of the border and bragged that he had been a member of the group that captured and burned the *Caroline* in 1837. New York authorities promptly arrested, jailed and indicted him for murder. The *Caroline* case had been slumbering for three years, and Van Buren would have preferred to leave it in that state; for there were compelling arguments on both sides and much more important issues to be resolved. He recognized at once that the McLeod case raised difficult, if not dangerous, questions just when relations between Great Britain and the United States were improving and the stage had been set for a final adjudication of the troublesome northeastern boundary. Although he thought the arrest and trial of McLeod was a state affair, he worked to make the strongest argument possible to what he anticipated would be a vigorous British protest.

The Administration was well prepared when Fox demanded McLeod's release asserting that the *Caroline* affair was a matter of state between Great Britain and the United States. Forsyth replied to Fox citing a wealth of precedents that the sovereignty of the United States and New York would be violated if Britain accepted responsibility, and if it did not, then it was purely a concern of New York.[3]

Congress had taken up the case by this time and demanded all relevant correspondence on the *Caroline* affair. Van Buren was happy to comply because the communications involving the *Caroline* on the American side presented as strong a case as possible for redress, short of going to war which no one desired. As he had with the Maine boundary, the Mexican Claims Commission and the Texas boundary, Van Buren had now staked out the American position, leaving any final settlement to the incoming Administration.[4]

Of more serious moment to the Administration politically was the collectorship in New York City. With Butler, who was now the federal district attorney in the City, keeping his eye on Hoyt's activities, Van Buren had felt reasonably secure the new collector would not turn out to be another Swartwout. Still, he worried about Hoyt, whose communications and reports to Woodbury were frequently unsatisfactory to the Secretary.[5]

When Hoyt, in anticipation of removal and prodded by Butler, began to settle up his accounts, a discrepancy of about $30,000 came to light in fees which Hoyt had never reported, claiming that under the subtreasury Act they were due him as part of the "emoluments" of his office. Woodbury thought otherwise as did his solicitor, Mathew Birchard, and the first auditor.

Hoyt may have been corrupt and his administration of the customs office extravagant, but he was an able lawyer as was, of course, Woodbury. The exchanges were sharp and adversarial, each man with his own interpretation of the issues. Van Buren turned to Silas Wright for advice on the merits of Hoyt's stand, and he, in turn, wrote Butler. The District Attorney, who himself felt somewhat defensive and embarrassed about the disagreement, assured Wright that Hoyt "has been right in principle, however hasty or disrespectful in manner. His integrity I believe unimpeachable." But Wright's intervention had alerted Butler, who began to press Hoyt about his interpretation of the "fees" he had withheld from the Treasury. By mid-February he had changed his mind about Hoyt's "integrity" and thought his position was a cover for private transactions with public funds.

After examining the documents Van Buren authorized a suit be instituted against Hoyt and his bondsman. Summarily he removed the Collector from office. The amount of alleged embezzlement was calculated to be $30,519.83, a mere trifle of the Swartwout fraud. Nor had Hoyt concealed this sum, but on the contrary, stoutly defended his right to it. Eventually he would win his case, but the whole affair should never have been allowed to develop. Van Buren, Butler and John Van Buren, who had interested himself in Hoyt's affairs, had permitted a public servant to engage in activities that, if not precisely unlawful, were decidedly an abuse of the public trust.[6]

With the exception of the Swartwout frauds, which may properly be charged to Jackson but for which he bore most of the blame, and the Hoyt affair, Van Buren could look back on an Administration largely free of corruption and scandal in high places. His subtreasury law was the first serious effort to define and clarify the responsibilities of federal officials who handled public funds, the first legislation which made it a felony for conviction in any such misappropriation.[7]

On the same day he removed Hoyt, Van Buren attended a memorial service for an old colleague, Philip P. Barbour, associate justice of the Supreme Court, who had died suddenly on February 24, 1841. As he listened to Chief Justice Taney deliver the eulogy, he pondered whether to appoint a successor, risking possible rejection and the same criticism that had been leveled at John Adams for appointing midnight judges, or should he leave the nomination to Harrison.

Van Buren had a healthy respect for the parliamentary tactics of

Henry Clay and if he was to make a nomination, he had no time to lose. Van Buren returned to the White House with his mind made up. He was indifferent to the criticism, and he would risk defeat. He would make the appointment and strictly on political grounds. Barbour's replacement must be a Virginian, a loyal Democrat, a judge who would stand as a bulwark against Henry Clay's American system and a politician who understood the temperamental Ritchie. There were several Virginians who fit this pattern, but none so well as Peter V. Daniel. It had been largely through Daniel's efforts that Virginia had voted for Van Buren in the election just passed. An ancillary but certainly contributing factor was that Daniel had five years of judicial experience as a federal judge for the eastern district of Virginia.

Before Barbour was buried, in fact the very afternoon after the service in the Supreme Court chamber, without consulting Daniel whether he would accept, Van Buren scribbled the few lines nominating him and sent Martin, with the note to the Senate. As expected, there was instant uproar, with Clay taking charge to delay and defeat confirmation. A first effort was to abolish the Virginia–North Carolina circuit uniting Virginia with Chief Justice Taney's Maryland circuit and North Carolina with Justice Wayne's Georgia circuit. A new western circuit would be created where there was a real need, but in the emergency it was conceived as a means of securing the support of western members. An added advantage, at least to Clay's thinking, was that even if Daniel's appointment was not districted out of existence, the proposal would call forth debate and delay. The bill succeeded in the Senate beyond Clay's wildest dreams by a top-heavy vote of 34 to 13 after a short, though spirited debate where a furious Clay heaped such opprobrium on William R. King of Alabama that everyone anticipated a duel. In the House, however, the Democrats stood fast and finally were able to table the bill effectively killing it. These maneuvers had taken up six of the nine days remaining before Congress must adjourn.

On the afternoon of March 2, the Daniel nomination came before the Senate. The situation was tense and dramatic enough to jam the galleries. Even many of the restless officeseekers who had been besieging the President-elect took time off to watch the spectacle as the clock ticked on, measuring out relentlessly the last hours of the Van Buren Administration. The failure of redistricting in the House had had its effect on the western Senators. When the motion came up to table Daniel's nomination, the Democrats held together and defeated it by four votes.

This was crucial, but Clay was not beaten yet. He lost another motion to refer the nomination to the judiciary committee; and while a roll call, the third, was being taken on his motion to postpone action, he counted the Whig and Democratic Senators present. If all the Whigs left the chamber there would not be a quorum. The word was spread, and when the roll call on postponement was rejected, he led all but two of the Whig Senators from the hall. By now, it was evening and when one of the Whigs who had remained on Clay's orders to demand a vote on whether a quorum existed, the Democrats were briefly thrown into disorder. Had it not been for the cool, collected Silas Wright, the Senate might have adjourned. He suggested the Senators remain in their seats, all business suspended while the clerk, the sergeant at arms and their messengers went out and scoured the city for absent Democratic Senators. In an hour or two, twenty-seven Senators, including one Whig, were in their seats, exactly a quorum. Should the Whig Senator, Smith of Indiana, leave, he would break the quroum, but then there would be no member present to ask for a determination. All he could do was remain and vote with four Democrats who opposed the nomination. Van Buren's midnight judge, Peter V. Daniel, was confirmed a few minutes before midnight, March 2, 1841.[8]

All was not work or even politics during the closing days of the Administration. Van Buren was as sociable as ever and the dinners went on as before. His fine spirits were so infectious at the Christmas dinner he presided over that even the dour Calhoun, a guest, engaged in frivolous small talk. Yet politics could not be kept out of conversation in a city where it was the only way of life; nor could Van Buren forgive all of his political foes. Those, like Clay or John Jordan Crittenden, his colleague from Kentucky, were always welcome at the White House. But for those of his opponents who had enjoyed his hospitality and then abused it by echoing the base slanders of political hacks like Congressman Ogle of the gold spoons lie, Van Buren did not conceal his disdain.[9]

After the first of the year, the city gradually filled up with expectant officeseekers; and when the President-elect arrived on February 9 to take up his temporary residence at Gadsby's, the slushy streets, taverns, oyster houses, hotels and lobbies of the capital were jammed with visitors, some, like Philip Hone, simply in Washington to witness the inauguration, but most voracious for any crumb of patronage that might come their way from the new Administration. Poor Harrison, feeling the weight of his sixty-eight years, could scarcely move about the city without being importuned by hordes of officeseekers.

The old gentleman called upon the President on February 9, the day after he arrived, for a pleasant half hour's chat. Van Buren, breaking precedent, returned the visit accompanied by his entire cabinet. He also entertained Harrison at a small dinner party and observing how tired he looked, offered to vacate the White House by February 20 so that Harrison might have more seclusion than he could get at the drafty barnlike Gadsby's with those smoking coal fires. The President-elect declined, but was touched at his predecessor's thoughtfulness.

Few of the Democrats were as civil and courteous as Van Buren. Marcy made no effort to hide his contempt for the "political rogues who enacted their parts 20 or 30 years ago and who as I supposed passed into obscurity or the grave are all here perambulating the streets in the cause of Tip'canoe and Tyler too." Marcy reserved his sharpest barb for Harrison who, on the eve of his inauguration, had regained his health, if not his nerves. "I have not seen old 'Tip', but all represent him as merry as a cricket," he wrote Wetmore, "careless of the future, garrulous in the display of obscene stories, thoroughly intent with the spirit of lechery. He has according to his own account devoted the seventy years he has lived to Venus and Mars."[10]

While the badgered Harrison was astonishing Washington's social circle by arising early in the morning and rushing off on foot to the central market to purchase his own groceries, Van Buren accepted the resignation of his cabinet to take effect at noon on March 4, 1841. Woodbury, who had been elected a Senator from New Hampshire, would stay on; the rest of his colleagues were planning their departure for home when the roads were dry enough after spring thaw to be passable. Although long sections of railroad track had been completed between Washington and New York City, there were still gaps requiring coach travel. The night before Harrison's inauguration Van Buren and his family moved out of the White House to stay with the Gilpins in their commodious house. Martin, Jr., who was quite ill at the time, was being nursed by Mrs. Blair. Van Buren probably would have left Washington immediately had it not been for concern about his son.

Kinderhook and the Van Ness mansion he had purchased a year and a half earlier for his retirement home seemed more alluring than ever and he was anxious to begin refurbishing the house and grounds.[11] Besides the fifty acres that went with the Van Ness estate, Van Buren purchased an additional 150 adjoining acres of farm land from Paulding's older brother, William.

He had never had a permanent home during his busy career though he owned several farm houses in Kinderhook and surrounding towns. Before the Van Ness property came on the market, he was planning to build what he described as "a cottage" on some spectacular real estate in Stuyvesant overlooking the Hudson and but a few miles west of Kinderhook. At fifty-eight and with the prospects of large family, as his sons married and grandchildren came along, Van Buren for the first time in his life looked forward to playing the role of patriarch.

When he had moved into the White House, his accommodations were grander, of course, but this was just the culmination of an entire life spent in temporary quarters, boarding houses, hotels, rented homes in Albany and in Washington. Though essentially a convivial person, Van Buren found the social obligations of the Presidency difficult without a hostess. Quickly it had become apparent he needed a housekeeper who could keep track of twenty or so servants, plan the elaborate state dinners and receptions, while overseeing the smaller gatherings with close friends and family he so much enjoyed. Harriet Butler and Mrs. William A. Duer had asked about among their friends and eventually found an experienced household administrator who met the President's needs.[12]

The White House had been a lonely place for Van Buren during his first year in office, despite his public face of charm and wit, especially to his female guests. Van Buren's social composure did not deceive one young lady who observed him closely. She thought he looked careworn and depressed, adding, "He goes out little and then seems to take little pleasure in society."[13]

The financial crisis was much on his mind; he worried about his son John who was in England. As Smith was at school in Pittsfield, Massachusetts, Van Buren depended upon Abraham, Martin, Jr., and Benjamin Butler for his family circle. It was strictly a male household where there was little relaxation for a weary President on a day-to-day basis beyond a bottle of Madeira or Sherry in his private drawing room with Abraham and Martin, Jr., relaying the Washington gossip Van Buren always relished. The high-minded Butler, a strict temperance man, may have disapproved, but he usually joined the group.

In the fall of 1838 an event took place in Van Buren's life that he had long desired. Abraham married Angelica Singleton, the younger daughter of a rich South Carolina planter related to many of the leading families of the South like the Cabells, the Prestons and the Stevensons. Angelica and her older sister Marion had met the President and his two sons at a White House dinner in March of 1838. The young ladies were attracted to both Abraham and Martin, Jr., whom Marion described in a burst of complimentary adjectives as "pleasant, unpretentious, unpretending, civil, amiable young men." Martin, Jr., frail and shy, resisted the charms of the southern belles; but not his older brother Abraham, who fell an easy victim to Angelica's charm. By late summer they were engaged, and in November they were married at Colonel Singleton's plantation.

Van Buren could not be present, but he was obviously delighted at the match and Angelica's even, sweet-tempered disposition. The bridal pair did not remain long in the White House. Both wanted to visit Europe and Abraham craved relief from the tedium of acting as his father's private secretary. The young couple left in the early spring of 1839, reaching London just before John Van Buren sailed for home. He had been delayed by illness and unusually severe weather conditions which made an Atlantic crossing a dangerous ordeal for passengers on the packets.[14]

The brothers met after almost two years of separation at the young couple's hotel. The Major and Angelica were in bed, but no sense of delicacy stopped the irrepressible John, who talked with his brother through the open bedroom door for an hour, the sole concession to privacy being that John sat in the dark in the entrance foyer. One wonders what Angelica must have thought, not just of John's informality, but of Abraham's too. After learning about matters at home, John was off to an exclusive London club where he danced, ate, drank and gambled far into the night. In the style of a Regency buck, he recorded this episode as he had all of his previous socializing for his father's amusement and perhaps his consternation.

> Went to Almack—all the world there, Ladies F. Cowper, W. Stanhope; presented to the latter, don't like her now I know her, she is insipid. Found old Lord Powerscourt dinner party there; Grand Duke and Prince William there. I danced with Breadalbane's sister . . . wound up at Crockford's.[15]

By the fall of 1839, all the younger Van Burens except John were at home. John was in Albany working hard to build up a law practice as much a victim of hard times as it was of his lengthy stay in Europe. Angelica now acted as hostess for the President and assisted by the housekeeper managed the White House establishment. For the Washington ladies, at any rate, Angelica's presence, her afternoon teas and dress balls, brightened the social season that had been rather dreary for them during the past two years. But Angelica became pregnant before the session adjourned and as was the custom remained in seclusion during the early months of the second session. When her baby daughter Rebecca was born in late March 1840, both mother and child were sick for several months. Little Rebecca never did recover. Van Buren's first grandchild died at the White House in the fall.[16]

Van Buren had long hoped his son John would marry and forsake the high-living bachelor life he was leading in Albany and in New York City. Though no prude about such matters, even among the adult males of his own family, he could not have approved of John's affair with a voluptuous Italian lady of easy virtue with the fanciful name of Maria Ameriga Vespucci. Claiming direct descent from the sixteenth-century explorer whose latinized first name came to be the accepted one for the continents of the western hemisphere, Ameriga had come to the United States ostensibly seeking a grant from Congress for the work of her ancestor. Van Buren received her kindly in Washington, but after he listened to her reason for visiting him, politely sent her on her way with one of his "noncommital" answers. The halls of Congress and the executive offices proved no more productive. Vespucci left for what she hoped would be a more fruitful tour of New York and Philadelphia. She cut quite a swath among the "fast crowd" of New York, where she met John Van Buren. Assuming the President's son would provide her with the proper entrée to pursue her absurd claim, she eagerly offered her abundant charms to an equally ardent John Van Buren. The liaison did not last long, though hints of it crept into James Gordon Bennett's scandal-mongering *New York Herald.*[17]

Despite weaknesses of the flesh, John Van Buren had for some years been rather deeply attached to Elizabeth, a daughter of Judge James Vanderpoel, his law teacher and an old friend of his father's.[18] As Van Buren made preparations for leaving Washington after Harrison's inauguration, he had the welcome news that John was engaged to marry Elizabeth in the summer and gave every indication that he was now ready to leave the gaming tables and the sporting life. Van Buren was also cheered to learn from Abraham that Angelica was pregnant again. By March 10 it was clear that Martin, Jr., was out of danger and while he was in no condition to travel, he was in good hands and could be safely left in Washington until he was well enough to make the trip north.

Van Buren called on the President to wish him well on March 11, 1841. Declining a dinner invitation at the White House the next day he, along with Silas Wright, boarded the cars for Baltimore. He planned to stop a day or so in that city, in Philadelphia, with Dickerson in Newark and at least three weeks with the Butlers in New York. As he explained to Marcy, he had to buy furniture for his new home in Kinderhook. Much of his old furniture he had sold in his many moves; and though John Van Buren had purchased some elegant glassware and other necessary items for his father in England, the Van Ness mansion was larger and grander than any of his former residences. Such standard household furnishings as carpets, beds, chairs, sofas, clothes presses and the like as he had in storage were totally inadequate, either too small or too few in number.[19]

Unquestionably, he craved a respite from political concerns; yet he spent most of his time at Gilpin's writing graceful notes of appreciation to loyal Democratic leaders in many of the more populous states, and in acknowledging resolutions from party members of state legislatures. He gave an emotional farewell tribute to the Democratic members of Congress. Of more political significance than these acknowledgements were two letters he took particular pains to write.

Without Van Buren's knowledge, Thomas Hart Benton had managed to have the Missouri legislature nominate him for reelection in 1844. Robert Morris, on behalf of the Democracy in New York City forwarded complimentary resolutions the party had adopted in a mass meeting. Whatever Van Buren may have thought about his political future, he realized at once that the Missouri nomination was premature. In a public letter he wrote Governor Reynolds of Missouri, a small masterpiece of political ingenuity, he declared he was not indifferent to the high offices in which he had served, a privilege and a great honor open to every American. "No one," he said, "can expect or should desire to be always in office under a government and institutions like ours." Without closing the door to a future nomination, he firmly closed the door on any speculation that he was already maneuvering for reelection; that he had an unquenchable desire for personal aggrandizement.

Here was the disinterested statesman speaking, not the perpetual officeseeker and chief spoilsman he had been pictured in the Whig Press. He responded with similar sentiments in his reply to the New York group, enlarging on the debt of gratitude he owed the citizens of his native state whose interests were ever close to his heart. In what seemed like an afterthought, he said he planned to be in New York City on March 23. This letter too was written for publication.

With such publicity Van Buren's return to New York took on the appearance of a triumphal procession rather than an escape from defeat. An admiring Marcy thought Van Buren's carefully orchestrated farewell was a marvel. "He played the last scene, always a very difficult one, with the greatest skill; he has been at fault in nothing," said that veteran of many a political campaign.[20] Van Buren was pleased with the attentions shown him on the way and he expected a decent turnout in New York. When he reached the New Jersey shore of the Hudson, however, a biting northeast wind blew in a heavy storm which burst in a torrent of rain while he was crossing the river. The storm, Van Buren assumed, would put a damper upon any public demonstration. It proved otherwise. To his astonishment the crowds under the pelting rain were, to his practiced eye, greater even than his carefully prepared state visit under sunny skies eighteen months before.

While the sodden multitude cheered him as he stepped ashore at Castle Garden, a bedraggled reception committee suggested he ride in a closed carriage. Quick-witted as always, he saw no reason why he should ride in dry comfort while his fellow citizens stood in the rain. Escorted by a troop of mounted militia, their plumes plastered against their helmets, and several hundred firemen, the ex-President and New York's senior United States Senator, both bareheaded, waved at the crowd. Wright's high-standing forelock, the delight of Whig caricaturists, flattened and streamed down the sides of his ample forehead, while Van Buren's bald dome glistened with water. Up Broadway the procession crawled along, their carriage obliged to stop several times before it could move through the dense throng that pressed about it. Turning south from 14th Street and east on Bleecker through the teeming slums of the City, south again through the Bowery to Tammany Hall, a plain five-story brick boxlike structure on the corner of Nassau and Frankfort near city hall. Here Robert Morris, the ex-Recorder, and Mayor Isaac Varian made short speeches of welcome and received an even shorter reply from a thoroughly drenched but eminently cheerful Van Buren.

The ex-President then made his way as quickly as the crowds would permit to Butler's house on Greene Street and Waverly Place near Gramercy Park. Evening had long since fallen by the time Van Buren and Wright changed into dry clothes and ate a late dinner. Both men presumably had fortified themselves against the chill, knowing well there was not a drop of liquor even for medicinal purposes in the Butler household. After dinner Wright went off to the City Hotel where he always stayed in New York, but the irrepressible Van Buren, bouyed up certainly by a reception beyond his fondest expectations, attended the Bowery Theatre where a visiting equestrian and acting group were presenting one of those potpourri performances of drama and circus then in vogue. Van Buren, who was recognized in his fifty-cent box seat, was repeatedly cheered from the pit and the gallery.[21]

For the next five weeks, Van Buren was a whirlwind of activity. Almost every evening he visited or received guests, attended operas, plays and other public entertainments, as much to be seen as to see and to enjoy the elaborate theatrical fare of New York. By day when he was not talking with old political friends, he was visiting cabinetmakers, upholsterers, shops that dealt in china and glassware, silversmiths and suppliers of farm implements. He kept abreast of events in Washington from the dozen or so newspapers he always scanned and a daily bundle of letters from the capitol. Instead of winding down the tensions and responsibilities that had taken their toll over the past four years. he at first seemed incapable of rest and relaxation. The quiet, orderly household of Harriet Butler's must have been subjected to great strain.[22]

In the midst of his activities came the news that President Harrison was seriously ill; he died on April 4.[23] If Van Buren had little respect for Harrison the man, he did for Harrison the President. When the Whig Mayor Isaac Varian and the Democratic majority on the city council in a display of nonpartisanship declared April 10, 1841, to be the day New York would observe official mourning, businesses all over the city closed down as did the theaters. The newspapers had published the procession route; so shops, houses, hotels and taverns bordering it were festooned with black crepe or some other funereal badge of mourning. At major crossings bunting interwined with crepe and with crude likenesses of the late President were suspended across the streets. Van Buren was conspicuous among the city's twenty-six distinguished citizens chosen as honorary pallbearers.[24]

On the morning of May 8, 1841, another stormy day, Van Buren boarded the day boat for Albany. Butler and his son Benjamin, Jr., were with him. Intermittent rain squalls drenched the deck of the day boat as it moved against the tide at a steady six knots, stopping every hour or so to discharge and take on passengers and freight at the villages and towns along the east and west banks of the river.

Van Buren had traveled on the Hudson as recently as November, but he had not viewed both banks of the river from the City to Albany since 1835. While other passengers remained below, he stood on the deck of the *Albany* braving the showers, eager to observe any changes in population, increases or decreases in factories and clearings for farm land. Very little escaped his attention—the traffic on the river, the number of wharves, any differences in economic activity over what he had remembered. At 1 o'clock, the showers ceased, the sun broke through and within an hour all was clear and bright and beautiful.

Butler and Van Buren arrived at Stuyvesant late in the afternoon where a large crowd of residents were gathered on the Point. Four marshals lined up the carriages, buggies, broughams, wagons and farm folk on foot and on horseback, and what Butler described as "a very respectable procession" escorted the two men to Kinderhook. Fortunately the warm weather of the past few days had dried up the muddy road so that the vehicles rolled along without delay, arriving at the village at dusk. Amid a pealing of bells from Kinderhook's three churches, and cannon salutes, Van Buren was met by its president, Mordecai Myers, who escorted him and Butler to the piazza of Stranahan's Hotel. There, under the illumination of dozens of torches, the ex-President responded briefly to the welcome of his fellow townsmen.[25]

He was then escorted to his new home about one mile to the

southeast of the village center where he was happy to find all there in readiness for his arrival. The stately Van Ness mansion had few furnishings, and those much worn, when he had purchased it a year before. Since then he had had the place thoroughly renovated and furnished.

In 1839, during his state visit, Van Buren, his son Smith and his widowed sister-in-law Catherine, had dinner there. The house, thought Smith, was "clean and no more." After dinner of fricassee and ham washed down with champagne, everyone tried to think of a name for the estate. "The present favorite," said Smith, "is the Locusts of which there are a great number about. The only objection is that the name is used in Cooper's *Spy* for one of his places." Because the front yard about 100 feet from the road was dotted with lofty pines and lindens, some one of Van Buren's friends or relations, possibly his son John, who had a smattering of German, suggested Lindenwald (linden forest), a rather pretentious name that appealed to Van Buren's knowledge of the immediate locale, yet imparted a sense of dignity, a family seat in the English or southern tradition.[26]

During his hectic month in New York at the Butlers he found time to send off instructions for improvements to the house itself, the grounds and outbuildings, many of which were in a sad state of repair. And as soon as he settled himself at Lindenwald, his days were spent directing carpenters, masons, painters and laborers. Crates of furnishings he had purchased in New York were arriving to be distributed throughout the house. "The size of the house and the length of time it has been unoccupied have added to expense and time of supervision," he wrote Woodbury in late July after seven months of restoration.

Prudent always in money matters, Van Buren did not spare his purse when it came to Lindenwald. Expensive French wallpaper brightened the plain plaster walls of the downstairs rooms. The finest Brussels carpets obtainable in New York covered the pine floors of the old mansion. He adorned the entrance hall with one of Colonel Earle's portraits of Jackson ("which is the best he took"), and in the dining room he hung "an excellent likeness of Mr. Jefferson which I have had the good fortune to procure."

On the main floor just off one of the parlors, he converted a spare room to an office, or more appropriately a library. During his years in Washington he had collected books and tracts on political subjects so that when he took up residence at Lindenwald he spent a good deal of time unpacking and arranging his library. Besides works of a practical or professional nature like these, Van Buren was acquiring novels, books of poetry, biographies and histories, primarily the work of American authors, Bancroft, Irving, Paulding, Bryant, to cite but a few of the inscribed copies that found their way to his library. Over the mantel in the library he placed a large steel engraving of Henry Clay, his mobile features, the hint of sly humor admirably caught by the artist and faithfully reproduced.

But the outbuildings, grounds and fields surrounding Lindenwald commanded Van Buren's principal attention. For he was determined to have a producing farm as soon as possible, and the approaches to the house, so long neglected, put in a presentable state for his own and his visitors' critical eyes. By the end of July, he had almost finished. The wild timothy, plantains, burdock, thistles and sumac in front of Lindenwald had been ploughed away, the earth manured, seeded, rolled into three acres or more of lawn gently falling away from the house on its slight rise. All the barns, stables, wood houses, poultry coops well behind the house, had been repaired or replaced. The tumbledown stone walls and rotted stiles were put in order. The mansion itself he had painted white.[27] On one side of the house, Van Buren built a glass-roofed conservatory, on the other were two ponds for ornamental fish. The semicircular drive from the Kinderhook road was thoroughly leveled and smoothed over with gravel.

Van Buren was very proud of the transformation and "barring the pressure upon my purse," he wrote Jackson, "the occupation has been pleasant enough and everybody that sees it says that I have made one of the finest places in the state." The kitchen garden that had been a cornfield when Van Buren purchased the property was now supplying him with provender. "I have had this season the first cabbage, first potatoes, beets and turnips and as early beans, corn, peas as we have grown in this neighborhood of flourishing farmers and splendid farms" he boasted.

That summer may have been the happiest interlude in Van Buren's life; it was certainly the sunniest.[28] In late June he went up to Albany for John's wedding to Elizabeth Vanderpoel and accompanied the happy wedding party to Lindenwald where the young folks organized picnics, walks along Kinderhook Creek, which flowed through Van Buren's property, fishing expeditions in the several small lakes near the mansion, music, singing and laughter at night after dinner, with plenty of champagne and other wines and sweet liqueurs. John had wanted all his brothers present at the wedding and urged Abraham and Angelica, who were at Colonel Singleton's, to make the trip north. But Angelica was six months pregnant, "prosperous and ponderous" as the Major described her condition, and the long trip at this stage was far too dangerous for her to undertake. He promised that they would be at Lindenwald as soon as the baby was born and Angelica could travel. Van Buren's second grandchild, a son, was born in May, and by the end of June the little family was at Lindenwald. The child, named Singleton, thrived in the cool, treeshaded house. "He is not large by any means but in perfect health," the Major wrote his father-in-law.[29]

The only family matter that bothered Van Buren during the summer was John's state of finances now that he was a married man. Though his son assured him his law practice was steadily improving, the younger Van Buren's speculations had left him with a debt of over $20,000. John had been frank with his father about his financial problems. After totting them all up he said, in his jocular fashion, "So you see, for a young gentleman about committing matrimony I am as badly off as my worst enemy could desire. One thing is certain Miss Elizabeth won't marry me for my money."[30] His economic prospects did not improve during the fall and he had to borrow $500 from Abraham on a ninety-day note to meet some pressing obligation. He talked of trying to obtain the attorney generalship, a lucrative post, but the Democratic legislative caucus had other plans. Van Buren drew back, refusing to meddle openly in state politics for what would obviously be seen as nepotism.[31] Busy and content in retirement, Van Buren had not lost his interest in politics, though at this time he was more of a bystander than an actual participant. He watched carefully the battle royal being waged between Tyler and the Whig majority in Congress headed by Clay.

Marcy, who was still in Washington on the Mexican Claims Commission, tried briefly to make some gains for himself and for the Democrats out of the rift by swinging the President over to the party. At first, Tyler welcomed Marcy's interest and support, and thus encouraged, the wily politician tried to gain control over the New York patronage all of which was still in Clay's hands. But he was quickly brought up short when Tyler ignored his suggestions about patronage and appointed John C. Spencer, anathema to all Democratic factions, to be secretary of war.[32]

If Van Buren had been annoyed at Marcy's flirtation with Tyler, he gave no sign, remaining conspicuously out of things, declining public dinners in his honor, replying graciously but innocuously to the resolutions from numerous Democratic committees in various parts of the country as local politicians tried in vain to draw him out on the next campaign.[33]

Yet his stance was not altogether apolitical. As the first chilling breezes from the northwest began gilding the tips of the lindens and touching the maples and oaks with crimson, Major Van Buren prepared to return with his family to South Carolina for the winter. Why not visit his son in the clement South when, as he knew, Kinderhook would be buried in snow and the Hudson frozen solid almost to the city? He had long wished to visit the Hermitage, had in fact planned such a trip in 1838 only to be dissuaded by James K. Polk who was running a close race for Governor and feared his presence in Tennessee would hurt his chances for election. Why not see some of the Southwest, visit the Hermitage and return home by the Northwest? Jackson's invitation was always open. An undeniable family visit combined with homage to Jackson would be innocent of politics. Or would it be? The answer would have to be a qualified negative.

At fifty-nine, Van Buren was still healthy and vigorous enough to brave the discomforts and the diseases of travel in underdeveloped and wild regions. He did want to escape the worst of the harsh winter months in the Hudson valley; and he did want to visit friends like Stevenson, Poinsett, Polk and, of course, Jackson, before the sick old man died, and he himself had lost the stamina for such an excursion. Still he knew that any trip, whatever personal reasons he might give for making it, would be construed as having a political bearing.

While he had not as yet made up his mind to face the uncertainties of securing the Democratic nomination in 1844, nor the burdens and responsibilities of the Presidency should he be nominated and elected, he had not ruled out a trial either. Much depended upon the November election in New York. If the Democrats lost or made a poor showing, it would shape his plans not just for the trip but for the future. If on the other hand there was a Democratic landslide, he would not discourage any groundswell in his behalf.[34]

Prospects looked favorable for the Democrats. They controlled the House but the Whigs had a solid majority in the Senate. Governor Seward had frightened conservatives and moderates in both parties by refusing extradition to Virginia of three blacks, one of whom was a fugitive slave. His Administration had also embarked on extensive programs of internal improvements that Flagg and Dix, highly respected in such matters, charged were impairing the credit of the state. But most political observers thought the Whigs would maintain their hold on the Senate. Seward's declining popularity, the inability of the Whigs to use the federal patronage effectively because of the party split in Washington and the continuing depression, provided the ingredients for a Democratic sweep which increased their majority in the House and restored their majority in the Senate.[35] With the legislature again under their control, the Democrats elected a complete slate of state officers. Flagg regained the comptrollership, Samuel Young, long identified with Van Buren, became secretary of state and a majority of his supporters received other important posts including the Canal board.[36]

Well before the legislature made these appointments, Van Buren had come to a decision after talking the matter over with Silas Wright, Dix and Croswell. A letter from Henry Horn, one of Van Buren's close political associates in Pennsylvania, had convinced him that he must act more positively if he were to be a serious candidate. Horn had written immediately after the defeat in November that some of Van Buren's friends had thought it expedient to open the campaign for the next presidential election. A majority, however, had deemed any such movement for Van Buren premature. In view of the New York election, Horn asked directly whether he and others should begin to organize the party in Pennsylvania. He alluded to other candidates who were in the field, an unmistakeable reference to Calhoun and to Buchanan. Van Buren's letter to the Missouri legislature, Horn said, was being cited as proof that he had taken himself out of the race for 1844.[37]

Van Buren now moved quickly to clarify his Missouri letter. In his reply to Horn, which was published widely, he repeated that neither he nor any public man should take the initiative in pushing for nomination. It was unbecoming, un-Jeffersonian and undemocratic for anyone to thrust himself forward simply to gratify his ambition. The party should select the person it thought best represented Democratic principles. Speaking of his Missouri letter he denied that it meant he would refuse a mandate from the party. His words were carefully chosen when he wrote, "Whilst such are the lights in which the subject is regarded by me, and whilst I shall most assuredly never take a single step with a view to be made a candidate, I have not said that I would decline the performance of any public duty . . . [if] the people of the United States should see fit to call me and which shall not be inconsistent with the station I have already held through their favor."[38]

The publication of the letter and the news that Van Buren was planning a southern trip suggested to some politicians that he was a candidate. Buchanan reacted as Van Buren assumed he would, by speaking of his unpopularity in Pennsylvania. "I think he would consult his lasting fame and the interest of the party," he wrote Campbell P. White, the New York politician and banker, "by remaining at his retirement near Kinderhook."[39] Other Democrats, for entirely different reasons, wanted Van Buren to remain at home. Albert Tracy, Marcy's friend, was uneasy when he read Van Buren's reply to Horn and learned of his plans to make a trip to the South and the West. Assuming that he was innocent of any political motives, Tracy wrote, "Our opponents will say it is an electioneering trip for place and power." Van Buren promptly disavowed any political motives and reassured Tracy that he "need have no apprehensions that I will do anything which can, with any show of reason, produce the feeling to which you allude and in respect to which your views are entirely right."[40]

Yet he could not ignore the fact at the time he wrote Tracy that more and more Democratic meetings had come out in his favor and that an increasing number of Democratic papers were talking of his renomination as a vindication of true Republican principles. Nor could he fail to notice the activities of would-be aspirants—Calhoun, Cass, Richard M. Johnson. Van Buren had sought to smoke out Calhoun's intentions before he had decided to become available for renomination. On behalf of the New York Lyceum, he had invited the South Carolinian to make the introductory lecture in the fall of 1841 and Calhoun, perhaps sensing motives other than complimentary, declined. In September he turned down an invitation from John Van Buren to attend a mass convention of the party in the eastern and southern counties of New York.[41]

The letter to Horn and rumors of a southern trip alerted Calhoun, if he needed alerting, to a Van Buren candidacy. Writing to Armistead Burt, one of his trusted lieutenants in Congress, he said if Van Buren were pushed for renomination, "it would be fatal to the cause," and with the tariff and abolition in mind, he amplified this comment. "In fact, no northern man ever has been, or ever will be, under existing circumstances, a reformer." Completely oblivious that he was doing Van Buren an injustice and making no attempt to conceal his own self-serving motives, he announced his candidacy. "If my friends should think my service will ever be of importance at the head of the Executive, now is the time," he said.[42]

Impatient, impulsive, Calhoun seemed completely unaware how imprudent an open campaign would be at this stage. The Whigs were in confusion; but Webster and Tyler might just, with the force of executive patronage, create a third party that would entice conservative Democrats, state bank factions in both parties, officeholders and malcontents everywhere who were mired in the fourth year of a grinding depression. As Van Buren understood, but Calhoun did not, the political situation was far too murky to make any firm commitments. Hugh Legaré, the abrasive observer of the passing scene, remarked, "Poor Calhoun! Nature certainly gave him great parts and experience has helped him but his intellectual organization is radically defective and he wants science. His glance is quick and penetrating but not broad . . . he comes to his conclusion always by the prior road and bends all facts to suit them."[43]

Van Buren could not pretend to Calhoun's intellectual powers, but in the game of politics he was the master player. Confirmation for his low, almost nonexistent profile came from one of his keenest political analysts in Washington, Robert McLellan, a member of the

New York congressional delegation. "Very little is being said in private regarding the next Presidential candidate," he reported. "The first man who pushes his head above water will be as dead as the oldest mummy of Thebes. The impression is this, let future events determine the matter."[44]

While his great rival was putting together his organization and opening himself up to attack from all quarters, Van Buren was following McLellan's advice and quietly planning his trip. Kemble visited him for a day in late November to see the estate and to dine. "I alluded to the probability of his reelection," he wrote Poinsett, "but he said he had nothing to ask, or to desire, that having held the highest office in the gift of the country, his curiosity as well as his ambition had been fully satisfied."[45]

Abraham and his family were in South Carolina. Smith, now married to Ellen King James, daughter of the immensely rich William James, had rented and was furnishing a house in Albany where he intended to spend the winter. John and his bride would also remain in Albany while he worked as he had never worked before to pay off his debts and keep up his extravagant style of living. With both of his new daughters-in-law pregnant, Van Buren happily looked forward to another increase in his family when he returned.[46]

Van Buren wanted Martin, Jr., now fully recovered from his severe illness, to accompany him, and he thought of Paulding as an enjoyable companion for the journey. After the Christmas holidays he invited his former secretary of the navy to make the trip, pointing out how pleasant the South would be at this time of year, the opportunity he would have to see Poinsett and Jackson. Paulding, who was always indecisive about any change in his daily routine, even short trips to New York City or to his numerous friends in the Hudson Valley, hesitated at what would be a monumental undertaking. He finally agreed and Van Buren set the date of departure for February 23, 1842, reassuring Paulding that "I shall accept no dinners and will make all your speeches for you that can't be avoided." Martin, Jr., decided to remain at Lindenwald where he could supervise the winter maintenance of the estate. Van Buren contracted with a Mr. Marquette to manage the farm on shares.[47]

Having settled affairs at Lindenwald and agreed with Paulding on their date of departure, Van Buren planned the route with great care. They would spend one day in New York City, one day with Gilpin in Philadelphia, bypass Washington and Richmond, heading straight for Poinsett's rice plantation on the Peedee River.[48]

With very little publicity Van Buren and Paulding made a quick, uneventful trip, arriving at their destination just ten days after leaving Philadelphia. Poinsett's home, "The White House," which he described as "only a humble cottage" was one of the most charming, well-managed plantations in South Carolina. A botanist of professional stature and a landscape gardener in his own right, Poinsett had personally laid out and planted the grounds of the White House with varieties of exotic shrubs and flowers neither of his visitors had ever seen. In the lush, semitropical setting of the Peedee River valley, the small mansion was bordered and shadowed from the hot sun by huge magnolias. Lawns reached down to the slow-moving river, broken here and there by flower beds that were ablaze with the colors of late spring. Even Poinsett's slave quarters were models of their kind. Each whitewashed cabin with its own garden plot was well-spaced from its neighbor. Though rice planting was hard and disagreeable work, Poinsett's some sixty slaves appeared content, certainly well-fed and clothed.[49]

Politically, this tidewater section between the Peedee and the Wackamaw had opposed nullification. Its leading men were deeply suspicious of Calhoun. Van Buren and Paulding were elected to the exclusive Planters' Club where old South Carolina families, the Izards, Allstons, Pettigrews, Sparkmans and Tuckers, entertained them sumptuously and made it obvious that they were among political friends. Paulding and Van Buren spent a week with Poinsett, leaving March 12 for Colonel Singleton's plantation near Statesburg in Sumter County.[50] Awaiting them was the Major and his family, Andrew Stevenson, who was recovering from a serious illness, and various Singleton relatives.

There followed another round of social activities, only this time among some of Calhoun's closest friends and political associates, Governor James Hammond, James Chesnut, Francis Pickens, ex-Governor John P. Richardson. Even the dyspeptic George McDuffie put in an appearance. Any political differences that may have existed went unmentioned as these prominent South Carolinians went out of their way to be cordial. "You can scarcely conceive a happier or more harmonious party," declared Stevenson's wife Sarah. "The gentlemen hunt and fish without any other result than the pleasure and benefit of air and exercise." As at the White House, the visiting northerners enjoyed themselves thoroughly and were sorry to leave the Singleton plantation, its fine fare and what Paulding called its "diabolical old Madeira, enough to corrupt a saint and convert an anchorite or teetotaler into a glutton and wine bibber."[51]

But their schedule was a tight one since they were following Jackson's advice and going south through Columbia to Mobile, New Orleans and up the Mississippi east to Nashville. The more direct overland route through the mountains of eastern Tennessee was apt to be impassable in the spring, Jackson warned.

Van Buren had written Jackson advising him of their departure date and that they would be in South Carolina during mid-March. What was the best route to Tennessee, he asked? Jackson's reply caught up with Van Buren while he was at Colonel Singleton's and forced a change of plans that added 2,000 miles to their original itinerary. It also changed the character of the trip because the ex-President could not travel through so much of the South and West or pass through population centers like Mobile, New Orleans and Memphis without pausing for receptions which would increasingly become political in nature.[52] Besides Jackson's letter, an invitation from Henry Clay was forwarded to Colonel Singleton's. Clay hoped Van Buren and Paulding would visit him on their return home. "Come directly to Ashland, and make it your headquarters," the Whig leader insisted. "We shall be glad to see you and give you a hearty welcome everywhere."

Van Buren responded graciously but with some caution, knowing all too well Clay's carelessness about his communications and how easily his remarks might be misconstrued. He was still determined, if at all possible, to play down any appearance of a political junket. Van Buren was not about to become one of McLellan's "Mummies of Thebes." "My movements after leaving here are not definitely settled," he wrote Clay, "and will have to be governed by circumstance of which I am not now fully advised." He hoped he would be able to visit Ashland, though not before early May.[53]

After the travelers left South Carolina, they took steamboats whenever possible. But whether by steamboat or by hired carriage, the news of their arrival at any given village or even hamlet preceded them. Mail began piling up at many a waystation with invitations to public dinners, barbecues, receptions, all in Van Buren's honor. He spent a good deal of his free time in penning courtly replies, but true to the promise he made Paulding, he declined all public festivities. Van Buren always had a brief speech ready for any group assembled to greet him along the way. The pace was strenuous, the living conditions primitive in the backwoods of Georgia and Alabama. Paulding suffered from diarrhea intermittently and from malaria when they reached the Alabama River. Not so Van Buren, who seemed to thrive on hardship; and though he made dozens of speeches and greeted thousands of people, his health remained remarkably good, his spirits resilient, his courtesy and good humor unfailing.

What amused him most was the gaping wonder of the farm folk and village dwellers who came to see him. Thoroughly conditioned by Whig stump speakers and the Whig press, they expected an effeminate little eastern "swell." "The rustical and barbarous Corydons," in Paulding's phrase, who stared at Van Buren with curiosity akin to disbelief, reached its highest point when he arrived in Nashville. The Whigs were better organized in Tennessee than they were anywhere in the Southwest. During the campaign against Van Buren, he had been so thoroughly vilified and caricatured that the crowds were astonished at his appearance. "Instead of a dwarf Dutchman, a little dandy who you might lift in a bandbox," Jackson wrote Blair, "the people found him a plain man of middle size, plain and affable."[54]

The Old Hero was quite ill, but his welcome was warm and his faculties as keen as ever. Jackson insisted on accompanying Van Buren to a huge public meeting in Nashville where, gray of face and in obvious pain, he shook hands for several hours and never missed introducing his distinguished visitor to any local politician or influential citizen he spied in the crowd. While at the Hermitage, probably at Jackson's urging and certainly that of Benton, whose many letters of encouragement he received along the way, Van Buren decided to revise his itinerary again. He would visit Polk at Columbia, Clay at Lexington and Richard M. Johnson at Louisville in Kentucky, then on to St. Louis, passing into Illinois from Missouri and going north to Chicago, returning home by way of Detroit, Cleveland, Erie, Buffalo.[55]

After the West discovered that Van Buren was not what the Whig press and speakers had made him out to be, the trip became a personal triumph, almost completely free of partisanship, Whigs and Democrats mingling together for a chance to greet the man who was supposed to be "used up." What had begun primarily as a personal tour had become another example of Van Buren's "wizardry." Though he had not planned it as such, he was quick to grasp the opportunity presented. When he arrived at home on July 28, 1841, he was the most conspicuous "noncandidate" in the nation. He had been on the road for over five months, had traveled over 7,000 miles, and besides being an ex-President, he found himself a national celebrity. "No gun powder popularity this," said an exultant Benton, "no machinery or secret springs have been set in motion to produce the result and it comes to a man whose life has been entirely civil."[56]

« CHAPTER 27 »

An Active Retirement

Van Buren, still healthy and vigorous, parted with an exhausted Paulding at Albany in late July. He spent a few days with his two sons and their families. Ellen Van Buren had just presented him with his second granddaughter who was christened Ellen James after her mother. Elizabeth was expecting any day now and gave birth also to a daughter three days after Van Buren left Albany for Kinderhook. "She weighed nine and half pounds at birth which is considered a light child," wrote the exuberant father, "has dark blue eyes, regular features and a very fine forehead. . . . we all agree to name it after my mother—was her name Anna or Hannah?"[1]

Van Buren was glad to be home and even happier that his farm was in such a flourishing condition. Martin, Jr., and Mr. Marquette had done an excellent job. After looking over the acres of vegetables, ripening grain, the fruit trees, the stock, he concluded he would have ample supplies for his own wants and a surplus for sale which would pay at least the wages of the gardeners and the laborers on the many fences and stone walls.[2]

Hard times still persisted; yet Van Buren's financial position was a comfortable one. Before he left on his tour, he reckoned his net worth exclusive of the Lindenwald estate, at $75,906. Most of this sum he had lent on good security and anticipated an income from $5,000 to $7,000 for the year 1842–43, out of which must come the expenses of his southern trip.[3]

Van Buren would maintain for himself and Martin, Jr., a household staff of six, who received monthly wages and their room and board. Felix, his chef, was paid $12 a month as was his coachman. His butler, who doubled as a waiter, and his valet who also helped serve guests at meals, were each paid $8 a month. Two household maids received $6 a month. His total annual cash outlay for household servants amounted to $648 a year. He also allocated a sum of

$200 over a six-month period in wages for help on the farm and garden above what income he expected from surplus produce.

Van Buren set a good table and expected to entertain extensively. The farm and orchard would produce fruit, vegetables in season, poultry, pork, beef, mutton and dairy products. He budgeted $670 for the delicacies that would grace his table, like game, fish, shellfish, oysters, and for groceries. His wine and champagne bill he estimated at $120, newspapers and books $150, other incidental expenditures brought his total outlay for six months to $3,252 which roughly balanced his expected income for that period.[4]

Van Buren's budget would be put to a severe test; for no sooner had he settled himself after his tour than the visitors began to come and not just for the day (some did, of course) but for a week or a fortnight. Richard B. Gooch, one of Ritchie's numerous relatives and a friend of Smith's and Martin, Jr., who arrived at Lindenwald twelve days after Van Buren's return, found his host entertaining Dr. Bethune, a rotund, wine-bibbing clergyman from Philadelphia, Abraham, Angelica, their baby and nurse, Martin, Jr., Smith and his family and servants. A New York City man, Jonathan Kent, who one of Van Buren's sons confided to Gooch was worth "a million," and the young Maine journalist, J. L. Stevens, completed the group.

Van Buren and some of his guests were absent fishing when Gooch arrived but returned with a fine catch in mid-afternoon. His three daughters-in-law joined the men for dinner which began at the now fashionable hour of 6:00 p.m. There were a dozen courses with wines and champagne for the ten diners. They were at table for three hours. After dinner, liqueurs and brandies were served in the drawing room, the ex-President dominating the conversation with a description of his trip. On a map he traced the route and claimed, his eyes twinkling, to have shaken 200,000 hands. No sooner had Gooch left than the Gilpins arrived from Philadelphia, then the Wrights en route to Washington and Flagg from Albany. A week later the Singletons were guests, while through it all Van Buren's sons, their families, nurses and servants remained at the mansion, which, as spacious as it was, must have been filled to overflowing with family and guests. It was no wonder that Van Buren remarked to Poinsett on October 1, 1842, that since his return he had not been able to leave Lindenwald except for a brief visit to the state fair at Albany.[5]

Yet he seemed to thrive on all the comings and goings and was even then planning to spend the winter in New York after visiting Edward P. Livingston at his nearby family estate, Clermont Manor. Van Buren, while acting the gracious host, managed the farm so well that he made a profit of $700 on his apples and pears alone for the year 1842, proudly explaining to Poinsett he had told his neighbors, all successful farmers from whom he had bought all of his staples the previous year, he now could help them if they fell short of hay, oats and potatoes.[6] Commenting on Van Buren's farm management and on the man himself, Paulding was filled with admiration:

> The same practical good sense the same sober, consistent and judicious adaptation of means to ends which has carried him successfully through every stage of his political life is discoverable in his system of farming. . . . He is always sanguine of success because he prepares the necessary means, and never sinks under disappointment because he feels it is not his fault that he did not succeed. Hence the good fortune which his enemies ascribe to cunning intrigue and which has acquired him the appellation of the Magician, is nothing more than the natural and just result of the joint qualities of his head and heart.[7]

Paulding's sketch of Van Buren's character omitted his tireless industry and his stamina. For somehow during an autumn crammed with social activities and management responsibilities, Van Buren found time to cope with his mounting correspondence while directing with great care his campaign for renomination.

At the moment his cause for immediate concern was the state of the party in New York. Before he left on his journey he was worried about the task ahead of the Democrats "to build up again the broken fabric of the state credit and to pay its debts." Outwardly confident they would achieve this objective, he was well informed by Flagg and Dix that the party itself might prove more of an obstacle than the Whigs who had created the problem.

When the Whigs gained control of the state in 1838, they began a program of increased state support for an overly ambitious system of internal improvement which not only enlarged existing canals but projected new ones, roads and railroads. Governor Seward in his message to the 1841 legislature downplayed the debt of over $17 million which his Administration had run up. With complete aplomb, he recommended doubling that amount to finish canal construction and meet the state's obligation to the New York and Erie railroad then being built across the state at a cost of over $36 million. If this were not enough, the Governor blandly asked the legislature to appropriate funds for a railroad connecting the St. Lawrence River with Lake Champlain and another from Albany to New York City.[8]

Seward's plans for New York dwarfed even those of De Witt Clinton. Unlike that great proponent of the Empire State, the Whig governor affected little concern for deficit spending which, for the times, was on a gigantic scale. The state's credit was already so badly impaired that its bonds were selling at a discount of 20 percent below par value. Just to meet current obligations the Seward Administration had gone into the money market for short-term loans at high interest rates.

This was the situation when Flagg was reelected comptroller and Young made secretary of state, while Michael Hoffman, who had been elected to the house in the Democratic sweep, became chairman of its Ways and Means Committee.

Eight days after taking office, Flagg had analyzed the state's finances and issued his report, a document that startled the legislature. There was an immediate cash problem of over $600,000 that had to be covered within two months. Rejecting the money market as a temporary and costly expedient, the new comptroller recommended the sum be taken from the general fund and a one mill tax be levied on the assessed value of all real property in the state. Meanwhile, a strict policy of retrenchment must be adopted. On March 7, 1842, Hoffman had a bill ready that embodied Flagg's and his own ideas.[9]

The one mill tax that would raise enough money in 1842 to repay a part of the money borrowed from the general fund would be continued until 1846. In 1843 and subsequent years one half of the money thus raised would be used to liquidate completely the debt to the general fund. The remainder was to be paid into the depleted canal fund for maintenance of the existing system. Hoffman's bill mandated a bond issue of $15 million at 7 percent for twenty years to wipe out all temporary indebtedness incurred for the various new canals. After Flagg's report, these sections of Hoffman's bill met with no opposition from the Democrats and even had considerable Whig support.

But the section dealing with the indefinite postponement of further public works created immediate dissension. Too many Democrats in the legislature had profited from Whig extravagance and were even now deeply involved in canal and railroad schemes. Many stood to lose heavily on their speculation in railroad ventures or in land purchases along feeder canals if retrenchment was to be the policy. Debate was heated. Hoffman and his radical contingent not only had to fight off the Whigs but Democrats like Daniel Dickinson who had invested heavily in the New York and Erie railroad stock and in other enterprises whose worth was tied to that venture. At the height of the debates, the party's candidate for governor in 1838, William C. Bouck, appeared in Albany to fight the bill. A member of the Canal board for nineteen years, he headed a potent faction which wanted to develop further the state's inland waterways.[10]

In the Senate, where the Democrats had a majority of only two, the bill passed because it attracted some Whig support. Bankers in New York and Albany were obviously worried about the several millions of dollars they had lent to the state over the short term and the millions of dollars in long-term state bonds they held. Those who had bought them when they sank well below par stood to reap substantial profits should the Flagg-Hoffman bill become law, while those who were original holders, mainly the banks, would recoup their paper losses. Even if Governor Seward were disposed to veto the bill, supporters were numerous enough in the legislature to override. When the bill became law, New York 5 and 6 percent bonds rose steadily over a two-month period until they reached par value.[11]

Flagg and Hoffman had rescued the credit of the state and placed its fiscal policies on a sound track. But they had, of necessity, dealt harshly with entrepreneurs like Bouck and Dickinson and the interests they represented.

There was a good deal of tension in party councils when Van Buren returned from his tour. As he looked over the prospects of the party, he thought Silas Wright was the logical man to head off a possible breach, if he could be persuaded to accept the gubernatorial nomination.[12] Radicals like Flagg and Hoffman were Wright's personal and political friends, and he was their choice for governor. Moderates and conservatives like Marcy and Croswell were equally enthusiastic. Van Buren stayed aloof though he must have been disappointed when Wright declined to run.

The radical leaders then tried to block Bouck's nomination only to discover that paradoxically the Whig program had done much to alter the shape of the coalition of small eastern farmers, mechanics and city bosses that Van Buren had found to be a winning combination. The population density of the rural counties was shifting inexorably west. No longer did the older river counties of the east enjoy the supremacy in state affairs they once had. Economically, the western counties had profited from Seward's expensive projects, Democrats as much if not more than Whigs.

In eastern cities and towns, business-minded Democrats, Olcott, Marcy, his friend Wetmore, the orthodox Tammany braves and sachems in New York City, chafed under the continuing depression. Many of them still attributed hard times to the subtreasury system even after its repeal. Michael Hoffman believed passionately that if Bouck were nominated and elected there would be "a continuation of Whig wastefulness and extravagance. If the State must be ruined, no more fit operators can be selected."[13]

The radicals, to their disgust, soon learned they could not sidetrack Bouck. At the convention in Syracuse on September 7, 1842, they accepted his nomination and that of Daniel Dickinson for lieutenant governor with deep misgivings. The Whig convention, meeting on the same day, dropped Seward for Lt. Governor Luther

Bradish, not so much for his attitude on internal improvements as for his more moderate stand on slavery.

Bouck, with his thatch of swept-back white hair, his close-set black eyes and his pinched features, looked for all the world like Dickens's Anthony Chuzzlewit whose face was "so sharpened by the wariness and cunning of his life, that it seemed to cut a passage through the crowded room." Bouck was not as shrewd as he appeared, but in reality had all the virtues and most of the prejudices of the practical farmer, which he was. Dickinson was of an entirely different cut. A Connecticut Yankee who had settled in Binghamton, an able lawyer and far better educated than Bouck, he was no less devoted to the interests of his region.

Everyone, including the Democrats themselves, was surprised at their landslide victory. Bouck triumphed over Bradish by over 22,000 votes, though his majority would have been less had the Abolitionists not run a third party ticket. In the congressional race, the Democrats elected twenty-five of New York's thirty-four Congressmen. They captured both houses of the legislature; and in the euphoria of victory, radicals, conservatives and moderates conveniently forgot their past differences. If the radicals distrusted Bouck and Dickinson, they had nothing to complain about in the distribution of state officers. Flagg was reelected comptroller by a unanimous vote in caucus. Samuel Young was again elected secretary of state.[14]

Shortly after the election Bouck wrote Van Buren a cordial letter asking advice for his message to the legislature. Seeing an opportunity to use the governor's message as a vehicle for his own stand on national affairs, he accepted the offer. The result was a thirteen-page manuscript, most of which was a defense of his Administration. On New York affairs, Van Buren was brief and moderate. He attacked the Whigs for their extravagance, pointing out that the people had repudiated their policies. Internal improvements were certainly essential to prosperity and growth, but outlays for them should never at any given time become a burden on the state's resources. Van Buren favored retrenchment in expenditures for the time being and a temporary halt in the financing of all public works until New York's credit was restored to the healthy state that existed under the last Democratic Administration.

Van Buren sent the document off to Marcy for criticism, asking him to share it with Flagg, then deliver it to the Governor-elect. Marcy read over Van Buren's suggestions and replied brusquely that Bouck's message was nearly completed and would be devoted primarily to state affairs. He did not think the Governor-elect would use any of the material and in any event he felt Bouck should see it before Flagg.[15] The letter reinforced doubts Van Buren had harbored about

Marcy over the past four years. Its style, its tone and its substance were scarcely that of a loyal friend and enthusiastic supporter.

As Marcy predicted, Bouck made precious little mention of national affairs; the few sentences he devoted to them bore no resemblanace to Van Buren's comments nor any reference to the policies of his Administration. The Governor's remarks on public works leaned more toward Seward's view than to those of the radical Democrats or even moderates like Dix or Van Buren.[16] Van Buren kept his chagrin to himself, while he watched the maneuvering of his rivals for the Democratic presidential nomination.

He was satisfied that his trip had greatly increased his popularity, had dispelled many of the myths that had sprung up around him and had sown confusion among his opponents. Calhoun belittled the effects of the trip, but clearly he was worried. Buchanan's budding candidacy was uprooted completely in Pennsylvania. Tyler, who seemed to be leaning toward the Democracy, began tentatively to form his own party on a patronage base. Marcy, who had been active in trying to recruit Tyler, understood his predicament. "The fact is there is a lion in the way," said Marcy, "and that lion is the Pres'ts' suspicion that the leading Democrats of New York mean to bring forward VB in the next Presidential election."[17] Calhoun was far and away Van Buren's strongest contender and his managers were able and persistent. Francis Pickens, Franklin Elmore, Robert Barnwell Rhett, editor of the *Charleston Mercury,* Dixon Lewis of Alabama, and Romulus Saunders of North Carolina helped plan his campaign. R. M. T. Hunter of Virginia and Governor James Hammond of South Carolina were accomplished wire-pullers. Their strategy was to balance the Calhoun ticket with a northern man who enjoyed a national reputation and had a regional following.

Before Van Buren left on his trip, Hunter and Hammond had put feelers out to Wright and when politely rebuffed they turned to Woodbury. The ex-secretary, now Senator from New Hampshire, was interested, but he had ambitions of his own for first place. Without refusing to back Calhoun, he acknowledged his "endowments" and if Van Buren was not a candidate, he looked forward to "unanimity in the selection."[18]

Woodbury had been carefully and stealthily building up a following among the anti-Van Buren politicians of New England. By March, 1842, the movement had gained enough impetus to include Isaac Hill, the ex-collector of Boston, David Henshaw, and what remained of the Conservative party in the other New England states.

The problem facing the Senator during the unseasonably cold spring of 1842 was Tyler. Woodbury's organization was peculiarly vulnerable to the President's patronage powers. Henshaw, Hill, and the Ingersolls in Connecticut, on whom Woodbury must depend and who had once trumpeted Jacksonian principles, had long since thought only of the power and perquisites of office. The talented, politically unreliable Caleb Cushing, John C. Spencer and Abel P. Upshur of Virginia were in the process of dividing the country up, removing Whigs and Democrats from office and replacing them with officeholders presumably loyal to Tyler.[19] By the summer of 1842, Calhoun's lieutenants were cautiously seeking an alliance with Tyler whose "Administration party" was detested by the Whig and Democratic organizations. Impractical, stubborn, easily influenced by men like Cushing and Spencer, Tyler had neither a public policy of his own nor any winning political formula other than the patronage. Newspaper support was feeble despite the liberal use of government subsidies through official notices.

The President had permitted only two Whig measures to escape his veto and not before he had wrung compromises from Congress that angered the Whigs and disgusted the Democrats. Tyler had sanctioned repeal of the Subtreasury Act, but rejected any attempt on the part of the Whig majority to reestablish a central banking system. In its place he presented a plan of his own which came to be known as the Board of Exchequer. Modeled closely on Britain's system of public finance, the Board regularized the emergency issues of Treasury notes and their circulation without interest as a kind of national bank note.[20] The Exchequer bill, which reflected the near desperate condition of the Treasury, was condemned unsparingly in Congress. Benton for the Democrats, Clay and Mangum for the Whigs, led an opposition which killed the bill in the committees of both Houses. The other measure Tyler sponsored was again prompted by the condition of the Treasury. Under the terms of the Compromise Tariff of 1833, revenue had diminished each year and what with the persistence of the depression, the government found itself unable to balance its budget.

Of the $12 million loan Harrison's extra session had authorized, the Treasury had only been able to market at par about $6 million, a severe reflection on the nation's credit. With a projected deficit for 1842 of more than $14 million, the Administration gained an extension of time for redeeming its outstanding indebtedness and the reissue of another $5 million in Treasury notes.[21] Desperate for revenue, Tyler finally agreed to a tariff bill that would supplant the Act of 1833. As the Whigs in Congress originally conceived it, the tariff would be a protective one with the revenue feature incidental. Clay and other Whig leaders were anxious, however, that these higher rates would not create a surplus and force a reduction. Included in their tariff bill was distribution to the states of all revenue the government received from the sale of public lands. Tyler was willing to approve the higher rates but not distribution, and the Whigs finally had to accept his caveat. Needless to say, they were further estranged from the President because of his intransigence on the distribution issue, while the Democrats, many of whom had been put in an awkward position, absolutely closed their ranks against overtures of the Administration party, patronage or not.[22]

Calhoun, of course, was the exception. The tariff was not only a compelling economic issue but a means of embarrassing Van Buren. He would attempt to undercut his rival through Wright, whom he assumed would have to support the tariff. Wright recognized the danger; whichever way he voted on this controversial question it would be construed as an indicator of Van Buren's position.[23]

On August 17, 1842, the Democrats met in caucus to determine the party line on the tariff bill after Tyler's veto. Wright, sensing Calhoun's eyes upon him, felt it essential to explain his position. The Treasury was bare, revenue was urgently needed and obviously there were only two sources for funds: sales of the public lands and duties on imports. The continuing depression had cut deeply into purchases of public lands. Revenue from that quarter was far too small for the kind of immediate and substantial relief the Treasury must have. An increase in duties was essential to maintain the public credit and keep the nation from drifting further into debt. Internal improvements in the states had come virtually to a standstill. Many of the northwestern states faced bankruptcy; his own state of New York had had to impose drastic remedies to escape the same fate. In rhetoric unusual for Wright, he warned his southern brethren that if the question remained open, "the Democrats of New York would be forced by their constituents to separate from them." As for himself, he was an avowed advocate of free trade, but not in these perilous times.[24]

Despite Wright's bold words, he was in a quandary, recognizing clearly that the tariff issue was a political minefield for Van Buren. The final version of the tariff came up for a vote in the Senate on August 26, 1842. As he explained to Flagg, "We are now upon the third tariff of the House, a horrible bill, am miserably perplexed to know how to vote myself but as we ought to pass some bill and as this one will kill distribution, I shall vote for it, and shall do it knowing that it will make a great noise and produce great dissatisfaction both in the city of New York and at the South."[25]

A few days later as Wright was preparing to leave for home, he was still worried that he had acted unwisely, but consoled himself that his intentions were honest. An exultant Calhoun, who had spoken vehemently against all the tariff bills on the Senate floor, was certain Wright's vote had finished off Van Buren in the South. "The whole is much like what happened in 1828," he said.[26]

During the cabinet upheavals, the harsh debates over banks and tariffs, the patronage forays of Tyler's Administration party, Van Buren remained silent. His letter to Henry Horn stood as the only statement of his position. There were, of course, "spontaneous" meetings calling for his nomination in various parts of the country, but no response from Lindenwald.

By late summer, many of his friends in Washington were becoming increasingly concerned at his reticence, his apparent refusal to commit himself. What they feared most was the momentum of the Calhoun campaign, the effort of Calhoun's organization to present him as the only defender of correct Republican principles. Jefferson's heir "has been going on without balk of any kind," said a worried Thomas Hart Benton. ". . . Everything here is intrigue and looking to the presidential election."

Bluntly, he warned Van Buren to give positive direction to his campaign or "you will soon find yourself what Crawford was in 1824, and it behooves your friends here to begin in time to counteract the schemes against you." What the impatient Benton had not considered carefully enough was the Tyler factor. Until Van Buren could be sure the Administration party had run its course and no longer stood a chance of splitting the Democracy, he would wait and let Calhoun have his head.[27]

Benton's view, conditioned by the closed society of Washington, made Calhoun look stronger than he was. One of the fruits of Van Buren's trip was that he now had a clear picture of the party in the South and West. He knew who its most active and able workers were, not just in the states, but in the counties, towns and even many of the villages. Very little escaped his observant eye; the best editors, the most popular newspapers, what areas were in need of newspapers and could support them. In short, he recognized that the apparatus for a nationwide campaign only required his signal to be set in motion.[28] Calhoun, on the other hand, had not traveled north of Washington for over twenty years; and he had never personally visited the West, knowing it only through its representatives in Congress, a fact of which he became tardily aware when his correspondents and party workers in the field warned him about his chances of nomination in a national convention of the party.

Van Buren waited while the imponderables worked their way out and then he would make his move in New York, in Pennsylvania and, he hoped, in Virginia. As the year came to a close, Tyler's party showed signs of running down despite lavish use of patronage. Woodbury had finally decided he could not gain the nomination and was moving with glacial slowness toward an arrangement with Calhoun.[29] Lewis Cass, whom Tyler had kept on as minister to France, had been led to believe he had a chance. Resigning his post, he made the perilous winter crossing of the Atlantic. A tepid reception accorded him in Boston and New York very nearly ruled him out of the running, though undismayed he began to pick up strength in the West.[30]

Still, Van Buren kept a wary eye on New York City where members of the Loco-Foco factions like Fitzwilliam Byrdsall, Levi Slamm and Townshend Lawrence maintained their ties with Calhoun. A newcomer to the politics of the city, Joseph H. Scoville, was writing editorials and public letters complimenting Calhoun in Park Godwin's *New Era,* seeking to swing Tammany over to Calhoun because of Wright's vote on the tariff. Scoville was a clever young journalist who ingratiated himself with the mercantile community, and who relished political intrigue. "Our aim," he wrote Calhoun, "is to lull all suspicion and let our VB friends suppose they are *all* right while we make sure that we are."[31]

But if Scoville had rounded up some disaffected merchants, Loco-Focos, and shipowners and formed what he thought was a secret committee, he was deluding himself. Van Buren had too many knowledgeable friends in Tammany and among the Loco-Focos who were reporting fully and accurately on Calhoun's activities in the city.[32]

Van Buren was adhering to his original proposition of a draft from the party, carefully cultivated, of course, but without his own direct involvement or any obvious organization. The only move he made on his own responsibility was an effort to keep Jackson from being drawn into the contest. His concern for Old Hickory's precarious health was genuine enough; yet more to the point, he was seeking vindication on his own terms, not just as a pale embodiment of Jackson's popularity.[33]

Calhoun, whose capacity for self-delusion seemed limitless, made just the right move in validating Van Buren's campaign strategy. Over the protests of the far-sighted Francis Pickens, he had the South Carolina legislature nominate him for the Presidency and draw up a series of resolutions as a general platform for his campaign.[34] Before the news of this precipitous action reached the North and the West, Van Buren's informal network of party workers was ready with nominations of its own. The first and most important event was a mass meeting in Philadelphia on January 7, 1843.

Organized by the irrepressible Henry Horn, the committee invited Jackson, Silas Wright and Benton to attend. All declined, but in what had become customary procedure, they gave a political gloss to their regrets. Wright's reply was worthy of the Magician at the height of his powers. In admitting Van Buren was his candidate, he felt it was nonetheless inappropriate for him, occupying the position he held as a Senator from New York and a personal friend of the ex-President, to influence by his presence the Democracy of another state. Let the Pennsylvanians exercise free choice without any outside interference. He did, however, declare that Van Buren had been struck down for standing by fundamental principles of the party. The inference was clear. In supporting Van Buren, Democrats were not backing a man, however deserving, they were upholding the integrity of their party. Jackson's letter was a direct and flattering espousal of Van Buren and his policies. Donelson ghosted the letter, which included a postscript that could be used as a toast if the situation called for one.[35]

Two days after the Philadelphia meeting, the Democrats of Indiana held their state convention, which nominated Van Buren and passed a series of resolutions asking his opinion on various public issues. Van Buren hesitated. Most of the questions dealt with measures or policies on which his views were widely known. He could add a few comments that would cover recent developments and refer the Indiana Democrats to his published opinions. Wright and Flagg advised him to do just that and avoid being accused of electioneering. But after weighing the situation carefully, Van Buren decided he should not only address recent issues, but polish and integrate all of his policies while setting them in a theoretical context that would proclaim unalloyed Jeffersonian principles.

On only three issues did he break new ground. He opposed treasury notes or government obligations in any form being used as a circulating medium; similarly he was against a constitutional amendment curtailing the veto power of the President; and he most definitely stigmatized distribution of any surplus that might accrue in the Treasury from the sale of public lands.

The tariff was the most politically sensitive question he was asked to comment on. Van Buren treated it carefully and ingeniously. He agreed with the resolution of the Indiana convention that a tariff must discriminate in its rates among the various items imported, but its essential purpose must be for revenue which incidently would offer protection to American industry. A discrimination feature would, within the limits imposed, permit sliding scales adjusted for maximum benefit of one region at minimum cost to another. His explanation of how this would work was clear; the plan was practical and carefully calculated to reinforce his position supporting the revenue principle, yet would not alienate either southern planters or deprive northern manufacturers of any chance to compete with the more advanced technology of British industry.[36] Van Buren's approach to the tariff question was aimed specifically at the border states, where industry was beginning to take hold, but where plantation agriculture still predominated. In particular, he had Virginia in mind because of its importance to his campaign, and because it was here that he would measure his strength in the South against Calhoun's.

Like Van Buren, Calhoun had recognized the vital importance of Virginia. The Old Dominion exercised a moral force over the slave states; yet as a border state and a western state, its influence was broadly national in scope. Since his break with Jackson, Calhoun had maintained his ties with Virginians of the Old Republican stripe—Littleton Tazewell, ex-Governor David Campbell, William Smith, the young, scholarly James A. Seddon. With Hunter representing the States' Rights Whigs as Tazewell did the old Republicans, Calhoun had gradually increased his hold and drew together these two factions within the major parties. His support of the subtreasury had cost him Whig support and drove conservatives like William C. Rives into its ranks. But when Van Buren's "Reply" was published, Calhoun's candidacy had made so much progress in Virginia, that Ritchie could not take a stand for Van Buren without imperiling his influence and that of his paper. Wright, who was looking after Van Buren's interests in Washington, had sought in vain to draw Ritchie out. Ex-Senator William H. Roane who was acting as his intermediary explained his position was "a delicate and difficult one."

After the Philadelphia and Indiana nominations, Wright, always fearful that he be misunderstood, especially by the sensitive Virginians, wrote Roane he had taken special pains "that Mr. Van Buren should have no clique of busy electioneers here." He had managed to keep the New York delegation silent on the Presidency, no mean feat in the rumor-laden capitol. "The truth is," said Wright, "we have not made a motion and any results which are seen are not of our production, but, as we believe, proceed from the deep prevading sense of justice and principle of the democracy of the country." His preference for an early convention, he explained to Roane, was not because he wanted Van Buren nominated, but because he wanted the question settled before Congress met and wasted its time in endless intrigues over the nomination rather than attending to the public business. The Democrats, he was certain, would control both Houses, and the onus of inaction must not fall upon them for the benefit of the Whigs. Honest and sensible as both Roane and Ritchie knew him to be, this full and free declaration from Wright and presumably Van Buren cleared the air in Richmond. The "Reply" together with Wright's letters convinced the Junto that Van Buren was the ideal candidate, whose motives were pure. As for Wright himself, he felt most uncomfortable in his lonely position as Van Buren's spokesman making decisions on his own. "I am attempting to do a very large business here on your capital but have no one to consult with," he complained.[37]

Calhoun and his advisors thought differently. They had finally come to the conclusion that Van Buren's diffidence was a clever political dodge; that he had been scheming for renomination ever since his defeat in November of 1840. Beset by organization problems and feeling very keenly their weakness in the free states, they came up with two suggestions they felt would strengthen their position. They would postpone the date for the national convention that had been held since 1832 in the early summer eighteen months before the national election. Scoville in New York, Hunter, Rhett, Lewis and Calhoun himself now favored its being held in May of 1844, or just six months before the election. Another year would give them time to perfect their organization and perhaps woo the President into their ranks as champions of the South and of Old Republicanism.

A further realistic recognition of Calhoun's minority position outside of his native region was Rhett's idea that delegates be chosen by districts, not by state conventions. Decked out in democratic trappings, when first broached in Congress and in the *Charleston Mercury*, the district idea would gain delegates for Calhoun in states where Van Buren could be expected to hold the majority.[38] Thus, even in New York they could pick up some delegates. In the City, Scoville, the old Loco-Foco leaders and their merchant allies, like Byrdsall and H. N. Salomon, were working hard for Calhoun; they would carry at least four wards and as many delegates. The same held true in Massachusetts. Bancroft and Morton still controlled the Party; but Henshaw and other conservative Democrats, if they could not oust the Van Buren leadership, could gain some of the delegates. Woodbury and Hill could be counted on in New Hampshire. As for Connecticut, Calhoun had long maintained personal and political ties with fellow alumni of Yale in the New Haven area. John M. Niles and Gideon Welles, who controlled the party, were close to Van Buren, but under a district plan Connecticut would elect some Calhoun delegates.

Roane warned Wright about the movement, and Blair came out against it in the *Globe*. The *Enquirer*, however, maintained its silence.[39] Other Virginians were less discreet. Besides the highly respected Roane, Peter V. Daniel, Van Buren's midnight appointment to the Supreme Court, Ritchie's sensible former partner Claibourne W. Gooch, Jefferson's grandson and namesake, Thomas Jefferson Randolph, and the hard-drinking, popular George C. Dromgoole, challenged the Calhoun position in eastern Virginia. James McDowell,

Benton's brother-in-law, vigorously contested it in western Virginia. the "Reply" was powerful ammunition in their hands, and they used it effectively.[40]

The first trial of strength between the two candidates and their respective positions came in the Virginia convention on March 2, 1843. Calhoun and his district plan were overwhelmingly defeated. Bitterly disappointed, he singled out Ritchie for special condemnation, declaring it had been his fixed determination since 1829 to divorce Virginia from the South and make the State a satellite of New York.[41]

For the next several weeks, conventions and mass meetings all over the nation made nominations and passed resolutions. After the Virginia convention all were looked upon as tests of strength among the various candidates. Van Buren had a distinct majority in northern and western states; but Cass, whose candidacy had been considered of no consequence, showed unexpected strength; and even Richard Mentor Johnson attracted some attention in his own state of Kentucky, in Arkansas and in western Pennsylvania. Buchanan kept up a pretense of candidacy but privately conceded to Van Buren.

To the beleaguered Calhoun forces, postponement of the convention and election of delegates by district became all the more imperative.[42] Van Buren, whose only organization in Washington consisted of Wright, understood Calhoun's position but found himself entrapped in his own lofty posture of awaiting the call of the party. He could not maintain his independent stand, yet at the same time tamper with the date for the convention. He must be above it all. Wright expressed this attitude clearly. "The friends of Mr. Van Buren," he wrote Roane, "believe if his nomination would not be made at one time, it would not at another. . . . if I supposed there would be any time between this and the election when the spontaneous voice of a fair convention would not select him as the candidate, that would satisfy me that he ought not to be nominated."[43]

For the sake of the party harmony and his own consistency of purpose, Van Buren was willing to go further and bow to Calhoun's date, though he stood firm against district nominations. Through Wright he let it be known he would leave the decision on the date of the convention to a majority of the states.[44]

While the Calhoun organization was slowly and painfully coming together, Virginia again dealt it a heavy blow. Unwisely, it entered candidates in the State election against the regular Democratic nominees, only to lose every contest. Still, neither Hunter, who was one of the casualties, nor Rhett was ready to give up and neither was Calhoun. Again Ritchie was the object of disappointment, anger and scorn. "I know Ritchie protests against it," said Calhoun, "but I have not the least faith in his protests. That he acts in concert with Albany I hold certain."[45] With Virginia lost to Van Buren and probably Pennsylvania, too, Tyler seemed the only resource left to Calhoun's advisors, and the patronage all important. There was still a chance to weaken Van Buren in his home state and perhaps capture the Party in New England.

Tyler had stepped up his patronage offensive and, either through ignorance or guile, had placed Calhoun men in positions where they could inflict the most harm on Van Buren.[46] Henshaw went into the Cabinet as Secretary of the Navy. After being rejected three times as Secretary of the Treasury, Cushing returned to Massachusetts and made a thorough sweep of the Customs House and other federal offices in Massachusetts replacing Whigs with nominal supporters of Tyler who were in reality secret adherents of Calhoun.[47] Webster's friend, Edward Curtis, remained unscathed as Collector in New York, though not before giving assurance he would support Tyler in forming a Presidential party. Like Henshaw and Barnes in Massachusetts, he, too, began secretly shifting toward Calhoun after Webster left the Cabinet and his influence with the Whig party went into a precipitous decline.

Officeholders everywhere knew there was no future with Tyler and they could expect little mercy from Van Buren or Clay. Calhoun was the only alternative, and he showed surprising strength in the New York City Democratic convention that met on August 29, 1843, to select delegates for the state convention to be held at Syracuse the following week. Had it not been for the timely appearance of Butler who managed to hold enough skittish delegates behind Van Buren, Calhoun would have carried the day. As it was, with Butler's eloquence and magnetism, the district plan was rejected by only one vote.[48]

With Marcy presiding, the convention at Syracuse displayed an air of harmony on the surface when it unanimously endorsed Van Buren for the Democratic nomination. By a decided majority it approved the traditional convention plan for electing delegates. The convention also pledged the New York Democracy to accept the majority will of the party in the state as to the date when the convention would take place, which, in effect, meant May 1844.[49] Defeat in New York seems to have dealt the mortal blow to the Calhoun organization and to Calhoun's ambitions. He offered no objection when, after pondering the wreck of his campaign, he finally decided not to permit his nomination at Baltimore, the site selected for the convention.

In an address he prepared during December 1843, he announced his decision to have his name withdrawn as a candidate at Baltimore.

His address betrayed personal as well as political differences between himself and Van Buren; but the central committee in Washington and in Charleston pruned out all the specific references to the tariff and to Van Buren, though it retained his conditions, extreme enough, which any candidate nominated at Baltimore must accept or forfeit Calhoun's support.[50] To Van Buren, Calhoun's address bespoke not just the anger of defeat but a probable division in the party.

Behind the rhetoric lurked the most contentious question of all—Texas. For months he had known that Calhoun's faction was reserving annexation as an issue, popular in the slave states but widely opposed in the North. Calhoun had not been openly identified with Texas as yet because the President had made it his own policy. Whatever the fate of Tyler, Van Buren was quite certain Calhoun's delegates to Baltimore would introduce the question, if for no other reason than to embarrass him and if possible deprive him of the nomination.

Texas was a subject that he had thought about long and earnestly for months, ever since Thomas Gilmer, then a Virginia Congressman and now Tyler's secretary of the navy, had written a public address urging annexation almost a year earlier. Van Buren did not have to be reminded by Roane, that a community of interest existed between the Tyler Administration and Calhoun's group on annexation. Roane's information came from John Letcher, an editor of a Van Buren paper in Abingdon, Virginia. "The Calhounites have won the first trick in securing postponement," Letcher had written of the convention, "and unless I am greatly deceived, are preparing to stack the deck." Alluding to the Gilmer address, he said that a member of the Hunter committee had sent a copy to Jackson and asked him whether he favored annexation. Jackson had replied with a resounding affirmative. The letter, according to Letcher who had seen it, "will be published before the Baltimore convention and Van Buren asked for his opinion."[51]

Besides Letcher's letter, Van Buren heard from Andrew Stevenson, who opposed the tactic in the midst of a presidential campaign. But the Calhoun connection with Texas was openly established in early October 1843, when Hunter gave a speech at a public dinner in Richmond advocating annexation. "Being forewarned of the game to be played," wrote Roane, "I hope you will be able to trump their winning tricks."[52]

The year had been an eventful one for Van Buren and it closed with reasonable prospects of success at Baltimore. Yet there were many imponderables. Ever the realist in politics, he understood the gravity of the Texas question which could bring a possible war with

Mexico and most certainly thrust to the forefront the deeply divisive issue of slavery. At home the party was far from being the disciplined machine he had created, though it gave him a unanimous vote of confidence at Syracuse instructing its delegates to vote for his nomination. If the Calhoun-Tyler forces threatened a split at Baltimore, the Hoffman-Flagg forces in the New York legislature were determined to drive Governor Bouck and his conservative wing to the wall.

It had taken all of Van Buren's influence to hold the party together. The task had been particularly distasteful because the feud involved his own family and his closest friends. Among the Radicals was John Van Buren, who for some years had been the chief organizer of the "young men" in the party and the undisputed leader in Albany. Samuel Young, Flagg and Hoffman, the Radical leaders, had been conspicuously loyal to Van Buren since their first appearance in politics.

Van Buren always enjoyed the company of Flagg, with his ready wit, his penetrating intelligence, his orderly mind. Hoffman was a welcome companion, a canny political strategist, whose judgment he respected, a man brimming with nervous energy, whose conversation he found amusing when it spilled over into picturesque denunciations of his political foes, real or imagined. Though he had had differences with the irrepressible Young over the years, Van Buren was never happier than when sharing a bottle of Madeira with this scarred veteran of many a political combat. And, of course, he valued Young's political links with the restless Loco-Focos in New York City.

Honest and capable, they spurned special interest groups, which hoped to enrich themselves through favorable legislation. In their approach to state affairs, they, like Van Buren himself, believed in careful management, balanced budgets, the prudent administration of the state's immense resources for orderly development. Though not opposed to manufacturing, banking and commerce, they set themselves grimly against what they regarded as speculation, particularly if it involved the credit of the state.[53] Only to a degree did their outlook differ from that of Van Buren, but this slight difference was all-important.

As dogmatic Jeffersonians, they tended to slant their views more distinctly than he in favor of an economy and a society based upon self-sustaining farm communities. Unlike Van Buren, they did not assign proper weight to the economic changes occurring all around them. They understood the importance of cheap, efficient transportation, but in terms of support for agriculture. Canals and turnpikes rather than railroads claimed the higher priority. Yet neither they nor Van Buren viewed the state canal system, in the same way as Governor Bouck, who saw it as a means of internal development and consequent increase of land values along its routes.

Flagg and Hoffman believed that the state should not support any means of transportation until there was sufficient population density and surplus production to warrant a safe investment. To them, Bouck and the conservatives were land speculators rather than sincere proponents of public works. They could see little difference between the Governor's program and that of his Whig predecessor, which had burdened the state with debt, enriched a few at the expense of the many.

In their zeal, Van Buren realized they were doing Bouck an injustice. The charge that he was interested in raising property values along the state waterways was a valid one; but he was no mere speculator. He was just as interested in fostering farm communities as they were; and where Flagg's career had been in journalism and accounting, Hoffman's and Young's in law, Bouck was and would remain a successful farmer to the end of his life. Agricultural development meant more to him than to his antagonists. Bouck would use the resources of the state to promote farming, not, as Flagg and Hoffman would have it, farming to promote the resources of the state. With such a personal philosophy, Bouck was much more receptive to the idea of public assistance for railroads and for turnpikes. Van Buren held broader views than either faction about the future of the Empire State, and he perceived that the differences separating them were not as fundamental as either side thought.

When state finances were put in proper order and income from the Erie Canal and its feeders began to produce a healthy surplus in the general fund, then Van Buren would encourage a cautious recognition of internal improvements. The convenience, the speed and flexibility of railroads, the fact that they could operate all year round were not lost on him. He, too, was at heart a farmer, but he did not foreclose, as Young, Flagg and Hoffman had, a future where manufacturing would be of great significance.[54]

What Van Buren feared most about the apparent schism in the party was that these slight differences separating the two factions would breed personal antagonism. They would feed upon themselves and congeal into warring camps long after the original sources of the quarrel had faded away. Politics, as he knew from his own experience, had a way of perpetuating differences after they had ceased to become issues.[55]

Shortly after the legislature convened in January 1843, Governor Bouck was already being criticized for bestowing several political plums on his immediate family. The *Argus* excited the suspicion of the Radicals when it did not take the Governor to task for flagrant nepotism. What really disturbed them, however, was its praise of the section in Bouck's message on internal improvements. The Radicals, who had not forgotten the *Argus*'s temporizing on the subtreasury plan and now certain of its unreliability, resolved to deprive it of the state printing it had enjoyed in the past.

They had purchased some months before the opening of the legislative session a struggling daily, the *New York Democrat.* Renamed the *Albany Atlas,* the radicals placed John Van Buren's brother-in-law, James M. French, and William Cassidy, a crusading editor of Irish birth and fiery opinion, in charge. What these young men may have lacked in the practical side of journalism they more than made up with their zeal and their determination to purge the party of what they regarded as its evil elements.[56]

With the help of the Whigs, who threw their weight first one way then the other, as best suited their partisan purposes, the *Argus* won out in the contest over the state printing. The Radicals, however, in the course of debate, indelibly stamped it as the Governor's paper, spokesman for officeholders and speculators or "Hunkers," a corruption of a Dutch verb that carried a pejorative ring of greed about it. As the Radicals saw it, the success of the *Argus* was a victory for the bank conservatives. In the language of Flagg, it was "a decided demonstration in their favor and they are loud in their denunciation of radicalism and all those doctrines Jefferson approved."

The protracted struggle over the printing, which lasted ten days, left some deep wounds in the party, which the bristling posture of the *Atlas* widened. Van Buren looked on with growing anxiety and a feeling of helplessness he had never experienced before. At his home in snowbound Canton, a gloomy Silas Wright felt the jealousies were now so intense that there was no hope for reconciliation.[57]

Van Buren sought a refuge of sorts in drafting the address the Democratic state convention would make to the party and to the country. The convention was months away but the address had to be approved by all factions before the legislature adjourned and its members dispersed at the end of April. He had finished a lengthy draft by the first week in April and forwarded it to Albany where an informal committee composed of Flagg, John and Smith Van Buren, Young, a newcomer to the party ranks, Preston King from Silas Wright's district, and Wright himself went over the draft, cutting, polishing and with what Flagg described as wry faces, accepting a resolution praising Governor Bouck.[58]

During the summer of 1843, Van Buren and Wright had succeeded in bringing a sense of unity to the party. Young agreed to step aside and permit Marcy, who was acceptable to the Hunkers, to be chosen chairman of the convention. The convention itself went ahead according to plan, except the Radicals made the mistake of excluding Marcy from the list of delegates to Baltimore. Though the sturdy old war horse professed indifference, he was hurt.[59]

After his rejection as a delegate, he wrote a long, anguished letter to Van Buren asking him to discipline the faction. In his reply, Van Buren declared he could not influence the policy of the *Atlas* and even if it were possible he would not do it. "Such is the fact," he said, "and if under these circumstances any of my friends think it right to make the course of that paper grounds for alienation of feeling towards myself they must do it." He would regret any break in personal relations and hoped there would be none, but politically he was indifferent even if it cost him the Presidency.

Van Buren admitted "that the relationship in which John stands to Mr. French naturally gives the impression of which you speak and I regret the circumstances most sincerely. He knows as well as you do my earnest desire that he should stand aloof from the dispute among our friends . . . and I hope he will do so, but if he does not, I cannot help it." Van Buren wanted to make sure Marcy knew as much about the radical side of the differences as he obviously knew of the conservative position. He sent Marcy's letter to Smith with a note to John asking him to read it and then give Marcy a full explanation.[60]

Van Buren's candid reply to his equally candid letter eased Marcy's state of mind. "Those who oppose the Governor," Marcy confided to his diary, "have a strong motive for electing Van Buren to the Presidency and would restrain themselves until that matter was concluded. Those who support Bouck would willingly sacrifice Van Buren, but not if it meant the Governor's defeat. Each party has a strong motive to keep the peace."[61] Marcy's observation would have been more accurate had he used truce instead of peace. In defeat, the Whigs had succeeded far beyond any reasonable expectation in dividing their opponents.[62]

« CHAPTER 28 »

The Hammet Letter

The last day of February 1844, gave portents of an early spring, bright and beautiful, with scarcely a breeze to ruffle the great flag on the Capitol. The weather could scarcely be better for an outing. And what must have seemed the happiest of coincidences, a special excursion was planned for an elite group on that very day aboard the nation's newest warship, the U.S.S. *Princeton.*

What quickly captured the public imagination was the *Princeton*'s unorthodox armament. Besides its twenty-four forty-two pounder's mounted broadside, it carried two wrought-iron smooth bore, 12-inch cannon capable of firing a 212-pound shot over three miles. One dubbed the *Peacemaker,* the other, the *Oregon,* they were to be fired on a trip down the Potomac for the President, other dignitaries and their guests. The 200 guests who boarded the *Princeton* about noon were all in a holiday mood. Even the widower President was relaxed, perhaps more in anticipation of the company of the shapely young Julia Gardiner, a New York belle, whom he was soon to marry and whose presence along with her father he had personally arranged.

The *Princeton*'s three powerful engines drove it smoothly as the picturesque shores of Maryland and Virginia glided by. At a point just below Mount Vernon, the captain, Robert Stockton, altered its course for the return trip. Guns and machinery had worked perfectly, but Stockton hesitated when Secretary of the Navy Gilmer asked for one more shot from the *Peacemaker.* He explained to Gilmer that the gun was overly heated and there was risk in another firing. The Secretary was insistent and Stockton reluctantly ordered another round.

President Tyler remained below with Julia and the champagne Stockton had provided for his guests, but Secretary of State Upshur, Gilmer, and Calhoun's friend and coadjutator Virgil Maxcy were among those who decided to witness the final firing. When the executive officer pulled the lanyard, the gun exploded throwing chunks of iron that weighed several hundred pounds each to the left crushing all those who were standing there. Upshur, Gilmer, Maxcy and a half dozen others were killed instantly. What had started as a holiday trip ended as a public tragedy.[1]

For Tyler, whose embattled administration had yet to develop a program, the tragedy on the *Princeton* was more than a personal tragedy. Upshur, to be sure, was not a man of any particular talent or experience, but he was deeply involved in secret negotiations whose successful outcome Tyler counted on to refurbish his image. The round-faced, ruddy Virginia judge had never served in Congress; nor had he been a conspicuous public figure even in his own state of Virginia when Tyler selected him to head the Navy Department and managed to have the refractory Senate confirm his appointment. As head of a small Department, fourth in rank among the cabinet posts, Upshur's duties were more ceremonial and social than policymaking. His brief tenure of office had scarcely prepared him for the burdens of the State Department, senior among the cabinet posts, to which Tyler promoted him when Webster resigned in May 1843.

Gilmer, the other cabinet casualty, was an abler man. A States' Rights Whig, he had served a term as governor of Virginia, like Tyler himself; but unlike the President he had formerly associated himself with the Democrats who elected him to the Twenty-eighth Congress. After the Senate refused to confirm David Henshaw to the post Upshur had vacated, Tyler appointed Gilmer to be secretary of the navy. As he had during Henshaw's brief tenure, the President relied on Gilmer to improve his position with Democrats of a conservative mind, like Marcy, with whom Gilmer enjoyed both political and personal rapport. More than any other member of his patchwork cabinet, the President was counting on Gilmer to fashion a new coalition that would reelect him in 1844. Upshur and Gilmer both were counseling the President on what had become his consuming ambition, the annexation of Texas. It had been, after all, Gilmer's public letter favoring annexation that created a brief flurry of political speculation a year before. Now Tyler was again faced with cabinet replacements, just when secret negotiation with Texas and Mexico seemed to be making progress.

The accident on the *Princeton* bore heavily on another individual, John C. Calhoun. Although he had formally withdrawn as a presidential candidate, he had by no means given up possible alternatives should the Baltimore convention, due to convene in three months, be unable to make a nomination. And there was always the possibility that he could gain the President's blessing and form a third party with himself at its head. Upshur and Gilmer had been leading members of his faction in Virginia before they went into the cabinet. Maxcy, an important member of his campaign organization, was the center of Calhoun's support in Maryland and through his legion of congressional contacts kept up the cause in Washington. "I really know not how our loss is to be repaired in Virginia and Maryland," said Rhett to Calhoun.[2]

Like Tyler, Calhoun was also counting on the Texas issue to interject a new component in the campaign. All signs indicated that Clay would oppose immediate annexation because this seemed the safer course for the many disparate and diverse interests that were supporting him and the Whig party. Van Buren remained silent, but clearly as a northern man he must consider sectional opposition to southwestern expansion particularly if it meant additional slave territory. Calhoun's ambition, even his personal desire to destroy the political careers of Clay and Van Buren, were not the only, nor the most important, motives for his espousal of annexation. He sincerely believed that the acquisition of Texas was essential for the safety and the prosperity of the South within a Union that was steadily moving toward the abolition of slavery.

When Calhoun and his lieutenants thought Van Buren would not be a candidate, they had begun carefully and cautiously to build up annexation as an issue for the campaign against Clay. They initiated the Gilmer letter and were well satisfied with Jackson's response to it. It had not been by chance that Gilmer went into the cabinet when Upshur was promoted to the State Department. Calhoun now had two devoted followers in the Administration, one in its most sensitive post. In addition to Gilmer and Upshur, who were members of Tyler's inner council, Calhoun could count on a third man who was close to the President, his old collaborator Duff Green. The accident on the *Princeton* may have cost Tyler a mediocre negotiator and a valued counselor, but it deprived Calhoun of his covert influence over a discredited Administration and forced him either to ally himself with it openly or abandon his carefully planned objectives. Tyler all along had stood athwart the South Carolinian's compulsive ambition. The President needed no persuasion from Gilmer, Upshur and Duff Green to make annexation a major policy of his Administration. For he, too, recognized its popular appeal in the South and the West. And he was bending every effort for his own reelection, if not as the Democratic nominee, then as the head of a third party.

The President was probably aware that his most trusted advisors were Calhoun men, but he must have thought he could rely on their loyalty to him. Unknown to the President, Upshur was in constant touch with Calhoun who was only too happy to provide advice and guidance. Upshur had scarcely been confirmed when Maxcy and Rhett descended upon him and discussed annexation.[3] A week later came a letter from Calhoun on the subject which prompted a correspondence that laid the basis for the Tyler Administration's policy toward Texas. "There can be no doubt, I think," responded Upshur, "that England is determined to abolish slavery throughout the American continent and islands if she can." He thought that the Administration would do nothing to prevent abolition in Texas because the North would not support this policy. Texas would become another Florida as it had been during the Seminole War, a sanctuary for fugitive slaves and warlike Indians. In his view, adjoining slave states would invade the Republic on their own responsibility, and he trembled for the Union at this dread eventuality. "What then," he asked Calhoun, "ought we to do? Ought we not to move immediately for the admission of Texas into the Union as a slaveholding state? Should not the South demand it, as indispensable to their security?" Rhett and Maxcy had done their work well. Upshur, backed up by Gilmer and Green, had no difficulty bringing the President around to their position, a course which Calhoun had outlined.[4]

Evidence of one sort or another indicating Britain's interest in Texas had been accumulating in the State Department for some time. That the British were trying to bring about a treaty of peace between Texas and Mexico was clear; that it hoped to increase its trade with an independent Texan republic was likewise obvious. But that it had further designs on Texas, annexation or abolition of slavery as official policy was largely in the imaginations of the American chargé to Texas, William S. Murphy, and the confidential agent Tyler sent to England, none other than the inimitable Duff Green, whose private letters went to the President and to Calhoun.

Lord Aberdeen, the British foreigh minister, made indiscreet remarks to a committee of delegates from a world convention of abolitionists that met in London the previous summer. He spoke informally and in general terms but included Texas among those regions of the world he hoped would be free of slavery. Informal comments like these from high-placed Britons made their way to Calhoun and to Upshur. Both Green and Murphy embroidered their letters with comments that the British were determined to eliminate slavery and that Texas claimed a high priority.[5] That was the situation when the *Peacemaker* exploded on the *Princeton.*

A distraught Tyler, his secret negotiations with Texas temporarily blocked by President Sam Houston for reasons of his own, his chief negotiator dead, grasped at the only man he felt could bring annexation to a successful conclusion, John C. Calhoun. One week after the explosion on the *Princeton,* Tyler appointed Calhoun secretary of state. He was promptly confirmed. Whether he would accept or not was another matter. Calhoun had twice rejected Tyler's overtures, but he was then a presidential candidate. It was reasonable to assume he would take the post on the basis of grave national need—Texas annexation and the settlement of the Oregon border with Great Britain—provided Tyler would explain satisfactorily his position on the tariff. At least this is what Calhoun's old colleague Senator George McDuffie thought after an interview with Tyler. And the Senator wrote as much to Calhoun who received the letter, news of the *Princeton* tragedy and word of his probable appointment all on the same day.

Calhoun remained in a quandary. He realized how desperately Tyler needed his stature and prestige to refurbish his Administration. An arrogant man, certain of his intellectual gifts, he had little taste for subordinating himself to the politically discredited Tyler. It was one thing to remain in the background with his close associates manipulating the President; it was quite another to involve himself openly and directly with Tyler's waning fortunes. But the Texas issue was all-important to the future of the South as he saw it; and the way parties stood, he might still manage to bring the President and his patronage to his aid in a third party movement. With his usual mixture of high-minded motives and personal ambition this complex person finally agreed to accept.[6]

Calhoun's great ability and drive quickly became apparent when he cleared up all the details outstanding between the two governments and had a treaty of annexation formally signed between Texas and the United States on April 12, 1844. Negotiations were secret, but no one in Washington had any doubt that a treaty would be presented to the Senate any day.[7] Van Buren prepared himself to make public his position on what had become in the space of three months a national debate.

For a year he had noted and weighed accounts of mass meetings demanding immediate annexation in the South and the West. Although alive to the mounting tempo of the Texas question during 1843, he had hoped it would not be drawn into politics, especially presidential politics. No man in the country recognized as clearly as he the perils immediate annexation posed to the Union and to the long-range objectives of American foreign policy. That this critical question should be decided impulsively at the eleventh hour in the heat of a presidential contest he felt was unwise because it reflected deeply on the character of the nation, carried with it the possibility of war with Mexico and, of course, threatened the party and his renomination at Baltimore.

Texas and Oregon had become catchwords in the lobbies of the capitol, the parties, the boarding houses and hotels where politicians gathered. Tyler refrained from any mention of annexation in his third message to Congress. But everyone in Washington had been aware that behind the President's carefully chosen words, Secretary of State Upshur was actively negotiating for the annexation of the Lone Star Republic. This knowledge had its effect on the Washington establishment and tended to weaken Van Buren, whom, it was conceded, would have difficulty with the issue in the North. Silas Wright found, to his dismay, he could not even count on unanimity in the New York delegation. He kept a wary eye on Rhett because he thought he was tampering with the New Yorkers, a sage observation.[8]

Robert J. Walker, the diminutive, intriguing Senator from Mississippi, an annexationist of long standing, added immeasurably to the excitement. In early February 1844, he published a long letter in which he ingeniously argued that the admission of Texas would benefit every section of the Union and even help solve the slavery question by drawing off the bondsmen from the Old South and dispersing them over the immense new territory.[9] Walker's letter instantly made Texas and annexation the major issue of the campaign. That the letter was aimed at damaging Van Buren was quite obvious to Wright and others of his watchdogs in Congress; for Walker had concluded Van Buren could not beat Clay and was quite open in expressing this view.[10]

While Walker's letter and Calhoun's withdrawal from the presidential race were still the prime subjects of conversation and editorial opinion in the capital, an enterprising group of Democratic Congressmen called a caucus to discuss preferences for the Baltimore nomination. They scheduled the meeting for the early evening of March 5, 1844. That very afternoon, Robert Tyler, the President's son and private secretary, handed Wright a message from Tyler requesting an interview at the White House after dinner. Concerned that the caucus might get out of hand if he were not present and perhaps detecting some purpose in keeping him away, Wright took what for him was the unprecedented step of declining, though he assured young Tyler if the meeting was over early enough he would hurry to the White House. The caucus that he had feared would be yet another intrigue against Van Buren's nomination, did not turn out that way, but when it adjourned, the hour was too late for a meeting with Tyler.

Several months earlier Smith Thompson had died leaving a vacancy on the Supreme Court that by custom should go to New York. Tyler had thought of Van Buren and Marcy but had been dissuaded from offering either man a seat on the bench. Nor had Wright been receptive when sounded out at the President's express direction. Tyler had not given up and Wright was quite certain why he had summoned him. He was prepared to cite personal reasons for dampening any intimations the President might make that he accept a Supreme Court appointment. But he had not expected Tyler, whom he visited early the following morning, to make a concrete offer. "I was really embarrassed because I knew that my determination upon the point had before been made to him," he wrote Van Buren.[11]

Though he did not say so, it had been a difficult decision for Wright to make. He was weary beyond belief at the constant demands made upon him to explain the anti–Van Buren material appearing under pseudonyms in the *Madisonian,* now Tyler's paper, the *Spectator,* speaking for Calhoun, and the *Daily National Intelligencer* for Clay.[12] Despite this remorseless barrage, Wright did not feel quite as vulnerable as he had the previous year because he had finally put together an informal council. Its collective wisdom eased his mind to a degree. The little group consisted of Preston King, a Congressman from his own district, John Fairfield, now a Senator from Maine, Lemuel Stetson, another New York Congressman, and Thomas Hart Benton. All were experienced politicians, but not one, not even Benton with all his presence and his stature in Congress, could match the short, fat, unprepossessing King in sorting out fact from rumor and forming accurate judgments. His was one of the keenest political minds among the younger members of the party.[13] Still, the responsibility was largely Wright's, and being a naturally modest man he felt awkward in his position of leadership. More often than not he sought refuge in solitary drinking.

The Court would have removed him at one stroke from all these distasteful labors, from the compulsion to answer dozens of letters a day, the burden of making stump speeches for the coming campaign, blessed relief from the feuds of New York politics. A man of modest means, the salary and lifetime tenure of a justice represented financial security. The whole frantic pace of his life for the past ten years would fade away. He would be able to spend more time with his wife, a shy person of uncertain health who disliked Washington intensely. Yet Wright was too honorable and too self-effacing a person to indulge his own private welfare at the expense of his principles and his friendship with Van Buren.

It was all the more to Wright's credit that he sacrificed his personal well-being for a man and a cause when he did. The full effects of the Walker letter and Calhoun's declension were just now being felt in Washington. Those who made it their business to manufacture public opinion in the politically overheated capital were combing through both documents and taxing the Van Buren men with hypothetical questions which concealed traps for those who relaxed their vigilance.

The master of Lindenwald, whose nomination seemed a certainty a few months before, who had either pledged or committed to him Democratic delegates from three-fourths of the states, now found he had an uphill battle on his hands holding his restless followers in place. Knowledgeable politicians sensed his growing difficulties. These misgivings were not just confined to the "Croakers" as Wright referred to them, but Democrats whose loyalty was considered beyond reproach. Members of the old Family faction in Pennsylvania responding to Walker's covert whispering campaign began doubting whether Van Buren could defeat Clay; whether indeed the party could survive if he were nominated. Old friends like Richard Rush, Charles J. Ingersoll and George M. Dallas joined with Walker in speculating about other candidates. Aspirants who had been counted out now were being discussed seriously. The picturesque, perpetual candidate, Richard M. Johnson, dusted off his red vest, let the press know he was in favor of Texas and considered himself the ideal choice at Baltimore. Lewis Cass was ready to accept all of Calhoun's conditions, and as everyone knew he had always favored annexation. The gaunt figure of Calhoun remained in the shadows, his followers discreetly keeping his name before the public. Buchanan, after thoroughly surveying the political ground and seeing a possibility for a deadlock, made sure he was not lost to public view. Even Tyler, whose party was made up almost exclusively of officeholders, redoubled his efforts to deprive Van Buren of the nomination and ride into triumph on the success of his negotiations with Texas.[14]

New words and phrases reflecting a change of mood in Washington began cropping up; "reannexation," "immediate annexation," "availability" sprinkled speeches in Congress, salted the gossip and appeared with increasing frequency in the editorials of the *Madisonian,* the *Spectator* and even of the staid *Richmond Enquirer.* From Washington the exchanges carried them far and wide over the country. The press of the two major candidates, taking its cue from the *Globe* and the *Daily National Intelligencer,* remained silent, as if all the political hubbub was beneath them. Blair, at heart an annexationist, remained loyal to Van Buren. Gales and Seaton, who always had been opposed to the acquisition of Texas, found no hardship in keeping Clay's name out of it.

Van Buren sympathized with Wright and his "council," but he was well aware of his friend's strain of pessimism and his tendency to exaggerate the perils of any given situation. Moreover, Wright and his associates were isolated in Washington and subject to its unique pressures. He needed a more objective observer who could sort out the fact from the rumor and whose ability and stature were respected in the capital. His mind naturally turned to Butler, who met all of these qualifications and who in addition favored annexation, though on certain terms. Fortunately, Butler had been invited to attend a series of lectures to be held in Washington in early April under the auspices of the recently established National Institute.

When Van Buren learned Butler was planning to visit Washington, he persuaded him to combine his scientific and cultural interests with a political errand. Why not have a talk with Tyler, or one of his cabinet, excepting Calhoun, and with Walker; compare notes with Wright and his "council"; and then make a visit to the Hermitage where he could bring Van Buren's views before Jackson and see the Old Hero for the last time? Butler, who needed a respite from his expanding law practice and his worries over his brother's possible bankruptcy, needed little convincing. Accompanied by his eldest son and Samuel Jones Tilden, whose political sense matched that of Preston King, Butler arrived in Washington on April 4 and put up at his old boarding house, Mrs. Fuller's. The Institute lectures formed an excellent cover for Butler's mission, but after it was observed that he had had two lengthy discussions with John Nelson, Tyler's attorney general, and with Robert Walker, the rumor mill in Washington for once correctly assumed his presence had something to do with Van Buren and Texas.

Butler conferred first with Nelson with whom he laid down certain prerequisites of a technical nature before annexation could be accomplished. Apart from raising these points, Butler's primary concern, and ostensibly Van Buren's, was that the government avoid any possibility of a war with Mexico. "We must keep our character unsullied in the eyes of the world," he said, emphasizing that it was indispensable to secure the consent of Mexico or public opinion in the North would not support annexation. The treaty of cession should be as simple as possible. "Above all," warned Butler, "it should not mention any stipulation or guarantee for the protection of slavery or slaves." Butler raised these same points with Walker in what he described as a "full and free conversation."

The annexationist leader seemed to accept Butler's position and added one of his own which had become current among southern Congressmen. The annexation issue would defeat Clay and elect Van Buren if he came out squarely for it. Butler assured him that it was not that clear-cut at all. Fixing the little Senator with his dark brown eyes that had swayed many a jury with their flashing intensity, Butler said it was highly delicate question which could be easily mismanaged and recoil upon its advocates. Besides the slavery question, what of the Texas debt? What of its boundaries that would have to be surveyed and stipulated? These practical questions must have made Walker wince. For, if he were not directly involved with the speculators in Texas lands and obligations, he knew that many of his colleagues were. Much of the enthusiasm for annexation might disappear if it were known the Congress would not assume the Texan debt and confirm Texan titles to real estate after annexation. There is no doubt the Senator gave no direct answers; yet on the whole Butler was satisfied he had made an impression. Whether Nelson and Walker would communicate his points to Secretary of State Calhoun and to the President and if so, whether these two individuals would accept any of them, was of course a highly debatable matter.[15]

Before he received any information from Butler, Van Buren had made up his mind how he must treat the Texas issue. After he had weighed the reaction to Walker's letter and judged the probable consequences of Calhoun heading the State Department, he knew he had to make his views public. He had before him an increasing stack of letters from Wright and others of his supporters in Congress detailing the pace with which pro-annexation sentiment was engulfing southern and southwestern members of the party in Congress.

Similarly, his correspondence from northern and northwestern friends, his extensive reading of the northern press, convinced him immediate annexation was unpopular, even feared in some quarters, as a conspiracy of slave owners to control the destiny of the Union. Immediate annexation as a majority of southern Democrats were demanding could break up the party and imperil the nation.

He had just concerns that any treaty Calhoun had a hand in would make matters worse in the North because it would inevitably contain clauses protecting slavery. Van Buren had not forgotten Calhoun's celebrated proslavery resolutions of January, 1838. The South Carolinian, in one of the most forceful speeches of his long career and posing as the defender of the South, had insisted that his region would never tolerate any attempt to interfere with slavery in the District of Columbia and in the territories. He had specifically included Texas in case it should become a part of the nation in his arguments.[16] Butler's warning to Nelson and Walker that the subject of slavery must be omitted from any treaty with Texas came directly from Van Buren.

Van Buren was also concerned about the moral consequences of a war with Mexico for the acquisition of territory that nation still regarded as its own. If hostilities with a weaker neighbor resulted in such a massive land grab and that land were used, as he confidently expected, for the expansion of slavery and slave states, what would the world think of the vaunted merits of democracy? Freedom, equality, human rights, representative government, all those great

Jeffersonian principles embodied in the Declaration, would they not be held up as a hollow mockery? On a more practical plane, would the people of the free states stand idly by and permit this vast extension of slave territory? How many new slave states could be carved out of Texas?[17] Yet Van Buren recognized that the northern states were not united on annexation. Certainly the party in the north had not as yet understood its implications for the future. If anything, the popular tide was running in favor of immediate acquisition. He was in agreement with Walker's attitude, as reported by Butler, that if he accepted immediate annexation, he would win the nomination and probably the election. Still, he came to the conclusion that he would rather risk nomination and election than bend before a surge of popular enthusiasm concocted by speculators, proslavery sectionalists and ambitious politicians who would play fast and loose with the nation's character and future. With these thoughts in mind, Van Buren began his fateful letter of reply to one William H. Hammet, an obscure, one-term Democratic Congressman from Mississippi, who, as an unpledged delegate to the Baltimore Convention, had been one of the many who wrote asking his opinion.[18]

In his billowing scrawl, Van Buren covered page after page, which Martin, Jr., carefully copied out in his more legible hand after his father had thoroughly edited his draft. It was an elaborate document—a state paper really—that covered seventy-two manuscript pages and when published took five full columns of the *Globe*'s tiny, tightly packed print.

Van Buren began his reply with an historical analysis of the problem in which he took pains to show how he, as secretary of state some six years before the Texan War of Independence, had under Jackson's direction sought to purchase the territory. In great detail, step by step, he reviewed the history of negotiations first with Mexico, then with Texas and Mexico jointly. In doing so, he made it plain that he had never been opposed to annexation and was not now, if the state of war between Texas and Mexico was resolved and sentiment of the country as expressed by a majority of Congress supported the measure. Van Buren dismissed the argument that if the United States did not now annex Texas, European powers would. He could not imagine Texans who were American emigrants and who shared a common heritage of independence willingly turning themselves into subjects of a colonial power. Should, however, a colonial power seek to acquire Texas, he pledged himself if he were in office, to resist any such move even if it meant a declaration of war. But the crux of his opposition to immediate annexation was fair play in American dealings with Mexico. It would be a travesty of international law and of comity among nations if the United States, having sworn in solemn treaties to maintain good and peaceful relations with its southern neighbor, unilaterally ignored these treaties as mere scraps of paper.

Van Buren made a plea for restraint, while pressing for a peaceful settlement before any steps be taken for annexation. He implied Mexican recognition of Texan independence would be accomplished in the near future through diplomatic means and concluded that if he were President he would not be "influenced by local or sectional feeling" but would act for what he regarded the benefit of the entire nation. If Mexico persisted in fruitless efforts to reconquer Texas and resumed the War, he assured Hammet, he would, upon a request from Texas, submit the question of annexation to Congress and abide by its decision.

The letter was too long, too studied, for popular consumption, much less understanding. But it was a courageous and statesmanlike view. Van Buren had arrived at his conclusions on his own, though to some extent he relied upon Butler's information and advice.[19] Obviously, he hoped Butler would convince Jackson to come out against immediate annexation. In his letter to Hammet he ventured an opinion that Old Hickory would never trifle with the honor and pledges of the nation. But he also knew Jackson believed Texas was once American territory bartered away by John Quincy Adams in the treaty negotiated with Spain for the acquisition of Florida.

For Jackson, it was not immediate annexation but "reannexation" of what had once belonged to the United States. In his ardent desire to expand the nation, he was not inclined to examine the fine print of treaties or be bound by provisions which Adams and the Senate had, in his opinion, no right to conclude. But apparently Van Buren had not made himself clear to Butler and to Blair. Either they misunderstood him or he had changed his mind at the last moment. The *Globe* came out with an editorial on April 15 in favor of reannexation as soon as possible. Butler at the Hermitage led Jackson to believe Van Buren favored the same policy.[20] The Senate Committee on Foreign Relations now had the treaty in hand. And copies were being circulated among the Senators.

Abraham and family were at Lindenwald when his father was putting the finishing touches on the Hammet letter. As he had done on many occasions in the past, Van Buren discussed the document with his sons who also helped with the editing.

Abraham, entrusted with carrying his father's reply to Hammet, left Lindenwald on the 24th, and delivered the letter and accompanying note to Wright on 8:00 p.m. April 27.[21] Van Buren's note gave

Wright complete discretion as to its contents and its publication. Wright, who had studied the treaty and the documents, read Van Buren's letter carefully in that context and then went to bed without making up his mind. Early the next morning he read it again and decided it was a "good letter." At breakfast he told his "council" he had Van Buren's letter on Texas. Would they accompany him to his office in the capitol where each could read it and give his opinion? All agreed. After each member had read the letter, reserving his opinion until all had studied its contents, there was complete concurrence that not a word should be changed and that it should be published as soon as possible.

Wright and Abraham Van Buren, gathering up Senator Allen of Ohio, then sought out Benton, who read the letter, and at Wright's request the others present went over it again. All thought it was excellent and agreed to its immediate publication. The morning papers were being hawked on the street corners. Wright picked up a copy of the Saturday *Intelligencer* as he was walking to the *Globe* office and was astounded when he read on the editorial page a long letter from Clay opposing annexation. Since he had complete control over the timing of publication, he might have paused and let the Clay letter become the chief topic of conversation in the capital. A more politically astute course would have been to wait at least until Monday. Opinion on the issue would have been more clearly developed. Clay and the Whigs would have absorbed the full attention and attack of the immediate annexationists. In view of Calhoun's Packenham letter that connected immediate annexation with slavery, Van Buren's carefully balanced and comprehensive answer would have appeared to better advantage. Compared with Clay's categorical rejection of annexation and Calhoun's insistence on immediate annexation coupled with slavery, Van Buren provided a middle ground.

But Wright reasoned that the public had a right to know at once Van Buren's views. Perhaps without thinking through the immediate political impact, he actually welcomed the coincidence that the views of opposing party chiefs should appear on the same day and that these opinions should appear only five days after the treaty had been submitted to the Senate along with Calhoun's array of what he saw as self-serving documents. To him, Van Buren's letter seemed so reasonable and so skillfully drawn up as to defuse an explosive issue. He was absolutely convinced it demanded immediate publication, and he hurried to the *Globe* office where he was told the typesetter had to start composition no later than 4:00 p.m. if it were to appear that evening; otherwise the letter would not be published until Monday evening, April 29. Wright's intention was to have Hammet read the letter at the *Globe* office. By the time he found Hammet there was not enough time for this. He prevailed upon the nervous Mississippi Congressman to have the leter published without his reading it. That night the long-awaited letter appeared in the *Globe*.[22]

« CHAPTER 29 »

Rejection at Baltimore

After church on April 28, 1844, Silas Wright listened carefully to the excited Congressmen and officeholders who gathered in groups on the street. The Illinois, Indiana and Ohio representatives he thought were the most complaining. On his way home he espied Walker's uncle, Secretary of War William Wilkins of Pennsylvania, Robert and John Tyler, Jr., Charles Jared Ingersoll and even Calhoun, buttonholing Senators.

Blair, who had been painfully sick for weeks, had the Van Buren letter read to him; and he too thought its statement of the case, its propositions and its tone well adapted for a great and potentially divisive public question.[1] On Monday morning, the *Globe* carried a long letter from Benton much along the lines of the Hammet letter but far more biting in its allusion to political conspirators among whom everyone recognized Calhoun as the chief instigator, with Tyler the willing accomplice.[2]

Blair, like Benton, persisted in seeing the treaty and immediate annexation as a Calhoun plot against Van Buren. The letter to Packenham, so explicit on slavery and Texas, convinced the recuperating editor of Calhoun's base intent. It was all a cabal to break up the party and the Union in Blair's fevered mind. Calhoun had "Duff Green in London in the pay of Tyler to work upon the British government to give pretexts and serve support to the conspiracy," Blair wrote Jackson. "Upshur and Tyler were, we now learn, the tools of Calhoun throughout the whole period since his retirement from the Senate."[3]

Though Blair was up to the mark about Calhoun's means, he was wrong about the ends Calhoun had in mind. Calhoun may have derived grim satisfaction at what he knew would be the results of the Hammet letter—Van Buren's candidacy irretrievably ruined in the South and the Southwest. But his primary object had not been to defeat Van Buren's nomination. He had been in favor of annexation since Texas declared its independence; and as the years followed, as abolition sentiment increased, not just in the free states but in Europe, he came to regard Texas as a bastion for the protection of the South, its prosperity, its culture, however tainted it might be by slavery, an institution he considered irreversible. Where Blair went wrong was in assuming Calhoun's motives to be purely partisan.

Had Van Buren come out for immediate annexation, Calhoun would have been compelled, if not to support him, at least not to enter the lists against him. Where Blair misread Calhoun he likewise misread Jackson. Old Hickory's motives for immediate "reannexation" were roughly the same as Calhoun's, except for the linkage of slavery with the question. Jackson believed that the immediate acquisition of Texas was a matter of self-interest, of protection from potentially hostile foreign powers. "In the southwest," said Jackson, "Texas is viewed as absolutely necessary for the defense of New Orleans and to keep foreign influence from tampering with our Indians and slaves in war and in peace. We cannot bear that Great Britain should have a canady [*sic*] on our west as she has on our north."[4]

The old man was saddened at the turn of events. He had counted on Van Buren's election and the triumphant vindication of their shared principles in 1844. Now he saw no chance and only hoped a sound Democrat who favored immediate reannexation would be selected to run. That Democrat, he made abundantly clear to Blair, was not to be Tyler or Calhoun, whose Packenham correspondence he had read with irritation and dismay. "How many men of talents want good common sense," he said.[5]

For the next several weeks, Van Buren in Kinderhook, and Wright in Washington, watched the passion for Texas sweep the country. In the capital it was particularly intense; as delegates to the Baltimore convention began appearing, members of Congress and officeholders seized upon them arguing that the Hammet letter made Van Buren "unavailable." Worried about the integrity of northern delegations once they reached Washington, Wright urged Bancroft and other leaders in New England to meet with the New York delegation in New York City so they could plan common action.[6]

Virginia, which had been counted solidly for Van Buren until his Hammet letter, reversed itself with Ritchie, William H. Roane and Thomas Jefferson Randolph taking the lead in asking Van Buren to withdraw. The *Enquirer* removed Van Buren's name from its masthead and asserted repeatedly that no instructions bound Virginia's delegates.[7] Van Buren might have expected as much from the nervous Richmond editor who had long supported westward expansion. But it must have been galling indeed ,to have Jefferson's grandson turn against him and particularly Roane, who had so recently alerted him that Texas would be an issue. At that time Roane had set himself firmly against annexation, referring to Texas as "the Botany Bay for all the human refuse of the southwest." Now the Virginians were talking about Cass as the most "available candidate."[8]

Wright watched with disgust as a combination of immediate annexationists headed by Walker made common cause with Calhoun's clique. Even in Massachusetts there were delegates secretly opposed to Van Buren but instructed to vote for him. In Vermont, Connecticut, Rhode Island and New Jersey the same feeling prevailed. The Walker-Calhoun group in Congress argued that delegates would not violate their instructions if they voted for the two-thirds rule, and if they stood by Van Buren for one or two ballots they would have discharged their obligation. On the other hand, should Van Buren be nominated, South Carolina and other southern states would desert the party and in alliance with Tyler, who had given notice he would stand for reelection, deliver the nation to Clay and the Whigs.[9]

Wright kept Van Buren well aware of these movements. From Blair and Donelson he had learned of Jackson's negative reaction to the Hammet letter. Donelson had written him enclosing a copy of Jackson's letter to the *Nashville Union* endorsing immediate reannexation. Jackson had gone through the documents and agreed with Tyler that Texas faced a clear and present danger from British expansionists.[10] Jackson concluded his letter with a favorable reference to Van Buren but took him to task for failing to understand the probability of foreign intervention if annexation were delayed.[11]

The Jackson letter distressed Van Buren. He could not believe that anyone "who desires the annexation only when it can be made consistently with what is due to the peace and honor of the country can find anything to object to in my letter." The only point he had not made was reference to slavery "by name." But then neither had Jackson. In what for Van Buren was an unusual burst of petulance, he dwelt on "the opposition and persecution to which I have been exposed" for what he had said and done to uphold the rights of the South and now to be accused of being its enemy. He would not, however, as Clay was doing, modify his stand on Texas. "Self respect," he said, "forbids me from adding another word in regard to it." He asked Butler, who headed the New York delegation, to have a long talk with Donelson, who was a member of the Tennessee delegation. Perhaps Jackson had misunderstood his position, but he warned Butler not to commit him to any "Codicils and additional explanations." Besides this request, he wrote Butler another confidential note on the same day.

Should Butler find that the Democracy at Baltimore simply wanted to "propitiate those who entertain personal prejudices ag't. me, decide in favor of Mr. Wright for the Presidency" then he must immediately contact the Senator and use all means in his power to have him accept. "Nothing," said Van Buren, "could occur that would give me more pain than his refusal under such circumstances."[12]

The political anarchy that was prevailing in Washington reached a frenzy as the day approached for the convention to assemble. Several Congressmen, the most prominent being Cave Johnson, Polk's manager for the vice-presidential nomination, wondered what to do should the opposition deadlock the convention. Would Wright agree to have his name placed before the convention if it became necessary to drop Van Buren?

Wright was shocked to the depths of his being. He shared Van Buren's political philosophy, his views on annexation and on all other public questions. The thought of his name being substituted for Van Buren's purely for personal reasons was not just deeply repugnant to his political values, but even worse, a betrayal of his public trust.

Under no circumstances would he permit himself to be placed in nomination at Baltimore. Wright was most emphatic on this point with Cave Johnson. But in order to make doubly sure there would be no mistake, he wrote a long letter declining absolutely any compromise arrangement involving his name and gave it to Judge Fine, a neighbor and the delegate from his home district, with instructions to read it to the convention should that be necessary.[13]

Bancroft was either unable or unwilling to have the New England delegates meet with those from New York as Wright had suggested. Butler, however, did arrange a caucus of most of the New York delegates at the Astor House on May 22.[14] Their mood was not optimistic as they debated what they should do if delegates from other states violated their pledges and nominated a candidate they could not accept. Butler convinced the delegates to await circumstances and vote as a unit for Van Buren against the two-thirds rule, which he thought would be the deciding issue. Pessimistic, he left with Cambreleng, Kemble and others of the thirty-five man New York delegation for Barnum's Hotel in Baltimore, their convention headquarters.[15]

Yet with all these forebodings, Van Buren forces had by no means given up. Blair, now recovered, his fighting instincts aroused, was counterattacking in the *Globe,* directly naming Calhoun and Tyler as the chief instigators in a political plot to defy the will of the party. Nor did the *Globe* spare Cass and other candidates. Ransacking the paper's file of back issues and begging letters from Van Buren supporters in Congress, Blair accumulated and published correspondence of Cass, Richard M. Johnson and other "available" candidates which, when placed in the current context, supported his conspiracy thesis. Benton thundered on the Senate floor, in the lobbies and elsewhere. Wright quietly but earnestly spent a part of each day seeking out dubious delegates, explaining the Hammet letter and urging them to vote against the two-thirds rule. In Virginia, Dromgoole and James McDowell remained faithful and fought hard for Van Buren against Cass who was emerging as the Junto's candidate. The counterattack had a significant effect in holding the line. Wright began to feel Van Buren's strength was increasing. "You may not be nominated," he wrote Van Buren, "but you must not withdraw. It would destroy the party."[16]

When Bancroft reached Washington on May 22, he found the city crammed with what he called "disorganizers," men in his own delegation like Rantoul, who urged him to support the two-thirds rule. After the high-strung historian talked with Wright and others, he made up a list of reliable delegates by state and by individual delegate. He was confident Van Buren would have a majority on the first ballot whatever mode the convention chose for its balloting; but he was skeptical about the two-thirds rule. "On this the battle will turn," he wrote Van Buren, "and the point may be severely contested.[17]

Butler arrived in Baltimore on Friday evening, May 24, and canvassed heads of other delegations. He was agreeably surprised to find a much stronger support for Van Buren than he had anticipated. The next day he went over to the Egyptian Saloon of the Odd Fellows' Hall on Gay Street, the convention center, and was satisfied that everything possible was being done to advance the cause. The auditorium, the largest in Baltimore but poorly designed with a small, dark gallery and abominable acoustics, was already a scene of confusion, crowded with delegates, officeholders, idlers and local politicians. Butler managed to make his way through the disorderly, noisy throng and search out delegates he recognized for brief chats. From this cursory sounding, he concluded Van Buren would be nominated on the first ballot, "if we are not handcuffed by a two-thirds rule."[18] But that was just what the Walker-Calhoun men were working desperately to bring about.

Promises were being made thick and fast to any and all candidates. They probed for weak spots in the northern delegations. Michigan, Pennsylvania, New Hampshire, Virginia and Tennessee were of particular attention. Rantoul, a Woodbury backer, worked over the New England delegations. The split in the Tennessee delegation had been recognized and special efforts were made to expand it. They exploited Cass's strength in his native state of Michigan and rounded up pockets of support for Richard Mentor Johnson, including the old naval hero, Commodore Stewart, in their effort to garner votes for the two-thirds rule and then secure the nomination of a compromise candidate.

Complicating Butler's problem was that a convention assembled to nominate Tyler was meeting at the same time only a few blocks away. Delegates to that convention, distinguished by their bright buttons sporting the Texas emblem, "The Lone Star," moved in and out of the motley crowd at the Egyptian Saloon, distributing handbills emblazoned with their slogan, "Reannexation of Texas—Postponement is Rejection." The Walker-Calhoun men were keeping in touch with the Tylerites and using effectively the threat of another party in the field that Calhoun could support if Van Buren were nominated. They had previously and successfully worked on the complicated structure of the Pennsylvania delegation to undermine Van Buren. Buchanan, whose hopes had lain dormant for months, had seen possibilities that he might be a compromise candidate. The shrewd, politically amoral Senator assured the Van Buren managers of his loyalty while he secretly agreed to a Calhoun proposal that he use his influence in support of the two-thirds rule. Instructed for Van Buren, the Pennsylvania delegation was a mass of uncertainty.[19]

Among those who spoke with Butler at Barnum's Hotel before the convention met were Marcus Morton, Bancroft, Cave Johnson and Donelson, whom Van Buren had specifically asked Butler to contact in order to clarify the Hammet letter. They were all quite certain the two-thirds rule would pass and that Van Buren would not be nominated. Could Wright be prevailed upon to change his mind and accept the nomination with Polk as his running mate? Butler, who had Van Buren's confidential notes in his pocket must have been receptive; yet he was not convinced that the two-thirds rule would prevail, and until that test was made, he would not commit himself.

From 10 o'clock Monday morning, May 27, until noon all manner of humanity continued to pack itself into the Egyptian Saloon. It was soon jammed to overflowing. In the gallery that O'Sullivan described as black as the hole of Calcutta, there was no standing room. The unruly mass made it virtually impossible for anyone to be heard, much less for the convention to be organized.

Robert John Walker, who had been the busiest operator in Washington and at Baltimore, had counted on the pandemonium to rush through the two-thirds rule before the New Yorkers had concentrated their forces. In order to do this he had to have his own presiding officer but one who would not arouse the suspicions of Butler. Walker selected the Calhounite, Romulus Saunders, a bearish man with a loud voice and an overbearing manner, to elbow his way through the throng, mount the stage and nominate the presiding officer. Walker's choice for this key post was one of his legion of

Pennsylvania connections, Hendrick Wright, who was obscure enough in political circles as not to have any known allegiance to a particular faction. The Mississippi Senator guessed that the Van Burenites would not oppose him because of the crucial position Pennsylvania held.

Saunders' figure, presence and especially his voice from the rostrum brought a momentary silence in the hall. With a few terse remarks demanding prompt organization, he nominated Wright as presiding officer and without waiting for a second insisted on a vote. Butler could have interposed at this point and nominated his man, Governor Henry Hubbard of New Hampshire; but unwilling to disturb Pennsylvania and having no reason at hand to oppose Wright, he and his delegates joined the roar of acclamation that greeted Saunders's proposal. No sooner had Wright assumed the chair than Saunders was again on his feet moving the two-thirds rule and citing as precedent its use in the conventions of 1836, 1840. Before Saunders had finished, Cave Johnson, standing on a chair, roared out a challenge. There were many present who were not duly accredited delegates. The Chair must appoint a Credentials Committee before any more votes were taken. Walker had anticipated just such a turn of events and had the Chair accede to Johnson's request. Wright selected a known delegate from each state and then entertained a motion to adjourn until four in the afternoon.

A man of honorable intent, Butler was ill-prepared at first for the ward boss tactics of the Walker-Calhoun men. He had been associated with politics in New York for a score of years and was thoroughly accustomed to heated debate in caucus and the discipline of party drill. But as boisterous as the Bucktails had been in their heyday, they had never deliberately practiced deception on a scale like this. Now alert to Walker's methods, a born fighter when aroused, an eloquent speaker, the spare, tense Butler regrouped his forces and prepared for a showdown. When the convention reassembled and the Credentials Committee made its report that showed 266 accredited delegates present from every state except South Carolina, he was as ready as he ever would be. No sooner had the report been received than Butler was on his feet with a motion which somehow he managed to put to the disorderly throng. He proposed a Committee on Rules. Then the raucus Saunders gained the floor to debate the motion. Interest and intensity stilled the uproar of the convention, still scarely more than a loosely disciplined mass meeting.

Saunders bellowed out the familiar arguments for the two-thirds rule—a reliance on precedent and an assertion that a bare majority in the convention did not constitute the will of the party. Walker followed, speaking vehemently along similar lines but dwelling on the danger of the convention making a minority nomination which would surely damage the party and, if it resulted in a secession of the southern wing, smash it irretrievably. When the little man with the black wig and the hoarse voice finished, he was treated to a thunderous ovation.

Butler was now thoroughly aroused, angry at the bluster of Saunders, the insinuations of Walker; he must also have realized he had been outmaneuvered from the outset. His emotions at a high pitch, he attacked their propositions, interlacing his argument with rhetorical flourishes, now scorning the pretensions of Walker and Saunders, now appealing with tears in his eyes to fair play from the South, now adumbrating the Jeffersonian principle that the will of the majority was the cornerstone of democracy. Three times, carried away by his discourse and the empathy he could feel surging up from the crowd, he leaped in the air. The crowd loved it, interrupting him time and again with applause. No one candidate, he admitted, could receive a two-thirds vote, and should the convention adopt the rule he implied many of the delegates would be violating their instructions. Further, the rule, if imposed, would either result in no nomination or force the majority to accept the will of the minority. As Walker had threatened secession, Butler hinted a withdrawal of the Van Buren wing. It was a magnificent one-man display and it almost succeeded. If Butler had forced a vote at the close of his speech while he still held the convention in his hands, he might have beaten Walker and Saunders. But the effort had exhausted him, and he did not oppose adjournment.

At nine the next morning, the convention resumed with Butler's motion for making a Committee on Rules the order of the day. Saunders was ready with an amendment that would substitute for it the rules that governed the conventions of 1836 and 1840. As before, debate was animated, though less emotional than it had been in the previous session. Butler participated, but the burden of resisting Saunders's amendment fell to Samuel Medary, the articulate editor of the *Ohio Statesman.* Finally, after four hours, when both sides had run through the gamut of their arguments, the Van Buren men allowed a vote to be taken.

Again the hall became still as the chairman of the Maine delegation was recognized and cast the state's 9 votes against the motion. New Hampshire followed. Walker had hoped for a break here. Isaac Hill and Levi Woodbury were both opposed to Van Buren, but New Hampshire cast its lot with Maine. Massachusetts came next. George Bancroft arose and announced in his piercing voice the first break in the North. Seven delegates had gone for the rule and five voted against, a distinct setback for the Van Buren men who had worked a good part of the evening—without success—to counter Woodbury's influence. Vermont, Rhode Island and Connecticut added 8 more votes to the Saunders amendment. Butler cast the massive 35 votes of the New York delegation against the rule. But when New Jersey gave its 7 votes for the southern position, the reporters who were tallying in "the black hole of Calcutta," realized Van Buren had lost and in his own section, even without Pennsylvania which split 12 for the rule and 13 against. The rest of the tally was anticlimactic; delegates from many western states violated their pledges without even a show of opposition.[20]

Still without formal organization, the convention began balloting for the nomination at 4:00 p.m. Van Buren had 146 votes on the first ballot, a clear majority over all contenders but falling short of the two-thirds requirement. Most of the delegates honored their pledges, though Tennessee, which everyone thought would cast its ballot for Van Buren, gave its 13 votes to Cass. The Michigan dark horse had piled up an astonishing 84 votes emerging as the probable compromise candidate.[21] Cass continued to add votes as seven more ballots were taken and when the convention adjourned he had, it seemed, the nomination in his grasp with 123 votes to Van Buren's 99. If Cass's managers could strike a bargain with the supporters of Buchanan and Richard Mentor Johnson who between them controlled 43 votes, he would crack the Van Buren bloc.

That evening Butler learned that Johnson would withdraw in favor of Cass on the next ballot. He thought he could hold about 100 votes for Van Buren on the eighth ballot, but feared his increasingly restive delegates would bolt for Cass and nominate him on the ninth or tenth ballot. Only a decisive move would prevent a stampede. Now was the time to present Silas Wright. Butler spent the entire evening canvassing his delegates for Wright and came away confident he could secure his nomination.[22] At midnight, he went to bed but not to sleep. His mind was troubled at the callous way the southern delegates had brushed aside Van Buren, at the contempt so many delegates had shown for instructions, at all the noise, the confusion and the intriguers. But more, far more, disheartening was that he had read Wright's letter to Judge Fine. Butler wrestled with Wright's declension through the early hours of the morning and finally convinced himself that under new circumstances, particularly Van Buren's letter, Wright could be persuaded to change his mind. He would read the letter to the New York delegation at a morning meeting he had arranged, suggesting Van Buren's name, in accordance with his instructions, be withdrawn and Wright's placed in nomination. If Wright received a majority but could not gain two-thirds, Butler was prepared to demand the recision of the rule. If that failed, he would have the New York bloc withdraw from the convention.[23]

Promptly at 8:00 a.m., Butler met with the New York delegation at Barnum's. He read the Van Buren letter and while urging that Wright's name be presented, he took great pains to convince Judge Fine of the necessity. But the Judge could not be moved. If a single vote were cast for Wright, he would read his letter to the convention and advise it against Butler's proposal. His hopes dashed, a sorrowful Butler informed the delegations he had lined up for Wright that his name would not be presented. What then, they asked, should they do? Butler replied that they should stand by Van Buren "so long as New York kept [him] before the convention."[24]

A significant minority, headed by Bancroft, thought another candidate should be named or Cass would gain the nomination. Polk, he felt, had the best chance; he was not mixed up with the intriguers, his Jacksonian principles were as sound as Wright's or Van Buren's for that matter, and he was for immediate annexation. Bancroft's suggestion, innocent as it seemed, was the product of intense negotiations he had personally conducted with the Tennessee and other delegations while Butler was making his rounds the past evening. Their paths must have crossed, though Butler evidently had an arrangement that if Wright were nominated, Polk would be supported for second place.[25] Butler was not surprised at the mention of Polk's name; he agreed that Polk was a far better man than Cass and he agreed to have New Hampshire present his name. After the eighth ballot was tallied, Van Buren picked up 5 votes and Cass had lost 9. Polk had garnered 44.

At this point Butler decided that if Cass were beaten, Polk was the only chance. He informed the Chair that New York would leave the convention temporarily for a conference. The New Yorkers were bitter, and it took all of Butler's persuasive talent before they would agree that Van Buren be withdrawn and Polk supported. While the final details were being worked out, Cambreleng slipped away and informed Maine of New York's decision just as the ninth ballot got underway.[26] Maine cast 8 of its 9 votes for Polk.

To thunderous cheers from the floor the bandwagon was underway. Van Buren votes in New England switched to the Tennessean. When the New York delegates returned, the roll call had reached Virginia, which switched from Cass to Polk. William Roane was at the rostrum explaining the change and, seeing the New Yorkers returning, spoke feelingly of Virginia's deep regard for Van Buren and the many close personal and political friends he had in the Old Dominion. Butler replied for his delegation by withdrawing Van

Buren's name. All of the delegates would vote as they chose, but he would support Polk. Lt. Governor Dickinson, who was standing beside Butler, immediately cast all but one of New York's votes for Polk. It was all over, even the South Carolina observers declaring their support.

Back in his hotel room, Butler's nerves gave way. He lay on his bed and wept uncontrollably despite the presence of O'Sullivan, Gilpin and several other friends. His only consolation was that the convention had scotched the Walker-Saunders intrigue and the party remained intact. Later, Judge Fine suggested to Butler that Wright might take the vice-presidential nomination since it would not conflict with his sense of loyalty to Van Buren and did not involve his views on Texas. Butler grasped at the chance of retrieving something for New York's honor and pride from the defeat. Seeking out Walker and Roane, he made the proposition, which they promptly accepted as due the sacrifice New York had made for the sake of party harmony. The next day Wright was nominated.

Not long before the convention assembled, Samuel F. B. Morse had installed the world's first telegraph line between Baltimore and Washington. News of his nomination was promptly carried to Wright. Without a moment's hesitation, the Senator telegraphed his refusal to Baltimore; but after three more messages and three more emphatic rejections, he finally sent Preston King and Orville Robinson, another New York Congressman, by wagon to Baltimore.[27] They bore a letter which contained a categorical rejection that could not be misunderstood. The convention thereupon nominated George M. Dallas of Pennsylvania, a bitter political foe of Buchanan, one of Walker's relatives who had kept secret his opposition to Van Buren's nomination.

Wright had been an anxious observer of the events in Baltimore as the news clicked over the telegraph key which Morse himself was operating in the Capitol rotunda. He had also stayed up late Tuesday night for the written reports messengers were rushing to Washington. After the balloting on Tuesday he wanted Van Buren's withdrawal from the fray, his only difficulty being how the retreat could be managed with honor. As he wrote Van Buren, he was "satisfied that the desperation of the principal agents was such and their public commitments such, that if you were nominated they would do all in their power to defeat you in self justification."

Both Wright and Butler believed Calhoun was the principal agent of the Walker-Saunders effort to invoke the minority rule. He may have hoped for a convention deadlock and a consequent breakup of the party as Butler claimed. Gouverneur Kemble remarked to Poinsett how disappointed Calhoun's men were at Baltimore, "at a nomination taking place at all when they expected the convention to dissolve in a row."[28] Wright was also convinced that Calhoun wanted the convention broken up and the party, too; as was Justice Daniel, who said, "The bitterness and animosity of J. C. Calhoun towards everyone whom he has regarded as standing in the way of his march to power would create surprise with some."[29] Whether these were Calhoun's intentions or not, Wright and Daniel were correct in their prediction that if Van Buren were nominated the South would not support him.

Van Buren had been prepared for his defeat at Baltimore at least three weeks before the convention. He, too, at one point was willing to have the party break up if a minority refused to acquiesce to the will of the majority. Now, like Wright and Butler he endorsed Polk, called for party harmony, and like his two loyal lieutenants he was disgusted at the tactics used to deny him the nomination. But there was no more talk of New York standing firm and letting "others dissolve the party if they choose."[30] Two things he was certain about: he would never again voluntarily take an active part in party politics; and for the first time he realized how fragile the party had become, how near the surface and how dangerous sectional tensions were to the very existence of the Union.[31]

« CHAPTER 30 »

The Best Laid Plans

Van Buren took his rejection with equanimity. He thought well of Polk, though in all likelihood agreed with Kemble that he was just one of a dozen or more state politicians who were shrewd, industrious and loyal to the Democracy, but with little else to recommend them. As Kemble described the candidate, he "is not wanting in decision of character; but he is too much of the village and party politician, deficient in scope of mind and of narrow prejudice."

Van Buren knew of Jackson's and Felix Grundy's high regard for the short, trim man. He had been a careful observer of Polk's three campaigns for governor of Tennessee and had been upset enough to write him a stern letter reproaching the Democracy of the state for poor management when he went down to a second defeat after a grueling personal campaign in 1843.[1] He had no qualms about Polk's party orthodoxy; nor did he doubt his political importance in battling against odds to hold Tennessee for the party throughout the Southwest.

The information he received from Wright, O'Sullivan and Butler presented Polk as the only alternative to Lewis Cass. Van Buren may not have considered Polk a statesman, but he was an infinitely better choice than Cass, whom he viewed a trimmer in politics, an ambitious intriguing sort of fellow whose timidity on the great public measures of the Jackson Administration had earned him the contempt of his colleagues in the cabinet.

With some misgivings about the candidate's ability, Van Buren was too much of a party man not to give Polk his support. Though he had decided that he would never again seek public office, the party, after all, was more his personal creation than that of any other politician, even Jackson. There was another factor, a most important factor; Polk had been completely innocent of the deals, the intrigues of the Baltimore convention. Or was he? A niggling doubt remained that the candidate, a westerner not a southerner, was still a slaveholder. Would he fall under Calhoun's influence? Would he stand forth as a southern President seeking more concessions from the North on the extension of slavery? Wright expressed these concerns to Poinsett when he said, "Our Democracy will support faithfully the Baltimore ticket, if there be no more southern bad faith, or sectional issues raised to produce southern suspicion."[2]

Calhoun and his friends exulted at the defeat of the New Yorkers. Unlike Van Buren, whose loyalty to the party was beyond reproach, Calhoun had no such devotion.[3] Abolition and the growth of public opinion in the North for the containment of slavery had raised serious doubts in his mind whether the South and its peculiar institution could coexist with the free states. His more impetuous followers, like Rhett, Hammond and Lewis, were constantly hammering at the idea of disunion.[4] Calhoun was prepared to go his separate way as he had during the Jackson Administration, only now he commanded a much stronger political position in the South than a decade earlier. Polk was as sensitive to this fact of political life as was Van Buren and Wright. He could not win without Calhoun's support; nor could he without Van Buren's for that matter. But thus far the candidate with Jackson's and Bancroft's help in the North and Walker's and Cave Johnson's efforts in the South had managed to balance the pretensions of the two wings and ease mutual suspicions the Baltimore convention had engendered.[5]

Van Buren concluded that even with a united party in the nation, the race would be a close one, which meant that Polk had to carry New York and Pennsylvania in both of which the Democracy was badly split. Before the Baltimore convention he managed to bring the Hunkers and their opponents, now being called "Barnburners," into an uneasy alliance. But after the convention, the factions, free of restraint, began abusing each other with increasingly bitter remarks in their respective presses, the *Atlas* for the Barnburners and the *Argus* for the Hunkers. As before, Governor Bouck, Lt. Governor Dickinson and Edwin Croswell of the *Argus,* Flagg and Hoffman were at the center of the controversy. The newspaper quarrel that erupted again in Albany quickly spread throughout the state as the party press took one side or the other. Bouck, who always regarded himself as a moderate, was cast as the creature of those who stood to profit from a resumption of deficit spending for internal improvements, a front for all the entrepreneurial interests who would gain if the Flagg-Hoffman-Dix "stop and tax" policy were scrapped.[6]

In Washington, Silas Wright correctly predicted that the Barnburners would seek to use him as the means of toppling Bouck. His suspicions were confirmed at a mass meeting of Democrats in

New York, and he was determined when he reached home to "put the use of my name unequivocally to rest."[7]

Van Buren had taken no public stand on these divisive issues, but his sympathies had always remained with the *Atlas* and the Barnburners. He shared their "stop and tax" policy on internal improvements. His opposition to speculation was unyielding. Banks, too, he regarded with a well-founded suspicion, having noted their renewed efforts to secure legislation more favorable to their interests, particularly increases in their note circulation as the depression finally began to lift. Was it to be 1837 all over again? Not if he could help it!

Apart from these economic reasons, Van Buren had strong personal and political motives that moved him toward the Barnburners. His sons Smith and John were closely associated with the *Atlas* group. John had political ambitions and was obviously the power among the younger element of the Democracy in Albany, in New York City and in other towns and cities throughout the state where, like his father before him, his law practice took him. At the height of his powers, the younger Van Buren was now a family man. While he still enjoyed society, good food and good liquor, marriage had curbed his tendencies toward dissipation. He was also an able politician, not as clever and calculating as his father, nor possessed of his powers of observation. Rather than restraining himself as his father would have done in the heat of a political contest, John was impetuous in his zeal for party principles. He could be thoughtless, arrogantly disparaging toward those who disagreed with him. There seemed to be no middle way for John—either devotion or deep enmity.

Van Buren took great pride in his second son's accomplishments; and after he had definitely decided upon retirement, began devoting all of his skill and his contacts to John's advancement. Indulging a father's blindness to his son's weaknesses, he mapped out a career in public service that would parallel his own. In 1842 John Van Buren had sought the state attorney generalship but neither father nor son made a concerted effort for the appointment and neither was disappointed when they failed. Since then John had used the time profitably in building his image among the radicals; yet he followed his father's advice and curbed his pugnacious inclinations to do battle with the conservatives. Still John Van Buren's ties with the Barnburners were sufficiently well known to rule out a second bid for the attorney generalship, if Bouck were again the candidate and won the election in the fall. A Bouck reelection would adversely affect Flagg, too. Of course, a Whig sweep would definitely drive Flagg from office and certainly damage John Van Buren's promising political career.

Unwilling to exert any personal influence over the *Atlas* or the

Argus, Van Buren despaired of carrying the state for Polk, for John and for Flagg unless Silas Wright could be persuaded to run for governor. He was well aware that Wright utterly abhorred the feuds and factions of Albany, the enemies any governor would make, the problems and vexations of executive office. He had read not long after the convention Wright's notice in the *Argus* saying he was not a candidate for the governorship and requesting all his friends to abide by this wish.[8]

Wright was happily content with his seat in the Senate where, if he was upset at the vagaries of Washington politics, he was by and large protected from the vicious rough and tumble of the New York arena. Van Buren knew all this, knew the factions in the state would buffet his dear friend after his election, knew also, and at first hand, the rising tide of violence in the river counties as the tenant farmers, whom he had once defended when a young lawyer, strove to abolish the old manorial rights of the great estates. Any governor elected this year would have to make unpopular decisions containing this public unrest of "anti-rentism" before the slow process of law could ease their condition and confirm them in what they demanded as their right.

If Bouck ran again, the Whigs would take New York and probably install Henry Clay in the Presidency. To avert this calamity, Wright must run and win—for the reconstruction of New York's Democratic party, for honest conservative government at home and for the restoration of the state's prestige in the nation. Not the least of Van Buren's thoughts bearing on the matter was a public statement from Bouck that he would withdraw in favor of Wright and only Wright. The Senator had preserved his neutrality in the factional fight without giving offense to either side. But Bouck and his advisers knew of Wright's desire to remain in the Senate and counted on his refusal to run for governor.

Wright stopped off at Lindenwald in mid-June on his way home from Washington. Van Buren was hospitable as always. John A. Dix, who had come down from Albany, joined the two men over Madeira after dinner in the capacious study. Local politics was touched upon but all studiously refrained from making any comment on the Baltimore convention. The next morning, after Dix left, Wright told Van Buren of his decision to publish another notice that he hoped would close all doors to his being named a candidate for governor. Though Van Buren was taken aback at Wright's declaration, he concealed his disappointment and set out to convince the Senator he must forego his personal wishes and stand by the party. He had, he said, anticipated Wright's stand, but cautioned him against adding anything more to the notice already published in the *Argus.*

Still talking, Van Buren led the way outside through the meadows beyond the lawn and along one of the pathways that cut through the woods of the estate. Wright was disturbed when his host said "the situation was a very delicate and important and responsible one; that [I] was in great danger of carrying my practice of declination to excess; that the temper of the Democracy was to win the Presidential election at any sacrifice and that it must and would be done; that I should easily be suspected of indifference upon that point."[9] Reverting to state affairs, Van Buren repeated what Wright had all along suspected. His defeat at Baltimore had broken the thin wafer sealing the factions together in New York. If Bouck were to stay in the race, the party would become hopelessly divided and the state would go to the Whigs. Wright was the only man who was acceptable to both sides.

Van Buren's wealth of arguments, his understanding of his friend's reluctance, his carefully pitched appeals to a higher duty, a duty which he had instilled in his associates during the heyday of the Regency, made a deep impression. After all, the unassuming Wright felt he owed his entire political career to this self-made country squire with his charming, winning ways, a man who had occupied the highest office in the state and the nation, yet had not permitted the nomination of Polk, a far lesser figure than himself, to shake his faith in the party, his resolution that it "must and would" win in November.[10]

Van Buren's sacrifice seemed far greater, far more noble to Wright than his petty concerns, his wife's dislike of Albany, his own mistrust of his executive abilities. All of these mixed emotions, these personal and political concerns, he voiced to Van Buren, who knew his man well and was prepared to answer them in a way that would make the most impression. He admitted that the next two years would not be easy ones in Albany; nor, he believed, would they be any more pleasant in Washington. Indeed, they might be far more difficult than Wright thought, especially if Clay won the Presidency. Whatever objections Wright might have, he "could never get the public to consider them nor to make allowances if [he] acted publicly upon them."[11] A troubled Wright left Lindenwald with a promise from Van Buren "that he would do nothing to try to make me a candidate for Governor."

Van Buren may have been true to his word but only in a highly technical sense. Three weeks later while Wright was in Vermont visiting his family, Smith, John and John A. Dix attended a Fourth of July mass meeting of the Columbia County Democrats in Hudson. Among the patriotic resolutions adopted was one proclaiming the virtues of Silas Wright and demanding that he become the party's candidate for governor.[12] When Wright arrived home and waded through the accumulated mail, he was not surprised that the coincidence of his and Dix's visit to Lindenwald and Dix's speech at the Columbia County meeting had been noted and the inference drawn in the Hunker papers. The three men had decided on this strategy to push Bouck out of the race, "that the thing was all arranged at Kinderhook." It never occurred to Wright that Van Buren may have broken his promise, and he seems to have thought he had convinced the Magician he was not to be a candidate.[13]

Remembering their conversation, however, he did not draft another letter, but simply wrote Croswell and asked him to reprint his original notice. As the basis for any further editorial comment, Wright added that his position remained unchanged and he had not authorized the use of his name in any circumstance.[14]

The Hunkers were placed in an untenable position. Governor Bouck could neither push for renomination nor withdraw until Wright's position was certain. From all sides the harried Senator was being forced to make the public declaration he dreaded; for if he refused as he so ardently wished, he would be deserting his friends, Flagg, Dix and his revered mentor, Van Buren. If he accepted, he would be immediately injected into the factional fight and find himself opposed to men like Marcy and Croswell, close associates of many years standing, politicians whom he still respected. Wright finally wrote a letter for the *Saint Lawrence Republican,* categorically refusing to accept "a nomination under any circumstances." Instead of sending it directly to the editor, he enclosed a copy to Preston King in nearby Ogdensburgh, thinking it only polite that his small circle of close friends should know of his decision before they read about it in the paper. His excessive thoughtfulness proved his undoing.

The very next day, King was on his doorstep with the letter and a lengthy remonstrance against his blanket refusal to be a nominee. There folowed a week of intense negotiations during which Wright's friends reiterated the fact that he was not free to make his own decisions in such political matters. He was "public property," they said, and "the Democratic Party had a right to use [him] as it thought proper." Wright countered that it was not the party but only a portion of the party making these demands upon him.[15] But his defense was really no defense at all because he was aware and his friends were aware that however he might phrase his position barring a categorical refusal, he would be nominated. Bouck would have to support him or be accused, and properly so, of bad faith. Just before the convention, he capitulated, permitting Judge Fine and other friends to let his name go before the convention. He had three-fourths of the delegates on the first ballot which was made unanimous on the motion of a Hunker delegate.[16]

Van Buren had deliberately sacrificed one of his closest friends; yet he did not see it as that but rather as part of a well-laid plan that would unify the party in New York and crush the resurgence of what he believed was the old Conservative organization of Tallmadge and Rives. He understood that he was loading a heavy burden on Wright, and he felt badly about it. Yet if Clay won the state there would be only the remembrance of a painful interlude while the Barnburners would gather in a rich harvest of party loyalty.[17] Should Wright carry the State and the election for Polk, Flagg would be secure, and John Van Buren would become attorney general, the first step to the White House. A grateful Administration, he assumed, would assist the Barnburners in crushing the heresy. Wright would find Albany a much more serene environment than Washington, where the Texas and Oregon issues, abolition, "Southern rights," quite probably a war with Mexico, possibly with Great Britain, too, not to mention tariff and banking issues, all threatened.

Under the leadership of Flagg, the state committee made every effort to conciliate Bouck, approving more nominations of Hunkers for seats in the legislature than their numbers warranted. The effect was immediately apparent in generating campaign enthusiasm. Van Buren watched with satisfaction as Hunkers and Barnburners labored together for the election of Wright and Polk.[18]

With a united party rallied behind Wright and Flagg's admirable direction of the elaborate campaign plan Samuel Tilden had drawn up, the contest was still in doubt to the very end. Wright's opponent, Millard Fillmore, proved a popular figure who was also acceptable to the warring Whig factions. He lacked Wright's stature and intellect, but as a four-term Congressman, he was well known in the state and was particularly strong in his home counties of the west. Had it not been for Wright and James Birney's small Liberty party of abolitionists, Clay would certainly have carried the state. Wright's crucial importance to the ticket was emphasized in his majority over Fillmore of some 10,000 votes while Polk's popular vote was less than half that number. As the returns came in from other states, it became obvious to all that New York had elected Polk. Silas Wright and Martin Van Buren deserved all of the credit.

The President-elect, self-conscious and sensitive, could have hoped for victory under different circumstances. It was mortifying enough to be the dark horse candidate, the object of the disparaging question the Whigs everywhere had hurled at him during the campaign. "Who is James K. Polk?" Unable even to carry his own state, he had become President at the sufferance of others, a fact that could not be ignored when his own party press made Wright the hero of the election, the man who could have been President but put his loyalty before his ambition.[19]

Obviously New York must be recognized in the cabinet, whose make-up Polk began considering as soon as his election was certain. Just how he would reward the state was another matter, and contrary to his later justification of his actions, he had a reasonably accurate though biased picture of the factions in New York which influenced his policy on appointments.

Shortly after Polk's nomination, William L. Marcy, ever hungry for public office and despairing of the Supreme Court judgeship that still remained vacant, had opened a correspondence with Polk. In his many letters to Tennessee, Marcy described the state of New York politics, exaggerating the strength of the Hunkers and implying that if Wright were elected he would have to remain at the helm or the party would dissolve.[20] The *Atlas* and the *Evening Post* speaking for the Barnburners added their weight to the demands that Wright must serve out his term as governor. Their attitude was backed up in letters Judge Fine wrote Cave Johnson and Francis Pickens.

Polk, of course, had to consult with Jackson, who, along with Wright and Van Buren, had made him president. Accompanied by Aaron Brown, a Tennessee Congressman and close friend, he visited the Hermitage and exchanged views with Jackson who was clearly dying but whose mental faculties, decided opinion and fierce prejudices were as keen as ever.[21] Nothing apparently was decided about the cabinet except geographic distribution. Buchanan was the subject of discussion because of his support for immediate annexation and the importance of the state he represented. Wright's name came up, if not from Polk or Brown, surely from Jackson whose friendship he valued. They discussed the problem of Calhoun, whom Jackson denounced freely. Polk assured him that no Calhoun, Benton or Tyler clique would control his Presidency and that every prospective cabinet member must disavow any presidential aspirations.

Evidently Jackson pushed Wright for the State Department, but acknowledged his great abilities in the financial area and would have been satisfied if he were given the Treasury, accounted second in rank among the cabinet members. Polk did not want Wright in the cabinet at all because like so many men unsure of their prestige and their ability, he feared the New Yorker would dominate his Administration. After the Hermitage conference he pondered the major appointments, and while he was reasonably confident Wright would not accept any cabinet post, he finally offered him the Treasury Department. Then, apparently concerned that the second position would offend New York and mindful perhaps of Jackson's advice, he gave Wright the option of the State Department. Reflecting on this alternative offer, Polk worried that Wright might accept State and rewrote the letter offering the Treasury Department.[22]

As congratulatory letters piled up on his table at Canton, the doleful Governor-elect sought to comfort his saddened wife. "In my house," he wrote Flagg, "the feeling is very different. We are no less grateful for the national result than our neighbors but there are personal considerations very far from joyful or triumphant."[23]

To Polk's relief, Wright confirmed what his associates were saying. He declined the offer saying he had pledged himself during the campaign to serve out his term as governor. Polk did not receive Wright's declination until January 4, and assuming Wright had consulted Van Buren before turning down the Treasury, tardily realized he had been remiss in not writing the ex-President. He immediately wrote a long letter to the master of Lindenwald, which in a deeply apologetic vein assured him that he was no wise responsible for the decision at Baltimore. Exuding deference and flattery, Polk asked his advice on the cabinet appointment he had reserved for New York, mentioning he had offered the Treasury Department to Wright, the "one individual about whom my mind was definitely made up, without reservation," but whose declension he had just received. Polk begged Van Buren to have Wright reconsider, and this time he added the option of the State Department. If Wright was adamant, who would Van Buren suggest for either position? He wrote Wright a similar letter asking him to reconsider, adding he would welcome suggestions for the entire cabinet. Van Buren and Wright received their letters on the same day, January 17, 1845, again delayed because of Polk's indecision. Wright promptly dispatched his letter to Van Buren. "Is there a doubt that I should name to him [Polk] Butler for State and Flagg for Treasury?" he asked.[24]

Van Buren studied Polk's letter. Beyond the delay in arrival which must have aroused some suspicion, he was pleased with what seemed the sincere offer of assigning either State or Treasury to New York. But the inference was plain; the President-elect preferred New York to have the Treasury, and this preference coincided with Van Buren's wishes, too. "He is fit for anything," said Van Buren of Wright, "but the Treasury is the very one to which his great talents are peculiarly adapted." He approved also Wright's declination of the post. If Polk wanted Wright to remain in New York, so did Van Buren.

Candidly, he spoke of the divisions in the party, explaining to Polk that in order to carry the state many of ex-Governor Bouck's friends had been conciliated by being nominated and elected to the legislature. During the campaign, however, Bouck had acted very badly. He had opposed Wright's nomination at the last moment, despite his public declarations to the contrary, forcing the convention "to lay him aside by a direct vote." The composition of the legislature and the activities of the Hunkers with Bouck at their head made it absolutely necessary that Wright's firm hand be at the helm of the state and of the party. What Van Buren did not say was that without Wright's influence the nominations of John Van Buren and Flagg for state offices would both be in jeopardy, for the Hunkers claimed half of the positions. An added concern about John was his emotional state at the time. He was in deep mourning for the sudden death of his wife not two months before. Van Buren thought the duties and prestige of the attorney general's office would lift his son's spirits, renew his zest for life.[25]

Although Van Buren recommended Butler for the State Department, he was far more emphatic in his backing of Flagg for the Treasury. In fact, he ruled out the Treasury for Butler as not compatible with his training or interest and made it evident that it would require some persuasion before he could be induced to accept State. He did declare, however, that Butler's views on Texas were close to those of Polk and reminded him that Butler had drafted the Texas plank at the Baltimore Convention.[26] For personal, political and practical reasons, Van Buren was prepared for New York to accept the second post. Flagg, a superb political tactician, was more conservative than he in matters of banking, internal improvements and monetary policy. But the sturdy, vibrant little moneyman would take care that New York's interests were well protected in Washington. As a second choice for the Treasury, Van Buren named Cambreleng, who, as Chairman of the House Ways and Means Committee, had worked closely with Polk when he was Speaker. Should New England be represented, he would nominate either Senator John Fairfield from Maine or Governor Hammond of New Hampshire, though his recommendations were rather tepid. Then he spoke of Bancroft, who he thought "greatly over tops both in that respect. If he lived in either state I should not hesitate to give him the preference." After finishing his long letter to Polk, he gave Wright the gist of what he had said and had one of the servants carry the message to Albany. The next day he received Wright's note which confirmed, to his relief, what he had done.[27]

The month of January passed without hearing from the President-elect, no cause for immediate concern considering it was the time of year when everyone expected delays in the mails. But during the interval the factors that had in part governed Van Buren's letter had changed. On February 1, 1845, the Democratic caucus in Albany renominated Flagg unanimously and John Van Buren was named attorney general by the narrowest of margins, one vote over his Hunker opponent.[28]

Relieved of his anxiety for his son's public career and feeling less of an obligation for Flagg's future, Van Buren turned his attention again to national concerns; as he watched the debate in Congress over Texas, he changed his mind about Butler and the State Department. An urgent message went out from Lindenwald for Butler to visit him for a consultation. Though trying to extricate himself from his brother's financial problems and coping with the heavy load of law business, Butler hastened to comply. He spent two days with Van Buren, during which they discussed the Joint Resolution for Annexation Calhoun and Tyler had introduced into Congress the previous session when their treaty failed in the Senate. The resolution again before Congress required a simple majority from both houses rather than a two-thirds vote in the Senate for a treaty to be confirmed.

Butler, differing from Van Buren, Wright and Dix, favored the resolution. While he still harbored doubts about accepting a cabinet post, Van Buren was able to convince him that he must accept the State Department if it were offered.[29] After Butler left Lindenwald, Van Buren wrote three letters to Polk, on February 4, 11, and 15. This time he addressed them under cover to Dix in Washington. The first and most important letter assured the President-elect that Butler would not only accept the State Department but his views on Texas conformed as far as he could see with those of the new Administration. Butler, he added, would faithfully carry out the President's foreign policies once a decision had been made even if he disagreed with them.

Apart from local considerations, it had been the warmth of the Texas debate in Congress that had prompted him to urge Butler for the State Department. In his mind there was no public figure in the country save Butler who could handle annexation dispassionately and successfully. If the Republic should enter the Union simply on the grounds of extending slavery, there would be a dangerous reaction in the free states. He realized that in his preoccupation with New York politics he had neglected material concerns and now he sought to rectify his error, while still maintaining both his own dignity as an elder statesman and the prestige of his native state.

In the second letter he recommended that Donelson be considered for a cabinet post on the grounds of his knowledge of Texas affairs. Jackson's nephew had just returned from the Republic where he had conducted negotiations with President Houston for the Tyler Administration. This experience, in Van Buren's opinion, made him uniquely qualified to offer expert counsel. That Donelson's appointment would please Jackson was also a consideration, though a minor one, in making this eleventh hour recommendation. These two letters he left unsealed along with a note to Dix asking him to read them and show them to Blair before presenting them to the President.

Van Buren's third letter, which he did seal, introduced Dix to Polk and praised him extravagantly as the most able and reliable of the New York delegation. Included in the package were two more letters, one to Blair, the other to Bancroft. Both expressed his deepest fears about annexation. Were it not carefully handled, the issue would wreck the party and could bring about war, even disunion.[30]

Although Van Buren strongly urged Blair to use his influence with the President-elect in Butler's behalf, he did not mention him by name to Bancroft. In that letter, he discussed appointments generally and expressed a belief that since the secretary of state would be conducting the negotiations with Texas, he should come from one of the free states. He also thought Bancroft should consult freely with Dix. "You know," he said, "that in his judgement, purity and discretion we place unlimited confidence."[31]

Van Buren's first letter of January 4, 1845, reached Polk before he left Tennessee. After studying its contents, he was relieved that Van Buren agreed with him on Wright and seems to have considered the endorsement of Butler for State, hedged as it was with doubts about his acceptance, to have cleared the way for offering the premier position to another state without giving offense to New York or to the partisans of the ex-President.

Polk must have noticed that neither Van Buren nor Wright had included Marcy in any of their recommendations; and with what he had gleaned from the letters he had been receiving about the condition of New York politics, he must have pondered the significance of the omission. Polk was especially pleased with Van Buren's high praise for Bancroft and settled on him at once for the New England appointment. The historian-politician, like Marcy, had begun corresponding with Polk shortly after the Baltimore convention. In describing the convention proceedings, Bancroft was not shy about his role in the nominations, an important one, which Gideon Pillow, Donelson and Cave Johnson, Polk's managers at Baltimore, confirmed.[32] At the same time, Bancroft kept his close ties with Van Buren and Wright intact. He visited Albany and Lindenwald during the campaign, and he wrote numerous warm, friendly letters that left no doubt where his allegiance lay.[33]

The arrival in Columbia of Van Buren's letter required another conference at the Hermitage, though by now Polk had pretty well settled the cabinet appointments in his own mind. Before he received Van Buren's letter, Polk had heard from Wright and others in New York recommending Flagg for the Treasury. Now, with Van Buren's strong endorsement before him, he decided tentatively to make that appointment. Jackson backed up his decision and approved of the entire list Polk had made up. Polk agreed with Jackson in his condemnation of Calhoun. He, too, mistrusted the champion of the South and decided he must go. Still, common sense dictated he test the temper of Washington before he decided on how he would deal with the South Carolinian.

Calhoun had emerged as the undisputed leader of the seaboard slave states south of Virginia and the rich cotton regions of Alabama and Mississippi. His frank defense of slavery, his untiring efforts to make the peculiar institution a matter of public and party policy had extended his influence as far north as Polk's own state. Nor was it any secret that Calhoun wished to remain secretary of state, at least until the Texas and Oregon issues were settled. During the past summer, Francis Pickens and other Calhoun supporters visited the Hermitage and Polk himself at his home in Columbia, all intent on ensuring Calhoun's tenure in the new Administration.[34]

Polk left Jackson with the impression that he would appoint Buchanan as secretary of state, Flagg treasury, Andrew Stevenson of Virginia war, Bancroft navy, Walker attorney general and Cave Johnson postmaster general. If anything, his cabinet list, despite Polk's determination to avoid cliques, inclined toward Van Buren—Flagg, Bancroft, Stevenson and Johnson all being his partisans. Walker was the only representative from the Deep South and the only member who may be said to represent Calhoun's influence. Northern by birth and upbringing, though a slaveholder, he opposed slavery and his tentative post, attorney general, was considered the least prestigious in the cabinet. Buchanan, of course, was anti-Van Buren, though Polk had no reason at this time to suspect the depths of his opposition to the New Yorker.

Polk arrived in Washington on February 14, 1845, after surviving two weeks of bad weather, tumultuous crowds, the ever-present lobbyists and officeseekers. By the time he finally settled down in his four-room suite at Coleman's Hotel, he had been taxed by the well-organized supporters of Buchanan, Walker and Marcy.

When several Van Burenite Senators visited Polk's crowded chambers during the next several days, they observed well-known advocates of these three aspirants surrounding him. One or two days after his arrival, Senator Dix managed to elbow his way to Polk's side, introduce himself and hand him Van Buren's letters.[35] Bancroft had consulted Dix and Blair as Van Buren suggested and knew about the Barnburners' substitution of Butler for Flagg, the State Department for the Treasury. He was ready to respond favorably should the President-elect ask his opinion. Polk did not.

Polk went ahead with his original plan and appointed Buchanan secretary of state, though not before he exacted a promise that if he should be a candidate for the Presidency, he would resign from the cabinet.[36] After rejecting Van Buren's suggestion of Butler for State, he decided he would not accept his and Wright's advice for Treasury either. New York would have to be content with a secondary post, the War Department. A union of Calhoun and Cass men, the same coalition that had wrested the nomination from Van Buren, were insisting that the South be awarded the Treasury in the person of Robert John Walker. Walker, though flawed as a public man due to questionable speculative ventures, was an able financier. The fact that he would be a makeweight against Buchanan and thus hold down cabinet intrigues to a minimum, was appealing to Polk's tidy, methodical mind that saw wisdom in narrowly constructed but symetrical checks and balances.

Yet Polk hesitated. He had found to his astonishment that New England Democrats with impeccable party credentials shied away from Bancroft, whom they suspected of drifting toward the conservatives of the Woodbury–Isaac Hill–David Henshaw stamp, the Hunkers of the Northeast.[37] Bancroft was not as popular in his native region as the President-elect thought, but he certainly enjoyed the respect and confidence of New York if Van Buren's phrase in his first letter to Polk was to be taken literally. Why not give the Treasury to Bancroft?

At one stroke he could accomplish three objectives: he would reassure the New York Barnburners that one of their own, even though a citizen of another state, would have the second position in the cabinet; he would free himself from any imputation that he was bowing to a clique, whether it be Van Buren's or Calhoun's; and New England would have a representative in the cabinet. It seemed all very logical; yet logic was no substitute for political realism, as Polk would learn in due course.

It was not until February 22, eight days after his arrival in Washington, that Polk finally answered. Polk began with a labored explanation of his decision to appoint Bancroft to the Treasury. Loaded with apologetic explanations for his delay in replying, Polk offered the War Department to New York and seemingly gave the choice to Wright and Van Buren. But even in recognizing New York with one of the minor posts, an insult on the face of it to the pretensions of the Empire State and its role in his election, he narrowed the choice to Butler or Marcy. Polk explained that he had asked Dickinson and Dix to recommend a New Yorker for War and that they jointly named Judge Sutherland, a man unknown to him and to the country at large.

That Dix should have agreed with Dickinson in recommending

Sutherland must have shocked Van Buren. For by this time he had evidence that Dickinson was part of the Marcy lobby. Somehow, Dickinson had taken advantage of Dix's inexperience with politics as played in Washington and had him commit a blunder that was embarrassing to Van Buren. In the same mail was a second letter from Polk dated February 25 stating that he had offered the post of secretary of war to Butler. Would Van Buren use his influence with Butler to have him accept?[38]

Van Buren usually kept his temper under control, but when he read Polk's two letters which arrived on the evening of February 26, 1845, he became furious. That the President-elect would use his own phrase praising Bancroft out of context as justification for appointing him to the Treasury Department was bad enough; but when Polk alluded to the Walker-Cass clique as the reason for rejecting Flagg, Van Buren could hardly restrain himself. He suspected wrathfully that Polk was counting on Butler's refusal. In his letter recommending Flagg to head the Treasury he had spoken of Butler's personal situation which might stand in the way of his accepting the State Department. The President-elect had not even mentioned Butler for the State Department, though he acknowledged he had read Van Buren's letters sent to Washington. Despite his emotional state, Van Buren was thinking clearly. His political relationship with Marcy had become much more distant as he watched him drifting toward Bouck and the Hunkers.[39] Though his sons John and Smith, his friends Flagg and Hoffman, had long since decided Marcy was not worthy of trust, Van Buren and Wright had continued their cordial relationship until after the state election. What had brought them to the breaking point was Governor Bouck's appointment of two Senators to replace Wright and Tallmadge.

There had been no reason for Wright to resign until a few days before his inauguration in January. But the Governor-elect, still overwhelmed at the abrupt change in his life and not as familiar with the party division as he ought to have been, made a grievous error. Bouck had offered to appoint Marcy to Tallmadge's unexpired term, which had only a few months to run, but he wisely declined. Then, with what must be considered malice aforethought, the Governor mentioned to Wright that Tallmadge had resigned.

Bouck was faced with a dilemma. Two of his Hunker henchmen, Lt. Governor Dickinson and Henry Foster, a bank lobbyist from Onondaga County and a former Congressman, coveted the appointment. Both men were self-centered politicians to whom position and place meant everything, and each was attacking the other with impunity, threatening yet another factional split. If Wright could be induced to resign, Bouck could seal off a breach in the Hunker ranks and satisfy the aspirations of Dickinson and Foster. Without a moment's thought, Wright neatly fell into the trap Bouck had prepared. He told Bouck that he did not think much could be accomplished in Congress during the three weeks or so before his inauguration. Bouck promptly declared that New York should be fully represented in the Senate when Congress convened. "If you think so Governor," said Wright. "I must resign for I do not intend to go to Washington." And with that remark went into Flagg's office and wrote out his resignation.

Wright never dreamed that Bouck might have Foster in mind for his replacement; nor did he pause to consider that his presence in Washington during the congressional session, however brief, might be important in protecting his own and the Barnburners' interests with the incoming Administration.

Bouck immediately appointed Dickinson and Foster to fill the two vacancies. The Barnburners could stomach Dickinson despite his devotion to deficit spending for internal improvements; he was, after all, a strict party man. But Foster was a renegade, a Hunker of the deepest dye, who had openly deserted his party in 1837 and had become a leading Conservative. In Congress he had been one of the bitterest opponents of the subtreasury system and had only returned to the Democracy because the Whigs made it plain they could do without his services.[40]

When the two men were appointed they came to an agreement that Dickinson would run for the full six-year term and Foster for Wright's unexpired term of four years. The composition of the legislature being what it was, Dickinson had a much better chance of being selected than Foster. This treacherous, petty politician, recognizing all too well his weakness, hit upon a scheme that might ensure his election.

As soon as he arrived in Washington, he sought out members of the Cass-Walker clique and spread the word that Wright and Butler were opposed to the admission of Texas on any terms. Moving about among the avid immediate annexationists, he managed to convince many that the Barnburners were antislavery men who would range themselves against the Administration on Texas whether it be a party question or not. He was particularly vehement against Flagg, who he implied was a crypto-abolitionist. The superheated atmosphere of Washington gave Foster's charges wide circulation. In far distant Tennessee, Old Hickory read about them in the *Nashville Union* and wrote one of his scorching letters to Van Buren deploring the attack on that "upright and valuable man Mr. Wright."

The legislature elected Dickinson to the Senate, and with Governor Wright's influence behind him, Dix narrowly defeated Foster.

But the trail of suspicion Foster had laid in Washington was well blazed when Polk arrived in February. And Dickinson kept it open as he and other conservative Democrats within the New York delegation pushed hard for Marcy.[41] Dix was the only prominent New York politician who tried to set the record straight. When asked whether Marcy would be acceptable to Van Buren and Wright for the War Department, he told Polk frankly that Marcy's appointment would wreck the party and grievously injure the Administration in New York. At the time Polk had before him a circular letter signed by ten Congressmen from New York recommending Marcy for the War Department. Polk also quizzed Gouverneur Kemble who was visiting Washington about his preference if the War Department went to New York. Kemble replied first Butler and if he refused, Marcy.[42]

While Polk claimed he anxiously awaited a reply from Lindenwald and was sending "twice a day to the [Post] office" for it, Van Buren moved rapidly to head off the Marcy appointment. "Deeming it indispensible to consult Mr. Wright before I replied," wrote Van Buren in a memorandum he made of the events, "I sent my son Martin to Albany the same night to give Mr. Wright time to reflect and give his opinion." The next morning Van Buren himself went to Albany and found Wright in complete agreement that Butler be persuaded to take the War Department. Smith was deputed to convey a letter from his father to Washington vouching for Butler's acceptance. Realizing that time was of the essence, Van Buren wrote out his reply to Polk at one sitting, Smith copying each page as it was drafted. In language unusually harsh for him, he took the President-elect to task for ignoring what he called the just claims of New York and could only believe "selfish" interests had misled him. Wright should have been offered the State Department "because it was due him and the pride of the State." He and Wright had, however, "cheerfully agreed to number two and proposed Flagg." While he thought well of Bancroft, the New York Democracy will be "insulted" at his being placed in the Treasury over Flagg. He advised that since Flagg was unacceptable, Butler be given the Treasury Department and Bancroft the Navy as Polk had originally intended. If this could not be done, then Butler was his and Wright's choice over Marcy for the War Department. "There is not time to consult Mr. Butler, but the exigency of the case is such and the damage of your making a fatal mistake in this state so immense," concluded Van Buren, "that I take it upon myself the responsibility of saying Butler will accept War if tendered. . . . If all fails by all means take Cambreleng."

Just in case there were any further delays, Van Buren sent another letter to Polk explaining briefly that Smith would leave for Washington on February 28 with his reply. This letter went out under cover to Dix on the morning boat, February 27, 1845. The next day Smith left on his mission.[43] Wright and John Van Buren, who now read Van Buren's letter to Polk, convinced him he had been too blunt, too severe in some of his phrases. The result was that Van Buren wrote again to the President-elect apologizing for the tone of his reply and assuring Polk that "neither Mr. Wright or myself have borne the slightest feeling of unkindness towards yourself in consequence of what has been done."[44]

In Washington, Polk, to his dismay, found that Bancroft's elevation to the Treasury aroused even more fierce opposition from New England Democrats than his appointment to the Navy Department. Vice-President-elect Dallas, still pushing for Walker, warned of dire political effects in Pennsylvania if he were not selected for the Treasury to counterbalance Buchanan. The disappointed New Englanders combined with the Cass-Calhoun-Walker forces brought very heavy pressure on Polk. For a time, he resisted. He appointed Walker attorney general only to be faced with such an intense barrage from all quarters that he began backing off and reconsidering the Navy Department for Bancroft.

Though a minority in the New York Democracy, the Marcy lobby, had made an impression on Polk. He did not wish to antagonize Van Buren and Wright, yet he honestly thought Marcy would be closer to his policies than Butler. And he seems to have been unaware that any serious political differences existed between Marcy and Van Buren despite Dix's explanations and warnings. After all, Kemble, who he knew was an intimate friend of Van Buren and Wright, had not thought so; nor had Bancroft. The politically astute move, however, was to offer Butler the War Department with the almost certain knowledge he would not accept unless Wright and Van Buren could change his mind.

On February 25 he sent a direct offer to Butler and on the same day informed Van Buren what he had done. Polk's offer came as a complete surprise to Butler, and after listening to his wife's vehement objections he declined the post, though he did indicate that he would accept the Treasury or the State Department. Should these appointments be already made and New York was to be represented by the War Department, he recommended Cambreleng. On February 27 while Van Buren was in Albany consulting with Wright and drafting his reply to Polk, Butler sent the youthful Tilden and the ebullient editor O'Sullivan to Washington bearing his letter of refusal.[45] They reached the capital that evening and went directly to Coleman's Hotel where Polk received them cordially. He read the letters, and when he had finished he said how deeply he regretted Butler's decision, adding that the other cabinet appointments were all settled. Perhaps they were in his own mind, but he was certainly rethinking his decision on Bancroft for the Treasury and still could have offered that post to Butler.

Throughout the conversation the President-elect took every advantage of expressing his deepest friendship for Van Buren and Wright. So apparently sincere were his oft-repeated words of praise for them that he fairly captivated O'Sullivan and even made a strong impression on the cautious Tilden. Polk complained that he had written Van Buren more than a week ago and had not as yet heard from him. He was extremely anxious for a reply, and it was only because time was becoming so short before his inauguration that he offered Butler the War Department without advice from Lindenwald and Albany. Although "all afloat again on the subject" Polk agreed with Tilden he would delay his decision a day or two longer hoping to hear from Van Buren before he acted. Altogether, he had gone out of his way to impress O'Sullivan and Tilden, young men he had just met. He even convinced them that he was thoroughly familiar with New York politics and favored the Barnburners, no mean feat with a man of such razor-sharp political instincts as Tilden.[46]

After they left, Polk wrote a long letter to Van Buren saying that he could not wait any longer and would appoint Marcy to the War Department. A defensive note came to the fore as Polk repeated his previous offers to New York and explained, as he had in his last letter to Van Buren and in his conversation with Tilden and O'Sullivan, that Flagg did not have the national reputation he required of his cabinet members. He mentioned how fiercely the New England Democrats were opposing Bancroft for the Treasury and said, in an obvious reference to Walker, they were backing a southern man in preference to him. He took notice of Butler's willingness to accept the State or the Treasury Departments, but tactlessly reminded Van Buren that in his letter of January 4 Van Buren had doubts about Butler's accepting the State Department and had ruled him out completely for the Treasury. Had he known earlier Butler would accept, he would have offered him either post. "It is now however too late," said Polk, very definitely placing the burden on Van Buren's shoulders. Since Polk had not as yet made up his mind on the Treasury, he was being deliberately evasive.

Still, he was uneasy about appointing Marcy. In a self-abnegating strain which must have cost Polk considerable mental anguish, he said, "I am fully sensible that I came into the office not only at an earlier period of life than any of my predecessors but perhaps with less of the public confidence in advance than any of them."[47]

Polk had need of using these soothing words. Marcy had never been recommended in any of his correspondence with Van Buren and Wright. Dix had condemned Marcy out of hand, while Governor Bouck, Edwin Croswell and other obvious opponents, at least of Van Buren in New York, had showered him with letters of recommendation for the man. He also knew that Bouck and the Hunkers had opposed Wright, and yet Wright had been nominated and triumphantly elected. He was too shrewd in politics not to have observed the connection between Marcy and the Hunkers. And he was not being truthful about the Treasury position. He could have offered it to Butler since he hinted to Van Buren that he was in the process of changing his mind about Bancroft. As he knew Butler would accept, he could have easily waited the one or two days for Van Buren's reply. But he had already, as Van Buren remarked, made up his mind that New York would have the third position and Butler's declination fixed Marcy in the place.[48]

Smith Van Buren, bearing the long-awaited reply, arrived in Washington on the morning of March 1. He had made all the connections and had there not been an accident in Baltimore, would have reached Washington the previous evening. Without pausing to book a room, the travel-stained, weary Smith went directly to Coleman's but was unable to gain admittance, though he sent in his card. He returned to his hotel, arranged for a room and ordered dinner. While he was dining Tilden and O'Sullivan appeared.

Much to his surprise, Tilden asked him if he had the letter and then related the substance of their interview. O'Sullivan suggested Smith write a note to Polk, saying he had an important letter for him from his father and asking him to set a time when it would be convenient for him to receive it. It was now late afternoon. Young Van Buren appeared at Coleman's a second time, gave the servant his note and after a short interval was told he could have an interview in a half an hour.

Precisely thirty minutes later, Smith reappeared only to be kept waiting for some time. Finally the servant threw open the door, and Smith found himself in an empty room. As he walked to the center of the chamber, a door that was slightly ajar opened and Polk strode in followed by a stranger whom he introduced as Aaron Brown from Tennessee. There was an awkward moment when Smith stood silent, not wishing to carry out his confidential mission with another person present. Nor did Polk take the initiative. Brown, recognizing the awkwardness, excused himself. When they were alone, Smith, who resembled his father in features and stature more than any of Van Buren's sons, apologized for troubling Polk on Sunday. Only his father's anxiety that he receive the letter as quickly as possible made the intrusion necessary and with that he handed him the letter. Polk seemed obviously embarrassed and turned the envelope over and over in his hands without breaking the seal. "I wish I had rec'd this last evening," he said. Smith told him of the accident at Baltimore which had delayed him. Polk continued to twist the letter this way and that and then said pointlessly, "I don't know what is in this letter." In all probability he wanted young Van Buren to leave while he read what he had good reason to think was bad news—Marcy was not as acceptable to Van Buren as he had hoped. But Smith, though a shy young man, was not about to be fobbed off when he knew his father would expect a full and accurate account of Polk's reaction. He said he knew what was in the letter because he had copied it and could give him "the substance in a word."

The President-elect may have been a scheming politician under very heavy pressure and unfamiliar circumstances, but he always observed the amenities. He could not be needlessly rude to anyone, much less Van Buren's son. With apparent eagerness, he agreed to hear Smith out. "The letter urged in the strongest terms the selection of Mr. Butler for the War Office and guaranteed his acceptance," said Smith. Polk, who was still holding the letter, hit it sharply with his hand and said excitedly, "It is too late."

With Smith standing in front of him he spoke rapidly and gave every appearance of deep concern as he related a plausible account of his actions—the delay in hearing from Van Buren, the two tenders of office, both declined, no one to consult. "All your Congressmen are new to me, your Senators are new to me . . ." Polk blurted out. He had to have a full cabinet before his inauguration or he would be considered indecisive. When Polk paused in his torrent of speech, Smith quietly interjected what Tilden had told him, that Polk would wait one or two days before making his decision. Polk did not deny this but said, "After my interview with those young men I found it necessary to decide and last night decided the matter." Everything would come all right, he was certain. "You father cannot doubt my friendship for him. I have battled in his ranks too long." As he talked, he twisted and turned the unopened letter in his hand.

Giving due allowance for Smith's bias, his evident weariness, his account has the ring of truth. Polk doubtless wanted Marcy rather than Butler because he thought Marcy was safer on slavery and Texas than Butler, less under the influence of Van Buren. But he surely did not want to alienate the Barnburners either. Smith's disappointment and his utter disbelief must have shown in his face. As reported to his father, he was so shocked at Polk's "complete capture by the rogues that I got out of his way as soon as possible and could scarcely say anything to him while with him."[49] Returning to his hotel, Smith learned that Bancroft would not have the Treasury. "The same northwestern delegations and Woodbury which have interfered heretofore," he wrote his father, "have been at him." Although Polk did not actually fix on Walker until the next day, he had made up his mind when Smith visited him. "The Treasury management you perceive," said Smith, "tells the whole story for New York."

Polk was distressed at Van Buren's letter which he read after Smith had left. "If I have committed an error, I can only say it was unintentional," he wrote Van Buren. "It pains me to think that you may suppose that I have acted unkindly towards you and your friends."[50] Van Buren's second letter, his "anodyne" as he called it, reached Dix on March 3. The Senator asked Smith to deliver it. This time young Van Buren had no difficulty in receiving a prompt audience. After greeting Smith warmly, he opened the letter and holding it in his hands asked Smith to read it over his shoulder. When he had finished, Polk turned around and said that "he felt relieved," but his honor being involved he could not recall his letter to Marcy.

At this remark, Smith gave vent to his feelings; and if he was not rude to the President-elect, he was certainly outspoken. He denounced Marcy, Bouck and Croswell roundly. Marcy had scarcely any following in New York, but this appointment would immensely strengthen a minority faction and "utterly paralyze the party in our state and prostrate the Administration and its friends." Startled at the vehemence of young Van Buren, Polk disclaimed any knowledge of Marcy's being so repugnant to Wright and Van Buren. He had relied on the advice of Kemble and Bancroft, both of whom had thought a Marcy appointment would be quite acceptable. Polk was not being entirely truthful because Kemble denied he ever gave such advice and in fact had in his possession evidence he had recommended strongly that Polk have Wright's concurrence before he offered Marcy any position. However he may have felt about Smith's outburst, the President-elect sought to mollify him, asking if he would write his father and say that "this error was one of the head, that it gave him more pain that it could give you."

He promised he would rectify any mistake he may have made. Van Buren and Wright would have complete control of New York patronage including an important foreign mission. After Smith left, Polk seems to have grasped fully the extent of his blunder. He was relieved at Van Buren's apologetic second letter and decided, rather impulsively it would seem, to follow up on what he had said to Smith about patronage. Tilden and O'Sullivan were summoned. Polk told them he had gone so far with Marcy he would stick with his decision; "even if the Heavens fell, he must fulfill" his commitment. But he said emphatically and specifically that Wright would have the New York City collectorship, the major post offices, district attorneys and all other important federal offices in the state. Tilden and O'Sullivan were suitably impressed, and even the angry and skeptical Smith seems to have softened his opinion. "The soundest judges here," he wrote his father, "think P. came here all right—but has been bedevilled since he arrived. To a large extent this is, of course, evident but not wholly so."[51]

After reading Smith's letters, both Wright and Van Buren decided the President had been playing falsely with them all along. Van Buren was mortified and Wright embarrassed at how the Hunkers had outmaneuvered them. Their prestige suffered a heavy blow at home and they knew it. Van Buren spurned the *carte blanche* on patronage and was indignant when Butler accepted his former office of district attorney for the Southern District of New York and Hoffman a lesser position without consulting him. Polk was at least as good as his word on rewarding the Barnburners with the patronage.[52]

Tilden and O'Sullivan were offered federal posts, but by this time they had felt the wrath of the Governor and the ex-President. Both reluctantly declined. Flagg spurned the offer of the collectorship. Next, the President asked through Butler whom they wished to designate for offices should removals be made, implying that he was in no hurry now to replace the hated Tylerite collector of New York, C.P. Van Ness.

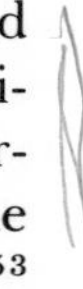

Butler sought to push Jonathan Coddington, whom Wright and Van Buren had settled upon for the collectorship. When the President reserved judgment, he sent Tilden to Washington for an interview. Careful by nature, Tilden was warned after he reached the capital that Polk had become extremely sensitive about coercion.[53] Tilden was on guard more than usual when he visited the President. But he presented the patronage problems of New York clearly, and quietly. Without hesitation he reminded the President of his blanket promise. Polk seemed enraged, denied he had made any such commitment and claimed that the Barnburners were trying to force his hand. He intended to remove Van Ness but in his own good time and would accept no further advice on the matter as to his replacement.[54]

Actually he was being advised by Marcy, who suggested Cornelius Lawrence, an elderly merchant who had for some years been president of New York's largest bank. He was not a Hunker, but Flagg, Van Buren and Wright certainly looked upon him with suspicion because of his financial connections. By now Van Buren had lost interest in the Polk Administration, which he was certain, after Blair was forced to sell out the *Globe* to Ritchie, was in the hands of "the rogues."

After his performance before Tilden, Polk lost all credibility among the Barnburners in New York and by so doing irretrievably weakened the Democratic party in the Empire State.[55] In his unrealistic absorption with his own personal prestige and image, Polk seemed determined to favor the weaker faction, not just in New York but also in New England. Nor did Calhoun's friends think much of Polk's appointment policy, though clearly it favored southern interests over northern.

Polk's last move was to offer Van Buren the mission to England after he had first offered it to Calhoun, who refused, then Pickens and Elmore with similar results. Van Buren, whose name George Bancroft had suggested, declined in a polite letter. But to make an offer fourth hand to a former President and still the most powerful political leader in the North, was so condescending as to be insulting.[56]

The Hunkers now held most of the major positions in New York and represented the Empire State in the cabinet. But Van Buren did not despair. His faction commanded by far the allegiance of the best and the brightest in the party. Since his repudiation in Baltimore and his further repudiation by the Administration, he was content to keep an eye on state politics and looked more and more to the younger men, his sons John and Smith, Tilden, Preston King and Dix. At age sixty-three, it was time to spend his remaining years pleasantly entertaining old friends, managing his estate, and perhaps enjoying some foreign travel and a book by the fireside of a cold winter evening.[57]

« CHAPTER 31 »

Free Soil, Free Labor, Free Men

In his suite of rooms at Julian's Hotel overlooking New York's Washington Square, Van Buren put the finishing touches on a manuscript he had been writing off and on for several months. It was an April morning in the year 1848; the manuscript was a draft that set forth in great detail the position of the Barnburners on the great political question of the day, the Wilmot Proviso. Composition involving much research and thought had occupied his time during the day and came as a welcome change to his routine. For this was the first time he was staying the entire winter in the City where he could indulge himself with Julian's famed cuisine, attend the theater which he always enjoyed, and talk politics with friends and associates.

The New York weather, though cold enough, was less severe than at Kinderhook and transportation south was more convenient if he chose to visit Gilpin in Philadelphia, or Blair at his country place, Silver Spring, but a few miles from the capital. His immediate family had all left Albany for the metropolis. Abraham was back in the army, an aide to General Taylor in Mexico. Angelica, who now had two sons, maintained a town house in the City. Smith and his small family were spending the season in a hotel nearby. And John, after his term as state attorney general, had found it more convenient to conduct his practice from the City than Albany. His office and residence on White Street was also near Julian's. Martin, Jr., his health slowly declining from incipient tuberculosis, lived with his father, though this winter, during the coldest months, he went to Washington where he reported regularly on the doings there.

Van Buren felt the need of his children and grandchildren more than ever. Unusually healthy for his sixty-five years, still maintaining his cheerful, optimistic outlook, he missed those friends of the past—Jackson, dead now for almost three years; his jocular companion over many a bottle of choice Madeira, Commodore Nicholson; but most of all he mourned Silas Wright, whose passing some eight months earlier had not only shocked and saddened him but left him with feelings of guilt. He had, after all, pressed his dear friend to run for governor, and along with others of his close political circle, believed the pressures of the office, the schism in the party, and the overwhelming defeat for reelection in 1846, had shortened his life. What had been gained for this sacrifice of a modest statesman, honest conscientous and able, above all loyal to his party and its principles?[1]

As far as Van Buren could see, the Polk Administration, if it had not directly undermined Wright, had at least lent credence to Wright's enemies, the enemies of true Democrats, all of whom remained secure in their federal offices long after Wright's downfall. These Hunkers had not even a decent respect for the memory of this great and good man in their journals and their talk. Rather, they rejoiced, or so it seemed to Van Buren, that another obstacle to Cass or Polk had been removed; another barrier to southern expansion lifted. Had not Wright supported that mischievous proviso of David Wilmot prohibiting slavery in territories the nation might gain from Mexico?[2]

Van Buren had read the papers, studied the speeches and debates in Congress with special attention to the expansion of slavery years before Wilmot broached it. One of his major objections to immediate annexation had been the extravagant boundary claims of Texas and the South's insistence on more slave territory in the Southwest. War with Mexico that he warned might occur in his Hammet letter was now a reality. Van Buren's political and public career was over; but apart from his personal hurt at the injustice done his associates and his state, he would be faithless to his trust as a concerned citizen of Jeffersonian principles if he did not speak out. When Wilmot moved his proviso in August of 1846, Van Buren began accumulating material on the subject and shaping his ideas. He had Martin, Jr., copy pertinent extracts on the problem of slavery from Madison's *Notes,* Elliot's *Debates,* the *Federalist Papers,* Wheaton's and Peters' *Reports* of Supreme Court decisions, newspaper articles and his own notes from the *Annals of the Congress, Register of Debates* and the *Congressional Globe.*

As the debate over slavery in the territories grew more intense, Van Buren carefully watched the increasingly strident demands from southern states that slavery must be extended into territories acquired from Mexico by right of conquest.[3] Now in the spring 1848, as the Whigs and the Democrats began preparations for the presidential contest, he noted with increasing concern that state conventions of his own party in Alabama, Virginia and Florida passed resolutions denying the power of Congress to make regulations prohibiting slavery in any territories either directly or indirectly through a territorial legislature. It was time he thought to organize all his material, adjust his thinking on the matter and inform the public as clearly and as forcibly as he could the position of the Barnburners on this matter of grave national import. Gradually the pile of manuscript rose on his table at Julian's until finally by April 1848, he had written over 15,000 words.

He had just completed the draft when Samuel Tilden appeared for a bit of political conversation. As the young man was preparing to leave, Van Buren, with a smile, handed the bundle of pages to him. "If you wish to be immortal," he said, "take this home with you, complete it, revise it, put it into proper shape and give it to the public." Tilden pleaded lack of time, a heavy burden of his own work, his health, a phobia with him, but he could not prevail over Van Buren's cheerful insistence. Tilden struck a compromise, however. He would go over half the manuscript if John Van Buren would help with the remainder and on that agreement they parted.[4]

Tilden may not at the time have wished to become "immortal," but after he read the manuscript he realized it was more than simply a political tract; it was that, of course, but it was so much more. Van Buren had written a constitutional history of the territories from their cession by states to the confederation government. He established beyond any possible doubt beginning with the Ordinance of 1787 and carrying his argument through the admission of a dozen states that Congress had under Article IV, Section 3, of the Constitution, exercised exclusive power in prescribing whatever conditions it thought proper for the governance of the territories. Citing Marshall in *McCulloch v. Maryland* and again in the *American Insurance Company v. Cantor,* which dealt with the territorial status of Florida, Van Buren showed how that great chief justice had defined "the rules and regulations" of Article IV.

The Constitution gave Congress powers over territories including the establishment of freedom or slavery. Even more emphatic than Marshall, he quoted Smith Thompson's majority opinion in the *United States v. Gratiot* that the Constitution vested this power in Congress without limitation. He brought Joseph Story forward with the most direct statement on the issue taken from his much-admired three-volume treatise on the Constitution. "The power of Congress over the public territory," Story had written, "is clearly exclusive and universal and their legislation is subject to no control but is absolute and unlimited."

Van Buren did not rest his case here, however. He extended his brief to a general indictment of slavery as the founders saw it, giving statements from Washington, Jefferson, Madison, Patrick Henry and George Mason, all of whom had condemned the institution and looked forward to its eventual abolition. Amplifying further the constitutional argument, he summarized and dissected the proslavery positions advanced by the legislatures of Virginia, Florida and Alabama, concluding with a ringing phrase that their "doctrine is therefore plainly stated . . . wherever the flag of the Union goes, it carries slavery with it; it overturns the local institutions, no matter how strongly entrenched in the legislation, the habits and affections of the people, if freedom be their fortunate condition and establishes in its place slavery."[5]

Van Buren had advanced well beyond the constitutional, legal and political limitations the South was attempting to impose on the free states when he reminded his slaveholder audience and more particularly the northern masses that slave labor and free labor could not coexist. He asserted it as a matter of history and of current experience:

> Free workers are unwilling to work side by side with negro slaves; they are unwilling to share the evils of a condition so degraded and the deprivation of the society of their own class; and they emigrate with great reluctance and in very small numbers to communities in which labor is mainly performed by slaves. With the exception of a few, and a comparatively very few, the white laborers, or in other words, the poor of those states where slavery is more extensively prevalent, are objects of commiseration and charity to the wealthy planter, and of contempt and scorn to the slaves.[6]

This manifesto, for it was nothing less, calmly set forth the entire Free-Soil issue not just in terms that would be understood in the free states, but in language and in argument that educated southerners could not casually thrust aside. Like Wright before him, he scouted the popular sovereignty idea that some politicians were talking about while affirming boldly the power of Congress over any territories to be acquired and refuting the argument that would be advanced ten years hence by his old friend, Chief Justice Taney, in the Dred Scott case. He recognized, however, that the proviso issue could split the party.

"On the part of the candidates for the Presidency, the Democracy of New York," he wrote, "had never made this a test question and felt called upon to apprise their southern brethren who persisted in doing so [what] would be the inevitable effect of such action."[7] Yet with the issue presented comprehensively and cogently, Van Buren did not hesitate to include some partisan remarks. There was an indirect indictment of the Polk Administration and its policy toward the party in New York. He made it clear that the Barnburners, after they first convened at Herkimer and later at Utica where they chose thirty-six delegates to the Baltimore convention, were the established party, the Hunkers, a seceding faction. Still, he even found words of praise for the Administration's Tariff Act and its passage of the independent treasury system that the Tyler Administration had repealed. Wherever it was appropriate, he took Clay to task for what he claimed was a deliberate obstruction of the war effort and his continuing advocacy of a national bank, high tariff, inflation and costly internal improvement programs at national expense.[8]

Van Buren's manuscript so impressed Tilden that he rushed over to White Street where John Van Buren read it. Far busier than Tilden, the younger Van Buren did not hesitate to push urgent work aside and take on part of the editorial chore. He proposed they divide up the manuscript, sharpen the points and in general polish the draft so that any well-informed reader would have no difficulty following the argument. In a few days they had completed the job to Van Buren's satisfaction.

How was it to be released—a series of letters or articles to the editor of a friendly newspaper like the *Evening Post* or the *Atlas,* a pamphlet to be broadcast over the land, an address in Congress that Preston King or Dix might deliver and publish in *The Appendix* of the *Congressional Globe*? They considered all these options and finally decided the maximum impact would be made if it were issued in the form of an "address" from the Democratic members of the legislature. On April 11, 1848, this, the first comprehensive statement of the Free-Soil position, its background in history, theory and practice, was read in the legislature and on the following day was published as an extra edition of the *Atlas* along with four resolutions that rang the tocsin of the Barnburners.[9]

The "address" was the culmination of two years of bitter intraparty conflict in which Van Buren and his circle continued their feud with the Polk Administration and the conservative Hunkers, political warfare which did little credit to either side. Had Butler accepted the War Department, and Flagg the collectorship of New York City in the first place, much of the factionalism would have disappeared and Wright's lot as governor would have been less burdensome. Had there been less pride and sensitivity on the one side, a more liberal and knowledgeable outlook on the other, the party in New York would have weathered its internal debate without inflicting upon itself wounds that would never heal. But Van Buren and Flagg himself indulging their resentment, rejected the President's gesture of conciliation.

From then on, Polk, through accident or design, not only reneged on the promise he had made to Smith Van Buren, and later to Tilden and O'Sullivan, but he let Marcy and Robert John Walker influence him on the New York patronage. Even the appointment of Wright's friend Ransome Gillet was downgraded through Walker's intervention from the promised post of solicitor of the treasury to the humdrum lesser position of register in the Department.[10]

The Hunkers got away with a lion's share of the New York patronage, and under Walker's direction used it effectively against Van Buren and Wright. Butler's office of district attorney for the Southern District and Hoffman's as surveyor of the port of New York had little or no leverage. Both were far outweighed politically when Polk selected Hunkers for the remaining attorney posts. Lawrence, whom Polk appointed collector over Wright's and Van Buren's candidate, was too close to Marcy and Bouck. If Lawrence made no overt use of the vast patronage at his command against Wright, he tolerated Walker's manipulation of it.

The Secretary of the Treasury's objective was to foreclose what he thought would be Wright's campaign for the Presidency. He seems to have been either unaware that the Governor had no interest in the succession, or believed Van Buren would overcome his denials and thrust him forward. Walker's candidates were his relative, Vice-President Dallas, or Lewis Cass, both of whom were "safe" on the slavery issue. The Walker campaign carried out through the collectors and the postmasters was to create an opinion in party ranks that the policies of Wright's administration were in direct opposition to those of Polk's. As Wright complained to the President, "It is unfortunately true that there has been an apparent conflict of object and purpose between your administration and mine from the commencement, so far as men in this state have been concerned. . . . every man holding an office under you has seemed to consider it his duty or his interest to oppose me."[11]

Although Wright had sought to close up the breach between the Hunkers and the Barnburners at the beginning of his Administration, he opened it wider. Late in the legislative session of 1845, a combination of Hunkers and Whigs passed a bill providing a resumption of expenditures for canal improvements and construction in direct violation to the Flagg-Hoffman "stop and tax" law of 1842. Wright promptly vetoed the bill and the legislature sustained him.

In his veto message, the Governor alluded to a constitutional convention that he hoped would place that law beyond the constant attacks of land speculators, canal contractors and other interested parties. This frank disclosure of policy completely alienated the Hunkers. Opposed to Wright on local matters, the Hunkers supported Walker's program to eliminate him as a presidential contender. Under political attack from the anti-Renters whose vigilante action he had put down in the interest of law and order and from the Hunkers and the federal officeholders, Wright was not hopeful about his reelection in 1846. He even had to endure the humiliation of seeing the Hunker chief Bouck be appointed, at Walker's instigation, Receiver of the public funds in New York City for the subtreasury. Nor had Wright proved a strong executive in forcing discipline on his own party, in making and carrying through a distinct policy. He had, as a contemporary observed, "a better faculty for making friends than for controlling them."[12] Yet even a Van Buren at the peak of his powers would have been sorely tried by the combination of politics and circumstance that beset the Governor during his two-year term of office.

In the summer of 1846, Secretary of State Buchanan, ever solicitous of his own political welfare, decided to see for himself how matters stood in New York. Wright was so explicit about his problems with the federal officeholders that Buchanan wrote a long letter to Polk saying he had been mistaken about the Barnburners. The conniving Secretary traced the troubles in New York to his enemy in the cabinet, Robert John Walker.[13] Buchanan's opinion made an impression on Polk. Already suspicious of the Hunkers, many of whom he knew had opposed his tariff and banking policies, Polk wanted more information about their activities at home.[14] Bancroft had just resigned from the Navy Department to accept the mission to England. On his way to Boston, would he stop over in Albany and relay his impressions to the White House? Wright had just been renominated for governor over bitter Hunker opposition, and Polk asked Bancroft to offer the assistance of the Administration in his reelection.

Bancroft's reception in Albany was scarcely enthusiastic. The triumphant Radicals inveighed against the interference of federal officeholders in the state convention, particularly the postmaster of New York City, Robert Morris, and told the President's emissary that they had fought their own battles alone against all enemies. *They* would judge Polk's offer "in its acts not its professions." Wright was far more cordial and sympathetic than his outspoken friends.[15]

The Governor described the opposition of the Hunker officeholders and was realistic in his assessment of his probable defeat in November. Bancroft went off to Lindenwald where he was made welcome despite Van Buren's growing skepticism about his political leanings. There he heard much the same story Wright had related. Convinced, Bancroft lost no time in acquainting Polk that a conspiracy against Wright existed. The President had, in the meanwhile, talked at length with Gillet and had begun to realize that his patronage policies in New York, far from creating a strong party free of cliques, was in the process of destroying what had been the most efficient party machine in the nation's most powerful, most populous state.[16]

On the advice of Buchanan, Wright shook off his lethargy and wrote lengthy letters to Cave Johnson and the President. With Johnson, he was completely candid, named the officeholders who were actively working against him and directly accused Walker of being the mastermind. Johnson was embarrassed when Wright told him all of his subordinates in the state, two of whom were Tylerite holdovers, were working against his reelection. Six days later, after talking with Bancroft, Wright wrote Polk a more general letter which nevertheless made it plain that federal officeholders directed from Washington were opposing the regular Democratic organization. For details he suggested Polk read his letter to Johnson.

The President took Wright's advice, and it would appear was genuinely shocked at the revelations it contained. He must have taken Johnson to task, for that worthy responded with an indignant letter to Wright hotly defending the Administration and accusing some of the Barnburners in New York's congressional delegation of pursuing "a cowardly as well as dishonest course . . . by acting in one way when no record of their acts could be preserved for the public eye and in another way when called to vote the ayes and noes."[17]

Polk also wrote a long, conciliatory letter to Wright assuming the blame and assuring him that "if he had made mistakes his intentions had been well meant." He would if possible seek out those persons who had acted against Wright "without my knowledge" and presumably discipline them.[18] And he did summon Walker, Johnson and Marcy to a meeting where, armed with Wright's charges, he demanded they have their minions in New York support Wright. The principal result of the President's stern injunction was a letter from Walker declaring his support of Wright and, what must have seemed the supreme irony to the beleaguered Governor, giving him permission to publish it.[19] This flurry of activity occurred about two weeks before the election and could not possibly have altered in any meaningful way the damage the officeholders had already done to Wright's image or even changed their posture without complete loss of credibility, which, of course, they were unwilling to do.

Van Buren had kept in touch with Wright and with politics in Albany and in Washington, but after Bancroft's visit in October took no active part beyond casting his vote in Kinderhook for the Democratic ticket. Troubled as he was at the condition of the party after Wright's defeat, he would give advice when asked, leaving active participation to the younger men. The death of Silas Wright on August 27, 1847, after only eight months of well-earned retirement came as a shock. A grief-stricken Van Buren wrote Blair that Wright's "whole heart was as intelligible as a book, he was a never ceasing source of admiration. . . . I would be less than human if I did not feel deeply the loss of so real a friend."[20] For the Barnburners the loss seemed irreparable. Whatever Wright's opinions had been, he was their choice for the presidential nomination and now they were left leaderless. His death occurred but a few weeks before the party convention in Syracuse, and among the sorrowful Barnburners a feeling of rage welled up against their adversaries.

Divisions over issues had been real enough but the Barnburners to a man felt the Hunkers had murdered Wright as surely as if they had used a dagger in the dark of night. Van Buren himself voiced their mood. "How difficult it would be to make the selfish and designing politicians of the day believe," he said, "that whilst they were using their greatest efforts to obstruct his advancement, his mind was never for a moment occupied with his own promotion." The Barnburners were in a vengeful mood when they met with the Hunkers in convention at Syracuse on September 7, 1847. This time the Hunkers had managed to gain a majority of the delegates.

From the very beginning when the chairman called the convention to order, the Barnburners were on the defensive. All they could do when the Hunker majority defeated Flagg's renomination for comptroller and John Van Buren's for attorney general was voice their angry disapproval. Order was finally restored only to descend to chaos again after the Hunkers tabled a resolution supporting the Wilmot Proviso. The next morning David Dudley Field, the brilliant New York City lawyer and a friend of John Van Buren, called the Proviso resolution up as an amendment to the platform. The Hunker chairman ruled him out of order, an action that immediately plunged the unruly gathering into another tumult.

A Barnburner delegate with a voice loud enough to be heard above the clamor, deplored the action, shouting that the convention was not doing justice to Silas Wright. A momentary hush came over the hall and a Hunker delegate took the opportunity of responding, his voice heavy with sarcasm, "It is too late, he is dead." This contemptuous remark touched off a vocal explosion that smashed the convention and the Democracy of New York. James Wadsworth, intimate friend of the Van Burens and a Barnburner to the core, leaped upon a table and roared out, "Though it may be too late to do justice to Silas Wright, it is not too late to do justice to his assassins." When the chairman managed to make himself heard after this retort, he adjourned the meeting. Before his ruling could take effect, the Barnburners in a body followed Wadsworth and John Van Buren out of the hall.[21]

During the remainder of the month, John and Smith Van Buren, Wadsworth and other Barnburner activists planned and organized a mass meeting to be held in Herkimer on October 26, 1847. Neither Van Buren nor Flagg sanctioned the movement or the meeting. Both men supported the Proviso in principle and would speak out for it when the issue became a matter of public policy, but as Van Buren explained, "to adopt it now as a party cry is another and very different thing." It would divide the party and place the Whigs in office. However much he deplored Polk's policies and distrusted the man, however much he detested the Hunkers for their treatment of Wright and their responsibility for the Whig victory in 1846, Van Buren was not prepared to accept a permanent Whig ascendency that he was certain would occur if the Barnburners clung to the Proviso. But neither he nor Flagg was able to restrain the activists. Departing from his usual course with his sons, Van Buren sought unsuccessfully to have the Herkimer meeting take a firm but moderate stand on the Proviso and temper John's "more violent utterances." Fearful that the Barnburners would be carried away and use his name, Van Buren was determined to keep clear of the meeting. He reminded Flagg that he had retired "from the bustle of politics and most resolutely determined to remain so." Since he could not control John, he relied on Flagg to protect him.[22] He had managed to persuade his restless son not to attend the meeting; but he could not prevent him from drafting the address, the resolutions and the call which appeared in the *Atlas,* the *Evening Post* and other radical sheets.

Yet if John Van Buren was not among the 4,000 citizens who converged on Herkimer to be addressed in the open fields outside the village, his fighting spirit was all pervasive. Under the direction of Wadsworth and Cambreleng, the meeting unanimously adopted the Wilmot Proviso resolution tabled at Syracuse and vigorously denounced the Hunker management of that convention.[23] Having advanced to the brink, the activists held back from the last irrevocable step, the nomination of a ticket. Van Buren's and Flagg's "scoldings" had made some impression.

The Herkimer meeting sent a thrill of anxiety through the leadership of the party. The part John Van Buren and Cambreleng played in the proceedings immediately connected Van Buren's name with the Proviso. Letters flowed in to Lindenwald from Washington asking if it were true that he was antislavery and whether he intended to be a candidate for the Presidency in 1848.[24] Van Buren took this opportunity to write a public letter disavowing any intention of running again for any office. On the Proviso issue he gave it as his personal, not political, opinion that he favored the exclusion of slavery from territory now free as a matter of principle, but thought the question was up to Congress when new territories were organized.[25]

With most of the Barnburners staying away from the polls, the radicals had the grim satisfaction of seeing the Whigs sweep the state electing twenty-three of New York's thirty-six representatives. The New York avalanche, along with Democratic reverses in other northern states, meant the Polk Administration would have a minority in the new Congress.

Undismayed, the Hunkers went ahead in late January 1848 with their convention at Albany where they appointed delegates to the forthcoming national convention in Baltimore. They also issued an address that condemned the Barnburners unsparingly as deserters who were using the Proviso issue as a pretext to revenge themselves on the regular organization, split the party and turn state and nation over to the Whigs. After reading their address, Van Buren knew that any hope he may have had for a reconciliation of the two factions was now beyond reach. And he did not seek to reason with his impetuous sons, especially John, who for some time had been the organizing spirit of the Barnburners.[26]

Before the Albany convention, John had drafted a call for an opposing convention at Utica and had the Democratic legislative caucus under Barnburner control issue it. Sweeping aside any criticism that the convention was not representative of the party, the Barnburners met as scheduled and adopted a platform and a slogan, named delegates to the Baltimore convention and set a date for a state convention to nominate state officers. They also took one more step, inviting like-minded men from other states to join with them, purge the Democracy of its evil elements and fight for "Free trade, free labor, free soil, free speech and free men." Describing the Utica delegates to Blair, John said, "More than thirty of the thirty-six are in every respect an honor to the state—probably my father is the first choice of all of them and Cambreleng, Kemble, Tilden, King, Beekman, Wadsworth, two sons of Governor Tompkins, etc., are of course his most intimate and confidential friends."[27]

All these events in New York created consternation among party leaders in Washington. Just before the Utica convention, Gouverneur Kemble, who was visiting the capital, attended Polk's regular weekly reception. Immediately after he shook hands with the President and moved into the throng, he was accosted by several members of the cabinet. The new attorney general, Nathan Clifford of Maine, and Cave Johnson were his most persistent questioners. Both men regretted the divisions of the party in New York and asked Kemble which faction he favored. Kemble stoutly denied any division and, while acknowledging he was a Barnburner, said the Hunkers were a small faction kept alive through the efforts of federal officeholders. Clifford and Johnson both affirmed that New York had to be carried in November. "Would it support Cass?" they asked.

Kemble replied "No!" "Buchanan?" "No!" "Any chance for Dallas?" "Doubtful!" He amplified this remark by saying that no member of the Polk Administration could carry the state because of its patronage policy.

Johnson, perhaps still smarting from the argument he had carried on with Wright just before the Democratic debacle of 1846 in New York, asked if the Barnburners "did not consider Lawrence, the Collector, a Democrat?" The blunt response was "No!" What about Hoffman and Butler? Kemble answered in the affirmative, but mincing no words added that Butler had been offered the War Department when everyone knew he would not accept. Said Kemble, "I had myself so stated to the President who had done more to distract and break up the party in New York than anybody else." A chopfallen Johnson grudgingly admitted there was some truth to that statement.

While they were talking, Polk came over and joined the conversation. He prevailed upon Kemble to remain after the reception was over. For the better part of an hour the President quizzed Kemble about the party in New York. Like Clifford and Johnson, he too wanted to know whether the Empire State would support Cass or Buchanan and received the same answer. "How could the divisions be healed?" he asked. Kemble could not advise him on that. Nor could he suggest who among the Democrats would gain the united support of the party. He thought the Whigs would carry the state. Woodcuts and lithographs of Taylor, he told the President, were prominently displayed in the bars of taverns which mechanics frequented.

Well before these developments, Preston King, an uncompromising advocate of the Proviso, had been in touch with other like-minded Democrats and Whigs who were willing to jump party traces on that issue alone. One of King's correspondents, the ambitious Salmon P. Chase of Ohio, though nominally a Democrat and a zealous antislavery man, was trying to coax William H. Seward away from the Whig establishment in New York.[28] The Utica platform had taken notice of these activities and welcomed them but was not yet ready to form any distinct organization. The Barnburners still maintained they represented the regular Democratic organizations and they declared explicitly that they would not make the Proviso a test for the presidential contest.

Van Buren was now taking a more active interest in the political scene. Against his better judgment he was permitting himself to be drawn more closely into the Barnburner cause. The pronouncements of several southern state conventions that summarily dismissed any power of government to prevent a citizen from taking his slaves anywhere within any of the states or territories had prompted him to reverse himself. These assertions must be answered, and he began preparation of a response.

Some weeks later he drafted a series of what he called "instructions" for the Barnburner delegates to the Baltimore convention. John Van Buren had induced him to do so. In their discussions since the Utica caucus, they had agreed that efforts would be made at Baltimore to challenge the credentials of the Barnburner delegates or otherwise frustrate their participation. Van Buren had given considerable thought to this possibility and what the delegates ought to do under various contingencies. A letter from John suggesting that if the Barnburners were admitted they would place him in nomination had spurred Van Buren to action. John had written that if his father were defeated as seemed almost a certainty, "assume the proviso is made the test of exclusion against a man who has said nothing on the subject. The delegates would then bolt the convention. . . ."[29]

His son's proposed tactic horrified Van Buren. In a long letter of advice he insisted that under no provocation or enticement should the delegates present his name to the convention. "Now it required a man of your talents and prolific imagination," he replied, "to have considered a more objectionable movement—of having the friends you were bringing forward upon their own responsibility but backed by a declaration from me that if nominated I would not decline. Nothing could be more cruel to me or more destructive to the stand you and your friends in the state have acquired."[30]

John Van Buren's letter convinced his father that he had better draw up some instructions for his guidance or for the guidance of the delegation. Assuming the Barnburners were admitted, Van Buren advised that they must emphasize they had no candidate. Nor should they raise the Proviso question even indirectly. The South had taken advantage of the extreme statements of Wilmot, Preston King and other radicals on the slavery issue to talk disunion and frighten the masses of Northern Democrats into line. Fortunately, the Barnburners in their speeches had followed Silas Wright's temperate approach and had, he thought, opened the eyes of party leaders in the free states to the southern strategy. But any rash act at Baltimore would destroy what had been accomplished and discredit the Barnburners as mere malcontents, schismatics bent on destroying the party for their own selfish and narrow aims.

Van Buren thought they should agree to all resolutions of the 1844 platform except the two-thirds rule. "If there shall be an apparent design to give the seventh resolution [approving the immediate annexation of Texas] a construction which will embrace slavery in the territories," Van Buren said, "they the [Barnburner delegation] will have no option but are bound to go against it. . . . If not, it may be best to let it pass in silence." On the burning question of which candidate to support, he ruled out Cass, Buchanan, Dallas and particularly Woodbury, who was now an associate justice of the Supreme Court. Van Buren would not, as he put, "have the Supreme Court contaminated by presidential intrigue."

He need not have instructed the delegates that they vote against Polk if he were placed in nomination, but he did so anyway. Should the President be renominated, one of the delegates ought to describe his treacherous role in undermining Wright and humiliating the real Democratic organization in New York. He must also declare that New York had refrained from informing the public of Polk's conduct because it would have been unpatriotic to criticize the President in war time. After this explanation the Barnburners must leave the convention. But if Polk were not renominated and if the remaining convention delegates, duly warned that the leading candidates could not carry New York, still persisted in nominating any one of them, Van Buren advised the Barnburner delegation to remain. However distasteful, the best course was submission to the majority will, disclaiming any responsibility for defeat in November.

While "the instructions" were supposed to be secret, they were inevitably leaked out. Barnburners all over the state heartily approved the spirit and the tactics. Their press, exaggerating the intent of the "instructions" probably to exert maximum pressure on regular party leaders elsewhere, hinted at the formation of a third party if the radical delegation were excluded or in any way hindered from full and free expression of their views at Baltimore.[31]

It quickly became apparent, even to the most obtuse observer of political events, that there would be a floor fight over the contesting New York delegations. Both factions remained adamant about any compromise. "Unless recognized as the true and the only representatives of the New York party," said Butler, speaking for the Barnburners, "our delegates will regard any proposal by the convention for admission of both sets of delegates or for the the admission of one half of each set . . . as equivalent to rejection—indeed more distasteful than the out and out admission of the Albany delegates as exclusive representatives of New York."[32]

Butler, of course, was voicing the prevailing sentiments of the Barnburners. As for Van Buren, though he had been their leading spirit and the cohesive force that bound them together, he still hoped that the various contingencies he had outlined would not occur at Baltimore. Only few of his most discreet friends and members of his immediate family knew just how important his work had been, how useful his advice behind the scenes. Yet he had freely given of his time, his wealth of experience in political matters, his great skill in maneuver, not because of any desire for office or personal preferment. Far from it! He was determined to remain in retirement, his public life over.[33] Nor had he acted for the political advancement of his family.

Rather, Van Buren had counseled the Barnburners and articulated their aims calmly, clearly and eloquently because he felt it was his duty as a disinterested citizen to defend principles he had prized during his long political career. He believed that if they were cast aside for place or for section, the Union was in peril. With his sense of history, his constant and careful analysis of social and economic trends, he knew that the Proviso was right. He knew also that the citizens of the free states did not understand clearly as yet the divisive character of slavery, not just to the major parties but to the nation itself. In the Barnburners he saw the instrument which would alert the public mind that his address, an appraisal and a justification of the spirit of the Proviso, was no radical doctrine sprung upon an unsuspecting people, but one older than the Constitution itself and supported by every Administration since Washington, the present one a sole exception. He had done his best, at first reluctantly, but when he saw the danger, he had moved rapidly and decisively.[34] After all the thought and labor he had given to the Barnburner position, Van Buren learned that it had gone for naught. The only contingency that came up at Baltimore was the fatuous ruling by the convention leadership to admit both contesting delegations from New York and award each delegate half a vote. New York remained silent throughout the voting for the nominee. After Cass went over the two-thirds mark on the fourth ballot and won the nomination, the chairman of the Barnburner delegation arose and after a short, sharp protest, left the hall followed by his delegation.

Van Buren had been prepared for this turn of events and knew it virtually guaranteed a Whig victory. The platform seemed innocuous enough, but the candidate was another matter. Although Van Buren had advised that the Barnburner delegation bow to the majority will on any candidate except Polk, he clearly had strong reservations about Cass. The nominee's letter to one of Polk's fellow townsmen and rivals in Tennessee politics, A.O.P. Nicholson, in September 1847 had openly opposed the power of Congress over the territories. His interpretation of the Constitution on this point was completely the reverse of Van Buren's. As Van Buren saw it, Cass had "stultified himself by giving the most exceptionable opinion on the subject of slavery in the territories" of any Northern candidate.[35]

But he was troubled at the determination of the bolting delegates to nominate an independent ticket at Utica a month hence. He agreed with Butler, who had written on May 29 doubting the expediency of any independent organization. "I am aware too," Butler said, "that no third party can long avoid either a merger in the Whig or Federal party . . . or else dissolution of its members into an isolated and disenfranchised individualism." Of the two alternatives he favored the latter because "we can start right in this forlorn hope, form a nucleus for the organization of a northern Democratic party, which within the next four years will control the votes of the northern states and bring the despots and ingrates and the south and their obsequious satellites of the north to their senses."

The day after he received this highly charged letter from Butler, he heard from Benton. Far from the sentiments he would have expected from this expansively vocal, independent Senator who had detested Cass for years and made no bones about it, the message it conveyed was one of acquiescence to the will of the party. Blair and Dix, he maintained, both agreed that it would be inexpedient to develop a separate organization in New York on the presidential question. "In the present state of things to say nothing of other reasons," Benton remarked, "we think it would have the effect of changing the aspect and position of the two parties before the nation. You would be considered the minority, the schismatics." Despite his own feelings about a third party and those of associates Benton, Blair, Flagg and Butler, Van Buren was unable to dampen the enthusiasm of the younger Barnburners for an out-and-out repudiation of the national ticket.[36]

Barnburner and Hunker alike awaited the outcome of the Whig convention in Philadelphia before either faction determined its campaign strategy. The Whigs too had deeper and more pervasive divisions, over the Proviso, than the Democrats. Their leading candidate, Zachary Taylor, a colorful war hero, was a citizen of Louisiana and a large slaveholder. Proviso Whigs, more numerous than their Democratic counterparts in the North, vowed they could not support Taylor. But there was one abiding difference which made for unity in the party. The Whigs as a national organization had been out of power for almost eight years. Most of them were so hungry for patronage and power they were able to rationalize their ideals. Just before the Philadelphia convention, John Van Buren appraised the opposition party with a realistic eye, "the conservatives [Whigs] are shockingly afraid of Taylor . . ." he said. "They have an empty stomach worse than nature abhors a vacuum and, when they see they must be honest or starve, painful as the alternative is, I think they will choose the former."[37]

Under the spurs of Preston King and Tilden, the Barnburners

had already taken steps to form a third party. Their elders managed to keep them committed to developing a state organization while remaining silent on the national ticket. Yet they continued to throw out enough hints to attract dissident Whigs and rebellious Democrats in other states.

Henry Stanton, an ex-Democrat, now a supporter of the abolitionist Liberty party, was quite certain they were approachable. In an astute observation to Chase, he added one condition, "if they felt assured they could head a general movement." At the same time he cautioned Chase who was forming a Free-Soil organization in Ohio that "they are so situated, they must lead off as 'Democrats' in the 'regular' way or perish."[38]

Any fears Chase and other radical Free-Soilers may have had that the Barnburners would place political revenge above their anti-slavery posture and confine their campaign exclusively to New York were allayed after the nomination of Taylor and Millard Fillmore in Philadelphia. John Van Buren assured Chase the Utica convention would name a "free territory" candidate. He also hinted that the Barnburners were ready to jump party traces and form a new organization. In Washington a disgruntled Marcy vented his spleen to his old friend Prosper Wetmore. "The Barnburners with little Ben [Butler], Randy John [Van Buren] a stolid CC [Cambreleng] at their head seem determined on mischief. They are a pretty set of pot knaves.[39]

There was no doubt that Free Soil was gaining momentum and shaking the leadership of both major parties throughout New England, New York, Pennsylvania and the northwestern states. Seward, who had not been happy about the Taylor nomination, worried about the Whig party in the North. "If the temper around me is at all like that of New England and Ohio and Indiana, what is to save us in those regions?" he asked Weed.[40]

By June 22, when the Barnburners met again at Utica, it was taken for granted they would launch a third party. John Van Buren had been at Lindenwald, and he had words of advice for the delegates. The convention should make a nomination for the Presidency but not for Vice-President. Ostensibly, the reason he gave was that Cass's running mate, William O. Butler's views on public policy were as yet unknown. The real reason was political, an attempt to attract regular Democrats who opposed Cass but would not back a third party. Similarly, the convention should not attack Fillmore, the Whig candidate for Vice-President. Avoid, if at all possible, a call for a national convention but instruct the Baltimore delegates to attend any Free-Soil meeting that might be called during the summer.

In offering these guidelines, Van Buren was speaking through his son and seeking to confine the movement to the radical Democrats of New York until the strength of the Free-Soil movement was more clearly developed.[41] He had an additional reason of a more personal nature to be cautious. For several months now he had been importuned from all quarters to offer himself as the Barnburners' candidate. He had tried to head off any premature movement while educating the Democracy of the North to the dangers that lay ahead if slavery were not contained. It would take time, he thought, before the party understood the issue was a real one, not simply an abstraction. In this effort he had succeeded and he had failed.

The Free Soil address he wrote and the patient and temperate explanations of the Barnburners in their public utterances, had had considerable effect in educating the people of New York and adjoining states as to what the Proviso meant to their future. But he had been unable to stem the rising excitement, particularly among the younger Barnburners for more decisive measures, more visible leadership. After the Baltimore convention, the demand for action became insistent; and as the political storm bore down upon him, Van Buren sought desperately to find an appropriate shelter.

Once he realized he could not slow the momentum of the younger men, that they were determined to nominate a candidate for the Presidency, he concluded that besides himself, there was only one other New Yorker of sufficient stature to command their respect, John A. Dix. The dignified Senator, however, opposed any nomination or any third party for that matter. All of this Van Buren knew, but he counted on Dix's loyalty to gain his acceptance. Accordingly, he wrote Dix one of his characteristic letters. All to no avail Dix; adamantly refused and was roundly criticized at Utica for his stand.[42]

Without waiting for a reply from Dix, Van Buren took the precaution of naming other possible nominees in what he described as "my last political letter" to Butler. After Dix, he proposed Addison Gardiner, a former lieutenant governor, Cambreleng, Colonel Samuel Young and Butler himself. Van Buren went on to outline what he thought must be done at Utica in words reminiscent of those his son John had previously written. He was very decided that the convention initiate a state movement only "without making any appeal for cooperation to other states. If the people of other states desire such a measure let them move on it and we will cooperate with them." Finally he advised that the nomination for Vice-President be deferred until the regular convention met in September.[43]

Next, Van Buren wrote a letter to the delegates which he enclosed in his instructions to Butler. Had he confined himself to his declaration that he would never again be a candidate for a public office, he would have spared himself much mental anguish. But for once

on a political matter of great personal concern to him, he let his emotions overcome his habitual caution. Van Buren assured the convention that he stood firmly for their objectives, roundly castigated the Baltimore convention and vehemently opposed any further extension of slavery. He even alluded to his opinion that Congress had the power to abolish slavery in the District of Columbia as well, of course, as in the territories. Without hesitation he sanctioned the refusal of the Barnburners to vote for Cass. "If no other candidates than those before the country are presented I shall not vote for President," he concluded.[44]

Some three hundred Barnburners poured into the flourishing town of Utica. They came from the east by the Utica and Schenectady railroad, from east and west by canal boat, from other remote areas by wagon, coach and even horseback. And as they saw their old heroes, like Colonel Young, stooped and white-haired but as fiercely independent as ever, Cambreleng, Butler, and among the younger men the impeccably dressed John Van Buren, voluble Preston King and reserved Tilden, they shouted themselves hoarse and then quenched their thirsts in the taverns.

The atmosphere in the crowded little hall was charged with emotion when Colonel Young, unanimously elected chairman, called for order. What the delegates did not know was that the evening before leaders of the bolting Whigs had met with the chief Barnburners in a caucus to determine whom they should nominate for the Presidency. John Van Buren took command of the proceedings and sought vainly to have Addison Gardiner made the candidate. When Van Buren's name came up, John insisted that his father would decline and offered himself as the candidate.[45] Nothing was decided, though the younger Van Buren finally had his way; and while no one could agree on a candidate everyone decided Van Buren should not "make the sacrifice of personal feelings involved in acceptance."

But on the following day, June 22, after Butler read Van Buren's letter to the convention, the pent-up emotion of the delegates broke loose. Before any measure of control could be imposed, the delegates nominated Van Buren with a mighty surge of approval so fervent that Tilden, Flagg and John Van Buren found it impossible to withdraw his name. Since they had gone against Van Buren's wishes, they went further and ignored his advice by nominating a vice-presidential candidate, Henry Dodge, governor of the new state of Wisconsin. They also adopted a platform that rehearsed the old grievances against the Baltimore convention and the Polk Administration and came out explicitly for free soil. Then in an obvious bid for votes from special interests, the convention adopted resolutions of support for internal improvements and labor reform, distinctly at variance with regular party canons. The Barnburners were making a clear break with the Democratic party.

On June 28, a few days after the last clamoring Barnburner left Utica, the Conscience Whigs, those who supported the Proviso, met in a mass meeting at Worcester, Massachusetts. There the assembled multitude listened to strident indictments of the South, of slavery and especially of its extension into the territories. The meeting took note of the Barnburners and their recent convention at Utica. Frankly looking to a coalition, the leaders at Worcester called for a convention to be held at Buffalo in September.[46]

When news of the Utica nomination reached Washington, Van Buren's small group of supporters ebbed away, Benton's defection being the most conspicuous. As for the master of Lindenwald, he was so upset that he upbraided John, who lamely explained his hands had been tied.[47] For a man who had made party discipline the keystone of his political career, the Utica nomination was almost a physical blow. As a citizen he would support whom he pleased, but as a public man he was situated differently; and as a Democrat, indeed a founder of the party, he had devoted himself to identifying and maintaining what he considered its just and true principles. Van Buren's political reputation meant much to him. Here in the twilight of his long and he hoped useful career in politics and public service, he found himself like De Witt Clinton being branded a schismatic. The very thought of political fellowship with Whigs and abolitionists filled him with revulsion. But there was still a chance some other nomination would be made at Buffalo; there was still a possibility the Barnburners would confine themselves to the state campaign, divorcing themselves entirely from the national ticket.

Van Buren's political instincts were unimpaired. If he found Whigs and abolitionists hard to accept as political allies, what must they think of him? As the summer wore on, it became increasingly evident that many of them mistrusted and disliked Van Buren as much as he detested them. Outside of New York, even the younger element of the radical Democracy were not enthusiastic about Van Buren. Many of them thought his past record of seeking to conciliate the South had shown a weakness and some even thought evidence of moral cowardice. Others concluded Van Buren was too old, too long a controversial political figure who had accumulated too many enemies. Chase was almost as much troubled by the Utica nomination as Van Buren himself. He thought Associate Justice John McLean, rather Whiggish in his politics, cautiously Free Soil in sentiment, would be a stronger candidate. Would the Barnburners attend the Buffalo convention now that they already had a ticket and a platform? If they attended would they support McLean rather thanVan Buren and some prominent New Yorker for the Vice-Presidency—Colonel Young or John Van Buren, for instance?[48] These were burning questions that not only bothered Chase, but most of the Free-Soil leadership. "Should Van Buren be the nominee, anti-slavery Whigs will have just cause to suspect the design is to favor the Democracy, and so on the other hand should the choice fall on [a] Whig," said Henry Stanton, expressing the Free-Soil dilemma. Amalgamation could not be accomplished without great difficulty.

Charles Sumner, a rising political mandarin from Massachusetts, was not so disturbed by old party enmities as he was with Van Buren himself. The Liberty party had nominated John P. Hale on an abolitionist ticket. Would it willingly subordinate itself to a Free-Soil amalgamation with Van Buren at its head? Sumner was doubtful, but he found some hope in Dodge's refusal to accept the Utica nomination for Vice-President.[49]

As for Van Buren, after some agonizing reflection, he decided he must accept the Utica nomination. He made his decision against his own personal wishes with the full knowledge that Free-Soil men in other states preferred some other candidate.[50] Van Buren could not desert his friends in New York any more than he could disavow his opposition to slavery in the territories. Yet he adopted the most neutral stand he could take when asked, if he were to be elected President would he veto a bill for abolition in the District of Columbia? Privately, Van Buren had long since decided he would approve such a bill, but he would not make any public comment on the matter.[51]

He was committed to head the New York Free-Soil–Democratic ticket and hoped to keep it that way. He would run as a Democrat and lose as a Democrat. But he would defeat Cass in the nation and the Hunkers in the state, and he would make Free Soil a national issue that sooner or later would be, he hoped, one the Democratic party would accept. As he reminded the cautious Tilden at a meeting in Lindenwald on July 4, 1848, "I had supposed that we wanted every man who was opposed to the extension of slavery. Would it not be well first of all to defeat General Cass, and show the pro-slavery party that they shall not invade free soil? To that end is not the vote of Gerritt Smith [abolitionist leader] just as weighty as that of Judge Martin Grover [a prominent Hunker]?" The Buffalo convention was not, after all, a Democratic convention. There would be no election of delegates to it from the ranks of the Barnburners. They would consider it, as Tilden said, "simply a voluntary assemblage of individuals, whose relations with each other are to be for the first time established." Speaking for the young Barnburners, Tilden wrote Chase that they stood by Van Buren. There could be no change in the Utica nomination even though the old leaders might think otherwise. "The Democracy of this State supports the cause and Mr. Van Buren," continued Tilden, "an organized party, having more than fifty presses, many of which are the longest established and most influential in the State."

Influenced by a strong upswing of Free-Soil sentiment throughout the North, the ambitious McLean toyed with the idea of accepting a Free-Soil nomination. But he eventually concluded the new movement would subside, and he refused to have his name placed in nomination at Buffalo. With McLean out of the picture, the Conscience Whigs, the Liberty party men and other disparate partisans united only in their opposition to the extension of slavery, felt they must accept Van Buren.[52]

The various groups that opposed each other and Van Buren in everything except the slavery issue may have been drawn together by necessity, but they were experiencing a steady falloff of early adherents. Both the Whig and the Democratic parties had strengthened their organizations. The Polk Administration mobilized the federal officeholders who were sowing dissension in Free-Soil ranks. Conscience Whigs and radical Democrats alike were shocked when Horace Greeley, so long a vociferous opponent of slavery, brought his influential *Tribune* behind Taylor and Fillmore. They were equally dismayed when Edward Everett, who sympathized with the Free-Soil movement, refused to endorse it.[53]

The disappointed northern followers of Henry Clay, who had seen their hero discarded at Philadelphia, began drifting back to the fold. Radical Democrats like Gideon Welles, while maintaining a covert sympathy for Free Soil, could not stomach the idea of amalgamation at Buffalo. He foresaw a motley group of "dissatisfied politicians—disappointed men—and ultra impracticable themselves who cannot conform their ideas to attainable ends."[54]

Van Buren, recognizing the unique character of the Buffalo convention, drafted a long letter, which Butler would read to the delegates. "While the convention will be very important," he said, "in one respect it will be wholly unlike any political convention which has been held in the United States. . . . It will be in a great degree composed of individuals who have all their lives been on different sides of public questions and politics, state and nation, and who still differ regarding most of these questions but who feel called upon to unite on one issue slavery extending into the territories."

He reminded the delegates he had not sought the nomination. At Utica it had been forced upon him. But he would not decline if the convention insisted on placing him at the head of the movement because, and he emphasized this point, he owed it to "the faithful Democracy of New York to sustain their position from which they were driven by the injustice of others."[55]

There was no lack of enthusiasm among the several thousand discordant spirits that descended upon Buffalo on August 8, 1848, the eve of the convention. A torchlight procession had been arranged; and the marchers, including a substantial number of Buffalo idlers, wound its way through the city's streets chanting Free-Soil slogans, many displaying crude representations of Van Buren, Jefferson, John P. Hale and David Wilmot.

A huge tent had been erected in the city park. And promptly at 8:30 on the sultry morning of August 9, the multitude made a mad rush for seats on the rough wooden stands inside. Within minutes the makeshift seats were occupied and every spot of open ground taken up. After considerable difficulty, the throng was brought under enough control to elect Charles Francis Adams President and to delegate its responsibility to a select committee of organization on what was known as the Ohio plan. Manageable in size, this group would be composed of six delegates at large from each state represented at Buffalo and three additional delegates from each congressional district. It was so designed as to give equal representation for the Liberty, the Democratic and the Whig parties.

The committee, elected Chase chairman and moved its deliberations to the county courthouse. Chase and King were the principal authors of the platform. Little disagreement arose over the Proviso resolution, though there was some argument over the tariff, which was resolved in such a way that would assure the people there would not be "a sudden and radical change in the mode of raising revenue and paying the national debt."[56]

The Committee on Resolutions had cast the net as widely as possible. Included in the platform was a demand for more federal expenditures on the improvement of rivers and harbors, cheaper postage for the masses, free land for actual settlers, a drastic reduction of federal officeholders. The business of the platform completed, delegates turned their attention to the nominations.

Barnburners had all along maintained that they were Democrats and that their Buffalo meeting was not a convention in the accepted sense, but purely a voluntary assemblage. As the campaign in New York moved on during the summer, however, they realized they would have to broaden their scope if they were to make the impact they desired. The speeches, too, as John Van Buren and Benjamin Butler electrified their audiences, charging the Hunkers with murdering Silas Wright, had developed almost a messianic frenzy among the Barnburners from which even Van Buren was not immune.

In his Buffalo letter he gave the Barnburner observers the official status of delegates; the meeting became in his words "The Convention." And in the substance of his letter he made it clear that the Barnburners were willing to merge their separate identity into the

Free-Soil party. No wonder prolonged cheering followed Butler's reading of Van Buren's letter and an ovation greeted his name when placed in nomination.

Hale was the only serious contender, but it came as no surprise when Henry B. Stanton read the committee Hale's letter offering to withdraw and work for the nominee of the convention. On an informal ballot taken immediately after the reading of Hale's letter, Van Buren had an absolute majority over all the candidates. Liberty party delegates, who had formed the most significant opposition, withdrew Hale's name and made the nomination unanimous. The committee would have preferred a western man for the vice-presidential nominee but could not agree on a candidate and finally decided on a party rather than a geographic division. Charles Francis Adams was chosen as Van Buren's running mate.[57]

The entire business had taken only three and a half hours. At 1:00 p.m. the committee reported to the mass meeting in the tent that the ticket would be Van Buren and Adams. The meeting approved it with thunderous acclamation. After the cheering had stopped, Chase took over the president's chair and read the resolutions. Cheering frequently drowned out his full clear voice, but when he came to the end and proclaimed "Free Soil, Free Speech, Free Labor and Free Men" as the campaign slogan, the tent was transformed into a frenzy of emotion as hundreds among the multitude wept openly. There were more torchlights in the streets that evening as the city of Buffalo paid tribute to those magic words.

Van Buren formally accepted the nomination two weeks after the Buffalo convention. When it was published in the *New York Evening Post* on August 24, the most captious critic among the Liberty party could not fault it. He was forthright in declaring he would not, as matters stood now, oppose the abolition of slavery in the District of Columbia. His insistence on the Proviso was bold and unequivocal. He fairly captured the heart of Joshua Leavitt, a leading figure in the Liberty party who was now referring to him as "our glorious old man in the battle against the slave power."[58] A different note was sounded from the White House when an overworked, embittered Polk charged the radical Democrats of New York with political treason: "Mr. Van Buren is the most fallen man I have ever known."[59]

« CHAPTER 32 »

"Patriotism Firm and Ardent"

The telegraphic dispatches from Hudson after election day and through the last week of November 1848 caused no surprise at Lindenwald. Taylor carried the state and was President-elect. Van Buren had not received one electoral vote. Hamilton Fish, the Whig candidate for governor, defeated John A. Dix, a reluctant Free-Soiler, and Reuben Walworth, a determined Hunker. The party in the State had split almost exactly in half. It was scant satisfaction to Van Buren that he had beaten Cass in New York and in Massachusetts, that he had tipped the scales to Taylor in two other states, and polled over 290,000 votes.

The only bright gleam in the campaign had been his son John's dazzling performance. His speeches, graceful, witty, studded with epigrams had been a source of great pride to his doting parent. John had given his all in a whirlwind campaign that took him through New England, New York and the Northwest. Loyalty to his family and to his friends, belief in the cause, were precisely the ideals Van Buren himself cherished. Second only to John in sacrificing himself on the hustings was Benjamin Butler, Van Buren's closest friend since the death of Silas Wright. But John Van Buren came home exhausted, his health impaired, and Butler had been summarily removed from his federal post.

To Van Buren "the forlorn hope," as he knew it would be, was an "exceedingly unpleasant but unavoidable sacrifice to the feelings and unsurpassed injuries of friends who had stood by me through evil and good report and had more than any others helped make me what I had been." When it finally came to upholding his convictions or adhering to party discipline, he accepted, albeit reluctantly, his convictions as had his son John, Butler and Dix, in the best tradition of the old Regency.[1] He felt the Free-Soilers had "accomplished all that was wished and more than we had any right to expect." Their moral impact on the two great parties in the North had been significant. The question of slavery in the territories became for a time the single, all-engrossing issue.[2] But what would be the reaction in the South? As Van Buren expected, Calhoun and his associates drew up a vigorous address asserting the right to establish slavery in the territories whether their inhabitants opposed it or not.

In a harsh attack on the Free-Soilers, he accused them of being abolitionists and demanded a strict Fugitive Slave Law. These passionate remarks and the sectional sentiment they displayed could not be taken lightly, though a substantial majority of southern Congressmen would not support the address and engaged in bitter debate with extremist minorities. This manifestation of moderation was reassuring, the turbulent short session of the Thirtieth Congress was not.[3]

Unable to rally a solid South behind his fervid appeal, Calhoun and his lieutenants managed to block the establishment of a territorial government over California and New Mexico. During the summer of 1849, their cause gained adherents when it became apparent that Taylor was coming under Free-Soil influence. California was the crux of the matter. Discovery of gold in that far western region set off a population rush. By the fall, the gold seekers and their support groups in California were clamoring for statehood. Yet not even a territorial government existed; and the miners, whatever their origins, were giving unmistakeable signs that there was no place for slavery in the gold fields or anywhere else in the region. New Mexico, sparsely populated, without any organized government, continued to be free as it had been under Mexican law.

Van Buren watched these events closely, especially the advance of the Free-Soil idea in the North. John Van Buren, after a month or so of comparative rest, had regained his former vigor. Clearing up his most urgent legal business he again plunged into politics. He was in constant touch with Chase, who had been elected to the United States Senate as a Free-Soiler, and other radicals on the slavery issue. On November 8 he delivered one of his finest and most radical speeches in a packed Faneuil Hall at Boston. His voice scarcely ever rising above a conversational pitch, he warned his rapt listeners of danger facing California and New Mexico. Vehemently, he insisted that his father had headed the Free-Soil party "neither out of resentment to the south nor to the national democratic party nor to Mr. Cass." In more extreme language than Van Buren would have counseled, his son called for the abolition of the slave trade and of slavery in the District of Columbia, in California and New Mexico. He opposed also the interstate slave trade, though he doubted the power of Congress to interfere with it. Emphasizing each sentence, he solemnly declared that no more slave states "ought to be admitted into this Union."[4]

While his son was electrifying northern audiences, Van Buren spent the winter in New York City part of the time incapacitated by severe attacks of gout. Yet he did not let his painful disorder interfere with his life-style. He curtailed his visits to the theater and dinner parties with friends; but he did not stint himself on the rich meals with wines he so enjoyed, and during the evening customarily polished off a glass or two of sherry or madeira.[5]

During his enforced leisure he devoted more time than usual to reading and to analyzing political and public issues. Like others of his generation in the North, Van Buren was becoming alarmed at the polarization of the nation. The call for a southern convention to be held at Nashville he viewed as a very serious matter; and he welcomed the compromise measures that settled some of the more vexing problems between the sections—the admission of California as a free state, the drawing of a permanent boundary for Texas and, that long-standing reproach to free institutions, symbolic as he knew it was, the abolition of the slave trade in the District of Columbia. He was, however, concerned about the Fugitive Slave Law, doubted its constitutionality and viewed it as an invasion of states' rights.

His son Martin and Blair, both of whom witnessed the great debate over the compromise measures, kept him fully informed of the politics and the logrolling that were involved in their passage. Friends like Kemble and Gilpin, who made frequent visits to Washington, relayed the gossip he relished along with shrewd observations of the political scene.[6] At home he advised his son John on state affairs, but otherwise took little active part in them. Free soil was a national issue, and the Taylor Administration seemed to be moving in that direction on California and New Mexico. Polk was dead, Cass discredited and both Barnburners and Hunkers would remain a hopeless minority in the state, bereft of patronage and of policy unless they composed their differences.

Internal improvements and deficit spending, the initial causes for division, were less compelling public issues than they had been a decade before. Railroads, canals, highways and banks were now not so dependent upon state support for their development and maintenance. The subtreasury system, which Marcy and Croswell opposed and the banking community had warned would spell ruin for the economy, had been in operation for almost five years and yet none of the dire predictions had materialized.

Venture capital from private sources was more than sufficient as the economy reacted to the increased government expenditures for the Mexican War. Gold from California and bumper cotton crops from the South bound for northern factories or for Europe were enriching the banks and the mercantile community of New York. The Corn Laws that protected British farmers had been repealed in 1846, opening up yet another expanding market for western foodstuffs that were shipped along the Erie Canal and the Hudson River for export from the City of New York.

All these factors, but particularly the political wilderness in which the Democrats had been wandering since 1846, made it possible for a working relationship to exist between the two factions without either side acknowledging a compromise of principle. Horatio Seymour, a moderate Hunker and, like his father, a suave politician, made overtures to John Van Buren, who saw the practical benefits of a united party on state affairs. Under John's leadership, the Barnburners attended the regular Democratic convention at Rome, New York, in September 1849. Henry B. Stanton, a Barnburner delegate, submitted a resolution in favor of the Proviso that very nearly destroyed the patient planning of Seymour and the bold tactics of John Van Buren. Displaying a sense of management worthy of his father, the younger Van Buren convinced his delegates that they could coalesce with their opponents, yet not sacrifice their ideals. "We are asked to compromise our principles," the younger Van Buren said to the convention. "The day of compromise is past; but in regard to candidates for state offices, we are still a commercial people. We will unite with our late antagonists. And we will hold them as we hold the rest of mankind—enemies in war, in peace friends."[7]

The coalition was a frail one and Seymour faced severe opposition from the more conservative, less malleable Hunkers. Croswell made no secret of his opposition. The vengeful Daniel Dickinson blamed Seymour when the coalition failed to gain a majority in the November election. The certainty that a Whig would replace him in the Senate may have overjoyed the Barnburners, but it infuriated Dickinson, who declared a personal vendetta on all moderate Hunkers. From now on Dickinson worked to build a prosouthern faction within the party, which soon interjected itself into the presidential contest under the slang term of the "Hards." The Barnburners and their Hunker allies became quickly known as "Softs."

In all of these maneuverings, Van Buren simply advised his son, and this only when asked.[8] Thus far he could find little fault with John's course. The compromise measures and the death of Calhoun deprived the extreme "Southern Rights" wing of the party of a platform and a leader. The Nashville convention, which had seemed so menacing in the fall of 1849, did not materialize. Demand for cotton, which had fallen off in 1849, was again on the increase; and with the prospect of higher prices, southern planters looked more to their fields than to politics. Kemble, who made a short visit to Washington early in 1851, found a dramatic change in personal relations. "I was greeted everywhere by the southern men," he wrote Van Buren, "with the utmost kindness in striking contrast to last year." The presidential cliques were forming, he reported, adding that he had left Marcy and Croswell at the capitol where they "will work for no good except it is to fall upon each other."[9]

Because he had been a third party candidate in 1848, Van Buren had no standing with the national organization, but his counsel was still sought and his influence felt through his son's leadership of the Softs in New York. Blair sounded him out on Benton as a presidential candidate; but John Van Buren, who long harbored serious doubts about Old Bullion, objected strenuously and his father had to agree that the embattled Senator from Missouri now nearing the end of his thirty years in the Senate was too controversial to gain enough support for the nomination.

Though father and son were as one in opposing Benton, they disagreed on alternative candidates. In the aftermath of the Compromise of 1850, Van Buren reexamined his earlier opposition to Woodbury. When the Justice publicly declared that the distractive slavery issue was settled, Van Buren chose to believe him. With what influence he commanded, he quietly lent his name to the Woodbury candidacy. Other ex-Barnburners, the more radical Preston King and the more conservative John A. Dix, began to think better of the Justice.[10] Blair, who had also come around to Woodbury, wrote John Van Buren that "the great point with us is to have Hunkerdom swallowed up. Woodbury of all others is most obnoxious to Cass, Buchanan, Marcy and the leaders of that class."[11]

In September, when it appeared that Woodbury would command the united support of the Democratic party in the North and be acceptable to the South, he died suddenly. "Let us know what we ought to do in the emergency," Dix wrote Blair, expressing a preference for Sam Houston but asking him to consult with Benton. Benton rejected Houston out of hand in one of his sweepingly oracular statements: "He has flaws in him that would make him fall to pieces in our hands." Blair suggested William O. Butler, Cass's running mate in 1848. Benton thought he would be the best candidate if New York supported him. Off went a letter from Silver Spring to John Van Buren, who sent a prompt reply. Butler would be acceptable if he pledged himself against extending slavery and would be "willing to work for the repeal of the fugitive slave act, returning the power to the states where it rightfully belongs."

The Softs were willing to take their chances without legislation on slavery in New Mexico and Utah and were confident that Free-Soil sentiment in Congress was strong enough to abolish the peculiar institution in that region. "Will Butler say he will not veto it?" asked John Van Buren. "If he agreed to all these conditions, he would have thirty of New York's thirty-six delegates."[12] Blair was too circumspect even to attempt gaining these assurances from Butler. Judging rightly that John's queries were not binding, Benton went ahead as Butler's chief manager and began to attract national attention until word leaked out that his candidate had made a deal with the Free-Soilers. The Kentucky Democratic convention which was then in session forced Butler to write a public letter approving a resolution that Congress had no power over slavery in the territories. The boom ended as abruptly as it had started.

"Butler's letter satisfied me that he is incompetent to fill the office," said John Van Buren to his father, "and I have written Blair that if he will stick with Butler I will stick with Marcy and I think that means the disposal of both." At his father's urging John began tossing out hints to Marcy, who was still a Hard, but who the elder Van Buren believed could be wooed and won.

Van Buren's judgment of Marcy's character and ambition was as sound as ever. He knew him to be a man of ability, whose talents qualified him for the Presidency. He knew him also as a man of soaring ambition; and though he would never excuse Marcy for his share in the downfall of Wright, Van Buren recognized the strength Marcy would bring to the Softs. He was, after all, popular throughout the Union, and he had shown in the past a strength of will Van Buren thought necessary to resist the extremists in the North and the South while working for the repeal of the Fugitive Slave Law.

Marcy accepted John Van Buren's offer of support with alacrity. At the party convention held in the fall of 1851 at Syracuse, he sealed the bargain, though not without some difficulty from the more extreme Softs and the fury of Daniel Dickinson with his small group of dedicated Hards. John Van Buren had hoped to lead the New York delegation to the national convention at Baltimore, but the state committee by a margin of two votes opted for the district rather than the convention plan for electing them. Dickinson and his Hards in New York City chose eleven delegates, Marcy gaining the rest—twenty-five.

Without a united delegation behind him, Marcy was gravely weakened; and John Van Buren was not disposed to work for a losing cause. If he could not control he would have none of it. Smith took his place on the delegation. The quixotic John was wearying of politics and of the role of the anonymous wire-puller; nor had he the patience of a Seymour, working step by step behind the scenes to unite the factions. Besides he was short of money and must spend more time and energy on his law practice.[13] As for his father, the pleasure of retirement in his declining years had always competed with his interest in politics. When his son decided not to make a contest with Seymour, Van Buren resumed his role as spectator.

There was little disappointment at Lindenwald when news arrived of Marcy's defeat and the nomination of Franklin Pierce. Van Buren had in the beginning counseled the approach to Marcy because he thought it would ensure a firm basis for the party in New York, with his son its undisputed leader. When Seymour emerged as the real source of power among the Softs, Van Buren's interest in Marcy's fortunes waned. The surprise nomination of Pierce, far from disturbing him, seemed almost providential. It had certainly been a rebuff to Seymour; it all but obliterated the Hards and Daniel Dickinson. Nor did it enhance Marcy's reputation at home. Should his son again shake off his political lethargy, he would be in a better position than before. He was not as enthusiastic as Blair, who called the Pierce nomination the "greatest triumph ever achieved over the jobbers"; but he was quite satisfied with the result.[14]

After he was assured that Pierce favored the Compromise and considered the slavery question settled, he wrote a public letter to a Tammany committee endorsing the nominee. Otherwise, he took no part in the campaign. His attention was now wholly concerned with the management of his farm, family affairs and with a history of political parties in the United States he had been thinking about for some time.[15]

Life at Lindenwald from the fall of 1848 to the spring of 1852 was the most tranquil period of Van Buren's busy life, though not unmarked by tragedy. In early 1849 Van Buren had made a proposition to Smith that if he and his family would live with him at Lindenwald, he would leave him the entire estate. The responsibilities of marriage and a family had not inspired this youngest son to pursue any definite career. Smith had been admitted to the bar in 1841 and had practiced for two years in New York City. After he married the heiress, Ellen James, however, he abandoned his struggling practice and moved to Albany, dabbled a bit in politics and lived as a gentleman of leisure.

A shy young man, he was warmhearted and charming, as were all of Van Buren's sons. But he had lived so long under the shadow of his father's fame and his brother John's brilliance that he underrated his own abilities. Blair, who was fond of Smith, described him as a person who was "afraid to display himself lest he may disappoint expectations. . . . he must by repeated efforts learn to shake off modesty and divers sorts of hypercritical anxieties which will embarrass his thoughts and stiffen, if not stifle him."[16]

Van Buren agreed with this fair estimate of his son's character, yet worried about Smith's lack of ambition, his apparent distaste for any worthwhile occupation. Smith did enjoy the rural pleasures of Lindenwald and shared his father's passion for farming. Why not become, as Van Buren put it, "heir apparent?" Van Buren explained to his friends that he wanted to keep Lindenwald in the family; not as he was quick to say "for any aristocratical purposes but because good treatment for my plants and trees which being the result of my own labors I have a strong attachment."[17] Smith was willing but insisted on remodeling the old mansion in the new fashionable mode of Gothic revival. The counterproposition disturbed Van Buren, who visualized great expense for no worthwhile end and great inconvenience to himself; Smith's plan was an ambitious one that would completely alter the appearance of his home and rearrange the interior he himself had altered to suit his comfort and his notions of elegance.

Smith engaged Richard Upjohn, the fashionable British-born architect who promptly went to work and drew up a detailed plan. Van Buren was appalled at the result, especially the costs he estimated for Upjohn's changes. The architect, with Smith's enthusiastic approval, would convert a fine mansion in the Federal style into a hybrid Renaissance Italian palazzo—with Gothic revival touches and ornate cornices on the dormers, a central gable on the right and a bell tower on the left. Upjohn retained the stately main entrance, where old Peter Van Ness used to sit smoking his long Dutch pipe behind the lower half of the Dutch double doors. He did, however, provide another entrance in the north wall of the house. Other changes expanded the size of Lindenwald to thirty-six rooms and corridor, one water closet, a bath and central heating. Only after Smith agreed to share a portion of the expenses was Van Buren persuaded to accept Upjohn's elaborate plan. He mused:

> What curious creatures we are. Old Mr. Van Ness built as fine a home here as any reasonable man could. Its stability unsurpassed and its taste of what was then deemed to be the best. William P. Van Ness had disfigured everything his father had done. I succeed him and pulled down without a single exception any erection he had made and with evident advantage. Now comes Smith and pulls down many things I had put up and made the alterations without stint. . . . what nonsense.[18]

On June 1, 1849, the work began. Van Buren moved to the village of Kinderhook where he engaged rooms for himself and where he expected to remain for a few months, then winter in New York City. His rooms in Kinderhook were comfortable enough and his manservant "faithful James" looked to his wants. After he had acclimated himself to his temporary surroundings he watched the work of demolition and reconstruction with interest. Van Buren had expected he would have to put up with temporary quarters for six months. But the work moved forward much more rapidly than his gloomy prediction. By August, he moved back to Lindenwald and invited Henry Clay, who was staying at Saratoga, to join him there for a visit.[19]

Smith's wife Ellen died suddenly in the late fall, casting a pall of gloom over the little family. Van Buren was very fond of his daughter-in-law and had looked forward to her gentle presence as the chatelaine of his refurbished home. He also worried about the effect on Smith, who was left a widower with three young children. But in the spring of 1850, all the work was done. Smith's spirits improved when everyone moved back to Lindenwald. His father soon became accustomed to the new arrangements and in fact enjoyed the creature comforts Upjohn had provided in his interior plans.[20]

Besides the pleasure of having Smith and his motherless children to ease the loneliness that had been his lot since the death of his wife, Hannah, over thirty years before, Van Buren had, of course, his farm, his books, friends and always politics to interest him. He enjoyed horseback riding for exercise, though unable to break a habit of a lifetime, he rode far too fast for a man of his age, causing concern to his family and friends. During the summer on pleasant days, usually with a friend or two, he fished for bass in the ponds on the estate, for trout in the streams and salmon in the Hudson.[21] Amid these pleasant moments of relaxation the old man enjoyed the merry hubbub of grandchildren—John's daughter, Anna, Abraham's two sons, Martin and Singleton, and Smith's children, all of whom spent the summer at Lindenwald.

Van Buren also occupied a part of his time with a serious work on his history of political parties in the United States. He had long had such a project in mind but something or somebody always seemed to require his attention when he was ready to begin. He was also thinking about an autobiography. Just before the carpenters, stonemasons and laborers descended on Lindenwald, he started retrieving letters he had written over the years to various friends and associates. Blair claimed his first attention because he was working on a biography of Jackson and had accumulated most of Old Hickory's papers at Silver Spring. He promised to send on all of Van Buren's letters to Jackson and suggested that Kendall, who, at one time had also thought of doing a biography of Jackson, would probably have additional correspondence.[22] It had been Thomas Hart Benton who spurred him to put aside some of the diversions and distractions and allocate a portion of his leisure to what at first he simply entitled "Inquiry."

Benton, who had been defeated for reelection to the Senate, was writing his personal history of the great political and public questions that had faced the nation since he entered Congress in 1821. He asked Van Buren to review his rough drafts as they were written. A labor of love, Van Buren found the work so absorbing that it rekindled his interest in his own recollections, his ideas about government and politics. Through the summer and fall of 1851 he read and made extensive comments on Benton's chapters. Old Bullion's stolid prose conjured up remembrances of the past, reminding him that he must set down his own impressions. Benton acted as a further goad, albeit a good-natured one. "Have you read Tocqueville?" he asked. "He is the authority in Europe and with the federalists here and will be with our posterity if they know nothing but what the federalists write." *Democracy in America* Benton admitted, "is a candid and intelligent appraisal. Do get it and read it and especially what he says of the contest with the Bank. . . . read it all and see what a figure we are to make if we do not write for ourselves." Van Buren, who may have read John O'Sullivan's extensive review of Tocqueville's book in the *United States Democratic Review,* took his advice. As he studied Reeve's translation of the persuasive French critic, his own thoughts about the history of political parties began slowly to take shape.

Martin, Jr., his health much restored by careful diet, abstention from wine and liquor, the restful atmosphere of Silver Spring and Ballston Spa, rejoined the family at Lindenwald.[23] He was more than happy to resume his old relationship with his father, that of a private secretary and helpmeet in his literary work. Van Buren did not pursue a systematic schedule either in his research or in his writing. Days, even weeks, would go by while he attended to other duties or other diversions. Many an evening when company was present Van Buren could not resist the pleasures of the whist table or lengthy conversations after dinner when the decanters were full, the topics redolent with frothy reminiscences or exciting events of the day.

Yet slowly he worked his way through dozens of volumes of political philosophy, biographies, printed volumes of letters, pamphlets, essays, speeches, his own voluminous correspondence and political tracts. Martin, Jr., copied out extracts that his father had marked out and was always ready to discuss themes for the essay format Van Buren decided on for his "Inquiry." He also prepared a chronological analysis of Jabez Hammond's *Political History of New York* for his father's contemplated autobiography.

So the days went pleasantly on for Van Buren, who, except for occasional attacks of gout, was remarkably healthy. The bouts of indigestion, the heavy colds and fevers that plagued him during his active public career seemed but unpleasant memories. Despite his active outdoor life, Van Buren put on weight as he grew older. But his walk was still elastic, his posture as ramrod straight as ever. He looked the stereotype of the country squire, with his ruddy complexion, a tribute to many a fine meal and rare bottle, his bald head fringed with carefully trimmed white hair his face fuller and smooth-shaven, his white sideburns perhaps a trifle thicker than in his youth. The penetrating blue eyes retained their luster and their mischievous twinkle at an apt remark.[24]

Unfortunately, Martin, Jr., on whom Van Buren had become quite dependent for his correspondence and his literary labors, again showed signs of his old illness during 1852. Slowly, the insidious ailment made inroads on a physique never robust. Nor would it now respond to the usual treatments—the waters at Saratoga and Ballston, the Virginia Springs—nor to any of the numerous remedies the best American physicians prescribed. Like other monied Americans whose families were faced with similar ailments, Van Buren accepted current medical advice that European physicians, in particular those in London and Paris, had achieved promising results in arresting tuberculosis. A change of climate and environment, it was thought, would also be beneficial.

Van Buren needed little persuasion to make a European trip. Though concern for his son's health was uppermost in his decision, he remembered how pleasant his previous journey had been in 1832 and how all too brief his stay. He had recently read Cooper's European travel books and was, of course, thoroughly familiar with Washington Irving's romantic sketches of European life and culture.

To Van Buren's delight, he found that the Gilpins and Kemble were planning a trip to England, Italy and France. After Martin was placed under the care of celebrated physicians in London and Paris, they would all meet at Rome where they would spend the winter of 1853–54. When this plan was decided upon, Van Buren devoted considerable time in talking with knowledgeable travelers, arranging his financial affairs, which included a letter of credit for $10 thousand with Barings, the British bankers most Americans favored.

Smith had the farm management well in hand; Flagg and Tilden were to handle his financial affairs, and son John and Butler any legal problems that might develop. President Pierce obligingly gave him a letter to Lord Aberdeen, Britain's Prime Minister, reminding that worthy of Van Buren's former eminence and asking any courtesies he might extend. Archbishop John Hughes of New York provided him a letter of introduction to Pope Pius IX.[25]

In Mid-May, when Van Buren arrived in London he renewed his old acquaintance with Lord Palmerston who was then home secretary in the Aberdeen ministry. As the first ex-President of the United States to visit England, he was received with high honors and entertained lavishly. When the brief stay in London ended, Palmerston, who had learned Van Buren's travel plans, alerted British diplomats in France and Italy to render whatever service they could. On June 22, 1853, Van Buren and Kemble bade goodbye to Martin, Jr., and made for Ireland. Another round of festivities Palmerston had arranged awaited them in Dublin and in the country estates of the Anglo-Irish nobility. The travelers crossed over to Scotland toward the end of the summer, then took passage over the North Sea to the Low Countries where they were lionized in Belgium and Holland. While at the Hague, Van Buren visited the state archives, where he made three pages of notes translating from the original Dutch on the early settlement of New York and his family's connection with the history of the colony. There was also a brief trip to the village of Buren. Here he found three individuals still bearing his name out of the 700 inhabitants. Conversing with them in fluent Dutch, which he had never forgotten, he discovered no trace of any ancestors.[26]

By late August, as Van Buren prepared to head south for Italy, an exhausted Kemble parted company and left for Paris. Van Buren's stamina, his energy and his delight with the lavish entertainments of high society had been too much for his friend. Had he eaten and drunk as much as Van Buren and took so little exercise Kemble complained, he "would have to live on bread and water." In Paris he found it necessary to put himself on a strict regimen, avoiding all table d'hôtes and restricting his diet to three dishes at dinner. Kemble and Van Buren had, as so often the case with close friends, proved to be bad traveling companions. "I was constantly in the situation of an old bachelor who had married a young wife," Kemble said apologetically, "and I think you had something of the same feeling towards me, but however strong the desire to accommodate in two persons travelling together, continued effort becomes irksome at last and as they are not married, they will be better friends by living apart."[27]

In early November, Van Buren arrived in Rome where he met the Gilpins and had an audience with the Pope who accorded him the honors of a visiting head of state. They spent over a month in Rome like typical American tourists viewing the antiquities in the city and the surrounding countryside, witnessing the colorful pageants and festivities of the Christmas holidays. Evenings were devoted to rounds of balls and galas. The Gilpins, like Kemble before them, could not stand the pace that Van Buren seemed to thrive upon and at the end of the winter season again he found himself alone. After the spring rains when travel south was more comfortable and less dangerous, Van Buren set off for Sorrento and rooms in the Villa Farangola the American minister in Naples had secured for him. On reaching Naples, King Ferdinand and Queen Maria Teresa needed no prompting from the British and American ministers to entertain Van Buren lavishly.

Van Buren must have known of the cruel and bigoted reign of King Ferdinand. It had been a common subject of conversation at London dinner parties since the publication of Gladstone's indictment of this petty tyrant's regime. Count Cavour surely broached that topic in his conversations with Van Buren when they visited together at Turin. Yet Van Buren made no mention in his letters home of conditions in the kingdom; nor did he ever say anything about the King except that he and the Queen had been very kind to him and in a reference some years later on to his residence "under the dominion of an absolute monarch."[28] Rather he seemed oblivious to the poverty, the squalor and the evidence of political oppression all around him. Most of the time he spent in Naples (when King Ferdinand was not feting him at Caserta) was with Lord and Lady Holland, who for the past three years had been leading a luxurious life and had established a salon for titled English expatriates like themselves, Neopolitan grandees and the few Americans of distinction who happened to be passing through. In early June, Van Buren left the surroundings of Holland House in Naples and the glittering tinsel extravagance of the court for Sorrento.

Nothing Cooper had written praising the village exceeded his expectations. He found his quarters in the villa cool, spacious, comfortable, and numerous servants ministered to his wants. The view from the Palazzo was spectacular, overlooking the broad sweep of the bay with the city of Naples in the far distance, Mount Vesuvius topped by its omnipresent plume of smoke on his right, the little villages and the dark green of orange and lemon groves interspersed with the lighter shades of olive, the gardens and flowers that swept up to his terrace.

Van Buren, whose iron constitution had finally succumbed to attacks of gout and whose thirst for European social life seems to have been slaked, temporarily at least, began a daily routine of early morning bathing in the warm water of the bay. Quickly he regained his usual good health and high spirits. He visited the partially excavated Roman cities of Pompeii and Herculaneum, the magnificent temple ruins at Paestum and in the evenings when he was not entertaining or being entertained, began writing his memoirs.[29]

Bundles of newspapers and packages of letters from home awaited him at Florence, at Naples and at Sorrento. The stark news most of them held contrasted vividly with the languid beauties of landscape that surrounded him. Blair's initial enthusiasm for Pierce had disappeared—no more chortling about the rout of the jobbers, but a smoldering anger over the President's weakness and his truckling to southern extremists. There was also the humiliation of Dix, twice offered positions, secretary of state and minister of France, only to have them withdrawn as Pierce bowed to a combination of southern pressure and the Hards in New York.

Greene Bronson, an intractable old Hard, he learned, was enjoying the emoluments of the New York collectorship and using his patronage to strengthen his faction's hold on the City. Marcy, though secretary of state, seemed powerless to influence the patronage policies of the President. If anything, the situation in New York was worse than it had been during the Polk Administration.[30] Nativism was on the rise, abolitionism seemed to be gaining rapidly and all manner of sectional outbursts were bruited about in Congress. There was good news too. His oldest son, Abraham, and family were planning a visit to Europe, along with John and his daughter. Martin was responding to treatment; his general health much improved.[31]

Aside from the pleasure this family information gave him, Van Buren found little in the state of the nation to ease his mind. And as spring took hold suddenly and brilliantly in that temperate climate, letters and papers he received became more ominous in tone.[32] The Kansas-Nebraska Act shocked Van Buren who saw in it the reopening of the slavery question he had thought settled by the Compromise of 1850. He correctly predicted the political and sectional turmoil that would follow Douglas's explicit repeal of the Missouri Compromise in the Act. "I felt, in all their force," he recalled, "the dangers to which our political fabric would be exposed by that act and mourned over its adoption . . . in exciting sectional animosities to a far more perilous height than they had ever reached before."[33]

Family affairs again took precedence over concern about the Union. The Van Burens had a glorious reunion, all present except Smith, at the lovely Swiss town of Vevey on the shores of Lac Leman. But before the end of their stay, Martin began again to show alarming symptoms indicating a return of his disease. It was decided that he should return to London by way of Aix-les-Bains for consultation with his physicians. With a heavy heart Van Buren saw his family go their separate ways, leaving him to continue on alone to Nice where he hoped Martin would join him.[34]

Van Buren, who had visited Nice before, found this Italian city with its French veneer most enjoyable, the weather pleasant and the large English colony agreeable company. He stayed at the Hotel Grand Bretagne "anxiously awaiting reports of Martin's condition and economizing with a vengeance." There he learned that Smith was to be married to Henrietta Irving, a great-niece of Washington Irving. "He of course thinks her perfect," Van Buren wrote Martin, "and she is certainly of a good breed, 25 years old." But the news from London was far from encouraging. Martin was not responding as he had before to the various treatments the London physicians prescribed. In mid-November, Van Buren became very worried about his son and urged him to leave chilly, damp London for the warm, brighter climate of the Côte d'Azur. He soon learned that Martin was too sick to make the long trip, and he changed his plans, arranging to spend the winter in Paris.[35]

Early in 1855, he reached Paris to find a desperately ill son awaiting him. Van Buren did all he could to make Martin comfortable, engaging the best medical treatment in the city and taking rooms which he described as "airey and pleasant as any in Paris." For once Van Buren did not conceal his emotions when he realized his son was dying. "We have . . . the best of servants and troops of obliging friends by which the painfulness of our situation is as much as possible lessened," he wrote, "but the last result is constantly before my eyes and cannot be disguised and I fear much longer delayed." Martin bore it all with "a steadiness and firmness seldom equalled, never exceeded," said the grief-stricken father, " a silent tear occasionally is all that is seen or heard by way of complaint." Three days after he wrote this letter to Blair, Martin died at age forty on March 19, 1855.[36] After his son's death, Van Buren remained in Paris during the spring months. In June he crossed over to London and sailed for home on June 20, 1855.

He found Lindenwald in excellent condition and again he resumed his daily rides, his work on his memoirs and on the history of political parties. As the summer wore on, he became increasingly despondent about the future of the Union. The Know-Nothing movement he found a deplorable negation of Jeffersonian principles and suspected it of being a façade behind which scheming Whig politicians lurked. Alarmed at the rapid organization of the Republican party espousing the same Free-Soil principles he had once articulated so clearly, Van Buren rationalized his opposition to it with his familiar dexterity. What had been, he thought, a salutary admonition to the southern Democracy in 1848 was now in 1856 a disruptive sectional movement of widespread appeal in the North. The growth of a similar organization, the ultra-Southern Rights wings of the party in the South posed a schism that could not be repaired. He felt he must restate his position on slavery and if anyone would listen to the oldest living ex-President, try to strengthen the voice of moderation.

The Republican candidate, John C. Frémont, he liked personally but considered far too young and inexperienced for the fearful responsibilities any President would face in 1857 if the Union were to be preserved. Before the election, Van Buren wrote Moses Tilden, older brother of Samuel, that slavery had figured in political contests since the beginning of the Republic, but never to the extent it had reached in the present campaign. "Now, for the first time in our history, one side and that the one in which we reside has undertaken to carry an election including the control of the Federal government, and against the united wishes of the other."

Van Buren was no admirer of Buchanan, but thought he was wise enough not to repeat the mistakes of Pierce. "I was so indignant against the repeal of the Missouri Compromise," he continued to Tilden, "that I could not do justice to the Kansas Organic Act as that was the instrument by which the outrage was perpetuated." Had Pierce acted firmly, however, and used all the force of the federal government to prevent outside interference for political purposes, Van Buren thought Kansas would have become "so decidedly free as to put an end to attempts to make it a slave state, the country would have been quiet, the party united, and he renominated." There was a certain plausibility to Van Buren's hindsight, but obviously he was unfamiliar with the complexities of the Kansas problem. After all, he had been in Europe during the early stages of the crisis and had gleaned his information primarily from the press and from a few selected correspondents who had decided opinions.

What he did recognize clearly was that "bleeding Kansas" had become a symbol of opposing cultures, of hostile ideologies and expanding nationalisms. Van Buren at seventy-five was just as perceptive as he had been in his prime. His belief that the Union faced dismemberment should the Democratic party lose the election was a practical one and was shared by the dying Thomas Hart Benton, who also supported Buchanan against his own son-in-law.[37] Other than a lengthy private letter to Tilden and one public statement to Tammany favoring Buchanan over Frémont, Van Buren took no part in the campaign and election. Still, he was relieved when, as he put it, the Democrats achieved a "hair-breadth escape" in November.[38] Buchanan's inaugural address, however, was not encouraging. The new President announced no policy on the slavery question but rather shifted the burden to the Supreme Court.

Van Buren was as startled as his friends among the Republicans at the direction and the scope of the Court's decision in the Dred Scott case, which Taney delivered just two days after Buchanan's inauguration. He interrupted his own writing to study carefully all the briefs (copies of which he received from Montgomery Blair, one of Scott's counsels), the opinions, and the legal, constitutional and partisan arguments that appeared in the press.[39] At first he could not accept one of Taney's central arguments that a free black residing in a state was not also a citizen of the United States. After further reflection, however, he decided that "the sense in which the word 'citizen' was used by those who framed and ratified the Federal Constitution was not intended to embrace the African race, whose ancestors were brought to this country and sold into slavery."[40] He admitted in extenuation, as he was bound to do in view of the fact that free blacks of property in his own state were certainly citizens of New York, the Chief Justice's opinion was open to conflicting interpretation. Ample precedents existed in declaring emphatically that a citizen of a state was also at least in some respects a citizen of the United States. Van Buren acknowledged that Justice Curtis's dissenting argument was "subtle and masterly," but he accepted Taney's contention that a black citizen of a free state, were he given status as an American citizen, would then "undoubtedly establish his own right under the Constitution to the enjoyment in a slave state of all the privileges allowed its own citizens" under Article IV, Section 2, of the Constitution.[41] He agreed with Taney that the South would never have ratified the Constitution had its delegates understood blacks wherever they resided enjoyed the same civil standing as whites in the slave states.

Where he disagreed with the Chief Justice was his ruling on the constitutionality of the Missouri Compromise. He felt strongly that this issue, pregnant with discord, should never have been raised. It was proper of course for the Court to hear the case on writ of error, but once it established that it had no jurisdiction the matter should have stopped there, though he recognized Taney was legally justified in going farther. It was "a grievous mistake," he wrote, "the question the court undertook to settle was political and had assumed a partisan character of great virulence." With a backward glance at the Federalists, Van Buren maintained that Jeffersonians had always opposed the interference of judges and clergymen in community affairs not specifically religious or legal. "Their want of sympathy, as a general rule, for popular rights is known throughout the world," he complained, "and in this country that repugnance received an enduring impulse from the unanimity with which the vast majority railed from the bench and the pulpit" in the late nineties.

When he made these comments, the Chief Justice was being subjected to violent criticism from Republican party presses and periodicals in the North. Some of the most scathing attacks had come from Van Buren's closest friends and associates of many years, anti-Nebraska Democrats like Benjamin Butler and Gideon Welles, both of whom were now leading Republicans. An essential factor in Van Buren's life was loyalty to one's friends but other equally compelling tenets were his realism and his political faith. The Chief Justice and President Buchanan, he thought, had indulged themselves in an unprofitable exercise that was distinctly opposed to "true republican" principles. Taney may have been attempting, as the Compromise of 1850 had attempted, to remove the slavery issue and, what must now be added, the racial issue, too, from partisan controvery.[42] But he had overstepped his judicial role whether for laudable ends or not.

Van Buren considered the question for almost a year before he began writing an essay on the decision. He finally concluded that the majority opinion of the Court had injured perhaps fatally the federal concept Madison had devised, and Jefferson and Jackson had developed through many perilous encounters. By now, Van Buren had seen the disastrous results of Buchanan's capitulation to southern extremists and the devastating impact of the Dred Scott decision on popular sovereignty. He had weighed and found completely wanting the fraudulent Lecompton Constitution the South was demanding Congress accept, thus forcing Kansas into the Union as a slave state against the majority will of its citizens.

On these points his discussion boldly reasserted his original contention that Congress had plenary powers over the territories. The language, dense and prolix even for Van Buren, placed the blame on the centralizing features of Federalist-Whig doctrine. He noted unfavorably that both the Chief Executive and the Chief Justice had been schooled in the tenets of the Federalist party and joined the Democrats "at comparatively advanced periods in their lives with opinions formed and matured in an antagonist school," and he singled out for particular criticism the most prominent advocates of Lecompton, Robert Toombs and Alexander Stephens, former Whig congressmen from Georgia. "I know neither personally," he wrote, "and never heard of either particularly, save as extreme partisans in the ranks of our opponents."[43] He closed his essay with a stern condemnation of judicial intrusion on political questions as utterly opposed to democratic principles. Van Buren composed the essay as he did most of his writing after his return from Europe, in bits and pieces over a period of time. If much of his argument was cast in a political mold, his discussion of the legal points involved was both lucid and comprehensive. His conclusions about the divisive character of the Dred Scott decision were eminently sound, his analysis of the case more perceptive than the opinions of most of his contemporaries in both parties. Finally he put the essay aside and never recast it. When his sons found it among his papers after his death, they published the essay as if it were an integral part of the *Inquiry,* which Van Buren never intended it to be. As for the author himself, he recognized his increasing difficulties in marshaling his thoughts, organizing his notes and putting them all together in a cohesive narrative. On a rainy day in Owasco, New York, Throop's home where he had gone for a fishing trip in the fall of 1858, Van Buren passed the time writing a twenty-two-page letter to his son Smith, literally pleading with him to devote more attention to arranging and editing his manuscripts. Van Buren, who had assumed Smith and his family would live with him at Lindenwald and be his literary executor as well as the second master of Lindenwald, was deeply disappointed when his son moved out of the mansion to be closer to the family of his new wife.

The old man was now alone again during the long winter evenings before he went to New York City for the coldest months of the year. He did not lack for company, surrounded as he was with relatives and friends from boyhood. What bothered Van Buren most was Smith's slow pace in working over his material. Van Buren had lost sight of the fact that Smith had his own family responsibilities and distractions.[44] The drudgery of deciphering Van Buren's sprawling hand and arranging his masses of unorganized material would have strained the patience, if not the skill, of the most devoted offspring. And Smith was not the kind of person who bore up well under steady application.

Despite Van Buren's complaints, Smith spent a good deal of time on the *Inquiry,* and as a result it is much better organized than the *Autobiography* on which Van Buren acted as his own editor. Neither work compares favorably with the political essays, letters and partisan tracts Van Buren wrote in the thirties and forties. Still, in substance, the *Inquiry* and the *Autobiography* are valuable sources for political and economic thought of the Jacksonian era. The *Inquiry* stands as the first major effort any American attempted to write a history of the political process in the United States from the founding of the Republic. Marred as it is by partisan bias, the *Inquiry* does explore the American experience in liberty and equality. Above all, it offers a credible defense against Tocqueville's concepts, as Thomas Hart Benton had called for when he was writing his own *Thirty Years' View.*

In many respects, the *Autobiography,* a loosely constructed partial account of Van Buren's life and career, is more useful than the *Inquiry* to the historian, though not to the general reader. The ponderous style, the digressions, the arguments that interrupt the narrative, make it tedious reading. No doubt Martin's illness and death, and Smith's casual attitude toward his father's work, left this valuable manuscript unfinished and poorly edited.

There are but a few references in passing to his own Administration, to the campaigns of 1836, 1840, and nothing on those of 1844 and 1848, though he clearly intended to fill in these great gaps. Van Buren's brief mission to England is documented much more fully than his two years as secretary of state or his four years as Vice-President. Here and there the diligent reader, if persistent, will uncover wonderful vignettes of historical episodes. The election of John Quincy Adams is told fully and entertainingly from the viewpoint of one of the losers. And often after pouring over a labored digression, one is rewarded with a sparkling observation of men and events.

John Randolph of Roanoke, "of annoyance to his opponents and of not a little uneasiness to his friends," suddenly emerges in all his eccentricities;[45] Andrew Jackson sits alone on a cold evening in 1829 in a darkened, drafty White House smoking his pipe before the fire; De Witt Clinton moves majestically and often deceitfully in and out of the narrative; Rufus King, his silver-buckled shoes and his clothes of an earlier era, spreads his napkin before him after dinner as he begins to speak on the topics of the day; the stalwarts of the Albany Regency in its heyday make plans, enjoy triumphs, and bring on themselves political disasters. They are limned in these pages as they appeared to an extraordinarily perceptive observer. Like any memoir written for the most part years after the events it describes the material must be used with care, particularly the disquisitions on the Bank war, the heroic portraits of Jackson and Jefferson, the bland innocence about his role in the Eaton affair. But the *Autobiography* with all its faults, is yet an impressive source, one of the most important legacies Van Buren left to future generations.

After 1858, Van Buren seems to have given up his historical writing completely. His health was beginning to fail. He had been through three serious accidents, in one of which he broke his left arm.[46] Fortunately, in his third accident (thrown ahead of his horse at a gallop), he somehow escaped the pounding hoofs and apart from a few scratches and bruises and a mild concussion, suffered no serious injury. As he playfully asked, "Does not this speak well of my skull?"[47]

Early in 1857, he had Butler rewrite his will, and in 1860 he made some further revisions. His affairs in order, Van Buren continued his round of activities as always, seemingly unaffected by the passage of time. But he had had to restrict his physical activities; debilitating attacks of gout and heavy colds sapped his accustomed energy. Twice he had to minister to his son John whose exuberant life-style had ravaged his once splendid physique. John's uncertain health worried his father and perhaps contributed to his own physical problems.[48]

His spirit and his understanding of events were unimpaired when the crisis came after the Democracy split at Charleston in 1860. The nominations of Breckenridge and Lane on a Southern Rights platform that warned openly of secession if the Republicans should carry the election in the fall roused the fighting instinct of the old Jacksonian.

Van Buren voted for Stephen A. Douglas in November and shortly afterward went to New York City for the winter. When South Carolina seceded from the Union in December 1860, he had long talks with Crittenden, who happened to be in the City at the time. Van Buren was still hoping his compromise plan, or one Thurlow Weed was circulating that was not substantially different, would prevail before more states left the Union and all efforts at composing differences between the North and the South were foreclosed.

As a last resort he proposed a constitutional convention. If such a body could not resolve the issue of slavery, he advised letting the South withdraw, but not until the question had been settled in the form prescribed. Until then public officers must execute the law throughout the Union and obey the Constitution.[49] For emphasis and for publicizing his views, he prepared a set of resolutions for the New York legislature in which he asserted that the Constitution was "a perpetual and irrevocable compact." Secession, he declared, "receives no countenance, much less sanction from the Federal Constitution and is founded neither on reason or justice."

Van Buren came down with a heavy cold after writing these resolutions, which rapidly developed into an attack of influenza. He had never been so sick for such a protracted period in his life; nor had his convalescence been so slow. By March, however, he was up and around, though far from his usual, sturdy self.[50] He found time to greet the new Lincoln Administration with a slashing attack on Buchanan's appeasement policies in a long letter to Blair. And he turned down a request from Franklin Pierce for a meeting of all the ex-Presidents in Philadelphia to call a Peace Convention. He might have agreed had Pierce contacted the Lincoln Administration beforehand and received "satisfactory assurances of their acquiescence." Van Buren was back in Kinderhook when the telegraph communicated the attack upon Fort Sumter in Charleston Harbor and its surrender to the Provisional Confederate troops of General Beauregard.

Like his neighbors and all of his relatives, he was swept up in the emotion of the hour. In a letter to Charles Jared Ingersoll, a recent guest, he said, "The disposition to give them [the Lincoln Administration] an earnest and vigorous support in the difficult struggle in which they are engaged, is every hour becoming more and more intense."[51] A few days later, a hastily formed Union committee in Kinderhook planned an evening meeting at Stranahan's Hotel supporting the President's call for 75,000 men to put down the insurrection. By now Virginia and North Carolina had left the Union; Tennessee was being overcome with secessionist sentiment; many citizens of Missouri, Kentucky and Maryland were uncertain in their allegiance. The committee, most of whose members differed from him in politics, met with Van Buren at Lindenwald and asked his advice about the proceedings. Worried about his health, he excused himself from actual participation but authorized the committee to give his unqualified support for the meeting. "The attack upon our flag," he said, "and the capture of Fort Sumter by the Secessionists could be regarded in no other light than as the commencement of a treasonable attempt to overthrow the Federal Government by military force."

Van Buren approved the call of the President on the states for the necessary means to suppress the rebellion. He was heartened at the response of the loyal states and gave his "earnest and vigorous support to the Lincoln Administration for its maintenance and the maintenance of the Union and the Constitution."[52] Old as he was and far from well, Van Buren called upon the Democracy of the Empire State to support the federal government. If he had faltered in his declining years, sacrificing some of his fundamental principles in a vain effort to save the Union, Van Buren's ringing endorsement of the war effort, his last public pronouncement, went a long way toward atonement.

That fall Van Buren went to New York for the winter and again was stricken with illness, this time pneumonia which kept him bedridden for several months during which he may have suffered some minor strokes. When Lawrence Van Buren met his brother at the Coahaxie Depot, he was shocked at his appearance. He could scarcely walk the few feet from the platform to the carriage awaiting him. And when he arrived at Lindenwald, he was put to bed. Thereafter his decline was rapid.[53]

Suffering from arteriosclerosis and from congestive heart failure, he drifted in and out of consciousness as the lovely summer days passed by. In early July it was obvious that the end was near, and the three sons hurried off to Lindenwald where they could be with their father in his last hours. At 2:00 a.m. on July 24, 1862, as thousands of his fellow countrymen were engaged in fierce combat, Van Buren died.

Distracted by military defeat and congressional opposition, virtually overwhelmed with unremitting labor, Lincoln still found time to add a passage of his own to the formal announcement of Van Buren's death. These few words were a just tribute to an American statesman who, with all his faults, his weaknesses, his little vanities, had made an impression of no little consequence on his state and his nation. "The grief of his patriotic friends," wrote Lincoln, "will measurably be assuaged by the consciousness that while suffering with disease and seeing his end approaching, his prayers were for the restoration of the authority of the government of which he had been head, and for peace and good will among his fellow citizens."[54] Van Buren would have approved the sentiment, one imagines, with a smile. After all, a crucial election campaign was going on in the Empire State; the race would be close. The Magician, whenever possible, had always offered proper respects for the departed with an eye on the election returns.

Abbreviations

BLP	Blair-Lee Papers, Princeton University
BMHS	George Bancroft Papers, Massachusetts Historical Society
BNYSL	Benjamin Butler Papers, New York State Library
BuP	Benjamin Butler Papers, Princeton University
CAL-CLEM	John C. Calhoun Papers, Clemson University
EMHS	Edward Everett Papers, Massachusetts Historical Society
FaLC	John Fairfield Papers, Library of Congress
FNYPL	A. C. Flagg Papers, New York Public Library
FO	Foreign Office Photostats, Public Records Office, UK
GLWC	Gideon Welles Papers, Library of Congress
HiSPA	Historical Society of Pennsylvania
HL	Huntington Library
JLC	Andrew Jackson Papers, Library of Congress (Micro)
JVBH	John Van Buren Hoes Collection
KNYHS	Rufus King Papers, New-York Historical Society
LC	Library of Congress
MLC	William L. Marcy Papers, Library of Congress
MHS	Massachusetts Historical Society
NA	National Archives
NYHS	New-York Historical Society
NYPL	New York Public Library
NYSL	New York State Library
OLCU	Thomas Olcott Papers, Columbia University
PNYSL	Henry Post Papers, New York State Library
RLC	W. C. Rives Papers, Library of Congress
TNYHS	John W. Taylor Papers, New-York Historical Society
UNC	University of North Carolina at Chapel Hill
UVA	University of Virginia
VBLC	Van Buren Papers, Library of Congress
VBMHS	Van Buren Papers, Massachusetts Historical Society

VBNYPL	Van Buren Papers, New York Public Library
VBNYSL	Van Buren Papers, New York State Library
VNNYPL	Van Ness Papers, New York Public Library
WOLC	Levi Woodbury Papers, Library of Congress

NOTES

Chapter 1. Beginnings

1. George S. Hillard, *Six Months in Italy,* 2 vols. (Boston, 1853), II, 160–62.

2. James F. Beard, ed., *The Letters and Journals of James Fenimore Cooper,* 6 vols. (Boston, 1960), I, 381–87; James Fenimore Cooper, *Gleanings in Europe, Italy: By an American,* 2 vols. (Philadelphia, 1838). I, 152–252; II, 2–45. Homer M. Byington to Walter Ferree, Apr. 11, 1972, Walter Ferree and George Franz, eds., *Van Buren Papers,* (National Historical Publications & Records Com.), Van Buren to Kemble, June 13; to John A. C. Gray, June 16, 1854, VBLC.

3. John C. Fitzpatrick, ed., *The Autobiography of Martin Van Buren,* Annual Report of the American Historical Association, 1918, 2 vols. (Washington, D.C., 1920), II, 7, 9.

4. Robert Johanssen, *Stephen A. Douglas* (New York, 1973), 143–45, 387; Van Buren, *Autobiography,* 8.

5. Van Buren, *Autobiography,* 9, 10; Pierre M. Irving, *The Life and Letters of Washington Irving,* 2 vols. (New York, 1869), I, 159.

6. John W. Barber and Henry Howe, *Historical Collections of the State of New York* (New York, 1841), 119; Franklin Ellis, *History of Columbia County New York* (Philadelphia, 1878), 223; Denis T. Lynch, *An Epoch and a Man: Martin Van Buren and His Times* (New York, 1929), 28; Edward A. Collier, *A History of Old Kinderhook* (New York, 1914), 386, 414.

7. Van Buren, *Autobiography,* 11, 13; Harriet A. Weed and Thurlow Weed Barnes, *The Life of Thurlow Weed,* 2 vols. (Boston, 1883), I, 162; Lynch, 57–59; Ellis, 74, 82, 221.

8. Van Buren, *Autobiography,* 14, 15, 19.

9. *Ibid.,* 14; John P. Van Ness to Van Buren, Jan. 6, 1803, VBLC. The date is wrong in the calendar of the Van Buren papers. See Elizabeth H. West, comp., *Calendar of the Papers of Martin Van Buren* (Washington, D.C., 1910), 12. The *Autobiography* is confusing on Van Buren's status. At one point he alludes to his being a clerk, at another he declares that he had already practiced law in Kinderhook. But the correspondence between Van Ness and Van Buren clearly indicates that the latter was a clerk during his stay in the city. See Van Ness to Van Buren, Jan. 6, 1802; May 26, Nov. 3, 1803, VBLC.

10. John P. Van Ness to Van Buren, Jan. 6, 1802, VBLC; only once in May of 1803 did he ask his brother for a loan; Van Buren to Van Alen,

May 26, 1803; Van Ness to Van Buren, Nov. 3, 1803, VBLC. In his *Autobiography,* Van Buren declares that except for a $40 loan from Van Ness which he promptly repaid, he had received nothing more. But see Van Ness's note for $100 payment at six months or one year. Van Buren, *Autobiography,* 14, 15; Van Ness to Van Buren, Nov. 3, 1803, VBLC.

11. Van Ness to Van Buren, Jan. 6, 1807, VBLC; Van Buren, *Autobiography,* 14; also, Van Buren to Smith T. Van Buren, Sept. 5, 1835, VBLC.

12. "Minutes of the Meetings of the Tammany Society or Columbian Order," Feb. 1, Mar. 4, May 11, 1801. NYPL.

13. See, for instance, Alfred Connable and Edward Silberfarb, *Tiger of Tammany* (New York, 1967); M. R. Werner, *Tammany Hall* (New York, 1928), for readable accounts of the early years of Tammany Hall.

14. Merrill Peterson, *Thomas Jefferson* (New York, 1970), 644–51; James Parton, *Life and Times of Aaron Burr,* 2 vols. (New York, 1857), II, Ch. XVI, *passim;* Jabez Hammond, *The History of the Political Parties of the State of New York,* 3 vols. (Syracuse, 1852), I, 134–44; Joseph Rayback, "Presidential Ambitions of Aaron Burr," in Hans Trefousse, ed., *Toward a New View of America: Essays in Honor of Arthur C. Cole* (New York, 1977), 54; Howard Lee McBain, *DeWitt Clinton and the Origin of the Spoils System in New York* (New York, 1907), 126–58.

15. Hammond, *New York,* I, 164–73; Peterson, *Jefferson,* 670; Van Ness to W. P. Van Ness, April 2, June 9, 1802, VBNYPL; Dumas Malone, *Jefferson and His Time,* 6 vols. (Boston, 1948–1981), IV, 396, 397.

16. Clinton's council of appointment did not remove all Federalists at this session; nor did it exclude all Burrites from office. The best patronage plums, however, went to the Clinton-Livingston faction. See Noble E. Cunningham, Jr., *The Jeffersonian Republicans in Power* (Chapel Hill, 1963), 43.

17. Van Buren, *Autobiography,* 15; Hammond, *New York,* II, 187–90; De Alva Stanwood Alexander, *A Political History of New York,* 3 vols. (New York, 1909), I, 122–27; see also James Cheetham, "A View of Aaron Burr's Political Conduct;" "Nine Letters on the Subject of Burr's Defections" (New York, 1803); William P. Van Ness, "Letters of Aristides" (New York, 1803), *passim;* Van Buren, *Autobiography,* 15.

18. Lynch, 73.

19. William Kent, ed., *Memoirs and Letters of James Kent L.L.D.* (Boston, 1898), 34, 35.

20. Van Buren to Van Ness, Mar. 13, 1804, VBLC.

21. Van Buren, *Autobiography,* 16.

22. *Ibid.,* 15–19.

23. In 1807, Van Buren became a counselor-at-law, which broadened his prospects. Despite the Van Ness and Van Schaack influence, the firm of Van Alen and Van Buren cleared $3,322 for the year 1804–05. Over $2,500 of their income for the year was in notes outstanding or unfinished business. Statement of Van Alen and Van Buren, Mar. 19, 1805, VBLC. Edward M. Shepard, *Martin Van Buren* (Rev. ed.). (Cambridge, 1899), 20. Ellis, *Columbia County,* 38–43; Van Buren, *Autobiography,* 22–24; Van Buren to Manor Committee, Apr. 28, July 28, 1811, VBLC.

24. Van Buren, *Autobiography,* 21, 22; James A. Hamilton, *Reminiscences of*

James A. Hamilton (New York, 1869), 42; George H. Humphrey, "Changes in the Practice of Law in Rochester," Ch. I, *The Rochester Historical Society Publication Fund,* Series IV (Rochester, 1925), 207.

25. Van Buren, *Autobiography,* 22, 25; Van Buren to Francis Stebbins, Aug. 19, 1811; to Manor Committee, Apr. 25, 1811, VBLC; Van Buren to Clinton, Apr. 15, 30, 1808; Apr. 9, 1810, Clinton Papers, Columbia Univ.

26. De Witt Clinton to Van Buren, May 29, 1806, VBMHS; Peter Irving to Van Ness, Oct. 24, 1808, VNNYPL; Van Buren to John E. Wool, May 21, 1811, VBLC.

27. Peter Irving to W. P. Van Ness, n.d., 1805, VNNYPL; Hammond, *New York,* I, 236–38.

28. Van Buren, *Autobiography,* 28; Elisha Jenkins to Van Buren, April 22, 1807, VBLC.

29. Hammond, *New York,* I, 262; Edward M. Shepard, *Martin Van Buren* (Boston, rev. ed., 1899), 45–46; Lynch *Van Buren,* 75.

Chapter 2. State Senator

1. *Albany Argus,* Apr. 19, 1830; Shepard, *Van Buren,* 21, 22; Lynch, *Van Buren,* 57, 164, 165; Harriet Allen Butler, MS. "My Father and Mother, Brothers and Sisters," B. F. Butler Papers, Kinderhook Historical Society; Laura C. Holloway, *The Ladies of the White House* (Philadelphia, 1884), 138, 140, 143, 317.

2. Ellis, *Columbia County,* 68, 155.

3. *Ibid.,* 195, 207; Collier, *Old Kinderhook,* 416; Barker and Howe, *Historical Collections,* 115–17.

4. Peter Irving to W. P. Van Ness, Oct. 24, 1808, VNNYPL; Alexander, *New York,* I, 168–69.

5. Dixon Ryan Fox, *The Decline of the Aristocracy in the Politics of New York* (New York, 1965), 41, 42; Hamilton, *Reminiscences,* 42; Adrian Hoffman Joline, *The Autograph Hunter and Other Papers* (Chicago, 1907), 66–78; Alexander, *New York,* I, 191.

6. Van Buren, *Autobiography,* 28–32; J. Van Ness to Van Buren, Apr. 12, 1812, VNNYPL J. Van Ness to Van Buren April 12, 1812, VBLC; Ellis, *Columbia County,* 154–55; Van Buren, *Autobiography,* 29–31.

7. *Ibid.,* 31, 32; W. P. Van Ness to Van Buren, Mar. 13, 1812, VBLC.

8. George W. Broom to Van Buren, Apr. 17, 1812, VBLC; J. Van Ness to Van Buren, Apr. 12, 1812, VBLC.

9. Van Buren, *Autobiography,* 32–35.

10. *Ibid.,* 38, 39; Richard Riker to Van Buren, July 21, 1812, VBLC.

11. Gorham A. Worth, *Random Recollections of Albany,* 3rd. ed. (Albany, 1866), 60; see also, William James Morgan, et al., eds., *Autobiography of Rear Admiral Charles Wilkes* (U.S. Navy, Washington, D.C., 1978), 215; C. R. Rosebury, *Capitol Story* (New York, 1964), 14–16.

12. Alexander, *New York,* I, 196; Van Buren, *Autobiography,* 40, 41.

13. Governor Tompkins was the deciding factor. He thought Van Buren too young and inexperienced to prosecute Solomon Southwick and David

Thomas, the chief lobbyists for the Bank of America; Van Buren, *Autobiography,* 37, 38, 39, 96.

14. Charles R. King, ed., *The Life and Correspondence of Rufus King,* 6 vols. (New York, 1899), V, 265–69. Hereafter cited as *King Correspondence;* Alexander, *New York,* I, 197; Hammond, *New York,* I, 314–16.

15. *King Correspondence,* V, 270.

16. *Ibid.,* 276–80; 288.

17. *Ibid.,* 291; Van Buren, *Autobiography,* 45.

18. The Federalist bank projectors of the Bank of America, in order to make their bank palatable to the people, to combat rival banking interests and to protect their supporters in the legislature, had agreed to pay the state an unprecedented sum for a bonus. Its charter included provisions binding it to pay $400,000 into the state school fund, $100,000 for the literature fund (state library) and an additional $100,000 if no other bank were chartered for the next 20 years. If this were not deemed ample, the projectors agreed to lend New York $1,000,000 at 5 percent interest for the building of canals (a project close to the heart of De Witt Clinton) and finally another $1,000,000 at 6 percent interest to the farmers for capital improvement; Alexander, *New York,* I, 191.

19. Charles Holt to Van Buren, Jan. 20, 1813, VBLC; *New York Evening Post,* Feb. 5, 1813; Van Buren, *Autobiography,* 45.

20. Charles Holt to Van Buren; Jan. 20, 1813, VBLC; Van Buren, *Autobiography,* 46; *King Correspondence,* V, 290; Richard Riker to Van Buren, Feb. 15, 1813, VBLC.

21. Van Buren, *Autobiography,* 46, 47.

22. Weed and Barnes, eds., *Thurlow Weed,* I, 43, 44, Alexander, *New York,* I, 193; Van Buren is not listed among original supporters such as John Taylor, Benjamin Knower, both violent anti-Clintonians; for evidence that Van Buren's support had a solid foundation, see *King Correspondence,* V, 291.

23. *Albany Argus,* Feb. 5, 1813; Van Buren, *Autobiography,* 44–55; Hammond, *New York,* I, 354, 355.

Chapter 3. War to the End

1. Van Buren, "To the Electors of the State of New York," Feb. 1814, VBLC.

2. *Ibid., Albany Argus,* Apr. 15, 1814.

3. Morris had asked in a savagely bitter letter, "Must we wait till the claws of human tigers rake our stinking bowels to look for a Heart?" Gouverneur Morris to Rufus King, Mar. 23, 1814, KNYHS.

4. Lynch, *Van Buren,* 115; Winfield Scott to Van Buren, Oct. 28, 1861, VBLC; Charles Elliott, *Winfield Scott, the Soldier and the Man* (New York, 1937), 143–44; see Van Buren, "Points of Evidence," Jan. 3, 1814; to A. J. Dallas, Jan. 5; to John Armstrong, Jan. 6; to John Smith, Jan. 6, 1814; Van Buren, "Charges and Specifications against William Hull," Mar. 23, 1814, VBLC.

5. Hammond, *New York,* I, 377, 378; Alexander, *New York,* I, 219.

6. Van Buren, *Autobiography,* 66, 67; Winfield Scott, *Memoirs of Lieutenant General Scott LL.D. Written by Himself,* 2 vols. (repr. New York, 1970), I, 49–66, 115.

7. William M. Holland, *The Life and Political Opinions of Martin Van Buren President of the United States* (Hartford, 1835), 103; Hammond, *New York,* I, 369–72; Van Buren, "Notes and Resolutions at the Capital in Albany April 14, 1814," VBLC; *Autobiography,* 49–52.

8. Taylor to Van Buren, Oct. 6, 1814, VBLC.

9. Hammond, *New York,* I, 377, 378; *King Correspondence,* V, 410, 410n.

10. Scott to Van Buren, Oct. 22, 1814, VBLC; *Albany Argus,* Sept. 3, 1814.

11. Van Buren, *Autobiography,* 55, 56; Burr to Van Buren, Oct. 5, 1814, VBLC.

12. An old adversary, William Stebbins, editor of the *Northern Whig,* filled the editorial columns with his violent attacks. The *Whig* advised Van Buren's acquaintances and relatives in Kinderhook and Hudson to break off all personal relations with such a political despot. Hudson *Northern Whig,* Oct. 1–14, 1814; Van Buren, *Autobiography,* 56.

13. *King Correspondence,* V, 426; *New York Evening Post,* Oct. 20, 1814.

14. Hammond, *New York,* I, 390.

15. Van Buren, *Autobiography,* 66–69, 71; Hammond, *New York,* I, 393; Nathan Sanford to Van Buren, Dec. 28, 1814, VBLC; see also Sanford to John W. Taylor, Feb. 28, 1815, TNYHS; Worth, *Albany,* 61, 392; Hammond, I, 392; *Albany Argus,* Feb. 24, 1815.

16. Dixon Ryan Fox, *Yankees and Yorkers* (New York, 1940), 152, 1819; Fox, *Decline of the Aristocracy,* 152; William Chazanoff, *Joseph Ellicott and the Holland Land Company* (New York, 1970), 116, 117; Butler to Van Buren. Dec. 2, 1854, VBLC; Harriet Allen Butler, MS. "My Father and Mother."

17. *Albany Argus,* Mar. 24, 1815; Fox, *Decline of the Aristocracy,* 154; Jones Fisk to John W. Taylor, Feb. 20, 1816, TNYHS.

18. In one of these skirmishes, Peter A. Jay, a Federalist leader, gained the floor while Van Buren was talking with a Republican assemblyman. Sarcastically, Jay alluded to their conversation by pointing to the assemblyman, but not acknowledging him by name or by the county he represented. In the course of his remarks, when the Speaker called him to order for a breach of decorum, Jay said, "I mean the gentleman who always speaks with the Attorney General at his elbow!"; Van Buren, *Autobiography,* 73.

19. Hammond, *New York,* I, 405, 411, 411n, 412; *King Correspondence,* V, 501; *Albany Argus,* Jan. 31, 1816.

20. Joseph G. Rayback, "A Myth Reexamined: Martin Van Buren's Role in the Presidential Election of 1816," Proceedings of the American Philosophical Society (Apr., 1980), Vol. 124, No. 2, 109–10.

21. Hammond, *New York,* I, 411. Hammond apparently asked Van Buren's opinion in a group situation or possibly with the New York delegation present because later he tried twice to have a private conversation with Van Buren about the succession and other topics, but did not find him in. Hammond to Van Buren, Jan. 23, 1816, VBLC.

22. Hammond to Van Buren, Jan. 23; Van Buren to Bibb, Jan. 29, 1816, VBLC.

23. Van Buren to Bibb, Jan. 23, 1816; Bibb to Van Buren, Feb. 5, 1816, VBLC.

24. Samuel R. Betts to Van Buren, Jan. 19; Rayback, "Myth Reexamined," 112; Van Buren to Bibb, Feb. 5, 1816, VBLC. Hammond, *New York,* I, 407–10.

25. Betts to Van Buren, Feb. 24, 1816, VBLC; *Albany Argus,* Mar. 5, 14, 1816; Taylor to Miss Jane Taylor, Mar. 17, 1816, TNYHS.

26. *King Correspondence,* V, 520; Nathan Sanford to Van Buren, Mar. 14; Betts to Van Buren, Mar. 16, 1816, VBLC. In his elegantly researched eminently persuasive article on Van Buren's role cited above, Professor Joseph Rayback has corrected many errors and idiosyncratic interpretations that have grown up about the campaign and election of 1816; Birdseye to Taylor, n.d., 1816, TNYHS.

27. King to Walcott, Feb. 24, 1816, KNYHS.

28. *Niles Weekly Register,* Feb. 17, 1816; *King Correspondence,* V, 521; Ambrose Spencer to John C. Spencer, Nov. 8, 1816, Ambrose Spencer Papers, NYSL.

29. Marcy to John Bailey, n.d., 1816, Alexander Washburn Coll., MHS; Van Buren, *Autobiography,* 73–77.

30. Spencer to John C. Spencer, Nov. 8, 1816, Ambrose Spencer Papers, NYSL.

31. James Emmott to King, Feb. 2, 1817, KNYHS; John Townsend to Taylor, Feb. 5, 1817, TNYHS.

32. John Townsend to Taylor, Feb. 12, 1817, TNYHS.

33. John P. Van Ness to W. P. Van Ness, Jan. 30, 1817, Van Ness Papers, NYHS; *Albany Argus,* Mar. 28, 1817.

34. Porter to Van Buren, Feb. 10, 1817, VBLC; Elisha Jenkins to John W. Taylor, Feb. 10, 1817; TNYHS Porter to Van Buren and Cantine, Feb. 13, 1817; Robert Swartwout to Van Buren, Feb. 26, Mar. 17, 1817, VBLC; Enos Throop, a Congressman from western New York, thought that Van Buren should not push the Porter candidacy. Everyone in his area was for Clinton, and moreover he said that Porter was a man of loose morals and dissolute habits; Enos Throop to Van Buren, Mar. 17, 1817, VBLC; Van Buren, *Autobiography,* 77–83.

Chapter 4. Mutiny in the Ranks

1. Butler to Harriet Allen, Jan. 20, Feb. 14, 1817, BNYSL; Ruggles Hubbard to Van Buren, Jan. 3, 1815; Samuel R. Betts to Van Buren, Feb. 24, 1817; Sanford to Van Buren, Jan. 23, 1819, VBLC; *Albany Argus,* May 3, 1818.

2. Van Buren, *Autobiography,* 89; Clinton to Post, Mar. 29, 1817, Apr. 6, 1817, PNYSL.

3. Elisha Jenkins to Jacob Brown, Feb. 16, 1818, Brown Papers, MHS; Hammond, *New York,* I, 449–51.

4. Van Buren to Gorham A. Worth, Apr. 27, 1818, VBLC.

5. Van Buren, *Autobiography,* 84, 85.

6. Hammond, *New York,* I, 451–53; Stewart Mitchell, *Horatio Seymour of New York* (Cambridge, 1938), 4–10; Van Buren, *Autobiography*, 144, 145; Frank Otto Gatell, "Sober Second Thoughts on Van Buren, the Albany Regency and the Wall Street Conspiracy," *Journal of American History* (June, 1966), Vol. 53, 29.

7. Smith Thompson to Van Buren, Aug. 30, 1819; Van Buren to D. E. Evans, Aug. 23, 1819, VBLC.

8. Hammond, *New York,* I, 447–60.

9. Clinton to Post, Apr. 6, 1819, PNYSL.

10. Hammond, *New York,* I, 475–76.

11. *Ibid.,* 478–82; *King Correspondence,* V, 102.

12. *Albany Argus,* Jan. 4–8, 1818; *National Advocate,* Jan. 5, 6, 1819.

13. *Albany Argus,* Feb. 5, 1819; *King Correspondence,* V, 199, 202, 203.

14. Van Buren to Gorham A. Worth, Nov. 26, 1818; Butler to Harriet Allen, Mar. 29, 1817; Feb. 18, 1818, BNYSL. Van Buren to Butler, Feb. 11, 1819, VBLC.

15. *King Correspondence,* V, 201; Hammond, *New York,* I, 507.

16. Van Buren to Worth, Apr. 27, 1819, VBLC.

17. Chazanof, *Ellicott,* 147–49.

18. Hammond, *New York,* I, 495, 496; Van Buren to D. E. Evans, May 15, 1819, VBLC; *Autobiography,* 91.

19. Smith Thompson to Van Buren, Nov. 30, 1818; Feb. 8, 1819; Nathan Sanford to Van Buren, Nov. 4, 1818, VBLC; Elisha Jenkins to Jacob Brown, Dec. 12, 1818, Brown Papers, MHS.

20. Van Buren to Worth, Nov. 26, 1818, VBLC; Clinton to Henry Post, Dec. 2, 1820, PNYSL; Hammond, *New York,* I, 497–98. Van Buren to James Tallmadge, Jr., Dec. 3, 1819, on circular signed by all the Bucktail Senators, Brock Coll., HL.

Chapter 5. The High-Minded Gentlemen

1. Marcy to Bailey, Dec. 13, 1818, Washburn-Bailey Papers, MHS; Henry Ammon, *James Monroe, The Quest for National Identity* (New York, 1971), 364; Thompson to Van Buren, Nov. 28; Dec. 3, 25, 1818; Jan. 23, 1819; Aug. 3, 1819; Dec. 5, 1819, VBLC; for Clinton's ambitions, see Clinton to Post, Aug. 12, 1819, PNYSL.

2. Ammon, *Monroe,* 365; Hammond, *New York,* I, 501–21.

3. Van Buren, *Autobiography,* 94–97; Tompkins to McIntyre, Aug. 6, 1819, VBLC; for details on the controversy and its final settlement in favor of Tompkins, see Hammond, *New York,* I, 501–21; *Albany Argus,* Aug. 3; Aug. 5, Nov. 9; Dec. 31, 1819.

4. W. P. Van Ness to Monroe, Aug. 16, 1819, Brock Coll. HL; Samuel R. Betts to Van Buren, Jan. 31, 1820; Van Buren to Tompkins, Jan. 17, 1820, VBLC; Van Buren, *Autobiography,* 94–96; *King Correspondence,* VI, 254; King to Van Buren, Jan. 31, 1820; Thompson to Van Buren, Jan. 28, 1820, VBLC; Van Buren to King, Feb. 2, 1820, VBNYHS; Ray W. Irwin, *Daniel D. Tompkins* (New York, 1968), 240–44.

5. Johnston Verplanck to Van Buren, Dec. 25, 1819, VBLC; W. C. Bryant, *A Discourse on the Life, Character and Writings of Gulian Crommelin Verplanck* (New York, 1870), 18; Fox, *Decline of the Aristocracy,* 213–15; John W. Taylor to Joel Keeler, Jan. 16, 1819, TNYHS; Thompson to Van Buren, Jan. 22, 1819; Nathan Sanford to Van Buren, Jan. 23, 1819, VBLC.

6. Van Buren to D. E. Evans, Nov. 17, 1819, VBLC; M. M. Noah to Van Buren, July 13, 1819, VBLC; James Hamilton to Van Buren, Dec. 31, 1818, VBLC. But he had to work carefully and skillfully if he were to convince the Bucktails of this notion; M. M. Noah to Van Buren, Dec. 19, 1819, VBLC; Mrs. Catharina V. R. Bonney, *A Legacy of Historical Gleanings* (Albany, 1874), 2 vols., I, 341; Van Buren to Noah, Dec. 17, 1819, VBLC.

7. Van Buren, "Fragment," Jan. 4, 1820, VBLC; *King Correspondence,* VI, 322, 323.

8. Henry F. Jones to Van Buren, Jan. 19, 1820; Van Buren to Jones, Jan. 21, 1820, VBLC. Van Buren wrote *Leonidas* during January 1820; see printed version in *Albany Argus,* Jan. 18, Jan. 25, 1820; B. F. Bartley to Van Buren, Oct. 8, 1820, VBMHS; Butler to My Dearest Wife, Mar. 16, 1820, BNYSL; Glover Moore, *The Missouri Controversy, 1819–21* (Lexington, Ky., 1966), 170, 178, 185; John W. Edmonds to Abraham Van Buren, Jan. 24; Van Buren to Tompkins, Jan. 17, 1820, in the hand of John W. Edmonds, VBLC; *King Correspondence,* VI, 322, 326; for the legislative debates and resolutions, see *Albany Argus,* Jan. 21, 1820.

9. Marcy to John Bailey, July 8, 1816; Aug. 3, 1817; Dec. 13, 1818; Washburne-Bailey Papers, MHS; Ivor D. Spencer, *The Victor and the Spoils* (Providence, 1959), 27–28.

10. *King Correspondence,* V, 192; VI, 216; *Albany Argus,* Dec. 14, 1819; Van Buren to Butler, Dec. 29, 1819; *King Correspondence,* VI, 245, 246, 248, 317; James A. Hamilton to Van Buren, Jan. 18, 1820; Van Buren to King, Feb. 9, 1820; to Evans, Dec. 9, 1819, VBLC; Van Buren to King, Feb. 9, 1820, KNYHS.

11. For the text of the "Address" see the *New York Evening Post,* Apr. 14, 1820; Van Buren, *Autobiography,* 96; Hammond, *New York,* I, 531, 532.

12. Van Buren to Noah, Dec. 17, 1819; Noah to Van Buren, Dec. 19, 1819; Van Buren to Isaac Q. Leake, May 28; Peter J. Hoes to Van Buren, June 29; "Memorandum," M. Van Buren, P. J. Hoes & Jesse Buel, attest I. Vanderpoel, July 28, 1820; to Leake, Aug. 1, May 28, 1820, VBLC. None of Van Buren's associates objected to making Cantine the editor, though he was a lawyer without any newspaper experience. For the business side of the firm, they hit upon a former banker, innocent of the printer's trade, Isaac Q. Leake.

13. Hammond, *New York,* I, 536; Van Buren to Charles E. Dudley, May 12, 1820, VBLC; William L. MacKenzie. *The Life and Times of Martin Van Buren* (Boston, 1846), 154; *King Correspondence,* VI, 331.

14. Butler to My Dear Harriet, June 2, 1820, BNYSL; Van Buren to Charles E. Dudley, May 10, 1820, VBLC.

15. Alexander, *New York,* I, 255; Clinton to Henry Post, Nov. 30; Nov. 27; Dec. 2, 1820, PNYSL.

16. Hammond, *New York,* I, 521.

17. Van Buren, *Autobiography,* 67–68. Why Woodworth acted as he did was the subject of much speculation at the time and for many years afterward. Clintonians freely charged that Van Buren had persuaded him to act as he had, an accusation that seems most unlikely. Van Buren actually disliked Woodworth, considering him a pliable man whose principles were erratic to say the least, and Woodworth had nothing to gain nor Van Buren anything to give for a special consideration. In all probability Woodworth differed from his Clintonian colleagues on a matter of detail, even though the assembly bill was both proper and constitutional. Hammond, I, *New York,* 546–49.

18. *King Correspondence,* VI, 355.

19. Clinton to Henry Post, Nov. 17, 19, 20, 25, 27, 1820, PNYSL.

Chapter 6. The Tammany Horse

1. *King Correspondence,* VI, 381; Hammond, *New York,* I, 553.

2. *Niles Weekly Register,* Dec. 2, 1820; *Albany Argus,* Nov. 20, 21, 22, 28, 1820; Clinton to Post, Nov. 25, 1820, PNYSL.

3. *Ibid.,* Nov. 17, 30, 1820.

4. MacKenzie, *Van Buren,* 165; B. F. Butler to My Dear Harriet, June 24, 1820; to Dearest Wife, June 28, 1820, BNYSL; Butler to Mrs. Butler, June 28, 1820, VBLC; Van Buren to King, Nov. 19, 1820, KNYHS.

5. Solomon Van Rensselaer to Post, Jan. 28, 1821, PNYSL; Rufus King to Van Buren, Dec. 1, 1820; Smith Thompson to Van Buren, Dec. 18, 1820, VBLC; Clinton to Post, Dec. 2, 1820, PNYSL; M. M. Noah to Van Buren, Dec. 29, 1820, VBLC; John W. Taylor to Mrs. Taylor, Dec. 24, 1820, TNYHS.

6. This letter in manuscript bears extensive editing. Obviously, Van Buren wanted to be clearly understood, but not offend either King or Van Rensselaer; Van Buren to King, Jan. 14, 1821, VBLC; *King Correspondence,* VI, 375–77; Bonney, *Legacy,* I, 360–62.

7. Bonney, *Legacy,* I, 362; MacKenzie, *Van Buren,* 167.

8. Bonney, *Legacy,* I, 303; Hammond, *New York,* I, 565–69.

9. MacKenzie, *Van Buren,* 167.

10. Bonney, *Legacy,* I, 354, 357, 358; Clinton to Post, Nov. 2; Dec., 1820. PNYSL. *Albany Argus,* Jan. 18, 26, 1821.

11. Bonney, Legacy, I, 361; Monroe to Van Buren, Jan. 7, 1821, VBLC; *King Correspondence,* VI, 384; Hammond, *New York,* I, 558; *Albany Argus,* Mar. 16, 1821.

12. Van Buren to King, Feb. 2, 1821, VBNYHS; Lynch, *Epoch and a Man,* 207; Hammond, *New York,* I, 562.

13. *King Correspondence,* VI, 384; Bonney, *Legacy,* I, 363; Taylor to Bartow White, Feb. 10, 1821, TNYHS.

14. King to Van Buren, Feb. 18, 1821, VBLC; see also King to John A. King, Feb. 21, 1821, *KNYHS; Van Buren to King, Jan. 14, 1821, VBNYHS; King Correspondence,* VI, 265.

15. Charles King to Rufus King, Dec. 19, 1820, KNYHS.

16. *King Correspondence,* VI, 413; Van Buren *Autobiography,* 105, 106. In accepting the nomination Van Buren said that he had declined many offers "because they came from political friends . . . ," Van Buren to Elisha Foot, June 16, 1821, VBLC.

17. *King Correspondence,* VI, 414.

18. Nathaniel H. Carter and William Stone, *Reports of the Proceedings and Debates of the Convention of 1821, assembled for the Purpose of Amending the Constitution of the State of New York* (Albany, 1821), 1, hereafter referred to as *Reports.*

19. Van Buren, *Autobiography,* 106, 107; *Reports,* 299.

20. *Ibid.,* 321; Uhlshoeffer to Van Buren, Sept. 21, 1821, VBLC.

21. Beardsley to Collins, Sept. 20, 1821; Van Buren to John A. King, Oct. 28, 1821, VBLC; Rufus King to Charles King, Oct. 19, 1821; to Dr. Sir, Oct. 21, 1821, KNYHS; King to John A. King, Oct. 4, 1821, KNYHS; *Reports,* 381.

22. *Ibid.,* 263.

23. *Ibid.,* 147.

24. Shepard, *Van Buren,* 78–87; see also Hammond, *New York,* II, 1–85 for a concise and reasonably objective reporting of the convention. For the animus against judges that permeated the Bucktail majority, see the speech of P. R. Livingston, *Reports,* 618–20.

25. Van Buren to Cornelius Miller, Nov. 20, 1821, VBLC; Rufus King to Charles King, Oct. 15, 1821, KNYHS; *King Correspondence,* VI, 422; King to Rufus King, Oct. 26, 1821, KNYHS.

26. Hammond, *New York,* II, 94.

Chapter 7. An Uncertain Regency

1. *King Correspondence,* VI, 452.

2. MacKenzie, *Van Buren,* 184.

3. Van Buren to C. E. Dudley, Dec. 15, 1821, VBLC.

4. Glover Moore, *The Missouri Controversy,* 139–44; John Munroe, *Louis McLane, Federalist and Jacksonian* (New Brunswick, 1973), 120, 121; Everett S. Brown, ed., *The Missouri Compromise and Presidential Politics from the Letters of William Plumer Jr.* (St. Louis, 1926), 25, 26, 33, 65; see Van Buren to Taylor, Nov. 25; Dec. 20, 1813; Feb. 15; Mar. 11; Apr. 1; Oct. 31, 1814, VBNYHS; John W. Taylor to M. Harris, Dec. 11, 1840, TNYHS; *King Correspondence,* VI, 356; Taylor to H. G. Spafford, Jan. 19, 1821, TNYHS; Charles Francis Adams, ed., *Memoirs of John Quincy Adams,* 12 vols. (Philadelphia, 1876), V, 439. Hereafter cited as *Adams Memoirs.*

5. Albert H. Tracy to Taylor, Sept. 21, 1821, TNYHS; *Adams Memoirs,* V, 431; Edward K. Spann, "John W. Taylor: The Reluctant Partisan" (unpub. Diss.), New York Univ., 1957, 168, 266–67.

6. Van Buren, *Autobiography,* 513.

7. See correspondence of Micah Sterling in Hemphill, *et al.,* eds., *Calhoun Papers* (Columbia, S.C., 1959–80), 13 vols., IV, 316, 317, 596, 597. Hereafter cited as *Calhoun Papers.*

8. Van Buren to Yates, Nov. 6, 1821, VBLC; Taylor to Richard Taylor, Dec. 12, 1821, TNYHS.

9. Munroe, *McLane,* 121; Adams, *Memoirs,* V, 437.

10. Plumer, 65–69; Charles Henry Ambler, *Thomas Ritchie, a Study in Virginia Politics* (Richmond, 1913), 79–81; *Adams Memoirs,* V, 523, 24.

11. Hammond, *New York,* II, 94; 75,422 votes for the revised constitution to 33,925 opposed.

12. Van Buren to Dudley, Dec. 15, 1821; Dudley to Van Buren, Dec. 21, 1821, VBLC.

13. Barnes and Weed, eds., *Weed,* II, 36; the *Monday Republican* lasted 18 months. Vol. I, No. 1 was published June 21, 1821, while the last issue came off the press in November 1822. Weed had worked as a printer for the *Argus* when Jesse Buel edited it in 1816. In 1821 the *Argus* again employed him, but Leake heard that Weed had worked for a Clintonian paper in Chenango and promptly discharged him, despite the fact that he was an excellent printer. Weed thought of appealing to Cantine, who in all likelihood would have reinstated him. But he did not, and went on to carve out an independent career of journalism and politics; *ibid.,* I, 85.

14. MacKenzie, *Van Buren,* 185.

15. Van Buren to Yates, Nov. 6, 1821, VBLC; Hammond, *New York,* II, 99; Bonney, *Legacy,* I, 373.

16. Van Buren, *Autobiography,* 125, 126; Bonney, *Legacy,* I, 364, 370–71, 374, 375. Van Buren's letter to Knower was written on Jan. 5, 1822, and is misdated in Bonney, *Legacy,* I, 374.

17. *Adams Memoirs,* V, 480. King to Charles King, Dec. 28, 1821, KNYHS.

18. *Ibid., Adams Memoirs,* V, 479, 480.

19. Bonney, *Legacy,* I, 373–75; King to Van Buren to Meigs, Jan. 3; Meigs to King and Van Buren, Jan. 4; Tompkins, King and Van Buren to Meigs, Jan. 4; Meigs to Tompkins, King and Van Buren, Jan. 4, 1822; Van Buren to Monroe, Jan. 5, 1822; M. J. Cantine to Van Buren, Jan. 6, 1822, VBLC.

20. *Adams Memoirs,* V, 581, 582; *King Correspondence,* VI, 441.

21. Bonney, *Legacy,* I, 377; Meigs to Tompkins and Van Buren, Jan. 7; Monroe to Van Buren, Jan. 7, 1822, VBLC.

22. Tompkins and Van Buren to Meigs, Jan. 7, 1822, MS. draft in Van Buren's hand; Michael Ulshoeffer to Van Buren, Jan. 13, 1822; Charles E. Dudley *et al.,* to Tompkins and Van Buren, Jan. 22, 1822, VBLC; *Albany Argus,* Jan. 12, 1822. On Jan. 24, 1822 the *Albany Register* gave the number of the New York Congressmen who had signed Van Rensselaer's recommendation; *Albany Register,* Jan. 24, 1822; *Albany Argus,* Feb. 1, 1822; Van Buren, *Autobiography,* 126; Ulshoeffer to Van Buren, Jan. 27, 1822, VBLC.

23. *King Correspondence,* VI, 456; Walter Patterson wrote Solomon Van Rensselaer on Jan. 20, 1822, "the delegation are much irritated at Martin Van Buren and the Vice President. Cambreleng has written a letter in defence of their conduct which will be published if more is said on the subject"; Bonney, *Legacy,* I, 387; Van Buren, *Autobiography,* 126.

24. *King Correspondence,* VI, 472; Van Buren to Dudley, Jan. 10, 1822, VBLC.

25. Ulshoeffer to Van Buren, Feb. 2, 1822; Knower to Van Buren, Mar. 4,

1822, VBLC; Van Buren to King, Feb. 18, 1822, KNYHS; Bonney, *Legacy,* I, 407; John A. King to King, Jan. 4, 1822, KNYHS.

26. *Adams Memoirs,* V, 431, 436, 451; Bonney, *Legacy,* VI, 402.

27. Van Buren to Gorham Worth, Feb. 15, 1822; Ulshoeffer to Van Buren, Feb. 17, 1822, VBLC.

28. *King Correspondence,* VI, 458.

29. Van Buren, *Autobiography,* 128, 129.

30. The account would not be settled until May 1824. Tompkins received from the Treasury $97,229 in two installments. See Public Acts, *Annals of Congress,* 18th Cong., 1st Sess.; 19th Cong., 1st Sess. See estimate of Gaillard in Thomas Hart Benton, *Thirty Years View,* 2 vols. (New York, 1897), I, 77; Rufus King to John A. King, Feb. 1822, KNYHS.

31. *King Correspondence,* VI, 452, 53; Van Buren to Worth, Mar. 16, 1822, VBLC.

32. *King Correspondence,* VI, 518; Van Buren, *Autobiography,* 126, 127; Ambler, *Ritchie,* 14, 15, 85–87.

Chapter 8. Harp on the Willows

1. *King Correspondence,* VI, 478; Clinton to Henry Post, Aug. 3, 1822, PNYSL; see, for instance, M. M. Noah to Van Buren, Mar. 12, 1822, VBLC.

2. *King Correspondence,* VI, 478; MacKenzie, *Van Buren,* 185.

3. Talcott to Van Buren, Feb. 7, 1822; Ambrose Spencer to Van Buren, Mar. 28, 1822, VBLC.

4. Van Buren, *Autobiography,* 113; Hammond, *New York,* II, 100; Ulshoeffer to Van Buren, Feb. 17; Mar. 19; Apr. 2, 1822, VBLC; *King Correspondence,* VI, 467.

5. See a contrary opinion of De Witt Clinton, who declared that Yates was "a weak man," had many enemies and would be prostrated in six months. Clinton to Henry Post, Aug. 29, 1822, PNYSL; Van Buren to Dudley, Oct. 15, 1822, VBLC.

6. Alexander, *New York,* I, 325, 326; Hammond, *New York,* II, 101–05; Barnes and Weed, eds., *Weed,* I, 8, 9, 103, 399, 400; II, 4, 470.

7. MacKenzie, *Van Buren,* 188; Van Buren, *Autobiography,* 114.

8. Van Buren to Yates, Dec. 10, 1822; *Autobiography,* 114, 115; Skinner to Van Buren, Feb. 15, 1823, VBLC; *Albany Argus,* Jan. 26, 1823; *New York Evening Post,* Jan. 30, 1823.

9. *National Advocate,* Jan. 8, 1822; John A. King to King, Jan. 11, 1823, KNYHS; *King Correspondence,* VI, 493.

10. Later Betts was nominated as a circuit court judge and confirmed. See MacKenzie, *Van Buren,* 190n.; *King Correspondence,* VI, 471; Clinton to Post, Aug. 29, 1822, PNYSL; *King Correspondence,* VI, 458; Bonney, *Legacy,* I, 402; Porter to Henry Clay, Sept. 30, 1822, Porter Papers, Buffalo Hist. Soc. In fact, the explanation for Noah's eccentric behavior in the matter of the old Judges was that they were partisans of Crawford.

11. Van Buren to King, May 31, 1822, KNYHS; *King Correspondence,* VI, 448; Cantine to Van Buren, Dec. 2, 1822; Van Buren to Hoes, Jan. 30,

1823, VBLC; MacKenzie, *Van Buren,* 190, 191; *King Correspondence,* VI, 504; Skinner to Van Buren, Feb. 15, 1823, VBLC.

12. Noah had wanted a share of the state printing, but unwilling to endanger his relations with the Regency, bowed out with good grace. Milton W. Hamilton, *The Country Printer, New York State 1785–1830* (New York, 1936), 17–20, 185, 266; Hammond, *New York,* II, 121–23; Roger Skinner to Van Buren, Feb. 15, 1823, VBLC; MacKenzie, *Van Buren,* 192; James F. Hopkins, *et al., The Papers of Henry Clay* (Lexington, Ky., 1959–73), 5 vols., III, 372. Hereafter cited as *Clay Papers.*

13. Van Buren, *Autobiography,* 513; Clinton to Post, Dec. 30, 1822, PNYSL; *Calhoun Papers,* VI, 259, 362, 364. *Adams Memoirs,* VI, 42; John Belohlovek, *George M. Dallas* (University Park, Pa., 1977), 6, 19, 21; Robert V. Remini, *Martin Van Buren and the Making of the Democratic Party* (New York, 1959), 50; Clay Papers, III, 313, 314; Thomas S. Rogers to Virgil Maxcy, Jan. 13, 1823, Galloway-Maxcy-Markoe Papers, LC; *Calhoun Papers,* VI, 165, 347, 357, 597; VII, 492, 515, 516, 532; *Clay Papers,* III, 300.

14. Van Buren to Johnston Verplanck, Dec. 22, 1822, VBLC; *King Correspondence,* VI, 507, 508; William Gwynn to Colonel George Peter, Mar. 31, 1823, Galloway-Maxcy-Markoe Papers, LC; Lynch, *Martin Van Buren,* 239, 240; Clinton to Post, Jan. 8; Apr. 13, 1823, PNYSL; *King Correspondence,* VI, 489.

15. *Clay Papers,* III, 211.

16. *King Correspondence,* VI, 252, 290, 494, 594; Chase Mooney, *William H. Crawford* (Lexington, Ky., 1974), 228–30.

17. Van Buren made several trips to Richmond during the winter and spring of 1823; Remini, *Van Buren,* 38.

18. *King Correspondence,* VI, 510, 511; Donald M. Roper, "The Elite of the New York Bar as seen from the Bench, James Kent's Necrologies," *The New York Historical Society Quarterly,* Vol. LVI, No. 3 (July 1972), 234, 235.

19. Rufus King expressed well Van Buren's dilemma in a letter he wrote his son John before the Thompson interview. "I am aware," he said, "that the presidential question will produce in our state as well as others much excitement. . . . He, Verplanck, seems to think that the leaders have less power over the Party, whose views he may oppose, than formerly: whether he judges correctly I am unable to predict, but there are many topics on which they may yet insist with much confidence."; *King Correspondence,* VI, 495.

20. See, also, more direct efforts of Thompson in William H. Crawford to Van Buren, Aug. 1, 1823, VBLC; Jacob Sutherland to Thompson, Mar. 10, 1823, Smith Thompson Misc. Papers, NYHS; Van Buren had a conference with Crawford just before he left Washington. Rufus King thought that he had avowed his support at that time, but if so it was a carefully guarded secret between the two men and not in Van Buren's style to make binding a compact of a political nature before he had personally surveyed the situation at home. King, "Memorandum," Apr. 7, 1823, KNYHS.

21. Jonathan Cramer to John W. Taylor, Jan. 6, 1823, TNYHS; *Albany Argus,* Jan.–Feb., 1823.

22. *Calhoun Papers,* VIII, 44, 45. Thompson to Van Buren, Mar. 25, 1823;

on Apr. 4 Van Buren wrote Thompson calling his attention to their conversation in Washington and declaring his opinion had not changed; Van Buren to my Dear Sir, Apr. 4, 1823 (draft copy), VBLC. Van Buren to King, May 31, 1822, KNYHS; *King Correspondence,* VI, 512–14, 516, 517, 520–25.

23. Thompson to Van Buren, Mar. 17, 21, 25, 1823; King, "Memorandum," Apr. 7, 1823, KNYHS; Van Buren to King, May 2, 1823, VBLC; Van Buren to Thompson, Mar. 30, 1823; King to Adams, Apr. 1; to Monroe, Apr. 2; to Van Buren, Apr. 4, 1823; Adams to King, Apr. 7, 1823; Van Buren to Thompson, Mar. 30; Thompson to Van Buren, Apr. 6, 1823; Van Buren to King, Apr. 12, 14, 1823; to Thompson, Apr. 15; Thompson to Van Buren, Apr. 25; July 11, 25, 1823; July 25, 1823, VBLC; *King Correspondence,* VI, 527.

Chapter 9. A Broken Politician

1. George W. Irving to William H. Crawford, July 14, 1823, Crawford Misc. Papers, LC; MacKenzie, *Van Buren,* 193.

2. *Calhoun Papers,* VIII, 44, 53–55, 58, 59, 72, 73; *King Correspondence,* VI, 527; Charles M. Wiltse, *John C. Calhoun, Nationalist* (Indianapolis; New York, 1944), 272, 274–75; Hammond, *New York,* II, 128–30; *King Correspondence,* VI, 529; Elliott, *Winfield Scott,* 235; Van Buren to Smith Thompson, June 4, 1823, VBLC.

3. *Calhoun Papers,* VIII, 70; Clinton to Post, July 11, 1823, PNYSL; Dix to John W. Taylor, July 25, 1823, TNYHS; George W. Irving to William H. Crawford, July 14, 1823, Crawford Misc. Papers, LC; Macon to Van Buren, May 9, 1823; Taylor to Van Buren, May 12, 1823; Crawford to Van Buren, Aug. 1, 1823; Van Buren to David E. Evans, Nov. 3, 1823, VBLC.

4. Van Buren, *Autobiography,* 142, 143.

5. *Clay Papers,* III, 475, 476; Van Buren, *Autobiography,* 145; *Clay Papers,* III, 486, 487; see also A. C. Flagg to Van Buren, Nov. 13, 1823, VBLC. Flagg had no preference.

6. Washington, D.C. *Daily National Intelligencer,* Sept. 3; Oct. 8, 20, 1824; Joseph Gales to Van Buren, Sept. 14, 1824, VBLC; Van Buren to Butler, Aug. 18, 1824, BuP.

7. Hammond, *New York,* II, 154; Van Buren, *Autobiography,* 146, 147.

8. *King Correspondence,* VI, 539; Mooney, *Crawford,* 241, 263–67.

9. Marcy to Van Buren, Jan. 11; Van Buren to March, Feb. 17; to Butler, Feb. 19, 1824, VBLC; Van Buren, *Autobiography,* 665, 666.

10. Van Buren to Butler, Apr. 22, 1824, VBLC; *King Correspondence,* VI, 564; Mooney, *Crawford,* 556, 557; Gallatin to Van Buren, Oct. 2, 1824, VBLC.

11. Van Buren to Butler, Feb. 15, 1824, VBLC; *Daily National Intelligencer,* Feb. 17, 1824; Butler to Van Buren, Mar. 27, 1824, VBLC; MacKenzie, *Van Buren,* 198, 199.

12. "If he [Governor Yates] says that he acted in consequence of a belief that the Constitution would be amended in season for the next election, does he not know that there is not a boy in the nation who does not know that that was impossible?" he asked. Van Buren to James Campbell, May 14, 1824, VBLC.

13. *Annals of Congress,* 18th Cong., 1st Sess., 74; Remini, *Van Buren,* 45; Hammond, *New York,* II, 142.

14. Van Buren, *Autobiography,* 181, 182; Joseph Gales and William W. Seaton to Van Buren, Sept. 3, 15, 1824, VBLC; Mooney, *Crawford,* 264, 265.

15. *Adams Memoirs,* VI, 365; *King Correspondence,* VI, 565, 566, 569, 571; Van Buren, *Autobiography,* 181, 182.

16. *Clay Papers,* III, 767; Van Buren to A. Dickens, May 30, 1824, VBLC.

17. Webster, who visited Jefferson the same year, gave a full contemporary description. See Charles M. Wiltse, ed., *The Papers of Daniel Webster,* 4 vols., series 1. (Dartmouth, N.H., 1974–1980), I, 370, 371.

18. Van Buren to Jefferson, June 8, 1824; to David E. Evans, June 9, 1824, VBLC; for the text of Jefferson's reply, see Martin Van Buren, *Inquiry,* Appendix, 425–36.

19. Barnes and Weed, eds., *Weed,* I, 108. *Clay Papers,* III, 768. Marcy, whose shrewd opinions on New York politics Van Buren respected, wrote in mid-February that "the embarrassing question about Governor begins to be agitated and I avow to you that the nomination of Gov. Yates is not an event to be put down as certainly to happen." Skinner to Van Buren, Dec. 30, 1823; Marcy to Van Buren, Feb. 15, 1824; Van Buren to Butler, Feb. 17, 1824, VBLC.

20. Hammond, *New York,* II, 166; Jacob Sutherland to Van Buren, Jan. 24; James Ganson to Van Buren, Feb. 22, 1824, VBLC; Butler to Van Buren, Mar. 11, 24, 27, 1824, VBLC.

21. Van Buren, *Autobiography,* 148; *Clay Papers,* III, 768; MacKenzie, *Van Buren,* 197; De Witt Clinton held Van Buren responsible; Clinton to Henry Post, July 2, 1824, PNYSL; Marcy to Van Buren, Jan. 11, 1824, VBLC; Van Buren, *Autobiography,* 144.

22. Barnes and Weed, eds., *Weed,* I, 109–11; Van Buren, *Autobiography,* 143, 144; *King Correspondence,* VI, 564–68; Hammond, *New York,* II, 164, 165; Van Buren, *Autobiography.* 148; Hammond, *New York,* II, 164, 165.

23. Van Buren, *Autobiography,* 143, 145, 148; Hammond, *New York,* II, 175–77; *Clay Papers,* IV, 17; Barnes and Weed, eds., *Weed,* I, 122–29; Remini, *Van Buren,* 73–76.

24. Weed makes a tortuous explanation of the failure of the Adams men to honor their commitment to Clay, which is in part a falsehood and in part a shifting of the blame to the Adams electors. Van Buren to William H. Crawford, Nov. 17, 1824, VBLC; Barnes and Weed, eds., *Weed,* I, 128, 129; see also Remini, *Van Buren,* 82, 83; *Albany Argus,* Dec. 3, 1824.

25. Van Buren, *Autobiography,* 149.

Chapter 10. A Quick Recovery

1. Van Buren, *Autobiography,* 149; Skinner to Van Buren, Dec. 1, 1824, VBLC.

2. Munroe, *Louis McLane,* 151–60; Van Buren to Butler, Dec. 27, 1824, VBLC.

3. Van Buren, *Autobiography,* 149–50; Hamilton, *Reminiscences,* 68; Van Buren to Mrs. John McLane, Feb. 6, 1825, VBLC.

4. Bonney, *Legacy,* I, 415, 256ff.; see also "Memoirs of a Senator from Pennsylvania, Jonathan Roberts," *Pennsylvania Magazine of History and Biography,* Vol. LXII, No. 3 (July 1938), 407.

5. Van Buren, *Autobiography,* 150–51, see also Nathan Sargent, *Public Men and Events in the United States,* 2 vols. (Philadelphia, 1875), I, 76, 77.

6. McLane to Van Buren, June 18; Oct. 17, 1824, VBLC; *Adams Memoirs,* VI, 474, 493; Robert V. Remini, *Andrew Jackson and the Course of American Freedom,* II (New York, 1981), 92–94.

7. Munroe, *McLane,* 183.

8. Van Buren, *Autobiography,* 150, 151; *Niles Weekly Register,* Feb. 12, 1825.

9. Van Buren, *Autobiography,* 151.

10. Munroe, *McLane,* 161, 172; John Bigelow, "De Witt Clinton as a Politician," *Harper's New Monthly Magazine* (March 1875), L, 298, 563.

11. John Van Buren, *MS. Diary, passim,* JVBH.

12. Van Buren, *Autobiography,* 151–53; Munroe, *McLane,* 179–86; Gaillard Hunt, ed., *The First Forty Years of Washington Society Portrayed in the Family Letters of Mrs. Samuel Harrison Smith* (New York, 1906), 170–81; McLane to Catherine McLane, May 18, 1824; Feb. 11, 20, 1825, McLane Papers, LC.

13. James D. Richardson, ed., *A Compilation of the Messages and State Papers of the Presidents 1789–1897* (Washington, D.C., 1900), 10 vols., II, 862–64, 865.

14. Van Buren, *Autobiography,* 153, 234–39.

15. As Daniel Webster wrote his brother Ezekiel, "Mr. Clay's ill judged card has produced an avowal or a sort of avowal which makes the whole thing look ridiculous." Wiltse, ed., *Papers of Daniel Webster,* II, 20; see, for instance, the *Albany Argus,* May 27, 1825; Lott Clark to Roger Skinner, Feb. 28, 1825, VBLC.

16. Hammond, *New York,* II, 183.

17. Ransome H. Gillet, *Life and Times of Silas Wright,* 2 vols. (Albany, 1874), I, 13–15, 18, 19; Stewart Mitchell, *Horatio Seymour of New York* (Cambridge, 1938), 73; John A. Garraty, *Silas Wright* (New York, 1949), 60; clippings *New York Journal of Commerce, Evening Post, New York Globe,* Sept. 28, 1848; Hoffman to Flagg, Dec. 15, 1827, FNYPL; Butler to Van Buren, Oct. 3, 1848, VBLC; MacKenzie, *Van Buren,* 131, 132.

18. *Albany Argus,* Oct. 28, 1826; Garraty, *Silas Wright,* 106.

19. Silas Wright, Jr. to A. C. Flagg, Feb. 1, 1825, FNYPL; *Albany Argus,* Feb. 4, 1825.

20. James R. Taylor to Dean Mather, Nov. 1, 1825, TNYHS; see also C. Vanderventer to Virgil Maxcy, Aug. 24, 1825, Galloway-Maxcy-Markoe Papers, LC; *Calhoun Papers,* X, 38, 39.

21. Hammond, *New York,* II, 205; *Niles Weekly Register,* Nov. 5, 12, 1825; Jan. 2, 1826.

22. Silas Wright, Jr. to A. C. Flagg, Nov. 20, 1825, NYPL; Van Buren to D. E. Evans, Aug. 11, 1825, VBLC; *Albany Argus,* Aug. 15; Nov. 10, 1825; *National Advocate,* Nov. 20–30, 1825; *New York American,* Nov. 12, 14, 16, 1825.

23. Silas Wright to A. C. Flagg, Nov. 18, 26, 28, 1826; Richard Hoffman to Flagg, Nov. 28, 1826, FNYPL; William C. Bouck to Van Buren, Nov. 17, 1826, VBLC; *Albany Argus,* Nov. 26, 1826.

24. The adjective "noncommittal" was first coined by Marcy in an editorial for the *Argus,* but attributed to Van Buren; Spencer, *Marcy,* 36.

25. Richardson, *Messages and State Papers,* II 868, 877, 878.

26. Van Buren to C. E. Dudley, Jan. 3, 1826, VBLC; Van Buren, *Autobiography,* 149.

27. *Ibid.,* 159; Michael Hoffman to Marcy, July 31, 1826, MALC; Edward Livingston to Van Buren, Nov. 30, 1825, VBLC.

28. Van Buren to Livingston, Dec. 7, 1825, VBLC; Van Buren, *Autobiography,* 159; Marcy to Van Buren, Dec. 11, 17, 1825, VBLC; Spencer, *Marcy,* 37, 42, 43.

29. James Tallmadge, Jr. to John W. Taylor, Mar. 4, 1826, TNYHS; see, for instance, *Albany Argus,* Apr. 3; Aug. 25, 1826; Wright's letter dated Apr. 4, 1826 was published in the *Albany Evening Journal,* Oct. 6, 1840. Croswell, too, had written on the same subject; Edwin Croswell to Van Buren, Apr. 3, 1826, VBLC.

30. Van Buren had been thinking about Rochester as a candidate for Lieutenant Governor as early as February 1826; Van Buren to D. E. Evans, Feb. 5, 1826, VBLC.

31. Van Buren to Evans, Aug. 6, 1826; to Dr. Sims, Oct. 3, 1826, VBLC; to Gulian C. Verplanck, Oct. 14, 1826, VBLC; *Clay Papers,* V, 743. "Every section of the party feels the inconvenience of this state of things and is desirous of restoring the party to its former integrity—and the nomination of Rochester has been made with an express view to such reunion," Porter wrote Clay. Even Van Buren, said Porter, "who you know has been attempting to drive a political bargain with Clinton and both of whom have, at length, become convinced that this act of perfidy to their respective friends, will not be tolerated, has found it necessary to retrace his steps and will give his support to Rochester against Clinton and Noah," *ibid.,* 763, 764.

32. P. N. Nicholas to Van Buren, Oct. 13, 1826; VBLC *New York American,* Oct. 7–10, 1826; *Albany Argus,* Oct. 14, 1826; Rudolph Bunner to Gulian Verplanck, Feb. 5, 1827, Verplanck Papers, NYHS.

33. *Clay Papers,* V, 876–77; *Albany Argus,* Oct. 21, 23, 1826.

34. Wright to Flagg, Nov. 18, 1826, FNYPL; Van Buren to P. N. Nicholas, Nov. 1826, VBLC; Van Buren to Cambreleng, Nov. 4; Marcy to Van Buren, Dec. 10, 1826, VBLC; Van Buren, *Autobiography,* 162; *National Advocate,* Sept., Oct., Nov., 1826; *Albany Argus,* Sept. 11, 15, 18–22, 1826; Hammond, *New York,* II, 232–35.

Chapter 11. With Muffled Oars

1. Marcy to Van Buren, Dec. 10, 1826; VBLC; Van Buren to Gulian Verplanck, Oct. 14, 1826; Van Buren to Benjamin Butler, Dec. 12, 1826; Marcy to Van Buren, Dec. 27, 1826, VBLC. "There is not the least doubt that everything Noah says against Adams," he wrote Hamilton, "does him [Adams] great good with our country Republicans who look upon Noah literally with abhorence. Is there not spirit enough in the great Democratic party of the great city of New York to establish a press in which honest men can con-

fide?" Hamilton, *Reminiscences,* 63; John C. Hamilton to Van Buren, Dec. 21, 1826, VBLC; Hamilton, *Reminiscences,* 64; John C. Hamilton to Van Buren, Dec. 21, 26, 1826, VBLC.

2. Hamilton, *Reminiscences,* 63; John C. Hamilton to Van Buren, Dec. 21, 1826; Marcy to Van Buren, Dec. 27, 1826, VBLC.

3. With an egregious want of tact, Adams cast doubt on the motives of the Senate when he left it up to that body to decide whether publication of documents involving the Panama mission was in the national interest and consistent with its own past usage. The Senate decided on a closed session after lengthy debate in both Houses, in which Van Buren made an elaborate speech. The mission was approved and funds appropriated to finance it. When the American ministers finally arrived at Panama they found that the Congress had never materialized because of differences among the new Latin American states. *Albany Argus,* Apr. 18, 1826; *Register of Debates,* 19th Cong., 1st Sess., 1825–26; Richardson, ed., *Messages and State Papers,* II, 884–86; III, 893; Benton, *Thirty Years View,* I, 65–66; Hamilton, *Reminiscences,* 63.

4. *Ibid.* Van Buren wrote Benjamin Butler that "The Jackson men being in the fold are, of course, looking out for weak points in the enemies' lines and are ready for the assault when opportunity offers. We of the Crawford school lay upon our oars and will not lightly commit ourselves except in defence of old principles." Van Buren to Butler, Dec. 25, 1826, VBLC.

5. Van Buren, *Autobiography,* 213–16; *Daily National Intelligencer,* Jan. 28, 1827; Van Buren, *Autobiography,* 216; Van Buren, "Notes," Jan. 23, 1827, VBLC. His support of most features of the bill went contrary to a powerful constituency, the merchants of New York. See Isaac Hone to Van Buren, Dec. 26, 1826, VBLC. The bankruptcy bill failed in the Senate and was sent back to a select committee of which Van Buren was a member. It was reported back with the ninety-third section omitted, but again failed in the Senate. *Albany Argus,* Feb. 8, 12, 1827.

6. *Daily National Intelligencer,* Apr. 19, 1827; Charles B. Sherman to James P. Tilden, Nov. 4, 1834; Tilden Papers, NYPL; Adrian Hoffman Joline, *The Autograph Hunter and Other Papers* (Chicago, 1907), 66.

7. Van Buren, *Autobiography,* 514.

8. Van Buren to D. E. Evans, Dec. 16, 1826; Jesse Hoyt to Van Buren, Jan. 11, 1827, VBLC; J. G. deRoulhac Hamilton, ed., "Letters to Bartlett Yancey," *The James Sprunt Historical Publications* (Chapel Hill, N.C., 1911), 61; A. Conkling to John W. Taylor, Jan. 27, 1827, TNYHS; Hammond, *New York,* II, 245–46; John W. Taylor to Charles Miner, Apr. 16, 1827, TNYHS.

9. Van Buren, *Autobiography,* 513, 514; Van Buren to Ritchie (copy), Jan. 13, 1827, VBLC.

10. *Ibid.*

11. Albert R. Newsome, ed., "Letters of Romulus Saunders to Bartlett Yancey," *North Carolina Historical Review,* VIII, Oct., 1931, 454, 457.

12. Hammond, *New York,* II, 247.

13. Van Buren, *Autobiography,* 205, 426, 427, 430, 481; Ambler, *Ritchie,* 106; E. B. Smith, *Magnificent Missourian: The Life of Thomas Hart Benton* (New York, 1958), 76, 77.

14. *Richmond Enquirer,* Dec. 14, 1826; Jan. 15, 16, 1827; *Clay Papers,* V,

227, 228, 1030, 1031; Van Buren, *Autobiography,* 210; *Calhoun Papers,* IX, 156, 157; Gales and Seaton charged that the vote on the congressional printing which went to the *Telegraph* "was a better test of the 'organization' than any other thing could well have been. . . . it was extorted, if we may say so, from reluctant hands by the power of that most potent of instruments, a secret sub-caucus." In its ironic conclusion, the *Intelligencer* said, "Mr. Van Buren himself wept over the 'dire necessity.' " *National Advocate,* Mar. 2, 1827; *Daily National Intelligencer,* Mar. 4, 7, 12, 20, 22, 29, 1827; *Albany Argus,* Mar. 27, 29, 1827; *Richmond Enquirer,* Mar. 6, 1827; Ambler, *Ritchie,* 109, 110.

15. Calhoun was working independently to improve his image with the North Carolinians, particularly with Bartlett Yancey. *Calhoun Papers,* X, 252, 253; Van Buren, *Autobiography,* 514, 515; Hamilton, ed., "Yancey Letters," 62; Newsome, ed., "Letters of Saunders to Yancey," 460, 461; *Raleigh Star,* Feb. 23, 1827.

16. Van Buren to Butler, Apr. 12, 1826, VBLC; Charleston *Southern Patriot,* Apr. 9, 1827; Van Buren to C. C. Cambreleng, Oct. 29, 1826, VBLC; Van Buren, *Autobiography,* 169; Munroe, *McLane,* 213, 224; James Hamilton, Jr., to Jackson, Feb. 16, 1827, JLC.

17. Mooney, *Crawford,* 302, 308; Crawford to Van Buren, Aug. 15; Dec. 21, 1827, VBLC; Van Buren, *Autobiography,* 368; William H. Crawford to Alfred Balch, Dec. 15, 1827, PNYSL; to Van Buren, Dec. 21, 1827, VBLC; William S. Hoffman, "Andrew Jackson and North Carolina," *Politics,* XXXX, *The James Sprunt Studies in History and Political Science* (Chapel Hill, N.C., 1958), 12, 13; MacKenzie, *Van Buren,* 200; *Raleigh Register,* May 4, 1827. For the political importance and family connections of the Junto, see Joseph H. Harrison, Jr.'s admirable article," "Oligarchs and Democrats: the Richmond Junto," *The Virginia Magazine of History and Biography,* Vol. 78, No. 2 (April 1970), 188–91; see also Harrison's "Martin Van Buren and His Southern Supporters," *The Journal of Southern History,* Vol. XXII, No. 4 (November 1956), 438–41.

18. Ambler, *Ritchie,* 111, 112; Van Buren, *Autobiography,* 514, 515; Ben Perley Poore: *Perley's Reminiscences of Sixty Years in the National Metropolis* (Philadelphia, 1886), 2 vols., I, 74, 75, 81; *Adams Memoirs,* VII, 272.

19. Van Buren to D. E. Evans, Feb. 11, 1827, VBLC; Van Buren to Aaron Burr, Jan. 14, 1825, VBNYHS; Van Buren to D. E. Evans, May 24, 1827, VBLC; Wiltse, ed., *Papers of Daniel Webster,* II, 206–8; Van Buren to Rev. Harry Crosswell, Dec. 10, 1825; see John Van Buren Autograph Album, March 30–Sept. 8, 1829, *passim,* copy in possession of Rufus King Duer, Essex, Conn; to John Van Buren, Jan. 19, 1826, VBLC; Van Buren, *Autobiography,* 174–76; Roper, ed., "Kent's Necrologies," 220; Van Buren was purchasing such properties; see "Indenture" between Tobias D. Van Buren and Harmon Vosburg and Martin Van Buren, Oct. 28, 1809; also see Trust Deed, Oct. 5, 1809, VBNYHS; see also indenture between Marcy and Van Buren of Sheriff's sale held in Mar. and Apr. 1826, dated July 16, 1828, recorded Oct. 20, 1836. Van Buren paid $60.90 for 600 acres of land in Oswego, VBMHS; see Van Buren to Cambreleng, Oct. 22, 1827, VBLC, in which he says "I wish to stay at your house (in New York City), but I cannot afford to pay $2 a day for a room, nor will it do for me to be stuck up in

the garret." Yet in the same letter he asked Cambreleng to draw off three dozen bottles of wine "so that I can give a dinner to all the young Jackson Bloods in the city."

20. MacKenzie, *Van Buren,* 201, 202; G. H. Barstow to John W. Taylor, Apr. 5, 1827; John Bailey to Taylor, May 2, 1827, TNYHS; Jabez Hammond to Van Buren, May 23, 1827, VBLC; Hammond, *New York,* II, 239; Barnes and Weed, eds., *Weed,* I, 210–40; Van Buren to Cambreleng, July 4, 1827; to Andrew Jackson, Sept. 14, 1827, VBLC.

21. See the *Rochester Telegraph,* June–July 1827; Van Buren to Cambreleng, June 22; July 4; Oct. 23, 1827, VBLC; *Albany Argus,* June 12–22, 1827; New York *Enquirer,* June 19–July 2, 1827; New York *Morning Courier,* June 15, 16, 1827.

22. Marcy to Van Buren, June 25; Van Buren to John Van Buren, Nov. 21, 1826, VBLC; Van Buren, *Autobiography,* 169–72; *Albany Argus,* July 11, 21, 1827.

23. Thomas Cooper to Van Buren, July 10, 1827, VBLC; J. S. Schermerhorn to Van Buren, July 15, 1827; Thomas Cooper to Van Buren, July 31, 1827, VBLC; *Albany Argus,* Aug. 13, 1827.

24. Hamilton, *Reminiscences,* 68; Van Buren, *Autobiography,* 239–42.

25. James Parton, *Life of Andrew Jackson* (New York, 1861), 3 vols., III, 132n; Van Buren to Jackson, Sept. 14, 1827, VBLC; John Spencer Bassett, ed., *The Correspondence of Andrew Jackson* (Washington, D.C., 1926–33), 6 vols., IV, 364, 365. Hereafter cited as *Jackson Correspondence; Calhoun Papers,* X, 308–9. On Clinton's ambitions see Colonel Arthur P. Hayne to Jackson, July 20, 1826, JLC; *Calhoun Papers,* X, 305, 309; Wiltse, ed., *Webster,* II, 184–84; Hammond, *New York,* II, 256; James Hamilton, Jr. to Jackson, Feb. 16, 1827, JLC; Alfred Balch to Van Buren, Nov. 27, 1827, VBLC; Van Buren, *Autobiography,* 243; Herbert Weaver, Paul Bergeron, Wayne Cutler eds., *Correspondence of James K. Polk* (Nashville, 1969–), 5 vols., II, 293, 294. Hereafter cited as *Polk Correspondence.*

26. Hammond, *New York,* II, 259.

27. Rives, the young Virginia Congressman, whom Van Buren had marked for greater things, sought to use his friendship with Ritchie and with Van Buren to push another loyal Crawfordite from Virginia, former Speaker Philip P. Barbour. Rives had not understood the complexities of the contest and the importance of having a Speaker who was not so emphatically identified with free trade as Barbour was. As Van Buren correctly prophesied, "Barbour will not do and that will be the correct opinion of seven-eights of our friends." Van Buren to W. C. Rives, Oct. 17, 1827; to Thomas A. Gilmer, Dec. 5, 1827, RLC; Francis Fry Wayland, *Andrew Stevenson* (Philadelphia, 1949), 74–77; Munroe, *McLane,* 224; John Randolph to Jackson, July 5, 1827, RLC; Ambler, *Ritchie,* 112, 113; Hammond, *New York,* II, 260; Silas Wright to Flagg, Dec. 13, 1827, FNYPL; Henry R. Storrs to Verplanck, Dec. 17, 1827, Verplanck Papers, NYHS.

Chapter 12. An Uncalculated Risk

1. Clinton had never fully recovered from a bout with pneumonia the previous fall. Dr. David Hosack, who visited him a week before his death, had been shocked by his appearance, "his anxious respiration, his anhelation upon the slightest motion—his livid countenance, his irregular and intermitting pulse—his swelling limbs, all indicated the dropsical and perhaps organic affliction of the heart and larger vessels pointed to the fatal issue, I thus confidently predicted." Hosack alerted Clinton's family and some close friends of his fears. David Hosack, *Memoir of De Witt Clinton* (New York, 1829), 130–31.

2. Jackson himself was disturbed enough to seek more information from Polk. Hamilton, *Reminiscences* (New York), 75; *Polk Correspondence,* II, 160, 161, 171; John J. DeGraff to Van Buren, Feb. 18, 1828; Rudolph Bunner et. al., to Van Buren, Feb. 19, 1828, VBLC; *Albany Argus,* Feb. 27, 1828.

3. Lot Clark to Van Buren, Apr. 10, 1828, VBLC; Silas Wright to Azariah C. Flagg, Dec. 20, 1827, FNYPL; Wright to Flagg, Feb. 18, 1828, FNYPL; Rives to Gilmer, Feb. 20, 1828, RLC; P. N. Nicholas to Van Buren, Oct. 13, 1826; Van Buren to Nicholas, Nov. 1826, VBLC.

4. T. Rudd to John W. Taylor, Feb. 26, 1828, TNYHS; Allan Nevins, ed., *The Diary of Philip Hone* (New York, 1936), 23; Flagg to Wright, Mar. 7, 1828, FNYPL.

5. Flagg to Wright, Mar. 7, 1828, PNYPL; Wright to Flagg, Dec. 20, 1827, FNYPL.

6. *Register of Debates,* 20th Congress, 1st. Sess., 884, 90; *Daily National Intelligencer,* Jan. 1828, *passim.* Levi Woodbury, who was edging into the Jackson camp at this time and was a keen supporter of a higher tariff on woolens, deprecated delay as weakening the manufacturers' case; Woodbury to Edward Cutts, Jan. 4, 1828, F. H. Merrill Coll., MHS.

7. Lemuel Williams to Virgil Maxcy, Apr. 1, 1828, Galloway-Maxcy-Markoe Papers, LC; Van Buren to Wright, May 2, 1826; Wright to Van Buren, May 15, 1828, VBLC; Wright to Flagg, Feb. 6, 1828, FNYPL; Wright to Flagg, Mar. 21; Apr. 7, 1828, FNYPL; Benjamin Knower to Van Buren, Jan. 27, 1828, VBLC; *Webster Papers,* II, 243, 244, 293, 294, 336–38; Wright to Flagg, Mar. 21, 1828, FNYPL; Garraty, *Silas Wright,* Chap. III, *passim.*

8. Van Buren to Wright, May 2, 1826; Wright to Van Buren, May 15, 1828, VBLC; Wright to Flagg, Feb. 6, 1828, FNYPL; *New York Assembly Journal,* 1828, 350, 351; Jesse Buel et. al., "Circular," Mar. 15, 1828; Hoffman to Flagg, Feb. 3, 1828; Wright to Flagg, Mar. 21, 1828, FNYPL; Smith Van Buren to Van Buren, Feb. 16, 1844, VBLC; Wright to Flagg, Dec. 20, 1827; Hoffman to Flagg, Apr. 27, 1828, FNYPL; Knower to Van Buren, Apr. 23, 1828, VBLC; for an explanation of the complex mixture of specific and ad valorem duties see Frank W. Taussig, *The Tariff History of the United States* (New York and London, 1931), 8th ed. 59–95, 98–102.

9. Levi Woodbury, one of the New Hampshire Senators, voted with Van Buren for the amendment. Robert V. Remini, "Martin Van Buren and the Tariff of Abominations," *The American Historical Review,* LXVIII, No. 4, July,

1958, 914–16. The bill, as amended, passed the Senate by a vote of 26 to 21. Taussig, *Tariff,* 100, 101. As Van Buren anticipated, the Tariff Act was popular in New York; Nevins, ed., *Diary of Philip Hone,* 2.

10. Wright to Flagg, Feb. 18, 1828, FNYPL; Cooper to Van Buren, Apr. 11, 1828, VBLC; *Calhoun Papers,* II, 370–74; Van Buren to Crawford, Feb. 15, 1828, W. H. Crawford Papers, LC; Buchanan to General George B. Porter, Jan. 22, 1829, Buchanan Misc. Papers, NYHS.

11. Remini, *Van Buren,* 166; Hoffman to Flagg, Dec. 21, 1827, FNYPL; Van Buren to Crawford, Nov. 14, 1828, VBLC.

12. Hammond, *New York,* II, 281; Robert V. Remini, *The Election of Andrew Jackson* (Philadelphia and New York, 1963), 59; Hamilton, *Reminiscences,* 65–69. Vincent J. Capowski, "The Making of a Jacksonian Democrat: Levi Woodbury, 1784–1831" (Unpub. Diss., Fordham Univ., 1966), 126.

13. William L. Marcy to Van Buren, Jan. 29, 1828, VBLC; William Russell to John W. Taylor, Feb. 21, 1828, TNYHS; *Rochester Telegraph,* Mar. 9, 1828; *Webster Papers,* II, 309; Barnes and Weed (eds.) *Weed* I, 254, 255.

14. Thomas Ritchie to Van Buren, Mar. 11, 1828; from time to time, Van Buren himself, on Apr. 4, 1828, for example, took the liberty of advising Coleman. He enclosed one of his speeches and suggested to the editor that "there is not sufficient honesty at the present moment in the Administration of the Federal Government to keep decent men in countenance." Van Buren to Coleman, Apr. 4, 1828, VBLC; *Albany Argus,* Mar. 1–15, 1828; *New York Enquirer,* Mar. 4, 1828; James L. Crouthamel, *James Watson Webb: A Biography* (Middletown, 1969), 17–21; Charles H. Brown, *William Cullen Bryant* (New York, 1971), 157–63; Hamilton, *Reminiscences,* 77, 78.

15. Hammond, *New York,* II, 277–80; William Coleman to Van Buren, Apr. 17, 1828, Wright to Van Buren, Dec. 7, 1828, VBLC.

16. Hoffman to Flagg, Apr. 4, 1828, FNYPL; Van Buren to Evans, Mar. 31, 1828, excerpts from *The Catalogue for Charles Hamilton Auction,* No. 12, 58, VBLC; Lot Clark to Van Buren, Apr. 10, 1828, VBLC.

17. *Register of Debates,* 20th Cong., 1st Sess., 314, 338; Richard Riker to Van Buren, Apr. 14, 1828, VBLC; *Albany Argus,* Apr. 20, 1828; William Coleman to Van Buren, Apr. 17, 1828, VBLC; Gustavus Myers, *The History of Tammany Hall* (New York, 1917), 72–75; Hamilton, *Reminiscences,* 78; MacKenzie, *Van Buren,* 205; William L. Marcy to Van Buren, n.d., 1838, Marcy Misc. Papers, NYHS; Clay to Porter, Jan. 14, 1828, Porter Papers, Buffalo Historical Society.

18. Hammond, *New York,* II, 284–85; Marcy to Van Buren, Mar. 9, 1828, Marcy Misc. Papers, NYHS; Van Buren to B. F. Butler, Jan. 22, 1828, VBLC; Henry Vanderlyn *MSS. Diary,* Feb. 14, 1830, NYHS; Flagg to Wright, Mar. 7, 1828; Hoffman to Flagg, Apr. 4, 1828, FNYPL

19. Randolph had written Van Buren in early August inviting him for a visit. Van Buren declined but in doing so opened up a discussion of college education for Martin, Jr. In his reply, Van Buren, perhaps influenced by John's expensive and wayward course at Yale, expressed an unflattering opinion of college education. Randolph agreed, "Of colleges you cannot have a poorer opinion than I have long entertained," said Randolph. "There is hardly a pin to choose between them. I think Master Martin's name will

carry him through the race of life." Randolph to Van Buren, Aug. 4; Oct. 13, 1828, VBLC. Henry B. Stanton, *Random Recollections* (New York, 1887), 32; Flagg to Wright, Aug. 24, 1828, FNYPL; *Albany Argus,* Sept. 29, 1828; Barber and Howe, *Historical Collections of the State of New York,* 194. Barnes and Weed, eds., *Weed,* I, 232, 233, 236, 237.

20. Wright to Van Buren, Dec. 7, 1828, VBLC; Silas Wright to Flagg, Aug. 1, 1828, FNYPL: Van Buren to Throop, July 16, 1828, VBNYSL; Albert H. Tracy to Weed, June 19, 1828, Seward-Weed Papers, Univ. of Rochester Library; Edward Everett to Joseph Blunt, Nov. 23, 1828, Everett Papers, MHS; Phineas Tracy to John W. Taylor, July 10, 1828, TNYHS.

21. Phineas Tracy to Smith Thompson, July 26, 1828, Thompson Misc. Papers, NYHS; Hammond, *New York,* II, 236.

22. Van Buren kept his eye on Southwick as soon as he came out openly for the anti-Masons. "They," he said, referring to the National Republicans, "are frightened to death by Solomon Southwick and the anti-Masonic free-booters." Van Buren to Butler, Mar. 15, 1828, VBLC; Worth, *Albany,* 72; Flagg to Wright, May 16, 1828, FNYPL; Van Buren to Butler, Mar. 15, 1828, VBLC. An infuriated Nathan Sargent accused Van Buren of arranging to have Southwick run on a third ticket. Writing Weed after Southwick's acceptance of the anti-Masonic nomination, he said, "They disgrace themselves and injure their country by such a course. Moreover, they are used by Van B. . . . As a people we are corrupt and unfit to govern ourselves." But the most Weed, who would have had the facts if any existed, would comment was that the National Republicans were "juggled" out of the election; Nathan Sargent to Weed, n.d., 1828, Seward-Weed Papers, Univ. of Rochester Library; Alexander, *New York,* II, 364.

23. Van Buren to N. F. Benton, Apr. 30, 1828, VBLC; Hammond, *New York,* II, 288, 289, Barnes and Weed (eds.), *Weed* I, 303–07. At this point Van Buren felt certain he would win. For he asked James A. Hamilton on Aug. 25, four days after his return from western New York and five days before Granger declined the anti-Masonic nomination, to "make this bet for me, or on our joint account, as you please, Viz. $500 that Thompson will be defeated, whoever our candidate may be, and $100 on every thousand of a majority up to 5,000 or if you can't do better, say $500 on the result and $50 on every thousand up to ten thousand." And in the same letter he begged Hamilton to raise more money in New York City than he had promised. "Our friends here (at best poor) will break down." Van Buren to Hamilton, Aug. 25, 1828, in Hamilton, *Reminiscences,* 78, 79. Hamilton placed the bets through William B. Lewis. Hamilton to Lewis, Oct. 3, 1828, Andrew Jackson-William B. Lewis Papers, NYPL. Hamilton did send $200 to Albany through John Targee, a prominent Tammany man for what political purposes, however, remain unknown; MacKenzie, *Van Buren,* 205–11. MacKenzie insinuates that Van Buren was the instigator of the Southwick nomination, but has no proof to offer.

24. Levi Woodbury to Van Buren, July 1, 1828, VBLC; Lemuel Williams to Virgil Maxcy, Apr. 1, 1828, Galloway-Maxcy-Markoe Papers, LC; Donald B. Cole, *Jacksonian Democracy in New Hampshire, 1800–51* (Cambridge, Mass., 1970), 60–70; Van Buren to Crawford, Feb. 15, 1828, Crawford Papers, LC;

James Wolcott, Jr. to Van Buren, Mar. 10, 1828, VBLC; Flagg to Wright, Apr. 13, 1828, FNYPL.

25. Barnes and Weed, eds., *Weed,* I, 307–8. On his western tour, the sharp-eyed Van Buren had read and was impressed with the style of a politically independent paper published in Rochester, the *Daily Advertizer,* and its weekly edition, the *Monroe Republican.* Anxious to promote his own and Jackson's campaign in the heart of anti-Mason country, he approached Luther Tucker, its editor, and solicited his support. Tucker, who was part owner, referred Van Buren's request to the *Advertizer*'s original proprietor, Henry C. Sleight, a prosperous printer and publisher of commercial papers in Jamaica, Long Island. Sleight answered with a decided negative. A brief correspondence between Van Burenites in Genessee County and Albany ended with an offer through Tucker to buy out Sleight's half ownership of the *Advertizer.* Sleight, whose real interests lay elsewhere, promptly agreed to part with his share for $1,500 down and the remainder in the form of a loan Tucker would pay off in installments. Van Buren, himself, and Charles Van Benthuysen, business manager of the *Argus,* endorsed Tucker's note which the Farmers' and Mechanics' Bank of Albany promptly discounted. Shortly after the transaction was completed on Sept. 15, the *Advertizer* came out for Jackson and Van Buren, prompting Weed's blatantly malicious circular. Tucker answered the charge the following day, pointing out that the transaction was purely a business matter. As an editor, he would support that party and those candidates he honestly believed held correct principles. Van Buren's loan had nothing to do with his editorial preferences and would be repaid with interest. John B. Mullan, "Early Masonic History in Rochester," *The Rochester Historical Society Publication Series,* Vol. VII (Rochester, New York, 1928), 7–21. Henry C. Sleight to Henry O'Reilly, Mar. 20, 1872, in *ibid.,* VI, New York, 1924. No reply was forthcoming. There was no need, the libel had already made its impact.

26. William L. Marcy to Van Buren, Mar. 9, 1828; MacKenzie, *Van Buren,* 204; Van Buren to Cambreleng, Sept. 8, 1828, VBLC.

27. Van Buren had kept his secret well. Such a close associate as Michael Hoffman was speculating about the gubernatorial candidate as late as July 30, in his correspondence with Flagg, who was likewise in the dark; Michael Hoffman to Flagg, July 30, 1828. There was some hint of a Van Buren ticket by late August, and there was mention of Throop along with several other candidates for Lieutenant Governor; Flagg to Wright, Aug. 24, 1828, FNYPL.

28. As John Vanderlyn, brother of the artist and a lawyer who lived in the town of Oxford, Chenango County, wrote in his diary after the election, "so prominent was the interest felt by all in relation to the canal that it overcame all petty prejudices and secured many thousand votes for Van Buren." Van Buren himself remarked, "the only four towns in Broome (a crazy county) have given me a unanimous vote, Viz. 1000 and the others it is supposed will not reduce that; John Vanderlyn, *MS. Diary,* Mar. 6, 1829, NYHS; MacKenzie, *Van Buren,* 205. Hamilton, *Reminiscences,* 79. A Matthew Warner from Lima, New York, had written Van Buren inquiring about Jackson's morals, his character and whether he was a practicing Christian

on Sept. 4. Van Buren enclosed Warner's letter to Hamilton, who answered on Sept. 18. The text which is reprinted in Hamilton's *Reminiscences,* 75, 76, bears the wrong date; see *Albany Argus,* Sept. 20; *New York Evening Post,* Sept. 19, 1828.

29. Van Buren to Butler, Sept. 10, 1828; Cambreleng to Van Buren, Sept. 18, 1828, VBLC.

30. This particular rumor had been floating around Washington since early spring. Henry Clay, who oddly enough, feared the Regency would repeal the law, warned Peter B. Porter to be on his guard for a sudden coup. Clay to Porter, Mar. 24, 1828, Porter Papers, Buffalo Historical Society. Considerable pressure had been exerted upon Van Buren to take the lead in changing New York's district election law to election at large. He had resisted all efforts. *Jackson Correspondence,* III, 397; William B. Lewis to Van Buren, Aug. 8, 1828; *National Intelligencer,* Sept. 11, 25; Oct. 4, 1828; *Richmond Enquirer,* Sept. 25, 1828, VBLC. He did write one other letter which was not for publication but for guidance of the desparate New England Jacksonians. "If it be true," he said to Benjamin Norton, one of the editors of the *Hartford Times,* "that Mr. Adams joined the Republican party to bring it down, it is gratifying to know that he has not been able to carry his scheme further than to federalize New England that only for a season." *Albany Argus,* Nov. 1, 1828; Van Buren to Norton, Oct. 1828, VBLC.

31. William B. Lewis to Van Buren, Aug. 8, 1828, VBLC; Van Buren to Cambreleng, Oct. 18, 1828, VBLC; M. M. Noah to Van Buren, Oct. 2, 1828, VBLC; Remini, *Van Buren,* 192, 193; *Albany Argus,* Sept. 16; Oct. 31, Nov. 9, 1828; *Rochester Advertizer,* Oct. 30; Nov. 3, 1828; Hammond, *New York,* II, 289–91; see also Richard P. McCormick, *The Second American Party System* (Chapel Hill, N.C., 1966), 118, 119.

Chapter 13. Governor Van Buren

1. *Albany Argus,* Nov. 24, 1828; Van Buren to Cambreleng, Nov. 7, 1828, VBLC.

2. William B. Lewis to Hamilton, Dec. 12, 1828, VBLC; John Van Fossen to John C. McLemore, June 9, 1828. McLemore sent it on to Jackson's nephew, Andrew Jackson Donelson, A. J. Donelson Papers, LC; Remini, *Election of Andrew Jackson,* 186–91. Remini has pointed out that Jackson would have received more electoral votes than his 178 to Adams 83, had all the states voted a general ticket. But conversely, had all the states voted by district, Adams's vote would have been substantially larger. Although Jackson's popularity counted, historians have tended to overemphasize this aspect, and have neglected the anti-Masonic vote which in New York held the balance of power in 20 counties. See *Albany Argus,* Nov. 21, 1828. Organization in the last analysis clinched the victory in which sectional interests played a large role.

3. Forsyth to Van Buren, July 25, 28, 1828; Balch to Van Buren, Nov. 27, 1828; Verplanck to Van Buren, Dec. 6, 1828; Wright to Van Buren, Dec. 9, 1828; Lewis to Hamilton, Dec. 12, 1828, VBLC.

4. Van Buren to Cambreleng, Dec. 17, 1828, VBLC.

5. Worth, *Albany,* 56.

6. Van Buren to Butler, Mar. 15, 1828, VBLC; Van Buren to Butler, Nov. 17, 1828, VBLC; Wright to Flagg, Dec. 8, 1828, FNYPL; Van Buren to Cambreleng, Dec. 9, 1828, VBLC; Wright to Flagg, Dec. 19, 1828, FNYPL; to Van Buren, Dec. 19, 1828, VBLC.

7. Hammond, *New York,* II, 303, 304; *Albany Argus,* Jan. 28, 1829; MacKenzie, *Van Buren,* 206, 207; Van Buren to John H. Suydam, Dec. 1828, VBLC; Wright to Flagg, Dec. 19, 1828, FNYPL; Garraty, *Wright,* 75, 76; Van Buren to Butler, Sept. 14, 1828, BuP.

8. Van Buren, *Autobiography,* 221, 222; Forman to Van Buren, Jan. 24, 1829; Dec. 17, 1828; Newbold to Olcott, Dec. 17, 1828, VBLC; Hammond, *New York,* II, 297–302. "You need not, I think, have any apprehension about the message," Van Buren wrote Hoyt, ". . . I shall do the best I can in whatever relates to my office and leave the result to Providence and the People." MacKenzie, *Van Buren,* 206.

9. Van Buren to James Auchincloss, Jan. 29, 1829, VBLC; John Vanderlyn, *MSS. Diary,* Mar. 6, 1829, NYHS; for the full text of the message, see the *Albany Argus,* Jan. 6, 1829; for comment on it from other papers in New York and elsewhere, see *ibid.,* Jan. 16, 1829; Thomas Ritchie to Van Buren, Jan. 31, 1829, VBLC; James A. Hamilton to Van Buren, Jan. 8, 1829, VBLC; Hammond, *New York,* II, 297.

10. Forsyth to Van Buren, July 25, 1828, VBLC; *Calhoun Papers,* X, 435, 436; Van Buren to Crawford, Feb. 15, 1828, Crawford Papers, LC; Crawford to Van Buren, Oct. 21, 1828; Van Buren to Crawford, Nov. 14, 1828, VBLC; Edward Everett to Joseph Blunt, Nov. 23, 1828, EMHS; Rives to William Gilmer, Dec. 5, 1828, RLC; Verplanck to Hoyt, Dec. 6, 1829, VBLC; Balch to Van Buren, Nov. 27, 1828, VBLC.

11. Van Buren to Cambreleng, Dec. 17, 1828, VBLC; Wright to Van Buren, Dec. 9, 1828; Lewis to Hamilton, Dec. 12, 1828; Hamilton to Van Buren, Jan. 1, 1829, VBLC; Van Buren to Cambreleng, Jan. 9, 1829, VBLC; Van Buren to Rives, Jan. 7, 1829, RLC, Hamilton, *Reminiscences,* 80.

12. Cambreleng to Van Buren, Jan. 1, 1829, VBLC; Hamilton, *Reminiscences,* 80–82.

13. Benton to Van Buren, Dec. 3, 1828, VBLC.

14. Hamilton, *Reminiscenses,* 88, 89; James Hamilton, Jr. to Van Buren, Jan. 23, 1829, VBLC; Hamilton, *Reminiscences,* 91, 92.

15. Edward Everett to My Dearest A., Feb. 15, 1829, EMHS; Hamilton to Van Buren, Feb. 12, 1829, VBLC.

16. There is some evidence that this move originated with Calhoun. As early as mid-November, William C. Rives wrote T. W. Gilmer, "Van Buren (I know) and most probably Calhoun, and all the rest of the candidates for the Presidential purple are in close correspondence with Ritchie and suing for his support." Rives to Gilmer, Nov. 13, 1828, RLC; Shepard, Van Buren's best biographer, accepts Rives comment as does James A. Hamilton; Shepard, *Martin Van Buren,* 180; Hamilton, *Reminiscences,* 90; Ambler, *Ritchie,* 127; Hamilton to Van Buren, Feb. 13, 1829, VBLC; *Richmond Enquirer,* Feb. 9–13, 1829.

17. Everett to Dearest A., Feb. 15, 1829, EMHS; Hamilton to Van Buren, Feb. 13, 1829, VBLC; *United States Telegraph,* Feb. 16, 1829.

18. Van Buren to Cambreleng, Feb. 1, 1829, VBLC; Jackson to Van Buren, Feb. 15, 1829, VBLC; Hamilton, *Reminiscences,* 93.

19. James A. Hamilton to Van Buren, Feb. 18, 1829; James Hamilton, Jr. to Van Buren, Feb. 19, 1829, VBLC; Munroe, *McLane,* 245–47; Van Buren to Jackson, Feb. 20, 1829, VBLC. The letter of acceptance shows much rewriting by Van Buren, Marcy and Butler; Hamilton, *Reminiscences,* 96–97.

20. C. P. Van Ness to Van Buren, Feb. 23, 1832, VBLC; Wiltse, *Calhoun,* II, 22, 23; Elias K. Kane to Van Buren, Feb. 19, 1829; Rudolph Brunner to Van Buren, Feb. 21, 1829; J. A. Hamilton to Van Buren, Feb. 27, 1829, VBLC; *United States Telegraph,* Feb. 26, 1829; *Albany Argus,* Feb. 28, 1829. That there was foundation for Van Buren to feel uneasy about Branch and Ingham can be seen in Rives's comment that they "hold principles highly favorable to the south." Rives to T. C. Gilman, n.d., RLC; Cambreleng to Van Buren, Mar. 1, 1829, VBLC; Amos Kendall to Francis P. Blair, Mar. 7, 1829, BLP. See Ritchie to M. M. Noah, Mar. 14, 1829; as did James A. Hamilton, who wrote "Be assured Calhoun is disappointed." Hamilton to Van Buren, Feb. 21, 1829, VBLC.

21. Hoffman to Flagg, Mar. 7, 1829, FNYPL; James Hamilton, Jr. to Van Buren, Mar. 5, 1829, VBLC.

22. For detailed coverage of Van Buren's administration, see *Albany Argus,* Jan.–Mar. 1829; Hamilton to Van Buren, Mar. 6, 1829, VBLC; see, for instance, Van Buren's notes on northeastern and northwestern boundaries of the United States, and letter of C. P. Van Ness to Gallatin, Jan. 3, 1829, VBLC; Cambreleng to Van Buren, Mar. 1, 1829, VBLC.

23. John Van Buren to Dear Father, Apr. 18; May 1829, JVBH; MacKenzie, *Van Buren,* 211; *Albany Argus,* Feb. 17, 19, 20, 1829; Mar. 13, 1829; James A. Hamilton to Van Buren, Mar. 6, 1829; Van Buren to Oliver Wiswall, Mar. 17, 1829; to Gentlemen, New York City, Mar. 23, 1829, VBLC; Van Buren, *Autobiography,* 221.

24. Marcy to Van Buren, May 6, 1829, VBLC.

Chapter 14. All the Coppers in the Mint

1. Hamilton, *Reminiscences,* 129; *Albany Argus,* Mar. 21, 1829; Van Buren, *Autobiography,* 228.

2. Nevins, ed., *Hone Diary,* 69–72; Walter Hugins, *Jacksonian Democracy and the Working Class* (Palo Alto, 1960), 11, 12; *Albany Argus,* Mar. 28, 29, 1829.

3. William B. Hatcher, *Edward Livingston* (Baton Rouge, 1940), 333, 334; Parton, *Andrew Jackson,* III, 220, 229; Van Buren, *Autobiography,* 229–30. Capowkski, "Woodbury," 203; Albert Gallatin had applied directly to Van Buren for the French mission on Mar. 4, 1829. Van Buren answered on Mar. 8 saying that he had sent Gallatin's letter to the President and that he would see Gallatin in New York "in about ten days." There is no acknow-

ledgment in the Jackson Papers; Gallatin to Van Buren, Mar. 4, 1829, VBNYHS.

4. Van Buren, *Autobiography,* 251.

5. *Ibid.,* 230; Munroe, *McLane,* 251, 252.

6. Wiltse, *Calhoun,* III, 12; Remini, *Jackson,* II, 192–94; Van Buren, *Autobiography,* 232.

7. William C. Rives to T. P. Moore, Mar. 18, 1829, VBLC; Van Buren, *Autobiography,* 251, 252; Tazewell to Jackson, Mar. 30, 1829, JLC; Van Buren, *Autobiography,* 251–59; see Van Buren to Livingston, Apr. 6, 1829; Livingston to Jackson, Apr. 17, 1829; Jackson to Van Buren, Apr. 20, 1829, JLC; Hamilton, *Reminiscences,* 138; Livingston to Jackson, May 3, 1829; to Van Buren, May 1, 1829, VBLC. The new secretary had regained the initiative after only two weeks at his new post, but what was more important he established warm personal and official relations with the President and with his small circle of advisors; Van Buren to W. C. Rives, May 5, 1829, VBLC; Munroe, *McLane,* 234.

8. *Ibid.,* 252, 257, 260. As he wrote Van Buren on April 15, "I was not altogether without my fears when I sent off my dispatches yesterday that Mr. B[errien] might not assent to the change." McLane to Van Buren, Apr. 15, 1829, VBLC; Hamilton, *Reminiscences,* 131.

9. Van Buren, *Autobiography,* 260; Van Buren to Jackson, Apr. 14, 1829, in Hamilton, *Reminiscences,* 131, 132. Indeed, he was probably happy with the appointment since it would ease Hughes's debts, a source of worry to the old man. John S. Pancake, *Samuel Smith and the Politics of Business* (Tuscaloosa, 1972), 102, 146; Woodbury to Van Buren, Apr. 27, 1829; *Jackson Correspondence,* III, 22.

10. Hamilton, *Reminiscences,* 138; Jackson to Van Buren, Mar. 31, 1829, JLC; *United States Telegraph,* Sept. 27, 1830; Ritchie to Van Buren, Mar. 27, 1829; W. S. Archer to Van Buren, May 6, 1829, VBLC; John Rutherford to Rives, May 22, 1829, RLC; Cole, *New Hampshire,* 86–88; MacKenzie, *Van Buren,* 214, 215; James A. Hamilton to Van Buren, Mar. 25, 1829; Van Buren to Rives, May 5, 1829, VBLC.

11. Jackson to Van Buren, Apr. 4, 1829; Van Buren to John Campbell, Apr. 17, 21, 1829, Ritchie to Van Buren, Apr. 13, 19, 1829, VBLC.

12. Van Buren, *Autobiography,* 418, 419; Randolph to Van Buren, May 2, 1830, VBLC.

13. See, for example, *Jackson Correspondence,* III, 268, 269, 268n.; Myers, *Tammany Hall,* 63, 76.

14. Hoyt to Van Buren, Apr. 11, 1829; Cambreleng to Van Buren, Apr. 15, 1829, VBLC; MacKenzie, *Van Buren,* 216; Van Buren to Cambreleng and Bowne, Apr. 1829; to Dudley, Apr. 20, 1829, VBLC.

15. Van Buren to Jackson, Apr. 23, 1829, Van Buren, *Autobiography,* 264, 265; Van Buren to Cambreleng, Apr. 25, 1829; Dudley to Van Buren, Apr. 25, 1829, VBLC.

16. Van Buren finally fell back on Nicholas Trist, who was a clerk when he took office, and who was married to one of Jefferson's granddaughters. Trist, a fussy individual, was not as able as either Hamilton or Butler, but he was conscientious, proficient in several languages, a Virginian and not likely to promote political jealousies which at this point Van Buren was most

anxious to avoid. Hamilton, *Reminiscences,* 103, 127, 129, 131, 137; Butler to Van Buren, May 5, 1829, VBNYSL; Van Buren to Richard E. Parker, May 11, 1829; Van Buren, *Autobiography,* 265; Van Buren to Cambreleng, Apr. 23, 1829, VBLC.

17. Van Buren to Cambreleng, Apr. 25, 1829, VBLC; Van Buren, *Autobiography,* 265–67.

18. Cambreleng to Van Buren, Apr. 28, 1829, VBLC.

19. See Cambreleng to Van Buren, Apr. 28, 1829, VBLC; Cambreleng to the Collector of the Port, Apr. 8, 30, 1829; McKenzie, *Van Buren,* 219, 220; Butler to Van Buren, May 6, 1829, VBNYSL; James A. Hamilton to Van Buren, June 19, 1829, VBLC.

20. Calhoun to Gouverneur, Mar. 30, 1830; Morton to Calhoun, Mar. 7, 1829; *Calhoun Papers,* XI, 7, 25, 26, 143; Baylies to William Smith, May 20, 1829, Baylies Misc. Papers, MHS.

21. Kendall to F. P. Blair, Jan. 9; Feb. 3, 1829, BLP; Michael Hoffman to Chancellor Walworth, Feb. 10, 1829, Hoffman Papers, NYHS.

22. McKenzie, *Van Buren,* 211; John W. Taylor to Jane Taylor, Jan. 3, 1830, TNYHS.

23. Van Buren to John Van Buren, June 25, 1830, VBLC; Edward Everett to Van Buren, Aug. 10, 1829, VBMHS.

24. John Van Buren to Van Buren, Apr. 8, May, 1829, JVBH; to Van Buren, Sept. 6, 1829; Butler to Van Buren, May 6, 1829, VBNYPL; Edward Everett to Van Buren, Aug. 10, 1829, VBMHS; James A. Hamilton to Van Buren, Aug. 3, 1829, VBLC.

25. John Niven, *Gideon Welles* (New York, 1973), 60–64; James Gordon Bennett to Duff Green, July 20, 1829, Green Papers, UNC; Hamilton *Reminiscences,* 129; *Calhoun Papers,* XI, 143, 144.

26. Kendall to Blair, Nov. 22, 1829, BLP.

Chapter 15. Intrigues and Intriguers

1. *Adams Memoirs,* VIII, 128, 129.

2. Van Buren to Throop, July 16, 1829, VBNYSL.

3. James Hamilton, Jr. to Van Buren, Nov. 16, 1829, VBLC; Van Buren to John C. Spencer, July 15, 1829, Misc. Papers, NYHS; to M. Younglove, Dec. 8, 1829, VBLC. In one of his last acts as governor, Van Buren appointed the capable, though cantankerous, John C. Spencer, as special prosecutor for the state in the Morgan affair. Spencer had opposed Van Buren for years, yet his was a logical appointment and he performed his duties with his accustomed diligence. From Washington the ex-Governor maintained a correspondence with Spencer who unwittingly kept him abreast of the anti-Masonic party in western New York.

4. *Albany Argus,* Nov. 13, 1829; Ambler, *Ritchie,* 15, 123; Gooch to Van Buren, Sept. 14, 1829, VBLC; see, for example, Gooch to George Thompson, Aug. 21, 1828, Gooch Papers, UVA; but Van Buren was careful to refrain from politics or public issues like the Bank of the United States; Gatell, "Sober Second Thoughts," 22, 23.

5. Ambrose Spencer, then a member of the House, thought it would come with the next presidential election in a struggle between the slave and the free states. "Most present appearances," he said, "indicate Mr. Calhoun is the candidate on the one side and Mr. Van Buren on the other." Although no friend of Van Buren, the ex-justice of New York's supreme court wondered "whether it be wise or politic to keep up prejudices against the man we may be compelled to support. . . ." Edward Everett to A. H. Everett, Dec. 29, 1828, EMHS; Spencer to William L. Stone, Ambrose Spencer Papers, NYSL; Bassett, ed., *Jackson Correspondence,* IV, 108–10, 110n.

6. Lyon C. Tyler, ed., *The Letters and Times of the Tylers* (Richmond, 1884), 2 vols., I, 423.

7. Levi Woodbury to Mrs. Elizabeth Woodbury, Mar. 6, 1830, WOLC.

8. *Jackson Correspondence,* IV, 68–72, 88, 123, 124; Van Buren, *Autobiography,* 352.

9. A disgruntled John Floyd of Virginia recognized the interaction, but had it completely reversed. See Floyd to Colonel John Williams, Dec. 27, 1830, Floyd Misc. Papers, LC; Richard B. Latner, "The Kitchen Cabinet and Andrew Jackson's Advisory System," *Journal of American History,* LXV, No. 2, Sept. 1978, 368–74; Remini, *Jackson,* II, Ch. 19, *passim;* A. H. Shepperd to Charles Fisher, Jan. 30, 1830, Fisher Family Papers, UNC; Henry R. Storrs to Peter B. Porter, Mar. 10, 1830, Porter Papers, Buffalo Hist. Soc. Richard P. Longaker, "Was Jackson's Kitchen Cabinet a Cabinet"? *Mississippi Valley Historical Review,* Vol. 44, (June 1957), 101–3.

10. Crouthamel, *Webb,* 31, 35; Hamilton, *Reminiscences,* 154; Van Buren, *Autobiography,* 398. Van Buren attributed the premature announcement to Noah "an editor proverbially imprudent." Van Buren, *Autobiography,* 398. Webb later suggested Van Buren as a suitable replacement for Calhoun; Crouthamel, *Webb,* 25, 30. Opposition papers kept the Van Buren candidacy before the public and this is just how Van Buren wanted it treated. Jackson to Coffee, Dec. 31, 1829, JLC.

11. *Register of Debates,* 21st. Cong., 1st Sess., I, 31–41; Sidney Nathanson, *Daniel Webster and Jacksonian Democracy* (Baltimore, 1973), 32–34; William W. Freehling, *Prelude to Civil War: The Nullification Controversy in South Carolina* (New York, 1968), 183–86; Parton, *Jackson,* III, 282; John S. Barbour to James Barbour, Mar. 1, 1830, P. p. barbour Papers, NYPL.

12. *Calhoun Papers,* XI, 117, 118, 122, 123, 143; Jackson to John Coffee, Apr. 10, 1830, JLC.

13. Van Buren, *Autobiography,* 409–13.

14. *Ibid.,* 414.

15. Benton, *Thirty Years' View,* I, 148, 149; Van Buren, *Autobiography,* 412–17; *United States Telegraph,* Apr. 14, 1830; *Daily National Intelligencer,* April 14, 15, 1830. The description and events of the dinner have been the subject of debate among participants and historians. Wiltse, in his monumental biography of Calhoun, declares Jackson made the toast with the adjective "Federal" included and that Hayne did not suggest the addition until after the toast was given. Benton, in his *Thirty Years View,* declares that 80 volunteer toasts in all were given and does not mention the Hayne episode. Van Buren states that his and the President's toasts were written out just before

they left for the dinner. Marquis James, in his biography of Jackson, follows Van Buren's account. Wiltse, drawing on Duff Green, a rather unreliable source, and upon a copy of the toast in Major Lewis's handwriting, disputes the Van Buren account. His evidence is at best circumstantial. Richard R. Stenberg's detailed analysis of the occasion, asserts that the dinner was not a nullification affair, but was made so through the machinations of Jackson and Van Buren. Their object, particularly Van Buren's, was to injure Calhoun politically. Stenberg, with some justice, is particularly critical of Benton's account and his own part in it. But Stenberg's sources are primarily from Calhoun or his group and from Webster and the National Republican press, each in his own way having an interest in attacking Jackson and Van Buren. W. R. Stenberg, "The Jefferson Birthday Dinner," *The Journal of Southern History,* IV, No. 3, Aug. 3, 1928, 334–45; for a vivid description of the dinner, see Remini, *Jackson,* II, 234, 237.

16. Van Buren to Rives, Apr. 6, 1830, RLC; Samuel Carson to Charles Fisher, Apr. 9, 1830, Fisher Family Papers, HisPA; Van Buren to Dr. Burnett, May 2, 1830, VBLC; Van Buren to Throop, Jan. 10, 1830, VBNYSL; Van Buren to Colonel Jack, Mar. 8, 1830, VBLC; *Albany Argus,* Apr. 8, 1830.

17. Gideon Welles in the *Hartford Times* directly charged Calhoun with being personally responsible for Hill's defeat. So biting was the indictment that Green complained to Welles's partner, John M. Niles, of misrepresentation, while admitting Hill believed Calhoun responsible. Green to Niles, May 17, 1830, Green Papers, UNC; Marcus Morton chided both the Administration and Calhoun for their action on Hill and Noah. Regarding Noah, he wrote Calhoun, "it was *wrong* to nominate him—*wrong* so long to postpone the nomination—*wrong,* the postponement was *wrong*—the rejection was wrong." Cole, *New Hampshire,* 89–90; *Calhoun,* X, 143, 144, 191, 192, 201, 202; Ritchie to Rives, Apr. 15, 1830, RLC.

18. Van Buren, *Autobiography,* Hamilton, *Reminiscences,* 195–98; Parton, III, 323, 324.

19. Van Buren, *Autobiography,* 376; Duff Green to Isaac C. Canby, May 14, 1830, Green Papers, UNC; Van Buren, *Autobiography,* 376–82; Parton, III, 326. Wiltse, *Calhoun,* II, 94, 95.

20. The veto message is in Richardson, ed., *Messages and Papers,* III, 1046–56.

21. Tyler, ed., *Tyler,* I, 411, 412; Crawford to Van Buren, Jan. 3, 1831, VBLC.

22. Richard Pollard to Rives, June 10, 1830, RLC; Van Buren, *Autobiography,* 320, 321, 324; Clay to the Honorable Mr. Washington, May 17, 1830, Clay Papers, J. P. Morgan Library; Jackson to Van Buren, June 26, 1830, VBLC; Thomas Ritchie to Archibald Ritchie, June 8, 1830, in "Letters of Thomas Ritchie," in *Randolph-Macon Historical Papers,* No. 2 (June, 1902), 148. Hereafter cited as *Ritchie Letters.*

23. Van Buren, *Autobiography,* 312–38; *Jackson Correspondence,* IV, 137–39. Madison generally agreed with the veto, but was not quite as restrictive on constitutional limitations as either Jackson or Van Buren. Madison to Van Buren, July 5, 1830, VBLC; Van Buren had asked his opinion on June 5.

24. William Stickney, ed., *Autobiography of Amos Kendall* (Washington, 1872),

370, 371; Bassett (ed.), *Jackson, Correspondence,* IV, 156; Green to Donelson, July 15, 1830, Green Papers, UNC; Van Buren to Cambreleng, June 25, 1830; to Rives, July 1, 1830, VBLC; William E. Smith, *The Francis Preston Blair Family in Politics* (New York, 1933), 2 vols., I, 60–61; Van Buren to Jackson, July 25, 1830, VBLC; Stickney, ed., *Kendall,* 372; Kendall to Dear Sir, Nov. 26, 1830, Kendall Misc. Papers, NYHS; Green to N. Green, Nov. 17, 1830, Green Papers, UNC. Blair was also corresponding with Jackson on political matters in Kentucky; *Jackson Correspondence,* IV, 166, 167, 174; Benton, *Thirty Years View,* I, 129; James A. Hamilton added his weight to the anti-Green forces in a letter Jackson received at the Hermitage.

25. Quoted by Lynn Marshall, "The Early Career of Amos Kendall, the Making of a Jacksonian," Unpub. Diss., U.C. Berkeley, 1962, 199; Van Buren, *Autobiography,* 568.

26. John A. Dix to Micah Sterling, Dec. 24, 1830, Dix Papers, Columbia University; Niven, *Welles,* 80–82; Calhoun to Virgil Maxcy, June 27; Aug. 6, 1830; *Calhoun Papers,* XI, 207, 208, 214, 215, 217, 218; *Albany Argus,* June 5; July 28; Aug. 31, 1830; Van Buren to Jackson, July 25, 1830, JLC; Alexander Hamilton to Maxcy, Aug. 8, 1830; Galloway-Maxcy-Markoe Papers, LC; Robert Y. Hayne to Van Buren, Oct. 13, 1830, VBLC; Joel Poinsett to Jackson, Oct. 23, 1830, Poinsett Papers, HiSPA; James Hamilton, Jr. to Van Buren, May 29, 1830; Crawford to Van Buren, May 31, 1830, VBLC; W. H. Rives to Rives, Apr. 30, 1830, RLC; the *Globe,* Jan. 22, 1831; Pollard to Rives, June 10, 1830, RLC; Van Buren, *Autobiography,* 377; the *Globe,* Jan. 22, 1831; though Grundy felt certain the breach was irreparable—"I have done all I could to bring them together," he said, "and have not succeeded. . . . If men want to differ, a peacemaker may make himself odious by interfering too much. . . . I told Mr. Calhoun that in a private quarrel between them, I had no concern, in a public one—my feelings would be with Gen'rl. Jackson. . . . he hoped [he] would not go to extremities as to offend their friends." Grundy to Daniel Graham, Jan. 24, 1831; to Eaton, May 17, 1831; Grundy and McGavock Letters, UNC. Rather than publish in the *Telegraph,* all agreed it should come out in the *Globe.*

27. Blair to Van Buren, Dec. 9, 18, 1830, VBLC; the *Globe,* Feb. 8, 1831.

28. Edward Everett to Mrs. Everett, Feb. 22, 1831, EMHS; James A. Hamilton to Van Buren, Feb. 3, 1831; to W. B. Lewis, Mar. 6, 1831, VBLC.

29. See, for instance, George Crawford to Blair, Mar. 27, 1831, BLP; Niven, *Welles,* 59, 61, 79–87. *United States Telegraph,* Feb. 25, 26, 1831.

30. See the *Globe,* Mar. 14; Apr. 1, 1831; *United States Telegraph,* Apr. 14, 15, 1831; James B. Gardiner to W. B. Lewis, July 5, 1830, *Misc. Papers,* GAR-GEI, LC.

31. "The quarrel between Calhoun and Van Buren, or rather Jackson, completely paralyzes the Administration. Calhoun has friends enough in the Senate to defeat any and every measure, by joining with Mr. Clay's friends," observed Edward Everett. Everett to Mrs. Everett, Mar. 2, 1831, EMHS; Van Buren to John Van Buren, Mar. 27, 1831, VBLC.

32. Van Buren, *Autobiography,* 401–8; Jackson to Van Buren, July 25, 1831, VBLC; the *Globe,* July 23, 1831; Ingham to Berrien, Apr. 20, 21, 26; May 4, 7, 1831, Berrien Papers, UNC.

33. See, for instance, William S. Archer to Van Buren, Mar. 12, 1831,

VBLC; Van Buren to Jackson, Apr. 11, 1831, JLC; see the nine-page manuscript draft in VBLC same date; Van Buren to Butler, Apr. 16, 1831, VBLC; Van Buren to Ritchie, Apr. 17, 1831, VBLC; Ritchie to Van Buren, Apr. 20, 21, 1831, VBLC.

34. Butler suggested four names of men who he thought should go into the cabinet and who were offered places, Livingston, McLane, Woodbury and Drayton, "a favorite with Mr. Cambreleng and as far as I can judge would be acceptable as Secretary of War." Butler to Van Buren, Apr. 22, 23, 1831, VBLC; Jackson was grateful when he learned that Hill applauded Woodbury's appointment. "This will unite all friends in New Hampshire," said Jackson; Jackson to Major Lewis, May 7, 1831, JLC; Duff Green to N. Green, June 1, 1830, Green Papers, UNC, Green and Woodbury were on good terms. Green to Woodbury, June 28, 1830, WLC; Van Buren, *Autobiography,* 577; Van Buren to Livingston, Apr. 9, 1831, VBLC; Van Buren, *Autobiography,* 704, 705. See, also, Jackson to Van Buren, Dec. 17, 1831, VBLC; *Albany Argus,* May 6, 1831; F. N. Poinsett to Jackson, Oct. 23, 1830, Poinsett Papers, HiSPA; Ambler, *Ritchie,* 131; Van Buren to Smith, Nov. 2, 1830.

35. *Calhoun Papers,* XI, 390, 391.

Chapter 16. Secretary of State

1. Van Buren, *Autobiography,* 445; John Van Buren, *MS. Diary,* 2, JVBH.

2. Edward Everett to Van Buren, Aug. 10, 1829, EMHS; Van Buren to John Van Buren, Mar. 27, 1831, VBLC; see, for instance, Van Buren to John Van Buren, June 25, 1830, VBLC.

3. Van Buren to John Van Buren, Apr. 1, 1829, VBLC; Van Buren to Rives, Jan. 7, 1832, RLC.

4. Cambreleng to Van Buren, Jan. 14, 1832, VBLC; Butler to Van Buren, Apr. 22, 1831, VBLC; Van Buren, *Autobiography,* 445–47; Van Buren to Harmanus Bleecker, Feb. 25, 1827, VBNYSL; see also Van Buren's speech on the West Indies Trade in the Senate, Feb. 24, 1827, in *Albany Argus,* May 10, 1827.

5. Van Buren, *Autobiography,* 260–62, 450; see, for example, Hamilton, *Reminiscences,* 149.

6. Van Buren to Samuel Smith, Apr. 28, 1829, Samuel Smith Papers, LC.

7. Samuel Smith to Van Buren, May 11, 14, 1829, Samuel Smith Papers, LC; Hamilton, *Reminiscences,* 142.

8. Cambreleng to Van Buren, Apr. 9, 1830, VBLC; see *Albany Argus,* Jan. 21, 1832. For debate on Van Buren's instructions to McLane, see *Executive Proceedings,* 22nd Cong., 1st Sess., 1310–36; Van Buren, "Notes," July 20, 1829, VBLC; Van Buren to Cambreleng, July 19, 1829, VBLC.

9. Hamilton, *Reminiscences,* 143; Van Buren to Donelson; to Cambreleng, July 27, 1829, VBLC; Hamilton, *Reminiscences,* 141, 142; Van Buren to Cambreleng, July 27, 1829, VBLC; *United States Telegraph,* July 17, 1829; Hamilton, *Reminiscences,* 143; James Hamilton, Jr. to Van Buren, Aug. 3, 1829, VBLC.

10. Van Buren, *Autobiography,* 251, 272, 273; Hamilton, *Reminiscences,* 142;

"General Instructions," July 20, 1829, *Register of Debates,* 22nd Cong., 2nd Sess., "Appendix," 210–17.

11. Van Buren claims in his *Autobiography* that all the Van Nesses, including Cornelius had been hostile to him and that he had nothing to do with the appointment. But the facts contradict his recollections. See Van Ness to Van Buren, Aug. 1, 1823; Van Buren to Van Ness, Aug. 15, 1823; Van Ness to Van Buren, Aug. 28, 1823, VBLC; *Autobiography,* 260.

12. *Richmond Enquirer,* Aug. 24, 1829; Hamilton, *Reminiscences,* 142; see Butler to Jackson, Aug. 11, 1829, VBLC; see also Van Buren "Notes on Texas" and Secretary to Mr. Poinsett, No. 30, Aug. 26, 1829, copy by State Department Clerk Aaron Vail, together with a draft commission for Poinsett to negotiate for a cession, VBLC; E. C. Barker, "President Jackson and the Texas Revolution," *American Historical Review,* July 1907, vol. 12, no. 4, 789–91; Jackson to Van Buren, Aug. 12, 1829; Jackson, "Notes for Instructions to Poinsett," Aug. 13, 1829; Donelson, "Project for the Acquisition of the Province of Texas," Aug. 13, 1829, VBLC; *Jackson Correspondence,* IV, 66, 79–82. For Poinsett's recall see Jackson to Van Buren, n.d., 1829; Van Buren to Poinsett, Oct. 17, 1829, VBLC; J. Fred Rippy, *Joel R. Poinsett* (Durham, 1925), Ch. IV, *passim;* Van Buren to Cambreleng, Aug. 25, 1829, VBLC.

13. Hamilton, *Reminiscences,* 143, 145; *House Exec. Docs.,* 22nd Cong., 1st Sess., Doc. No. 250, 69–75; Hamilton, *Reminiscences,* 146.

14. Jackson to Randolph, Sept. 16, 1829; Van Buren to Randolph, Sept. 17, 1829; Randolph to Jackson, Sept. 24, 1829, John Randolph Papers, LC; Richardson, ed., *Messages and State Papers,* III, 1006–7.

15. *Register of Debates,* "Appendix," 22nd Cong., 2nd Sess., 210–60; Rives to Van Buren, Sept. 8, 1830, VBLC; Rives to Van Buren, Feb. 14, 1831, VBLC; Van Buren to Rives, July 28, 1829, RLC.

16. *Jackson Correspondence,* IV, 133.

17. McLane to Van Buren, Mar. 22, 1830, *State Department Diplomatic Dispatches, Great Britain,* XXXVII, Record Group 49, NA; McLane to Cambreleng, Mar. 30, 1830, VBLC; Richardson, ed., *Messages,* III, 1043, 1044; Van Buren to McLane, June 29, 1830; *Dispatches Great Britain,* XXXVII.

18. F. Lee Benns, *The American Struggle for the British West India Carrying Trade, 1815–30* (Bloomington, 1923), Indiana University Studies, X, 185–88.

19. *Jackson Correspondence,* III, 73; Van Buren to Randolph, Sept. 17, 1829, Randolph Papers, LC; Randolph to Van Buren, May 2, 1830; Jackson to Van Buren, May 4, 1830; Van Buren to Randolph, May 6, 1830, VBLC; *Albany Argus,* June 26, 1830; Richard Rush to Mr. Fendall, June 23, 1830, Rush Papers, LC; Van Buren to Jackson, June 25, 1830; Van Buren to Thomas Ritchie, Nov. 5, 1830; William Leigh to Van Buren, Mar. 7, 1831; William S. Archer to Van Buren, Mar. 29, 1831, VBLC; Van Buren, *Autobiography,* 421–31; William Cabell Bruce, *John Randolph of Roanoke* (New York, 1922), 2 vols., I, 635–36, 646, 654, 656; II, 13, 14.

20. Van Buren was not long in discovering his mistake. Cornelius Van Ness, who as former governor of Vermont knew Preble well, wrote from Madrid that "the state of Maine has infamously abused all who have heretofore been concerned in that business, and . . . the general government

has now put the business into the hands of the man of its choice." Van Ness to Van Buren, Feb. 10, 1830; see also Preble to Van Buren, Jan. 17, 1831; Van Buren to Jackson, Sept. 28, Oct. 14; Van Buren to Livingston, Nov. 5; Jackson to Van Buren, Dec. 17; Van Buren to Livingston, Jan. 14, 1832, VBLC.

21. See "Instructions and Correspondence" in *House Exec. Docs.,* 22nd Cong., 1st Sess., Doc. No. 250, 68–95.

22. The original sum claimed was over $12 million, but this amount had been inflated for bargaining purposes and contained many claims of dubious validity. See Van Buren to Rives, July 20, 1829; *Register of Debates,* 22nd Cong., 1st Sess., "Appendix," 210, 214.

23. Van Buren to Abraham, John and Martin, Jr., May 24, 1831, VBLC; Hamilton, *Reminiscences,* 221; Van Buren to Serurier, May, 1831, VBLC; Van Ness to Rives, July 6, 1831, RLC; Van Buren to John Forsyth, June 8, 1831, VBLC; James Gordon Bennett, *MSS Diary,* June 21, 1831, NYPL; *Albany Argus,* July 11, 1831.

24. Hamilton, *Reminiscences,* 226, 227; Munroe, *McLane,* 290; See Edward Livingston, Department of State, Washington, to Martin Van Buren, in Vail's handwriting, "Commission," Aug. 1, 1831; Van Buren to Woodbury, Aug. 4, 1831, VBLC.

25. Van Buren to Livingston, Sept. 9, 1831, VBLC; Van Buren to Rives, Sept. 11, 1831, RLC; Munroe, *McLane,* 269; McLane to Van Buren, Aug. 10, 1831, VBLC.

26. Van Buren to Lord Palmerston, Oct. 11, 1831, VBLC; see also memorandum of expenses, Dec. 1831; Rives to Van Buren, Oct. 22, 1821, VBLC; Van Buren to Rives, Nov. 1, 1831, RLC.

27. Van Buren to Livingston, Sept. 14, 1831, VBLC; Washington Irving to Livingston, Sept. 22, 1831 in "Despatches from United States Ministers." Record Group 59. NA; Van Buren to Cambreleng, Oct. 14, 1831, VBLC.

28. Van Buren to Jackson, Sept. 28, 1831; to Livingston, Nov. 5, 1831, VBLC; Van Buren to Samuel Smith, Nov. 21, 1831, Samuel Smith Papers, LC; Van Buren to Rives, Nov. 1, 1831, RLC; Rives to Van Buren, Sept. 29; Van Buren to Jackson, Oct. 14, 29, 1831, VBLC; for a recapitulation of his short career as minister see Van Buren to Jackson, Mar. 9, 1832, VBLC.

29. Van Buren to Lady Holland, Dec. 1831, VBLC; John was not at all abashed by these *Grandes Dames.* He thought Lady Holland "a coarse strong-minded, masculine woman of whom everybody is afraid," himself excepted, of course. John Van Buren *MSS. Diary,* 21, JVBH. Tom Moore, the poet, visited Van Buren and left an interesting account of Van Buren's attitude toward himself and high society. Moore, who thought Van Buren "a well-bred and intelligent man" recorded him as saying: "If there is anything which rank and station cannot do in England, I have not found it out." Moore observed that anyone like Van Buren or himself who "lived with that class without naturally belonging to them was apt to be regarded with suspicion by [his] own equals, were naturally inclined to say 'oh yes, he is flattered by living with the great and therefore flatters them in return'." Van Buren agreed at the same time declaring that he was disgusted with "the perpetual struggle towards this higher region that was visible in those below it; all

trying to get above their own sphere and sacrificing comfort and temper in the effectual effort." One wonders whether Van Buren was speaking of himself or had a concealed distaste for American democracy with its emphasis on equality of opportunity. This extract from Russell's *Memoirs,* VI, "Journal and Correspondence of Thomas Moore" was copied out by Aaron Vail on Mar. 30, 1832 for Van Buren, VBLC.

30. Van Buren, *Autobiography,* 473–79; Van Buren to Livingston, Nov. 22, 1831, VBLC.

31. Rives to Van Buren, Oct. 26, 1831, VBLC; Van Buren to Rives, Nov. 1, 1831, RLC; Rives to Van Buren, Nov. 12, 1831, VBLC; see Van Buren to Rives, Jan. 2, 1832, RLC, for his summer plans; Jackson to Van Buren, Dec. 17, 1831; Hamilton to Van Buren, Dec. 23, 1831; Webb to Van Buren, Dec. 31, 1831; Cambreleng to Van Buren, Jan. 4, 1832, VBLC; *Jackson Correspondence,* III, 396.

32. Cambreleng to Van Buren, Jan. 4; James A. Hamilton to Van Buren, Feb. 12; Van Buren to Jackson, Feb. 20, 1832, VBLC.

Chapter 17. Minister as Martyr

1. Van Buren, *Autobiography,* 453, 454; Van Buren to Jackson, Feb. 20, 21, 1832; VBLC. Blair to Van Buren, Apr. 25, 1858, VBLC; Cambreleng to Van Buren, Feb. 5, 1832, VBLC. For debate on the confirmation, see *Executive Proceedings,* 22nd Cong., 1st Sess., 1310–36; John J. Crittenden to Clay, Feb. 23, 1832, Crittenden Papers, LC.

2. Van Buren, *Autobiography,* 455, 456; Van Buren to Jackson, Mar. 6, 1832, VBLC. Lord Palmerston went out of his way to show respect for the fallen minister. He arranged to have King William invite him for an informal country weekend at Windsor where he thoughtfully included among the guests Van Buren's close friend, Charles Vaughan, the British minister to the United States, who was still on leave from his post. Van Buren to Jackson, Mar. 28, 1832, VBLC.

3. Lowrie to Van Buren, Jan. 27, 1832, VBLC.

4. Charles Butler to Van Buren, Jan. 21, 1832; Marcy to Van Buren, Jan. 26; Forsyth to Van Buren, Jan. 28; Elijah Hayward to Van Buren, Jan. 30, 1832; Cambreleng to Van Buren, Feb. 4, 1832; James A. Hamilton to Van Buren, Feb. 12, 1832; Marcy to Van Buren, Feb. 12, 1832; Isaac Hill to Van Buren, Feb. 12, 1832; Van Buren to John Van Buren, Feb. 23, 1832, VBLC; Hamilton, *Reminiscences,* 237–43.

5. Blair to Van Buren, Jan. 28, 1832, VBLC; Ambrose Spencer to John W. Taylor, Jan. 31, 1832, TNYHS.

6. Marcy to Van Buren, Jan. 26, 1832, VBLC.

7. Van Buren to Jackson, Feb. 20, 1832, VBLC.

8. N. P. Tallmadge et al. to Van Buren, Feb. 10, 1832; Van Buren to Gentlemen, Mar. 14, 1832, in *Albany Argus,* Feb. 3, 4; Apr. 26, 1832. There seems to be an error in dating. Van Buren informs Jackson on Mar. 6 that he had received no news from the United States since Feb. 1st, and had sent his answer to the New York committee by the last packet, probably March

1st or 2nd. But Van Buren wrote two responses, one to the citizens of Albany on Mar. 14, and a much longer one to the legislative committee. It is likely that his letters missed the packet and that he changed the date. Van Buren to Jackson, Mar. 6, 1832, VBLC.

9. Washington *Globe,* Jan. 28, 29, 1832, *Richmond Enquirer,* Feb. 1, 2, 3, 1832; Van Buren to John Van Buren, Mar. 8, 1832; Van Buren to Jackson, Feb. 20; Mar. 6, 9, 1832; Cambreleng to Van Buren, Jan. 28, 1832, VBLC; see Washington D.C. *Globe,* Mar.–May, 1832, *passim; Jackson Correspondence,* IV, 417; Van Buren to Jackson, Mar. 9, 13; to Marcy, Mar. 14; to Wright, Mar. 14, 1832, VBLC.

10. Van Buren to Jackson, Apr. 20, 1832, VBLC; *Albany Argus,* May 22, 1832. William B. Lewis, if anyone, was the architect of the first Democratic convention. Parton in his *Life of Jackson* publishes his letter to Kendall suggesting this course, because so many candidates for the vice-presidential nomination had cropped up that he feared destructive rivalries and intrigues. A convention would settle the issue once and for all. Parton reads into this letter a prearrangement for Van Buren, but even the closest of readings does not bear out this contention. Parton, *Jackson,* III, 382, 383.

11. Van Buren, *Autobiography,* 590, 591; Jackson to Van Buren, Dec. 17, 1832, VBLC; P. M. Gibbes to Poinsett, May 2, 1832, Poinsett Papers, HiSPA; see Van Buren's *Autobiography,* 582–84, 586, 587, 590, 591 which contains lengthy correspondence between Lewis and Van Buren on the efforts to stop Van Buren's nomination. They were written long after the events they describe occurred, but seem reasonably accurate and certainly confident in tone; see also W. B. Lewis to Joseph A. Larwill, Mar. 15, 1839; Larwill to Lewis, Mar. 27, 1839, VBMHS.

12. Niles to Welles, May 22, 1832, Welles Papers, LC; *Albany Argus,* May 26, 1832.

13. Alexander, *New York,* I, 384, 386; see *Proceedings of a Convention of Republican Delegates . . . Nominating a Candidate for Vice President in 1832* (Albany, 1832); Morgan Dix, comp., *Memoirs of John Adams Dix* (New York, 1853), 2 vols., I, 110, 120, 121, 130–33. For a comprehensive and stimulating account of the conventions of 1832 and the campaign, see Robert V. Remini, "Election of 1832" in Arthur M. Schlesinger, Jr., ed., *History of American Presidential Elections,* 4 vols. (New York, 1971), I, 495–566.

14. Robert Lucas et al. to Van Buren, May 22, 1832; Van Buren to Lord Palmerston, May 31, 1832, VBLC; "Public Dinner in Honor of Martin Van Buren, June 25, 1832," VBMHS. On the day Van Buren left for Washington there were 12 deaths from cholera in New York City and 42 new cases. *Albany Argus,* July 7, 9, 1832; *Jackson Correspondence,* III, 461, 462.

15. Benton made a series of lengthy speeches against the Bank beginning on Feb. 1, 1831. The one referred to in the text was delivered during the debate over the recharter bill in the spring of 1832. See Benton, *Thirty Years' View,* I, 220–50; see also his remarks on the veto, which are vintage Bentonian, *ibid.,* 256–65; Van Buren, *Autobiography,* 566, 625; Carl B. Swisher, *Roger B. Taney* (New York, 1936), 178–98; Taney to My Dear Sir, Feb. 20, 1832, Taney Papers, LC; see veto message in Richardson (ed.), *Messages,* III, 1139–54; for Jackson's reference to the Bank, see *ibid.,* 1121; for McLane's Re-

port, see *House Exec. Doc., No. 3,* 22nd Cong., 1st Sess.; Jackson to Van Buren, Dec. 6, 1831, VBLC; Van Buren to Rives, Jan. 7, 1832, RLC. Van Buren was less charitable to McLane on the specific issue of the Bank than the President. And he was not surprised when Cambreleng wrote that Calhoun and Clay would force the recharter bill through Congress and thus force a presidential veto. Cambreleng to Van Buren, Feb. 5, 1832, VBLC; Van Buren, *Autobiography,* 625.

16. *Albany Argus,* June 22, 1819; Ellis, *Columbia County,* 178, 179.

17. Van Buren to Worth, Apr. 27, 1818, VBLC; Van Buren to C. E. Dudley, May 13, 1820, VBLC. See, for instance, *Albany Argus,* Apr. 14, 1814, where Van Buren voted with the majority in the state senate to strike out a banking proviso in a charter for the North America Coal Co., and refusing a charter to the Albany Commercial Bank. Both votes were 16 to 11 along party lines. He was most frequently accused of using his political power to protect the Farmers' and Mechanics' Bank from competition. And there may have been some truth in these assertions. But the bank's practice can be summed up in the notion that all things being equal, it would favor a party stalwart over a party opponent, general policy not just in Albany, but in New York City, Boston and elsewhere.

18. Clinton to John Pintard, n.d., 1820, Clinton Papers, NYPL.

19. Hammond, *New York,* II, 297, 301.

20. Hamilton, *Reminiscences,* 149–75; Robert V. Remini, *Andrew Jackson and the Bank War* (New York, 1967), 61–65; *Jackson,* II, 222.

21. Jackson to Van Buren, Nov. 1, 1830, with enclosure, VBLC; see Hamilton, *Reminiscences,* 190, 191; Richardson, ed., *Messages,* III, 1092.

22. Van Buren, *Autobiography,* 635, 663; Van Buren, *Inquiry,* 224–28; Van Buren, *Autobiography,* 510; to Jackson, Feb. 20, 21, 1832, VBLC; Van Buren, *Autobiography,* 627; Woodbury to Mrs. Woodbury, July 10, 1832, WOLC; Van Buren, *Inquiry,* 224–28; *Autobiography,* 621.

23. Van Buren, *Autobiography,* 501, 502, 541, 566–68; Hamilton, *Reminiscences,* 246; Van Buren to Throop, July, 1832, VBNYSL.

24. Henry Clay to Weed, Apr. 14, 1832; John C. Spencer to Weed, July 13; Sept. 21, 1832; Barnes and Weed, eds., *Weed,* II, 42–45; Frederick W. Seward, ed., *William H. Seward: An Autobiography,* 2 vols. (New York, 1891), I, 99; Glyndon Van Deusen, *William Henry Seward* (New York, 1967), 16–17; *Albany Argus,* July 26, 1832.

25. McKenzie, *Van Buren,* 229; Lot Clark to Marcy, Apr. 10, 1832, MALC; Butler to My Dearest Harriet, Mar. 20, 1832, Butler Papers, NYSL.

26. Van Buren to Marcy; to Wright, Mar. 14, 1832, VBLC; McKenzie, *Van Buren,* 324, 325; Clark to Marcy, Apr. 10, 1832, MALC; Butler to My Dearest Harriet, Mar. 20, 1832, Butler Papers, NYSL; William H. Seward to Weed, Aug. 19, 1832, Weed-Seward Papers, Univ. of Rochester Library.

27. Marcy to Knower, May 6, 1832, MALC; Garraty, *Wright,* 95; Marcy's biographer, Ivor D. Spencer, asserts that Marcy did not like the Senate and was ambitious for the governorship of New York. His major source is the notoriously unreliable *Courier and Enquirer,* and an account of Marcy's ambition for the office by James Gordon Bennett, then the *Courier and Enquirer's* Washington correspondent which Bennett published in his *Herald* on

Sept. 24, 1845, 23 years after the events described. Hammond, who had the means of knowing the situation at first hand declares that Marcy "was not over-anxious to put his popularity to the test." The correspondence of Van Buren, Croswell, Hoyt and Knower seems to bear out Hammond's contention. See Hammond, *New York,* II, 421. As late as mid-August, Marcy was still uncommitted. Seward to Weed, Aug. 19, 1832, Weed-Seward Papers, Univ. of Rochester Library.

28. Garraty, *Wright,* 91; Van Buren to Jackson, Sept. 20, 1832, VBLC; Van Buren to Cambreleng, July, 1832, VBLC; *Albany Argus,* Aug. 18, 1832.

29. Van Buren to Jackson, Aug. 29, 1832, VBLC; see also Seward to Weed, July 16, 1832. Seward agreed with Van Buren on the veto message, "Admirably prepared to produce effect with the people."

30. *Albany Argus,* Aug. 9; Oct. 22, 1832; David Henshaw to Hill, Aug. 29, 1832, Henshaw Misc. Papers, LC; Granger to Weed, Aug. 10, 1832, Seward-Weed Papers, Univ. of Rochester Library; Butler had written the sketch in 1830 and it was published in the *Cabinet and Talisman,* a short-lived annual. Van Buren to Gorham A. Worth, Jan. 30, 1830, VBLC; P. V. Daniel to Croswell, July 2, 1832; Croswell to Van Buren, July 29, 1832, VBLC; *Albany Argus,* July 28, 1832; *Jackson Correspondence,* III, 467, 468.

31. *Albany Argus,* Oct. 6, 1832.

32. William H. Seward to Weed, Aug. 19, 1832, Weed-Seward Papers, Univ. of Rochester Library; Van Buren, "Letter from Martin Van Buren in Reply to the Letter of a Committee Appointed at a Public Meeting Held at Shocco Springs, North Carolina, Owasco, Cayuga County, Oct. 4, 1832," VBLC; Van Buren, *Autobiography,* 562; Crouthamel, *Webb,* 45, 46; Hoyt to Van Buren, Aug. 22, 1832, VBLC; *Albany Evening Journal,* Oct. 10, 1832; McKenzie, *Van Buren,* 239, 240; Seward to Weed, Oct. 26, 1832, Weed-Seward Papers, Univ. of Rochester Library; Hugins, *Jacksonian Democracy,* 27, 28.

33. Hammond, *New York,* II, 424; *Albany Argus,* Nov. 24, 1832.

34. McLean to Taylor, Nov. 26, 1832, TNYHS.

Chapter 18. In Reference to a Possible Future

1. Nevins, ed., *Hone Diary,* 88; *New York Standard,* Mar. 6, 1832; *Albany Argus,* Mar. 11, 12, 1833. The whole ceremony could not have lasted more than 20 minutes.

2. *Ibid.,* Cambreleng to Van Buren, Dec. 18, 26, 29, 1832, VBLC.

3. Elias K. Kane to Van Buren, Jan. 2, 1833; Van Buren to Jackson, Nov. 13, 1832, VBLC.

4. Cambreleng to C. W. Gooch, n.d., 1833, Gooch Papers, UVA.

5. Forsyth to Van Buren, July 7, 1832, VBLC; Daniel to Van Buren, July 12, 1832, VBLC.

6. *United States Telegraph,* June 20–23, 1832.

7. C. W. Gooch to Cambreleng, Oct. 9, 1832, VBLC; *Richmond Enquirer,* Dec. 4, 1832; Ambler, *Ritchie,* 145, 146; *Calhoun Papers,* XI, 649–50; Tyler, ed., *Tyler,* I, 451–52; *Richmond Enquirer,* July–August, 1832; W. P. Elliott to

Van Buren, Sept. 5, 1832, VBLC; Kendall to Van Buren, Nov. 2, 19, 1832, VBLC.

8. Van Buren, "Notes on Internal Improvements," Nov. 18, 1832, VBLC; *Polk Correspondence,* I, 576, 577.

9. Taney to Van Buren, June 30, 1860, VBLC; Van Buren to Jackson, Nov. 13, 1832, VBLC.

10. Hoffman to Van Buren, Dec. 7, 9, 1832, VBLC; *Jackson Correspondence,* IV, 485–504.

11. John Van Buren to Van Buren, Jan. 2, 1833, JVBH. There was good reason for Van Buren's fears, see, for instance, Henry T. Shanks, ed., *The Papers of Willie P. Mangum,* 5 vols. (Raleigh, 1950–56), II, 14–16.

12. Jackson to Van Buren, Dec. 12, 1832; Cambreleng to Van Buren, Dec. 18; Jackson to Van Buren, Dec. 23, 25; Van Buren to Jackson, Dec. 27, 1832, VBLC. William G. Carleton, "Politics Aspects of the Van Buren Era," *The South Atlantic Quarterly,* Vol. 1., No. 2 (Apr. 1957), 168. For a succinct discussion of Poinsett's role in the nullification crisis, see Rippy, *Poinsett,* Ch. X.

13. Jackson to Van Buren, Jan. 13, 1833, VBLC. Benton to Van Buren, Dec. 16, 1832; Cambreleng to Van Buren, Dec. 18, 1832, VBLC; Van Buren to Jackson, Dec. 27, 1832, VBLC; Marcy to John A. Dix, May 1, 1832, Dix Papers, MHS; Marcy to Knower, May 6, 1832, MALC; *Jackson Correspondence,* V, 3.

14. Charles Z. Lincoln, ed., *Messages of the Governors of the State of New York,* 9 vols. (Albany, 1909), III, 420–21; *Jackson Correspondence,* IV, 508.

15. Dix, *Memoirs,* I, 133–37; *Albany Argus,* Jan. 25, 26, 1832, *Albany Evening Journal,* Jan. 25, 1833; Flagg to Wright, Jan. 21, 1833, FNYPL.

16. Van Buren, *Autobiography,* 549–53, 565; Hammond, *New York* II, 431–33; *Jackson Correspondence,* V, 12.

17. Van Buren to Cambreleng, Jan. 27, 1833, VBLC; Van Buren, *Autobiography,* 562–65, 549–53; *Albany Argus,* Feb. 1, 1833; Flagg to Wright, Feb. 13, 1833; Seward, *Autobiography,* I, 227, 228; Jackson to Van Buren, Jan. 25, 1833, VBLC.

18. *Ibid.,* see the *Albany Evening Journal,* Feb. 17, 1837, for a similar indictment of Van Buren and Wright. In a partisan editorial Weed claimed both men opposed the proclamation because of their inveterate states' rights doctrines. Noting that Van Buren was in Albany at the time Weed said he was "silent and dark."

19. *Albany Argus,* Feb. 26, 1833.

20. *New York American,* Feb. 6–15, 1833; *New York Weekly Commercial Advertizer,* Feb. 14; the *Globe,* Feb. 12; *Richmond Enquirer,* Feb. 11; *Albany Argus,* Feb. 15, 1833.

21. W. C. Rives to Ritchie, Jan. 5, 1833, RLC; see also Carleton, "Political Aspects," 168.

22. William R. King to Van Buren, Jan. 9; William H. Haywood to Van Buren, Jan. 10, 1833, VBLC. Jackson wrote Poinsett that if Congress failed to pass the "Force Bill" in time he would take matters into his own hands and call out the militia of the various states. Jackson to Van Buren, Dec. 25, 1832, VBLC; Silas Wright to Van Buren, Jan. 13, 1833; Van Buren to Jack-

son, Jan. 9, 1833, VBLC; Benton, *Thirty Years,* I, 303–08; *Jackson Correspondence,* IV, 5, 6, 11, 12.

23. John Y. Mason to George C. Dromgoole, Dec. 24, 1832; Jan. 8, 1833, Dromgoole Papers, UNC; Cambreleng to Van Buren, Dec. 26, 1832, VBLC; Hoffman to Flagg, Jan. 15, 1833, FNYPL.

24. Ritchie to Rives, Jan. 6, 1833, RLC; *Richmond Enquirer,* Jan. 17, 18, 1833; Wright to Dix, Feb. 4, 1833, Dix Papers, Columbia Univ., Randolph to Rives, Feb. 21, 1833, RLC.

25. Dickerson to Van Buren, Jan. 11, 1833, VBLC.

26. Cambreleng to Van Buren, Dec. 29, 1832, VBLC; Hoffman to Flagg, Jan. 4; McLane to Van Buren, Jan. 23, 1833, FNYPL; Cambreleng to Van Buren, Feb. 5, 1833, VBLC; see, also speech of Cambreleng in the House, Jan. 28, 1833, in *Albany Argus,* Feb. 22, 1833; Flagg to Wright, Jan. 21, 1833, FNYPL

27. Elias Pitts to S. J. Tilden, Mar. 18, 1833, Tilden Papers, NYPL; Hoffman to Flagg, Feb. 6, 1833, FNYPL; Wright to Van Buren, Jan. 29, 1833, VBLC; Marcy to Van Buren, Feb. 13, 1833, VBLC.

28. Hoffman to Flagg, Feb. 4, 1833; Nathaniel Macon to Van Buren, Mar. 2, 1833, VBLC.

29. Smith, *Blair Family,* I, 103–4; Hoffman to Flagg, Feb. 14, 1833, FNYPL; John W. Taylor to Mrs. Taylor, Feb. 16, 1833, TNYHS; Benton to Van Buren, Feb. 16, 1833, VBLC; John McLean to S. D. Ingham, Feb. 4, 1833, VBLC.

30. Van Buren to Jackson, Sept. 20, 1832; Jackson to Van Buren, Nov. 18; Nov. 25, 1832; Van Buren to Jackson, Nov. 29, 1832, VBLC; Van Buren to Wright, Feb. 20, 1833, VBLC.

Chapter 19. A Perilous Course

1. *Jackson Correspondence,* V, 214.

2. *Albany Argus,* Mar. 28, 1832; see confirmation of Van Buren's fears in Taney to Thomas Ellicott, Sept. 28, 1833, Taney Papers, LC; Gatell, "Sober Second Thoughts," 28, 29.

3. In Virginia, for example, Ritchie opposed removal of deposits; see Ritchie to Rives, Aug. 26, 1833; see an insightful evaluation of Kendall and Blair in Nathaniel Niles to Rives, July 23, 1833, RLC.

4. Munroe, *McLane,* 384; Stickney, ed., *Kendall,* 375; Van Buren, *Autobiography,* 601, 602.

5. Benton, *Thirty Years,* I, 291; Munroe, *McLane,* 384; Smith, *Blair Family,* I, 112–15; Parton, *Jackson,* III, 500; *Jackson Correspondence,* V, 42; Major Lewis, in a "Narrative" Parton reprints, says Blair visited Van Buren in Feb. 1833 and to his dismay found the Vice-President-elect decidedly opposed to removal. This may well be true, though Van Buren did not arrive in Washington until Feb. 27 and it seems doubtful if he would commit himself so bluntly without spending some time with such close advisors like Wright, Cambreleng and Tallmadge, among others. Parton, *Jackson* III, 503–5; *Jackson Correspondence,* V, 14; Stickney, ed., *Kendall,* 374.

6. *Jackson Correspondence,* V, 32–44, 55–56. James A. Hamilton declares that before Van Buren left Washington he had told Cass and McLane among others that he was opposed to removal. Any declaration of such a decided character about an issue which he knew the President held very strong views seems very much out of character with Van Buren. Hamilton's assertion should be discounted without any positive evidence to the contrary. See Hamilton, *Reminiscences,* 258; *Jackson Correspondence,* V, 75–101, 102–4. Jackson had thought of making McLane secretary of state as early as September 1832, and asked Van Buren what he thought of Silas Wright as his successor in the Treasury. Jackson to Van Buren, Sept. 16, 1832, VBLC; also, see Munroe, *McLane,* 357.

7. Van Buren, *Autobiography,* 592–94. Ari Hoogenboom and Herbert Ershkowitz, "Levi Woodbury's Intimate Memoranda of the Jackson Administration," *The Pennsylvania Magazine of History and Biography,* Vol. 92, No. 4 (Oct. 1968), 510–13.

8. Jackson to Van Buren, Nov. 25; McLane to Van Buren, Nov. 26, 1832, VBLC; Van Buren, *Autobiography,* 592–94.

9. Taney to Van Buren, Mar. 8, 1860, VBMHS; Van Buren, *Autobiography,* 594.

10. See, for instance, Whitney's 71-page argument for removal, which Jackson sent to Van Buren. Whitney to Jackson, Apr. 30, 1833, VBLC; see also John McFaul and Frank O. Gatell, "The Outsider Insider Reuben M. Whitney and the Bank War," *The Pennsylvania Magazine of History and Biography* (Apr. 1967), Vol. 91, No. 2, 115–43; Van Buren to Jackson, Apr. 21; May 15, 1833; Samuel C. Allen to Van Buren, May 13, 1833, VBLC.

11. Jackson to Van Buren, May 12, 19, 1833, VBLC; Fletcher M. Green, "On Tour with President Andrew Jackson," *New England Quarterly,* XXVI, No. 1, June 1963, *passim;* Van Buren to Jackson, May 16, 1833, VBLC.

12. Green, "On Tour," 218, 219.

13. *Jackson Correspondence,* V, 109; Nevins, ed., *Hone Diary,* 96.

14. Van Buren, *Autobiography,* 602.

15. Kendall to Van Buren, June 9, 1833, VBLC.

16. *Jackson Correspondence,* V, 110–28.

17. Hamilton, *Reminiscences,* 258; Gatell, "Sober Second Thoughts," 29–31.

18. *Jackson Correspondence,* V, 128, 129, 131–42.

19. Stickney, ed., *Kendall,* 379, 380; Jackson to Van Buren, July 30, 1833; Blair to Van Buren, Nov. 13, 1859, VBLC; *Jackson Correspondence,* V, 142, 143.

20. Van Buren to Jackson, Aug. 16, 1833, VBLC; *Albany Argus,* Aug. 12, 1833; Van Buren to Jackson, July 29, 1833, VBLC.

21. Hamilton, *Reminiscences,* 261; Van Buren to Jackson, Aug. 6, 1833, VBLC.

22. *Jackson Correspondence,* V, 142–43; Frank O. Gatell, "Secretary Taney and the Baltimore Pets: A Study in Banking and Politics," *Business History Review,* XXXIX, No. 2 (Spring 1965), 208, 209; Whitney to Donelson, June 10, 1833, Donelson Papers, LC; Jackson to Vice-President Van Buren, Aug. 12, 1833, VBLC; Hamilton, *Reminiscences,* 253.

23. Van Buren to Jackson, Sept. 4, 1833, VBLC; Hamilton *Reminiscences,*

260; *Jackson Correspondence,* V, 151–56; Van Buren to Cambreleng, July 22, 1833; Whitney to Jackson, Apr. 30, 1833; to Van Buren, Aug. 16, 1833, VBLC.

24. *Jackson Correspondence,* V, 179, 180–82.

25. *Ibid.*, Wright to Flagg, Aug. 8, 1833, FNYPL; *Albany Argus,* Aug. 14, 1833; *Daily National Intelligencer,* Aug. 15; New York *Daily Advertizer,* 12, 17, 30, 1833.

26. Stickney, ed., *Kendall,* 383; *Jackson Correspondence,* V, 156.

27. Olcott to Van Buren, Aug. 1833, OLCU; *Jackson Correspondence,* V. 182; Frank O. Gatell, "Sober Second Thoughts," June, 1966, 25–27.

28. Taney to Ellicott, Sept. 28, 1833, Taney Papers, LC; Stickney, ed., *Kendall,* 380, 381; *Jackson Correspondence,* V, 181, 182; Bray Hammond, *Banks and Politics in America from the Revolution to the Civil War* (Princeton, 1957), 419, 420.

29. In June Van Buren wrote John, "You must not draw on me for $1,000 for I have not got the money," Van Buren to John Van Buren, June 10, 1833; to David E. Evans, Sept. 7, 1833, VBLC; Evans owed Van Buren $5,000 at the time; Van Buren to Cambreleng, July 22, 1833, VBLC.

30. Wright to Flagg, Aug. 8, 1833, FNYPL; *Jackson Correspondence,* V, 160.

31. *Ibid.,* 152, 167, 187.

32. Blair to Van Buren, Aug. 17, 1833, VBLC; *Jackson Correspondence,* V, 150–65.

33. Wright to Van Buren, Aug. 28, 1833, VBLC; Croswell to Van Buren, June 7, 1833, VBLC; Bennett had made similar charges to Jackson. And he wrote Van Buren again rehearsing the same charges; Bennett to Van Buren, Sept. 25, 1833, VBLC; to Jackson, WOLC; to Van Buren, Oct. 4, 1833, VBLC; *Jackson Correspondence,* V, 179–82. Jackson heeded Van Buren's note of caution. On Sept. 26 he wanted Van Buren's advice on three banks Taney had chosen in New York City. Van Buren replied on Sept. 28, urging that a fourth bank be selected. Jackson to Van Buren, Sept. 26, 1833; Van Buren to Jackson, Sept. 28, 1833, VBLC.

34. Whitney to Van Buren, Aug. 16, 1833, VBLC; *Jackson Correspondence,* V, 183, 184; Van Buren to Jackson, Sept. 14, 1833, JLC; *Albany Argus,* Sept. 12, 1833.

35. *Jackson Correspondence,* V, 184, 187; Blair to Van Buren, Oct. 15, 1845, Blair Family Papers, LC; Nov. 13, 1859, VBLC.

36. *Jackson Correspondence,* V, 203–7; to Van Buren, Sept. 23, 1833, VBLC.

37. *Albany Argus,* Sept. 12, 31, 1833; Irving, *Life and Letters of Washington Irving,* II, 263; *Albany Argus,* Sept. 21, 1833; Van Buren, *Autobiography,* 606.

38. Parton, *Jackson,* III, 501–3; 528–31; Van Buren, *Autobiography,* 607, *Richmond Enquirer,* Oct. 2, 1833; see also *Albany Argus,* Aug. 13, 1833, quoting the *Charleston Mercury's* attack on Kendall specifically and the Administration in general.

39. Van Buren to Jackson, Sept. 26, 1833, VBLC; Irving, *Irving,* II, 62; Van Buren, *Autobiography,* 610, 611; Munroe, *McLane,* 403, 404; Duff Green to John Floyd, Nov. 20, 1833, Floyd Papers, LC.

40. Van Buren to Jackson, Apr. 21, 1833, VBLC; Van Buren, *Autobiography,* 606, 609; Hoogenboom and Hershkowitz, "Woodbury Memoranda," 513; Van Buren to Jackson, Aug. 8, 1833, VBLC.

41. *Albany Argus,* June 3, 1833; *Albany Evening Journal,* Oct. 24, 1833; *New York American,* Sept. 26, 27, 30, 1833; New York *Courier and Enquirer,* Sept. 30; *Albany Evening Journal,* Sept. 27, 30, 1833; *Albany Daily Advertizer,* Oct. 3, 1833; *Albany Argus,* Oct. 4, 1833; Van Buren to Jackson, Oct. 2, 1833, VBLC; see also New York *Evening Post,* Oct. 1, 1833.

42. Van Buren to Jackson, Oct. 2, 1833, VBLC; Jackson to Van Buren, Oct. 5, 1833, JLC; Van Buren to Mahlon Dickerson, Oct. 26, 1833, Dickerson Papers, New Jersey Historical Society; Abraham Van Buren to Van Buren, June 3, 1833, VBLC; Jackson to Van Buren, Oct. 5, 1833, JLC; Van Buren to Rives, July 22, 1833, RLC; Stickney, ed., *Kendall,* 331.

43. For positive reaction to Butler's appointment, see *Albany Argus,* Dec. 3, 1833, with favorable notices from the *New York American,* the *Daily National Intelligencer, New York Evening Star, New York Commercial Advertizer,* all opposition papers. Swisher, *Taney,* 144; Van Buren to Butler, Nov. 8, 1833; Jackson to Van Buren, Nov. 16, 19, 1833, VBLC; Butler to the President, Nov. 12, 1833, BuP.

Chapter 20. A Pinch of Snuff

1. *Register of Debates,* 23rd Cong., 1st Sess., 826–31.

2. *Ibid.;* Van Buren to John Van Buren, Jan. 29, 1834, VBLC; Benton, *Thirty Years,* I, 420; see, for instance, Blair to Jackson, Jan. 7, 1834, Blair Family Papers, LC; *Albany Argus,* Mar. 15, 1834.

3. Elam Tilden to John W. Edmonds, Tilden Papers, n.d., 1834, NYPL.

4. Elijah Whittlesey to John W. Taylor, Oct. 10, 1833, TNYHS; Nathanson, *Webster,* 49, 60.

5. Seward, *Autobiography,* I, 237.

6. Thereby earning the criticism of John Tyler who thought Clay's carping about Van Buren's absence had not been productive. Tyler, ed., *Tyler,* I, 482; *Register of Debates,* 23rd Cong., 1st Sess., 13, 22, 23, 27; Van Buren to Jackson, Oct. 12, 1833, with endorsement, VBLC; *Jackson Correspondence,* IV, 420.

7. Van Buren, *Autobiography,* 673–79.

8. *Register of Debates,* 23rd Cong., 1st sess., 42, 43; Jesse Hoyt to Van Buren, Dec. 19, 1833, VBLC.

9. Van Buren to John Van Buren, Jan. 29, 1834, VBLC.

10. Wiltse, ed., *Webster Papers,* II, 364, 365; Van Buren, *Autobiography,* 661.

11. *Ibid.,* 517–521, 663; Green to Floyd, Floyd Misc. Papers, LC, Nov. 30, 1833; *Albany Argus,* Nov. 1, 1833; Van Buren to John Van Buren, Jan. 24, 1834, VBLC.

12. See also Cambreleng to W. C. Bryant, Dec. 12, 1833, Bryant-Godwin Papers, NYPL; C. A. Wickeliffe to Crittenden, Jan. 15, 1838, Crittenden Papers, LC; Cambreleng to Gooch, Feb. 8, 1834, Gooch Papers, UVA.

13. James A. Hamilton to Van Buren, Dec. 30, 1833, VBLC; J. M. Ewing et al. to Crittenden, May 19; James Lane to Crittenden, May 30, 1834, Crittenden Papers, LC.

14. Henry Clay had been informed of Knower's problems and the strains that these imposed in the Regency. He asked Biddle to "accommodate Mr. Knower. It may be done with perfect safety. He is, you know, father-in-law of Governor Marcy and he belongs to that powerful interest in New York (he is indeed head of it) which is held by very loose bonds to the Regency and the desired accommodation would have the best of affects." Biddle apparently knew more about Knower's financial difficulties than Clay. The loan was not made. Clay to Biddle, Feb. 2, 1834, Biddle Papers, LC. Inevitably, Van Buren was drawn into the affair. Biddle and Clay had not forgotten that Knower had been high in Regency councils; nor his past connections with the Regency Farmers' and Mechanics' Bank on whose board Van Buren had once sat. These aspects of the past provided the ingredients of a circumstantial case. Without checking facts, the word went out to the opposition press that Van Buren had endorsed Knower's paper to protect his own speculations. So successful were these untruths in molding public opinion that Van Buren felt he had to deny them in important private circles. "You see the attempts which have been made and are making all over the Union to impress the belief that I am involved in consequence of the failure of Mr. Knower," he wrote Gorham Worth, cashier of Manhattan Bank in New York. "No sooner is one story put down than they put up another." Van Buren to Worth, Feb. 28, 1834; Spencer Stafford to Van Buren, Oct. 5, 1833, VBLC; Elias Pitts to S. J. Tilden, Oct. 18, 1833, TNYPL. Benjamin Butler to Thomas W. Olcott, Feb. 1, 1834, OLCU; Silas Wright to Flagg, Feb. 7, 1834, FNYPL; see also R. K. Colt to Biddle, Feb. 24, 1834, Biddle Papers, LC; for a vivid account of the "Panic Session," see Arthur Schlesinger, Jr., *The Age of Jackson* (Boston, Toronto, 1945), Ch. IX; Van Buren, *Autobiography,* 740–42.

15. *Register of Debates,* 23rd Cong., 1st Sess., 631, 746, 1207–9; Myndert Van Schaick to Van Buren, Mar. 12, 1834, VBLC.

16. Van Buren to John Van Buren, Feb. 10, 1834, VBLC; *Albany Argus,* Feb. 17, 1834; Van Buren to John Van Buren, Feb. 10, 1834, VBLC; see Butler to Olcott for a detailed refutation of Clay's charges, Mar. 20, 1834, OLCU; Flagg to Dr. Sir, Feb. 14, 1834, FNYPL; *Albany Argus,* Jan. 7, 1834.

17. James A. Hamilton to Van Buren, Dec. 20, 1833; Hoyt to Van Buren, Jan. 29; Feb. 4, 1834; Throop to Van Buren, Feb. 1, 1834; Morgan Lewis to Van Buren, Feb. 11, 1834, VBLC; *Albany Argus,* Jan. 28, 1834; Butler to Olcott, Jan. 27; Feb. 1, 8, 1834, OLCU. Thomas Govan, *Nicholas Biddle* (Chicago, 1959), Chap. 25; Reginald C. McGrane, *The Correspondence of Nicholas Biddle Dealing With National Affairs* (Boston and New York, 1919), 219–45. Van Buren to John Van Buren, Feb. 10, 1834; VBLC.

18. Taney to Van Buren, June 30, 1860, VBLC; Butler to Olcott, Jan. 27, 1834, OLCU.

19. Myndert Van Schaick to Van Buren, Mar. 12, 1834, VBLC; *ibid.;* James King to Van Buren, Mar. 14, 1834, VBLC.

20. *New York Evening Post,* New York *Courier and Enquirer, New York American,* Apr. 9–12, 1834; *Albany Argus,* Apr. 11, 1834.

21. Silas Wright to Flagg, Jan. 3, 1834, FNYPL; McKenzie, *Van Buren,* 246–48; Eldad Holmes to Van Buren, Feb. 7, 1834, enclosing "Proceedings and Resolutions of the Fourth Ward, New York." Throop to Van Buren, Feb. 1, 1834; clipping, "Our Country and our Country's Good. Democracy

Unconquered and Unconquerable." Proceedings and Resolutions of a public anti-Bank meeting. Philadelphia, Feb. 8, 1834. Van Buren to the Cordwainers of Philadelphia, Feb. 15, 1834; Van Buren to Robert Lucas, Feb. 17, 1834; Stephen Allen to Van Buren, Feb. 21, 1834, enclosing clipping on meeting at Tammany Hall, Feb. 19, 1834, VBLC.

22. McKenzie, *Van Buren,* 246; C. C. Gooch to Ritchie, Sept. 7, 1837; Van Buren to Rives, Feb. 17, 1834, RLC; Tyler, ed., *Tyler,* I, 484–86.

23. See, for instance, George M. Dallas to Rives, Mar. 10, 1834, RLC; Belovalek, *Dallas,* 54, 55.

24. See Garraty, *Wright,* 122, and Hammond, *New York,* II, 440, 441. Curiously, William H. Seward, the Whig leader in New York, had virtually the same opinion as Wright. In a letter to his father, Seward called the loan a "ruinous policy . . . which will embarrass and cripple the state for twenty years and sink its credit while it will swell the great corrupted fund already too large." Seward to Samuel Seward, Mar. 28, 1834, Samuel Seward Papers, NYSL; Ralph C. H. Catterall, *The Second Bank of the United States* (Chicago, 1903), 343; Wright to Flagg, Mar. 25, 1834, FYPL.

25. Butler to My Dearest H., Jan. 17, 1834, BNYSL; Van Buren, *Autobiography,* 724–47.

26. Benton, *Thirty Years,* I, 422–25; R. H. Wilde to Gulian Verplanck, May 1, 1834, Verplanck Papers, NYHS; Van Buren to Marcy, Mar. 31, 1834, VBLC; *Jackson Correspondence,* V, 278, 279; Benton, *Thirty Years,* I, 423, 428–33.

27. Van Buren, *Autobiography,* 200, 203, 737, 755, 792; Van Buren to My Dear Friend (Mrs. William C. Rives), Apr. 6; May 23, 1834, RLC.

28. Ralph M. Aderman, ed., *The Letters of James Kirke Paulding* (Madison, 1962), 61, 145, 147, 196. Van Buren paid Paulding $264.64 a month in rent. Barnes and Weed, eds., *Weed,* I, 392, 406; John Van Buren to Van Buren, Jan. 1835, JVBH.

29. *Ibid.;* Dix to Van Buren, Feb. 16, 1835, VBLC.

30. McKenzie, *Van Buren,* 250.

31. Butler to My Dearest H., Jan. 17, 1834, BNYSL.

32. James A. Hamilton to Van Buren, May 5, 1834, Van Buren to John Van Buren, June 22, 1834, VBLC; Govan, *Biddle,* 259.

33. *Jackson Correspondence,* V, 271, 272.

34. Livingston had recommended that Jackson in a special message to Congress place a total embargo at once on all French imports. James A. Hamilton to Van Buren, May 5, 1834, VBLC; Van Buren to Rives, May 12, 1834; Rives to Van Buren, May 15, 16, 1834, RLC; Munroe, *McLane,* 417, 418.

35. Taney to Van Buren, Apr. 9, 1860, VBLC; some allowances for age must be made in Taney's exposition of his debate with McLane. The Chief Justice was writing of events that occurred 26 years before; but they square in the main with Van Buren's impressions that are not based on the Taney letter; Van Buren, *Autobiography,* 612.

36. *Jackson Correspondence,* V, 271–72; Van Buren, *Autobiography,* 611–13.

37. Butler to My Dear Wife, June 29, 1834, BNYSL; Alvin Laroy Dudset, *John Forsyth, Political Tactician,* (Athens, Ga., 1962), 158–60, 162, 167–69.

38. Van Buren, *Autobiography,* 611–17.

Chapter 21. Mediator

1. Mangum to Major John Beard, Oct. 7, 1834, Fisher Family Papers, UNC.

2. Van Buren to Jackson, Oct. 18, 1834, VBLC; Van Buren to Woodbury, Nov. 6, 1834, WOLC; to Mahlon Dickerson, Nov. 7, 1834, Dickerson Papers, New Jersey Historical Society.

3. John Law to Van Buren, Aug. 20, 1833; Abner Harris to Van Buren, Sept. 3, 1833; David Disney to Van Buren, Dec. 8, 1834, VBLC; *Jackson Correspondence,* V, 291.

4. See *United States Telegraph,* July 3, 11, 1834; John Catron to Jackson, Mar. 21, 1834, JLC.

5. James Gordon Bennett to Van Buren, Sept. 25; Oct. 5, 1833, VBLC; *Jackson Correspondence,* V, 327, 328; Nathaniel Niles to Rives, July 23, 1833, RLC; J. W. James to Blair, June 14, 1834, Blair Family Papers, LC; *Jackson Correspondence,* V, 283–85; Leland W. Meyer, *The Life and Times of Colonel Richard M. Johnson of Kentucky* (repr. New York, 1967), 406, 407; Richard B. Latner, *The Presidency of Andrew Jackson* (Athens, Ga., 1979), 130, 131.

6. Hamilton, *Reminiscences,* 283; *Polk Correspondence,* II, 475: III, 130, 134, 154, 155, 384; *Jackson Correspondence,* V, 472.

7. Ratner, *Jackson,* 135, 194, 195; *Jackson Correspondence,* V, 273; Van Buren to Jackson, Nov. 5, 1834, VBLC.

8. *Ibid.,* 202; Van Buren, *Autobiography,* 728–30; Nathaniel Niles to Rives, Jan. 11, 1835, RLC; C. C. Gooch to Rives, July 25, 1833; Nathaniel Niles to Rives, July 23, 1833; Apr. 25, 1835, RLC.

9. Nathaniel Niles to Rives, July 23, 1833; Jan. 15, 1835, RLC; Rives to Van Buren, June 16, 1834, VBLC; Van Buren to Rives, July 19, 1834, RLC.

10. Rives to Van Buren, Oct. 14, 1834, VBLC; Jackson to Van Buren, Oct. 27, 1834, VBLC.

11. Van Buren to Rives, Oct. 23, 1834, RLC; Rives to Van Buren, Oct. 14, 1834; Van Buren to Jackson, Oct. 23, 1834, VBLC.

12. "I have been a strenuous advocate for forebearance heretofore (more than you are aware of) and will continue to be," Van Buren to Rives, Oct. 23, 1834.

13. Jackson to Van Buren, Oct. 27, 1834, VBLC.

14. He recommended the same policy to Donelson, who for reasons of his own had also queried Rives on the subject. Rives to Van Buren, Nov. 15, 1834 [copy], RLC; to Donelson, Nov. 13, 1834, Donelson Papers, LC; Elmendorf to Van Buren, Dec. 21, 1835, VBLC.

15. Jackson to Van Buren, Sept. 14, 1834, VBLC; Van Buren to Jackson, Oct. 13; Nov. 5, 1834, VBLC; Donelson to Rives, Nov. 13, 1834, Donelson Papers, LC; Van Buren to Rives, Nov. 15, 1834; Van Buren to Woodbury, Nov. 30; Dec. 4, 1834, WOLC; see drafts of the President's message in the Jackson Papers, Nov. 1834; that part dealing with France, is entirely in Van Buren's hand except the section on reprisals which Kendall drafted; Duckett, *Forsyth,* 170–72; Van Buren to Woodbury, Nov. 30; Dec. 9, 1834, WOLC.

16. Van Buren to Rives, Dec. 3, 1834, RLC.

17. *Ibid.*

18. Van Buren to Mrs. Rives, Dec. 12, 1834, RLC; Aaron Vail to Van Buren, Jan. 27, 1835; Livingston to Van Buren, Jan. 29, 1835.

19. Leverett Harris to Van Buren, Jan. 29, 1835, VBLC.

20. For the diplomatic correspondence see *Register of Debates,* 23rd Cong., 2nd Sess., 165–69, 275–77; see also Hatcher, *Livingston,* Ch. 16 *passim,* and Richardson, ed., *Messages and State Papers,* III, 1342–46, 1348–51.

21. Parton, *Jackson,* III, 573.

22. Nevins, ed., *Hone Diary,* 184.

23. Compare Jackson's message with Van Buren's manuscript draft of the message of Dec. 1835, JLC; compare also with Jackson's notes for the annual message, Dec. 7, 1835, in *Jackson Correspondence,* V, 577–79.

24. The *Globe, Daily National Intelligencer,* Dec. 8, 9, 1834; Elmendorf to Van Buren, Dec. 21, 1835, VBLC.

25. Parton, *Jackson,* III, 574–76.

26. See draft in Livingston's hand with some pages which Donelson copied in the Jackson Papers, Dec. 1835, JLC; the rest of the message, while upholding American honor, was sufficiently vague not to encourage any war party in France. Again, in the annual message to Congress, additional diplomatic correspondence was included to strengthen the American case at home and abroad. For a complete text of the message, see Richardson, ed., *Messages,* V, 1407–32.

27. Irving to Van Buren, Feb. 24, 1836, VBLC.

28. A. G. Hammond to Van Buren, Dec. 23, 1824, VBLC.

29. John B. Jentz, "The Anti-Slavery Constituency in Jacksonian New York City," *Civil War History,* Vol. 27, No. 2 (June 1981), 101; *Annals of the Congress,* 17th Cong., 1st Sess., 277.

Chapter 22. Partisan Leader

1. *Richmond Enquirer,* Apr. 16, 20; June 23, 1833.

2. *Calhoun Papers,* XII, 160–64.

3. *United States Telegraph,* Feb. 28, 1835.

4. Ritchie to Van Buren, Mar. 2, 1835, VBLC.

5. Wright to Joseph Watkins, Feb. 9, 1835, VBLC; Smith Van Buren searched the journals of the legislature and contemporary newspapers at his father's request. He found no evidence of Van Buren's involvement in the resolutions; Smith Van Buren to Van Buren, Feb. 16, 1844, VBLC; Benton to Major General Davis, Dec. 16, 1835, in *Albany Argus,* Jan. 10, 1835.

6. Wright to Ritchie, Mar. 10, 135, VBLC; *Richmond Enquirer,* Mar. 15, 1835; see also Joel Silbey, "Election of 1836," in Arthur M. Schlesinger, Jr., ed., *American Presidential Elections,* 4 vols. (New York, 1971), I, 589–91.

7. Van Buren to My Dear Friend, Apr. 1, 1835, RLC.

8. Butler to Olcott, Mar. 27, 1835, OLCU.

9. B. F. Butler, esq. of New York, Attorney General to Hugh Garland of Mecklenburgh, Virginia, Mar. 1835, VBLC; Butler to Rives, Apr. 4, 1835, RLC; Silbey, "Election of 1836," 378.

10. See Van Buren drafts of E. K. Kane to Ritchie, Mar. 10; Wright to Ritchie, Mar. 10, 1835, VBLC.

11. S. J. Tilden to Elam Tilden, Mar. 23, 1835, Tilden Papers, NYPL; Rives to Van Buren, Apr. 10, 1835, RLC; Van Buren to My Dear Friend, Apr. 1, 1835, RLC.

12. Rives to Woodbury, Apr. 8, 1835, WLC; Van Buren to Rives, May 5, 1835, RLC.

13. Belohvalek, *Dallas,* 57–58; Charles J. Ingersoll to Biddle, Feb. 9, 1832; Biddle to Samuel Breck, Mar. 1, 1834, Biddle Papers, LC.

14. *United States Telegraph,* July 13, 16, 23, 1835; Silas Wright to Dr. Sir, Feb. 28, 1835 [draft], VBLC; *United States Telegraph,* July 17, 1835; Klein, *Buchanan,* 105, 106.

15. David Petrikin to Van Buren, Feb. 11; Mar. 8, 1835; Charles Penrose to Van Buren, Mar. 15, 1835; David M. Brodhead to Van Buren, Feb. 23, 1835; Roberts Vaux to Van Buren, Jan. 24; Apr. 27; Sept. 16, 1835; Gilpin to Van Buren, Apr. 5, 1835; Richard Rush to Van Buren, Sept. 28, 1835, VBLC.

16. MacKenzie, *Van Buren,* 258, 262; George D. Strong to Van Buren, Dec. 23, 1834; Gideon Lee to Van Buren, Dec. 25, 1834, VBLC; *New York Times,* Dec. 22, 1834; Tilden to Tilden, Feb. 25, 1835, Tilden Papers, NYPL; Silbey, "Election of 1836," 583.

17. Blair to Van Buren, Aug. 26, 1834, VBLC; Stickney, ed., *Kendall,* 331–46.

18. Latner, *Andrew Jackson* 193–96; Polk to Jackson, May 15, 1835, Jackson Papers, LC; Taney to Van Buren, Mar. 25, 1835; Richard E. Parker to Van Buren, May 8, 1835, VBLC; Kendall to Caleb Butler, May 12, 1835, Kendall Papers, LC; Harriet Martineau, *Retrospect of Western Travel* 2 vols. repr. (New York, 1969), II, 156–57.

19. Taney to Rives, May 12, 1835, RLC.

20. R. A. Worrell to Duff Green, Green Papers, LC.

21. John C. McLemore to Jackson, Apr. 6, 1835, JLC; see John M. McFaul, *The Politics of Jacksonian Finance* (Ithaca, 1972), Ch. 4, *passim;* see also, Niven, *Welles,* Ch. 9, 128–31, 137–41.

22. See, for instance, anon. letter from Gallatin, Tennessee, in *Richmond Enquirer,* Mar. 30, 1835.

23. Jackson to Donelson, May 12, 1835, Donelson Papers, LC.

24. Apparently any citizen from Baltimore with the time and the interest could attend and be listed as an official delegate. New York had 42 delegates, Pennslyvania 60, New Jersey 71, Virginia 96; *Albany Argus,* May 25, 1835; Blair to Jackson, May 19, 1835, Blair Family Papers, LC; Latner, *Jackson,* 202, 203; Wright to Van Buren, May 22, 1835, VBLC.

25. Nathaniel Niles to Rives, Apr. 10, 25, 1835; R. C. Mason to Rives, May 28, 1835, RLC; William B. Slaughter to Rives, May 28, 1835, RLC; Charles Sellers, *James K. Polk, Jacksonian* (Princeton, 1957), 272–73; Meyers, *Johnson,* 310–11; Wright to Van Buren, May 22, 1835, VBLC; to Erastus Corning, May 24, 1835, Wright Papers, NYSL; Mason to Rives, May 28, 1835, RLC.

26. Van Buren to Rives, May 26, 1835, RLC; W. B. Slaughter to Rives,

May 28, 1835, RLC; see also Van Buren's draft of same date, VBLC; Rives to Van Buren, June 2, 25, 1835, VBLC.

27. Wright to Van Buren, May 22, 1835; Buchanan to Van Buren, May 21, 1835, VBLC.

28. *Albany Argus,* June 9, 15, 16, 1835.

29. Welles to Van Buren, Apr. 7, 1835, RLC; Dutee J. Pierce to Van Buren, Apr. 17, 1835, RLC; Bancroft to Van Buren, Jan. 8, 1835; see also Van Buren draft of letter for Silas Wright to William Foster, Jan. 3, 1835, VBLC; Bancroft to Marcy, Nov. 19, 1835, Marcy Papers, NYHS.

30. *Albany Argus,* Sept. 3, 5, 7, 11, 12, 15, 16, 22; Oct. 2, 1835. Van Buren's letter, written to William Schley, was sent from Owasco, New York, Sept. 10, 1835, VBLC.

31. Marcy to Van Buren, Nov. 22; to Van Buren, Dec. 3, 1835, VBLC; Wright to Marcy, Dec. 18, 1835, Marcy Papers, LC; Rives to Van Buren, Jan. 29, RLC; Wright to Flagg, Jan. 17, 1836, FNYPL; *Albany Argus,* Jan. 6, 1836.

32. Williams to Van Buren, Apr. 7, 1836; Van Buren to Williams, June 14, 1836, VBLC.

33. Van Buren to Bancroft, Aug. 17, 1836, BMHS; to Rives, Oct. 11, 1836, RLC; to James A. Hamilton, Sept. 28, 1836, VBLC; Marcy to Prosper Wetmore, July 25, 1836, MALC.

34. Caleb Cushing to Edward Everett, Everett Papers, NYHS; see, for instance, William H. Seward's astute analysis of the Whig dilemma; Seward to Weed, Feb. 15, 1835, WSR; Crittenden to J. L. Moorhead et al. Dec. 3, 1835, Crittenden Papers, LC.

35. C. W. Gooch to Rives, May 5, 1835, RLC.

36. Buchanan to Van Buren, Nov. 16, 1836, VBLC.

37. Cambreleng to Flagg, Nov. 21, 1836, FNYPL.

38. Van Buren to John Van Buren, Nov. 21, 1836, VBLC.

39. Taney to Van Buren, Sept. 16, 1834, VBLC; on patronage, see for instance, Van Buren to Jackson, Nov. 5, 1834, in *Jackson Correspondence,* V, 306, 307; Judge Richard E. Parker to Van Buren, June 18, 1835, VBLC.

40. Jackson to Van Buren, *ibid.,* 212.

Chapter 23. Be Careful of Cataline

1. N. P. Willis, *The Prose Works of N. P. Willis* (Philadelphia, 1849), 569.

2. *Niles' Weekly Register,* Dec. 26, 1836; W. R. Hallet to Van Buren, Jan. 12, 1837, VBLC; *Albany Evening Journal,* Feb. 13, 1837; Meyer, *Johnson,* 425–29.

3. Van Buren to Cambreleng, May 10, 1835; Memorandum, "Views on Public Lands, Banks and Paper Currency," Aug. 11, 1835; George H. McWorter to Van Buren, Nov. 1; Cambreleng, Nov. 2; Joseph D. Beers, Nov. 21; Roberts Vaux to Van Buren, Dec. 19, 1835, VBLC.

4. Van Buren to Jackson, Aug. 5; to Gorham Worth, Aug. 25; to John Van Buren, Nov. 22; Dec. 30, 1836; S. M. Skinner to Moses Tilden, Jan. 3, 1837; Worth to Van Buren, Mar. 12, 1837, VBLC; Wright to Flagg, Jan. 1837, FNYPL; Samuel Swartwout to Rives, Jan. 1837, RLC.

5. Nevins, ed., *Hone Diary,* 184, 185–90.

6. Lewis F. Linn to Van Buren, 1836, VBLC. The recipe was 1 quart fine hickory ashes, 1 teacup of soot, 1 gallon of boiling water, combine and take 1 full glass four times a day.

7. Van Buren, "Memorandum," 1837, VBLC.

8. Van Buren to John Van Buren, Dec. 30, 1836, VBLC; Butler to Van Buren, Feb. 15, 1837; Van Buren to Butler, Feb. 15, 1837, VBLC.

9. See, for instance, a retrospective comment of John Catron; Catron to A. J. Donelson, May 4, 1829, Donelson Papers LC; Capowski, "Woodbury," 13, 193; Hoogenboom and Hershkowitz, "Woodbury Memoranda," 512–15.

10. Wetmore to Marcy, Mar. 19, 1837, Simon Gratz Autograph Coll., HiSPA; James C. Curtis, *The Fox at Bay* (Lexington, Ky., 1970), 55, 56.

11. Elbert B. Smith, *Francis Preston Blair* (New York, 1980), 166; see also Kendall to Van Buren, Dec. 7, 16, 17; Wright to Van Buren, Dec. 19, 23; John C. Rives to Kendall, Dec. 21; Blair to Kendall, Dec. 24, 1842, VBLC; Samuel Lawrence to Amos Lawrence, May 1840, Amos Lawrence Papers, MHS.

12. Van Buren to John Van Buren, Dec. 30, 1836; Van Buren to Rives, Feb. 1, 1837, in Rives "Memorandum," Feb. 2, 1837, RLC; Richard E. Parker to Van Buren, Feb. 7, 1837, VBLC; Rippy, *Poinsett,* Ch. I, III, VI, VIII, X, *passim;* Van Buren to Poinsett, Feb. 4; Poinsett to Van Buren, Feb. 9, 1837, VBLC; Poinsett to Jackson, Oct. 23, 1830, Poinsett Papers, PHS; see an interesting profile of Poinsett in the *United States Democratic Review,* Vol. 1, No. 3 (Feb. 1838), 361–68, 443–56.

13. Buchanan to Daniel M. Brodhead, Dec. 18, 1836, Brodhead Papers, LC; Van Buren to Dallas, Feb. 16, 1837; Buchanan to Van Buren, Feb. 19, 28, 1837; Simon Cameron and Ovid Johnson to Van Buren, Feb. 24, 1837, VBLC.

14. When it was certain he had been elected, a group of friends and supporters in New York wished to present him with a carriage. Van Buren knew at the time that the party in New York had contributed funds for the construction of a magnificent phaeton whose oaken body and wheels were being fashioned from planks of the frigate U.S.S. *Constitution.* He hastened to place a damper on any similar gift for himself that might detract from Jackson's sense of his own self-importance. Never averse to a bit of political gamesmanship, he had Gideon Lee, one of the city's congressmen, write the well-wishers that he felt "a very strong and I fear an uncontrollable unwillingness to do or permit any act which may lead to ostentation or bear the slightest resemblance to the customs of the monarchies of the Old World." He tipped his hand, however, by saying that "on retirement from public to private life he would accept any testimonies of the kind with the most grateful sensations." Gideon Lee to James M. Mellon, Dec. 29, 1836, VBLC.

15. John Fairfield to Dear Ann, Dec. 18, 1836; to Dear Wife, Jan. 13, 1837, FaLC; Amos Kendall to Campbell P. White, Nov. 8, 1834, White Papers, NYHS; McFaul and Gatell, "Outsider Insider," 134; Vanderlyn to Garry Vanderlyn, Jan. 28, 1837, Vanderlyn Papers, NYHS.

16. Fairfield to Dear Ann, Jan. 13, 1837, FaLC.

17. The *Globe,* Mar. 6, 1837; Fairfield to Dear Wife, Feb. 17, 1837, FaLC; Willis, *Works,* 569–70; Richardson, ed., *Messages,* IV, 1530–47.

18. The *Globe,* Mar. 9, 1837; Charles Windrow Elliott, *Winfield Scott, The Soldier and the Man* (New York, 1837), 332.

19. Blair to Van Buren, Sept. 27, 1839; Van Buren to Smith Van Buren, Jan. 5, 1835, VBLC.

20. See, for instance, John Van Buren to James Wadsworth, Mar. 25; Apr. 3, 1834, James Wadsworth Papers, Univ. of Rochester; Worth to Butler, Mar. 12, 1837, BuP.

21. Nevins, ed., *Hone Diary,* 254.

22. In Albany, the Farmers' and Mechanics' Bank was all but stripped of its specie after sending $2 million to sustain the credit of the country banks in the west; see, for instance, Marcy to Woodbury, Apr. 9, 1837; Croswell to Butler, Mar. 13, 1837, BuP. John Brockenbrough to Woodbury, May 1, 1837; Woodbury to Brockenbrough, May 1, 1837, WOLC; Blair to Jackson, May 1, 1837, Blair Family Papers, LC; John Van Buren to Woodbury, May 9, 1837, WOLC.

23. Flagg to Van Buren, Apr. 10, 1837, VBLC.

24. N. P. Tallmadge to Van Buren, Mar. 15, 1837, VBLC; Rives to Van Buren, Apr. 8, 1837, RLC.

25. *Jackson Correspondence,* VI, 465–67, 476; Wright to Van Buren, May 13, 1837, VBLC.

26. The memorandum in the hand of Martin, Jr., was as follows: "Applications having been made for repeal or modification of the Treasury order of July, 1836, the President wants opinions in writing of Cabinet upon the following questions. (1) If suspended is it reasonable to expect loss of public funds from sale of public lands, (2) Does public interest require that order be rescinded or suspended by the President until the next session of Congress, (3) If order rescinded what substitute?" March 1837, VBLC; Mahlon Dickerson to Rives, May 2, 1837, RLC.

27. The *Globe,* Mar. 23, 1837; Rives to Niles, Mar. 26, 1837; Van Buren to Rives, Apr. 8, 1837, RLC; The *Globe,* Mar. 23, 1837; see Harry N. Scheiber, "The Pet Banks in Jacksonian Politics and Finance, 1833–41," *The Journal of Economic History* (June, 1963), XXIII, No. 2, 210.

28. See Morgan Lewis to Van Buren, Apr. 27, 1837, VBLC; the *Globe,* May 5, 1837.

29. Butler, "Draft replying to the "Merchant's Memorial," May 4, 1837, VBLC; Isaac Hone to Van Buren, May 3; Van Buren to Hone, May 3, 1837, VBLC; Cambreleng to Woodbury, Apr. 24, 1837, WOLC; *Albany Evening Journal,* May 10, 1837.

30. Cambreleng to Van Buren, May 6, 1837, VBLC; Hone, *Diary,* 258–59; Wright to Van Buren, May 25, 1837; Tallmadge to Rives, May 1, 1837, RLC; Cambreleng to Van Buren, May 10; Wright to Van Buren, May 13, 1837, VBLC. A more accurate observer was state senator William H. Seward; he wrote his father on the legislative action, "except things are now at the worst and any change must be for the better." The younger Seward added that political repercussions would be as extensive as the economics. "There is a marvelous change in public opinion." Seward to Samuel Seward, May 13, 1837, Samuel Seward Papers, NYSL. For an interesting discussion of political change and consistency, see William G. Carleton, "Political Aspects

of the Van Buren Era," *South Atlantic Quarterly* (Apr. 1951), Vol. 50, 112, 167–85.

31. Woodbury to George Newbold, May 12, 1836, Letter Book, WOLC.

32. Marcy to Woodbury, Apr. 9; John Brockenbrough to Woodbury, May 1; Woodbury to Brockenbrough, May 1, 3; John Van Buren to Woodbury, May 9, 1837, WOLC; even Jackson, whose near hysterical defense of the Specie Circular could be construed as harboring inner doubts about it, asked Woodbury to help bail out the Planter's Bank of Tennessee; Jackson to Woodbury, May 16, 23, 1837, WOLC; Van Buren, "Memorandum," May, 1837, VBLC.

33. Brockenbrough to Woodbury, May 1, 1837, WOLC; Nathaniel Niles to Rives, June 8, 1837; Van Buren to Rives, May 13, 1837. Finished drafts were not sent until May 25, 1837, VBLC; Rives to Tallmadge, May 31, 1837, RLC.

34. Wright to Van Buren, June 22, 1837, VBLC; Polk to Van Buren, May 29, 1837; Dutee Pearce to Van Buren, May 29, 1837; Van Buren to Niles, June 10, 1837; Niles to Van Buren, July 1, 1837; Wright to Van Buren, May 28, June 4, 1837, VBLC.

35. Ritchie "Letters," 223–25.

36. Van Buren to Gouge, June 21, 1837, VBLC. Report on the Finances, "Appendix," *Register of Debates,* 25th Cong., 1st Sess., 18. Woodbury reported that the Deposit Banks on Aug. 15 held $11 million and $34 million in notes and loans of other banks; *ibid.*

37. "Ritchie Letters," 223–26; *Richmond Enquirer,* Sept. 15, 1837; the *Globe,* July 7, 21, 1837; Tallmadge to Rives, June 28, 1837, RLC.

38. Marcy to Bancroft, June 17, 1837, BMHS; to Wetmore, July 20; Aug. 18, 1837, MALC; Carl Degler, "The Locofocos, Urban Agrarians," *Journal of Economic History* (Sept. 1956), Vol. 16, No. 3, 332, 333.

39. Flagg to Butler, July 12, 1837, VBLC.

40. See John M. Niles to Rives, June 8, 1837, RLC.

41. Marcy to Wetmore, Aug. 18, 21, 1837, MLC; Niles to Rives, June 8, 1837, RLC; "Ritchie Letters," 234–36; the *Globe,* July 1, 28, 1837.

42. *Albany Argus,* July 20, 1837; Van Buren to Ritchie, Aug. 11, 1837, VBLC.

43. W. B. Hodgson to Rives, July 13, 1837, RLC; Blair to Jackson, July 21, 1837, Blair Family Papers, LC.

44. *Polk Papers,* IV, 463.

45. Allen to Rives, Aug. 5, 1836, RLC; the *Madisonian,* Aug. 21, 23, 25, 1837.

46. *Ibid.,* Sept. 7, 11, 1837. From the Hermitage came a sure sign of the growing importance and the purpose of the *Madisonian,* "A viper in the hypocritical disguise of a *friend* to the Administration" was how Jackson described it. *Jackson Correspondence,* V, 508.

47. Butler to Dear Wife, Aug. 26, 1837, BNYSL; see Van Buren, "Notes," Sept. 1837; see also Poinsett, "Notes," Sept. 1837, VBLC.

48. The *Madisonian,* Sept. 12, 14, 1837; Taney to Van Buren, July 25, 1837, VBLC; *Jackson Correspondence,* V, 491–95; *New York Weekly Commercial Advertizer,* Sept. 3, 1837.

49. Hoyt to Van Buren, Aug. 23, 1837, VBLC.

50. Wright to Flagg, Sept. 2, 5, 1837, FNYPL; Weed to Seward, Sept. 1, 1837, Seward-Weed Papers, Univ. of Rochester; Blair to Jackson, Sept. 9, 1837; Jackson to Blair, Sept. 15, 1837; *Jackson Correspondence,* V, 509–13.

51. Henry Clay had been in the Senate during the special session of the 11th Congress, but he was not in Congress for the special session of the 13th Congress.

52. H. D. Gilpin to Butler, Sept. 8, 1857, BuP; for the text of the message and Woodbury's report, see "Appendix," *Register of Debates,* 25th Cong., 1st Sess., 1–50.

53. Initially, it would appear Van Buren favored treasury notes not bearing interest. At any rate, Cambreleng favored such a measure in the Ways and Means Committee and on the floor of the House. *Register of Debates,* 25th Cong., 1st Sess., 1250; also in Van Buren's message to the special session is a veiled reference to the receipt of specie-paying bank notes in payment of debts owed the government. Richardson, ed., *Messages,* III, 324–46. See also Amos Kendall, "Memorandum," Sept. 1837, Kendall Papers, MHS. Wrote Kendall, "If banks of circulation must be used none chosen that issue small notes and receive same. Deposits not permitted to be lent—pay banks for service."

54. *Congressional Globe,* 25th Cong., 1st Sess., 35–57.

55. *Ibid.*

56. Jackson to Blair, Sept. 27, 1837, Blair Family Papers, LC; for a thoughtful overview of the "Panic" and the special session, see Reginald C. McGrane, *The Panic of 1837* (Chicago, 1965, Phoenix Ed.), 91–154.

Chapter 24. Beset by Difficulties

1. Henry D. Gilpin to Dear Father, Sept. 25, 1837, Gilpin Papers, HiSPA.

2. Hammond, *New York,* II, 479.

3. Marcy to Wetmore, Nov. 9, 1837, MALC; Van Buren to Jackson, Nov. 18, 1837, VBLC.

4. Gary E. Moulton, *John Ross, Cherokee Chief* (Athens, 1978), 2–12, 72–91; Ronald N. Satz, *American Indian Policy in the Jacksonian Era* (Lincoln, Neb., 1975), 97–106.

5. Richardson, ed., *Messages,* IV, 1596–99; *Senate Exec. Doc.,* 25th Cong., 2nd Sess.; Marcy to Van Buren, Dec. 8, 1837, VBLC.

6. Arthur G. Staples, ed., *The Letters of John Fairfield* (Lewiston, Maine, 1922), 198; the *Globe,* Feb. 12, 1838; *Congressional Globe,* 25th Cong., 2nd Sess., 83–93, 150–52.

7. Wright to Flagg, Jan. 29, 1839, FNYPL; Polk to Jackson, Mar. 7, 1837, JLC; to Donelson, May 29, 1838, Donelson Papers; Sellers, *Polk,* 332–34.

8. See Van Buren, *Inquiry,* 140–44, 148, 149, 161, 222–26.

9. See, for instance, Barnabas Bates to Van Buren, Feb. 11, 1835, VBLC; Hoyt to Major Abraham Van Buren, Dec. 30, 1837, VBLC.

10. J. W. James to Blair, June 14, 1834, Blair Family Papers, LC; J. G. Harris to Bancroft, n.d., 1838, BMHS; Leonard James to Van Buren, Jan.

24, 1838, VBLC; Robert Rantoul to Woodbury, Oct. 10, 1837, WOLC. There were few Loco-Focos in Boston, and if the slogan "equal rights" was voiced, it was but a faint signal of distress from an ephemeral newspaper whose brief life was being squeezed out while the collector withheld government advertising and other sources of petty patronage which kept such sheets alive; Bancroft to Butler, June 6, 1836, BuP.

11. Bancroft to Van Buren, Jan. 10, 1830, VBLC; J. W. James to Blair, Jan. 24, 1838, BLP; Ralph Huntington to Van Buren, Jan. 27, 1838, Leonard Jarvis to Van Buren, Jan. 24, 1838, VBLC.

12. Morton explained his refusal in a cryptic note to Bancroft saying ". . . it is morally impossible for me to take the office. My situation is peculiar and better known to myself than others." Morton to Bancroft, Dec. 24, 25, 1837, BMHS; M. A. DeWolfe Howe, *The Life and Letters of George Bancroft* (New York, 1908), 2 vols., I, 224; J. W. James to Blair, Jan. 12, 1838, BLP; Bancroft moved very cautiously; see Bancroft to Woodbury, Apr. 16; July 16, 1838, WOLC; but Henshaw remained determined to regain his station. See Henshaw to A. Woodbury, July 19, 1838; Woodbury to Henshaw, July 22, 1838, WOLC; Woodbury to Bancroft, Dec. 20, 1837; Seth Thomas to Bancroft, Jan. 29, 1838, BMHS.

13. Russell B. Nye, *George Bancroft, Brahamin Rebel* (New York, 1945), 114–18; Bancroft to Woodbury, Mar. 12, 1838, WOLC.

14. Buchanan to Van Buren, n.d., 1838, VBLC.

15. Henry Horn, Feb. 19; J. N. Barker, Apr. 5, 9; John M. Read, Apr. 5; Muhlenburg, Mar. 12; George G. Leiper, Mar. 18; George F. Lehman, Apr. 1, 1838 to Van Buren, VBLC; Philip S. Klein, *President James Buchanan: A Biography* (University Park, Pa., 1962), 115; Richard P. McCormick, *The Second American Party System* (Chapel Hill, N.C., 1966), 144.

16. Jeromus Johnson to Woodbury, Dec. 21, 1837, WOLC; Van Buren to Jackson, Jan. 10, 1838, VBLC; W. C. Bryant to Van Buren, Dec. 22; Saul Alley to Van Buren, Dec. 22; Jabez Hammond to Van Buren, Dec. 24, 1837, VBLC.

17. Wright to Flagg, Jan. 29, 1838, FNYPL; Hoyt to Major Abraham Van Buren, Dec. 30, 1837, VBLC; Marcy to Wetmore, Nov. 13, 1838, MALC.

18. Van Buren to Butler, Dec. 12, 1838, VBLC.

19. Butler to Van Buren, Apr. 17; Van Buren to Jackson, Apr. 24; Jackson to Van Buren, May 1, 1838, VBLC.

20. For evidence of Woodbury's maneuvers, see Isaac O. Barnes, Feb. 12, 16, 18, 19, 1838 to Woodbury; Henshaw to Woodbury, Feb. 23, 1838; Woodbury to Henshaw, Mar. 30, 1838, WOLC.

21. Hammond, *New York,* II, 455; Jacob Sutherland to Van Buren, Apr. 9, 1838, VBLC.

22. Van Buren to Irving, Apr. 23, 1838; Butler to Van Buren, Apr. 30, 1838; Irving to Van Buren, Apr. 30, 1838, VBLC.

23. See Paulding's correspondence with Blair, Van Buren, James Henry Hammond, Lewis Cass and others in Aderman, ed., *Paulding,* Ch. IV.

24. Richard E. Parker to Van Buren, May 2, 1838; VBLC; Joseph H. Parks, *Felix Grundy* (Baton Rouge, 1940), 314–17; Grundy to McGavock, May 29, 1838, McGavock-Gonzales Papers, UNC; Parks, *Grundy,* 324; Butler hailed

the Grundy appointment as a "brilliant finale in all the doings and nondoings of the session." Butler to Van Buren, July 9, 1838; see also Van Buren to Jackson, July 22, 1838, VBLC.

25. Woodbury to Hill, June 9, 1838; to Dr. Sir, June 1838, WOLC.

26. Marion Singleton to Dear Mother, Mar. 23, 1838, Angelica Singleton Van Buren Papers, LC.

27. Peter B. Porter to Clay, Dec. 13, 1837, Clay Papers, LC.

28. Spencer, *Marcy,* 105, 106; Peter B. Porter to Clay, Nov. 12, 1837, Clay Papers, LC; the *Globe,* Dec. 23, 1837; Marcy to Wetmore, Dec. 22, 31, 1837, MALC.

29. See Sir Francis Bond Head to Henry S. Fox, Dec. 23, 1837, in which he complains that he has as yet received no reply from Marcy regarding MacKenzie's activities; William R. Manning, ed., *Diplomatic Correspondence of the United States, Canadian Relations,* 4 vols. (Washington, D.C., 1943), III, 406, 407.

30. See Fox to Palmerston, May 17, 1838, F.O. 5, Vol. 323, LC Photostat. "The Navy," Fox said, was "in an unusual state of backwardness, it was almost of neglect. The present Secretary of the Navy, Mr. Dickerson, is a feeble, superannuated and very inefficient public officer. The services of the Department have suffered much, during the last three years through his neglect, and want of vigour in enforcing discipline and subordination."

31. Scott, *Memoirs,* I, 307; for the text of the instructions, see *Exec. Docs.* in 25th Cong., 2nd Sess., No. 73, 74; see also Richardson, ed., *Messages,* IV, 1616, 1617, 1838; Poinsett to Scott, Jan. 12, 1837, Poinsett Papers, HiSPA.

32. Marcy to Wetmore, Jan. 22, 1838, MALC; see Scott to Poinsett, Feb. 22, 1838, Poinsett Papers, HiSPA.

33. Poinsett to Scott, Feb. 5, 1838, Poinsett Papers, HiSPA; John E. Wood to Marcy, Feb. 5, 1838, Marcy Papers, NYSL; Fox to Palmerston, Apr. 5, 1838, F.O. 5, Vol. 323, LC Photostat.

34. Richardson, ed., *Messages,* IV, 1618–21.

35. J. Franklin Jameson, ed., "Correspondence of John C. Calhoun," *American Historical Assoc. Annual Report* (Washington, 1900), 2 vols., II, 388; hereafter cited as *Calhoun Correspondence;* Staples, ed., *Fairfield,* 137.

36. Van Buren to Palmerston, draft, May 16, 1838, VBLC; Van Buren to Jackson, Apr. 29, 1838, VBLC.

37. John Van Buren, *MSS. Diary,* 26–57, JVBH.

38. Manning, *Canadian Relations,* III, 469–74; Fox to Palmerston, Nov. 24, 1838, F.O. 5, Vol. 323, LC Photostat.

39. Fox to Forsyth, Jan. 10, 1838, F.O. 5, Vol. 323, LC Photostat. Fox kept Whitehall well informed on the powers granted the Administration under the Neutrality Act. Fox to Palmerston, Jan. 21, Feb. 5, 1838, in *ibid.*

40. Edward Everett to Van Buren, Mar. 4, 1839, EMHS; Staples, ed., *Fairfield,* 24–26, 28, 35, 81–83.

41. For Harvey's proclamation see Manning, *Canadian Relations,* III, 482, 483; Fox to Palmerston, Mar. 7, 1834, F.O. 5, Vol. 331; *ibid.,* Apr. 5, 1838, F.O. 5, Vol. 333, LC Photostat.

42. Benjamin Butler to My Dear Wife, Feb. 23, 1839, BNYSL; Charles R. Sanderson, ed., *The Arthur Papers* (Toronto, 1943–57), 3 vols., I, 57–58.

43. Scott to Poinsett, Jan. 12, 1839, Poinsett Papers, HiSPA.

44. Scott, *Memoirs,* II, 332–34.

45. Staples, ed., *Letters of Fairfield,* 271; Scott, *Memoirs* II, 338–45; Fairfield to Van Buren, Feb. 22, 1839; "Memorandum," Feb. 27, 1839, signed H. S. Fox and John Forsyth, VBLC; see also Fox to Palmerston, Mar. 17, 1839, F.O. 5, Vol. 331, LC Photostat.

46. *House Docs.,* 25th Cong., 3rd Sess., 1838–39, Doc. No. 270; Manning, *Canadian Relations,* III, 61–67, 482, 483; Edward Everett to Van Buren, Mar. 22, 1839, VBLC.

47. Everett to Van Buren, Mar. 4, 1834, EMHS; Fairfield to Van Buren, Mar. 22, 1839, VBLC; Sir John Harvey to Fairfield, Mar. 7, 1839; Fox to Palmerston, Mar. 31, 1839, F.O. 5, Vol. 331, LC Photostat.

48. Van Buren to Stevenson, Apr. 1834; Stevenson to Van Buren, May 16, 1839, VBLC, Stevenson Family Papers, LC. Fortunately, the new Canadian Governor Sir George Arthur, who replaced Lord Durham, acted with moderation; out of the 160 Americans captured in the assault upon Windsor, 17 were executed, 20 were transported to Australia and the remainder pardoned unconditionally. Sanderson, ed., *Arthur,* II, Pt. 1, 27, 29, 152. "We shall yet have much trouble with Canadian affairs," Van Buren wrote Stevenson in June, 1839, "but with great prudence on both sides will ultimately be able to allay the great excitement still prevailing on the frontier." Stevenson to Van Buren, May 16; Memorandum for Mr. Forsyth, June 6; Fairfield to Van Buren, Nov. 22, 1839, VBLC; Van Buren to Stevenson, June 24, 1838, Stevenson Family Papers, LC; Fox to Sir George Arthur, Jan. 31, 1839, F.O. 5, Vol. 331, LC Photostat.

49. William R. Manning, Diplomatic Correspondence of the United States: Inter American Affairs 1831–1860 (Washington, D.C. 1932–1939), 12 vols. *Canadian Relations,* VIII, 422–25.

50. Greenhow to Poinsett, Aug. 14; Bustamente to Poinsett, Aug. 14, 1837, in Gilpin Papers, HiSPA; Curtis, *Van Buren,* 164.

51. Richardson, ed., *Messages,* IV, 1594–96; all of the pertinent documents he had Forsyth include in his report.

52. See, for instance, Fairfield's comments in Staples, ed., *Fairfield,* 156, 242.

53. Leo Hershkowitz, "The Land of Promise: Samuel Swartwout and Land Speculation in Texas 1830–1838," *New York Historical Society Quarterly,* Vol. XLVIII, No. 4 (Oct. 1964), 312, 313, 319; Richardson, ed., *Messages,* IV, 1484–88; Wharton to Houston, in Garrison, ed., *Diplomatic Correspondence of the Republic of Texas, American Historical Assoc. Annual Report,* 1907 (Washington, D.C., 1808), 2 vols., II, Pt. 1, 180; Van Buren to John Van Buren, Dec. 22, 1836, VBLC.

54. *Congressional Globe,* 24th Cong., 2nd Sess., 185–263, 270.

55. Aderman, *Paulding,* 201–3; Garrison, ed., *Texas,* II, Pt. 1, 180; Richardson, ed., *Messages,* IV, 1500–1; Garrison, *Texas,* 201.

56. *Ibid.,* 160, 208, 236–41.

57. The British Foreign Office kept a close watch on Mexican-American affairs with particular attention to Texas, and Henry Fox kept Whitehall well informed; see, for instance, Fox to Palmerston, Oct. 12, 1839, F.O. 5,

Vol. 249; Jan. 20, 1837, F.O. 5, Vol. 314, No. 1; Garrison, ed., *Texas,* II, Pt. 1, 245, 255, 256, 268.

58. Manning, *Texas,* XII, 11–13, 129–40; Van Buren Memorandum, Aug. 1837, VBLC; Curtis, *Van Buren,* 158–59; the *Globe,* Oct. 19, 1837; Garrison, ed., *Texas,* II, Pt. 1, 263, 266, 285.

Chapter 25. "Little Van Is a Used-up Man"

1. Croswell to Van Buren, May 11, 1838; Ritchie to Van Buren, May 1838, VBLC; *Adams, Memoirs,* XI, 516, 517: X, 25, 26; *Courier and Enquirer,* Mar. 25, 27, 1838.

2. *Albany Argus,* Sept., Oct., 1838; the *New York Evening Post,* Oct., 1838; Marcy to Wetmore, Oct. 19, 1838, MALC.

3. John Van Buren, *MSS. Diary,* 96, JVBH.

4. Van Buren to Woodbury, Aug. 5, 1838, WOLC; Cambreleng to Van Buren, Nov. 9, 12, 1838, VBLC; Flagg to Van Buren, Nov. 9, 1838; Hoyt to Woodbury, May 5, 1838, WOLC.

5. The *Globe,* Dec. 20, 1838; Richardson, ed., *Messages,* IV, 1706–10.

6. *House Exec. Docs.,* 25th Cong., 3rd Sess., Doc. 13, 1–31. In his quarterly returns, Swartwout magnified the accounts of bonds uncollected, reducing accordingly cash on hand and accounts unsettled. *House Reports,* 25th Cong., 3rd Sess., Doc. 313; Pt. II, 23–95, 99–141, 442–50, 467–83, 492–95. With Butler's assistance, an effective minority report was made which rather conclusively shifted the onus where it belonged to Swartwout and Price, demonstrating clearly that under the system of reporting and auditing accounts, an embezzler could easily conceal his frauds from the authorities. See also *Richmond Enquirer,* May 14, 1839; Matthew A. Crenson, *The Federal Machine* (Baltimore, 1975), 80–82; 262–85.

7. *Congressional Globe,* 25th Cong., 3rd Sess., 183, 197.

8. See, for instance, Woodbury to Van Buren, Aug. 9, 24, 25, 1838, VBLC; Van Buren to Woodbury, Aug. 25, 1838, WOLC.

9. *MSS.* Draft report of the secretary of the treasury, Oct. 21, 1839, VBLC; the Treasury had not issued notes to the full extent provided. Congress had authorized $19 million of which the Treasury had issued only $5 million thus far.

10. John W. Edmonds to Van Buren, June 21, 1839, VBLC; Buchanan to R. J. Walker, Aug. 1, 1839, Buchanan Papers, NYHS: Dickinson to Poinsett, July 13, 1839, Poinsett Papers, HiSPA.

11. John W. Edmonds to Van Buren, June 27, 1839; Van Buren to New York Democrats, July 2, 1839, VBLC; Seward to Thomas Tallmadge, June 30, 1839, VBLC; the *New York Evening Post,* July 3, 1839; the *Globe,* July 5, 1839; Kemble to Poinsett, July 5, 1839, Poinsett Papers, HiSPA.

12. Marcy to Wetmore, July 20, 1828, MALC; *Albany Argus,* July 26, 1839; Van Buren to Jackson, July 30; Jacob Sutherland to Van Buren, Aug. 2, 1839, VBLC; *Albany Evening Journal,* Sept. 4, 1839.

13. Woodbury wrote that the banks were selling treasury notes, a sure sign of tight money. Bank officials had promised not to sell the notes and

were breaking faith with the Administration; Woodbury to Van Buren, Aug. 8, VBLC; also, Hoyt to Woodbury, Aug. 10, 1839, WOLC.

14. Calhoun to Micah Sterling, July 26, 1838, Cal-Clem; *Calhoun Correspondence,* 427–29.

15. J. Garland to Rives, Jan. 1, 1840, RLC. "My personal relations with MVB remain as they were when you left and must continue so," wrote Calhoun to Virgil Maxcy, "till they can be changed consistently with honor and propriety. Politically, it is understood that I desire to see the Administration support themselves because they approach nearer to my principles than their opponents, but it is also understood that I am perfectly free to follow my own judgement and principles on all questions." Calhoun to Virgil Maxcy, Feb. 28, 1839, Galloway-Maxcy-Markoe Papers, LC; Francis Granger to Weed, Jan. 2, 1840, Seward-Weed Papers, Univ. of Rochester.

16. *Calhoun Correspondence,* 388–95. See also Wiltse, *Calhoun Nullifier,* 358–61.

17. Ralph W. Hidy, *House of Baring in American Trade and Finance* (Cambridge, 1949), 273–77.

18. *Ibid.,* 272; Hoyt to Woodbury, Aug. 26, 1839, WOLC; George Newbold to Olcott; to Woodbury, Aug. 28; Sept. 19, 1839, Campbell-Mumford Papers, NYHS.

20. Marcy to Wetmore, Oct. 14; Nov. 3, 1839, MALC.

21. Woodbury to Van Buren, Oct. 21; Dec. 2, 1839, VBLC; Richardson, ed., *Messages,* IV, 1767.

22. Marcy to Van Buren, Dec. 21, 1839, VBLC; John A. Dix to Van Buren, Jan. 2, 1840; *New York Evening Post,* Dec. 3, 1839; *Albany Argus,* Dec. 3, 4, 1839.

23. Franklin Elmore to General A. Bailey, Sept. 18, 1839, UNC.

24. *Calhoun Correspondence,* 435; *Polk Correspondence,* V, 343, 344; Blair to Jackson, Dec. 22, 1839; JLC.

25. *Jackson Correspondence,* VI, 43, 44; see also *Polk Correspondence,* V, 343, 344, 350.

26. *Congressional Globe,* 26th Cong., 1st Sess., 1839–40, 88–89.

27. See, for instance, Green to Andrew P. Calhoun, Feb. 28, 1840, South Caroliniana Library; Green to Calhoun, Jan. 2, 1840, Green Papers, LC; *Calhoun Correspondence,* 438–41; Ambler, *Ritchie,* 208–10.

28. *Calhoun Correspondence,* 437, 438, 444.

29. Van Buren, *Autobiography,* 749, 750; Van Buren to Jackson, Feb. 2, 1840, VBLC.

30. Charles Ogle, *The Regal Splendor of the President's Palace* (Boston, 1840), *passim.;* Robert G. Gunderson, *The Log Cabin Campaign* (Lexington, 1947), 102–5.

31. Kendall to Van Buren, Aug. 22, 1839, with enclosure, VBLC.

32. *Jackson Correspondence,* VI, 55; Calhoun to Maxcy, July 7, 1840, Galloway-Maxcy-Markoe Papers, LC; Klein, *Buchanan,* 113–14, 130, 131; *Jackson Correspondence,* VI, 59–60.

33. Wright to Flagg, Jan. 26, 1840, FNYPL; *Jackson Correspondence,* VI, 55; *Extra Globe,* No. 1, May 16, 1840; *Jackson Correspondence,* VI, 61; *Polk Correspondence,* V, 444, 454, 464, 476.

34. *Courier and Enquirer,* Dec. 31, 1839; Ritchie to Poinsett, May 25, 1840, Poinsett Papers, HiSPA; *Polk Correspondence,* V, 374–75.

35. *Niles Weekly Register,* LVIII, July 11, 1840; see also the *Madisonian,* June 16, 18, 1840; *Daily National Intelligencer,* July 24, 1840; Moses Dawson to Poinsett, July 24, 1840, Poinsett in Gilpin Papers, HiSPA; Samuel Medary to Van Buren, Aug. 18; Dickerson to Van Buren, Sept. 11, 1840, VBLC.

36. Van Buren to John B. Carey and others, July 31, 1840, VBLC; also published in the *Richmond Enquirer,* Aug. 7, 1840.

37. Rippy, *Poinsett,* 182; Scott to Van Buren, Apr. 11; Van Buren to Scott, Apr. 11, 1838, VBLC.

38. Jackson to Blair, June 4, 1838, Blair Family Papers; Van Buren to Jackson, June 17, 1838, VBLC; much to the disgust of the hungry speculators, see, for instance, Nathaniel Smith to R. M. Johnson, Aug. 13, 1838, VBLC.

39. Moulton, *John Ross,* 93–101; Satz, *Indian Policy,* 64, 65, 158, 159, 190–92, 215, 261, 262.

40. Scott, *Memoirs,* I, 320–30; see, for instance, "Appendix," *Congressional Globe,* 25th Cong., 2nd Sess., 536–38.

41. *Ibid.,* 373, 558–61.

42. Joshua Giddings, *The Exiles of Florida* (Columbus, Ohio, 1858), Ch. XII, *passim.; House Exec. Docs.,* 25th Cong., 2nd Sess., Doc. No. 285, 11–15, 20; 3rd Sess., *Exec. Doc.,* No. 225, 5–31; yet Van Buren intervened with projected removal of the Tribes of the Six Nations in New York. See "Remonstrances of Seneca Indian Council, Buffalo Creek Reservation," Oct. 2, 1837; and Council of Chiefs of the Senecas Residing at the Buffalo Creek Reservation to Van Buren, n.d., 1838, A. C. Parker Coll., HL; *Daily National Intelligencer,* Jan. 15, 25; Mar. 2, 1838. *Daily National Intelligencer,* Jan. 15, 24, 25; Feb. 9; Mar. 10, 1838; Grant Foreman, *Indian Removal.* New Ed. (Norman, Ok, 1953), 357–77.

43. Nevins, ed., *Hone Diary,* 300, 385, 434; *Adams Memoirs,* X, 256, 270.

44. Woodbury to Van Buren, Sept. 5, 1841, VBLC. As late as 1841, he was complaining bitterly about War Department demands. "I am almost in despair," he wrote Poinsett. Woodbury to Poinsett, Jan. 5, 1841, WOLC; see also *Albany Evening Journal,* Deb. 12, 1841 on the state of the Treasury.

45. Van Buren, *Autobiography,* 275, 279, 288, 289, 290, 293, 294.

46. Lewis Tappan to Benjamin Tappan, Apr. 24, 1840, with Aaron Vail "Notes," VBLC; Forsyth to Van Buren, Sept. 23, 1839, VBLC.

47. *House Exec. Docs.,* 26th Cong., 1st Sess., Doc. 185, 59–64, 67–69; *New Haven Herald,* Jan. 11–13, 1840; *The Liberator,* Jan.–Feb., 1840.

48. Klein, *Buchanan,* 131; Van Buren to Gilpin, Jan. 7; Gilpin to Van Buren, Jan. 8, 1840, VBLC; Buchanan to Porter, Jan. 8, 1840, Buchanan Papers, HiSPA; Calhoun to Thomas G. Clemson, Jan. 11, 1840, Cal-Chem; *Adams Memoirs,* X, 378, 379.

49. Van Buren to Butler, Sept. 17, 1840. BuP.

50. Wright to Flagg, Mar. 10, 1840, FNYPL; John Van Buren to My Dearest Father, June 30, 1840, VBLC; Wright to Tilden, July 3, 1840, Tilden Papers, NYPL; Wright to Van Buren, Aug. 20; Flagg to Van Buren, Sept. 24, 1840, VBLC.

51. Others could be purchased for a penny or two at newspaper offices. Jared Bell's *New Era* in New York City had quantities of documents and pamphlets on hand which he sold in bulk to the general public or to Democratic committees and Young Men's Associations when they ran short of available stock. See "Van Buren's Letter to the Committee of Elizabeth City, Virginia," *New Era* edition, N.Y., 1840. Ritchie's office in Richmond was another distribution center. And the *Enquirer* published a weekly special campaign paper, the *Crisis.*

52. Buchanan to Van Buren, Sept. 25, 1840; Jackson to Van Buren, Sept. 22, 1840, VBLC.

53. Benjamin Butler to Bancroft, May 1, 1838, BMHS. O'Sullivan pursued an independent course and occasionally rubbed, Democratic politicians the wrong way. Marcy was offended by an article and complained to Wright who, after reading the piece, went to the co-publisher of the *Review,* J. D. Langton, and read him a lecture on the sensitivity of New York politics. But these incidents were rare and the *Review* generally pursued a discreet line on controversial interparty squabbles. Marcy to Wetmore, Jan. 27; Langton to Wright, Feb. 11; Wright to Marcy, Feb. 16; July 4, 1840, MALC; Arthur M. Schlesinger, Jr., Orestes A. Brownson, *A Pilgrim's Progress* (Boston, 1938), Ch. III, *passim.*

54. Van Buren to Marcy, Apr. 22; Marcy to Wetmore, Apr. 26; to Van Buren, Apr. 26, 1840; Marcy to Wetmore, Aug. 11, 28, 1840, MALC.

55. Van Buren to Welles, May 17, 1843, Welles Papers, LC; Marcy to Wetmore, Aug. 28, 1840, MALC.

56. See, for instance, Dickerson to Poinsett, May 9, 1840, Poinsett in Gilpin Papers, HiSPA.

57. Marcy to Wetmore, Aug., 1840, MALC.

58. Gunderson, *The Log Cabin Campaign,* 253–56; *Adams Memoirs,* X, 346, 347, 352, 353, 355, 356.

59. Jackson to Van Buren, Nov. 24; to Blair, Dec. 18, 1840, VBLC; Van Buren to Harmanus Bleecker, Jan. 27, 1841, VBNYSL; Calhoun to Maxcy, Feb. 19, 1841, Galloway-Maxcy-Markoe Papers, LC. For a spirited defence of Van Buren's administration and a highly critical analysis of Jackson's, see James C. Curtis, "In the Shadow of Old Hickory: The Political Travail of Martin Van Buren," *Journal of the Early Republic,* Vol. I, No. 3 (Fall 1981), 249–67.

60. *Polk Correspondence,* V, 586–89; Van Buren to Butler, Dec. 21, 1840, VBLC; Cambreleng to Van Buren, Dec. 15, 1840, VBLC.

Chapter 26. Search for Vindication

1. Wright thought it would be impossible to "unite the Democratic party in 1844 upon any candidate but Van Buren and therefore [he] did not think [he] had the right to place [himself] beyond the reach of that party for that purpose." Van Buren considered this wise counsel and decided for the moment not to make any pronouncements about the future. Wright to Richard Roane, Feb. 28, 1843, Wright Papers, NYHS.

2. Richardson, ed., *Messages,* V, 1814–35.

3. He asked Gilpin to research English and American precedents in chancery and common law; European authorities on international law. His suggestions to Gilpin were pertinent and comprehensive, showing he had lost none of his skill in dealing with the technical aspects of an incident which would surely involve politics, the federal system and foreign relations. The works of Vatell, Bynkershock, Sir Leslie Jenkins, Woodson's *Lectures,* V, II; Hawkins' *Pleas of the Crown;* Foster's *Common Law,* and V.I of Kent's *Commentaries* were all on the President's list.

4. Marcy to Wetmore, Feb. 6, 1841, *MS.* "Legal Opinion and Authorization, copy Martin Van Buren, Jr. and Henry D. Gilpin," Dec. 1841, VBLC; *Congressional Globe,* 26th Cong., 2nd Sess. For British correspondence, see F.O. Vol. 5, 349, II, III. LC Photostats.

5. Jonathan J. Coddington to Van Buren, Dec. 7; W. C. Bryant to Van Buren, Dec. 8; Samuel Alley to Van Buren, Dec. 22, 1840, VBLC; Butler to Wright, Jan. 30, 1841; Jeromus Johnson to Woodbury, Dec. 21, 1837, WOLC.

6. Van Buren to Butler, Oct. 29; Nov. 15, 1839; Butler to Wright, Jan. 30, 1841; Butler to Gilpin, Feb. 16, 17, 1841, VBLC; *New York American,* Feb. 27; *New York Evening Post,* Mar. 1, 3, 1841; see, for instance, John Van Buren, *MS Diary,* 144. "What Glorious News from New York," wrote the younger Van Buren in 1839, "a Republican Mayor and twelve wards out of seventeen; 41,000 votes. Good Lord, how Jesse Hoyt must have worked."

7. *United States Statutes at Large* (Boston, 1846), V, Ch. XLI, 385–92.

8. *Congressional Globe,* 26th Cong., 2nd Sess., 213–16; Van Buren to Jackson, Mar. 12, 1841, VBLC; John P. Frank, *Justice Daniel Dissenting* (Cambridge, Mass., 1964), 154–60; *Richmond Enquirer,* Mar. 2, 6–11, 1841; *Albany Evening Journal,* Mar. 8, 1841.

9. Marcy to Wetmore, Dec. 25, 1841; Jan. 3, 37, 1841, MLC; Lynch, *Epoch and a Man,* 465.

10. Bonney, *Legacy,* I, 155–59; Marcy to Wetmore, Mar. 3, 1841, in F. H. Sweet's *Catalogue List,* 1937, item 85, NYPL.

11. Marcy to Wetmore, Mar. 12, 1841, MALC; Van Buren to Jackson, Mar. 12, 1841, VBLC; Van Buren to Woodbury, July 14, 1841, WOLC; Aderman, ed., *Paulding,* 352, 353.

12. Harriet Butler to Van Buren, June 19, 1838, VBNYSL.

13. Marion Singleton to Dear Mother, Mar. 23, 1838, Angelica Singleton Van Buren Papers, LC.

14. Gabriella Singleton to Mrs. Marion Debeaux, Oct. 5, 1838, Singleton-Debeaux Papers, South Caroliniana Library; John Van Buren *MS. Diary,* 98, 103, 104–8, JVBH.

15. *Ibid.,* 196.

16. Angelica Van Buren to Dear Mother, Nov. 11, 1840, Angelica Singleton Papers, LC. John Van Buren, *MS. Diary,* 76, JVBH.

17. *New York Herald,* Dec. 11, 1841; John Van Buren to James Wadsworth, Dec. 12, 1841, Wadsworth Papers, Univ. of Rochester; Aderman, *Paulding,* 314, 315; Beard, *Cooper,* III, 452, 454; Nicholas B. Wainwright, ed., *A Philadelphia Perspective, The Diary of George Sidney Fisher* (Philadelphia, 1967), 66, 67.

18. John Van Buren, *MS Diary,* 156, JVBH; Van Buren to Miss Vanderpoel, July 2, 1838, VBLC.

19. Harrison to John Forsyth, Mar. 11, 1841, VBLC; Marcy to Wetmore, Mar. 12, 1841, MALC; see, for instance, bill of sale, Davenport and Company, Longport, Staffordshire, June 25, 1839, VBLC.

20. Van Buren to Thomas Reynolds, Mar. 6, 1841, VBLC; also published in the *Globe,* Mar. 13, 1841; Van Buren to Robert H. Morris and others, Mar. 6, 1841, VBLC; published in the *Globe,* Mar. 20, 1841; Marcy to Wetmore, Mar. 12, 1841, MALC.

21. Van Buren to Jackson, Mar. 30, 1841, VBLC; Wright to Elam Tilden, Dec. 6, 1841, Tilden Papers, NYPL; *New York Evening Post,* Mar. 23, 24, 1841.

22. Van Buren to A. J. Donelson, Apr. 28, 1841, A. J. Donelson Papers, LC.

23. *New York Evening Post,* Apr. 6, 1841.

24. *Ibid.; New York Herald,* Apr. 11; *New York Evening Post,* Apr. 11, 1841; Philip Hone, *MS Diary,* Apr. 11, 1841, NYHS.

25. Butler to My Dear Harriet, May 11, 1841, BNYSL; Collier, *Old Kinderhook,* 259–65.

26. Smith Van Buren to Martin Van Buren, Jr., July 21, 1839, VBLC.

27. Van Buren to Woodbury, July 24, 1841, WOLC; Van Buren to Jackson, May 15; July 30, 1841, VBLC. William B. Hesseltine and Lary Gara, eds., "A Visit to Kinderhook," repr. *New York History,* April 1954. All the brick mantle pieces gave way to marble with ionic pediments. Lynch, *An Epoch and a Man,* 476.

28. Van Buren to Jackson, July 30, 1841, VBLC; Van Buren to Woodbury, July 24, 1841, WOLC; Van Buren to Jackson, Oct. 12, 1841, VBLC.

29. Abraham Van Buren to Van Buren, Apr. 23, 1841, VBLC; Angelica Van Buren postscript Abraham Van Buren to John Van Buren, May, 1841, JVBH; Abraham Van Buren to Richard Singleton, Oct. 20, 1841. Singleton-DeBeaux Papers, South Caroliniana Library.

30. John Van Buren to Van Buren, June 3, 1841, JVBH.

31. Van Buren did ask a confidential and discreet friend in the legislature to be on the lookout for any sentiment that might develop for John. But there would be no special favors, no direct intervention, from Lindenwald. John Van Buren to Van Buren, Oct. 12; Dec. 4, 1841, JVBH; John Hunter to Van Buren, Feb. 4, 1842, VBLC.

32. Thereafter, Marcy joined the Democratic critics of the hybrid administration. Marcy to Wetmore, Sept. 23; to George Newell, Sept. 27, 1841; Thomas Gilmer to Marcy, Sept. 28, 1844; Marcy to Wetmore, Oct. 3, 1841; Gilmer to Marcy, Oct. 7, 18, 1841, MALC.

33. See, for instance, New York Ninth Ward Convention to Martin Van Buren, Aug. 28; Van Buren to the Ninth Ward Convention, Sept. 4, 1841; Rodney H. Chipp and others to Martin Van Buren, Sept. 24; Van Buren to Chipp and others, Oct. 2; Joseph McChesney and others to Martin Van Buren, Oct. 1, 1841, VBLC.

34. Van Buren to Jackson, Nov. 9, 1841, VBLC.

35. Millard Fillmore to Weed, Sept. 23, 1841, in *Publications of the Buffalo*

Historical Society (Buffalo, 1907), XI, 224, 226; Wright to Elam Tilden, Dec. 6, 1841, Tilden Papers, NYPL.

36. Van Buren's discreet friend, John Hunter, a member of the legislative caucus found no opportunity to suggest John for attorney general. He watched carefully the opinion in the meeting for a chance to nominate "our young friend. However popular he may be personally with our friends," Hunter reported, "the drill out of doors had given a popular current to certain men which it was impossible to stem." Hunter to Van Buren, Feb. 4, 1842, VBLC.

37. Wright to Van Buren, Oct. 25, VBLC; Van Buren to Albert Tracy, Dec. 13, 1841, VBNYSL; Van Buren to Jackson, Oct. 12, 1841, VBLC; Horn to Van Buren, Nov. 13, 1841, VBLC.

38. Van Buren to Horn, Nov. 13, 1841, VBLC.

39. Buchanan to White, Dec. 17, 1841, White Papers, NYHS.

40. Tracy to Van Buren, Nov. 24, 1841, VBLC; Van Buren to Tracy, Dec. 13, 1841, VBNYSL.

41. *Calhoun Correspondence,* 485, 493, 494.

42. *Ibid.,* 495–97; Gerald Capers, *John C. Calhoun, Opportunist* (Gainesville, Fla., 1960), 204–6; Poinsett to Kemble, Nov. 24, 1841, Gilpin Papers, HiSPA; to Van Buren, Nov. 28, 1841, VBLC; P. V. Daniel to Van Buren, Dec. 16, 1841, VBLC; Silas Wright to Elam Tilden, Feb. 13, 1842, Tilden Papers, NYPL.

43. See Nathanson, *Webster,* 181–87; Hugh Legaré to George T. Bryan, Aug. 11, 1838, Legaré Papers, South Caroliniana Library.

44. Robert McLellan to Van Buren, Feb. 6, 1842, VBLC.

45. Kemble to Poinsett, Nov. 28, 1841, Gilpin Papers, HiSPA; Thurlow Weed to J. Watson Webb, Jan. 15, 1842, Weed Papers, NYSL.

46. John Van Buren to Van Buren, July 31, 1842, JVBH; Van Buren to Jackson, July 30, 1842, VBLC.

47. Aderman, ed., *Paulding,* 312–19; Van Buren would supply a house for Marquette's family, all seed, farm equipment, stock, wagons and teams. In return he would receive two-thirds of the farm's output, Marquette one-third for his services; Van Buren MS Draft Contract with Mr. Marquette, 1841, VBLC.

48. Aderman, ed., *Paulding,* 319; Van Buren to Jackson, Feb. 7; Francis Pickens to Van Buren, Feb. 9, 1842, VBLC; Wright to Poinsett, Feb. 17, 1842, Poinsett Papers, HiSPA.

49. Adolph B. Benson, ed., *America of the Fifties, Letters of Frederika Bremer* (New York, 1924), 106–8.

50. See "The Secretary's Records of the Planters' Club," I, Mar. 3; Van Buren to Allston, Mar. 26, 1842, Grundy-McGavock Papers, UNC; Poinsett to Kemble, Mar. 14, 1842, Gilpin Papers, HiSPA.

51. Sarah Stevenson to Dear Brother, Mar. 14, 1842, Singleton-Debreaux Papers, South Caroliniana Library; see, for instance, John Slidell to Van Buren, Mar. 10, 1842, VBLC; Aderman, ed., *Paulding,* 327.

52. See, for instance, John Slidell to Van Buren, Mar. 10, 1842, VBLC.

53. Jackson to Van Buren, Feb. 22; Van Buren to Jackson, Mar. 26, 1852, VBLC; Clay to Van Buren, Mar. 17, 1842, VBMHS; Van Buren to Clay, Mar. 26, 1842, Clay Papers, LC.

54. *Jackson Correspondence,* VI, 152.

55. See, for instance, *Nashville Union,* Apr. 28, 1842, clipping in VBLC; Van Buren to Martin Van Buren, Jr., May 1, 1842, VBMHS; Benton to Van Buren, Apr. 14, 17; June 3, 1842, VBLC; Van Buren to Jackson, May 27, 1842, VBLC.

56. Henry Clay to Crittenden, June 3, 1842, Crittenden Papers, LC; Benton to Van Buren, June 8, 1842; Thomas Ford to Van Buren, Sept. 11, 1843, VBLC.

Chapter 27. An Active Retirement

1. John Van Buren to Van Buren, July 31, 1842, JVBH.

2. Van Buren to Jackson, July 30, 1842, VBLC.

3. Van Buren, Memorandum "Funds and Disbursements for 1842" Jan., 1842, VBLC.

4. Van Buren, "Probable Expenses, 1 May 1842 to 1 Jan. 1843," N.D., VBLC.

5. Richard B. Gooch to Mrs. Gooch, Sept. 10, 1842, Gooch Papers, UVA; Gilpin to Van Buren, Sept. 15, 1842, VBLC.

6. Van Buren to Poinsett, Oct. 1, 1842, Gilpin Papers, HiSPA.

7. Aderman, ed., *Paulding,* 353.

8. *Docs. of the Assem. of the State of New York,* 65th Sess. (Albany, 1842), IV, No. 61, 13–14.

9. *Ibid.,* 71st Sess., I, No. 4, 10–11.

10. *Albany Argus,* Feb. 16; Mar. 8, 1842.

11. Hammond, *New York,* III, 268–85.

12. Wright to Van Buren, Aug. 2, 1842, VBLC.

13. Hoffman to Flagg, June 30, 1842, FNYPL.

14. *Albany Argus,* Sept. 8, 9, 18, 1841; Hammond, *New York,* III, 308–24.

15. Van Buren to Marcy, Dec. 3, 1842, MALC; Van Buren, "Suggestions for Governor Bouck's Message of Jan. 3, 1842," VBLC; Marcy to Van Buren, Dec. 10, 1842, VBLC.

16. The debt in January, 1843, stood at over $23 million and as the Governor admitted, Canal tolls and other sources of state income were declining each year by almost a quarter of a million dollars. *Docs. of the Assem. of the State of New York,* 71st Sess., 1848 (Albany, 1848), I, 10, 11.

17. Isaac O. Barnes to Woodbury, Nov. 17, 1842, WOLC; Benton to Van Buren, June 8, 1842, VBLC; Bancroft to Van Buren, Sept. 28, 1842, in *Mass. Hist. Soc. Proc. 1908–9,* Vol. 42 (Boston, 1909), 391–93; hereafter cited as *Van Buren-Bancroft Correspondence;* Marcy to Wetmore, Oct. 3, 1841, MALC.

18. Pickens to Calhoun, Oct. 18, Nov. 6, 1841, enclosing a letter from Franklin Elmore, Cal-Clem.

19. Cole, *Jacksonian Democracy,* 216, 217; Hill to Cushing, July 2, 1842, Cushing Papers, LC; Woodbury to Isaac O. Barnes, Mar. 2, 1842, WOLC; Upshur to Duff Green, July 11, 1842, Green Papers, LC; Jacob Johnson to Cushing, Mar. 23, 1842, Cushing Papers, LC; William H. Roane to Van Buren, Sept. 11, 1842, VBLC; W. C. Preston to Waddy Thompson, Aug. 29, 1842, South Caroliniana Library; Capowski, "Woodbury," 195; Charles

Sellers, "Election of 1844" in Schlesinger, ed., *American Presidential Elections,* I, 755.

20. See "Memorial" from Citizens of New Orleans, May 1842, in Cushing Papers, LC.

21. Benton, II, 373–95.

22. Mangum to Clay, June 15, 1842, Henry Clay Papers, LC; "Appendix," *Congressional Globe,* 27th Cong., 1st Sess., 222–24, 364, 473–75.

23. Calhoun to James H. Hammond, Sept. 24, 1842, misdated in *Calhoun Correspondence,* 489–93.

24. W. Selden to Bancroft, Aug. 20, 1842, BMHS; *Calhoun Correspondence,* 552–4.

25. Wright to Flagg, Aug. 26, 1842; Wright to Flagg, Aug. 29, 1842, FNYPL.

26. *Calhoun Correspondence,* 516; James A. Hammond to Calhoun, Nov. 17, 1842, Cal-Clem.

27. Benton to Van Buren, Aug. 16, 1842, VBLC; Charles W. March to Cushing, July 22, 1842, Cushing Papers, LC; see also Wright to William Holdridge, Aug. 24, 1842, Wright Misc. Papers, NYHS.

28. See, for instance, John Law to Woodbury, June 21, 1842, WOLC.

29. *Calhoun Correspondence,* 516–17; Gilpin to Van Buren, Dec. 14; John Letcher to Benton, Dec. 15, 1842, VBLC; Charles H. Woodbury to Woodbury, Aug. 20, 1842, WOLC; John Goldsmith to RMT Hunter, Aug. 24, 1842, Cal-Clem.

30. See clipping from the New Jersey *Emporium and True American,* Dec. 7, 1842, VBLC; Gilpin to Bancroft, Dec. 12, 1842, BMHS; to Van Buren, Dec. 14, 1842, VBLC; *Van Buren-Bancroft Correspondence,* 397, 398.

31. Scoville to Calhoun, Nov. 8; Dec. 29, 1842; R. B. Rhett to Calhoun, Oct. 3, 1842, Cal-Clem. "You are doubtless aware," wrote Rhett, who had just returned from a visit to New York, "of the secret organization in your favor and altho the system of denying the very thing [we] are accomplishing is not consonant with our views of the fitness of things its effort has been very formidable."

32. Augustus Vanderpoel to Van Buren, Aug. 29, 1842, VBLC.

33. Van Buren to Jackson, Jan. 5, 1843, VBLC.

34. Pickens to Calhoun, July 19, 1843, South Caroliniana Library.

35. Wright to Gentlemen, Jan. 2, 1843, in the *Globe,* Jan., 3, 1843; Jackson to William H. Boyce and other Committee, Dec. 18, 1842, JLC.

36. Paulding, whose assessments of political doctrine were especially acute, praised "the Reply" or "confession of faith" as he called it. The sections on the tariff and the veto he thought were particularly apt, "and as to the appointing power, its liability to abuse, is an objection that may be brought with equal force against every Executive prerogative." Aderman, ed., *Paulding,* 333. "Mr. Van Buren's Reply to the Democratic state convention of Indiana, Feb. 14, 1843," VBLC. "The Reply" made the strongest impression where it was meant to be made in Virginia.

37. William H. Roane to Wright, Feb. 9, 14, 1843, Wright to Van Buren, Feb. 27, 1843, VBLC; to Roane, Feb. 25, 1843, Wright Misc. Papers, NYHS.

38. Wright to Van Buren, Jan. 27; Feb. 19, 1843, VBLC; Scoville to

Woodbury, Jan. 13, 1843, WOLC; Scoville to Calhoun, Jan. 15, 1843, Dixon Lewis Papers, U. of Texas; Franklin Elmore to Fitzwilliam Byrdsall, Sept. 9, 1843, James Hammond Papers, LC.

39. Roane to Wright, Feb. 9, 14, 1843, VBLC; Niven, *Welles,* 196, 201–3.

40. P. V. Daniel to Van Buren, July 6, 1843, VBLC.

41. Calhoun to Hunter, Apr. 2, 1843, Cal-Clem.

42. Buchanan to Campbell P. White, July 27, 1843, Buchanan Misc. Papers, NYHS.

43. Wright to Roane, Feb. 25, 1843, Wright Misc. Papers, NYHS.

44. *Calhoun Correspondence,* 539; Wright to Van Buren, Feb. 19, 1843, VBLC.

45. Maxcy to Hunter, Mar. 23, 1843, Cal-Clem; Calhoun to Hunter, May 9, 16, 1843, Calhoun Papers, LC; Rhett to Hunter, May 12; A. M. Scott and others to Ritchie, May 9; Ritchie to Scott, May 16, 1843, Hunter Papers, UVA; Calhoun to Hunter, May 16, 1843, Hunter-Garnett Coll. UVA.

46. See, for instance, *Albany Atlas,* Mar. 29, 1843; Sellers, "Election of 1844," 758–60.

47. *Van Buren-Bancroft Correspondence,* 402, 404, 405; Isaac O. Barnes to Daniel D. Brodhead, Apr. 29, 1843, Brodhead Papers, LC; Charles Henry Ambler, ed., "Correspondence of Robert M. T. Hunter," *Twelfth Report of the Historical Report of the American Historical Association for the Year 1916,* Manuscript Commission, Dec. 30, 1914." II, (Washington, D.C., 1916), 59; P. R. George to Isaac Barnes, Feb. 14; Mar. 9; to Cushing, Apr. 28, 1843, Cushing Papers, LC; Winfield Scott to Crittenden, July 8, 1843; Nathan Sargent to Crittenden, Sept. 30, 1843, Crittenden Papers, LC; Butler to Marcy, Aug. 30, 1843, MALC.

48. Van Buren-Bancroft *Correspondence,* 410, 411; Nelson Waterbury to Tilden, Aug. 30, 1843, Tilden Papers, NYPL; Butler to Marcy, Aug. 30, 1843.

49. *New York Evening Post,* Sept. 7–9, 1843.

50. Calhoun's Address was dated Dec. 21, 1843, but not published until Jan. 24, 1844; *Albany Evening Journal,* Feb. 5, 1844.

51. Letcher to Ritchie, Sept. 23, 1843, VBLC.

52. Andrew Stevenson, Oct. 8; Roane to Van Buren, Oct. 17, 1843, VBLC.

53. See, for instance, P. S. Root to Marcy, Aug. 18, 1843, MALC.

54. In 1827, Silas Wright, then a state senator, sponsored a bill which passed the legislature and became law, lending $500,000 to the Delaware and Hudson Canal Company. In 1829, during his short term as governor, Van Buren approved a further loan to the same company of $300,000; *Albany Evening Journal,* Jan. 14, 1844.

55. See Marcy's view which was similar, Marcy, *MS* Diary, Sept. 12, 1843, MALC.

56. *Albany Atlas,* Feb. 1, 14, 22, 1843; see advertisement for the *Atlas,* Sept., 1843, VBLC. It published a tri-weekly and weekly edition as well as a daily. Marcy to Wetmore, Jan. 15, 1843, MALC.

57. Flagg to Tilden, Jan. 23, 1843, Tilden Papers, NYPL; Wright to Van Buren, Apr. 10; Marcy to Van Buren, Jan. 27, 1843, VBLC.

58. Flagg to Van Buren, Apr. 12, 1843, VBLC.

59. Marcy to Wetmore, Sept. 8, 1843, MALC; Flagg to Van Buren, Sept. 4, 1843, VBLC; Marcy to Wetmore, Sept. 8, 1843, MALC; Hoffman to Flagg, Nov. 5, 8, 1843, FNYPL.

60. Van Buren to Marcy, Sept. 11, 1843, MALC.

61. Marcy *MS* Diary, Sept. 12, 1843; see also Butler to Marcy, Sept. 15, 1843, MALC.

62. Van Buren to Flagg, Apr. 22, 1844, VBLC.

Chapter 28. The Hammet Letter

1. Benton, II, 68, 507; see also vivid descriptions in the *Globe,* Mar. 1, 2, and *The Madisonian,* Mar. 2, 1844; Henry A. Wise, *Seven Decades of the Union* (New York, 1971), 226–27.

2. Rhett to Calhoun, Mar. 5, 1844, Cal-Clem.

3. *William and Mary Quarterly,* Series II, XVI, No. 4 (Oct. 1, 1937), 554–57.

4. *Ibid.;* see also Capers, *Calhoun,* 206, 217, 218; Winfield Scott to Crittenden, Oct. 4, 1843, Crittenden Papers, LC; Claude H. Hall, *Abel Parker Upshur, Conservative Virginian 1790–1844* (Madison, Wisc., 1964), 86, 93, 94, 197, 198.

5. *Calhoun Correspondence,* 545–46, 555, 559, 560; Green to Calhoun, Oct. 18; Pickens to Calhoun, Nov. 23, 1843, in R. P. Brooks and Chauncy S. Boucher, eds., "Correspondence Addressed to John C. Calhoun, 1837–1849," *Annual Report of the American Historical Association,* 1929 (Washington, D.C., 1929), 188–90, 191–92; Manning, ed., *Inter American Affairs* XII, 299–313, 319–23.

6. Rhett to Calhoun, Mar. 5, 1844, Cal-Clem; *Calhoun Correspondence,* 576–77.

7. As early as Mar. 18, before Calhoun's acceptance of the post of secretary of state, Blair was certain Tyler had reached an agreement with the government of Texas; Blair to Van Buren, Mar. 18, 1844; Butler to Mrs. Butler, Apr. 6, 1844, BNYSL; Wright to Bancroft, Apr. 8, 1844, BMHS.

8. Richardson, ed., *Messages,* V, 2111–15; Wright to Flagg, Jan. 20, 1844, FNYPL; Bouchard and Brooks, eds., *Calhoun Correspondence,* 214, 215.

9. The *Globe,* Feb. 5, 1844.

10. James P. Shenton, *Robert John Walker: A Politician from Jackson to Lincoln* (New York, 1961), 35.

11. Wright to Van Buren, Mar. 6; to Marcy, Jan. 25; Marcy to Wetmore, Feb. 16; Gorham A. Worth to Marcy, Feb. 18; Washington Hunt to Marcy, Feb. 24, 1844, MALC; Wright to Van Buren, Mar. 1, 6, 1844, VBLC.

12. Gillet, *Wright,* II, 1544–46.

13. Wright to Van Buren, Mar. 22; Apr. 29, 1844, VBLC.

14. Upshur to Duff Green, Feb. 26, 1844, Green Papers, LC.

15. Butler to Mrs. Butler, Apr. 6, 11, 1844, BNYSL; to Van Buren, Apr. 6; Wright to Van Buren, Apr. 6, 8, 1844; to Bancroft, Apr. 8, 1844, BMHS.

16. *Calhoun Correspondence,* 386–92; "Appendix," *Congressional Globe,* 25th Cong., 2nd Sess., 55, 60, 63, 65, 70–74.

17. Wright to Van Buren, Apr. 2, 6, 1844, VBLC.

18. After an extensive tour through the Southwest, Polk wrote Butler that Van Buren was certain to win if he declared for immediate annexation; Polk to Butler, Apr. 24, 1844, BuP; Wright to Van Buren, Apr. 11, 1844, VBLC.

19. Butler to Van Buren, Apr. 6; Wright to Van Buren, Apr. 6, 8, 11, 1844, VBLC; a portion of the letter shows the influence of Theodore Sedgewick, Jr.'s articles under the pseudonym of "Veto" in the *Evening Post.* Sedgewick had emphasized the prospect of a dozen slave states being carved out of the territory. *New York Evening Post,* Apr. 1844; Van Buren to Hammet, April 20, 1844, VBLC.

20. Butler to Mrs. Butler, Apr. 20, 1844, NYSL; the *Globe,* Apr. 15, 1844; *Jackson Correspondence,* V, 283; *Senate Exec. Docs.,* 28th Cong., 1st Sess., Doc. 341, *passim.*

21. Wiltse, *Calhoun,* III, 170–171; *Senate Exec. Docs.,* 28th Cong., 1st Sess., Doc. No. 341; *New York Evening Post,* Apr. 27, 1844.

22. Wright to Van Buren, Apr. 29, 1844, VBLC; *The Globe,* Apr. 27, 1844.

Chapter 29. Rejection at Baltimore

1. Wright to Van Buren, Apr. 29, 1844, VBLC.

2. The *Globe,* Apr. 27, 29, 1844; Wright to Van Buren, Apr. 29; Kendall to Van Buren, Apr. 29, 1844, VBLC; Elbert B. Smith, *Francis Preston Blair* (New York, 1980), 152–55.

3. *Jackson Correspondence,* VI, 281.

4. *Ibid.,* VI, 283–87.

5. *Ibid.*

6. Wright to Bancroft, May 21, 1844, BMHS.

7. William H. Roane to Van Buren, Apr. 30; Ritchie to Van Buren, May 5, 1844, VBLC; T. J. Randolph to Ritchie, May 4; David Hubbard to Ritchie, May 18, 1844; Ritchie Misc. MSS., LC.

8. Roane to Van Buren, Oct. 17, 1843, VBLC; Ritchie to Randolph, May 4, 1844, Ritchie Misc. MSS, LC; Van Buren to Butler, May 22; Bancroft to Van Buren, May 25, 1844, VBLC.

9. Wright to Van Buren, May 6, 10, 11, 12, 26, 1844, VBLC.

10. Donelson to Van Buren, May 16, 1844, VBLC.

11. Clipping, Jackson to the Editors of the *Nashville Union,* May 13, 1844, VBLC.

12. Van Buren to Butler, May 20, 1844, VBLC.

13. Gillet, II, 1534–37; Johnson to Polk, Apr. 30, 1844, Polk Papers LC.

14. Edwin S. Croswell to Van Buren, May 24, 1844, VBLC; Aaron Vanderpoel to Van Buren, May 22, 1844, VBLC.

15. Kemble to Abraham Van Buren, May 18, 1844.

16. The *Globe,* May 11, 16, 17, 1844; Kendall to Van Buren, May 13; Wright to Van Buren, May 13, 20, 1844, VBLC.

17. *Van Buren-Bancroft Correspondence,* 428, 429.

18. O'Sullivan to Van Buren, May 27; Butler to Van Buren, May 27, 1844, VBLC; Butler to Mrs. Butler, May 27, 1844, BNYSL.

19. *New York Tribune,* May 29, 1844; S. Croswell to Croswell, May 26, 1844, VBLC.

20. O'Sullivan to Van Buren, May 27, 1844, VBLC; the two-thirds rule carried by 30 votes, 20 of which came from New England and New Jersey, another 13 from Pennsylvania. Had they voted as pledged, the Van Buren forces would have defeated the Saunders amendment by ten votes; O'Sullivan to Van Buren, Mar. 38, 1844, VBLC.

21. *Albany Evening Journal,* May 30, 31, 1844; Henry Simpson to Van Buren, May 31, 1844, VBLC; Simpson charges that Buchanan caused the split in the Pennsylvania delegation.

22. Four Michigan delegates, for instance, violated their instructions from the outset; A. Teneyck to Van Buren, June 4, 1844, VBLC; S. Croswell to E. Croswell, May 26, 1844; Butler to Van Buren, May 31, 1844, VBLC.

23. Ibid., See Ambrose Spencer on Cass. He wrote Rev. William B. Sprague, "I have known since 1824 that this gentleman was very aspiring, ambitious, as to the means of ascending the ladder." Spencer to Sprague, May 30, 1844, Spencer Papers, NYSL.

24. Butler to Van Buren, May 31, 1844, VBLC.

25. *Ibid.*

26. Neither Butler nor Bancroft made any specific mention of a conversation, but Bancroft wrote Polk that he argued until midnight with the Ohio and New York delegations. Bancroft spent much of his time talking with Medary of Ohio and Gouverneur Kemble; unquestionably both men relayed his plan to Butler; Bancroft to Polk, July 6, 1844, BMHS.

27. Kemble to Poinsett, Nov. 23, 1844, Gilpin Papers, HiSPA; Wright to Marcy, Sept. 13, 1844, MALC; P. V. Daniel to Van Buren, June 11, 1844; Auguste Davesac to Van Buren, May 30, 1844, VBLC.

28. O'Sullivan to Van Buren, May 29, 1844; Butler to Van Buren, May 31, 1844, VBLC.

29. Wright to Flagg, June 8, 1844, FNYPL.

30. J. R. Livingston to Van Buren, May 28; Van Buren to Gansevoort Melville, June 3, 1844, VBLC.

31. O'Sullivan to Van Buren, May 29; Butler to Van Buren, May 31; Wright to Van Buren, June 2, 1844; Kendall to Van Buren, Oct. 13; Van Buren to Butler, May 20, 1844, VBLC; the *Globe,* May 28–30; June 6, 1844; *Daily National Intelligencer,* May 28–30, 1844; *Albany Evening Journal,* May 30, 31, 1844; Wright to Poinsett, June 26, 1844, Poinsett Papers, HiSPA.

Chapter 30. The Best Laid Plans

1. Kemble to Poinsett, Nov. 12, 1844, Gilpin Papers, HiSPA. Van Buren to Polk, Dec. 27, 1843, Polk Papers, LC.

2. Wright to Poinsett, June 18, 1844, Poinsett Papers, HiSPA.

3. Polk to Donelson, June 26, 1844; Buchanan to Donelson, July 17, Donelson Papers, LC; Ingersoll to Van Buren, Sept. 10, 1844, VBLC.

4. See, for instance, James H. Hammond to Calhoun, May 10, 1844, Cal-Clem; Calhoun to Henry Workman Connor, July 3, 1844, Henry Workman Connor Papers, LC.

5. Jackson to Butler, June 24, 1844, BuP; *Calhoun Correspondence,* 592–94, 601–03.

6. See *Albany Atlas Extra,* Dec., 1846, VBLC, for a remarkably able, historical outline of the two positions.

7. Wright to Marcy, Sept. 3, 1844, MALC.

8. Lucius E. Chittenden, *Personal Reminiscences* (New York, 1893), 14; *Albany Argus,* May 8, 1844.

9. Wright to Marcy, Sept. 13, 1844, MALC.

10. *Ibid.*

11. *Ibid.*

12. *Albany Atlas,* July 5; *Albany Argus,* July 18, 1844.

13. *Ibid.*

14. Gillet, *Wright,* II, 1555–57; Wright to Dix, July 14, 1844, Dix Papers, Columbia Univ.

15. Gillet, *Wright,* II, 1555–57; Garraty, *Wright,* 299–301.

16. *Ibid.,* Gillet, Wright, II, 1558–1565.

17. Van Buren to Jackson, Sept. 16, 1844, VBLC.

18. Wright to Dix, Feb. 14, 1845, Dix Papers, Columbia Univ; Van Buren to Polk, Jan. 18, 1845, VBLC.

19. Samuel Medary to Van Buren, Nov. 16, 1844; John Bragg to Van Buren, Nov. 24, 1844, VBLC; Butler to Donelson, Nov. 8, 1844, Donelson Papers, LC.

20. Polk to Marcy, July 9; Marcy to Wetmore, July 20; Nov. 10, 18, 1844, MALC; Marcy to Polk, Sept. 11, 1844, Polk Papers, LC; *Albany Argus,* Nov. 12, 1844; Sellers, *Polk Continentalist,* 178, 179.

21. Jackson to Van Buren, Oct. 2, 1844, VBLC.

22. Sellers, *Polk Continentalist,* 178; Gillett, *Wright,* II, 1631, 1632.

23. Wright to Flagg, Nov. 12, 1844, FNYPL.

24. Gillett, *Wright,* II, 1633–34; Polk to Van Buren, Jan. 4, 1845; Wright to Van Buren, Jan. 17, 1845, VBLC.

25. Hammond, *New York,* III, 515; John Van Buren to Van Buren, Jan. 21, 1845, JVBH; Kemble to Poinsett, Nov. 22, 1844, Gilpin Papers, HiSPA; Abraham Van Buren to Van Buren, Nov. 28, 1844, VBLC; John Van Buren to Van Buren, n.d., 1845, JVBH; Van Buren to Flagg, Dec. 17, 1844, VBLC.

26. Van Buren to Polk, Jan. 18, 1845; copy in the hand of Martin Van Buren, Jr., VBLC; Van Buren's original in his own hand is in the Polk Papers, LC. Robert John Walker to Butler, June 11, 1844, BuP.

27. Van Buren to Polk, Jan. 18, 1845, Polk Papers, LC.

28. Butler to Van Buren, Jan. 16, 1845, VBLC; Hammond, *New York,* III, 525.

29. Butler to Van Buren, Feb. 15, 1845, VBNYSL; Polk to Butler, Feb. 25, 1845, BuP.

30. Van Buren to Polk, Feb. 10, 11, 15, 1845, Polk Papers, LC; to Blair, Feb. 11, 1845, Blair Family Correspondence, LC; Van Buren, "Memorandum," filed with Van Buren to Polk, Jan. 18, 1845, VBLC.

31. Dix to Van Buren, Feb. 18, 1845, VBLC; *Van Buren-Bancroft Correspondence,* 437–439.

32. Polk to Bancroft, Jan. 30, 1845, BMHS; Bancroft to Polk, July 6, 1844; Polk to Bancroft, July 20, 1844, BMHS.

33. *Van Buren-Bancroft Correspondence,* 430–32; 434–36.

34. Hammond to Calhoun, June 7, 1844, James H. Hammond Papers, LC; *Calhoun Correspondence,* 465–71; Boucher and Brooks, eds., *Calhoun Correspondence,* 243, 244; Armistead Burt to Hammond, Mar. 2, 1845, James H. Hammond Papers, LC.

35. John M. Niles to Gideon Welles, Feb. 14, 1845, Welles Papers, LC; Dix to Van Buren, Feb. 18, 1845, VBLC.

36. Dix to Van Buren, Feb. 18, 1845, VBLC; Smith Van Buren to Van Buren, Mar. 2, 1845, VBLC.

37. Marcus Morton to Benjamin Tappan, Jan. 6, 1845, Tappan Papers, LC.

38. Polk to Van Buren, Feb. 22, 25, 1845, VBLC; Dickinson was actually pushing very hard for Marcy at the time; Samuel Nelson to Marcy, Mar. 1; Dickerson to Marcy, Mar. 1, 1845, MALC.

39. Marcy to Bancroft, Jan. 4, 1845, BMHS.

40. Hammond, *New York,* III, 525.

41. Wright to Dix, Feb. 15, 1845, Dix Papers, Columbia Univ.; Jackson to Van Buren, Feb. 10, 1845, VBLC; Lemuel Stetson to Flagg; Orville Robinson to Flagg, Dec. 31, 1845; Garraty, *Wright,* 343, 344; Marcy to Bancroft, Jan. 4, 1845, BMHS; Marcy to Wetmore, Feb. 25, 1845; Samuel Nelson to Marcy, Mar. 1, 1845, MALC.

42. Marcy to Wetmore, Feb. 25, 1845, MALC. In a letter he wrote to Polk and in a memorandum of a conversation he had with the President-elect, Kemble said the Marcy lobby did not represent the majority opinion of the party in New York. He said he advised Polk that Marcy was not as close to Wright as Butler and Flagg and that he ought to have the Governor's concurrence before he appointed him to any office; Smith Van Buren to Van Buren, Mar. 3, 1845, VBLC.

43. Van Buren to Polk, Feb. 27, 1845; Mar. 1, 1845, VBLC.

44. The matter of dates in this correspondence is cloudy. Van Buren in his memorandum says he received Polk's two letters on the evening of the 26th. He then says he alerted Polk to Smith's mission in a letter he sent "the next day," but either then or later inserted the "28th." If his note to Polk went out the 27th, Polk would have known Smith would arrive on Mar. 1 or Mar. 2 at the latest. And even if it were posted on the 28th, the President would have received it on the evening of Mar. 1, assuming no delay in the mail and giving Polk notice his letter of the 22nd had not miscarried. One wonders that if Polk were so anxious about hearing from Van Buren why he did not send his letter of Feb. 22 by special messenger. There is no evidence that Polk received Van Buren's letter of the 27th or 28th. Van Buren to Polk, Mar. 1, 1845; "Memorandum," Mar. 1845, VBLC. In an endorsement on his copy of this second letter, Van Buren wrote that the enclosed was sent as an "anodyne."

45. Polk to Van Buren, Feb. 24; Butler to Van Buren, Feb. 28, 1845.

46. Smith Van Buren to Van Buren, Mar. 2, 1845; O'Sullivan to Van Buren, Mar. 1, 1845, VBLC; Tilden to Butler, Mar. 1, 1845, BuP.

47. Polk to Van Buren, Mar. 1, 1845, VBLC.

48. Polk to Van Buren, Feb. 25, 1845, VBLC; Polk to Marcy, Mar. 1, 1845, MALC.

49. Smith Van Buren to Van Buren, Mar. 2, 1845, VBLC.

50. Polk to Van Buren, Mar. 3, 1845, VBLC.

51. Smith Van Buren to Van Buren, Mar. 4, 1845, VBLC.

52. Tilden to Van Buren, Mar. 29, 1845, VBLC; Wright to Van Buren, Mar. 38, 1845, VBLC; Van Buren and Wright to O'Sullivan, Mar. 15, 1845, VBLC.

53. Butler to Polk, Apr. 8, 1845, Polk Papers, LC; Marcy to Wetmore, May 12, 1845, MALC; Polk to Butler, May 3, 1845, BuP; Butler to Polk, May 6, 1845, Polk Papers, LC; Tilden to O'Sullivan, May 31, 1845, Tilden Papers, NYPL.

54. *Ibid.*, Bancroft to Butler, June 9, 1845; Butler to Bancroft, June 11, 1845, BuP; Cambreleng to Poinsett, Nov. 25, 1845; Poinsett Papers, HiSPA; Blair to Van Buren, Apr. 11, 1845; see Van Buren to Blair, Apr. 16, 1845, BLP; John Van Buren to Blair, Apr. 19, 1845, JVBH.

55. Dix to Butler, June 16, 1845, BuP.

56. See, for instance, Hammond to Calhoun, June 7, 1844, James H. Hammond Papers, LC; *Calhoun Correspondence,* 1027–29, 1043; Sellers, *Polk Continentalist,* 389, 390; Polk to Butler, May 5; Butler to Van Buren, May 7, VBLC; Butler to Bancroft, May 8, 1845, Butler Papers, NYSL; Van Buren to Bancroft, May 12; Flagg to Van Buren, May 16, 1845, VBLC.

57. Van Buren to Worth, Oct. 25, 1845, VBLC.

Chapter 31. Free Soil, Free Labor, Free Men

1. Van Buren to Gorham A. Worth, Jan. 2, 1848, VBLC; Van Buren to Blair, Nov. 29, 1847, VBLC; Aderman, ed., *Paulding,* 477, 478; Van Buren to Blair, Nov. 18, 1846, Blair Family Papers, LC.

2. Certainly Croswell shed few tears for Wright, see Croswell to James A. Seaver, Sept., 1847, NYSL.

3. Van Buren to Flagg, Oct. 12, 1847, VBLC; John M. Niles to Van Buren, Jan. 20, 1848, VBLC.

4. John Bigelow, ed., *Letters and Memorials of Samuel J. Tilden* (New York, 1908), 2 vols., I, 49, 50; "Address of the Democratic Members of the Legislature of the State of New York," Apr. 12, 1848, VBLC.

5. *Ibid.*

6. *Ibid.*

7. *Ibid.*

8. *Ibid.*

9. Bigelow, ed., *Letters and Memorials,* 50; *Albany Atlas Extra,* Apr. 12, 1848.

10. Butler to Van Buren, Nov. 6, 1847, VBLC; Gillet, II, 1697, 1788.

11. Hammond, *New York,* III, 557–61, 569, 685, 686; Gillet, II, 1657.

12. See a thoughtful, though eulogistic profile of Wright in Benton, *Thirty Year's View,* II, 700–702; Hammond, III, 569.

13. Buchanan to Polk, Sept. 5, 10, 1846, Polk Papers, NYPL; Garraty, *Wright,* 373.

14. Kemble to Poinsett, Oct. 10, 1846, Gilpin Papers, HiSPA.

15. Bancroft to Polk, Oct. 4, 1846, Polk Papers, NYPL; Gillet, II, 1790, 1791.

16. Bancroft to Polk, Oct. 4, 1846. Polk Papers, NYPL; Gillet, 1697–1711.

17. *Ibid.,* 1657–60, 1670–71, 1697–1711.

18. *Ibid.,* 1670.

19. In a conversation with Buchanan on Nov. 6, 1846, Polk attributed the defeat "to the bad faith of that portion of the Democratic party in New York opposed personally to Gov. Wright, called the Old Hunkers." He said gravely, "This faction should hereafter receive no favors at my hands if I know it." Milo M. Quaife, ed., *The Diary of James K. Polk, During His Presidency 1845–1847* (Chicago, 1910), 4 vols., II, 218.

20. Van Buren to Blair, Sept. 1, 1847, Blair Family Papers, LC.

21. Stanton, *Random Recollections* 159–60; *Albany Atlas,* Oct. 12, 13, 15, 1847.

22. Van Buren to Flagg, Oct. 12; Flagg to Van Buren, Oct. 13, 1847, VBLC.

23. Van Buren to Flagg, Oct. 12, 1847; John Van Buren to Wadsworth, Oct. 22, 24; Van Buren to Blair, Nov. 6, 1847, VBLC.

24. See, for instance, P. V. Daniel to Van Buren, Nov. 1; James W. Taylor to Van Buren, Nov. 2; G. M. Dallas to Van Buren, Nov. 2; Van Buren to Daniel, Nov. 13, 1847, VBLC.

25. Flagg to Dix, Nov. 13, 22, 1847, Dix Papers, Columbia Univ; John Van Buren to Van Buren, Nov. 13, 1847, VBLC.

26. *Albany Argus,* Feb. 23; *Albany Evening Journal,* Feb. 23, 24, 1848.

27. John Van Buren to Blair, Feb. 22, 1848, Blair Family Papers, LC.

28. Gouverneur Kemble "Memo," Jan. 1848, Charles Lyon Papers, UNC.

29. F. S. Hamlin to Chase, May 20, 1845, Chase Papers, HiSPA; Seward to Chase, Dec. 9, 1845; Feb. 2, 15, 18, 1846; Chase to My Dear Sir, Feb. 15; H. B. Stanton to Chase, Mar. 17, 1847, Chase Papers, LC; Preston King to Chase, Aug. 16, 1847, Chase Papers, HiSPA; John Van Buren to Van Buren, Apr. 30, 1848, VBLC.

30. Van Buren to John Van Buren, May 3, 1848, VBLC.

31. Van Buren had been receiving information that the President was seeking renomination; see, for instance, Van Buren, "Memo. of Instructions to New York Delegates, Baltimore, 1848," VBLC; John M. Niles to Van Buren, Apr. 18, 1848, VBLC. Joseph G. Rayback, *Free Soil, the Election of 1848* (Lexington, Ky., 1970), 181; this Barnburner campaign just prior to the convention did have considerable impact on the party. The address also aroused both opposition and praise, and since its publication on Apr. 12 had been circulated widely all over the country: see *Albany Atlas,* Apr. 12–30, 1848.

32. Butler to Ransome H. Gillet, May 8, 1848, VBLC.

33. Van Buren to J. Willson, May 9, 1848, VBLC.

34. Aderman, ed., *Paulding,* 481.

35. Van Buren to Samuel Medary, May 12, 1848, J. P. Morgan Library.

36. Butler to Van Buren, May 29, 1848, VBLC; Benjamin Tappan to Chase, May 20, 1848, Chase Papers, HiSPA; Butler to Van Buren, May 29.

37. Benton to Van Buren, May 20; Butler to Van Buren, May 30, 31, 1848, VBLC. Peter Cagger, who had taken over John Van Buren's leadership of the Democratic young men in Albany, announced the formation of a committee opposing Cass and Butler that denounced the activities of federal officeholders in the state and the expansion of slavery; Cagger to Van Buren, June 3, 1848, VBLC.

38. John Van Buren to William Collins, May 15, 1848, William Collins Papers, NYSL; Henry B. Stanton to Chase, June 6, 1848, Chase Papers, LC.

39. Marcy to Wetmore, June 10, 1848, MALC.

40. Seward to Weed, June 10, 1848, Seward-Weed Papers, Univ. of Rochester; Morgan Dix, *Dix,* I, 234.

41. John Van Buren to Butler, June 16, 1848, VBLC.

42. Dix to William Cullen Bryant, June 1, 1848, Bryant-Godwin Papers, NYPL; Flagg to Van Buren, June 11, 1848, VBLC; Van Buren to Dix, June 20, 1848, VBLC; Dix to Butler, June 19, 1848, BuP.

43. Van Buren to Butler, June 20, 1848, VBLC.

44. Van Buren to Samuel Waterbury and others (draft), June 20, 1848, VBLC; New York *Evening Post,* June 24, 1848; Chase to John Van Buren, misdated, June 19, 1848, Tilden Papers, NYPL.

45. John Van Buren to Van Buren, June 26, 1848, VBLC; Addison Gardiner to John Van Buren, June 24; John Van Buren to Van Buren, June 26, 1848, VBLC.

46. *New York Tribune,* July 1, 1848; Rayback, *Free Soil,* 212.

47. Blair to Van Buren, June 20, 1848, VBLC; John Van Buren to Van Buren, June 26, 1848, VBLC.

48. Chase to John Van Buren, June 19, 1848, misdated, Tilden Papers, NYPL.

49. B. F. Angel to Tilden, June 25, 1858, Tilden Papers, NYPL; Stanton to Chase, July 3; Sumner to Chase, July 7, 1848, Chase Papers, LC.

50. Arden Walter to Van Buren, July 3, 1848, WOLC; Chase to John Van Buren, June 19 (misdated), 1848, Tilden Papers, NYPL; William F. Kentzing to Van Buren, June 28, 1848, VBLC; John P. Curran to Weed, July 1, 1848, Weed-Seward Papers, Univ. of Rochester.

51. Preston King to Van Buren, July 12; Adams to Van Buren, July 16; Van Buren to C. F. Adams, July 24; Gamaliel Bailey to Van Buren, Aug. 2, 1848, VBLC.

52. Chittenden, 16; Tilden to Chase, July 18, 1848, Tilden Papers, NYPL; Flagg to Dix, July 20, 1845, Dix Papers, Columbia Univ. See, for instance, E. S. Hamlin to Chase, Aug. 1, 1848, Chase Papers, HiSPA; Hale to Chase, June 14, 1848, Chase Papers, HiSPA.

53. See Tilden's letter in *New York Evening Post,* Dec. 23, 1848; Everett to Sumner, Aug. 4, 1848, EMHS.

54. Welles *MS Diary,* Aug. 7, 1848, HL; quoted in Niven, *Welles,* 230.

55. Van Buren to New York delegates, Buffalo Convention, Aug. 2, 1848, VBLC.

56. Chase to My Dear Taylor, Aug. 14, 1848, Chase Papers, HiSPA.

57. *Ibid.*, Aug. 15, 1848.

58. Leavitt to Chase, Aug. 21, 1848, Chase Papers, HiSPA.

59. Quaife, *Diary,* III, 74, 509; IV, 65, 67. For the sincerity of Van Buren's convictions at this time, see Van Buren to Edward Coles, Oct. 1, 1848, Edward Coles Papers, Princeton; see also his parting remarks as Lucius Chittenden remembered them on leaving Lindenwald on July 5, 1848, in his *Personal Reminiscences,* 17.

Chapter 32. "Patriotism Firm and Ardent"

1. Isaac N. Arnold, *The Life of Abraham Lincoln* (Chicago, 1885), 105. A member of the Regency during its heyday said of Van Buren, "patriotism firm and ardent—a consciousness of a regard for what was abstractly right and a persistence in adhering to his friends . . . were in fact the main motive power with him." John Van Buren to Chase, Aug. 30; Chase to James H. Briggs, Sept. 12; Amos Jewett to Chase, Oct. 9; Benjamin Welch to Chase, Oct. 16; Jacob Brinkerhoff to Chase, Oct. 18, Chase Papers, HiSPA; Van Buren to Andrew Beaumont, Feb. 10, 1852; James W. Edmonds to Dr. Sir, June 29, 1866, VBLC.

2. Van Buren to James C. Holmes, Jan. 6, 1849, in the *New York Evening Post,* Jan. 6, 1849; to Blair, Dec. 1, 1849, Blair family Papers, LC.

3. Blair to Van Buren, Jan. 17, 1849, VBLC; Dix to Butler, Feb. 3, 1849, BuP.

4. John Van Buren to Chase, Feb. 23, 1849, VBLC; "Speech of John Van Buren," *New York Evening Post Extra,* Nov. 8, 1849, VBLC; S. P. Chase to Butler, July 26, 1849, BuP; John M. Niles to Van Buren, Dec. 7, 1849, VBLC.

5. Van Buren to Gorham A. Worth, Oct. 31, 1849, VBLC; Kemble to Poinsett, Feb. 15, 1850, Poinsett Papers, HiSPA; Van Buren to Blair, Dec. 14, 1850, VBLC.

6. Blair to Van Buren, Jan. 17, 27; Dec. 3, 1849; Martin Van Buren, Jr., to Van Buren, May 21; Blair to Van Buren, July 15–16, 20, Aug. 1, 3, 1850, VBLC.

7. Stanton, *Random Recollections,* 165.

8. Van Buren to Gorham A. Worth, Oct. 30, 1850, VBLC.

9. Kemble to Van Buren, Feb. 16, 1851, VBLC.

10. John Van Buren to Van Buren, Jan. 2, 1840, JVBH.

11. John Van Buren to Van Buren, Mar. 4; Blair to John Van Buren, Mar. 24, 1851, VBLC.

12. Dix to Blair, Sept. 6, 1851; Blair to John Van Buren, Sept. 14; John Van Buren to Blair, Sept. 17, 1851, BLP; John Van Buren to Van Buren, Sept. 19, 1851, VBLC. Feb. 25, 1852, JVBH; to Blair, Feb. 28, 1852, BLP, Princeton Univ.; Preston King to John Van Buren, Feb. 25, 1851, VBLC.

13. See a valuable collection of 22 letters from Marcy and Dickinson to Judge Thomas A. Osborne, printed in the *New York Sun,* Aug. 22, 1886.

14. Blair to Van Buren, June 26; Gilpin to Van Buren, June 6, 1852, VBLC; Van Buren to Chase, July 7, 1852, VBLC; Van Buren to Blair, Oct. 8, 1852, Blair Family Papers, LC.

15. Roy F. Nichols, *Franklin Pierce* (Philadelphia, 1948), 2nd ed., 192; Van Buren to Gentlemen, July 1, 1852 in *New York Evening Post,* July 7, 1852.

16. Blair to Van Buren, Mar. 30, 1849, VBLC.

17. Van Buren to Gorham A. Worth, Apr. 9, 1849, VBLC.

18. Anon: "Lindenwald, Home of Martin Van Buren, Association for the Preservation of Lindenwald." New York, 1937, *passim;* Collier, *Old Kinderhook,* 376–78, 498; George D. Berndt to author, Oct. 13, 1981. Blair to Van Buren, Mar. 30; Van Buren to Gorham A. Worth, Apr. 4, 1849. Smith advanced $7,400 of the cost; Van Buren "Last Will and Testament," June 8, 1860, VBLC.

19. Van Buren to Enos T. Throop, June 1, 1849, VBNYSL; Clay to Van Buren, Aug. 9, 1849, VBLC. Van Buren wrote Clay on Aug. 4.

20. Van Buren to Throop, June 1, 1849, VBNYSL; to Gorham A. Worth, Oct. 31; Blair to Van Buren, Nov. 11, 1849, VBLC.

21. Van Buren to Kemble, Jan. 5, 1852, William Lyon Chandler Papers, UNC; see Paulding to Van Buren, July 21; Aug. 6; Oct. 4, 1849, in Aderman, ed., *Paulding,* 462, 468; Blair to Van Buren, Nov. 12, 1850, VBLC.

22. Blair to Van Buren, July 20, 1849, VBLC.

23. Benton to Van Buren, Sept. 25, 28; Oct. 22, 1851, VBLC; see *The United States Democratic Review,* Vol. I, No. 1; John Van Buren to Van Buren, Jan. 20, 1851, VBLC; Blair to Van Buren, Sept. 30, 1849, VBLC.

24. Van Buren, *Autobiography,* 9, 595; *Inquiry,* Chs. IV, V, *passim.;* Van Buren to Smith Van Buren, Mar. 24, 1857; Oct. 7, 1858, VBLC; Chittenden, *Personal Reminiscences,* 15.

25. Van Buren to Kemble, Jan. 15, 1852, William Lyon Chandler Papers, UNV; Blair to Van Buren, Sept. 30, 1849; John Van Buren to Van Buren, Jan. 20, VBLC; Martin Van Buren, Jr. to Blair, Apr. 14, 1850, BLP; Van Buren to Butler, May 20, 1831; Pierce to Lord Aberdeen, Apr. 15, 1853, VBLC.

26. Lord Palmerston to Van Buren, May 26; June 21, 1853; Lord Clarendon to Van Buren, Oct. 1, 1853, VBLC; Van Buren to Lord Palmerston, June 21, 1853, VBLC; Van Buren, *Autobiography,* 10 n; Van Buren misdated his visit to Holland. Lord Palmerston to Van Buren, May 26, 1853, VBMHS; Van Buren, "Notes," 1853; Van Buren to Lord Palmerston, June 21, 1853; Van Buren to My Dear Judge, Sept. 7, 1853, VBNYHS; Jules Van Braet to Van Buren, Aug. 6, 1853, VBLC.

27. Kemble to Van Buren, Oct. 25, 1853, VBLC.

28. Van Buren to Blair, Sept. 30, 1854, Blair Family Papers, LC; to Martin Van Buren, Jr., Nov. 1, 1854, VBLC; Van Buren, *Inquiry,* 355; Cavour to Van Buren, Oct. 1, 1853, VBLC.

29. Van Buren to Kemble, June 13, 1845, VBLC; *Autobiography,* 9.

30. John Van Buren to Blair, Mar. 21, BLP; Marcy to Tilden, Oct. 16, 1853, Tilden Papers, NYPL; to Blair. Oct. 11, 1853, BLP.

31. John Van Buren to Blair, Oct. 11, 1853; Blair to Van Buren, Nov. 27, 1853, BLP; John W. Ludlow to Van Buren, Nov. 23, 1853, VBLC.

32. John P. Beekman to Van Buren, Dec. 2, 1853, VBLC.

33. Postscript of John Van Buren on Anna Van Buren, to Van Buren, Mar. 18, 1854, JVBH; Van Buren, *Inquiry,* 355.

34. Benjamin Butler to Van Buren, Dec. 2, 1854, VBLC; Van Buren to Martin Van Buren, Jr., Sept. 19, 1854, VBLC.

35. Van Buren to Martin Van Buren, Jr., Oct. 28; Nov. 1, 12, 20, 1854, VBLC.

36. Van Buren to Blair, Mar. 16, 1855, Blair Family Papers, LC.
LC.

37. Benton to Van Buren, Feb. 17, 1857; VBLC; Bigelow, ed., *Letters of Tilden,* I, 119–21.

38. Van Buren, *Inquiry,* 375.

39. Van Buren to Montgomery Blair, Jan. 31, 1857, Blair Family Papers, LC.

40. Van Buren, *Inquiry,* 355–57.

41. *Ibid.,* 357.

42. *Ibid.,* 362–65.

43. *Ibid.,* 370, 371.

44. Van Buren to Butler, July 27, 1857, Butler Papers, LC; to Smith Van Buren, Oct. 7, 8, 1858, VBLC.

45. Van Buren, *Autobiography,* 205.

46. Gilpin to Van Buren, June 7; Benton to Van Buren, Oct. 27, 1856, VBMHS; Van Buren to Blair, Jan. 31, 1857, Blair Family Papers, LC.

47. Van Buren to Blair, June 15, 1856, Blair Family Papers, LC.

48. Van Buren to Mr. Bronson, Oct. 15, 1860, VBNYSL; to Throop, July 15, 1861, Throop-Martin Papers, Princeton Univ.; to Alfred Goodwin, Nov. 11, 1861, VBLC.

49. Van Buren, "Draft Resolutions for the New York State Assembly, July 1, 1860," VBLC.

50. Van Buren to Kellogg, Percy and Merrit, Esq., Feb. 20, 1861, VBLC; to Blair, Mar. 1, 1861, Blair Family Papers, LC.

51. Van Buren to Ingersoll, Apr. 27, 1861, VBLC.

52. Van Buren to John Haberton, Nov. 31, 1861, VBLC.

53. Van Buren to Throop, Jan. 12; Feb. 9, 1862, Throop-Martin Papers, Princeton; to Flagg, May 28, 1862, VBLC.

54. Lawrence Van Buren to Blair, Oct. 23, 1862, BLP; *The New York Times,* July 25; *New York Evening Post,* July 25, 1862; NA "Record Group 645," Roy Basler, ed., *The Collected Works of Abraham Lincoln* (New Brunswick, 1952), 8 vols., index, V, 340, 341; Washington, D.C. *Evening Star,* July 24, 1802; *Daily National Intelligencer,* July 25, 1862.

Index